Robots and Autonomous Machines for Agriculture Production

Robots and Autonomous Machines for Agriculture Production

Editors

Jin Yuan
Wei Ji
Qingchun Feng

MDPI • Basel • Beijing • Wuhan • Barcelona • Belgrade • Manchester • Tokyo • Cluj • Tianjin

Editors

Jin Yuan
Shandong Agriculture University
Taian, China

Wei Ji
Jiangsu University
Zhenjiang, China

Qingchun Feng
National Research Center of Intelligent Equipment for Agriculture
Beijing, China

Editorial Office
MDPI
St. Alban-Anlage 66
4052 Basel, Switzerland

This is a reprint of articles from the Special Issue published online in the open access journal *Agriculture* (ISSN 2077-0472) (available at: https://www.mdpi.com/journal/agriculture/special_issues/robots_autonomous_machines).

For citation purposes, cite each article independently as indicated on the article page online and as indicated below:

LastName, A.A.; LastName, B.B.; LastName, C.C. Article Title. *Journal Name* **Year**, *Volume Number*, Page Range.

ISBN 978-3-0365-8376-1 (Hbk)
ISBN 978-3-0365-8377-8 (PDF)

Contents

About the Editors

Jin Yuan

Jin Yuan is a professor and doctoral supervisor at the School of Mechanical and Electrical Engineering, Shandong Agricultural University, and a deputy director of the Artificial Intelligence Branch of the Chinese Society of Agricultural Machinery. He is a postdoctoral fellow at Shanghai Jiaotong University and engaged in scientific cooperation research at the Norwegian University of Science and Technology. He has presided over more than 20 scientific research projects in the fields of agricultural robotics, intelligent agricultural equipment, and precision agriculture, such as the National Natural Science Foundation of China, the National Key Research and Development Program, etc. He won two National Science and Technology Advancement Prizes in 2009 and 2011, and was awarded the Third Youth Science and Technology Prize of the Chinese Society of Agricultural Machinery in 2012. He has published more than 60 academic papers, and more than 50 invention patents have been authorized.

Wei Ji

Wei Ji, Ph.D., is a professor in the School of Electrical and Information Engineering at Jiangsu university. He received B.Sc. and M.Sc. degrees in Electrical Engineering from China University of Mining and Technology, Xuzhou, China in 1999 and 2002, respectively, and a Ph.D. degree in Electrical Engineering from Southeast University, Nanjing, China in 2007. Since 2007, he has been with the School of Electrical and Information Engineering, Jiangsu University, Zhenjiang, China. His current research interests include agricultural robots, automatic control and navigation technology for agricultural machinery and equipment, comprehensive automation control systems for production processes, and non-standard robot control systems.

Qingchun Feng

Qingchun Feng is the Director of the Agricultural Robotics Laboratory at the National Intelligent Equipment Engineering Technology Research Center. He is engaged in research on agricultural robots. His achievements include five new product certifications, an Excellent Scientific Research Achievement Award from higher education institutions, the Wu Wenjun Artificial Intelligence Science and Technology Award, and the China Agricultural Science and Technology Award. He has led eight research projects, including the National Natural Science Foundation and National Key Research and Development Program. He has published 35 academic papers in SCI/EI journals, obtained 15 invention patents, and co-authored 1 international monograph. He was selected as Beijing Nova, Beijing City Outstanding Talent, and Youth Talent in the Chinese Association for Science and Technology Think Tank. Additionally, he serves as a member of the IFAC Agricultural Automation Technical Committee.

Preface to "Robots and Autonomous Machines for Agriculture Production"

Global agriculture faces critical pressures, including an aging population, rising production costs, and labor shortages. An important alternative solution for these challenges, robots and autonomous machines represent a high-level application of smart agriculture, which is based on a precise and resource-efficient approach to sustainably achieving higher efficiency and quality in agricultural production. On the one hand, robotics and autonomous machines continue to spread into various new agricultural scenarios, and on the other hand, technologies such as deep learning and machine learning are increasingly being used in agricultural production. By exploring the diverse methodologies employed in addressing such challenges, this Special Issue aims to advance the field and improve the efficiency of agricultural production through robotic and autonomous innovations.

In order to investigate the advancements in robots and autonomous systems for agriculture, by using modeling, detection, and control technologies and emphasizing their potential in precision farming, crop protection, crop harvesting, etc., we have organized this Special Issue, "Robots and Autonomous Machines for Agriculture Production". The issue has a total of 26 papers which cover a wide range of agricultural operations, including cotton planters, maize planters, apple harvesting, shrimp peeling, rice phenotyping, pest control, bale collection, pineapple processing, garlic seeding, agricultural film collecting, lettuce growth modelling, egg microcrack detection, forage pushing, fungus harvesting, and jujube pruning. In terms of the research field, the issue focuses on robotics and its related application research, such as the following areas: soft gripper design, autonomous robots, humanoid field-phenotyping robots, apple detection, manipulator motion planning, dairy robots, vineyard spraying robots, fungus harvesting robots, and orchard visual navigation. In addition, it also focuses on intelligent agricultural machines in different scenarios of seeding, crop monitoring, agricultural products, etc.

In summary, this Special Issue highlights different approaches in the development of agricultural robots and intelligent agricultural machines in several agricultural application scenarios for scene and object perception, intelligent decision support methods, and operational mechanisms and their control. It is expected that the insights gained from this Special Issue will be useful to researchers in the field of agricultural robots and autonomous machines.

Jin Yuan, Wei Ji, and Qingchun Feng
Editors

agriculture

MDPI

Editorial

Robots and Autonomous Machines for Sustainable Agriculture Production

Jin Yuan [1,*], Wei Ji [2,*] and Qingchun Feng [3,*]

1 School of Mechanical and Electronic Engineering, Shandong Agriculture University, Taian 271018, China
2 School of Electrical and Information Engineering, Jiangsu University, Zhenjiang 212013, China
3 National Research Center of Intelligent Equipment for Agriculture, Beijing 100097, China
* Correspondence: jyuan@sdau.edu.cn (J.Y.); jiwei@ujs.edu.cn (W.J.); fengqc@nercita.org.cn (Q.F.)

Citation: Yuan, J.; Ji, W.; Feng, Q. Robots and Autonomous Machines for Sustainable Agriculture Production. *Agriculture* **2023**, *13*, 1340. https://doi.org/10.3390/agriculture13071340

Received: 2 June 2023
Revised: 6 June 2023
Accepted: 29 June 2023
Published: 1 July 2023

The global agriculture faces critical pressures, including an aging population, rising production costs, and labor shortages. As an important alternative solution for those challenges, robots and autonomous machines represent a high-level application of smart agriculture, which is based on a precise and resource-efficient approach that attempts to sustainably achieve a higher efficiency in the agricultural production with an increased quality. On the one hand, robotics and autonomous machines continue to expand in various new agricultural scenarios, while on the other hand, technologies such as deep learning and machine learning are increasingly used in agriculture, and their application in various scenarios of agricultural production has become more in-depth. By exploring the diverse methodologies employed in addressing such challenges, this Special Issue aims to advance the field and improve the efficiency of agricultural production through robotic and autonomous innovations.

In order to investigate the advancements in robots and autonomous systems for agriculture, using modeling, detection, and control technologies, emphasizing their potential in precision farming, crop protection, crop harvesting, etc., we have organized this Special Issue "Robots and Autonomous Machines for Agriculture Production (RAMAP)". The Special Issue of RAMAP has a total of 26 papers [1–26], and papers were submitted from eight countries: Spain, Italy, Germany, Brazil, China, Sweden, Czech Republic and Croatia. Moreover, the Special Issue covers a wide range of agricultural operations, including cotton planters [15], maize planters [2], apple harvesting [3,7,11], shrimp peeling [4], rice phenotyping [6], pests control [9], bales collection [8], pineapple processing [10], garlic seeding [12], agricultural film collecting [13], lettuce growth modelling [14], egg microcrack detection [17], forage pushing [18], fungus harvesting [24] and jujube pruning [25]. In terms of the research field, the Special Issue not only focused on robotic and its related application research, such as soft gripper design [3], autonomous robot [5], humanoid field-phenotyping robot [6], apples detection [7], manipulator motion planning [11], dairy robot [18], vineyard spraying robot [20], fungus harvesting robot [24] and orchard visual navigation [26], but also refers to intelligent agricultural machines in different scenarios on seeding [1,2,12,15], shrimp peeling [4], recycling film collecting [13], crop and monitoring [14], and agricultural products [16,17,23].

Agricultural robots are multi-degrees-of-freedom autonomous operation machines used in agricultural production, with perception, decision-making, control and execution capabilities, mainly including information perception systems, decision-making systems, operation actuators, that is sensing, decision making and execution. Overall, most of the papers in the Special Issue of RAMAP were grouped into four categories: sensing for the crop or machine system [1,4,7,9,10,12,16,17,22,23,26], methodological studies for decision-making and control [2,8,11,14,18,20,24], designs related to intelligent machinery execution [1,3,13,15,25] and systematic solutions [5,6,19,21].

Generally, agricultural robots first need to sense the operating environment, the operating object and the state of the robot itself, and to provide the panoramic data related to the operating process to the agricultural robots to complete the operating tasks.

The first category has eleven papers under the following sub-heading: Intelligent sensing for the crop or machine system [1,4,7,9,10,12,16,17,22,23,26]. Currently, a large number of studies focus on deep learning techniques, which have shown their superb impact on robotic sensing applications, as reflected in this issue. Some papers utilized improved YOLO-based [7,23,26], CNN-based [12,16] or RCNN-based [9] methods to developed a detection model for operating target recognition or performance evaluation from the RGB images. To achieve more accuracy, faster and compacter models may be popular due to the cost-effective and feasiblity with low-computing platforms. The paper by Liu et al. [10] proposed a 3D localization algorithm to fuse the depth information based on multiangle image matching and YOLOv5 detection information. Some papers utilized the manual features combined with machine learning, such as the adaptive recognition boundary model [4], density-based lightning connection clustering [22], random forest [17], etc., to achieve target detection, due to a small training dataset or more efficient features.

For intelligent agricultural machines, Bai et al. [1] designed a monitoring system for the sowing quality of cotton precision planters, to realize the real-time monitoring of the cotton precision seeding operation processes and improve the intelligence level of cotton precision planters.

Generally, intelligent decision-making and intelligent control systems aim at deep fusion of perception information, cognitive reasoning, predictive planning, and coordinated control of agricultural robot perception and execution subsystem operations, which is the core element of agricultural robots.

The second category has seven papers under the following sub-heading: Methodological studies for decision making and control [2,8,11,14,18,20,24]. Three studies focus on the optimization of motion planning for robots. The paper by Latif et al. [8] optimized path planning approaches using a new autonomous articulated concept vehicle with neighborhood reach capabilities (AVN). The paper by Liu et al. [11] proposed a time-optimal rapidly exploring random tree (TO-RRT) algorithm to reduce the obstacle avoidance effect and increase picking efficiency of the manipulator. The paper by Yang et al. [24] proposes a multi-objective optimization algorithm of the multi-arm cooperative harvesting trajectory to improve the harvesting efficiency.

A novel method [14] for predicting the dynamic growth of leafy vegetables based on the in situ sensing of phenotypic and environmental data of batches is proposed to predict the dynamic fresh weight of substrate-cultivated lettuce grown in a solar greenhouse under normal water and fertilizer conditions. A model predictive control (MPC)-based approach [20] for vineyard spraying was presented to adapt to different vine row structures and suitable for real-time applications. Additionally, a control system [2] for an electrically driven precision maize seeder based on the CANopen protocol was designed. An obstacle avoidance strategy [18] based on the improved artificial potential field method is proposed for an autonomous navigation pusher robot.

The third category has seven papers under the following sub-heading: Designs related to intelligent machinery execution [3,13,15,25]. Zhang et al. [25] designed a pruning manipulator with five degrees of freedom for jujube trees. It is of reference value to solve the problems of poor working conditions and the labor intensity of manually pruning jujube trees. Chen et al. [3] developed a fin ray structure-based soft gripper mechanical model and its real-time servo-driven control strategy to reduce the potential danger of damage to the apple pericarps during robotic harvesting. Yu et al. [15] designed a cotton seeder duckbill welding robot to improve the automation, welding efficiency, and welding quality of duckbill welding of cotton seeds.

The final category has four papers under the following sub-heading: Systematic solutions [5,6,19,21]. Emmi et al. [5] presented an architecture to integrate the different components of an autonomous robot that provides access to the cloud, taking advantage of the services provided regarding data storage, scalability, accessibility, data sharing, and data analytics. Huang et al. [6] presents a new in-field interactive cognition phenotyping paradigm, and a humanoid robot equipped with image-acquiring sensory devices is designed containing an intuitive remote control for field phenotyping manipulations; subsequently, an attentional residual network (AtResNet) is proposed for rice tiller number recognition. The paper by Vasconcelos et al. [19] proposed a demo of agricultural field image data acquisition with a low-cost autonomous robot.

Precision agriculture, which addresses the spatial and temporal variability of soils and crops to reduce agricultural inputs and improve agricultural production reporting, varies greatly in implementation from country to country. Vrchota et al. [21] evaluated precision agriculture technologies' practical use in agricultural enterprises in the Czech Republic, which is a reference for the development and implementation of precision agriculture technology and equipment in each country.

In summary, this Special Issue highlights different approaches in the development of agricultural robots and intelligent agricultural machines in several agricultural application scenarios for scene and object perception, intelligent decision support methods, and operational mechanisms and their control. It is expected that the insights derived from this Special Issue will be useful to researchers related to the field of agricultural robots and autonomous machines.

Author Contributions: Conceptualization, J.Y.; investigation, W.J. and Q.F.; writing—original draft preparation, J.Y.; writing—review and editing, W.J. and Q.F.; All authors have read and agreed to the published version of the manuscript.

Funding: This material is based upon work that is supported by the National Natural Science Foundation of China (Grant No. 52275262).

Acknowledgments: We would like to sincerely thank all of the authors who submitted papers to the Special Issue of *Agriculture* entitled "Robots and Autonomous Machines for Agriculture Production", the reviewers of these papers for their constructive comments and thoughtful suggestions, and the editorial staff of *Agriculture*.

Conflicts of Interest: The authors declare no conflict of interest.

References

1. Bai, S.; Yuan, Y.; Niu, K.; Shi, Z.; Zhou, L.; Zhao, B.; Wei, L.; Liu, L.; Zheng, Y.; An, S.; et al. Design and Experiment of a Sowing Quality Monitoring System of Cotton Precision Hill-Drop Planters. *Agriculture* **2022**, *12*, 1117. [CrossRef]
2. Chen, J.; Zhang, H.; Pan, F.; Du, M.; Ji, C. Control System of a Motor-Driven Precision No-Tillage Maize Planter Based on the CANopen Protocol. *Agriculture* **2022**, *12*, 932. [CrossRef]
3. Chen, K.; Li, T.; Yan, T.; Xie, F.; Feng, Q.; Zhu, Q.; Zhao, C. A Soft Gripper Design for Apple Harvesting with Force Feedback and Fruit Slip Detection. *Agriculture* **2022**, *12*, 1802. [CrossRef]
4. Chen, X.; Xiong, Y.; Dang, P.; Tao, C.; Wu, C.; Zhang, E.; Wu, T. A Real-Time Shrimp with and without Shells Recognition Method for Automatic Peeling Machines Based on Tactile Perception. *Agriculture* **2023**, *13*, 422. [CrossRef]
5. Emmi, L.; Fernández, R.; Gonzalez-De-Santos, P.; Francia, M.; Golfarelli, M.; Vitali, G.; Sandmann, H.; Hustedt, M.; Wollweber, M. Exploiting the Internet Resources for Autonomous Robots in Agriculture. *Agriculture* **2023**, *13*, 1005. [CrossRef]
6. Huang, Y.; Xia, P.; Gong, L.; Chen, B.; Li, Y.; Liu, C. Designing an Interactively Cognitive Humanoid Field-Phenotyping Robot for In-Field Rice Tiller Counting. *Agriculture* **2022**, *12*, 1966. [CrossRef]
7. Ji, W.; Pan, Y.; Xu, B.; Wang, J. A Real-Time Apple Targets Detection Method for Picking Robot Based on ShufflenetV2-YOLOX. *Agriculture* **2022**, *12*, 856. [CrossRef]
8. Latif, S.; Lindbäck, T.; Karlberg, M.; Wallsten, J. Bale Collection Path Planning Using an Autonomous Vehicle with Neighborhood Collection Capabilities. *Agriculture* **2022**, *12*, 1977. [CrossRef]
9. Li, Y.; Feng, Q.; Lin, J.; Hu, Z.; Lei, X.; Xiang, Y. 3D Locating System for Pests' Laser Control Based on Multi-Constraint Stereo Matching. *Agriculture* **2022**, *12*, 766. [CrossRef]
10. Liu, A.; Xiang, Y.; Li, Y.; Hu, Z.; Dai, X.; Lei, X.; Tang, Z. 3D Positioning Method for Pineapple Eyes Based on Multiangle Image Stereo-Matching. *Agriculture* **2022**, *12*, 2039. [CrossRef]
11. Liu, C.; Feng, Q.; Tang, Z.; Wang, X.; Geng, J.; Xu, L. Motion Planning of the Citrus-Picking Manipulator Based on the TO-RRT Algorithm. *Agriculture* **2022**, *12*, 581. [CrossRef]
12. Liu, J.; Yuan, J.; Cui, J.; Liu, Y.; Liu, X. Contour Resampling-Based Garlic Clove Bud Orientation Recognition for High-Speed Precision Seeding. *Agriculture* **2022**, *12*, 1334. [CrossRef]

13. Liu, J.; Liu, X.; Jiang, Y.; Zhou, X.; Zhang, L.; Wang, X. Research on the Adaptability of High-Performance Film for Full Recycling to the Curl-Up Film Collecting Method. *Agriculture* **2022**, *12*, 1051. [CrossRef]
14. Liu, L.; Yuan, J.; Gong, L.; Wang, X.; Liu, X. Dynamic Fresh Weight Prediction of Substrate-Cultivated Lettuce Grown in a Solar Greenhouse Based on Phenotypic and Environmental Data. *Agriculture* **2022**, *12*, 1959. [CrossRef]
15. Ren, Y.; Guo, W.; Wang, X.; Hu, C.; Wang, L.; He, X.; Xing, J. Design and Test of Duckbill Welding Robot for Cotton Seeder. *Agriculture* **2023**, *13*, 31. [CrossRef]
16. Shi, C.; Cheng, Y.; Zhang, C.; Yuan, J.; Wang, Y.; Jiang, X.; Zhu, C. Wavelet Scattering Convolution Network-Based Detection Algorithm on Nondestructive Microcrack Electrical Signals of Eggs. *Agriculture* **2023**, *13*, 730. [CrossRef]
17. Shi, C.; Wang, Y.; Zhang, C.; Yuan, J.; Cheng, Y.; Jia, B.; Zhu, C. Nondestructive Detection of Microcracks in Poultry Eggs Based on the Electrical Characteristics Model. *Agriculture* **2022**, *12*, 1137. [CrossRef]
18. Tian, F.; Wang, X.; Yu, S.; Wang, R.; Song, Z.; Yan, Y.; Li, F.; Wang, Z.; Yu, Z. Research on Navigation Path Extraction and Obstacle Avoidance Strategy for Pusher Robot in Dairy Farm. *Agriculture* **2022**, *12*, 1008. [CrossRef]
19. Vasconcelos, G.J.Q.; Costa, G.S.R.; Spina, T.V.; Pedrini, H. Low-Cost Robot for Agricultural Image Data Acquisition. *Agriculture* **2023**, *13*, 413. [CrossRef]
20. Vatavuk, I.; Vasiljević, G.; Kovačić, Z. Task Space Model Predictive Control for Vineyard Spraying with a Mobile Manipulator. *Agriculture* **2022**, *12*, 381. [CrossRef]
21. Vrchota, J.; Pech, M.; Švepešová, I. Precision Agriculture Technologies for Crop and Livestock Production in the Czech Republic. *Agriculture* **2022**, *12*, 1080. [CrossRef]
22. Wang, Y.; Geng, C.; Zhu, G.; Shen, R.; Gu, H.; Liu, W. Information Perception Method for Fruit Trees Based on 2D LiDAR Sensor. *Agriculture* **2022**, *12*, 914. [CrossRef]
23. Xu, B.; Cui, X.; Ji, W.; Yuan, H.; Wang, J. Apple Grading Method Design and Implementation for Automatic Grader Based on Improved YOLOv5. *Agriculture* **2023**, *13*, 124. [CrossRef]
24. Yang, S.; Jia, B.; Yu, T.; Yuan, J. Research on Multiobjective Optimization Algorithm for Cooperative Harvesting Trajectory Optimization of an Intelligent Multiarm Straw-Rotting Fungus Harvesting Robot. *Agriculture* **2022**, *12*, 986. [CrossRef]
25. Zhang, B.; Chen, X.; Zhang, H.; Shen, C.; Fu, W. Design and Performance Test of a Jujube Pruning Manipulator. *Agriculture* **2022**, *12*, 552. [CrossRef]
26. Zhou, J.; Geng, S.; Qiu, Q.; Shao, Y.; Zhang, M. A Deep-Learning Extraction Method for Orchard Visual Navigation Lines. *Agriculture* **2022**, *12*, 1650. [CrossRef]

agriculture

Article

Design and Experiment of a Sowing Quality Monitoring System of Cotton Precision Hill-Drop Planters

Shenghe Bai [1,2], Yanwei Yuan [1,2,*], Kang Niu [2], Zenglu Shi [3], Liming Zhou [2], Bo Zhao [2], Liguo Wei [2], Lijing Liu [1,2], Yuankun Zheng [1], Sa An [2] and Yihua Ma [2]

1 College of Engineering, China Agricultural University, Beijing 100083, China; baishenghe@caams.org.cn (S.B.); liulijing@caams.org.cn (L.L.); zhengyuankun@caams.org.cn (Y.Z.)
2 The State Key Laboratory of Soil Plant and Machine System Technology, China Academy of Agricultural Mechanization Sciences Group Co., Ltd., Beijing 100083, China; niukang@caams.org.cn (K.N.); zhouliming@caams.org.cn (L.Z.); zhaobo@caams.org.cn (B.Z.); weiliguo@caams.org.cn (L.W.); ansa@caams.org.cn (S.A.); mayihua@caams.org.cn (Y.M.)
3 College of Mechanical and Electrical Engineering, Xinjiang Agricultural University, Urumqi 830052, China; b20213070561@cau.edu.cn
* Correspondence: yyw@caams.org.cn; Tel.: +86-135-2232-6652

Abstract: To realize the real-time monitoring of the cotton precision seeding operation process and improve the intelligence level of cotton precision planters, based on automatic color matching detection technology and visualization technology, this study designs a monitoring system for the sowing quality of cotton precision planters. The monitoring system is based on the double-silo turntable type cotton vertical disc hole seed metering device as the research carrier, and is composed of a missed seeding monitoring module and a visualization module. Among them, the missed seeding monitoring module includes an incremental rotary encoder, color code electric eye color fiber optic sensor, color code sensor amplifier, etc.; the visualization module includes data acquisition module, industrial computer, and so on. The missing seeding monitoring module is installed on the seed spacer of the cotton precision seed metering device. It uses Labview software for graphical programming and is equipped with a multi-functional industrial computer. It realizes the monitoring of parameters such as the number of sowings, the number of missed sowings, the speed of the hole seeder, the forward speed of the machine, and the sowing area. The results of the bench test and field test of the sowing monitoring system showed that the accuracy rate of the system's broadcast monitoring was over 93%, and the accuracy rate of missed broadcast monitoring was over 91%. The system solved the technical problem that cotton film-laying and sowing were not easy to detect. It could accurately detect the quality of cotton sowing in real time and meet the actual requirements of sowing monitoring.

Keywords: cotton precision planter; cotton seeds; broadcast monitoring; missed broadcast monitoring; sowing quality

Citation: Bai, S.; Yuan, Y.; Niu, K.; Shi, Z.; Zhou, L.; Zhao, B.; Wei, L.; Liu, L.; Zheng, Y.; An, S.; et al. Design and Experiment of a Sowing Quality Monitoring System of Cotton Precision Hill-Drop Planters. *Agriculture* **2022**, *12*, 1117. https://doi.org/10.3390/agriculture12081117

Academic Editors: Jin Yuan, Wei Ji and Qingchun Feng

Received: 19 June 2022
Accepted: 11 July 2022
Published: 29 July 2022

1. Introduction

Cotton precision sowing is the key to realize mechanized cotton planting. Sowing quality directly affects crop growth and yield. Among them, missed sowing is an important factor affecting sowing quality [1,2]. Therefore, breaking through the sowing quality monitoring technology is a research hotspot in the current sowing field. It can provide key support for technical development, such as for real-time adjustment of sowing amount, real-time reseeding of missed sowing, and machine operation management systems. This has important practical significance for improving the informatization of cotton precision sowing operations and promoting the quality development of mechanized sowing operations [3,4].

At present, it is in a closed state during the sowing process. The sowing quality cannot be directly monitored by human senses alone [5–7]. Therefore, it is particularly important to develop a seeding quality monitoring system. The research and application of precision seeder monitoring systems abroad began in the 1940s. Nowadays, the research on sowing quality monitoring technology is mature. The monitoring devices matching with seeders are widely used [2]. Precision Planting of the United States developed a 20/20 SeedSense monitoring system using WaveVision particle sensors. It could monitor the seeding quality in real time, and had the function of automatically correcting the seeding performance [8]. The Sistema Full Semina precision seeding system developed by MC electronic in Italy could realize the sowing monitoring of large and medium-sized seeds [9]. The John Deere Precision Planter was equipped with a Seed Star monitor. Statistics and analysis were carried out in various graphics, so that the operator could grasp the seeding quality information in real time. It uploaded data to the information center to provide data support for subsequent operations [10]. For different crop seeds, based on photoelectric method, image recognition method, piezoelectric method, capacitive method, etc. [11–14], different forms of monitoring systems have been developed abroad to monitor the seeding process in real time. Foreign seeding monitoring equipment has been commercialized and has good performance. However, it is expensive and not suitable for domestic general seeding tools and working environment conditions.

The domestic research on sowing monitoring system started relatively late. However, scholars were also actively exploring and developing a planting monitoring system suitable for the actual situation in our country. Che Yu et al. [2] designed an infrared monitoring system for seeding quality. It could monitor the sowing count, missed sowing, and outage of sowing, and the monitoring accuracy rate could reach more than 95%. Sun et al. [15] adopted the non-blind area anti-dust monitoring technology of non-point source, which improved the adaptability and monitoring accuracy of the no-tillage planter monitoring system to the dusty environment. Zhou et al. [16,17] developed a series of seed metering performance monitoring systems based on the dielectric properties of seeds and using capacitance detection technology; it realized the seeding detection of corn, rice, and cottonseed. Ding et al. [18–22] realized real-time monitoring of seeding frequency and total amount of seeding based on the characteristic analysis of the collision signal between seeds and piezoelectric films. Based on machine vision and BP neural network technology, Tan et al. [23,24] realized the precise monitoring of the seeding amount in the holes, with an average accuracy rate of 94.4%. To sum up, the existing monitoring methods of sowing parameters mainly included photoelectric monitoring, capacitive monitoring, and high-speed camera monitoring. Among them, the photoelectric monitoring method was the most widely used, with the advantages of low cost, reliable performance, and easy maintenance [25]. The machine vision method could solve the problems of low measurement accuracy and low degree of automation. However, the system was relatively complex and the cost was high, which was not suitable for field production applications [26–28]. The capacitive type was simple and economical, easy to maintain, and capable of non-contact real-time measurement. However, it had weak anti-interference and unstable performance, so it was difficult to apply it to field agricultural production activities [29].

At present, the cotton precision seeders generally adopt the method of hole seeding. Different from the common seed metering device structure and seed metering method, it belongs to "zero-speed seeding" (the instantaneous speed of the seeds falling into the seedbed is close to zero relative to the ground) [30,31]. The above monitoring technology is difficult to use directly. Therefore, to realize real-time monitoring of the sowing quality of cotton precision planters, a method for monitoring sowing parameters based on a color-coded electric eye color fiber optic sensor is proposed. Labview is used to build and develop a seeding quality monitoring system, and bench tests and field performance tests are conducted. This method is expected to improve the quality of cotton precision sowing operations and meet the actual production needs of cotton.

2. Materials and Methods

2.1. Research Carrier

The double-storage rotary disc type cotton vertical disc hole seed metering device is used as the research carrier of the sowing quality monitoring system. It is mainly composed of a moving plate, a seed-taking plate, a seed-casting bin assembly, a seed spacer, and a core plate. The structure is shown in Figure 1. The working area is divided into 5 areas: a seed filling area, seed clearing area, transfer area, seed transfer area, and seed casting area. The specific working principle can be found in reference [32]. In the transit area, the installation position of the color-coded electric eye color fiber optic sensor is determined based on the structure of the seed spacer. It is necessary for it to face the hole of the seed tray.

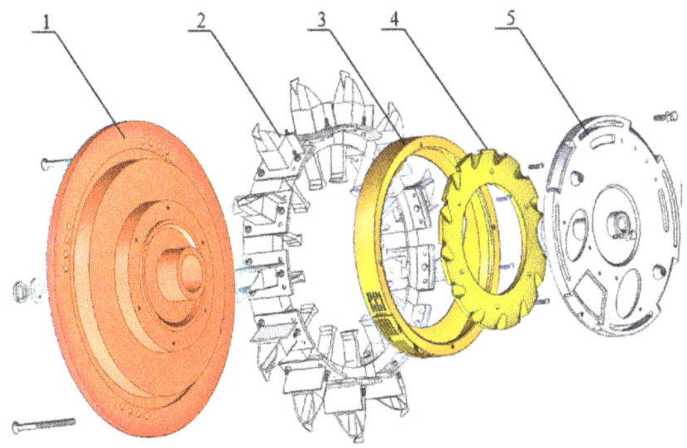

Figure 1. Structural diagram of double-bin rotary disc type cotton vertical disc hole seeding and metering device: (1) moving plate, (2) seed hopper assembly, (3) seed Spacer, (4) seed tray, and (5) core plate.

When the seed-taking tray enters this area, the seeds in the socket will slide along the inner wall of the socket into the seeding cavity composed of the seed-taking tray and the seed spacer. The color-coded electric eye color fiber optic sensor collects the RGB color of the cotton species, compares the RGB value of the reference color, and identifies the color. It obtains the seeding amount by calculating the change in the number of pulses; the seeding distance is obtained by multiplying the time interval between the two adjacent pulses identified by the forward speed of the implement. Missing seeding is obtained by comparing the actual seeding grain spacing with the expected grain spacing, and an alarm is given. After the seeds in the socket hole pass through the color-coded electric eye color fiber optic sensor, they slide along the inner wall of the socket hole into the seed rowing cavity composed of the seed taking plate and the seed spacer. After entering the seeding area again, the seeds in the warehouse slide into the duckbill along the spacer sleeve. After the mouthpiece is opened, the cotton seeds fall into the seed hole to complete the seeding operation. This provides the basis for the design of the subsequent seeding quality monitoring system.

2.2. Monitoring System Design

The seeding quality monitoring system mainly includes two modules: a missed seeding monitoring module and a visualization module. Among them, the missing-seeding monitoring module mainly realizes the real-time monitoring of the seeding amount and the missing-seeding situation, while the visualization module mainly realizes the visualization of the monitoring results of seeding quality. The system structure diagram is shown in Figure 2.

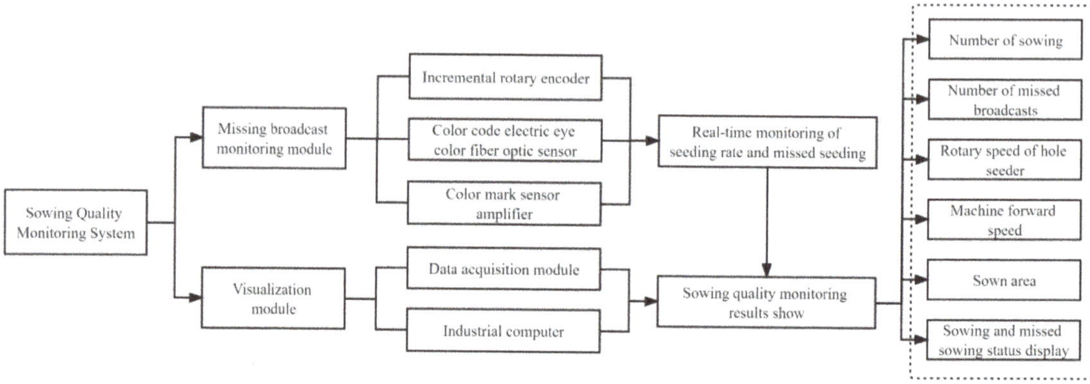

Figure 2. Structural diagram of monitoring system.

2.2.1. Hardware Design

(1) Missing broadcast monitoring module

The missing broadcast monitoring module includes an incremental rotary encoder, color code electric eye color fiber optic sensor, color code sensor amplifier, shielded cable, and so on. Cotton precision planters generally use hole seeding. The cotton seed slides directly into the duckbill from the seeding cavity. After the mouth is opened, it falls into the seed hole to complete the seeding operation. This greatly limits the detection space. That is, higher requirements are placed on the sensor installation location and detection method. Therefore, this study chooses the color-coded electric eye color fiber optic sensor (BV-501-RGB) and adopts the color automatic matching detection method. By comparing the two different RGB colors of the seed tray and the cotton seed, the light source is selected as three primary colors and a 4-element LED body. It solves the problem that cotton seeds are not easy to detect. The surface of the color-coded electric eye color fiber optic sensor is in the form of a cylindrical thread with a diameter of 6 mm. It is installed on the seed spacer and located in the transfer area, which requires facing the hole of the seed tray, as shown in Figure 3. It solves the problem that the sensor is not easy to install. It is connected to the color mark sensor amplifier through the shielded cable. The detection distance is 0~50 mm. It can adjust the detection height to control the spot size, ranging from 3 to 10 mm. It has the characteristics of a large light-receiving surface and multi-point collection, which can effectively reduce the misjudgment caused by uneven color.

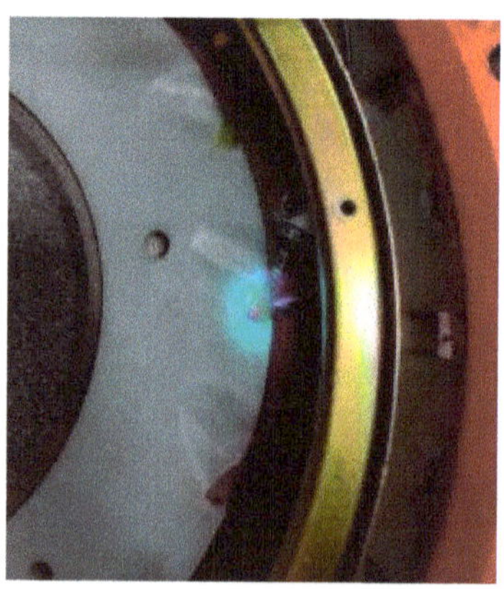

Figure 3. Sensor installation diagram.

The above sensors monitor whether there are cotton seeds in the hole of the seed-taking pan, and feed back the information of the sowing quality. Among them, the color-coded sensor amplifier is set in advance by aligning the color of the cotton with the color-coded electric eye color fiber optic sensor. The color is memorized: if it is within the set color threshold range, it is considered as normal seeding. A seeding status signal is sent.

An incremental rotary encoder (AB two-phase, 100 pulses) is installed on the moving plate. It is connected with the data acquisition module through the shielded wire, which has strong anti-interference and high cost performance. The rotating grating disc and the optocoupler generate a counting pulse signal that can identify the direction. It is transmitted to the industrial computer to calculate the speed of the hole seeder (that is, the forward speed of the machine).

(2) Visualization module

The visualization module includes a data acquisition module, industrial computer, etc. Among them, the data acquisition module is mainly used to receive the pulse signal collected by the incremental rotary encoder and the color mark sensor amplifier. It is uploaded to the industrial computer through RS485 communication. The industrial computer uses the Labview software to build the system, and the data is processed and displayed. The industrial computer adopts a multi-function integrated machine. The interface of the sowing quality monitoring system of the cotton precision seeder is shown in Figure 4. It can visually display parameters such as the number of seeds, the number of missed seeds, the speed of the hole seeder, the forward speed of the machine, and the sowing area. It realizes real-time display, alarm, and storage of data such as alarm information and data monitoring information. The system is equipped with a variety of detection modes suitable for various targets. It has the advantages of good handling of workpiece motion and vibration, and can meet the actual requirements of the sowing quality monitoring system.

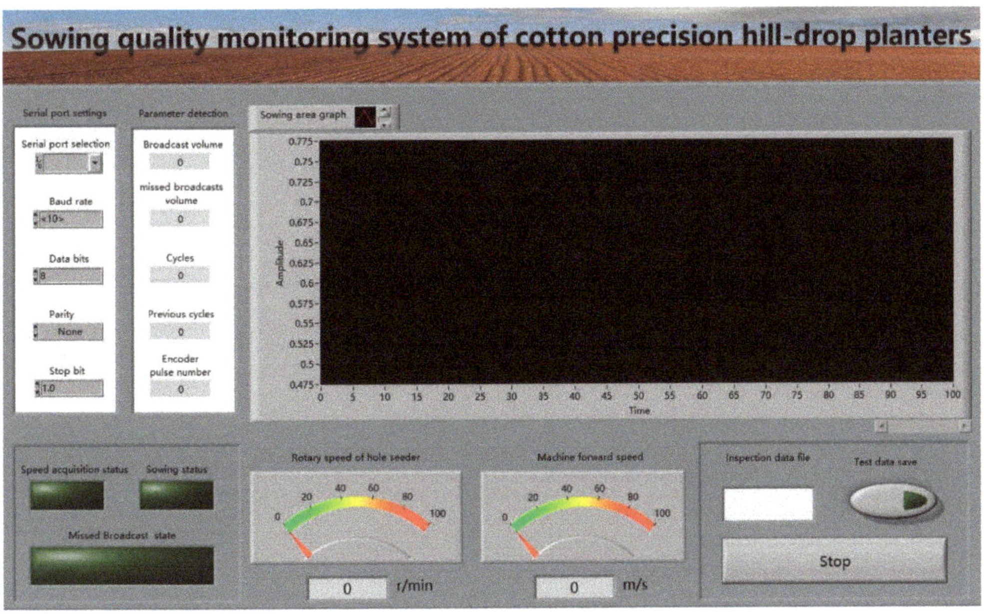

Figure 4. The interface of sowing quality monitoring system of cotton precision planter.

2.2.2. Monitoring System Software Design

This software program uses Labview software graphical programming to monitor parameters such as the number of sowings, the number of missed sowings, the speed of the hole seeder, the forward speed of the machine, and the sowing area. They are displayed on the corresponding controls on the interface in real time. It has the functions of alarming and storing data such as alarm information and data monitoring information (as shown in Figure 4). The software provides quick real-time understanding of cotton planting and missed planting status. It can solve the technical problem that cotton film-laying and sowing is not easy to detect, and effectively improve the operation quality and work efficiency of cotton sowing.

The system software flow chart is shown in Figure 5. Before the system starts, it performs an initial configuration and then sets system parameters. It obtains each parameter value through the corresponding protocol analysis. When starting the system, the color-coded electric eye color fiber optic sensor collects the RGB color of the cotton species. At the same time, the color mark sensor amplifier compares the reference color RGB value and the identification color to form a pulse signal. Incremental rotary encoders generate direction-identifiable counting pulse signals through rotating grating discs and optocouplers. They are transmitted to the industrial computer through the data acquisition module, and the forward speed of the machine and the seeding amount are calculated. Combined with the machine advance speed feedback data, it obtains the seeding grain distance (multiplied by the time interval between the two adjacent pulses identified by the implemented advance speed). Then, through the comparison between the actual seeding grain distance and the expected grain distance, the missed seeding can be judged, and the real-time alarm of missed seeding can be realized. Using Labview software, the industrial computer receives and processes sensor data in real time, and visualizes and monitors its parameters.

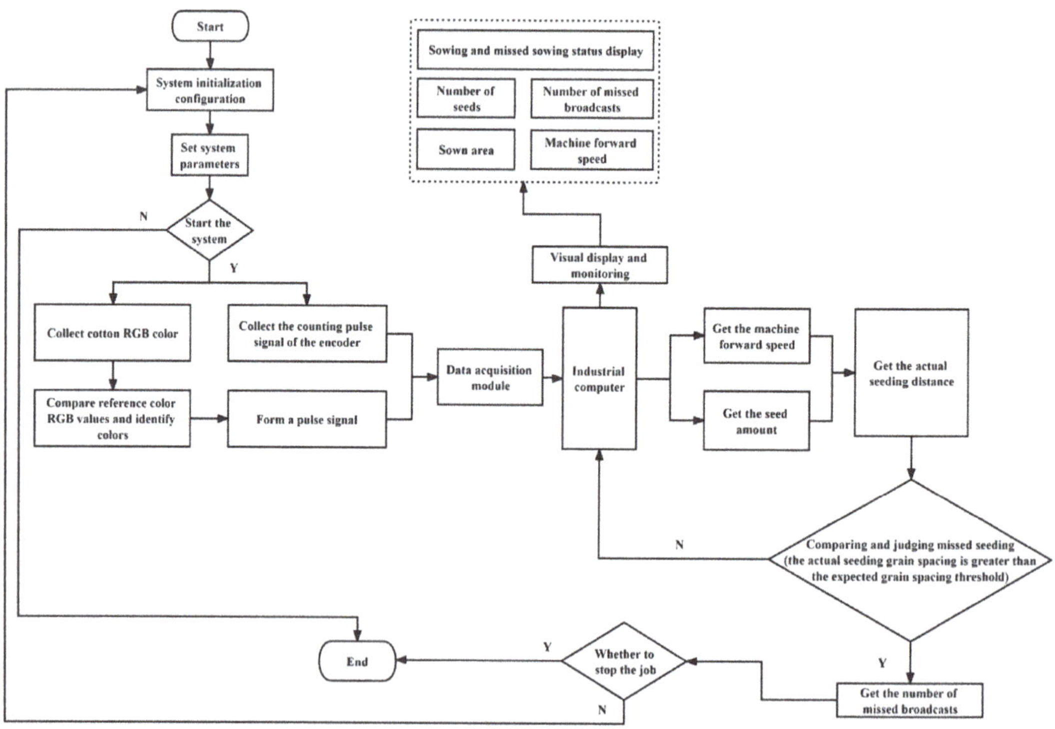

Figure 5. Program flow diagram of software system.

2.3. *Test Equipment and Methods*

2.3.1. Bench Test

In this study, the cotton seeds of Xinluzao 78 were used as the experimental material. Its thousand-grain mass was 84.40 g, and the moisture content was 4.45%. The average values of the maximum length, maximum width, and maximum thickness of the cotton seeds were 8.52, 4.50, and 4.48 mm, respectively. The standard deviations were 0.43, 0.32, and 0.41, respectively. The test equipment included a double-silo turntable test bench for the cotton vertical disc seed metering device, a stepper motor controller, and a cotton sowing quality monitoring system. To evaluate the accuracy of sowing quantity and missed sowing monitoring of the cotton sowing quality monitoring system, a bench test of the sowing quality monitoring system was designed. The test bench is shown in Figure 6.

Figure 6. Test bench of the monitoring system.

The purpose of the experiment was to evaluate the accuracy and stability of the sowing quality monitoring system of cotton precision planters. In the experiment, a double-silo turntable type vertical disc hole seeding and metering device was used. It set different rotational speeds for seed metering monitoring tests. The missed-seeding monitoring module was installed on the seed spacer. An inoculation bag was used to collect cotton seeds that had passed the missed-seeding monitoring module. By changing the speed of the seeding disc, the amount of seeding per unit time could be adjusted. In order to ensure that the seed meter works within the normal range of seed metering frequency, the rotation speeds were set to 20, 25, and 30 r/min [32]. It recorded the seed number displayed by the industrial computer. The actual number of cotton seeds in the inoculation bag was obtained by a manual method to calculate the actual sowing rate. It was compared with the monitoring value of the monitoring system to evaluate the accuracy of the system's sowing detection. The test was repeated 3 times, and the results were averaged to avoid errors.

2.3.2. Field Test

In order to investigate the stability and monitoring accuracy of the seeding monitoring system under field working conditions, a field trial of seeding monitoring was carried out in Tiemenguan City, Xinjiang on 20 March 2022. The test equipment was: a John Deere 804 tractor, Tiancheng 2MBJ-2/12 cotton precision film laying planter (Xinjiang Tiancheng Agricultural Machinery Manufacturing Co., Ltd., Tiemenguan City, Xinjiang, China, working width 4.4 m, laying 12 rows of film holes, weight 1300 kg), vertical disc hole seeding metering device, cotton sowing quality monitoring system, etc.

Before the sowing monitoring test, the influence of light conditions on the missed sowing monitoring module was examined [4]. No seeds were placed in this box, leaving the planter in a stationary state in the field. It connected the industrial computer with the sensing device and turned on the monitoring system. It was tested under conditions of sunlight, artificial lighting, and artificial blocking of natural light. The test results showed that under normal light conditions in the field, the monitoring system count was always 0. The normal light conditions in the field had no effect on the work of the monitoring device. In order to further test the effect of field monitoring of the monitoring system, an appropriate amount of cotton seeds were put into the seed box. It installed the 12-channel missing-seeding monitoring modules with different channel addresses and different channel numbers on the 12-channel cotton precision seed metering device. The tester held the cotton sowing monitor and showed that the industrial computer was sitting in the co-pilot position.

During the test, the sowing quality monitoring system of the cotton precision seeder was powered on. The data were cleared, and then the machine was started to move forward (Figure 7a). The speed of the seeding plate was controlled by the forward speed of the implement. This made the John Deere 804 one gear faster with a small throttle speed (about 2 km/h, the diameter of the hole seeder was 420 mm) as the base speed, which translated into a seeding disc speed of about 25 r/min. The rotation was controlled by adjusting the motor. The rotation speed was set to 3 levels, which were 20, 25, and 30 r/min. The forward distance of the machine was 100 m and the number of rotations was 75.8. There were 15 holes in each circle, with a total of 1137 holes. Monitoring of seeding and missed seeding was recorded. By looking at the cotton seeds discharged into the soil and counting them manually (Figure 7b), the actual seeding amount was counted. Compared with the results obtained by the monitoring system, this was used to evaluate the work stability and reliability of the monitoring system under field work. The test was repeated 3 times, and the results were averaged to avoid errors.

(a) (b)

Figure 7. Field sowing monitoring test site: (**a**) prototype field test and (**b**) artificial counting.

3. Results and Discussion

3.1. Bench Test

The seed breakage rate ranged from 0.10 to 0.20%, which was negligible. The monitoring results of the sowing amount at different rotational speeds are shown in Table 1 and Figure 8. In order to further evaluate the accuracy of the system's monitoring of the missed seeding amount, the amount of cotton seeds in the seed picking area of the hole seeder was artificially reduced at each rotation speed, so as to increase the missed seeding amount. According to the above method, the actual missed seeding amount on the seed bed belt was obtained. It was compared with the monitoring value of the monitoring system to evaluate the accuracy of the system's missed broadcast detection. Table 2 and Figure 9 show the monitoring results of the leakage amount at different rotational speeds.

Table 1. Sowing rate monitoring results of the seed meter at different speeds.

Seed Plate Speed/(r/min)	No.	Actual Broadcast Volume/Piece	Monitor Broadcast Volume/Piece	Accuracy Rate of Broadcast Monitoring/%
20	1	300	296	98.67%
	2	294	291	98.98%
	3	304	300	98.68%
	4	297	302	98.51%
	5	299	305	98.03%
	6	302	306	98.69%
25	1	306	298	97.39%
	2	300	294	98.00%
	3	298	304	98.03%
	4	302	296	98.01%
	5	302	308	98.05%
	6	303	295	97.36%
30	1	312	304	97.44%
	2	294	286	97.28%
	3	294	306	96.08%
	4	308	315	97.62%
	5	299	293	97.93%
	6	305	297	97.38%

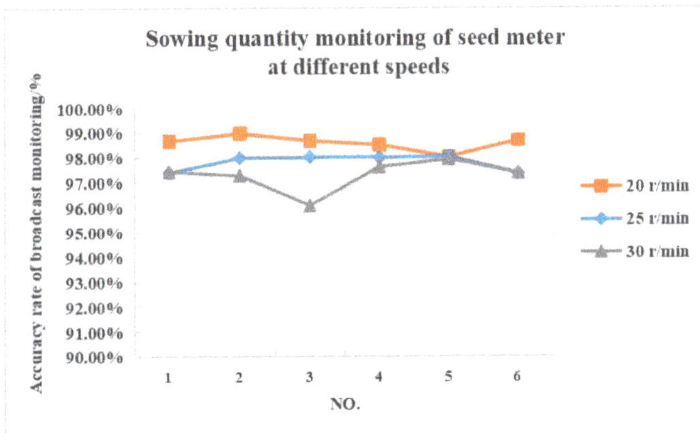

Figure 8. The sowing rate monitoring curve of the seed meter at different speeds.

Table 2. Missed broadcast monitoring results of the seed meter at different speeds.

Seed Plate Speed/(r/min)	No.	Actual Missed Broadcast/Piece	Monitor Broadcast Volume/Piece	Missed Broadcast Monitoring Accuracy/%
20	1	303	295	97.36%
	2	292	285	97.60%
	3	307	300	97.72%
	4	295	302	97.84%
	5	297	306	97.06%
	6	302	309	97.73%
25	1	305	295	96.72%
	2	303	294	97.03%
	3	299	310	96.45%
	4	301	291	96.68%
	5	297	308	96.30%
	6	303	295	97.36%
30	1	315	302	95.87%
	2	299	287	95.99%
	3	293	306	95.75%
	4	303	315	96.19%
	5	299	285	95.32%
	6	308	295	95.78%

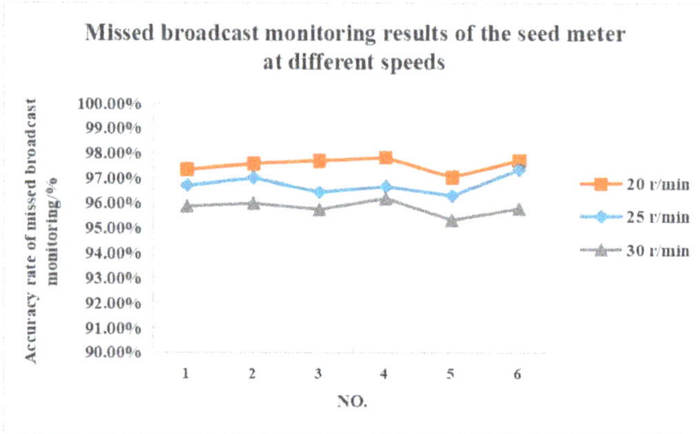

Figure 9. The missed broadcast monitoring curve of the seed meter at different speeds.

During the whole test process, there was no white crash screen and no wireless data transmission failure. It could be seen from Table 1 that with the increase of the rotation speed of the seed metering plate, the accuracy rate of sowing monitoring did not decrease significantly. The monitoring accuracy rate was above 97%. The reason for the slight decrease in accuracy was that the higher seeding speed would cause some smaller cotton seeds not to be fully filled into the hole, so that they could not be monitored by the sensor. However, the overall sowing rate monitoring accuracy of the system met the needs of cotton precision sowing monitoring. It can be seen from Table 2 that with the increase of the rotation speed of the seed metering plate, the monitoring accuracy of the missed seeding quantity decreased. The reason was that the high rotation speed of the seeding disc caused the seed flow to collide with the seeding tooth disc many times. This produced a small number of broken seeds, which caused the sensor to falsely detect. However, the monitoring accuracy of the system could still reach more than 95%. The monitoring accuracy of missed broadcasts met the requirements of the monitoring system. The above results show that the monitoring system could more accurately judge the seeding amount and the degree of missing seeding for the stable working seed meter. This could be used for evaluation and reference of sowing quality.

3.2. Field Test

The seed breakage rate ranged from 0.10 to 0.20%, which was negligible. The results of field sowing monitoring are shown in Table 3 and Figure 10. The field test results showed the following: The accuracy rate of broadcast monitoring was maintained above 93%. The accuracy rate of missed broadcast monitoring was maintained above 91%. The seeding monitoring system worked stably and reliably in the field under the rotating speed of the seeding disc that met the seeding requirements. Compared with the traditional monitoring method, the system could detect the seeding device of the hole seeder more accurately, and find the missed seed in time. This effectively saved the amount of cotton seeds, greatly reduced the cost of sowing, and met the requirements for monitoring the sowing status of cotton precision hole seeders.

Compared with the results of the bench test, the monitoring accuracy of the overall system was slightly reduced during the field test. The reason was that the cotton seeds used in the field test had not been selected manually. Some broken seeds and debris were included. The machine vibrated during operation. Moreover, a lot of dust in the field adhered to the color fiber optic sensor probe and affected the optical fiber transmittance. This subsequently calls for enhancement of the monitoring performance of the sensor and optimization of its installation position to ensure more stability. A condenser lens or a self-cleaning dust removal device should also be added to maintain a good light transmittance of the sensor and improve the environmental adaptability and operational reliability of the system. In the future, high-precision Beidou positioning technology and mobile Internet technology can be combined to provide support for the field sowing map, missed sowing state map, and variable reseeding prescription operations. The parameters of the seeder would also be automatically set, making the system function more in line with the actual production situation and suitable for different operating scenarios.

Table 3. Field sowing monitoring test results.

Seed Plate Speed/(r/min)	No.	Actual Broadcast Volume/Piece	Actual Missed Broad-casts/Piece	Monitor Broadcast Volume/Piece	Monitor the Amount of Missed Broad-casts/Piece	Accuracy Rate of Broadcast Monitoring/%	Accuracy Rate of Missed Broadcast Monitoring/%
20	1	1137	25	1107	26	97.36%	96.15%
	2	1137	24	1110	23	97.60%	95.83%
	3	1137	28	1111	27	97.72%	96.43%
	4	1137	22	1162	21	97.84%	95.45%
	5	1137	18	1171	19	97.06%	94.74%
	6	1137	30	1163	31	97.73%	96.77%
	7	1137	25	1121	24	98.59%	96.00%
	8	1137	21	1155	22	98.44%	95.45%
	9	1137	24	1111	25	97.71%	96.00%
	10	1137	28	1109	27	97.54%	96.43%
	11	1137	24	1168	23	97.35%	95.83%
	12	1137	29	1110	28	97.63%	96.55%
25	1	1137	33	1089	31	95.78%	94.55%
	2	1137	29	1187	27	95.79%	93.75%
	3	1137	35	1189	33	95.63%	94.29%
	4	1137	27	1187	25	95.79%	93.33%
	5	1137	22	1090	23	95.87%	94.74%
	6	1137	40	1075	37	94.55%	93.00%
	7	1137	27	1195	29	95.15%	93.75%
	8	1137	29	1081	27	95.07%	93.10%
	9	1137	29	1195	30	95.15%	96.00%
	10	1137	35	1088	33	95.69%	94.29%
	11	1137	35	1196	33	95.07%	94.29%
	12	1137	35	1090	37	95.87%	94.05%
30	1	1137	43	1078	40	94.81%	93.02%
	2	1137	35	1080	32	94.99%	93.75%
	3	1137	43	1200	40	94.75%	92.09%
	4	1137	40	1205	37	94.36%	92.50%
	5	1137	35	1074	27	94.46%	92.11%
	6	1137	48	1070	32	94.11%	91.67%
	7	1137	38	1077	36	94.72%	94.74%
	8	1137	35	1199	32	94.83%	93.10%
	9	1137	40	1068	43	93.93%	93.02%
	10	1137	46	1069	45	94.02%	92.00%
	11	1137	44	1199	41	94.83%	93.18%
	12	1137	42	1070	44	94.11%	91.49%

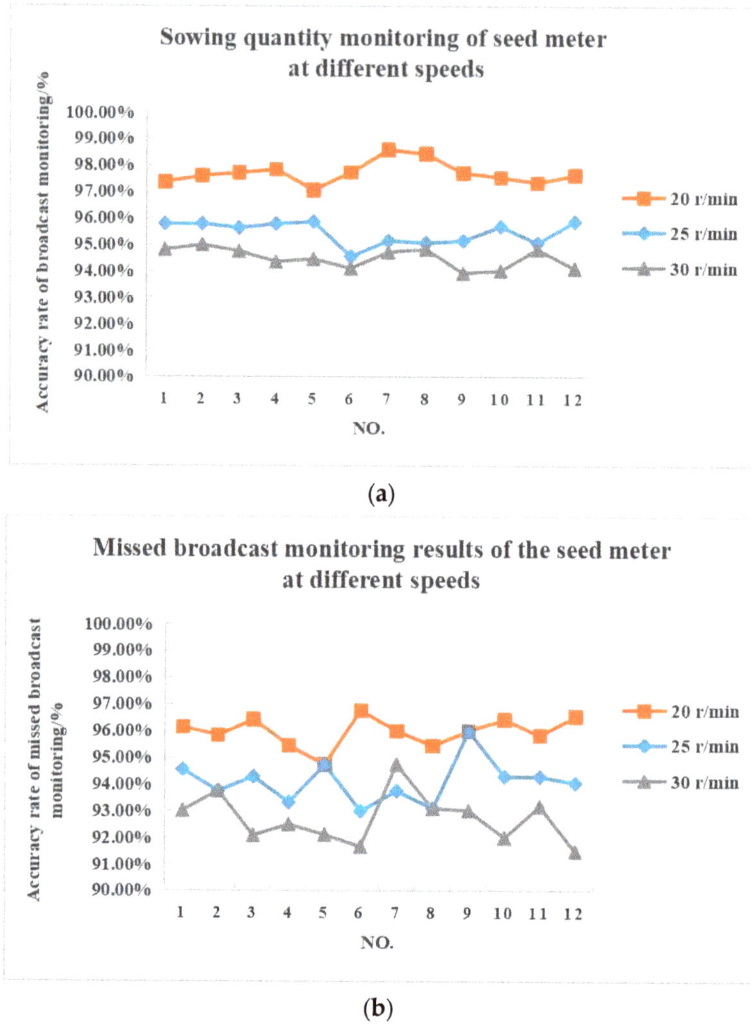

Figure 10. The field sowing monitoring curve of the seed meter at different speeds: (**a**) sowing rate monitoring and (**b**) missed broadcast monitoring.

4. Conclusions

In this study, a planting quality monitoring system suitable for cotton precision planters was designed. The bench performance test of sowing quantity monitoring and missing sowing detection was carried out on the monitoring system. A field test was carried out on the 2MBJ-12 cotton precision film laying planter to verify the reliability of the system. The bench test and field test of the sowing quality monitoring system of the cotton precision planter showed that the sowing quality monitoring system worked stably and reliably. In the bench test, the accuracy of sowing quantity monitoring was no less than 97%, while the accuracy of missing sowing monitoring was no less than 94%. In the field test, the accuracy rate of sowing rate monitoring was no less than 93%, while the accuracy of missed sowing monitoring was no less than 91%. This improved the work quality and work efficiency of cotton sowing. It met the requirements for monitoring the sowing quality of cotton precision hole seeders.

Author Contributions: Conceptualization, Y.Y. and B.Z.; methodology, S.B. and K.N.; software, S.B. and Z.S.; validation, S.B. and Y.Z.; formal analysis, K.N. and S.B.; investigation, S.A. and Y.M.; resources, Y.Y. and B.Z.; data curation, S.B. and L.W.; writing—original draft preparation, S.B.; writing—review and editing, K.N. and L.L.; visualization, Y.Z. and L.Z.; supervision, Y.Y. and B.Z.; project administration, Y.Y. and K.N.; funding acquisition, K.N. and L.Z. All authors have read and agreed to the published version of the manuscript.

Funding: This work was mainly supported by the Autonomous Region Regional Collaborative Innovation Special (Science and Technology Aid to Xinjiang Program) Project of China (2021E02055).

Data Availability Statement: Data are contained within the article. The data presented in this study can be requested from the authors.

Conflicts of Interest: The authors declare no conflict of interest.

References

1. Qiu, Z.M.; Zhang, W.P.; Zhao, B.; Ji, J.T.; Jin, X.; He, Z.T. Design and Test of Operation Quality Monitoring System for Small Grain Electric Seeder. *Trans. Chin. Soc. Agric. Mach.* **2019**, *4*, 77–83.
2. Che, Y.; Wei, L.G.; Liu, X.T.; Li, Z.L.; Wang, F.Z. Design and experiment of seeding quality infrared monitoring system for no-tillage seeder. *Trans. Chin. Soc. Agric. Eng.* **2017**, *33* (Suppl. S1), 11–16.
3. Wang, Z.M.; Pei, J.; He, J.; Zhang, M.H.; Yang, W.W.; Luo, X.W. Development of the sowing rate monitoring system for precision rice hill-drop drilling machine. *Trans. Chin. Soc. Agric. Eng.* **2020**, *10*, 9–16.
4. Ding, Y.C.; Chen, L.Y.; Dong, W.J.; Wang, W.C.; Liu, X.D.; Wang, K.Y.; Liu, W.P. Design and experiment of the sowing monitoring system for a wide-width rapeseed planter. *Trans. Chin. Soc. Agric. Eng.* **2021**, *13*, 38–45.
5. Hu, J.P.; Lu, L. Design of the image monitored control system for the magnetic type precision seeder. *Trans. Chin. Soc. Agric. Mach.* **2006**, *11*, 88–91.
6. Wang, T.J. *Experimental Study on the Monitoring Device for Missing Seeding of Rice Direct Seeding Machine*; Shenyang Agricultural University: Shenyang, China, 2017.
7. Liang, K.; Luo, H.Y.; Shen, M.X.; He, R.Y.; Zhang, L. Review and prospect for the detection technology of paddy seeding quality. *Acta Agric. Zhejiangensis* **2010**, *2*, 253–257.
8. 20/20 SeedSense Monitoring System. Available online: http://www.precisionplanting.com/Products/WaveVision/Default.aspx (accessed on 20 January 2022).
9. MC Electronics. Sistema Full Semina. Available online: https://www.mcelettronica.it/it/prodotti/semina/semina-a-righe/fotocellula-blockage-pro-seeder_256c28.html (accessed on 8 February 2022).
10. John, D. Monitoring and Documentation. Available online: http://www.deere.com/en_US/parts/parts_by_industry/ag/seeding/monitoring/monitoring.page (accessed on 19 December 2021).
11. Hadi, K.; Hossein, N.; Bahram, B.; Hossein, B.; Iraj, E. A practical approach to comparativedesign of non-contact sensing techniques for seed flow rate detection. *Comput. Electron. Agric.* **2017**, *142*, 165–172.
12. Karayel, D.; Wiesehoff, M.; Ozmerzi, A.; Muller, J. Laboratory measurement of seed drillseed spacing and velocity of fall of seeds using high-speed camera system. *Comput. Electron. Agric.* **2006**, *2*, 89–96. [CrossRef]
13. Leemans, V.; Destain, M.F. A computer-vision based precision seed drill guidance assistance. *Comput. Electron. Agric.* **2007**, *59*, 1–12. [CrossRef]
14. Navid, H.; Ebrahimian, S.; Gassemzadeh, H.R.; Mousavi, M.J. Laboratory evaluation of seed metering device using image processing method. *Aust. J. Agric. Eng.* **2011**, *2*, 1–4.
15. Sun, Y.J.; Shen, J.X.; Dou, Q.Q.; Li, Q.L.; Chen, G.; Sun, Y.T. Design and test of monitoring system of no-tillage planter based on Cortex-M3 processor. *Trans. Chin. Soc. Agric. Mach.* **2018**, *8*, 50–58.
16. Zhou, L.M.; Wang, S.M.; Zhang, X.C.; Yuan, Y.W.; Zhang, J.N. Seed monitoring system for corn planter based on capacitance signal. *Trans. Chin. Soc. Agric. Eng.* **2012**, *13*, 16–21.
17. Zhou, L.M.; Li, S.J.; Zhang, X.C.; Wang, S.M.; Yuan, Y.W.; Dong, X. Detection of seed cotton mass flow based on capacitance approach. *Trans. Chin. Soc. Agric. Mach.* **2014**, *6*, 47–52.
18. Ding, Y.C.; Wang, X.L.; Liao, Q.X. Method of real-time loss sowing detection for rapeseed precision metering device based on time changed window. *Trans. Chin. Soc. Agric. Eng.* **2014**, *24*, 11–21.
19. Ding, Y.C.; Zhu, K.; Wang, K.Y.; Liu, X.D.; Du, C.Q. Development of monitoring device for medium and small size seed flow based on thin surface laser- silicon photocell. *Trans. Chin. Soc. Agric. Eng.* **2019**, *8*, 12–20.
20. Li, M.; Liu, X.H.; Ding, Y.C.; Liao, Q.X. Loss sowing test technology and equipment of rapeseed pneumatic precision seeder based on the seeding frequency. In Proceedings of the Annual Conference of Chinese Society of Agricultural Engineering, Chongqing, China, 22 October 2011; pp. 299–304.
21. Ding, Y.C.; Yang, J.Q.; Zhu, K.; Zhang, L.L.; Zhou, Y.W.; Liao, Q.X. Design and experiment on seed flow sensing device for rapeseed precision metering device. *Trans. Chin. Soc. Agric. Eng.* **2017**, *9*, 29–36.

22. Ding, Y.C.; Zhang, L.L.; Yang, J.Q.; Zhu, K. Sensing device improvement and communication design on sowing monitoring system of precision planter for rapeseed. *Trans. Chin. Soc. Agric. Eng.* **2018**, *14*, 19–26.
23. Tan, S.Y.; Ma, X.; Wu, L.L.; Li, Z.H.; Liang, Z.W. Estimation on hole seeding quantity of super hybrid rice based on machine vision and BP neural network. *Trans. Chin. Soc. Agric. Eng.* **2014**, *21*, 201–208.
24. Zhao, Z.B.; Liu, Y.C.; Liu, Z.J.; Gao, B. Performance detection system of tray precision seeder based on machine vision. *Trans. Chin. Soc. Agric. Mach.* **2014**, *Z1*, 24–28.
25. Xie, C.J.; Yang, L.; Zhang, D.X.; Cui, T.; Zhang, K.L. Seeding parameter monitoring method based on laser sensors. *Trans. Chin. Soc. Agric. Eng.* **2021**, *3*, 140–146.
26. Wu, Z.Q.; Liu, J.J.; Yang, X. The design of JPS-16 seeder test-bed. *J. Agric. Mech. Res.* **2011**, *10*, 59–62.
27. Zhang, H.N.; Ma, Y.F. Research on performance testing method of machine sowing based on image processing. *Meas. Control. Technol.* **2015**, *2*, 44–47.
28. Chen, J.; Bian, J.; Li, Y.M.; Zhao, Z.; Wang, J.L. Performance detection experiment of precision seed metering device based on high speed camera system. *Trans. Chin. Soc. Agric. Eng.* **2009**, *9*, 90–95.
29. Zhou, L.M.; Zhang, X.C. Monitor system of precision seeder based on capacitive sensor. *J. Agric. Mech. Res.* **2009**, *11*, 37–39.
30. Tong, C. Zero-speed seeding technology and its theoretical design. *Mech. Res. Appl.* **1995**, *1*, 16–18+25.
31. Yuan, Y.W.; Bai, H.J.; Fang, X.F.; Wang, D.C.; Zhou, L.M.; Niu, K. Research Progress on Maize Seeding and Its Measurement and Control Technology. *Trans. Chin. Soc. Agric. Mach.* **2018**, *9*, 1–18.
32. Zhang, X.J.; Chen, Y.; Shi, Z.L.; Jin, W.; Zhang, H.T.; Fu, H.; Wang, D.J. Design and experiment of double-storage turntable cotton vertical disc hole seeding and metering device. *Trans. Chin. Soc. Agric. Eng.* **2021**, *19*, 27–36.

agriculture

MDPI

Article

Control System of a Motor-Driven Precision No-Tillage Maize Planter Based on the CANopen Protocol

Jincheng Chen [1], Hui Zhang [1], Feng Pan [1], Mujun Du [2] and Chao Ji [1,*]

[1] Institute of Mechanical Equipment, Xinjiang Academy of Agricultural and Reclamation Science, Shihezi 832003, China; shznkycjc@163.com (J.C.); zhanghxj2021@163.com (H.Z.); 17609922366@163.com (F.P.)
[2] Heilongjiang dewo Science & Technology Development Co., Ltd., Harbin 150030, China; dumujun@163.com
* Correspondence: jicobear@163.com

Abstract: To reduce the cost of machinery and manual operation, greatly improve the efficiency of maize sowing, and solve the problems of slow sowing speed, unstable operation quality, and the difficult monitoring of the sowing process of traditional seeders, a control system for an electrically driven precision maize seeder based on the CANopen protocol was designed. In this system, an STM32 is used as the main controller, and the vehicle terminal is used to set the operating parameters, such as the spacing of sowing plants and the number of holes in the metering plate. The GPS receiver is used to collect the forward speed of the tractor. An infrared photoelectric sensor is used to monitor the working state of the seeder. In this study, tests were conducted on different evaluation indices. The results showed that the detection accuracy of the photoelectric sensor reached 99.8% and the fault alarm rate reached 100%. The qualified rate of sowing was more than 91.0%. Based on indoor test results, the qualified rate was higher when the grain spacing was larger. The field test showed, in terms of the seeding performance, that the control system had good stability. When the grain spacing was set to 20 cm and the operating speed was 6~12 km/h, the qualified index was more than 89% and the reseeding index was less than 1.93%. The variation in sowing performance between different monomers was small, and the seeding performance was good. The control system helps to improve the performance of the seeder.

Keywords: precision planter; motor-driven; CANopen protocol; photoelectric sensor; no-tillage

Citation: Chen, J.; Zhang, H.; Pan, F.; Du, M.; Ji, C. Control System of a Motor-Driven Precision No-Tillage Maize Planter Based on the CANopen Protocol. *Agriculture* **2022**, *12*, 932. https://doi.org/10.3390/agriculture12070932

Academic Editors: Jin Yuan, Wei Ji and Qingchun Feng

Received: 30 May 2022
Accepted: 24 June 2022
Published: 28 June 2022

1. Introduction

Maize is the largest food crop in China and occupies an extremely important position in the whole agricultural planting system. In the new era, with the rapid development of China's economy, the actual demand for corn has increased greatly [1,2]. Changes in corn supply and demand have a great influence on maintaining national food security and stabilizing the grain market and supply [3]. In recent years, with the major management mode of agricultural production gradually developing to a large-scale and intensive direction, to effectively ensure the cultivated area and grain production task and the completion of sowing operations in high-yield periods, higher quality requirements have been proposed for maize precision sowing technology [4–6]. Facing the higher cost-savings and efficiency requirements of farmers and the more urgent demand for agricultural time, the deficiencies of ground wheel drive are increasingly prominent [7,8]: (1) Low operation efficiency. At present, the operation of precision planters in China is still at a low-speed level of 6~8 km/h. (2) Unstable sowing quality. Under high-speed operation, it is easy to bump and slip, resulting in a series of problems, such as missed sowing, reseeding, and poor uniformity of plant spacing. (3) It is hard to monitor the sowing process [9]. Traditional machines and tools operate in a closed environment, requiring auxiliary personnel to follow the machine and observe, which is not only labor-intensive and costly but also easily causes personal injury, and the observation results make it difficult to eliminate the influence of human subjective factors. Sensor-based electronic metering systems can minimize t' lacunae of mechanical metering systems. The application of electronic seed metering a control systems in planters is required for better seed uniformity in the field [10].

Sound and sustainable agriculture without electronics is inconceivable today, as el tronic systems are used to reduce farm inputs, protect the environment, secure farm income, and produce high-quality products [11]. In the last few decades, a number of active seeding control and detection systems have been proposed to solve the above-mentioned problems. Yuan et al. [6] used prescription operation maps and GPS information, combined with speed, to drive a servo motor seed space and achieved precision planting that could be steplessly adjusted from 10 to 20 cm. Yang et al. [12] designed a mechatronic driving system. Compared to the mechanical driving system, the advantage of the mechatronic driving system is noticeable, especially when the forward speed is more than 11 km/h. Anil et al. [13,14] developed an electromechanical drive system (EMDS) for seed metering units of a classic single-seed planter to attain uniform seed spacing. EMDS realizes the recommended optimal seeding rate; the possibility for fast and simple setting, synchronization, and real-time control; the ability to work at higher speeds; single movement; and the control of each metering unit. The dynamic relationship model between the speed of the tractor and the speed of the metering plate is established to ensure the accurate matching of the tractor time and the seed entry to better realize seed spacing consistency. Ding et al. [15] proposed a control system of a motor-driven precision maize planter based on GPS speed measurements. At the same plant spacing and operating speed, the variation coefficient of the GPS velocity measurement method is smaller than that of the encoder velocity measurement method. At a high speed of 12 km/h, the average qualified GPS index is 14.32% higher than that of the encoder. This shows that the GPS velocity measurement method is more suitable for high-speed operation. Li et al. [16] resolved the problem that GPS receivers cannot meet the requirement of precision seeding at low speed based on a Kalman filter.

Variable-rate seeding (VRS) technology can adjust the seed input according to regional soil differences, ensure the most suitable plant density, make full use of nutrients and moisture in the soil, and exert the maximum yield potential in specific soil regions, thus significantly increasing yield and reducing cost. He et al. [17,18] developed a low-cost VRS control system based on a controller area network (CAN) bus and developed a compensation algorithm for seeding lag (CASL) that could decrease the seeding lag distance immensely. The developed VRS control system was capable of flexibly expanding planter rows and independently controlling each row's seeding rate. Ding et al. [19] developed a variable rate planter row-unit driver for maize. The overall test results of the row-unit driver confirmed that it could realize the functions of seed metering, seeding quality detection, and CAN communication with the main controller.

To improve the seeding uniformity of a maize planter, He et al. [20] designed a GPS-based turn compensation algorithm to offset the seeding rates of planter units. Field experiments indicated that a four-row planter equipped with the developed turn compensation control system had seeding accuracies (above 97%) and seeding coefficients of variation (below 1.52%) values better than those of a noncompensation planter under equivalent working conditions. To find the problem of seeding blockage and missing seeding in time, Meng et al. [21] developed a monitoring system to solve the phenomenon of maize precision seeding machines in operation and to improve the economy and efficiency of seeding. Xie et al. [22] conducted a study testing the accuracy of the sensor to monitor the seeding parameters of a precision metering device under different seeding speeds and seeding spacings. Improving the accuracy of the sensor's monitoring of the seed passing frequency is of great help in improving the seeding monitoring accuracy under the conditions of high seeding speed and small seeding spacing. Xie et al. [7] developed a precision seeding parameter monitoring system based on laser sensors Field tests showed that the average monitoring error of the seeding quantity was less than 1%, and the average monitoring error of the seeding qualified rate was less than 1.5%. The monitoring system could trigger an alarm in time when the seeder had a missing seed fault.

In summary, most field experiments involving the seeder use four-row or six-row mechanical seeders. For the eighteen-row air suction seeder, in this study, an electrically driven precision sowing control system based on the CANopen protocol was designed, and a circuit board integrating the motor drive and sowing quality detection was developed. A seeding parameter dictionary with the CANopen protocol was constructed. A separable trapezoidal integral proportional integral derivative (PID) control algorithm was used to match the tractor speed and motor speed. In this paper, the performance of the control system was evaluated by laboratory bench and field tests.

2. Materials and Methods

2.1. System Components

The proposed maize precision planter system consisted of a monitoring subsystem and a mechanical device system. As shown in Figure 1, the required hardware components included a 12 V DC power supply, an on-board computer with a CAN bus (eMT3070B, Weintek Technology Co., Ltd., New Taipei, China), an in-house-designed integrated controller based on STM32F103VET6, an infrared monitoring sensor (Shandong Zhucheng Dilico Automotive Electronics Co., Ltd., Weifang, China), an inertial and satellite navigation module (WTGPS-200 WitMotion Shenzhen Co., Ltd., Shenzhen, China), brushed DC motors, and in-house-designed motor speed measurement modules. The mechanical part included a reducer, a planter plate, and a seed tube. The motor was used as an intermediary to integrate the control system and mechanical part.

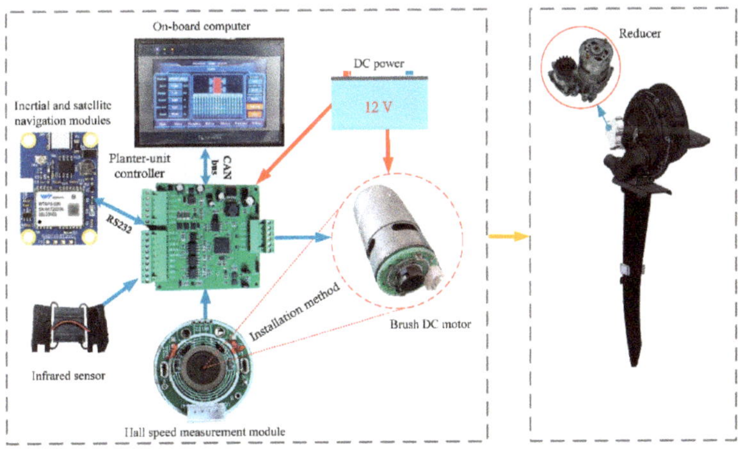

Figure 1. Planter monitoring system.

In detail, the on-board computer communicated with the controller via a CAN bus and was used for setting the seed spacing, current threshold, and width; monitoring the various working states of the system (such as the motor current and rotational speed); and controlling the start and stop of a single motor. To reduce field wiring, in this paper, the controller was integrated with the motor drive and CAN communication, which was mounted on each planter unit, to expand flexibly based on the planter row number and to adjust the motor speed to achieve the desired seeding rate. In this study, speed acquisition was performed through inertial and satellite navigation modules with a velocity accuracy of 0.05 m/s, and bidirectional credit guaranteed communication with the controller through RS232. A brush DC motor was utilized to drive the seed meter at a desired speed, and a hall speed measurement module for the brushless DC motor was developed, nested on the shaft side of the motor, and a pulse was generated by an interaction with the magnetic ring on the motor shaft. Additionally, a photoelectric sensor with a large field of view, high sensitivity, and strong dust resistance was installed on the seed tube to monitor the state of falling seeds. A circuit schematic diagram of the system with STM32F103VET6 as the main controller is shown in Figure 2.

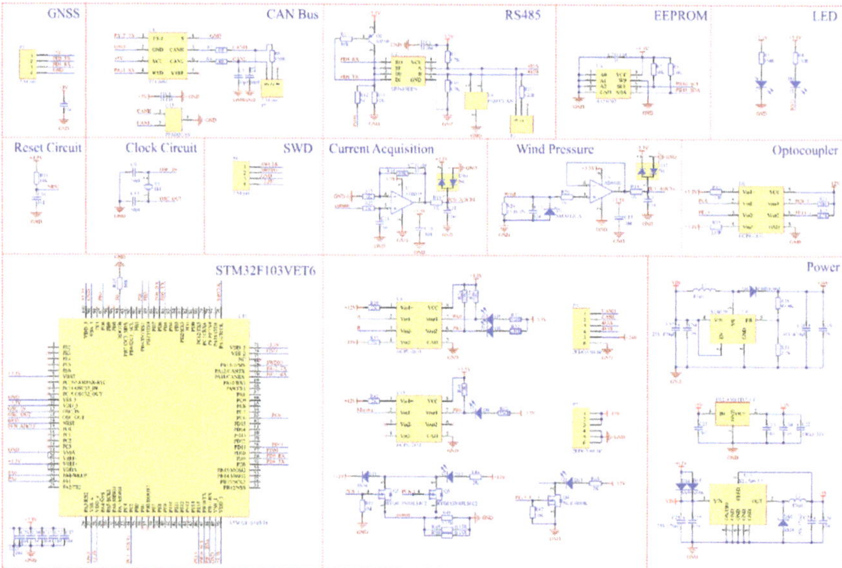

Figure 2. Schematic diagram of the system circuit.

2.2. *CAN Bus with CANopen Protocol*

The exchange of data packets in the system was based on a CAN bus. However, ISO 11,783 was specifically designed for tractor control system development [23–25]. In China, fewer products have been developed using subprotocols, especially in sensors and on-board computers. The on-board computer (eMT3070B) used in this paper had a CAN interface and conformed to the CANopen protocol. To test the designed system as soon as possible and to enhance the scalability at the present stage, the CANopen protocol was used as the basic protocol. CANopen is a high-level communication protocol based on the controller area network. It includes a communication subprotocol and a device subprotocol and has often been presented in embedded systems and industrial controls [26]. The CANopen protocol usually consists of three parts: a user application layer, an object dictionary, and communication. The core part is the object dictionary, which describes the relationship between the application object and the CANopen message. The user application layer in this paper refers to the application interface downloaded to the eMT3070B using EasyBuilder Pro development software provided by Weintek. Figure 3 shows the partial display interface design of the monitoring software for the eighteen-row seeding. In the communication layer, considering the field working environment, the well-established TJA1050 chip was selected as the transceiver of the CAN bus. This chip can work normally even with electrostatic interference and in voltage-mutating and high-noise environments and communicates with electricity.

To be stable, reliable, and controllable, the CANopen network needs to be set up with a network management master (NMT-Master) that controls the start and stop of all nodes. Communication between the on-board computer as the NMT host and the NMT slave via the NMT network management message is responsible for the layer management, network management, and ID distribution services. NMT management involves six states of a CANopen node following power-up: initializing, application reset, communication reset, preoperational, operational, and stopped. The NMT management state transition diagram is shown in Figure 4.

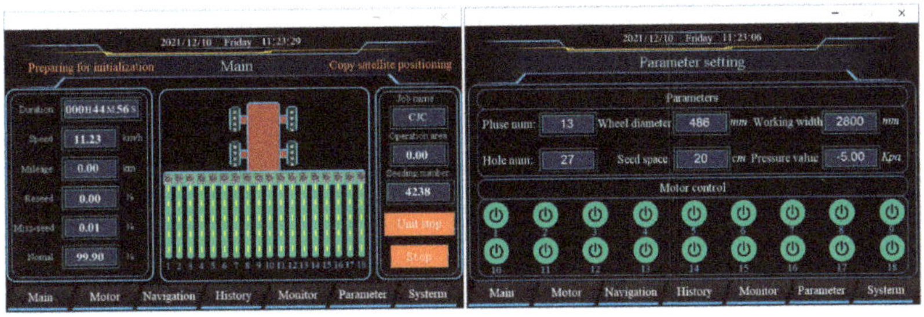

Figure 3. Partial human–computer interaction software interface.

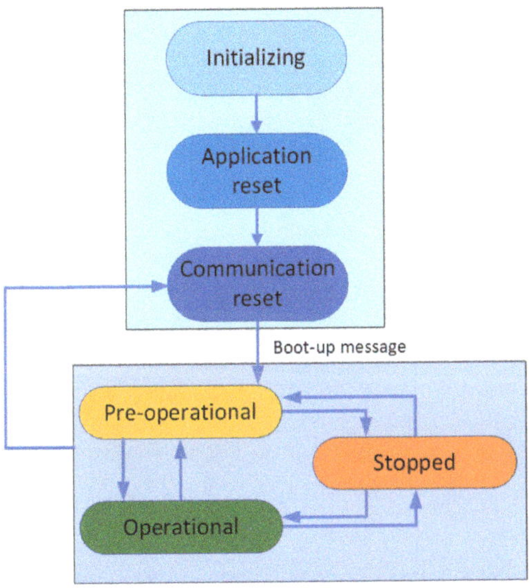

Figure 4. NMT management state transition diagram.

The object dictionary is the most important part of a device specification. It is an ordered set of parameters and variables, including all parameters of device description and device network state. The CANopen protocol uses an object dictionary with 16-bit indices and 8-bit subindices, and all parameters of the device can be accessed through the object dictionary. The parameter object dictionary of the system is defined in the 2000H–5000H (H represents hexadecimal system) index region according to the CANopen CiA 301 document. Real-time data use the process data object (PDO) for asynchronous one-way transmission without a node response. The service data object (SDO) is mainly used for the parameter configuration of slave nodes in the CANopen master station. Service validation is the largest feature of an SDO, generating a response for each message to ensure the accuracy of data transmission. The CAN bus system in this paper consisted of a master node and eighteen slave nodes with 104 object dictionaries. Partial object dictionary descriptions are shown in Table 1. The CAN bus data transmission mode is shown in Figure 5.

Table 1. Partial object dictionary description.

Parameter	Number of Bits	Transport Types	Dictionary Index	
			Indices	Subindices
Operating speed	32	PDO	2000	00
Operating area	32	PDO	2003	00
Motor status	8	PDO	2004	00
Seeding status	8	PDO	2005	00
Seeding number	32	PDO	2006	00
Miss-seeding rate	32	PDO	2007	00
Replay rate	32	PDO	2008	00
Seed spacing	16	SDO	2001	00
Motor control	8	SDO	2009	00
Working width	16	SDO	200A	00
...	...	...	...	...

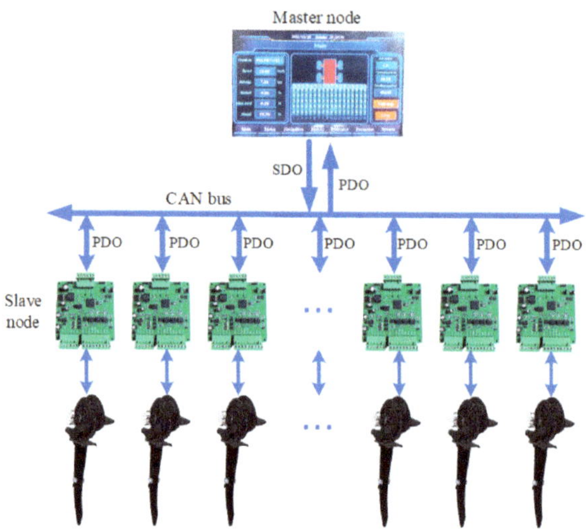

Figure 5. CAN bus data transmission mode.

2.3. *Motor Speed Matching Operation Speed*

To ensure the uniformity and qualified rate of seed spacing, it is very important to establish the dynamic matching relationship between the motor speed and the tractor speed. The rotational speed of the motor is determined by the tractor speed, the number of holes in the seeding plate, the transmission ratio from the reducer to the seeding plate, and the seeding distance. Accordingly, the required rotary speed of the planter unit can be calculated as:

$$R = \frac{1000VI}{36\ X_{ref}N} \tag{1}$$

where R is the motor speed (r/s), V is the tractor speed (km/h), I is the transmission ratio from the reducer to the seeding plate, X_{ref} is the setting seeding distance (cm), and N is the number of holes in the seeding plate. For a well-processed seeding cell, I and N are fixed values. X_{ref} is set based on the agricultural technology. Therefore, the tractor speed is the most critical factor affecting the sowing quality.

2.4. *Speed Acquisition and Motor Control*

GPS speed measurement is not affected by the structure of the seeder and surface conditions and can provide a variety of data, including latitude and longitude, heading angle, and elevation. Compared with other velocity measurement methods, such as encoders, it has great advantages. WTGPS-200 is a high-performance vehicle-mounted integrated navigation system for vehicle navigation. When the signal accuracy of the GNSS system is reduced or if the satellite signal is lost, the WTGPS-200 system uses pure inertial navigation technology without the aid of odometer information. It can also independently carry out high-precision positioning, velocity measurement, and attitude measurement for vehicle carriers over a long time. The accuracy of 0.05 m/s can meet the requirements of the GBT6973-2005 single-seed (precision) seeder test method. The controller obtains GPRMC frames conforming to the NMEA0183 protocol by RS232. Figure 6 shows the GPRMC frame format with fifteen fields. Field 0, as the frame head, represents the beginning of a frame, field thirteen is the frame data validation, and the frame ends with CR/LF. Field one to field twelve represent the data fields, in which field seven represents the speed value. Therefore, the seventh field in a frame can be extracted to obtain the speed.

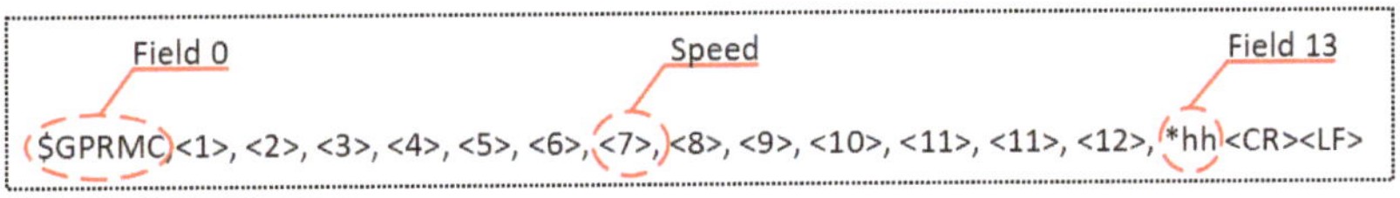

Figure 6. GPRMC frame format.

The real-time motor speed was controlled via the pulse width modulation (PWM) signal generated by the STM32 chip's internal timer. The PWM mode could generate a signal whose frequency was determined by the TIMx_ARR register, and the duty ratio was determined by the TIMx_CCRx register. The duty ratio could be adjusted to control the motor speed at a certain PWM frequency. Limited to the computing power of the chip used, more complex intelligent control algorithms are not adopted, such as adaptive PID [27], particle swarm optimization algorithm [28], fuzzy PID Control Algorithm [29,30], and ant colony optimization [31]. On the other hand, the experimental results indicated that the motor speed showed a linear relationship with the duty ratio. Therefore, closed-loop control can be carried out by the PID control algorithm [32]. PID control is a closed-loop control method based on deviation, which can eliminate the deviation between the target speed and the actual speed of the motor in the adjustment process. In discrete PID control, the realization of integration is the rectangular addition calculation in the case of infinite subdivision. In the discrete state, the time interval is large enough, and the accuracy of rectangular integration appears to be lower in some cases. To minimize the difference, the rectangular integration was changed into trapezoidal integration to improve the calculation accuracy. Introducing the trapezoidal integral into the incremental PID algorithm modifies the formula as follows:

$$\Delta v(k) = K_p(e(k) - e(k-1)) + K_i\frac{e(k) + e(k-1)}{2} + K_d(e(k) - 2e(k-1) + e(k-2)) \quad (2)$$

where $\Delta v(k)$ is the adjustment value, K_p is the proportional coefficient, K_i is the integral coefficient, K_d is the differential coefficient, and $e(k-1)$, $e(k)$, and $e(k-2)$ are the last three deviations. Figure 7 shows an analysis of the bench test data. The optimum motor speed control could be achieved when K_p was 4.15, K_i was 1.2, and K_d was 0.

The theoretical motor speed calculated by Formula (1) is the target value; the rotor position sensor measures the speed signal as a feedback value. The theoretical calculation of the target speed does not consider the influence of external factors. However, due to the factors of actual operation, such as zero drift of the speed sensor, error of DC motor speed measurement, and the efficiency of mechanical transmission, the error of the control parameters ($e(k)$) is affected. Therefore, setting a threshold variable, t, does not perform the PID algorithm when the deviation is less than the absolute value of the threshold. Experimental results showed that the control precision was best when the absolute value of threshold t was 0.15. On the other hand, if a system always has a uniform direction deviation, infinite accumulation and saturation can occur, which greatly affects the system performance. To solve the problem of integral saturation, the PID algorithm anti-integral saturation was introduced. The idea is to determine whether the control, $C(k-1)$, of the previous moment has exceeded the limit when calculating $e(k)$. If $C(k-1) > C_{max}$ (C_{max}: sets the TIMx capture compared to the register maximum value), only negative deviations are accumulated; if $C(k-1) < C_{min}$ (C_{min}: sets the TIMx capture compared to the register minimum value), only positive deviations are accumulated. This avoids the control quantity from staying in the saturated zone for a long time. The PID control algorithm is shown in Figure 8.

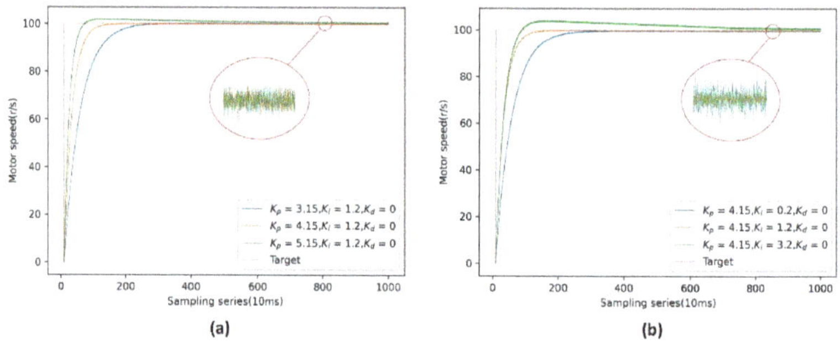

Figure 7. Data analysis curve of different PID parameters: (**a**) response curves under different K_p conditions and (**b**) response curves under different K_i conditions.

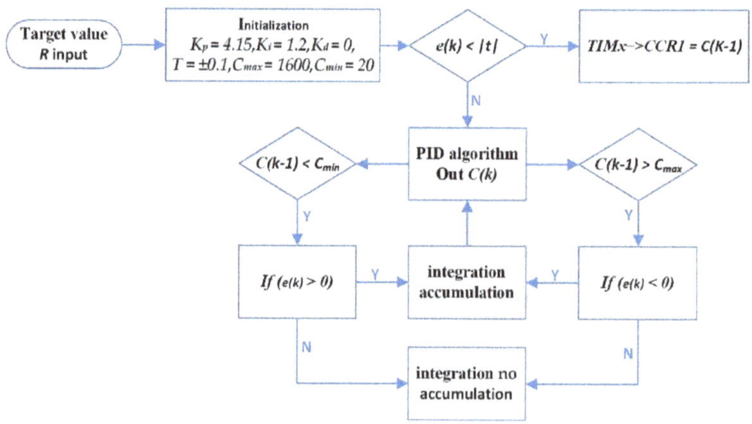

Figure 8. PID control algorithm.

2.5. Sowing Monitoring

To realize the real-time monitoring of the quality of maize no-tillage precision seeding operations, a seeding monitoring system based on reflective infrared photoelectric induction was designed. The monitoring probe used an infrared emitting diode and a photodiode as the signal transmitting and receiving ends. During the seeding operation, corn seeds were separated into single seeds from the seed metering device, dropped into the seed guiding tube, and were finally discharged into the soil through the lower seed guiding mouth. Among the working components involved in the seeding process, the structure of the seed guiding tube was the simplest and the closest to the seed dropping point. Therefore, mounting the seed monitoring probe on the seed guiding tube was preferred.

According to GB/T 6973-2005, the ratio of actual adjacent seed spacing, X (cm), to theoretical seed spacing, X_{ref} (cm), is the benchmark for evaluating the quality of seed metering. In addition to field measurements, the actual seed spacing is generally estimated by multiplying the tractor speed, V (km/h), of the seeder by the interval time, T (ms), between adjacent seeds. The forward speed, V (km/h), of the seeder can be obtained by the pick-up circuit. Therefore, the comparison between the actual seed spacing and the theoretical seed spacing can be converted to a numerical comparison between the actual adjacent seed falling time interval, T (ms), and the theoretical time interval, T_0 (ms). According to the standard, if $X > 1.5X_{ref}$, it is judged as a miss-seeding, and if $X \leq 0.5X_{ref}$, the seeding is judged as a reseed. For the convenience of system calculation, the judgment basis is converted to the relationship between the tractor speed, V (km/h), and the theoretical distance, X_{ref} (cm). If $VT > 54X_{ref}$, the seeding is judged as a miss-seeding. If $VT \leq 18X_{ref}$ is judged as a reseeding and if $18X_{ref} < VT < 54X_{ref}$, the seeding is a quality seeding. When a fault (miss-seeding or reseed) occurs, an alarm is triggered. Figure 9 shows three different states of falling seeds in the seed tube. Figure 10 shows the seed condition monitoring process.

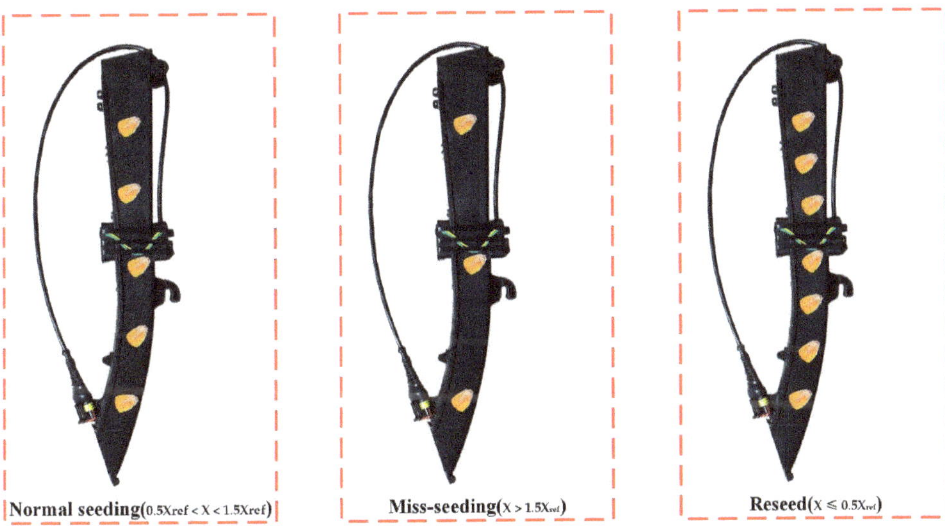

Figure 9. Judging the state of falling seeds in the seed tube.

2.6. Performance Test of the Seeder Monitoring and Control System

To verify the performance of the seeder monitoring and control system, laboratory bench tests and field tests were conducted. These tests included photoelectric sensor detection performance tests, abnormal alarm rate reliability tests, motor dynamic speed response tests, and statistical analyses of real-time sowing monitoring parameters.

The related tests were carried out on the JPS-12 seed metering device performance test bench (Bona Technology Co., Ltd., Harbin, China). The test materials were Xinyu No. 9 hybrid maize seeds produced by the Crop Research Institute of Xinjiang Academy of Agricultural Reclamation Sciences. The moisture content was 9.10%, the purity was 98.75%, and the thousand-grain weight was (274.22 ± 2.52) g. We randomly measured 300 seeds, and the shape was horse tooth, and the length, width, and height were 10.04 ± 1.06 mm, 7.45 ± 0.86 mm, and 5.50 ± 1.01 mm, respectively.

The seeding unit motor drive control system and experimental test setup are shown in Figure 11. The metering device was an air suction seed metering device produced by Precision Planting Company in the United States. The diameter of the metering plate was 4.5 mm, and the number of seed holes was 27. The DC motor was an NC3SFN-6035-CVC carbon brush variable-resistance brush DC motor produced by Transmotec, Sweden. The working voltage was 12 V, the current was 5.6 A, the rated speed was 10,700 r/min, and the stall torque was 446.8 mN·m. The motor reducer was a three-stage gear reducer developed by Devo, Heilongjiang Province, and the deceleration ratio was 82.8125. The power output gear of the DC motor reducer engaged with the external gear of the seeding plate.

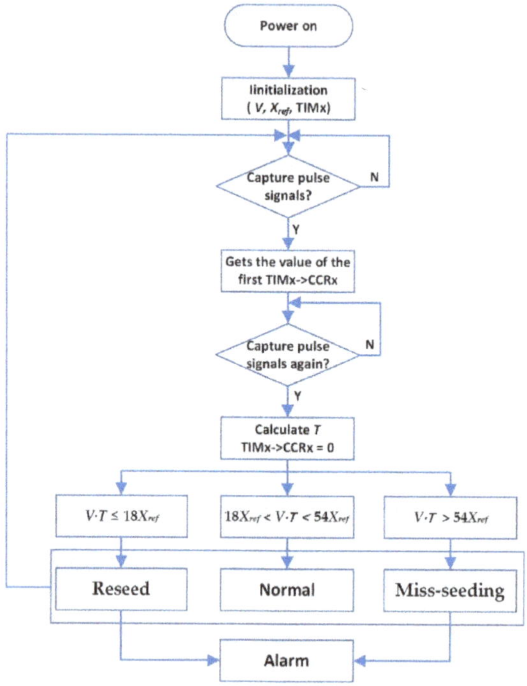

Figure 10. Seed condition monitoring process.

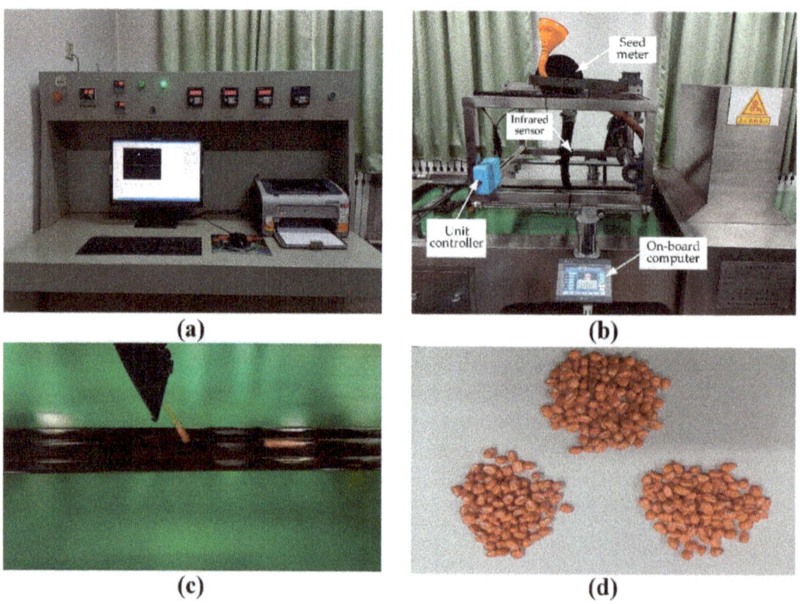

Figure 11. Seeding parameter monitoring on the JPS-12. (**a**) Control cabinet; (**b**) test bench; (**c**) seeding; (**d**) data statistics.

Since no interface can obtain the real-time speed on the JPS-12 test bench, to obtain the real-time operating speed of the seedbed belt as much as possible to simulate the field environment, ten groups of magnetic steel were installed on the inner side of the seedbed drive roller. NPN constant open all-pole Hall sensors were used in pulse signal detection. Figure 12 shows the installation position of the magnetic steel and the Hall sensors. The dynamic speed of the seedbed could be calculated according to Formula (3) after the signal of the speed pulse was collected by the Hall sensor.

$$V_b = \frac{\pi d n}{m T_c} \tag{3}$$

where V_b is the speed of the seedbed belt (m/s), d is the roller diameter (mm), n is the number of pulses in the T_c cycle, m is the number of magnetic steels, and T_c is the count cycle (ms).

Figure 12. Schematic diagram of seedbed belt speed detection.

The field experiment was conducted in Xiangshui County, Yancheng City, Jiangsu Province, on 17 February 2022, using a dual row with an eighteen-row seeder developed by Devo, Heilongjiang Province (Figure 13). To explore the influence of different operating speeds on seeding performance, the negative pressure of the fan output was adjusted to 4.5 kPa, the grain spacing was set to 20 cm, and the operating speeds were changed to 8 km/h, 10 km/h, and 12 km/h. To explore the effects of different grain spacings on sowing performance, the operating speed was 8 km/h, and the grain spacings were changed to 15 cm, 20 cm, and 25 cm. At the same time, we explored the differences in sowing performance parameters between different planting units. The grain spacing data were obtained by manual measurement.

Figure 13. Precision electric seeder and monitoring system test. (**a**) Eighteen-row maize precision electric seeder and monitoring system; (**b**) label seed position; (**c**) seed spacing measurement.

According to GB/T 6973-2005, the qualified index, QI, reseed index, RI, missing index, MI, and coefficient of variation, CV, were calculated as evaluation indices of sowing quality.

$$\mathrm{QI} = \frac{n_1}{N'} \times 100\% \tag{4}$$

$$\mathrm{RI} = \frac{n_2}{N'} \times 100\% \quad (5)$$

$$\mathrm{MI} = \frac{n_0}{N'} \times 100\% \quad (6)$$

$$X = \frac{\Sigma(n_i X_i)}{n_2} \quad (7)$$

$$\sigma = \sqrt{\frac{\Sigma(n_i X_i)^2}{n_2} - X^2} \quad (8)$$

$$\mathrm{CV} = \sigma \times 100\% \quad (9)$$

where N' is the total number of normalized intervals, n_0, n_1, and n_2 are the missing numbers ($X_i \in (1.5, +\infty]$), the qualified number ($X_i \in (0.5, 1.5]$), and the replay number ($X_i \in [0, 0.5]$), respectively, n_i and X_i are the grain spacing number and interval median in the i(th) interval, respectively, and X and σ are the mean and standard deviation of the sample, respectively. At the same time, these indicators were evaluated according to the NY/T 1143-2006 standard provided by the Ministry of Agriculture of China. Table 2 shows the main performance indices of the precision seeder.

Table 2. Main performance indices of the precision seeder.

Index	Indicators		
	Seed Spacing ≤10 cm	**Seed Spacing >10 cm~20 cm**	**Seed Spacing >20 cm~30 cm**
Qualified index	≥60.0	≥75.0	≥80.0
Reseeding index	≤30.0	≤20.0	≤15.0
Missing index	≤15.0	≤10.0	≤8.0
Coefficient of variation	≤40.0	≤35.0	≤30.0

3. Results and Discussion

3.1. Photoelectric Sensor Monitoring Performance and Real-Time Online Monitoring Test

To test the performance of the photoelectric sensor, the numbers of monitored corn grains at speeds of 6, 8, 10, and 12 km/h were tested in the laboratory and the field. When the speed reached the set value, the test seeder monomer was started by the virtual button on the on-board computer, and the seeder monomer was stopped at a random time. The number of corn seeds collected in containers fixed below the metering tube was manually counted. The statistical results indicate that the photoelectric sensor monitoring performance was quite good, and there were no differences between the laboratory and field monitoring data. Table 3 shows the statistical results of the monitoring data and actual data. The average monitoring accuracy was 99.8%.

To test the reliability of the system fault alarm, two kilograms of corn seeds were added to each sowing monomer. In the initial stage of operation, the metering tube was in normal planting, and the system did not send alarm information. When the seed box was empty, the system was checked to determine whether the alarm was prompted and whether the corresponding sowing monomer was shown in the vehicle terminal. According to the same method, during the normal seeding period, the seeding tube was artificially blocked at a given time, and the system blocking alarm rate was checked. The test results of fifty trials showed that the fault alarm rate was 100%.

The statistical analysis of the performance index data is shown in Figure 14. It can be seen from the chart that each evaluation index was basically similar at different speeds and substantially exceeded the standard (NY/T 1143-2006).

Overall, the qualified rate was higher when the grain spacing was larger. It was also found that when the speed was 12 km/h, the qualified rate decreased compared with the other speeds and the missed rate increased. The reason is that with the increase in the speed of the seedbed belt, the sliding degree of the seedbed pulley relative to the seedbed belt increased, resulting in inaccurate speed measurement. When the speed was 8 km/h and 10 km/h, the consistency of the indices is good, and the difference was significant when the speed was 6 km/h and 12 km/h.

Table 3. Statistical results of monitoring data and actual data.

Site	Speed (km/h)	Monitoring Value			Actual Value		
		A_m	B_m	C_m	A_a	B_a	C_a
Laboratory	6	245	304	258	245	305	258
	8	275	236	459	275	236	460
	10	236	389	321	236	389	321
	12	335	286	462	335	287	462
Field	6	365	428	303	365	428	303
	8	274	354	482	275	355	482
	10	382	298	407	382	299	408
	12	473	387	496	473	385	496

Table 4 shows the correlation analysis between the factors and performance indicators. There was a strong correlation between two factors (V and X_{ref}) and the seed distribution uniformity index (CV, QI, RI, and MI). In addition, the statistical values describing the correlation between various factors and performance indicators show that there was a strong correlation between QI, RI, MI, and V: QI decreased with an increase in V and RI, and MI increased with an increase in V; there were significant correlations between QI and CV and between RI and MI. The CV, RI, and MI decreased with increasing QI, and the correlation between MI and QI was the strongest. The coefficient of determination was 0.983, and the level of visibility was far less than 0.01.

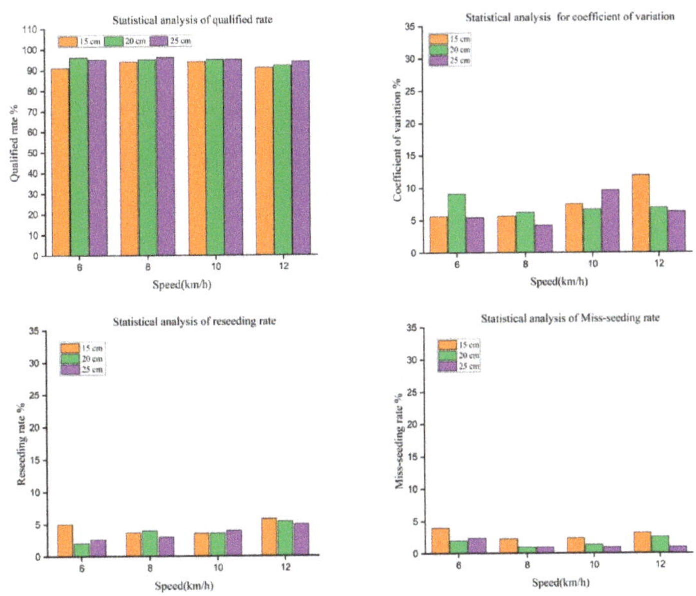

Figure 14. Statistical analysis of the performance index data.

Table 4. Correlation analysis between factors and performance indicators.

Items	V/(km/h)	X_{ref}/mm	CV/%	QI/%	RI/%	MI/%
V/(km/h)	1.000 [a] \ [b]	0.000 1.000	0.527 0.145	−0.791 * 0.011	0.738 * 0.023	0.843 ** 0.004
X_{ref}/mm	0.000 1.000	1.000 \	−0.369 0.329	0.527 0.145	−0.264 0.493	−0.474 0.197
CV/%	0.527 0.145	−0.369 0.329	1.000 \	−0.733 * 0.025	0.717 * 0.030	0.633 0.067
QI/%	−0.791 * 0.011	0.527 0.145	−0.733 * 0.025	1.000 \	−0.867 ** 0.002	−0.983 ** 0.000
RI%	0.738 * 0.023	−0.264 0.493	0.717 * 0.030	−0.867 ** 0.002	1.000 \	0.833 ** 0.005
MI/%	0.843 ** 0.004	−0.474 0.197	0.633 0.067	−0.983 ** 0.000	0.833 ** 0.005	1.000 \

Note: Because the prior uncertainty is a positive correlation or negative correlation, the double tail test was chosen; descriptive statistics of sample data were used to calculate the average and variance; and the visibility of the output results must be marked. When the visibility level reaches 0.05, the upper right corner uses '*'; when the visibility level reaches 0.01, the upper right corner uses '**'. [a] represents the coefficient of determination; [b] represents the p value, namely, the level of dominance.

3.2. Differences in Seeding Performance among Different Planting Units in the Field

The test results of the seeding performance at different operating speeds are shown in Table 5. When the operating speed was 8 km/h, the seeding performance was excellent. The qualified index of single seeding was 94.14%, the reseeding index was 1.72%, and the missing seeding index was 4.14%. With the increase in the operation speed, the reseeding index always maintained a certain level. However, due to the insufficient wind pressure of the fan and the irregular bounce of the seed when landing, the missing seeding index increased significantly, resulting in a decrease in the seeding accuracy. When the operating speed was 10 km/h, the seeding qualified index was reduced to 91.48%, and the leakage index was increased to 7.46%. When the operating speed was 12 km/h, the seeding qualified index was still greater than 90%.

Table 5. Results of the seeding performance at different operating speeds.

Items	V = 6 (km/h)		V = 8 (km/h)		V = 10 (km/h)		V = 12 (km/h)	
	No. 2	No. 7	No. 2	No. 7	No. 2	No. 7	No. 2	No. 7
Average distance (cm)	19.38	19.24	19.10	19.40	20.30	19.18	21.50	20.31
QI/%	93.17	93.40	94.14	94.53	91.48	91.94	90.35	90.01
RI/%	1.38	1.86	1.72	1.14	1.06	0.80	1.02	1.93
MI/%	5.45	4.74	4.14	4.33	7.46	7.26	8.63	8.06
Standard deviation	5.32	6.73	4.59	5.46	9.65	8.41	10.01	9.35

Note: No. 2 and No. 7 represent the second and seventh sowing planting units, respectively.

The seeding performance test results at different seed spacing settings are shown in Table 6, and the operating speed remained 8 km/h. With the increase in the seed spacing, the qualified index decreased and the reseeding index and leakage index increased. For a grain spacing of 15 cm, the average qualified index of two single seedlings was 93.34%, the average reseeding index was 2.09%, and the average missing seeding index was 4.57%.

When the operating speed was within the range of 6~12 km/h or the grain spacing was set to 15~25 cm, there was no significant difference in the seeding performance between the No. 2 and No. 7 planting units. These differences may be caused by factors such as the processing technology of the planting unit mechanical mechanism, mechanical vibration, and measurement error. Therefore, it is considered that the variability of the seeding performance between the monomers is small. In summary, when the quality of seeds and soil preparation meets the agronomic requirements of sowing, the electric drive seeding control system designed in this study meets the requirements of precision sowing under high-speed working conditions.

Table 6. Seeding performance for different driving modes.

Items	X_{ref} = 15 cm		X_{ref} = 20 cm		X_{ref} = 25 cm	
	No. 2	No. 7	No. 2	No. 7	No. 2	No. 7
Average distance (cm)	14.43	14.15	19.93	19.28	23.10	23.40
QI/%	91.79	92.25	93.19	93.98	94.12	94.71
RI/%	2.65	3.01	2.12	1.89	1.72	1.14
MI/%	5.56	4.74	4.69	4.13	4.16	4.15
Standard deviation	7.61	8.26	6.32	5.78	4.59	5.46

Note: No. 2 and No. 7 represent the second and seventh sowing planting units, respectively.

3.3. Discussion of the Results

Based on the control system of the electric drive precision seeder, laboratory bench tests and field tests were carried out. Its performance indicators tended to be consistent, which also fully illustrated the system reliability. The bench test explored the effects of different operating speeds and grain spacing on the seeding performance indices. At present, many scholars have carried out electric drive seeding experiments, and their working performance has been greatly improved compared with the traditional mechanical seeders. The performance indicators involved in this study are similar to those used in previous studies. Due to the differences in the environment and mechanical structure, the qualified rate of sowing in the field was lower than that of the bench test.

Since the test bench is designed for a traditional mechanical seeder, the influence of the seedbed vibration and slip ratio of the seedbed belt during high-speed operation has not been fully considered, thus affecting the test results to a certain extent [33]. In the field experiment, previous researchers mostly used 4-row or 6-row seeding machines for experiments. In this study, an 18-row air suction precision seeding machine was used. Due to the increase in seeding monomers, the airflow of the fan was unstable at high speeds, resulting in insufficient pressure during high-speed operation and a slight decrease in the seeding qualified index; seeding monomers on both sides of the seeding machine was a common malfunction. Nevertheless, more than 90% of the qualified rates fully met the actual work requirements. In the selection of photoelectric sensors, based on previous studies, a rectangular infrared radiation surface was selected, which greatly improved the sensing area of the photoelectric sensors and reduced the blind area. The high sensitivity of the sensor increased the fault alarm rate.

The sensors used in the system and the electronic components used in the design circuit are commonly used in the market. Compared with the laser detection sensor used in [7], the photoelectric sensor has a high value for practical application, and the monitoring performance was better than the laser detection performance; compared with the expensive LiDAR used in [10], the system used the common satellite acquisition module and achieved good data acquisition and control effect through certain filtering algorithms. The CAN bus control method greatly reduces the difficulty of field wiring. The brush DC motor is easier to control and lower cost than the brushless DC motor used in [17,18]. Usually, brushless motors perform better than brush motors.

In order to prevent electrostatic interference to the system, the electrostatic shielding circuit was specially designed in the circuit, which improved the anti-interference properties and robustness of the system. In practical field applications, a shielded twisted pair is used in CAN bus transmission, and terminal resistance is connected to the transceiver end. At the same time, CAN bus through the data link layer and physical layer has achieved high bus data security and bus stability; the correctness of data transmission is ensured by establishing a CANopen object dictionary. The above measures enhance the robustness of the system to subsystem faults and electromagnetic interference. Overall, whether in economic cost or system performance, the system was suitable for agricultural machinery operation.

On the other hand, the acquisition accuracy of the tractor speed directly affects the sowing quality. Although the accuracy of GPS can meet the current operating requirements, once the GPS signal, as the only acquisition speed, is affected, it seriously affects the operating quality. In the future, multisensor information fusion technology will be used to compensate for the speed signal to ensure that the speed measurement accuracy can still be maintained under sensor fault and interference conditions to ensure the consistency of the operation quality in a complex working environment.

4. Conclusions

A control system of an electrically driven precision maize seeder based on the CANopen protocol was designed and developed. A circuit board with motor drive and sowing performance detection was integrated. The matching model of vehicle speed and seeding plate speed was established through the PID control algorithm. Terminal monitoring software for real-time monitoring of sowing parameters was designed. According to the GB/T 6973-2005 standard, the evaluated parameters were the following: photoelectric sensor detection performance, fault alarm rate, qualified rate, reseeding rate, and missed rate. The following conclusions can be drawn:

(1) In terms of photoelectric sensor detection performance, there was not a large difference between the indoor bench tests and field tests with dust pollution, and the detection accuracy reached 99.8%. This also shows that the sensor has a strong penetration ability and a large radiation detection surface. The fault alarm function of the system was accurate and timely, and the fault alarm rate reached 100%.

(2) Based on the indoor test results, the qualified rate was higher when the grain spacing was larger. It was also found that when the speed was 12 km/h, the qualified rate decreased compared with other speeds, and the missed rate increased. When the speed was 8 km/h and 10 km/h, the consistency of the indices was good, and the difference was significant when the speed was 6 km/h and 12 km/h. Overall, the qualified rate of sowing was more than 91%. At the same time, the correlation of the seeder index parameters was analyzed: there were strong correlations between QI, RI, MI, and V; QI decreased with increasing V; RI and MI increased with increasing V; and CV, RI, and MI decreased with increasing QI. Furthermore, the correlation between MI and QI was the strongest. The coefficient of determination was 0.983, and the level of visibility was far less than 0.01.

(3) Based on field test results, the seeding performance results showed that the control system has good stability. When the grain spacing was set to 20 cm and the operating speed was 6~12 km/h, the qualified index was more than 90%, and the reseeding index was less than 1.93%. The variation in sowing performance between different monomers was small, and the seeding performance was good, which can provide a reference for the development and design of high-speed precision corn seeders.

Author Contributions: Conceptualization, J.C., C.J. and M.D.; methodology, J.C. and F.P.; software, J.C.; validation, J.C., C.J. and H.Z.; formal analysis, J.C. and M.D.; investigation, J.C., C.J. and H.Z.; resources, C.J. and M.D.; data curation, J.C. and C.J.; writing—original draft preparation, J.C.; writing—review and editing, J.C. and C.J.; visualization, J.C. and H.Z.; supervision, C.J.; project administration, C.J. and M.D.; funding acquisition, C.J. and M.D. All authors have read and agreed to the published version of the manuscript.

Funding: This research was funded by the Key Scientific and Technological Projects in Key Areas of the Xinjiang Production and Construction Corps (No. 2020AB011) and the National Natural Science Foundation of China (NSFC) (No. 32101634).

Institutional Review Board Statement: Not applicable.

Informed Consent Statement: Not applicable.

agriculture

MDPI

Article

A Soft Gripper Design for Apple Harvesting with Force Feedback and Fruit Slip Detection

Kaiwen Chen [1,2], Tao Li [2,*], Tongjie Yan [1,2], Feng Xie [1,2], Qingchun Feng [2], Qingzhen Zhu [1] and Chunjiang Zhao [3,*]

1 School of Agricultural Equipment Engineering, Jiangsu University, Zhenjiang 212013, China
2 Intelligent Equipment Research Center, Beijing Academy of Agriculture and Forestry Sciences, Beijing 100097, China
3 National Engineering Research Center for Information Technology in Agriculture, Beijing 100097, China
* Correspondence: lit@nercita.org.cn (T.L.); zhaocj@nercita.org.cn (C.Z.)

Abstract: This research presents a soft gripper for apple harvesting to provide constant-pressure clamping and avoid fruit damage during slippage, to reduce the potential danger of damage to the apple pericarp during robotic harvesting. First, a three-finger gripper based on the Fin Ray structure is developed, and the influence of varied structure parameters during gripping is discussed accordingly. Second, we develop a mechanical model of the suggested servo-driven soft gripper based on the mappings of gripping force, pulling force, and servo torque. Third, a real-time control strategy for the servo is proposed, to monitor the relative position relationship between the gripper and the fruit by an ultrasonic sensor to avoid damage from the slip between the fruit and fingers. The experimental results show that the proposed soft gripper can non-destructively grasp and separate apples. In outdoor orchard experiments, the damage rate for the grasping experiments of the gripper with the force feedback system turned on was 0%; while the force feedback system was turned off, the damage rate was 20%, averaged for slight and severe damage. The three cases of rigid fingers and soft fingers with or without slip detection under the gripper structure of this study were tested by picking 25 apple samples for each set of experiments. The picking success rate for the rigid fingers was 100% but with a damage rate of 16%; the picking success rate for soft fingers with slip detection was 80%, with no fruit skin damage; in contrast, the picking success rate for soft fingers with slip detection off increased to 96%, and the damage rate was up to 8%. The experimental results demonstrated the effectiveness of the proposed control method.

Keywords: apple harvesting; soft gripper; Fin Ray effect; finite element analysis; constant-pressure feedback; slip detection

Citation: Chen, K.; Li, T.; Yan, T.; Xie, F.; Feng, Q.; Zhu, Q.; Zhao, C. A Soft Gripper Design for Apple Harvesting with Force Feedback and Fruit Slip Detection. *Agriculture* **2022**, *12*, 1802. https://doi.org/10.3390/agriculture12111802

Academic Editors: Jacopo Bacenetti and Tao Cui

Received: 27 July 2022
Accepted: 27 October 2022
Published: 29 October 2022

1. Introduction

Harvesting is an important element of orchard production since it has a brief window period, high labor intensity, and high labor volume. The high labor cost in the harvesting stage limits the fruit industry's development. With this backdrop, fruit-picking robots have become a hotspot for study in related fields [1,2]. Researchers have completed several projects and made significant progress in important technologies such as robot perception and positioning [3,4], system integration [5], and efficient harvesting end effector design.

As a critical step in robotic harvesting, grasping determines the picking effect directly. During harvesting, the traditional robotic rigid clamping mechanism has issues: high requirements for fruit positioning [6] and easy damage to the apple pericarp [7,8]. In practical applications, it not only required the grippers to be dexterous, light, stable, and reliable to grasp but also to ensure that the appearance of the fruits is not damaged, to prevent harming commerciality. As a result, research on non-destructive harvesting end grippers for safe, reliable, and stable gripping is an important topic for harvesting robots with a promising application.

To lower the fruit damage rate, the soft gripper technology is attracting more and more researchers' attention. Some researchers [9–11] used soft materials on the surface of the fingers to increase the gripper flexibility and, hence, prevent damage to grabbed objects. However, due to the rigid support of the fingers' main body, it is also easy to cause different degrees of damage to the fruit pericarp. Furthermore, the structure is more complex, and the grasping stability is insufficient.

The soft structure gripper has a high adaptability, wide range of variability, and excellent working ability for gripping objects that are susceptible to damage [12,13].

Shepherd et al. [14] proposed the PneuNet (pneumatic mesh) structure, a bending multi-cavity pneumatic soft actuator. The soft gripper [15–17] designed by Whiteside's group has the characteristics of minimal pressure bearing, large deformation, and flexible movement. However, the end contact force is limited, and the stability is insufficient when grasping objects. A vision-equipped six-finger soft harvesting gripper [18] can identify the type and maturity of fruits and vegetables, and it can softly grab fruits and vegetables based on their shape but only for tiny fruits. Muscato et al. [19] created a soft citrus harvesting gripper out of spirally organized rubber sheets that had a strong wrapping capacity for gripping things but that lacked rigidity.

German bionics researcher Leif Kniese accidentally discovered the "Fin Ray effect" in 1997 [20], which was later widely employed in the study of robotic soft grippers [21,22]. Fin Ray soft fingers are highly compliant and can take greater loads than other soft constructions. Thanks to its superior grabbing stability, the Fin-Ray-effect-inspired grippers have received extensive attention from researchers.

However, the basic finger structure is not optimal for soft grippers, and studies have recently increased the gripping force by improving the finger structure [23–26]. Crooks et al. [23] proposed a multi-material structure gripper with a higher grabbing weight, but the fabrication method for this multi-material structure is quite tricky. Basson et al. [24] varied the slope and curve of the cross beams in a Fin Ray finger and analyzed the stress and displacement on the improved finger through simulation. However, the effects of other variables have not been fully tested. Shin et al. [25] analyzed the changes in stress and displacement when the finger touched an object by varying the number of cross beams, the front beam slope, and the slope of the cross beams. Elgeneidy et al. [26] developed a soft finger that could handle fragile objects by varying the angle and number of cross beams. Nevertheless, whatever structure maximizes the Fin-Ray finger gripping force while causing no damage to the object has yet to be determined.

Although it can greatly avoid fruit damage due to grasping by using the soft fingers, it is not sufficient to rely solely on the soft structure to ensure the gripper's lossless grasping. The gripper's lack of a force feedback system makes it unable to collect the contact state information between fingers and gripping items, which may cause damage due to excessive gripping force or slippage owing to insufficient gripping force.

Some researchers added force sensors to the fingers of soft grippers [27–31]. The sensing system is simple, but the sensor deforms with soft fingers, which has a great influence on the accuracy. When directly embedding force sensors through the manufacturing process but the cost is large and the universality is low due to its sophisticated driving scheme and manufacturing method [32,33]. Some researchers [34–36] estimated the contact force by substituting the force perception model from finger deformation by vision. Belzile et al. [37] used the quasi-static analysis method to calculate the contact force generated by the gripper, which realizes the internal force perception without the use of additional force sensors, but the solution process and control algorithm are complex.

In addition to preventing fruit damage due to excessive gripping force, slip detection is also an important factor due to the rough surface of the fingers [38,39]. Some studies use multi-axis or more force sensors to monitor the static friction coefficient between the finger surface and the object [40,41] or to detect vibration caused by sliding between the two contact surfaces using piezoelectric phenomenon [42], time–frequency conversion technique [43], or filtering [44] to accomplish slip detection. However, the sensors are dependent on the working environment, and utilizing more sensors to gather more tactile information would not only dramatically raise the cost but will also place a significant load on the gripper structure and control system. Some recent studies employ tactile data for training, and neural networks can predict item sliding [45,46], as well as physical parameters such as temperature, electromagnetism, light intensity, and acceleration to predict slippage [47]. Liu et al. [48] introduced a novel design of the GelSight Fin Ray gripper, which used a vision-based tactile sensor for tactile reconstruction, orientation estimation, and slip detection. But it is difficult to grasp heavier objects due to the design of its hollowed-out finger. Nonetheless, these technologies are rarely used on harvesting grippers.

To solve the above problems, this work proposes a novel soft harvesting gripper with flexible adaptive envelope, force feedback, slip detection, and other features. To design the Fin-Ray finger structure in such a way that the gripping force is high enough to ensure it is sufficient to successfully separate fruits from stems, the influence of various parameters of the Fin Ray structure on the gripping force and deformation of the finger was investigated through simulations, as the basis for the design of the soft gripper structure. The following are the main contributions:

(1) A new three-finger force feedback soft gripper for the apple harvesting robot is proposed. The relationship between the gripping force, the pulling force, and the servo torque was established to achieve the constant-pressure flexible clamping of fruits. Then the sensing system of the soft gripper was implemented by using the servo's feedback information instead of adding additional sensors, making the structure of the gripper simpler and less costly.

(2) A force feedback gripper dynamic control approach with slip detection is presented. The relative location of the fruit and the gripper is detected in this manner by incorporating a distance sensor, which makes the gripper structure and calculation simple. When the fruit slippage occurs, the servo output torque is adjusted in real time to reduce fruit harm using the feedback information.

In addition, to provide a theoretical basis for the design of the gripper, some mechanical properties of apples are given in the experiments.

Remark 1. *It should be clarified that the force feedback system and slip detection are two main contributions in this paper. To provide a stable mechanical design of the gripper as a study basis for these two points, we also analyze the structural parameters of Fin Ray fingers by the finite element analysis method.*

2. Structural Design of a Soft Gripper with Three Fingers

2.1. Finite-Element Analysis of Finger Structure with the Fin Ray Effect

The harvesting gripper's finger mechanism uses a triangular Fin-Ray soft finger component, which has a passive compliance quality and can implement an envelope while clamping spherical items. The general construction of the finger consists of the front and rear beams, cross beams, and base, as shown in Figure 1a. The front beam comes into contact with the fruits, and the front and rear beams are linked by cross beams. These cross-beam support rods are the foundation of Fin Ray fingers. Because of the presence of these crossbeam support rods, the Fin Ray structure can withstand greater loads than conventional flexible constructions.

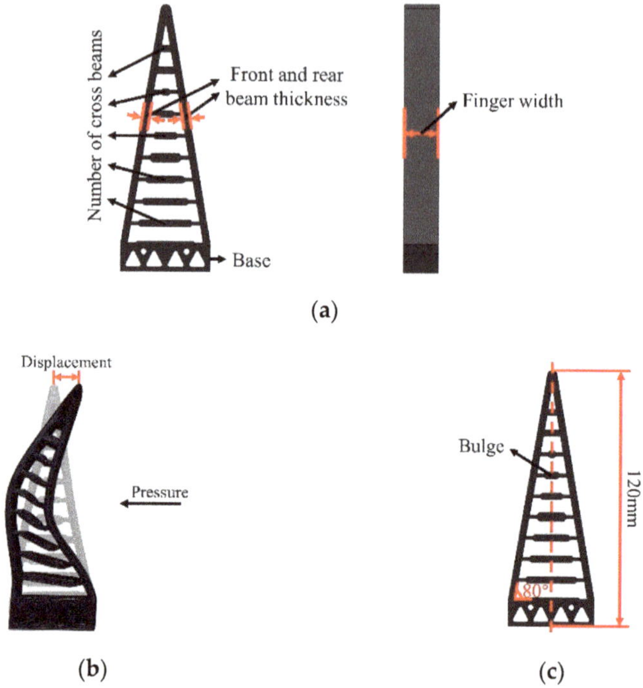

Figure 1. Characteristics of the Fin Ray finger: (**a**) basic components; (**b**) displacement of the fingertip; (**c**) structure of the Fin Ray finger.

2.1.1. Pre-Preparation of the Simulation Experiment

The finger gripping force must be sufficient to improve the grasping stability. Furthermore, the pressure per unit area of the pericarp should be small enough to guarantee that the fruit pericarp remains intact. As a result, the finger gripping force and the bending degree are two critical criteria. The finger gripping force can ensure clamping stability, while the finger bending degree can assure clamping stability and safety by increasing the contact area between the fingers and the fruits. The stress of the Fin Ray finger during deformation is complicated by making the mathematical modeling difficult. As a result, using the simulated tests, this research investigates the effect of the front and rear beam thickness, the finger width, and the number of cross beams on the finger gripping force and bending degree, as shown in Figure 1a. In the simulation experiment, the contact stress between the finger and apple is used to characterize the gripping force, and the displacement of the fingertip is used to characterize the bending degree, as shown in Figure 1b.

A single finger adopts a symmetrical structure; the total length of the finger is 120 mm, and the front beam and the rear beam are each at an angle of 80° to the base. The cross beams are parallel to the base; the distance is equal, and the thickness of the cross beams is 1.40 mm. The little bulges are designed on the cross beams to increase the rigidity and strengthen the load capacity, as shown in Figure 1c.

The TPU 95A [49] was chosen as the finger material. The TPU soft material is a hyperelastic nonlinear material with isotropic properties throughout the stress process. Furthermore, because the bending deformation of the soft finger is a nonlinear large deformation, the Yeoh model can better represent its material properties [50]. The strain energy density function W can be written as follows:

$$W = \sum_{i=1}^{N} C_{i0}(I_1 - 3)^i + \sum_{k=1}^{N} \frac{1}{D_k}(J - 1)^{2k}, \quad (1)$$

where N is the order of the model; I_1 is the deformation tensor; C_{i0} and D_k are the material constants; J is the volume ratio. When TPU is regarded as the incompressible material, $J = 1$.

The strain energy density function in the form of the binomial parameters is usually used [51], and the typical binomial parameter form of the Yeoh model is

$$W = C_{10}(I_1 - 3) + C_{20}(I_1 - 3)^2. \tag{2}$$

The fitting curve of the stress and strain of the TPU 95A was obtained through the uniaxial tensile test, as shown in Figure 2. The material parameters obtained after processing and analysis are shown in Table 1.

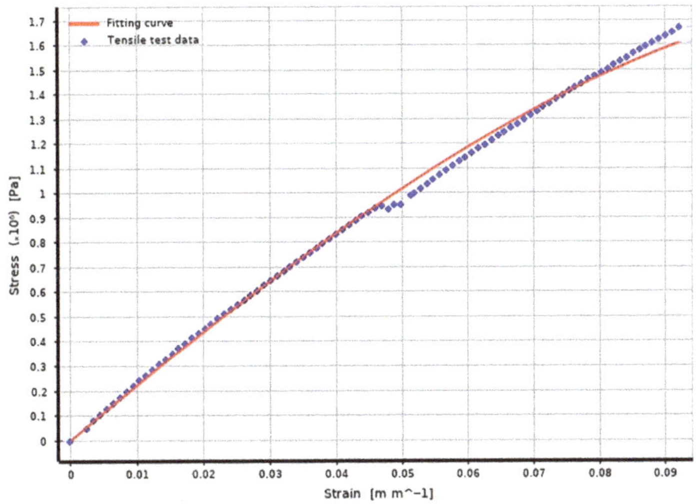

Figure 2. Strain–stress curve of the tensile test and fitting using the Yeoh model (TPU95A).

Table 1. Mechanical property parameters of materials.

Materials	Density (kg/m^3)	Young's Modulus (MPa)	Poisson's Ratio	C10 (MPa)	C20 (MPa)
PA12	1010	1900	0.4	—	—
Apple [52]	840	5	0.35	—	—
TPU 95A	1200	—	—	3.7358	−11.88

Because the contact stress between the three fingers and the fruit is the same, the contact between a single finger and the fruit can be considered to reduce the quantity of simulation calculation, to simplify the analysis.

During the simulation, the center of the bottom plate of the gripper is kept aligned with the center of the fruit at a distance of 65 mm [49].

2.1.2. Influence of Geometric Parameters on Contact Stress and Fingertip Displacement

Each geometric parameter has a varied effect on the contact stress and fingertip displacement. All other parameters were held constant to compare their changes when the given parameters were altered, and the influence of the given single parameter on them was gradually optimized.

First, the influence of the thickness of the front and rear beams was analyzed. The stress increases dramatically as the thickness increases, while the fingertip displacement decreases, as shown in Figure 3a.

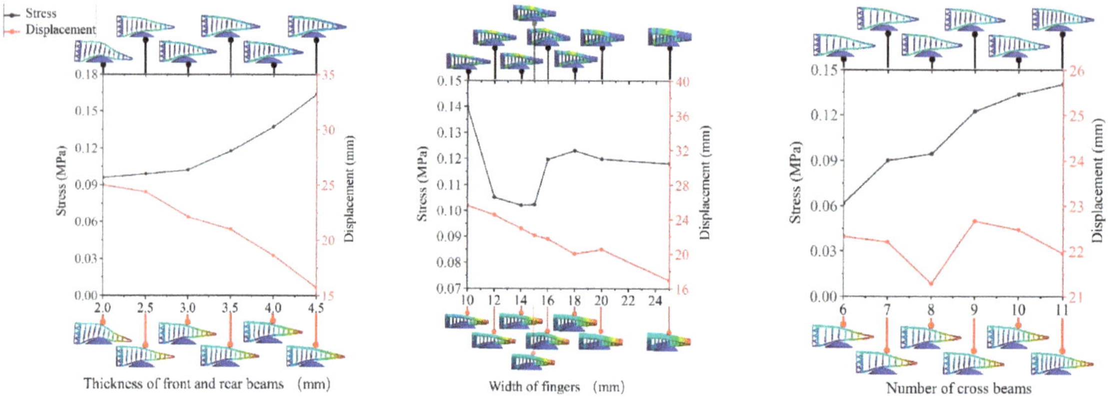

Figure 3. Changes in stress and displacement according to three factors: (**a**) thickness of front and rear beams; (**b**) width of fingers; (**c**) number of cross beams.

When the thickness of the front and rear beams is 2 mm, the stiffness of the finger after contact with the apple cannot be guaranteed, resulting in a small gripping force and easy fruit slip; when the thickness is 4.5 mm, the stress of the material itself will greatly limit its bending deformation and reduce the contact area between the fingers and fruit. At the same time, because excessive stress might cause fruit damage, the thickness of the front and back beams should not be too tiny or too large. When the thickness is 3.5 mm, the downward trend of the fingertip displacement becomes stronger as the thickness increases, while the upward trend of stress tends to be soft. As a result, selecting a thickness of 3.5 mm for the front and rear beams not only meets the requirement of the increasing gripping force but also allows fingers to make good contact with the fruits.

The effect of the finger width was then investigated. With the increase of the width, the fingertip displacement diminishes. However, the stress does not follow a constant pattern, as shown in Figure 3b. When the width is 10 mm, the stress and fingertip displacement is the greatest. This is because the finger width is excessively narrow, resulting in a limited contact area between the finger and the apple and high contact stress acting on the apple surface, which is easily damaged. Although the degree of the finger bend is greater when the finger is thin, it also results in insufficient grasping stiffness and fruit slide. When the width is 25 mm, the contact area between fingers and fruit increases, but its structure affects its bending, and it is not suitable for collecting fruits in the complex growing environment. When the width is 16 mm, as the width continues to increase, the fingertip displacement decreases dramatically and the stress tends to be flat. As a result, the best finger width is set to 16 mm in this study.

Finally, the number of beams was taken into account. Because the cross beams are the primary components that influence the stiffness of fingers, the number of cross beams has a substantial impact on the Young's modulus of the fingers [25]; hence, the distribution of the cross beams may have a major impact on the gripper performance. In distribution, there are several combinations of the cross beams. For the sake of simplicity, just the simplest equidistant parallel arrangement of the cross beams was considered in this study. Change the thickness of the front and rear beams to 3.5 mm, the width of the fingers to 16 mm, and change the number of cross beams. As the number of cross beams grows, so does the stress, and the fingertip displacement declines first and subsequently increases. When the number of beams is 9, the fingertip displacement reaches the maximum and then decreases again, as shown in Figure 3c. As a result, one selects nine as the optimal number of beams.

According to the results of the aforementioned analysis, the thickness of the front and rear beams has the greatest influence on the contact stress and fingertip displacement among the three geometric parameters. It is mostly because the finger surface is in direct contact with the fruit, and the thickness of the front and rear beams has a direct impact on the stiffness of the fingers. The structural parameters of the Fin Ray fingers are extremely complex, and this study just considers the most basic scenario. As a result, the best structural parameters are as follows: the thickness of the front and rear beams is 3.5 mm, the width of the fingers is 16 mm, and the number of beams is 9.

2.2. Overall Design of the Soft Gripper

The overall structure of the three-finger soft gripper for apple harvesting built with optimized Fin Ray fingers is shown in Figure 4a. It can be divided into three parts: the driving and sensing part, the transmission part, and the grasping part for clamping objects. The driving part is performed by a servo with torque and position feedback. To measure the relative distance between the gripper and the fruit, a distance sensor is mounted on the servo installation side of the gripper bottom plate. The transmission part is primarily accomplished by a slider, and the rocker mechanism was composed of a rocker, a connecting rod, a moving plate, and guide rods, as shown in Figure 4b. The servo rotates to drive the moving plate to move up and down. Because the fingers and their connectors are connected with the moving plate through the support rods, the fingers will move with the moving plate moving up and down, as shown in Figure 4c.

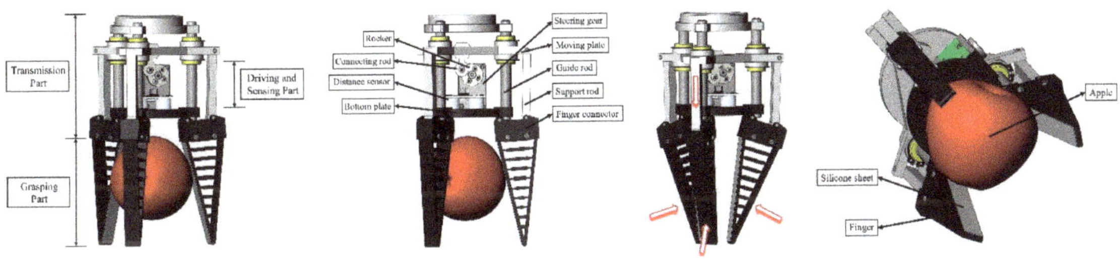

(a) (b) (c) (d)

Figure 4. Overall design of the soft gripper: (**a**) overall structure; (**b**) details of transmission and the driving and sensing part; (**c**) the designed gripping mechanism; (**d**) details of the grasping part.

In the grasping part, three Fin-Ray finger units are evenly distributed around the bottom plate of the gripper disc, connected with the transmission mechanism by the finger connectors to drive the Fin Ray fingers. A silicone pad is attached to the surface of each finger to increase the contact friction between the finger and the fruit, which ensures the clamping stability, as shown in Figure 4d.

At the initial position, the finger connectors are inclined outward at a certain angle relative to the bottom plate. Because the bottom of the fingers is connected in parallel with the bottom of their connectors, and the finger has a triangular symmetrical structure, the clamping range of the gripper is expanded.

3. Kinematic Mechanics Analysis of a Soft Gripper

The driving force begins with the servo, travels through the slider and rocker mechanism, multi-link mechanism, and Fin Ray soft structure, and eventually acts on the gripped fruit. In conclusion, the static analysis of the rigid multi-link mechanism and the soft finger structure was performed to acquire the gripping force on the fruit surface. Simultaneously, the relationship between the gripper pulling force and the gripping force was investigated in connection with the pulling harvesting method. Because the three fingers are symmetrically arranged relative to the bottom plate of the gripper, and the structure is the same. Furthermore, the servo output torque operates on the center of the moving plate, and the movement process and stress situation are comparable. As a result, the stress analysis of the direct contact between fingers and fruit begins with a single finger, making the analysis procedure simpler.

3.1. *Force Analysis of Rigid Multi-Link*

The basic structure and motion principle of the soft gripper is shown in Figure 5a,b. The force acting on the fruits of the Fin Ray structure can be equivalent to a single concentrated force in the analysis of the rigid multi-link (the analysis of the soft Fin Ray structure will be discussed below). The servo drives the rocker to rotate counterclockwise when grabbing, the moving plate to travel down along the guide rod, and the support rod to move. Following that, the support rod drives the finger connector to rotate around the joint FF, resulting in the envelope-gripping movement of the finger.

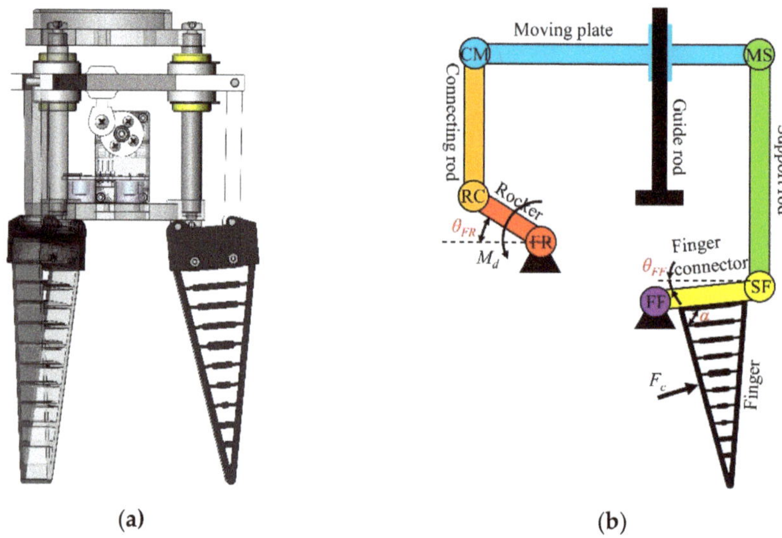

Figure 5. Motion schematic of the gripper: (**a**) physical model; (**b**) kinematics model.

In the figure: M_d is the servo output torque; θ_{FR} is the angle between the rocker and the horizontal direction; θ_{FF} is the angle between the finger connector and the horizontal direction; α is the angle between the front and rear beams of fingers and the base.

Because of the light weight of each moving part of the rigid multi-link, the gravity and inertia force during the movement of the gripper can be ignored.

The mechanical analysis of the multi-link mechanism is performed under static equilibrium conditions. The connecting rod is vertical to the moving plate at the time of initial contact. Their angle does not alter much when the rocker is rotated. To make the calculation easier, the difference is negligible. Among the multi-link, the connecting rod is a two-force member, and the moving plate is employed to assess the force, as shown in Figure 6. Therefore, one has

$$F_{CM} = F_{MS}, \tag{3}$$

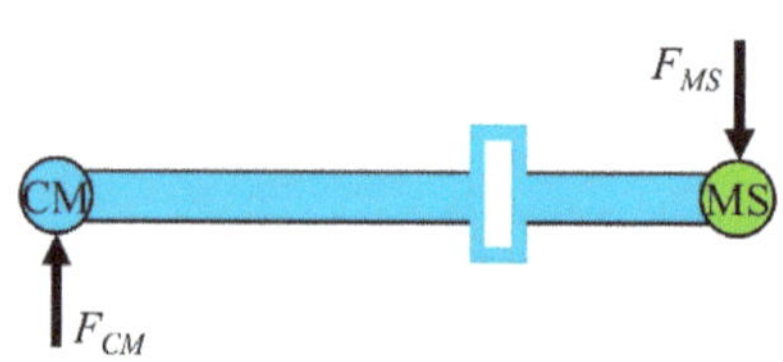

Figure 6. Force analysis of the moving plate.

For which, F_{CM} and F_{MS} are in the opposite direction. F_{XY} is the force of member X applying to member Y. To simplify the analysis, the sliding friction between the moving plate and the guide rod is negligible.

Thus, the support rod is a two-force member. Figure 7a shows the force analysis of the finger and its connector. The closing force triangle shown in Figure 7b can be obtained according to the geometric conditions for the equilibrium of the plane-intersecting force systems.

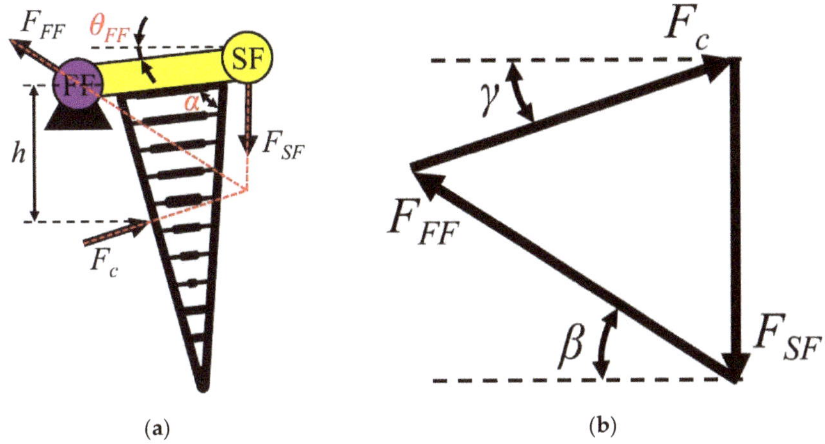

Figure 7. Force analysis of the finger and its connector: (**a**) force diagram; (**b**) closing force triangle.

To maintain the force balance of the finger and its connector, one obtains

$$F_c \cos\gamma = F_{FF}\cos\beta, \tag{4}$$

$$F_c \sin\gamma + F_{FF}\sin\beta = F_{SF}, \tag{5}$$

where F_c is the contact reaction between the finger and fruit, that is the finger gripping force; γ is the angle between F_c and the horizontal direction; β is the angle between F_{SF} and the horizontal direction.

According to Equations (4) and (5),

$$F_c = \frac{1}{\sin\gamma + \cos\gamma\tan\beta}F_{SF}, \tag{6}$$

where

$$\gamma = \frac{\pi}{2} - \alpha + \theta_{FF}, \tag{7}$$

$$\tan\beta = \frac{h}{L_{Fc}\cos\theta_{FF}\sin^2(\alpha-\theta_{FF})} - \frac{1}{\tan(\alpha-\theta_{FF})}, \tag{8}$$

where h is the distance from the center of fruit to the bottom plate of the gripper; L_X is the length of component X, that is L_{Fc} is the length of the Finger connector, and L_R is the length of the rocker.

To obtain the relationship between the servo torque M_d and the gripping force F_c, the rocker is taken as the forced object, and the force situation is shown in Figure 8. The moment balance at joint FR is

$$F_{CR}L_R\cos\theta_{FR} = M_d. \tag{9}$$

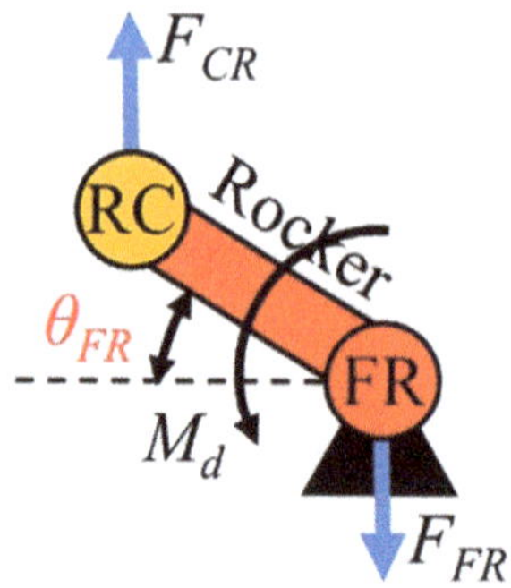

Figure 8. Force analysis of the rocker.

According to the force characteristics of the two-force members,

$$F_{CR} = F_{MS} = F_{FS} = F_{SF}, \tag{10}$$

where F_{FS} and F_{SF} are in opposite directions.

From Equations (6), (9), and (10), one can obtain

$$F_c = \frac{1}{\sin\gamma + \cos\gamma\tan\beta} \cdot \frac{1}{L_R \cos\theta_{FR}} \cdot M_d. \tag{11}$$

3.2. Contact Force Analysis between Soft Finger and Fruit

When the finger comes into contact with the fruit, it creates an adaptable envelope, and the contact area expands. The flexible deformation of the Fin Ray structure makes the mechanical analysis difficult. Therefore, to facilitate the calculation, the fruit is simplified as a regular sphere. Aiming at the picking method for pulling fruits, a simplified single-finger plane stress model is given in Figure 9.

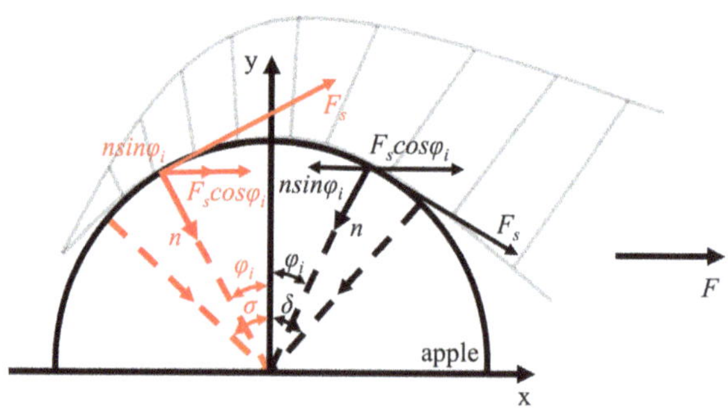

Figure 9. Plane force model of fruit (the forces are shown in red when the load divisions on the x-axis are in the same direction as the x-axis and black in the opposite direction).

The contact between the finger and the fruit is divided into two areas with angles of δ and σ, with the y-axis as the limit. The positive touching pressure of the fruit is simplified as a uniform load; the size is n; the unit is N/m, and the directions all point to the center of the circle, whose angle with the y-axis is φ_i (i = 1, 2, ... , m). F_s is the static friction force generated by the positive pressure of the finger on the fruit. When pulling the fruit, the positive pressure on the fruit and the component force of the static friction force generated along the x-axis direction are the main forces to ensure the stability of grasping. Specify that the direction of the force is positive along the positive x-axis. To obtain the resultant force in the x-axis direction F, one has

$$\begin{aligned} F &= \int_{-\delta}^{\sigma}(F_s \cos\varphi + n\sin\varphi)\cdot rd\varphi \\ &= r\cdot[F_s(\sin\delta+\sin\sigma)+n(\cos\delta-\cos\sigma)], \end{aligned} \tag{12}$$

where

$$F_s = \mu \cdot n. \tag{13}$$

In the Equation, μ is the maximum static friction coefficient between the finger and the pericarp of the fruit; r is the radius of the fruit.

Therefore, from Equations (11) and (12), the relationship between the resultant force F and the positive touching pressure on the fruit can be obtained,

$$F = rn\cdot[\mu(\sin\sigma+\sin\delta)+(\cos\delta-\cos\sigma)]. \tag{14}$$

The relationship between the equivalent single concentrated force F_c and the uniform load n in the rigid multi-link force analysis above is

$$F_c = \int_{-\sigma}^{\delta} n\cdot rd\psi = nr\cdot(\delta+\sigma). \tag{15}$$

According to Equations (11), (14), and (15), the relationship between the servo torque M_d and the resultant force F can be obtained as

$$F = \frac{\mu\cdot(\sin\sigma+\sin\delta)+(\cos\delta-\cos\sigma)}{L_R\cdot(\sin\gamma+\cos\gamma\tan\beta)\cdot\cos\theta_{FR}\cdot(\delta+\sigma)}\cdot M_d. \tag{16}$$

4. Soft Gripper Control Method for Slip Detection and Constant-Pressure Feedback

During the actual grasping, the gripping force is f_c, which is the same magnitude as the force F_c but in the opposite direction, and the pulling force is the resultant force in the x-axis direction F. From Equations (11) and (16), the relationship between the gripping force f_c of the gripper, the pulling force F, and the servo torque M_d can be obtained, as shown in Figure 10a. Therefore, when the fruit detachment force F_d is determined, the driving torque required for fruit detachment can be calculated according to the diameter of the fruit, thereby setting the servo output torque M_d. Simultaneously, it is possible to conclude that the gripping force f_c on the fruit surface at this time. To ensure constant pressure acting on the surface of the fruit, f_c should not be greater than the maximum pressure F_m that the pericarp of the fruit can withstand.

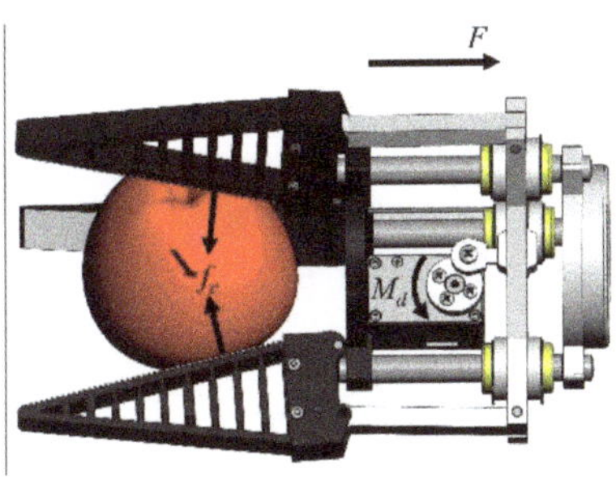

(a)

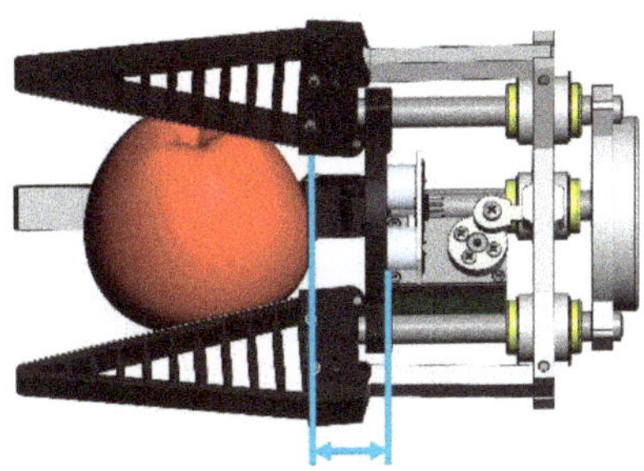

(b)

Figure 10. Model of the soft gripper control method: (**a**) the relationship among f_c, F, and M_d; (**b**) Relative position detection between fthe ruit and the gripper.

In addition to the fruit damage caused by the excessive gripping force of the fingers, which also includes bruises and scratches caused by the relative sliding between the fruit and the fingers, as shown in Figure 11. Therefore, to avoid the slippage between the fruit and the finger during harvesting, this paper detects the relative position between the fruit and the gripper by integrating a distance sensor to assess the fruit slippage and minimizing the damage caused by fruit slippage during harvesting, as shown in Figure 10b.

Figure 11. Damage to the fruit: (**a**) scratches; (**b**) bruises.

Combined with the constant-pressure feedback state, the specific implementation steps of the soft gripper control method for slippage detection are as follows:

4.1. Control Method of Constant-Pressure Feedback

The required servo output torque M_d can be obtained by identifying the diameter of the target fruit. To ensure that f_c is less than F_m at this time, the output torque must be adjusted further. When the f_c obtained at this time is greater than F_m, it should be ensured that the maximum torque can be output while the fruit is safely held. From Equation (11), let f_c equal F_m at this point to obtain the critical torque M_m of safe clamping, which is set as the servo's output torque. The gripper control method of the constant-pressure feedback is shown in Figure 12.

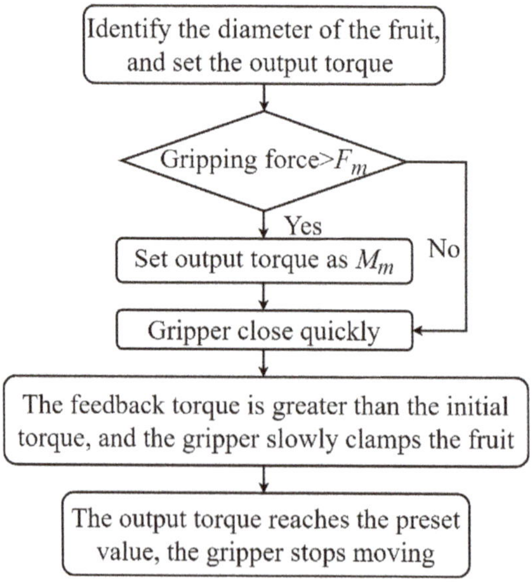

Figure 12. Gripper control method of constant-pressure feedback.

During the no-load closing motion of the gripper, the servo output torque is stable at an initial torque. When the feedback torque of the servo is greater than the initial torque, the finger and the fruit seem to be in contact. To avoid fruit damage due to the impact of the dynamic load, close the gripper quickly to reach the contact position before touching the fruit, and slowly close the gripper after the finger is in contact with the fruit. When the feedback torque reaches the preset value, it is assumed that the fruit has been grasped, and the servo stops rotating.

In contrast to the sensor system embedded in the finger, the servo with feedback information is used as the driver to ensure constant-pressure contact between the finger and the fruit, simplifying the structure of the soft harvesting gripper and facilitating the fruit harvesting in complex growth environments.

4.2. Control Method of Slip Detection

Fin Ray soft fingers have great advantages in dealing with the problem of fruit unilaterally damaged by extrusion. The cross beams act as rigid support rods to ensure the stiffness of the fingers while also allowing the fingers to adaptively wrap the entire fruit, preventing fruit damage due to the stress concentration.

However, it is difficult to ensure that relative slippage between the fingers and the fruit does not occur during the fruit detachment process. Because of the rough silicone pads attached to the surface of the fingers, the sliding friction force between the fingers and the fruit is relatively great when there is relative slippage between them, and it is easy to cause bruises and scratches on the fruit pericarp. As a result, effectively avoiding relative slippage is essential to ensure that the fruit is not damaged. The condition of the relative slippage, which causes the fruit damage, is complicated and will not be discussed in this paper.

A slip detection method is proposed for the designed soft gripper, which obtains the fruit position in real-time through the distance sensor. One believes that when the relative slip distance between the fruit and the fingers ΔL reaches L_s, the fruit tends to slip off, as shown in Figure 13. At this time, the output torque can be increased on the premise of ensuring that the maximum gripping force F_m is not exceeded, and the fruit can be clamped to prevent further sliding; if the relative slip distance ΔL can still reach L_s after increasing the output torque, clamping and pulling the fruit will increase the risk of damage, such as bruises and scratches. It means that the fruit is difficult to harvest at this point, and it is considered a harvesting failure, and the soft gripper is released. Controlling the gripper to perform the aforementioned operations n times, if harvesting fails all n times, give up picking this fruit. The slip detection control method is shown in Figure 14.

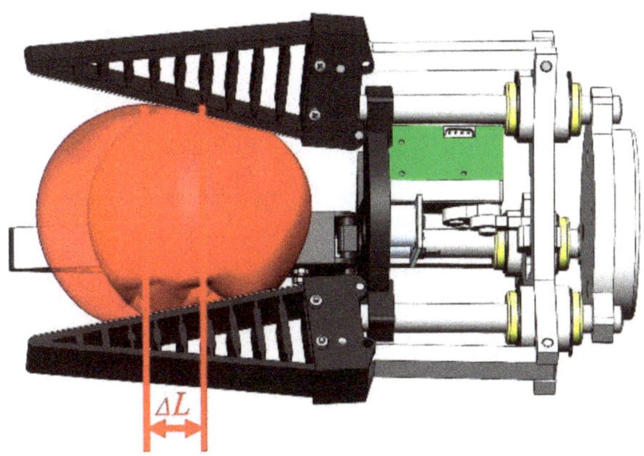

Figure 13. Slipping trend of fruit.

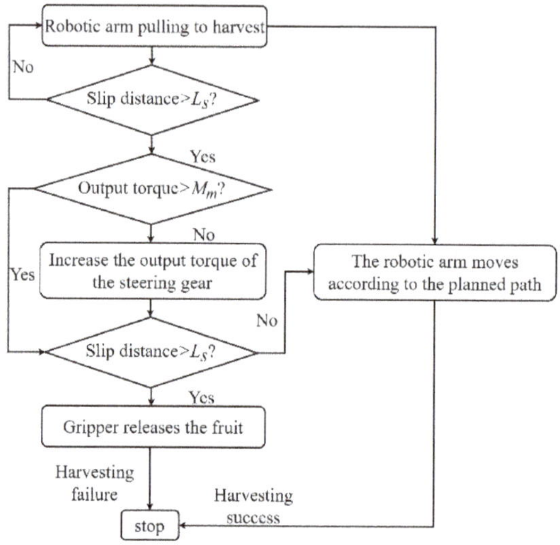

Figure 14. Gripper control method of slip detection.

Although the risk of harvesting failure is increased by the method proposed above, it does not cause damage to the fruit, and the fruit after harvesting failure can still be harvested manually without affecting its economic value or reducing economic losses.

5. Test and Analysis

5.1. Test Analysis of the Mechanical Properties of Apple

The Model E43 of MTS Exceed® Electromechanical Test Systems was used to conduct the relevant tests to obtain the relevant mechanical properties of the apples as the basis for the design of the gripper in this study. The range is 100 N, and it has a force and displacement sensor. Yantai Fushi apples were chosen as the test samples during the experiments.

In our study, a silicone pad is attached to the surface of the finger to improve the grasping performance by increasing the friction of the fruit's surface. To measure the maximum static friction coefficient μ between the silicone pad and the fruit, the pressure F_n was applied to the fruit through Model E43, and a silicone pad was pasted on the upper indenter and lower support, respectively. The tensile force of horizontally pulling the fruit was measured with a tension meter, as shown in Figure 15, and the horizontal pulling force F_p was measured from the beginning of the fruit slippage.

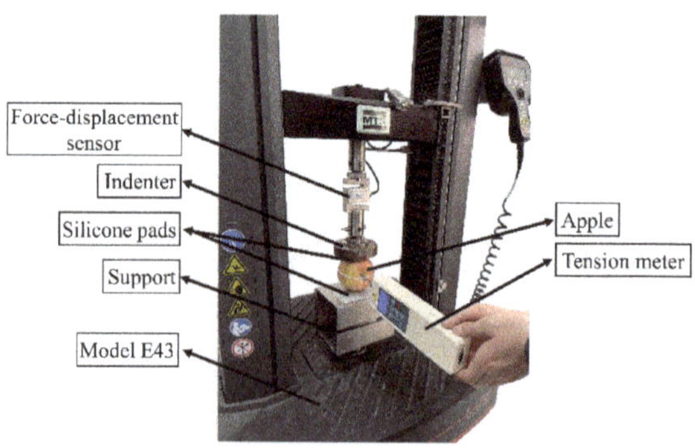

Figure 15. Diagram of mechanical properties test device.

Ignoring the apple's weight, it can be obtained from the static balance of the apple,

$$F_p = 2\mu \times F_n. \tag{17}$$

The test results and fitting function are shown in Figure 16, $R^2 = 0.92$. Therefore, $\mu = 0.8$ can be obtained.

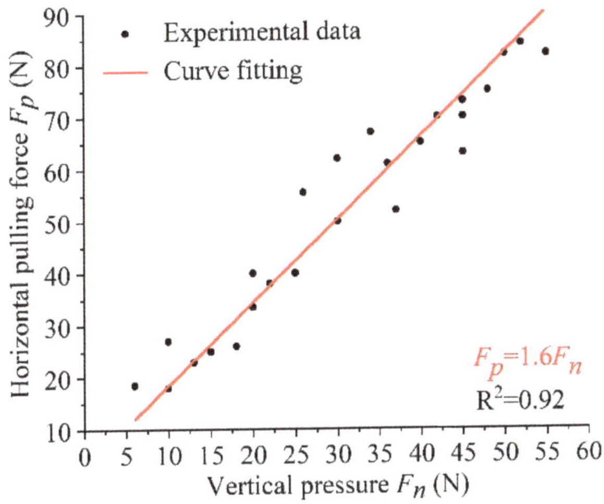

Figure 16. Test data and fitting curve.

To obtain the detachment force F_d required for fruit detachment, the apple was fixed on the support and kept still; then one end of the branch was fixed with the collection of the Model E43 and pulled axially. When the fruit branch was broken through the force sensor, the maximum pulling force was recorded. The experimental results are shown in Figure 17. The experiment used twenty apple samples with diameters ranging from 65 mm to 95 mm.

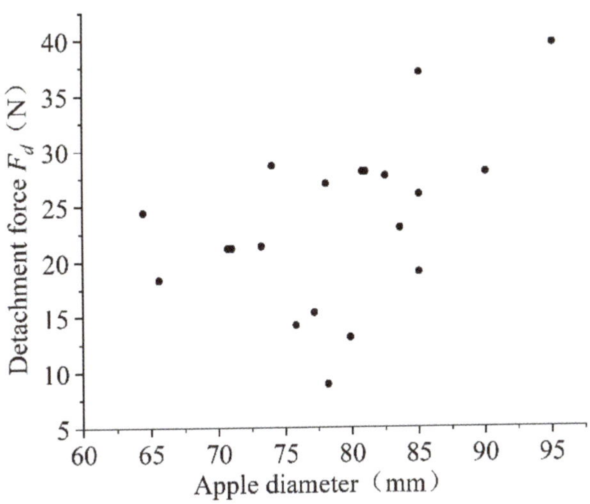

Figure 17. The influence of apple diameter on detachment force.

The results show that F_d is distributed between 8.88 N and 39.6 N. F_d generally increased as the apple diameter increased, but a small portion showed an irregular distribution. This could be because fruits with larger stem diameters have more connection force between branches and apples, necessitating more detachment force. At the same time, in the report of Bu [53] et al., the detachment force is much greater when the natural growth angle of the fruit is obtuse than when it is acute, as shown in Figure 18. In this experiment, we did not pay too much attention to the relation of detachment force to stem diameter and fruit growth angle. The test results were consistent with those of Bu [53] et al.

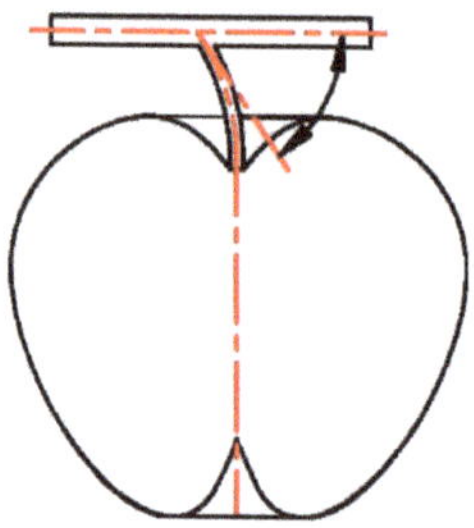

Figure 18. The natural growth angle of apple.

To avoid damaging the apple pericarp due to excessive gripping force, the maximum pressure F_m that the fruit pericarp can withstand must be known. We make a rectangular apple sample block of 10 mm × 10 mm × 20 mm near the apple's pericarp, place it on the middle of the support of the Model E43, and apply a load to the apple sample until it is destroyed. The force–displacement relationship during the apple-sample compression experiment was recorded, and the results are shown in Figure 19.

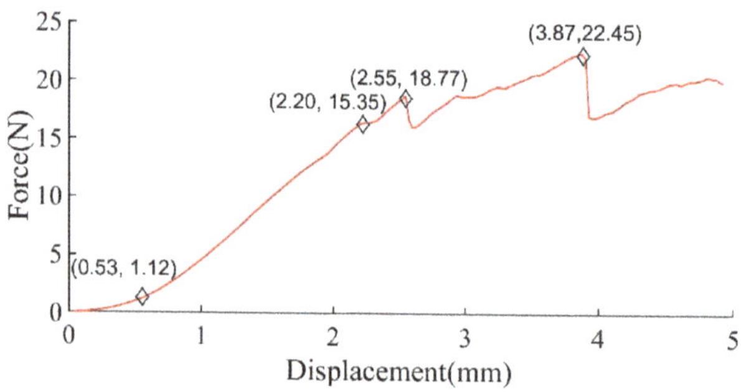

Figure 19. Force–displacement curve of the apple samples.

It can be seen that, once the force reaches 15.35 N, it remains almost unchanged with a one-stage displacement increase. This demonstrates that, when the force reaches 15.35 N, the apple begins to tend to plastically deform. According to the energy principle of the apple damage proposed by Schoorl [54], the damage volume of the apple is proportional to the energy it absorbs. To reduce the amount of energy transmitted to the apples during harvesting, set the maximum pressure F_m that apples can withstand to 15.35 N. The test results were consistent with those of Grotte [55] et al.

5.2. Gripping Force Verification Experiment

The rated torque that the servo can provide in this soft gripper is 12 kg·cm (1.2 N·m), assuming that the maximum torque that a single finger can provide is 0.4 N·m. L_{Fc} is 28 mm; L_R is 12 mm; h is 65 mm; α is 80°, and μ is 0.8. According to the structural design of the gripper, θ_{FF} is between −12°~15°, and θ_{FR} is between −30°~53°. Given that the fruit radius r varies, σ is customarily between 0° and 40°, and δ is traditionally between 0° and

25°. In the test, the diameter of the apple sample is about 90 mm. At initial contact, the finger and the fruit can be regarded as point contact. θ_{FF} is generally around 10°, and θ_{FR} is generally around 30°, as shown in Figure 20a. According to Equation (11), the maximum initial gripping force f_c of a single finger is approximately 15.34 N. The output torque M_d is little as the first contact, so the contact force between the finger and the fruit is far less than the maximum initial gripping force. When the gripper continues to close, σ and δ become larger, γ becomes smaller, and β becomes larger, so the finger gripping force f_c becomes larger, as does the pulling force F. At full contact, θ_{FF} is typically around −12°, and θ_{FR} is typically around −30°, as shown in Figure 20b. Therefore, the maximum gripping force f_c of a single finger is about 16.21 N. On the basis of Equation (16), the maximum pulling force F of a single finger is about 14.18 N, resulting in the maximum pulling force of the entire gripper being approximately 42.55 N.

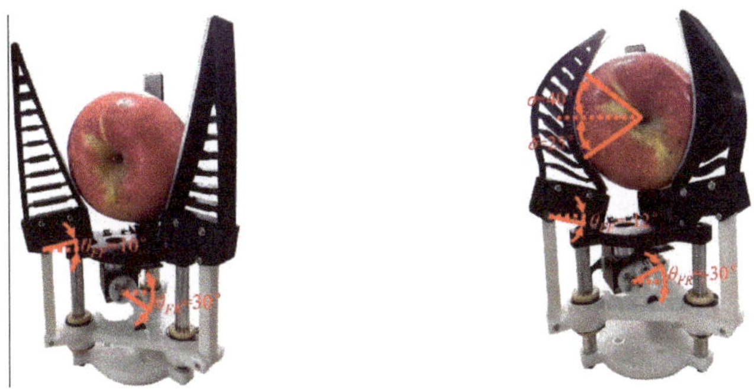

(a) (b)

Figure 20. Contact model of soft gripper: (**a**) initial contact; (**b**) full contact.

According to the above test results, the detachment force when pulling to harvest the fruit is about 8.88 N–39.6 N, indicating that the designed gripper's maximum pulling force meets the detachment requirement.

The gripping force resulting from the adaptive bending deformation of the soft fingers in contact with the fruit surface, which was measured by a thin-film pressure sensor (RP-L TDS REV C.) mounted between each finger and the silicone pad, as shown in Figure 21a. The RP-L type soft thin-film pressure sensor was composed of polyester film, high conductive material, and pressure-sensitive material. It converts the pressure acting on the thin-film area of the sensor into a change in resistance.

The test started when the finger made contact with the apple, and the output torque of the servo increased by 0.2 kg·cm (0.02 N·m) each time until it reached the rated torque of 12 kg·cm (1.2 N·m). To compare the difference in the gripping force of the finger on the surface of the fruit when the diameter of the fruit changes, apples with diameters of 70 mm, 80 mm, and 90 mm were chosen for the test, as shown in Figure 21b. In each test, the pressure output by the sensor and the servo torque was recorded, as shown in Figure 22.

As can be seen from the figure, there is an approximate positive relationship between the gripping force of the soft finger and the servo torque, and the image fits the theoretical curve well. Furthermore, it can be found that the effect on the gripping force is not very significant when the diameter of the fruit changes. Therefore, the finger output force during picking can be controlled by adjusting the servo output torque.

Nevertheless, the single-finger gripping force at a torque up to 1.2 N·m for the fruit diameter of 90 mm does not reach the theoretically calculated maximum value, which is probably due to the lack of accuracy from the thin-film pressure sensor.

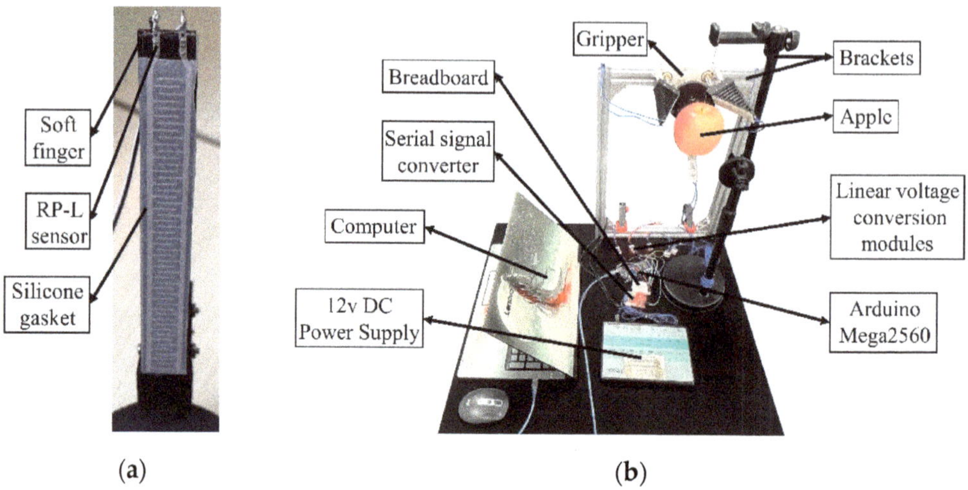

Figure 21. Experimental structure diagram: (**a**) finger with a RP-L sensor; (**b**) experimental platform.

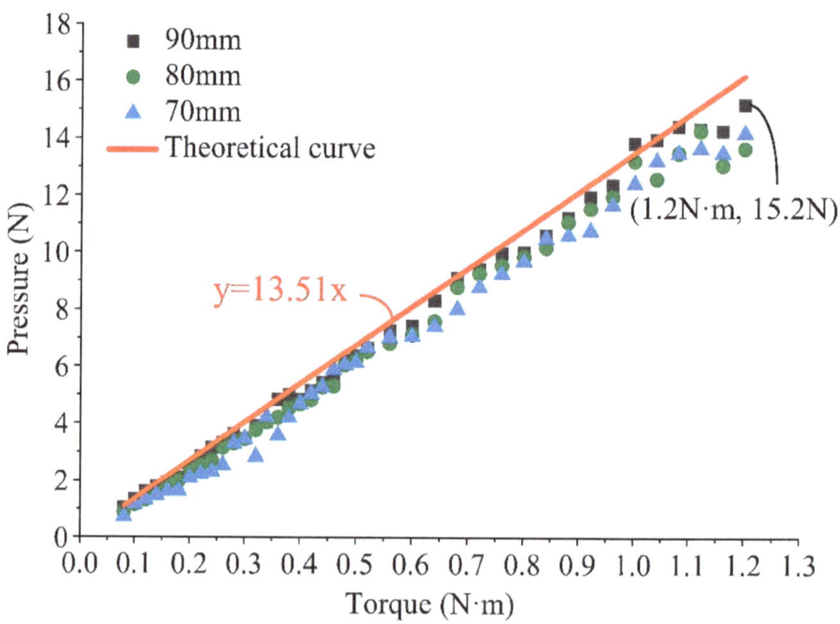

Figure 22. Relationship between torque and gripping force.

5.3. Test Analysis on the Harvesting Performance of the Soft Gripper

During the grasping and harvesting tests, the soft gripper was fixed on Franka, a seven-axis robotic arm with a high-sensitivity force-control performance, as shown in Figure 23. The tests were carried out in an orchard located in Changping Distrct, Beijing.

Figure 23. Experimental scene in the orchard.

5.3.1. Feasibility Test Analysis of Constant-Pressure Feedback System

To ensure that the finger gripping force is less than 15.35 N, the servo output torque is set to control the maximum gripping force f_c. Assuming that the detachment force required for fruit detachment is 40 N, it can be obtained from Equation (16) that the required output torque is 10.25 kg·cm (1.025 N·m). A single finger's grasping force f_c is 13.89 N, which is not harmful.

Therefore, a grasping comparison test was performed to verify the improvement of the soft gripper's safe grasping performance by the force feedback system. In this experiment, a total of 20 apple samples with no damage on the fruit skin were selected and divided into two groups of ten apples each. In the first set of experiments, the force feedback system was turned on, and the clamping test was performed on each apple. The clamping process followed the logic of the flowchart in Figure 12, and the clamping posture is shown in Figure 24a. After the gripper has completely and stably grasped the apple, hold it still for 5 s before releasing the fruit. In the second set of experiments, all experimental conditions were the same except that the force feedback system was turned off. As there is no output torque control, the clamping will stop until the servo reaches the locked rotor torque, and the clamping posture is shown in Figure 24b. The contact area on the fruit was marked after each release, and the fruit was then stored at the same constant temperature for 7 days. After taking them out, make a note of the damage on the apple surface's contact area. The radius of the damaged area was less than 10 mm for slight damage and greater than 10 mm for serious damage.

(**a**) (**b**)

Figure 24. Gripper attitude with force feedback on and off: (**a**) force feedback on; (**b**) force feedback off.

Observing the apple surface, the contact area of the apples clamped by the gripper with an open force feedback system was not damaged, so the damage rate was 0%; however, the apples were clamped by the gripper with a closed force feedback system. On the other hand, the slight damage rate was 10%, and the severe damage rate was 10%; the specific pericarp damage is shown in Figure 25. The experimental results show that activating the constant-pressure feedback system improves the soft gripper's safe grasping performance and effectively ensures non-destructive fruit grasping.

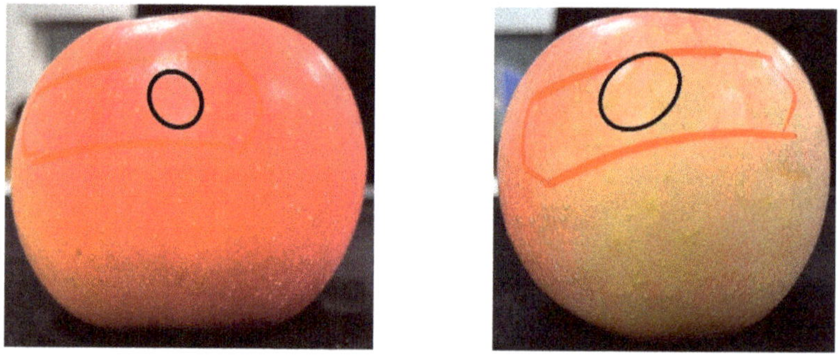

(**a**) (**b**)

Figure 25. Specific damage to apple pericarp: (**a**) slightly damaged; (**b**) severely damaged.

5.3.2. Test Analysis of Harvesting Success Rate and Apple Damage Rate

We carried out picking experiments to verify the stability and safety of the soft harvesting gripper designed in this paper. The harvesting process followed the logic of the flowcharts in Figures 12 and 14 with the force feedback system on. The soft finger length is 120 mm, while the effective gripping length is 100 mm. In the tests, L_s was set to 10 mm. To grab and separate the fruit, the soft picking gripper was controlled by Franka's arm with a pulling speed of 2 mm/s.

First, we analyzed various situations that occurred in the fruit harvesting process with the fruit slip detection turned on. The process began with the gripper approaching the fruit and ended with the fruit being harvested. The condition of the fruit slip and the change in the servo output torque for the three situations of no obvious slip, first slip, and second slip was recorded afterwards, as shown in Figure 26.

The figure shows that, even if the fruit did not slip for the first time, there would be a slight relative movement to the finger during harvesting, which might be due to the fingertip not being completely in contact with the fruit. After the fruit slipped slightly, the fingertip and the fruit made complete contact, providing adequate support for the apple. It was also conceivable that the measurement distance was floating within the accuracy range due to a lack of sensor accuracy. When the fruit slipped for the first time, the occurrence time was approximately 10 s, implying that the gripper pulled the fruit 2 cm. At this point, the fruit branch was completely straightened, and sufficient force was required to detach it from the branch; if the fruit slipped for the second time, it proved that it was not enough to harvest the fruit under the premise of safe harvesting; in addition, further harvesting might damage the fruit.

It can be ascertained that, during fruit harvesting, the stable servo output torque can ensure that the fruit does not break free due to the gripper loosening.

To further verify the effectiveness of the gripper harvesting, the tests for the three cases of rigid fingers and soft fingers with or without slip detection under the gripper structure of this study were carried out, as shown in Figure 27. For each group of the experiments, 25 apples with completely undamaged pericarps were selected. The picking situation and fruit harvesting damage were observed and recorded. The experimental results are shown in Tables 2–4.

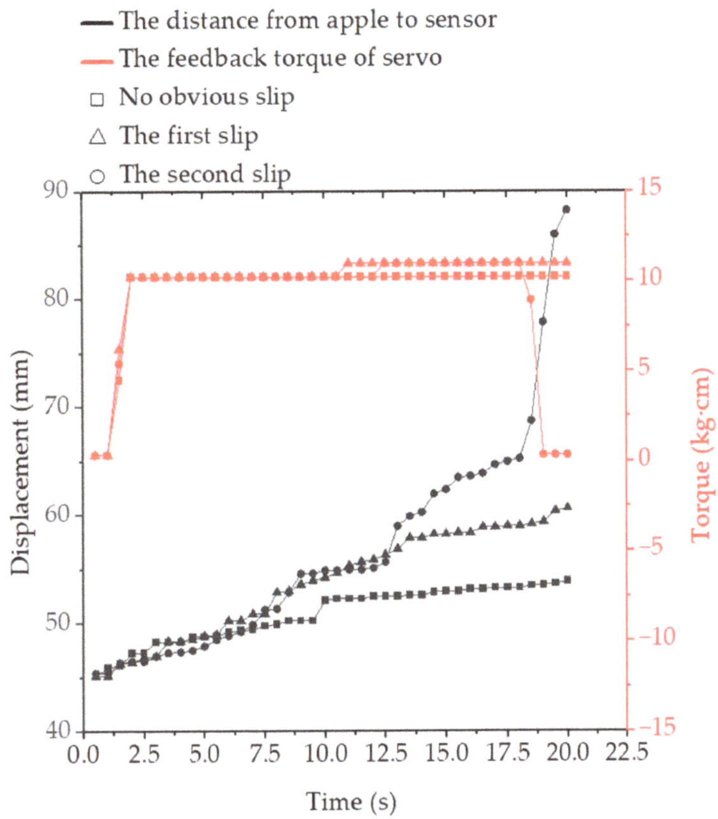

Figure 26. The slip condition of the fruit and the change of the output torque with time.

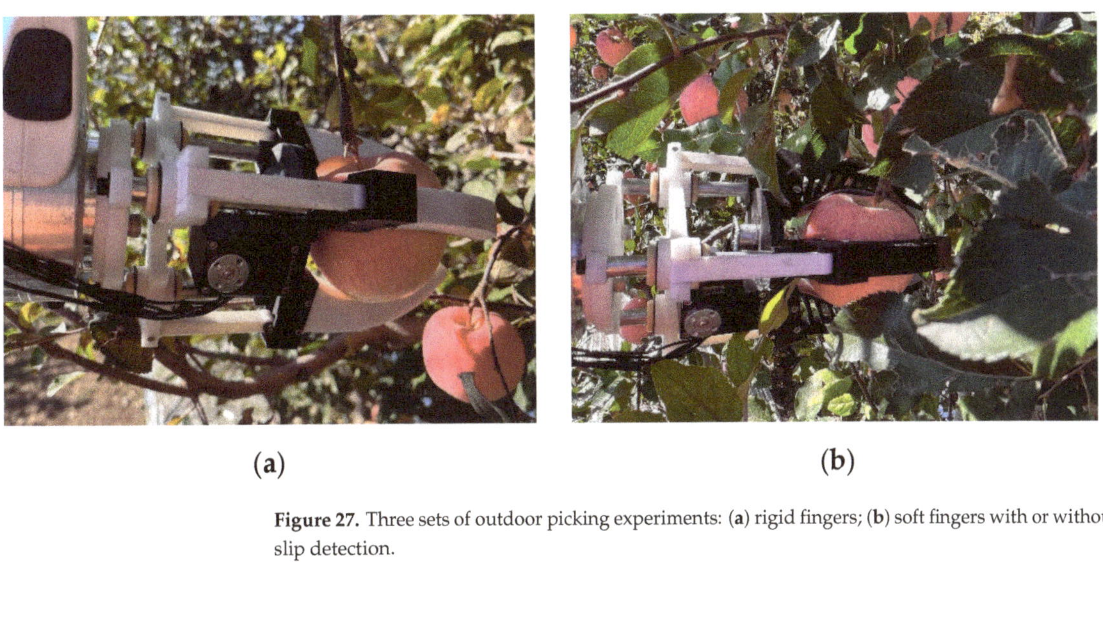

(a) (b)

Figure 27. Three sets of outdoor picking experiments: (**a**) rigid fingers; (**b**) soft fingers with or without slip detection.

Table 2. The harvesting situation of rigid fingers.

Average Diameter (mm)	Average Mass (g)	Number of Visible Slippage [1]	Number of Picking Success	Number of Picking Damage	Number of Slippage Damage
82.7536	236.748	7	25	4	3
		Damaged Fruit Characteristics			
Fruit Diameter (mm)	**Fruit Mass (g)**	**Visible Slippage or Not?**	**Picking Success or Not?**	**Picking Damage or Not?**	**Damage Causes [2]**
83.04	245.7	Yes	Yes	Yes	Slippage
86.37	280	Yes	Yes	Yes	Slippage
88.86	262.5	Yes	Yes	Yes	Slippage
90.23	278	No	Yes	Yes	Grasping

[1] When the slip detection is turned off, we define the visible slippage as the fruit that is about to slide to the fingertips of the gripper or that has already broken from the gripper (the same as below). [2] Because the fruit damage due to slippage in the gripper and due to grasping are quite different in character, we can distinguish them more easily (the same as below).

Table 3. The harvesting situation of soft fingers without slip detection.

Average Diameter (mm)	Average Mass (g)	Number of Visible Slippage	Number of Picking Success	Number of Picking Damage	Number of Slippage Damage
83.7548	232.724	9	24	2	2
		Damaged fruit characteristics			
Fruit Diameter (mm)	**Fruit Mass (g)**	**Visible Slippage or Not?**	**Picking Success or Not?**	**Picking Damage or Not?**	**Damage Causes**
82.35	235.5	Yes	Yes	Yes	Slippage
86.66	260.3	Yes	Yes	Yes	Slippage

Table 4. The harvesting situation of soft fingers with slip detection.

Average Diameter (mm)	Average Mass (g)	Number of First Slippage	Number of Second Slippage	Number of Picking Success	Number of Picking Damage
84.2252	242.932	13	7	20	0
		Second Slippage Fruit Characteristics			
Fruit Diameter (mm)	**Fruit Mass (g)**		**Second Picking Success or Not?**	**Picking Damage or Not?**	**Damage Causes**
82.32	226.5	**Second Picking after First Failed Harvesting [1]**	Yes	No	—
83.31	233.5		No	No	—
84.45	226		No	No	—
84.65	256.1		Yes	No	—
86.19	266.6		No	No	—
90.19	279.4		No	No	—
91.11	309.8		No	No	—

[1] With slip detection on, the second slippage of the fruit means that the picking has failed. At this point, each fruit was picked twice; it implies that the fruit has failed, and the next fruit would be chosen if both pickings failed.

Comparing Tables 2 and 3, the picking success rate for the rigid fingers is 100%, with a damage rate of 16%, while the success rate for the soft fingers is 96%, and the damage rate is 8%, both of these have the silicone gasket applied to the surface. This shows that the optimized Fin-Ray soft fingers in this paper are able to reduce the fruit damage better. At the same time, we can see that visible slippage of the fruit was common in both cases and that most of the damage occurred during the fruit slippage in the gripper. In the rigid fingers experiment, three fruits were damaged by slippage and one by grasping, which also shows that the rigid support structure is prone to fruit damage despite the flexible silicone gasket applied to the surface. In the soft fingers experiment, both damaged fruits were caused by slippage. Therefore, the effective control of the fruit slip in the gripper is essential to reducing the risk of fruit damage.

Comparing Tables 3 and 4, although the picking success rate dropped to 80% with slip detection on, there was no fruit damage. It turns out that the soft gripper with slip detection can effectively reduce fruit damage. Despite the fact that the harvesting success rate will decrease, the fruit will not be harmed, and its economic value will not be impacted after manual harvesting. In addition, we can see from Table 4 that 13 fruits made the first slippage, and in 7 of them, the second slippage occurred, further demonstrating the prevalence of fruit sliding during picking. Although five fruits failed in the second picking, no fruit were damaged, which indicates that the proposed control method for slip detection is effective in preventing damage to the fruits.

According to the above experimental results, the proposed Fin-Ray soft harvesting gripper with force feedback and fruit slip detection enables stable and non-destructive fruit picking. Notably, to improve the harvesting lossless rate, it is necessary to sacrifice some harvesting success rates by detecting slippage between the fruit and the fingers.

Remark 2. *It should be noted that the experimental results of the outdoor harvesting could be regarded as the effect of combining both force feedback and slip detection on the basis of the optimized harvesting gripper.*

6. Conclusions

This paper presents a three-fingered apple-harvesting soft gripper with constant-pressure feedback inspired by the Fin-Ray effect. First, the structural parameters of the single-soft-finger model were optimized using finite element analysis, and the influence of different Fin Ray finger structural parameters on the contact stress and fingertip displacement was investigated. The optimal structural parameters of the single soft finger were proposed: the front and rear beam thickness is 3.5 mm; the finger width is 16 mm, and the number of beams is 9. A three-fingered apple harvesting soft gripper was designed based on the above-optimized fingers. The determined gripper structure's mathematical model was then statically analyzed, and the relationship between the gripping force, the pulling force, and the servo torque was obtained. Therefore, the finger output force during picking can be controlled by adjusting the servo output torque.

We also propose a dynamic control method for detecting fruit slip during apple harvesting by integrating a distance sensor in this study. The maximum static friction coefficient between the finger and the apple, the detachment force of the apple, and the damaged condition of the apple were obtained through an experimental analysis of the apple's mechanical properties, which provides a theoretical basis for the gripper design. In indoor experiments, the results show that the servo output torque has an approximately linear relationship with the contact pressure between the fingers and the apple, and it is suitable for all sizes of apple. In the outdoor orchard experiments, turning on the constant-pressure feedback system can improve the safe grasping performance of the soft gripper, which can effectively ensure non-destructive fruit gripping. Comparing the tests for the three cases of rigid fingers and soft fingers with or without slip detection, the optimized Fin-Ray soft fingers in this paper are able to reduce the fruit damage better, and opening the slip detection can effectively avoid fruit damage. Furthermore, the stable output torque of the servo can ensure that the fruits do not break free due to the gripper loosening during harvesting.

In this study, we believe that the soft harvesting gripper is not only suitable for harvesting apples but also for harvesting some other fruits (e.g., tangerine and kiwi) and vegetables (e.g., tomato) and can provide an application reference. It has a high degree of adaptability and can effectively avoid fruit damage by adjusting the servo output torque.

However, our current research work still has some limitations. On the one hand, the complex structure of the Fin Ray fingers needs further investigation, and we will conduct more in-depth and detailed research on it in the future. On the other hand, the theoretical analysis and design of the gripper are only for the single-pulling fruit harvesting method in this study, which has significant limitations. This is only the first step in our exploration. In future work, combined with the optimal method and the posture of the fruit harvesting, the harvesting method combining gripper rotation and pulling will be studied.

Author Contributions: Conceptualization, K.C. and T.L.; methodology, K.C.; software, K.C.; validation, K.C., T.L., Q.F. and Q.Z.; formal analysis, K.C., F.X. and T.Y.; investigation, K.C.; resources, Q.F.; data curation, K.C.; writing—original draft preparation, K.C.; writing—review and editing, T.L.; visualization, K.C.; supervision, T.L. and Q.F.; project administration, C.Z.; funding acquisition, Q.F. and T.L. All authors have read and agreed to the published version of the manuscript.

Funding: This research was funded by Beijing Science and Technology Plan Project (Z201100008020009); Construction Project of Beijing Key Laboratory of Agricultural Intelligent Equipment Technology in 2022 (PT2022-37); BAAFS Innovation Capacity Building Project, grant number KJCX20210414; Beijing Key Laboratory of Intelligent Equipment Technology for Agriculture, Beijing 100097, China; the Priority Academic Program Development of Jiangsu Higher Education Institutions (Grant No. PAPD-2018-87); the Jiangsu Postdoctoral Sustentation Fund [Grant No. 2020Z378].

Institutional Review Board Statement: Not applicable.

Data Availability Statement: Not applicable.

Acknowledgments: The authors would like to thank the editor and the anonymous reviewers for their valuable suggestions to improve the quality of this paper.

Conflicts of Interest: The authors declare no conflict of interest.

References

1. Li, T.; Qiu, Q.; Zhao, C.; Xie, F. Task planning of multi-arm harvesting robots for high-density dwarf orchard. *Trans. CSAE* **2021**, *37*, 1–10.
2. Liu, C.; Feng, Q.; Tang, Z.; Wang, X.; Geng, J.; Xu, L. Motion Planning of the Citrus-Picking Manipulator Based on the TO-RRT Algorithm. *Agriculture* **2022**, *12*, 581. [CrossRef]
3. Li, T.; Feng, Q.; Qiu, Q.; Xie, F.; Zhao, C. Occluded Apple Fruit Detection and Localization with a Frustum-Based Point-Cloud-Processing Approach for Robotic Harvesting. *Remote Sens.* **2022**, *14*, 482. [CrossRef]
4. Li, T.; Yu, J.; Qiu, Q.; Zhao, C. Hybrid Uncalibrated Visual Servoing Control of Harvesting Robots with RGB-D Cameras. In *IEEE Transactions on Industrial Electronics*; IEEE: Manhattan, NY, USA, 2022.
5. Qingchun, F.; Wei, C.; Yajun, L.; Bowen, W.; Liping, C. Method for identifying tomato plants pruning point using Mask R-CNN. *Trans. Chin. Soc. Agric. Eng.* **2022**, *38*, 128–135.
6. Bontsema, J.; Hemming, J.; Pekkeriet, E.; Saeys, W.; Edan, Y.; Shapiro, A.; Hočevar, M.; Oberti, R.; Armada, M.; Ulbrich, H. CROPS: Clever robots for crops. *Eng. Technol. Ref.* **2015**, *1*, 1–11. [CrossRef]
7. Russo, S.; Ranzani, T.; Liu, H.; Nefti-Meziani, S.; Althoefer, K.; Menciassi, A. Soft and stretchable sensor using biocompatible electrodes and liquid for medical applications. *Soft Robot.* **2015**, *2*, 146–154. [CrossRef]
8. Deimel, R.; Brock, O. A novel type of compliant and underactuated robotic hand for dexterous grasping. *Int. J. Robot. Res.* **2016**, *35*, 161–185. [CrossRef]
9. Ma, L.; Yang, W.; Wang, C.; Chen, X.; Xue, C.; Lin, Y.; Liu, A. Structure Design and Experiment of the End-effector for Apple-harvesting Robot. *J. Agric. Mech. Res.* **2009**, *12*, 65–67.
10. Li, X.; Yu, J. Design of an Underactuated End-effector for an Apple Picking Robot. *J. Mech. Transm.* **2015**, *39*, 74–77.
11. Yao, X.; Wang, X.; Zhang, L.; Li, H.; Dai, L.; Lu, G. Flexible Robot Hand Based on Slider and Rocker Mechanism. *Trans. Chin. Soc. Agric. Mach.* **2021**, *52*, 396–405.
12. Paek, J.; Cho, I.; Kim, J. Microrobotic tentacles with spiral bending capability based on shape-engineered elastomeric microtubes. *Sci. Rep.* **2015**, *5*, 10768. [CrossRef] [PubMed]
13. Sun, Y.; Yap, H.K.; Liang, X.; Guo, J.; Qi, P.; Ang, M.H., Jr.; Yeow, C.-H. Stiffness customization and patterning for property modulation of silicone-based soft pneumatic actuators. *Soft Robot.* **2017**, *4*, 251–260. [CrossRef] [PubMed]
14. Shepherd, R.F.; Ilievski, F.; Choi, W.; Morin, S.A.; Stokes, A.A.; Mazzeo, A.D.; Chen, X.; Wang, M.; Whitesides, G.M. Multigait soft robot. *Proc. Natl. Acad. Sci. USA* **2011**, *108*, 20400–20403. [CrossRef] [PubMed]
15. Martinez, R.V.; Branch, J.L.; Fish, C.R.; Jin, L.; Shepherd, R.F.; Nunes, R.M.; Suo, Z.; Whitesides, G.M. Robotic tentacles with three-dimensional mobility based on flexible elastomers. *Adv. Mater.* **2013**, *25*, 205–212. [CrossRef] [PubMed]

16. Polygerinos, P.; Lyne, S.; Wang, Z.; Nicolini, L.F.; Mosadegh, B.; Whitesides, G.M.; Walsh, C.J. Towards a soft pneumatic glove for hand rehabilitation. In Proceedings of the 2013 IEEE/RSJ International Conference on Intelligent Robots and Systems (IROS), Tokyo, Japan, 3–7 November 2013; pp. 1512–1517.
17. Bartlett, N.W.; Tolley, M.T.; Overvelde, J.T.; Weaver, J.C.; Mosadegh, B.; Bertoldi, K.; Whitesides, G.M.; Wood, R.J. A 3D-printed, functionally graded soft robot powered by combustion. *Science* **2015**, *349*, 161–165. [CrossRef]
18. Fruit-Picking Robot Solves Automation Challenge. Available online: https://www.theengineer.co.uk/content/news/fruit-picking-robot-solves-automation-challenge (accessed on 7 July 2022).
19. Muscato, G.; Prestifilippo, M.; Abbate, N.; Rizzuto, I. A prototype of an orange picking robot: Past history, the new robot and experimental results. *Ind. Robot. Int. J.* **2005**, *32*, 128–138. [CrossRef]
20. Nestorovic, T.; Struhar, V. Designing a personal assistance application using wizard of Oz methodology. In Proceedings of the Annals of DAAAM for 2011 and 22nd International DAAAM Symposium "Intelligent Manufacturing and Automation: Power of Knowledge and Creativity", Vienna, Austria, 23–26 November 2011; pp. 715–716.
21. Crooks, W.; Vukasin, G.; O'Sullivan, M.; Messner, W.; Rogers, C. Fin ray® effect inspired soft robotic gripper: From the robosoft grand challenge toward optimization. *Front. Robot. AI* **2016**, *3*, 70. [CrossRef]
22. Wilson, M. Festo drives automation forwards. *Assem. Autom.* **2011**, *31*, 12–16. [CrossRef]
23. Crooks, W.; Rozen-Levy, S.; Trimmer, B.; Rogers, C.; Messner, W. Passive gripper inspired by Manduca sexta and the Fin Ray® Effect. *Int. J. Adv. Robot. Syst.* **2017**, *14*, 1729881417721155. [CrossRef]
24. Basson, C.I.; Walker, A.; Bright, G. Testing flexible grippers for geometric and surface grasping conformity in reconfigurable assembly systems. *South Afr. J. Ind. Eng.* **2018**, *29*, 128–142. [CrossRef]
25. Shin, J.H.; Park, J.G.; Kim, D.I. A universal soft gripper with the optimized fin ray finger. *Int. J. Precis. Eng. Manuf.-Green Technol.* **2021**, *8*, 889–899. [CrossRef]
26. Elgeneidy, K.; Lightbody, P.; Pearson, S.; Neumann, G. Characterising 3D-printed soft fin ray robotic fingers with layer jamming capability for delicate grasping. In Proceedings of the 2019 2nd IEEE International Conference on Soft Robotics (RoboSoft), COEX, Seoul, Korea, 14–18 April 2019; pp. 143–148.
27. Becedas, J.; Payo, I.; Feliu, V. Two-flexible-fingers gripper force feedback control system for its application as end effector on a 6-DOF manipulator. *IEEE Trans. Robot.* **2011**, *27*, 599–615. [CrossRef]
28. Chin, L.; Yuen, M.C.; Lipton, J.; Trueba, L.H.; Kramer-Bottiglio, R.; Rus, D. A simple electric soft robotic gripper with high-deformation haptic feedback. In Proceedings of the 2019 International Conference on Robotics and Automation (ICRA), Palais des congres de Montreal, Montreal, QC, Canada, 20–24 May 2019; pp. 2765–2771.
29. Kultongkham, A.; Kumnon, S.; Thintawornkul, T.; Chanthasopeephan, T. The design of a force feedback soft gripper for tomato harvesting. *J. Agric. Eng.* **2021**, *52*. [CrossRef]
30. Hao, Y.; Liu, Z.; Liu, J.; Fang, X.; Fang, B.; Nie, S.; Guan, Y.; Sun, F.; Wang, T.; Wen, L. A soft gripper with programmable effective length, tactile and curvature sensory feedback. *Smart Mater. Struct.* **2020**, *29*, 035006. [CrossRef]
31. Zhou, H.; Wang, X.; Kang, H.; Chen, C. A Tactile-enabled Grasping Method for Robotic Fruit Harvesting. *arXiv* **2021**, arXiv:2110.09051.
32. Truby, R.L.; Wehner, M.; Grosskopf, A.K.; Vogt, D.M.; Uzel, S.G.; Wood, R.J.; Lewis, J.A. Soft somatosensitive actuators via embedded 3D printing. *Adv. Mater.* **2018**, *30*, 1706383. [CrossRef] [PubMed]
33. Bilodeau, R.A.; White, E.L.; Kramer, R.K. Monolithic fabrication of sensors and actuators in a soft robotic gripper. In Proceedings of the 2015 IEEE/RSJ International Conference on Intelligent Robots and Systems (IROS), Congress Center Hamburg, Hamburg, Germany, 28 September–2 October 2015; pp. 2324–2329.
34. Zhu, H.; Li, X.; Chen, W.; Li, X.; Ma, J.; Teo, C.S.; Teo, T.J.; Lin, W. Weight imprinting classification-based force grasping with a variable-stiffness robotic gripper. *IEEE Trans. Autom. Sci. Eng.* **2021**, *19*, 969–981. [CrossRef]
35. Xu, W.; Zhang, H.; Yuan, H.; Liang, B. A compliant adaptive gripper and its intrinsic force sensing method. *Trans. Robot.* **2021**, *37*, 1584–1603. [CrossRef]
36. De Barrie, D.; Pandya, M.; Pandya, H.; Hanheide, M.; Elgeneidy, K. A Deep Learning Method for Vision Based Force Prediction of a Soft Fin ray Gripper Using Simulation Data. *Front. Robot. AI* **2021**, *8*. [CrossRef]
37. Belzile, B.; Birglen, L. A compliant self-adaptive gripper with proprioceptive haptic feedback. *Auton. Robot.* **2014**, *36*, 79–91. [CrossRef]
38. Zhou, J.; Zhu, S. Slippage detection in gripping fruits and vegetables for agricultural robot. *Trans. Chin. Soc. Agric. Mach.* **2013**, *44*, 171–176.
39. Zhang, W.; Yuan, L. Design of the Control System of Fruit and Vegetable Flexible Grasping of Agricultural Robot Based on Slip Detection. *J. Agric. Mech. Res.* **2017**, *6*, 228–232.
40. Okatani, T.; Nakai, A.; Takahata, T.; Shimoyama, I. A MEMS slip sensor: Estimations of triaxial force and coefficient of static friction for prediction of a slip. In Proceedings of the 2017 19th International Conference on Solid-State Sensors, Actuators and Microsystems (TRANSDUCERS), The University of Tokyo, Tokyo, Japan, 18–22 June 2017; pp. 75–77.
41. Zhang, T.; Fan, S.; Zhao, J.; Jiang, L.; Liu, H. Multifingered robot hand dynamic grasping control based on fingertip three-axis tactile sensor feedback. In Proceedings of the 11th World Congress on Intelligent Control and Automation, Shenyang, China, 29 June–4 July 2014; pp. 3321–3326.

42. Xin, Y.; Tian, H.; Guo, C.; Li, X.; Sun, H.; Wang, P.; Lin, J.; Wang, S.; Wang, C. PVDF tactile sensors for detecting contact force and slip: A review. *Ferroelectrics* **2016**, *504*, 31–45. [CrossRef]
43. Agriomallos, I.; Doltsinis, S.; Mitsioni, I.; Doulgeri, Z. Slippage detection generalizing to grasping of unknown objects using machine learning with novel features. *IEEE Robot. Autom. Lett.* **2018**, *3*, 942–948. [CrossRef]
44. Romeo, R.A.; Oddo, C.M.; Carrozza, M.C.; Guglielmelli, E.; Zollo, L. Slippage detection with piezoresistive tactile sensors. *Sensors* **2017**, *17*, 1844. [CrossRef] [PubMed]
45. Li, J.; Dong, S.; Adelson, E. Slip detection with combined tactile and visual information. In Proceedings of the 2018 IEEE International Conference on Robotics and Automation (ICRA), Brisbane, Australia, 21–25 May 2018; pp. 7772–7777.
46. Su, Z.; Hausman, K.; Chebotar, Y.; Molchanov, A.; Loeb, G.E.; Sukhatme, G.S.; Schaal, S. Force estimation and slip detection/classification for grip control using a biomimetic tactile sensor. In Proceedings of the 2015 IEEE-RAS 15th International Conference on Humanoid Robots (Humanoids), Seoul, Korea, 3–5 November 2015; pp. 297–303.
47. Romeo, R.A.; Zollo, L. Methods and sensors for slip detection in robotics: A survey. *IEEE Access* **2020**, *8*, 73027–73050. [CrossRef]
48. Liu, S.Q.; Adelson, E.H. GelSight Fin Ray: Incorporating Tactile Sensing into a Soft Compliant Robotic Gripper. In Proceedings of the 2022 IEEE 5th International Conference on Soft Robotics (RoboSoft), Edinburgh, UK, 4–8 April 2022; pp. 925–931.
49. Bu, L.; Hu, G.; Chen, J. Assessment of Grasp Ability for An End-effecter with Fin-ray Structure. In Proceedings of the Journal of Physics: Conference Series, Digital Meeting, 10–14 October 2021; p. 032030.
50. Lee, C.; Kim, M.; Kim, Y.J.; Hong, N.; Ryu, S.; Kim, H.J. Soft robot review. *Int. J. Control Autom. Syst.* **2017**, *15*, 3–15. [CrossRef]
51. Qian, S.; Lu, Y.; Yang, X. Overview of selection and parameter determination for hyperelastic constitutive model of rubber material. *Rubber Sci. Technol.* **2018**, *16*, 5–10.
52. Ahmadi, E.; Barikloo, H.; Kashfi, M. Viscoelastic finite element analysis of the dynamic behavior of apple under impact loading with regard to its different layers. *Comput. Electron. Agric.* **2016**, *121*, 1–11. [CrossRef]
53. Bu, L.; Hu, G.; Chen, C.; Sugirbay, A.; Chen, J. Experimental and simulation analysis of optimum picking patterns for robotic apple harvesting. *Sci. Hortic.* **2020**, *261*, 108937. [CrossRef]
54. Schoorl, D.; Holt, J. Impact bruising in 3 apple pack arrangements. *J. Agric. Eng. Res.* **1982**, *27*, 507–512. [CrossRef]
55. Grotte, M.; Duprat, F.; Loonis, D.; Piétri, E. Mechanical properties of the skin and the flesh of apples. *Int. J. Food Prop.* **2001**, *4*, 149–161. [CrossRef]

agriculture

MDPI

Article

A Real-Time Shrimp with and without Shells Recognition Method for Automatic Peeling Machines Based on Tactile Perception

Xueshen Chen, Yuesong Xiong, Peina Dang, Chonggang Tao, Changpeng Wu, Enzao Zhang and Tao Wu *

College of Engineering, South China Agricultural University, Guangzhou 510642, China
* Correspondence: wt55pub@scau.edu.cn

Abstract: Accurate and automatic real-time recognition of shrimp with and without shells is the key to improve the efficiency of automatic peeling machines and reduce the labor cost. Existing methods cannot obtain excellent accuracy in the absence of target samples because there are too many species of shrimp to obtain a complete dataset. In this paper, we propose a tactile recognition method with universal applicability. First, we obtained tactile data, e.g., the texture and hardness of the surface of the shrimp, through a novel layout using the same type of sensors, and constructed fusion features based on the energy and nonstationary volatility (ENSV). Second, the ENSV features were input to an adaptive recognition boundary model (ARBM) for training to obtain the recognition boundary of shrimp with and without shells. Finally, the effectiveness of the proposed model was verified by comparison with other tactile models. The method was tested with different species of shrimp and the results were 88.2%, 87.0%, and 89.4%, respectively. The recognition accuracy of the overall, shrimp with shells and shrimp without shells verified the generalizability of the proposed method. This method can help to improve the efficiency of automatic peeling machines and reduce the labor cost.

Keywords: shrimp; automatic peeling machines; tactile perception; recognition

Citation: Chen, X.; Xiong, Y.; Dang, P.; Tao, C.; Wu, C.; Zhang, E.; Wu, T. A Real-Time Shrimp with and without Shells Recognition Method for Automatic Peeling Machines Based on Tactile Perception. *Agriculture* **2023**, *13*, 422. https://doi.org/10.3390/agriculture13020422

Academic Editors: Jin Yuan, Wei Ji and Qingchun Feng

Received: 24 October 2022
Revised: 14 January 2023
Accepted: 30 January 2023
Published: 10 February 2023

1. Introduction

The shrimp industry is a key sector of the fishing industry [1]. Research on equipment for the automated processing of shrimp is important because manual processing not only leads to low productivity and high production costs but also reduces the quality of shrimp products [2,3]. The typical process used by shrimp peeling equipment is to first remove the head of the shrimp, followed by the shell, by squeezing it through a roller sleeve [4,5]. The automated recognition of shrimp with and without shells must be explored because existing automatic peeling machines are not perfect and require the secondary manual recognition of shrimp with shells.

Machine vision is widely used as a nondestructive detection technique for the quality evaluation and body measurement of shrimp [6–8] Some scholars have implemented shrimp detection tasks by extracting color, shape, and texture features from images and combining them with machine learning models [9–11]. Deep learning, which can automatically learn the feature representations of original image pixel data without relying on specific features, has achieved great success in the field of image recognition [12,13]. Zhang et al. proposed a YOLOv3 multisource fish detection framework based on multiscale fusion and identified fish bodies in fish images based on a CenterNet target detection network with an average accuracy of 90.2% [14]. Conrady et al. constructed a sea bream recognition model based on a mask region-based convolutional neural network (R-CNN) with good accuracy [15]. However, the visual method can recognize samples that are similar to the training samples [16]. As there are more than 2000 shrimp species, it is difficult to obtain a comprehensive sample dataset. In addition, its processing is mainly in the form of video [17], which limits its application in shrimp identification because of its long training time and high equipment requirements.

Tactile sensing is another form of perception that ignores the influence of shrimp species. Tactile sensing recognizes and detects the objects to be measured by analyzing the tactile time-series signals of these objects [18,19]. It is widely used in different fields owing to its high processing speed and recognition accuracy for objects with large force differences [20–22] Wang's team and Zhang's team applied principal component analysis (PCA) to reduce the dimensionality of tactile signals and recognize different objects by machine learning methods [23,24]. Keser's team and Qin's team used the discrete wavelet transform (DWT) method to generate feature vectors of tactile sample signals and then implemented the classification of tactile signals [25,26]. In the abovementioned studies, most of the tactile data on the object being tested are homogeneous, and whole or partial features are directly extracted for recognition by manual experience. However, the tactile data obtained from the surface of the shrimp is inhomogeneous, and shrimp with shells have complex and variable shell attachment sites, making it difficult to obtain accurate experimental results.

In this paper, we propose a method to identify shrimp with and without shells by tactile sensation. First, we use two sensors of the same type to obtain tactile data on the texture and hardness of the shrimp surface, and construct fusion features based on energy and nonstationary volatility (ENSV). Then, based on the feature distribution of the ENSV, an adaptive recognition boundary model (ARBM) is constructed. Finally, we verify the feasibility and generalizability of the proposed method. The main contributions of this study are as follows.

(1) This is an attempt to identify shrimp with and without shells using a tactile method to address the problem of the non-universality of existing recognition methods because of the large number of shrimp species.
(2) A physically meaningful ENSV-ARBM tactile signal processing scheme is proposed to amplify the tactile differences between shrimp with and without shells and reduce the effect of uncertainty in the recognition of shrimp with and without shell samples.
(3) The proposed method can meet the requirements of automatic peeling machines for accurate recognition of different species of shrimp in real time, which helps to improve the efficiency of automatic peeling machines and reduce the labor cost.

2. Materials and Methods

2.1. Experimental Setup

In this study, a tactile sensor was developed. When the tactile sensor slides across the surface of an object, it senses the surface texture and hardness information of the object and transmits signals over time through two sensing cells. The tactile sensor consists of four carbon fiber plates (Zesheng Carbon Fiber Products Factory, Zhongshan, China) and two piezoelectric film polyvinylidene fluoride (PVDF) sensors (Jiangmen Antai Electronics Co., Ltd., Jiangmen, China). The fabrication of the tactile sensor proposed in this study is simple, as shown in Figure 1a.

Four carbon fiber plates were offset and stacked in turn. When the tactile sensor touches an object, it amplifies the vibration features to show the tactile features of the object. Two piezoelectric film PVDF sensors with a copper block embedded in each end increases the visibility and recognizability of the tactile signal.

One piezoelectric film PVDF sensor (Sensor A) is horizontally installed in the middle of four carbon fiber plates. The copper block extends out of the carbon fiber plate and is in a suspended state. In this manner, the piezoelectric film PVDF sensor can obtain the surface texture information when the object is touched by the tactile sensor. The other piezoelectric film PVDF sensor (Sensor B) is installed in the middle of the longest carbon fiber plate. The copper block faces downward along the carbon fiber sheet. In this manner, the piezoelectric film PVDF sensor can obtain the hardness information when the object is touched by the tactile sensor. The material specifications of the tactile sensor are listed in Table 1.

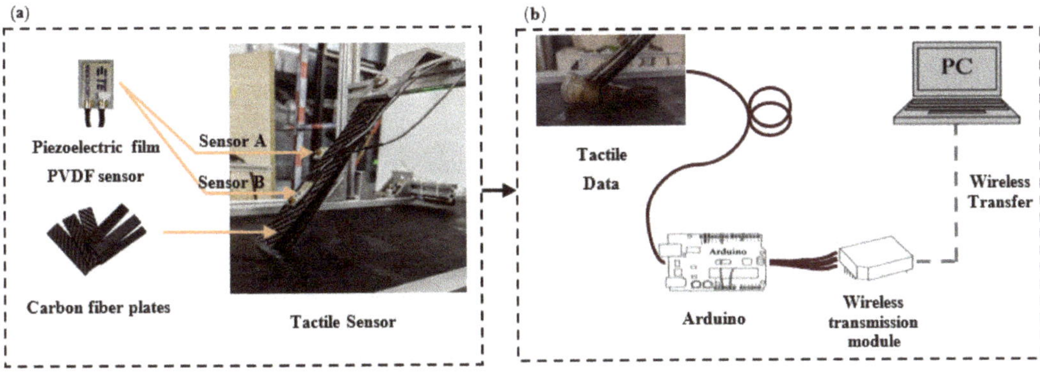

Figure 1. Schematic diagram of tactile sensor. (**a**) The physical diagram of tactile sensor; (**b**) diagram of experimental setup of tactile time-series acquisition.

Table 1. Material and structural parameters of the tactile sensors.

Material Type	Parameter	Structure Type	Parameter
Carbon fiber plates (Length/mm × Width/mm × Thick/mm)	150 × 30 × 1	Extended length of transverse piezoelectric film PVDF sensor/mm	10
Piezoelectric film PVDF sensor (Length/mm × Width/mm × Thick/mm)	20 × 10 × 1	Tilt angle of the sensor/°	60
Copper blocks (Bottom area/mm^2 × Height/mm)	2.25π × 3	Carbon fiber plates offset distance/mm	3

The conveyor belt speed is fixed, the carbon fiber plate of the tactile sensor scans the surface of the shrimp. Sensor A captures information about the texture of the shrimp's body by vibrating as the carbon fiber plate comes into contact with the shrimp. Sensor B, which is bent by the force created by the contact, captures information about the hardness of the shrimp. Shrimp with shells generally have a hard and rough surface, whereas shrimp without shells have a soft and smooth surface. This differential data of the shrimp's body surface is obtained through the use of two sensing units.

The experimental setup for tactile time-series acquisition is shown in Figure 1b. First, the shrimp were transported by a conveyor belt. When the shrimp pass the laser sensor, the data from the tactile sensor were acquired. Then, the Arduino (Shanghai Longzhan Information Technology Co., Ltd., Shanghai, China) collected the output signals of the tactile sensor. These were transmitted to the Bluetooth module and wirelessly transmitted to the computer in real time for processing and analysis. The data visualization interface is based on the LabVIEW software for computing. Finally, the obtained tactile signals were processed in a MATLAB (mathematical tool) environment.

2.2. *Data Processing*

Tactile time-series data were obtained from shrimp with and without shells. First, discrete tactile data were preprocessed by theoretical waveform analysis. Second, the ENSV features were extracted from the preprocessed tactile data. Finally, the ENSV was input into the ARBM to obtain the recognition models of shrimp with and without shells.

2.2.1. Tactile Signal Acquisition and Preprocessing

The tactile sensors described in Section 2.1 were used to acquire tactile data from the shrimp. All samples were placed on a conveyor belt moving at a speed of 0.1 m/s for tactile data acquisition. Taking into account the distance between the end of the tactile sensor and the laser sensor, data acquisition starts 2000 ms after the laser sensor is activated to analyze the data efficiently and reduce storage space. The sampling frequency of the analog signal of the tactile sensor was set to 1300 Hz, which is twice of that of the tactile data frequency to ensure that the tactile data does not overlap in the frequency domain. To obtain the complete tactile sensing process of the shrimps, the data capacity of one sample was set to 5000 data points (2500 data points each for the Sensor A and Sensor B). The acquisition ends after 5000 data were collected for each sample.

The raw signal plot is shown in Figure 2a. The blue waveform represents the data acquired by Sensor A, i.e., the shrimp surface texture information. The red waveform represents the data acquired by Sensor B, i.e., the shrimp hardness information. During the dynamic process of data acquisition, the raw output signal contains a DC component, which leads to a nonzero starting signal and different starting values for the two sensing cells.

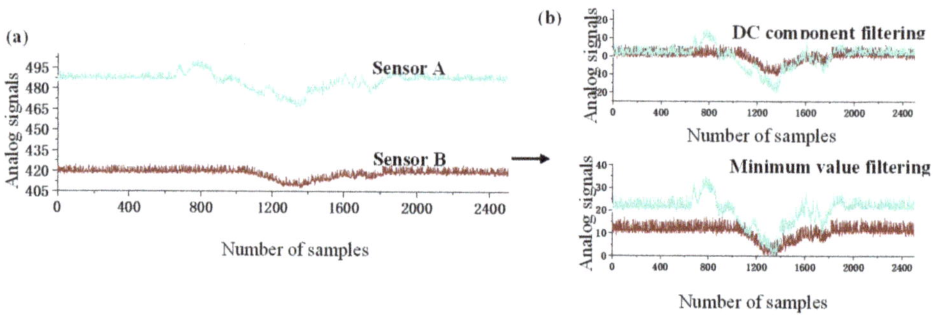

Figure 2. Data preprocessing process diagram. (**a**) Waveform diagram of the original tactile signal; (**b**) Waveform diagram of the tactile data after preprocessing.

When the energy features were extracted, the direct calculation of energy features would result in large energy values for each segment. When the nonstationary volatility was extracted, the direct calculation of nonstationary volatility of tactile signals would result in small nonstationary volatility values for each segment. Both cases affect the recognition accuracy of the sensor. Therefore, it is necessary to filter the DC components of the signal. However, when the DC component is filtered, the tactile signal will contain data less than 0, which leads to errors in the calculation of nonstationary volatility values. The data after preprocessing are shown in Figure 2b.

To reduce the interference of DC components in feature extraction, two tactile signal preprocessing methods were used. These are the direct filtering of the DC components from tactile signals when extracting energy features, and the minimum value filtering of tactile signals when extracting nonstationary volatility features. The specific raw signal processing is expressed as Formulas (1) and (2).

$$SC_N = TS'_N - \overline{TS'_N} \tag{1}$$

$$SM_N = TS'_N - \min\left(TS'_N\right) \tag{2}$$

where SC_N is the filtered DC component signal, N is the number of sampling points per sensing cell (N = 2500), TS'_N is the original tactile signal, $\overline{TS'_N}$ is the average value of the raw tactile signal, SM_N is the minimum value of the filtered signal, and $\min\left(TS'_N\right)$ is the minimum value of the raw tactile signal.

Tactile signals are directly used to train the model to recognize different objects by undergoing a complex learning process that ignores the detailed feature information about the surface texture and hardness [27]. The segmentation of the preprocessed tactile signal can tap into the details of the tactile signal, and reducing the signal length to process at each instant while keeping the signal characteristics [28]. The sliding window method was used to segment the data with a certain step size to ensure data continuity after segmentation.

The effects of data segmentation are shown in Figure 3a,b. The number of segments is calculated by

$$i = \frac{N - Sl + Ss}{Ss} \tag{3}$$

where i is the number of segments (i is an integer), N is the length of the preprocessed data, Sl is the window data length, and Ss is the sliding step of the segments. The window data length and sliding step length were set as 50 and 10, respectively.

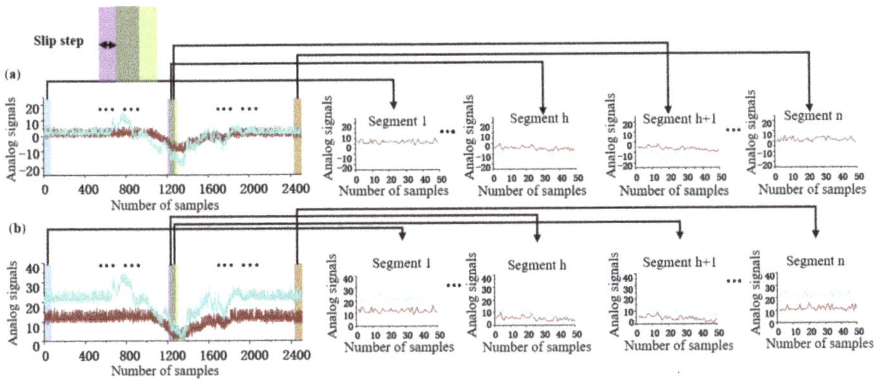

Figure 3. Schematic diagram of tactile data segmentation. (**a**) Schematic diagram of tactile signal segmentation with DC component filtering; (**b**) schematic diagram of tactile signals segmentation with minimum value filtering.

2.2.2. ENSV Features Extraction

Machine learning techniques combined with feature extraction methods can improve the recognition accuracy as well as speed-up the training process. We selected the ENSV features as the feature vector for the recognition of shrimp with and without shells. The energy feature in ENSV characteristics can well reflect the changes of the force on the sensor, and the nonstationary volatility feature can make the sensor more clear in the force process. The fusion of the two features can reduce the interference of invalid information and amplify the tactile differences between shrimp with and without shells.

First, the energy features of each segment were extracted after DC component filtering. Second, the nonstationary volatility features of each segment were extracted after minimum filtering. Finally, the energy and nonstationary volatility features were fused to obtain the identification feature vector. The feature extraction process is shown in Figure 4.

In the process of acquiring tactile signals, there is a difference in the blocking force between the tactile sensor and shrimp with and without shells. The surface of shrimp without shells is smooth and soft, producing a small blocking force. In contrast, the surface of shrimp with shells is rough and hard, producing a large blocking force. To describe the process of changing force on the tactile sensor as it slides across the shrimp surface, we extracted the energy of each segment as a feature after DC component filtering. The calculation formula is expressed as (4). The effect is shown in Figure 4a.

$$E_i = \frac{\sum_{n=(i-1)Ss}^{(i-1)Ss+Sl} (Sq_i)^2}{Sl} \tag{4}$$

where E_i is the average energy of each segment. Sq_i is the tactile data of each segment after DC component filtering, (n = 1, 2, … , 2500). n is the sequence number of the sampling point. The formula of the segmented energy feature vector is expressed as (5).

$$E = (E_1, E_2 \ldots E_i) \tag{5}$$

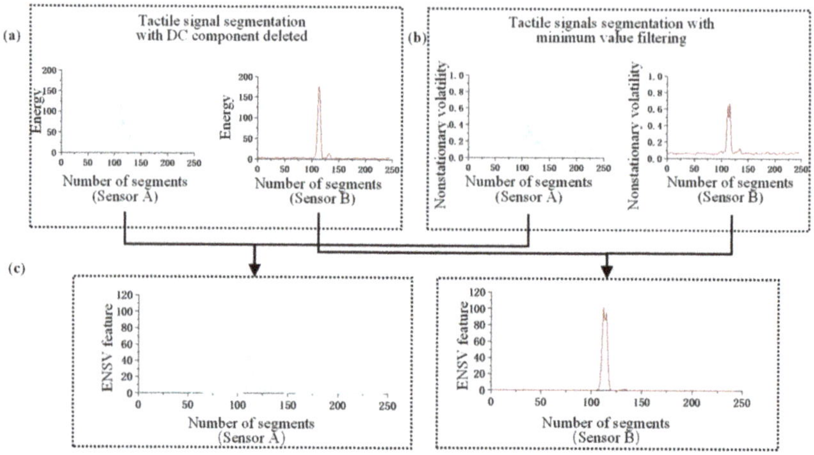

Figure 4. Feature extraction process diagram. (**a**) The result of the energy features of each segment; (**b**) the result of the nonstationary degree of fluctuation of each segment; (**c**) the result of feature fusion.

When the tactile sensor is not in contact with the shrimp, the tactile sensor data is stable. When the tactile sensor is in contact with the shrimp, it deforms and vibrates, and tactile signals produce nonstationary volatilities. This type of volatility differs from that of a stationary signal. We extracted the nonstationary volatility of each segment as a feature after minimum value filtering. The effect is shown in Figure 4b.

Ideally, for stationary volatility data, the sum of squares of any two tactile data points is equal to two times the square of the initial value. Let Sw_n be the value of any sampling point in a segment after the minimum value is deleted. Sw_{n+m} is the value of exploring m sampling points backwards from the nth sampling point. These are expressed as

$$Sw(n) = (Sw_n)^2 + (Sw_{n+m})^2 \tag{6}$$

where Sw is a function that varies with the sampling point n, represented as $Sw(n)$. In an ideal case, the Sw is constant for stationary data. The mathematical expectation of the Sw in a certain segment is

$$\text{ESw} = \frac{\sum_{n=(i-1)Ss}^{(i-1)Ss+Sl} Sw(n)}{N} \tag{7}$$

The relative mean square deviation (σW_i) of $Sw(n)$ and its mathematical expectation ESw is

$$\sigma W_i = \frac{\sqrt{E\left\{[Sw(n) - ESw]^2\right\}}}{ESw} \tag{8}$$

σW_i increases with the degree of nonstationarity. If the data are stationary under ideal conditions, then $\sigma W_i = 0$. The degree of nonstationary volatility of the feature vector is expressed as

$$\sigma W = (\sigma W_1, \sigma W_2, \ldots, \sigma W_i) \tag{9}$$

This feature amplifies the textural and hardness characteristics of the tactile sensor during contact with the shrimp, and reduces the data interference in the noncontact state. The expression of this feature is provided in (10) and illustrated in Figure 4c.

$$V = E \odot \sigma W \tag{10}$$

2.2.3. ARBM Construction

As mentioned in the introduction, shrimp with shells have complex shell attachment sites. Therefore, we propose an ARBM to solve this problem. First, the ENSV feature vectors were pretrained using a back-propagation (BP) neural network fitting model, as shown in Figure 5. Then, the feature vectors of shrimp samples with and without shells were assumed to be located in different circular regions, and the center of each class was calculated, as shown in Figure 6a. Finally, the radius of the recognition boundary was obtained by training. The recognition boundary of shrimp without shells was retained, while that of shrimp with shells was discarded. The shrimp with and without shells are located outside and inside the boundary, respectively, as shown in Figure 6b.

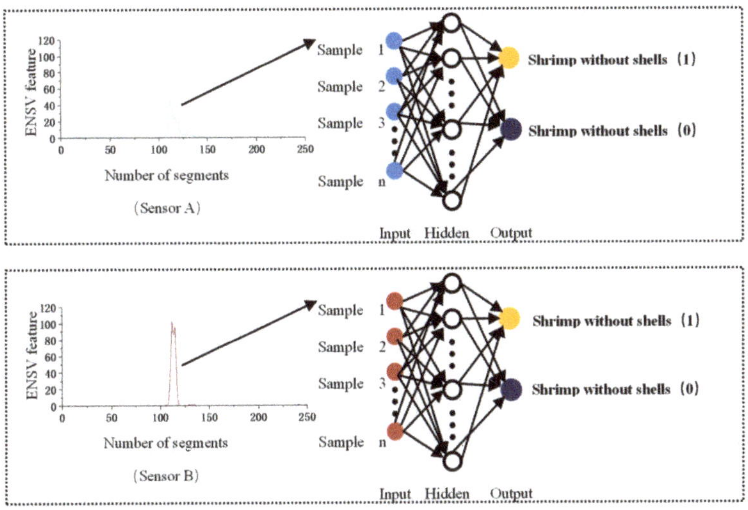

Figure 5. Pretraining flow chart.

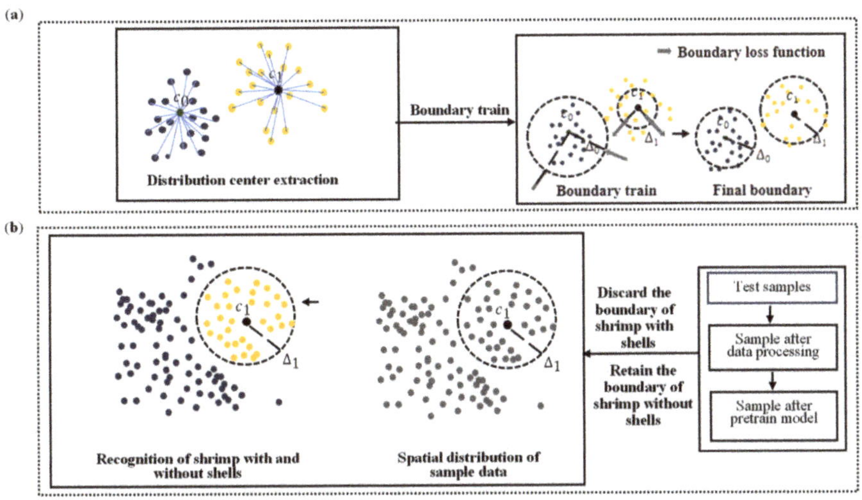

Figure 6. Boundary training and recognition schematic. (**a**) The boundary training schematic; (**b**) the recognition schematic.

Pretraining

We pretrained the model using shrimp with and without shells as prior knowledge. This enables more respective clustering of the ENSV feature vector distributions of shrimp with and without shells. Artificial neural networks were used to classify the feature dataset [29]. In general, BP neural networks do not have strict data distribution requirements. These can automatically transform the initial "bottom" feature representation into the "top" feature representation through multilevel and nonlinear transformations [30]. This part uses ENSV as prior knowledge for the pretraining process. The data from Sensor A and Sensor B are trained separately. The number of neural nodes in the input layer corresponds to the number of segments of the samples. The model input is the ENSV feature vectors extracted from the sample. The number of neurons in the hidden layer is 10. The number of neural nodes in the output layer corresponds to whether the shrimps have shells (1 for shrimp with shells and 0 for shrimp without shells), as shown in Figure 5.

Boundary Training

In this section we input the data to the pretrained neural network fitting model. Then the fitted values of shrimp with and without shells from different sensors are obtained. Place the values of Sensor A and Sensor B in the same two-dimensional coordinate system. The flow chart of boundary training and recognition is shown in Figure 6.

Pretraining uses the ENSV features of Sensor A and Sensor B as input quantities and shrimp with and without shells as output quantities. This process groups shrimp by their class and separates different classes. To make data computation more efficient and improve real-time processing, we use a circular boundary defined by only two parameters (radius and cluster center) to simplify the data analysis. Before training the recognition model, the centers of the feature vector distributions of shrimp with and without shells must be determined. Shrimp with shells are one class, while shrimp without shells are another. The sample dataset of a class is treated as a cluster, and the cluster centers are determined by calculating the mean feature vector of each cluster.

$$c_k = \frac{1}{|D_k|} \sum_{(Q_j, Y_j) \in D_k} Q_j \tag{11}$$

where $D_k = \left\{ (Q_1, Y_1), \ldots, \left(Q_j, Y_j \right) \right\}$ is the set of ENSV and its label. For shrimp with and without shells, $Y_j = 1$ and $Y_j = 0$, respectively. D_k is the number of sample sets marked as the same class.c_k is the cluster center.

Define Δ_k as the radius of the recognition boundary relative to the center of the circle c_k. The ENSV should satisfy the following constraints:

$$\forall Q_j \in D_k, \|Q_j - c_k\|_2 \leq \Delta_k \tag{12}$$

where $\|Q_j - c_k\|_2$ represents the Euclidean distance between Q_j and c_k. The SoftPlus activation function was used to map the radius and radius parameters.

$$\Delta_k = log\left(1 + e^{\widehat{\Delta_k}}\right) \tag{13}$$

where Δ_k is the cluster radius and $\widehat{\Delta_k}$ is the radius parameter.

On the one hand, it is hoped that the recognition boundary can surround most shrimp with and without shells. On the other hand, it is also hoped that the boundary of the circle is not too far from the center of the cluster. Therefore, the following boundary loss function is adopted.

$$\mathcal{L}_b = \frac{1}{M} \sum_{j=1}^{M} \left[\delta_j \left(\left\| Q_j - c_{Y_j} \right\|_2 - \Delta_{Y_j} \right) + (1 - \delta_j) \left(\Delta_{Y_j} - \left\| Q_j - c_{Y_j} \right\|_2 \right) \right] \tag{14}$$

where M is the total number of shrimp samples with and without shells, and Y_j is the label of the ith sample. δ_j is defined as

$$\delta_j = \begin{cases} 1, \|Q_j - c_k\|_2 > \Delta_{Y_j} \\ 0, \|Q_j - c_k\|_2 \leq \Delta_{Y_j} \end{cases} \tag{15}$$

Then, the radius parameter $\widehat{\Delta_k}$ is optimized using a random gradient descent:

$$\widehat{\Delta_k} = \widehat{\Delta_k} - \eta \frac{\partial L_b}{\partial \widehat{\Delta_k}} \tag{16}$$

where η is the learning rate of the boundary parameters. The $\frac{\partial L_b}{\partial \widehat{\Delta_k}}$ is calculated by

$$\frac{\partial L_b}{\partial \widehat{\Delta_k}} = \frac{\sum_{j=1}^{M} \delta'(Y_j = k) \cdot (-1)^{\delta_j}}{\sum_{j=1}^{M} \delta'(Y_j = k)} \cdot \frac{1}{1 + e^{-\widehat{\Delta_k}}} \tag{17}$$

If $Y_j = k$, then $\delta'(y_j = k) = 1$; if $Y_j \neq k$, then $\delta'(y_j = k) = 0$. In this way, the learned radius parameters not only surrounds most shrimp with and without shells, but also avoids the cluster centers of each type.

After learning the center and recognition boundary radius of shrimp with and without shells, we discarded the boundary of shrimp with shells and retained that of shrimp without shells. This is because shrimp with shells have different shell attachment areas, resulting in a wider spatial distribution of feature vectors for tactile recognition. In contrast, the feature vectors of shrimp without shells are relatively fixed. The presence of interference samples affects the recognition accuracy when training is conducted using only shrimp samples without shell samples. In addition, the maximum number of shrimp with shells must be recognized to ensure the effectiveness of the industrial production process.

During the test, the distance between the test sample and the class center of shrimp without shells was calculated. When the distance is less than the radius of the boundary of the shrimp without shells, it is judged as shrimp without shells; otherwise, it is judged as shrimp with shells.

3. Results and Discussion

Two experiments were conducted to evaluate the performance of the proposed method in recognizing shrimp with and without shells. In one experiment, the species Macrobrachium rosenbergii was selected as the training sample, and the trained ARBM was compared with the proposed tactile recognition model. In the other experiment, five different shrimp species were selected for testing, and the trained ARBM model was compared with the vision model. The overall recognition accuracy for shrimp with and without shells is the performance evaluation index expressed as (18)–(20).

$$AT = \frac{TP + TN}{TP + TN + FP + FN} \tag{18}$$

$$AS = \frac{TP}{TP + FN} \tag{19}$$

$$AP = \frac{TN}{FP + TN} \tag{20}$$

where AT is the overall recognition accuracy, TP is the number of correct recognitions of shrimp with shells, TN is the number of correct recognitions of shrimp without shells, FP is the number of incorrect recognitions of shrimp with shells, FN is the number of incorrect recognitions of shrimp without shells, AS is the recognition accuracy of shrimp with shells, and AP is the recognition accuracy of shrimp without shells.

3.1. Compare Different Tactile Recognition Models

To verify the validity of the ENSV-ARBM, we selected headless Macrobrachium rosenbergii shrimps as our experimental samples. The samples of Macrobrachium rosenbergii had a length of 8.2–9.8 cm and a weight of 35.2–40.1 g. Five hundred (500) shrimp with shells and another 500 without shells were examined.

First, the speed of the conveyor was fixed at 0.1 m/s. The shrimps passed the tactile sensor at specific time intervals, which must be longer than the time required to fully acquire the tactile sensations of a shrimp. After each shrimp passes the tactile sensor, the corresponding tag is manually recorded and the tactile data is saved. The experiment is conducted in MATLAB 2022a 64-bit (MATLAB, 2022a) platforms using a 2.7 GHz notebook computer with an Intel(R) Core (TM) CPU and 8 GB RAM. The samples of the tactile recognition of shrimp are shown in Figure 7. The device described in Section 2.1 was selected for data collection. Tactile data were collected from all experimental samples (i.e., 500 shrimp with shells and 500 without shells). Finally, 70% of shrimp with and without shells were randomly selected as the training set, 15% as the validation set, and 15% as the test set.

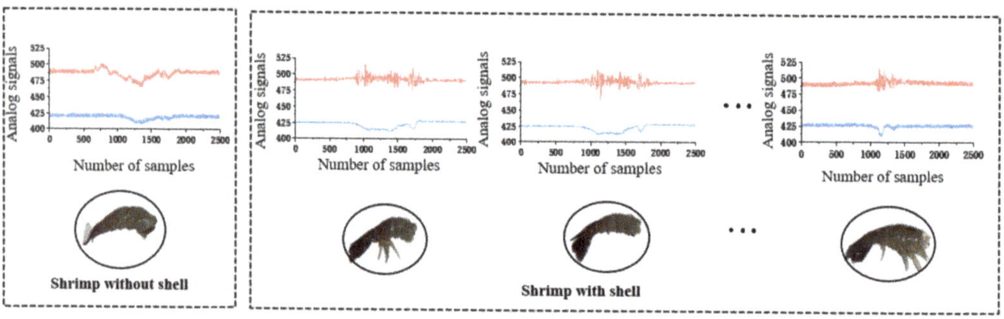

Figure 7. Plot of raw data of shrimp with and without shells.

The identification of shrimp with and without shells is based on the difference in their waveforms. When a tactile sensor scans a shrimp with shells, the grooves on its body cause the sensor to produce a more pronounced jitter and oscillation signal. As the surface area of the shrimp shell increases, the duration of the oscillation signal generated by the sensor decreases. Sensor B, located on the outermost carbon fiber plate, detects a certain protruding waveform due to the increased hardness of the shrimp's body. On the other hand, when a tactile sensor scans a shrimp without shells, the friction gradually increases as it scans the shrimp's smooth and soft body. As a result, the waveforms obtained from Sensors A and B on the surface carbon fiber plate are smoother and contain less energy. The results of the comparisons with tactile perception methods proposed in the literature are listed in Table 2.

Table 2. Comparison of the results of the proposed scheme with other tactile methods.

REF	Methods	AT %	AS %	AP %
[23]	PCA-KNN	74.0	77.3	70.7
[24]	PCA-SVM	72.7	66.7	78.7
[25]	DWT-KNN	72.7	73.3	72
[26]	DWT-ELM	77.4	78.7	76.0
Our proposed model	ENSV-ARBM	88.7	85.3	92.0

The statistical results in Table 2 show that ENSV-ARBM method has the highest *AT*, *AS*, and *AP* of 88.7%, 85.3%, and 92.0%, respectively. The ENSV feature is a fusion of the energy and nonstationary volatility features, in which the energy feature reflects the dynamic changes in the force during the tactile process, and the nonstationary volatility feature extracts the fluctuating data during the contact between the sensor and shrimp surface. The fusion of these two features can effectively amplify the differences in the surface texture and hardness between shrimp with and without shells as well as mask invalid data to identify the physically significant features of both types of shrimps. The ARBM is a recognition model based on the spatial distribution of the data, which enables the secondary classification of shrimp with and without shells. The model uses the distribution boundary of the sample space of shrimp without shells as a classification boundary in the presence uncertainty regarding the attachment surface of shrimp samples with shells.

The results in Table 2 demonstrate the effectiveness of the ENSV-ARBM-based tactile recognition of shrimp with and without shells. The overall recognition rate is better than that of the other tactile recognition algorithms. Shrimp without shells are smooth and soft to the touch, whereas shrimp with shells are rough and hard to the touch. By effectively extracting the texture and hardness tactile features of different shrimp body surfaces, the recognition accuracy of shrimp with and without shells can be improved. In addition, the stable tactile data boundaries of shrimp without shells reduces the problem of the complex shell attachment locations of shrimp with shells.

3.2. Compare Different Vision Recognition Models

To verify the generalizability of the proposed ENSV-ARBM for the tactile recognition of shrimp with and without shells, we selected five different shrimp species for comparative experiments using the machine vision approach and the tactile approach described in Section 3.1. These include Panulirus argus, Macrobrachium rosenbergii, Penaeus chinensis, Oratosquilla oratoria, and Metapenaeus ensis, as shown in Figure 8. There were 100 shrimp with shells and 100 shrimp without shells for each species.

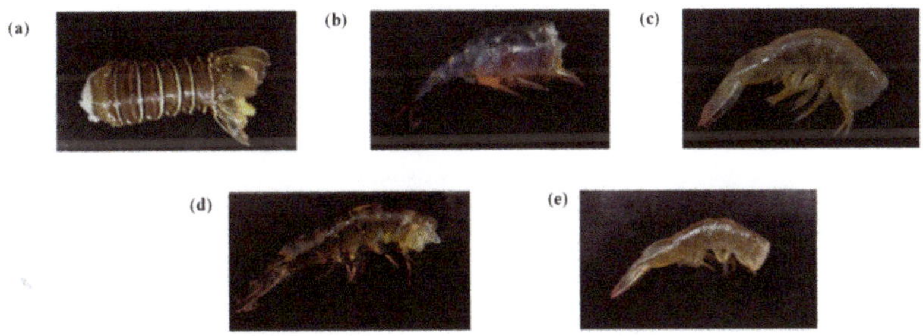

Figure 8. Photos of five different shrimp species. (**a**) Panulirus argus; (**b**) Macrobrachium rosenbergii; (**c**) Penaeus chinensis; (**d**) Penaeus japonicus; and (**e**) Metapenaeus ensis.

After decapitation, we measured the size and weight of the shrimp samples using a ruler and an electronic scale, respectively. The samples of Panulirus argus had a length of 13.1–15.9 cm and a weight of 69.3–72.5 g. The samples of Macrobrachium rosenbergii had a length of 8.2–9.8 cm and a weight of 35.2–40.1 g. The samples of Penaeus chinensis had a length of 10.7–12.8 cm and a weight of 33.3–39.8 g. The samples of Oratosquilla oratoria had a length of 11.3–13.1 cm and a weight of 35.4–42.5 g. The samples of Metapenaeus ensis had a length of 7.3–8.4 cm and a weight of 28.7–32.0 g.

In the area of tactile recognition, we selected the ENSV-ARBM-based tactile recognition method for our experiments. The trained model in Section 3.1 was selected to test the five different shrimp species. In the area of visual recognition, we used an industrial camera (HIKVISION) with a CMOS sensor as the data source for visual recognition. The sensor size is 22.3 mm $\times$ 14.9 mm, the effective pixels are 18 million, and the acquired image resolution is 2928 $\times$ 3904 (pixels). To test the fairness of the assessment, samples of 500 shrimp with shells and 500 without shells were photographed along the conveyor belt. The image information obtained was fed into the YOLOv3 and R-CNN frameworks for training purposes. The trained model then was applied to test the recognition of five different shrimp species with and without shells. The experimental results are listed in Tables 3 and 4.

Table 3. Comparison of results between the proposed scheme and other tactile methods.

Species \ Indicators	ENSV-ARBM AT %	ENSV-ARBM AS %	ENSV-ARBM AP %	PCA-KNN AT %	PCA-KNN AS %	PCA-KNN AP %	PCA-SVM AT %	PCA-SVM AS %	PCA-SVM AP %	DWT-KNN AT %	DWT-KNN AS %	DWT-KNN AP %	DWT-ELM AT %	DWT-ELM AS %	DWT-ELM AP %
(a)	89.5	89.0	90.0	73.5	77.0	70.0	73.0	69.0	77.0	74.0	74.0	74.0	78.5	79.0	78.0
(b)	89.0	86.0	92.0	76.5	77.0	76.0	74.5	72.0	77.0	73.0	73.0	73.0	77.5	78.0	77.0
(c)	88.0	87.0	89.0	76.5	78.0	75.0	71.0	67.0	75.0	72.0	73.0	71.0	79.0	79.0	79.0
(d)	87.5	85.0	90.0	74.5	75.0	74.0	74.5	71.0	78.0	71.0	71.0	71.0	77.5	78.0	77.0
(e)	87.0	88.0	86.0	75.5	75.0	76.0	72.5	70.0	75.0	71.5	72.0	71.0	77.0	77.0	77.0
Mean	88.2	87.0	89.4	75.3	76.4	74.2	73.1	69.8	76.4	72.3	72.6	72.0	77.9	78.2	77.6

Table 4. Comparison of results between the proposed scheme and other vision methods.

Species \ Indicators	ENSV-ARBM AT %	ENSV-ARBM AS %	ENSV-ARBM AP %	YOLOv3 AT %	YOLOv3 AS %	YOLOv3 AP %	R-CNN AT %	R-CNN AS %	R-CNN AP %
(a)	89.5	89.0	90.0	91.5	89.0	94.0	90.5	89.0	92.0
(b)	89.0	86.0	92.0	99.0	99.0	99.0	93.0	95.0	91.0
(c)	88.0	87.0	89.0	77.0	89.0	65.0	76.0	80.0	72.0
(d)	87.5	85.0	90.0	83.5	72.0	95.0	80.0	70.0	90.0
(e)	87.0	88.0	86.0	75.5	84.0	67.0	74.5	85.0	64.0
Mean	88.2	87.0	89.4	85.3	86.6	84.0	82.8	83.8	81.8

From Table 3, we can see that our proposed tactile perception method is better compared to other tactile methods. From Table 4, in terms of the average overall recognition accuracy, the ENSV-ARBM-based tactile recognition method exhibited the best performance for the *AT*, *AS*, and *AP* with 88.2%, 87.0%, and 89.4%, respectively. This was followed by the YOLOv3. R-CNN exhibited the worst performance for the *AT*, *AS*, and *AP* with 82.8%, 83.8%, and 81.8%, respectively. With regard to the recognition accuracy for each shrimp species, the vision recognition methods for the Macrobrachium rosenbergii and Panulirus argus were better than the tactile recognition approach.

Macrobrachium rosenbergii and Panulirus argus, with and without shells, were visually distinguished. The Penaeus chinensis and Metapenaeus ensis have transparent shells; hence, the visual recognition method misidentified shrimp with shells as shrimp without shells when the shells were attached to the tail. For Penaeus japonicus, both shrimp with and without shells showed a black color; hence, the visual method misidentified shrimp without shells as shrimp with shells.

The ENSV-ARBM-based tactile recognition method identifies whether shrimps have shells mainly through the dynamic changes in the texture and hardness of the shrimp surface. Although the flesh and shell of different shrimp species have different forms, textures, and colors, the variations in texture and hardness are similar. The machine vision training samples must be comprehensive, whereas the tactile method only identifies the physical features of texture and hardness of shrimp with or without shells; hence, it has better universality. The experimental results demonstrate the universality of ENSV-ARBM-based tactile recognition and provides good results for the recognition of different shrimp species.

4. Conclusions

A tactile recognition method based on ENSV-ARBM is proposed to address the problem of recognizing shrimp with and without shells. The method uses Macrobrachium rosenbergii shrimp as our training samples. We obtained tactile information by a self-designed tactile sensor and achieved the recognition of shrimp with and without shells by the proposed ENSV-ARBM method.

The effectiveness of the proposed ENSV-ARBM approach to tactile perception was verified by comparison with other tactile models. The method we propose amplifies the tactile differences between shrimp with and without shells. In addition, it overcomes the problems of uneven distribution of tactile data and the lack of shrimp samples to enable the recognition of shrimp with and without shells. The method was tested with different species of shrimp and the results were 88.2%, 87.0%, and 89.4% for AT, AS, AP, respectively, which verified the generalizability of the proposed method.

In conclusion, it has an advantage over other methods in the universality of different species of shrimp recognition and can meet the requirements of real-time and high-accuracy recognition for embedded devices. The method can provide an effective solution for automatic peeling machine to recognition shrimp with and without shells.

Author Contributions: Conceptualization, X.C. and Y.X.; methodology, X.C., T.W. and Y.X.; software, Y.X.; validation, X.C., Y.X. and P.D.; formal analysis, Y.X.; investigation, P.D.; resources, E.Z.; data curation, C.T.; writing—original draft preparation, Y.X.; writing—review and editing, X.C.; visualization, C.W.; supervision, X.C.; project administration, X.C.; funding acquisition, X.C. All authors have read and agreed to the published version of the manuscript.

Funding: This research was funded by Guangdong Province key areas of research and development program, grant number 2021B0202060002.

Institutional Review Board Statement: Not applicable.

Informed Consent Statement: Not applicable.

Data Availability Statement: The data presented in this study are available on request from the corresponding author. The data are not publicly available since future studies are related to current data.

Conflicts of Interest: The authors declare no conflict of interest.

References

1. Chang, Z.Q.; Neori, A.; He, Y.Y.; Li, J.T.; Qiao, L.; Preston, S.I.; Liu, P.; Li, J. Development and current state of seawater shrimp farming, with an emphasis on integrated multi-trophic pond aquaculture farms, in China—A review. *Rev. Aquac.* **2020**, *12*, 2544–2558. [CrossRef]
2. Zhao, L.T.; Xi, M.J.; Wang, W.; Chu, X.; Zhao, X.; Yuan, Y.; Liu, S.Q. A review of technology development process of a typical shrimp peeling machine. *J. Adv. Mech. Des. Syst. Manuf.* **2015**, *9*, JAMDSM0013. [CrossRef]
3. Zhang, X.H.; Zhao, Q.L.; Zhao, Y.D.; Wang, Z.H.; Yi, J.G. Parameter and pretreatment condition optimization of dual rollers extrusion peeling device for Litopenaeus vannamei. *Editor. Off. Trans. Chin. Soc. Agric. Eng.* **2014**, *30*, 308–314. [CrossRef]
4. Yi, J.G.; Ma, J.Y.; Wang, Z.H.; Zhang, X.H.; Yang, S.H. Status and Development of Shrimp Processing Equipment. *Mod. Agric. Equip.* **2020**, *41*, 2–11.
5. Zhang, X.H.; Zhao, Q.L.; Wang, Z.H.; Yi, J.G.; Wang, J. Parameter optimization experiment of adjustable five rollers shrimp peeling machine. *Trans. Chin. Soc. Agric. Eng.* **2016**, *32*, 247–254.
6. Hosseinpour, S.; Rafiee, S.; Mohtasebi, S.S.; Aghbashlo, M. Application of computer vision technique for on-line monitoring of shrimp color changes during drying. *J. Food Eng.* **2013**, *115*, 99–114. [CrossRef]
7. Hong, H.M.; Yang, X.L.; You, Z.H.; Cheng, F. Visual quality detection of aquatic products using machine vision. *Aquac. Eng.* **2014**, *63*, 62–71. [CrossRef]
8. Hong, C.; Liu, Z.H.; Wang, X.Q.; Gao, L.; Zheng, Z.Y.; Xu, Z.L.; Jun, Z. Construction of completeness recognition method for shrimp (*Litopenaeus vannamei*) based on morphological characteristics. *J. Food Saf. Qual.* **2021**, *2*, 8666–8673.
9. Hu, W.C.; Yang, C.Y.; Huang, D.Y. Robust real-time ship detection and tracking for visual surveillance of cage aquaculture. *J. Vis. Commun. Image Represent.* **2011**, *22*, 543–556. [CrossRef]
10. Lee, D.J.; Xiong, G.M.; Lane, R.M.; Zhang, D. An Efficient Shape Analysis Method for Shrimp Quality Evaluation, 2012 12th International Conference on Control, Automation. In Proceedings of the 2012 12th International Conference on Control Automation Robotics & Vision (ICARCV), Guangzhou, China, 5–7 December 2012; pp. 865–870.
11. Zhang, D.; Lillywhite, K.D.; Lee, D.J.; Tippetts, B.J. Automatic shrimp shape grading using evolution constructed features. *Comput. Electron. Agric.* **2014**, *100*, 116–122. [CrossRef]
12. Liu, Z.H.; Jia, X.J.; Xu, X.S. Study of shrimp recognition methods using smart networks. *Comput. Electron. Agric.* **2019**, *169*, 104926. [CrossRef]

13. Liu, Z.H. Soft-shell Shrimp Recognition Based on an Improved AlexNet for Quality Evaluations. *J. Food Eng.* **2020**, *266*, 109698. [CrossRef]
14. Zhang, L.; Huang, L.; Li, B.B.; Chen, X.; Duan, Q.L. Fish School Counting Method Based on Multi-scale Fusion and No Anchor YOLO v3. *Trans. Chin. Soc. Agric. Mach.* **2021**, *52*, 237–244.
15. Conrady, C.R.; Er, S.; Attwood, C.G.; Roberson, L.A.; Vos, L. Automated detection and classification of southern African Roman seabream using mask R-CNN. *Ecol. Inform.* **2022**, *69*, 101593. [CrossRef]
16. Ramik, D.M.; Sabourin, C.; Moreno, R.; Madani, K. A machine learning based intelligent vision system for autonomous object detection and recognition. *Appl. Intell.* **2014**, *40*, 358–375. [CrossRef]
17. Kiyokawa, T.; Tomochika, K.; Takamatsu, J.; Ogasawara, T. Fully Automated Annotation With Noise-Masked Visual Markers for Deep-Learning-Based Object Detection. *IEEE Robot. Autom. Lett.* **2019**, *4*, 1972–1977. [CrossRef]
18. Chen, X.S.; Mao, Y.Y.; Ma, X.; Qi, L. A Tactile Method for Rice Plant Recognition Based on Machine Learning. *Sensors* **2020**, *20*, 5135. [CrossRef]
19. Chen, X.S.; Mao, Y.Y.; Xiong, Y.S.; Qi, L.; Jiang, Y.; Ma, X. Intra-row weed density evaluation in rice field using tactile method. *Comput. Electron. Agric.* **2022**, *193*, 106699. [CrossRef]
20. Iskarous, M.M.; Thakor, N.V. E-Skins: Biomimetic Sensing and Encoding for Upper Limb Prostheses. *Proc. IEEE* **2019**, *107*, 2052–2064. [CrossRef]
21. Luo, S.; Mou, W.X.; Althoefer, K.; Liu, H.B. iCLAP: Shape recognition by combining proprioception and touch sensing. *Auton. Robot.* **2019**, *43*, 993–1004. [CrossRef]
22. Niu, H.S.; Li, H.; Gao, S.; Li, Y.; Wei, X.; Chen, Y.K.; Yue, W.J.; Zhou, W.J.; Shen, G.Z. Perception-to-Cognition Tactile Sensing Based on Artificial-Intelligence-Motivated Human Full-Skin Bionic Electronic Skin. *Adv. Mater.* **2022**, *34*, 2202622. [CrossRef] [PubMed]
23. Wang, S.A.; Alessandro, A.; Perla, M.; Fulvio, M.; Giorgio, C. Fabric Classification Using a Finger-Shaped Tactile Sensor via Robotic Sliding. *Front. Neurorobotics* **2022**, *16*, 808222. [CrossRef] [PubMed]
24. Zhang, Z.; Zhou, J.; Yan, Z.; Wang, K.; Mao, J.; Jiang, Z. Hardness recognition of fruits and vegetables based on tactile array information of manipulator. *Comput. Electron. Agric.* **2021**, *181*, 105959. [CrossRef]
25. Keser, S.; Hayber, S.E. Fiber optic tactile sensor for surface roughness recognition by machine learning algorithms. *Sens. Actuators A-Phys.* **2021**, *332*, 113071. [CrossRef]
26. Qin, L.H.; Yi, Z.K.; Zhang, Y.L. Enhanced surface roughness discrimination with optimized features from bio-inspired tactile sensor. *Sens. Actuators A-Phys.* **2017**, *264*, 133–140. [CrossRef]
27. Kawazoe, M.; Kosemura, Y.; Miki, N. Encoding and presentation of surface textures using a mechanotactile display. *Sens. Actuators A-Phys.* **2017**, *261*, 30–39. [CrossRef]
28. Soualhi, M.; Nguyen, K.; Medjaher, K. Pattern recognition method of fault diagnostics based on a new health indicator for smart manufacturing. *Mech. Syst. Signal Process.* **2020**, *142*, 106680. [CrossRef]
29. Bakker, T.; Wouters, H.; Asselt, K.; Bontsema, J.; Tang, L.; Müller, J.; Straten, G. A vision based row detection system for sugar beet. *Comput. Electron. Agric.* **2008**, *60*, 87–95. [CrossRef]
30. Ruan, X.; Zhu, Y.; Li, J.; Cheng, Y. Predicting the citation counts of individual papers via a BP neural network. *J. Informetr.* **2020**, *14*, 101039. [CrossRef]

agriculture

Article

Exploiting the Internet Resources for Autonomous Robots in Agriculture

Luis Emmi [1,*], Roemi Fernández [1], Pablo Gonzalez-de-Santos [1], Matteo Francia [2], Matteo Golfarelli [2], Giuliano Vitali [3], Hendrik Sandmann [4], Michael Hustedt [4] and Merve Wollweber [4]

1 Centre for Automation and Robotics (UPM-CSIC), 28500 Arganda del Rey, Madrid, Spain
2 Department of Computer Science and Engineering (DISI), Alma Mater Studiorum-University of Bologna, 40127 Bologna, Italy
3 Department of Agricultural and Food Sciences (DISTAL), Alma Mater Studiorum-University of Bologna, 40127 Bologna, Italy
4 Laser Zentrum Hannover e.V., Hollerithallee 8, 30419 Hannover, Germany
* Correspondence: luis.emmi@car.upm-csic.es

Abstract: Autonomous robots in the agri-food sector are increasing yearly, promoting the application of precision agriculture techniques. The same applies to online services and techniques implemented over the Internet, such as the Internet of Things (IoT) and cloud computing, which make big data, edge computing, and digital twins technologies possible. Developers of autonomous vehicles understand that autonomous robots for agriculture must take advantage of these techniques on the Internet to strengthen their usability. This integration can be achieved using different strategies, but existing tools can facilitate integration by providing benefits for developers and users. This study presents an architecture to integrate the different components of an autonomous robot that provides access to the cloud, taking advantage of the services provided regarding data storage, scalability, accessibility, data sharing, and data analytics. In addition, the study reveals the advantages of integrating new technologies into autonomous robots that can bring significant benefits to farmers. The architecture is based on the Robot Operating System (ROS), a collection of software applications for communication among subsystems, and FIWARE (Future Internet WARE), a framework of open-source components that accelerates the development of intelligent solutions. To validate and assess the proposed architecture, this study focuses on a specific example of an innovative weeding application with laser technology in agriculture. The robot controller is distributed into the robot hardware, which provides real-time functions, and the cloud, which provides access to online resources. Analyzing the resulting characteristics, such as transfer speed, latency, response and processing time, and response status based on requests, enabled positive assessment of the use of ROS and FIWARE for integrating autonomous robots and the Internet.

Keywords: precision agriculture; autonomous robots; artificial intelligence; IoT; cloud computing

Citation: Emmi, L.; Fernández, R.; Gonzalez-de-Santos, P.; Francia, M.; Golfarelli, M.; Vitali, G.; Sandmann, H.; Hustedt, M.; Wollweber, M. Exploiting the Internet Resources for Autonomous Robots in Agriculture. *Agriculture* **2023**, *13*, 1005. https://doi.org/10.3390/agriculture13051005

Academic Editors: Jin Yuan, Wei Ji, Qingchun Feng and Massimo Cecchini

Received: 16 March 2023
Revised: 17 April 2023
Accepted: 29 April 2023
Published: 2 May 2023

1. Introduction

The year 2022 ended with more than 8 billion inhabitants of the world. Most governments understand that feeding this vast and growing population is one of the significant challenges they must face in the coming years. Some associations have predicted that food production will need to increase by 70% to feed the entire population in 2050 [1]. In developed countries, cultivated land is close to its maximum output; therefore, the solution is oriented toward optimizing the available resources. Many different cultural and technological methods for increasing crop yield are being used. Some improve crop yields, but at the extra cost of increasing environmental pollution and the carbon footprint. These side effects are unacceptable in many industrialized nations, such as those in the European Union, which is committed to using sustainable methods.

Precision agriculture leverages technologies to achieve those objectives and avoids undesired effects. PA is a concept for farm management founded on observation, measurement, and response to crop variability [2]. It assembles different methods to manage variations in a farm to enhance crop yield, improve commercial profit, and guarantee eco-environmental sustainability. PA uses current information and communication technologies (ICT), automation, and robotics to monitor crop growth, predict the weather accurately, perform optimal irrigation, apply fertilizers smartly, manage weeds and pests accurately, test soil quality precisely, etc.

Since the late 1980s, precision agriculture techniques have been introduced step by step in the agricultural production sector, integrating the following:

- Sensors to acquire geolocated biodata of crops and soil, e.g., nitrogen sensors, vision cameras, global navigation satellite systems (GNSS), etc.
- Computers for analyzing those data and running simple algorithms to help farmers make simple decisions (applying or not applying a given process, modifying a process application map, etc.).
- Actuators in charge of executing the decisions (opening/closing valves, altering a trajectory, etc.) for modifying crops. As an actuator, we consider the agricultural tool, also called the agricultural implement, and the vehicle, manually or automatically driven, to move the tool throughout the working field and apply the farming process.

The integration of subsystems onboard robotic vehicles started in the late 1990s. Some illustrative examples, based on retrofitting conventional vehicles, are the autonomous agricultural sprayer [3], which focuses on achieving a pesticide spraying system that is cheap, safe, and friendly to the environment, and the autonomous orchard vehicles for mowing, tree pruning, and training, spraying, blossoming, and fruit thinning, fruit harvesting, and sensing [4], both deployed in the USA. In Europe, we can find the RHEA fleet (see Figure 1a), consisting of a fleet of three tractors that cooperate and collaborate in the application of pesticides [5]. Regarding robotic systems based on specific structures designed for agriculture (see Figure 1b), we can remark on LadyBird in Australia, intended for the valuation of crops using thermal and infrared detecting systems, hyperspectral cameras, stereovision cameras, LIDAR, and GPS [6], and Vibro Crop Robotti in Europe, built for accurate seeding and mechanical row crop cleaning [7]. These robots were integrated around computing systems based on centralized or elementary distributed architectures to handle a few sensors and control unsophisticated agricultural tools.

In addition to those developments, related technologies have evolved drastically in recent years, and now sensors can be spread throughout the field and communicate with each other. This is possible because of the Internet of Things (IoT). This computing concept describes how to cluster and interconnect objects and devices through the Internet, where all are visible and can interact with each other. IoT defines physical objects with devices (mainly sensors) and includes processing power, software applications, and other technologies to exchange data with other objects through the Internet.

Moreover, computers can run artificial intelligence (AI) algorithms, considering AI as the ability of a machine (computer) to emulate intelligent human actions. The application of AI to agriculture has been focused on three primary AI techniques: expert systems, artificial neural networks, and fuzzy systems, with significant results in the management of crops, pests, diseases, and weeds, as well as the monitoring of agricultural production, store control, and yield prediction, for example [8].

AI techniques are also applied to provide vehicles with autonomy; therefore, autonomous agricultural robots leverage this technology. AI-based vision systems can fulfill the following roles:

- Detecting static or dynamic objects in their surroundings.
- Detecting row crops for steering purposes.
- Identifying plants and locating their positions for weeding are clear examples of the current use of AI techniques in agricultural robotics [9].

(a)

(b)

Figure 1. (**a**) Agricultural robots based on retrofitted conventional vehicles (RHEA fleet); (**b**) agricultural robots designed on purpose (Courtesy of AgreenCulture SaS).

Another technology that has evolved in the last decade is cloud computing, defined as the on-demand delivery of computing services, mainly data storage and computing power, including servers, storage, databases, networking, software applications, artificial intelligence methods, and analytics algorithms over the Internet. The main objective of cloud computing systems is to provide flexible resources at adapted prices. A cloud computing system allows the integration of data of different types, loaded from many sources in batch and real-time. In particular, the integration can be based on georeferenced data in the precision farming area. Data can range from trajectory data to images and videos related to fields and missions and any sensors installed on the autonomous robot.

Cloud computing allows the use of services available in the cloud (computing, storing, etc.), with increasing advantages provided by big data techniques. Many agricultural applications of big data technologies have already been introduced in agriculture [10] and should be present in future robotic systems.

This article presents an architecture to integrate new technologies and Internet trends in agricultural autonomous robotic systems and has two main objectives. The first objective is to provide an example of designing control architectures to connect autonomous robots to the cloud. It is oriented toward robot designers and gives significant technical details. The second objective is to disclose to farmers the advantages of integrating the new technologies in autonomous robots that can provide farmers with significant advantages regarding (i) data storage, which is a secure and efficient way to store, but also access and share, data, eliminating the need of physical storage and, thus, reducing the risk of data loss; (ii) scalability, which allow the farmers to expand or reduce their storage needs, efficiently optimizing their resources, and (iii) analytics services, which allow a farmer to analyze their own data to make informed decisions taking advantage of the AI tools available on the cloud. These are general advantages of using the cloud, but autonomous robots have great potential for collecting data and must facilitate communicating those data to the cloud.

To base the architecture on a specific example, the integration of a laser-based system for weed management is considered. Thus, Section 2 presents the material, defining the robot's components, and the methodology, detailing the system's architecture. Section 3 then introduces the experiments to be assessed and discussed in Section 4. Finally, Section 5 summarizes the conclusions.

2. Materials and Methods

This section first describes the components and equipment integrated for building the autonomous robot used to validate and assess the proposed integration methodology. Second, the methods for the integration of components are detailed.

2.1. System Components

2.1.1. Main Process Loop in PA Autonomous Robots

The autonomous systems used for precision agriculture generally follow the structure of an automatic control loop that consists of the following (see Figure 2):

- Selecting the references for the magnitudes to be controlled, i.e., defining the desired plan.
- Measuring the magnitudes of interest.
- Making decisions based on the measured and desired values of the magnitudes (control strategy).
- Executing the decided actions

In our application, the selecting references are made with the smart navigation manager (mission planner), the measures of the magnitudes of interest are performed with the perception system and the IoT sensor network, the decisions are made with the smart navigation manager (smart operation manager), and the actions are executed with the agricultural tool and the autonomous robot that move the implement throughout the mission field. In addition, our system also takes care of the interaction with the cloud and the operator. In our proposed integration method, these components are grouped into modules, as illustrated in Figures 2 and 3. These modules are as follows.

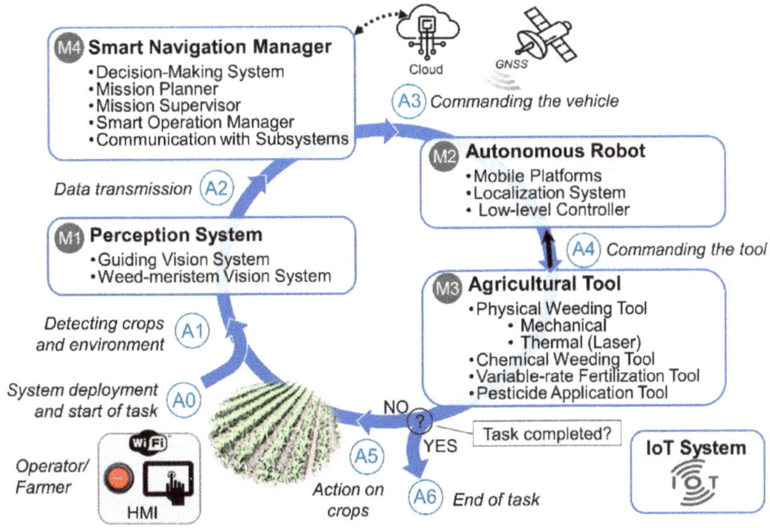

Figure 2. Components of a precision agriculture robotic system and main information flow.

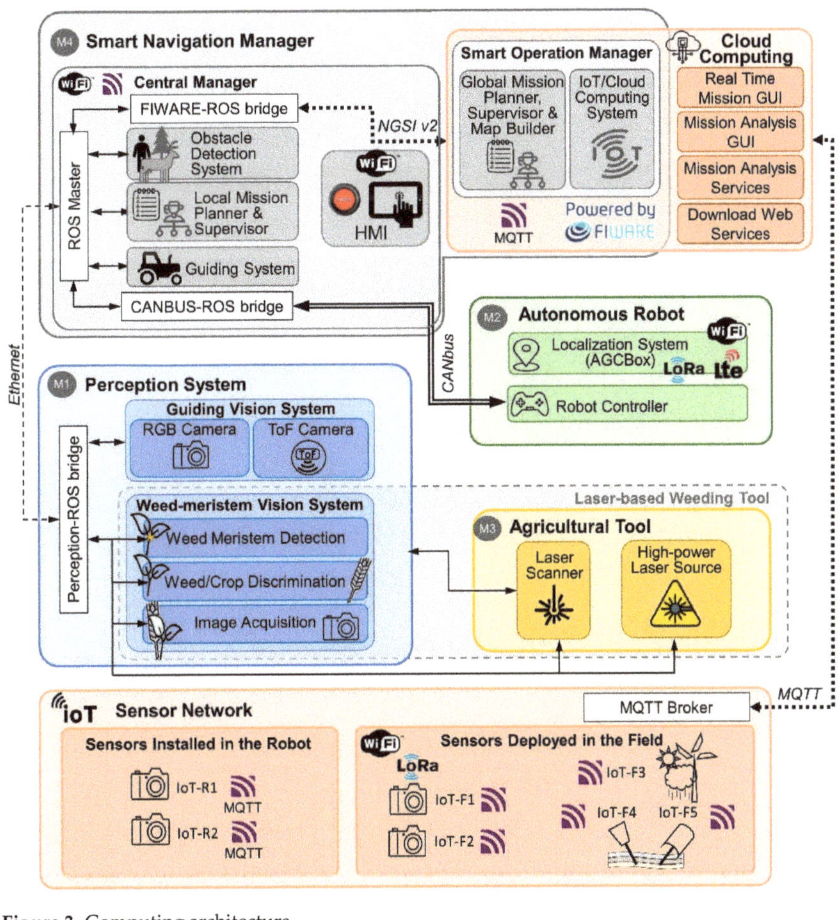

Figure 3. Computing architecture.

2.1.2. Agricultural Robot

A manually driven or autonomous vehicle is essential in agricultural tasks to perform the necessary actions throughout the working field. In this case, we use a compact mobile platform based on a commercial vehicle manufactured by AgreenCulture SaS, France. This is a tracked platform, and, thus, it operates as a skid-steer mechanism. The track distance can be adapted to the crop row space. Equipped with an engine or batteries, the platform can follow predefined trajectories at 6 km/h with a position accuracy of ±0.015 m using a global positioning system (GPS) based on the real-time kinematic (RTK) technique. This mobile platform is illustrated in Figure 4a.

Figure 4. (**a**) Mobile platform (AgreenCulture SaS) and (**b**) autonomous laser weeding system.

2.1.3. Perception System

A perception system is based on computer vision algorithms that obtain, process, analyze, and understand images and data from the environment. With these inputs, the system produces numerical and symbolic information for making decisions. The perception system for this study consists of the following systems:

- Guiding vision system: This system aims to detect static and dynamic obstacles in the robot's path to prevent the robot tracks from stepping on the crops during the robot's motion. Furthermore, it is also used to detect crop rows in their early growth stage to guide the robot in GNSS-denied areas [8]. The selected perception system consisted of a red–green–blue (RGB) wavelength vision camera and a time-of-flight (ToF) camera attached to the front of the mobile platform using a pan-tilt device, which allows control of the camera angle with respect to the longitudinal axis of the mobile platform, x. Figure 4 illustrates both cameras and their locations onboard the robot.
- Weed–meristem vision system: The system is based on 3D vision cameras to provide the controller with data on crops and weeds. These data are used to carry out the main activity of the tool for which it has been designed: weed management, in this case. For example, the perception system used in this study consists of an AI vision system capable of photographing the ground and discriminating crops from weeds in a first step using deep learning algorithms. In the second step, the meristems of the detected weeds are identified. Figure 3 sketches this procedure.

2.1.4. Agricultural Tools

Agricultural tools focus on direct action on the crop and soil and rely on physical (mechanical, thermal, etc.) or chemical (pesticides, fertilizers, etc.) foundations. This study used a thermal weeding tool based on a high-power laser source that provided lethal laser doses to be deployed on the weed meristems using scanners.

An AI video system provided the positions of the weed meristems. Indeed, this specific solution physically integrated the AI vision system, the laser scanner, and the high-power laser source into the laser-based weeding tool component. The video frames acquired with this system were sent to the central controller at a rate of 4 frames/s. After the mission, all stored images were sent to the cloud.

2.1.5. The Smart Navigation Manager (SNM)

This manager is a distributed software application responsible for driving the autonomous robot and coordinating all other modules and systems. The SNM is split into (i) the smart operation manager and (ii) the central manager, which also includes the human–machine interface (HMI).

Smart Operation Manager (SoM)

The smart operation manager is a human–computer interaction module that can acquire, process, and deliver information based on computer algorithms and is devoted to assisting farmers in making accurate, evidence-based decisions. The SoM is specialized for laser weeding technology, the tool selected for this study.

Data management is performed through the Internet using FIWARE. Data access control is provided via a virtual private network (VPN) to secure data transfer to/from the cloud. The visual dashboard will also be available on the HMI for field operations. Through the dashboard, the operator will also interact with the robot.

The smart operation manager is allocated in the cloud. It contains the global mission planner and supervisor, the map builder, and the module for managing the IoT and cloud computing system (see Figures 3 and 5). The hardware of the SoM relies on a cluster of 10 servers.

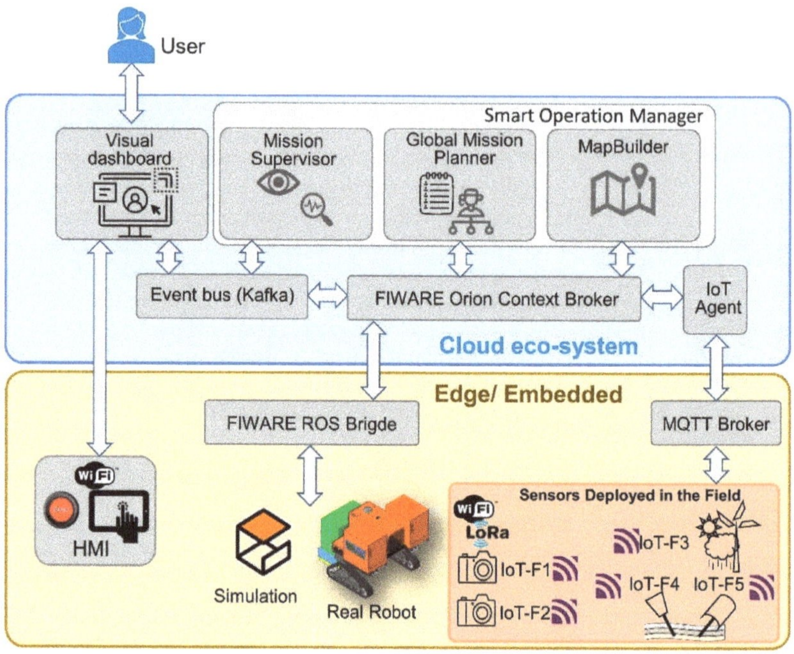

Figure 5. Cloud computing modules/containers.

(a) Global Mission Planner

A planner is a software tool responsible for computing the trajectories of the vehicle and an a priori known treatment map. The planner obtains some types of information from the Internet, including the following:

- Map information according to the data models on the Internet;
- Other information provided by third parties, such as weather forecasts;
- Data models to create maps for accessing already known treatment maps (sets of points in the field) which commonly originate from third-party map descriptions (Google Earth; Geographic Information System (GIS); GeoJSON.io, an open standard format to represent geographical features with nonspatial qualities).

Regarding robot location, two types of systems are envisaged, as follows:

- Absolute location based on GNSS: GNSS integrates several controllers for line tracking and is based on Dubins paths [11];
  - Relative location based on RGB and ToF cameras, LIDAR, and IoT sensors: These methods are based on different techniques for navigation in the field and navigation on the farm, such as hybrid topological maps, semantic localization and mapping, and identification/detection of natural and artificial elements (crops, trees, people, vehicles, etc.) through machine learning techniques.

(b) Global Mission Supervisor

A supervisor is a computational tool responsible for overseeing and monitoring the execution of the mission plan while helping the farmer (operator) manage potential failures. Most supervisor systems are designed around two actions: fault detection and fault diagnosis. The supervisor executes the following actions:

- Receiving alarms from the system components (vehicle, sensors, weeding tool, etc.).
- Detecting faults in real-time.
- Executing diagnosis protocols.

- Collecting all available geo-referred data generated by every module onboard the robot. The data are stored in both the robot and the cloud.

(c) Map Builder

A map builder is an application used to convert maps based on GeoJSON into FIWARE entities. Its main function is to support farmers in using the robotic system in a simple, reliable, and robust way by giving the robot enough information a priori (e.g., farm schema and boundaries, field locations and shapes, crop types, and status). This module takes advantage of the data models created by the FIWARE community to represent the farm and other environments digitally, where they have been conditioned to be adapted to robotic systems and especially oriented to navigation [12]. The design of the Map Builder allows the user to accomplish the following:

- Select the field in GeoJSON.IO, an open-source geographic mapping tool that allows maps and geospatial data to be created, visualized, and shared in a simple and multiformat way.
- Assign essential attributes to comply with FIWARE. These attributes are those based on the farmer's knowledge. They can include static (i.e., location, type, category) and dynamic (i.e., crop type and status, seeding date, etc.) attributes.
- Export in * GeoJSON format. The map obtained will be imported for extracting the information required to fill in the FIWARE templates, which include the farms and parcel data models, and other elements in a farm, such as buildings and roads.

This conversion makes it easier to connect the robot to the cloud by standardizing data. These data, after processing, constitute a source for the design of processes with the robot, and its storage and subsequent analysis can provide forecasts of future events in the field or behavior of the robot.

(d) IoT System

This study integrates an IoT sensor network to collect data from the following:

- The autonomous vehicle: The data and images acquired with IoT sensors onboard the vehicle are used to monitor and evaluate performances and efficiency and to identify the effects of treatments and traffic on surfaces.
- The environment: Data acquired with IoT sensors deployed on the cropland are used to (i) monitor crop development and (ii) collect weather and soil information.

Two IoT sets of devices are used in our study, as follows:

- Robot–IoT set: It consists of two WiFi high-definition cameras installed onboard the autonomous robot (IoT-R1 and IoT-R2 in Figure 3). The cameras are triggered from the cloud or the central controller to obtain a low frame rate (approximately 1/5 sec). The pictures are stored in the cloud and are used to monitor the effects of the passage of the autonomous vehicle; therefore, they should include the robot's tracks.
- Field–IoT set: It consists of the following (see Figure 3):
    - Two multispectral cameras (IoT-F1 and IoT-F2) placed at the boundary of cropped areas to obtain hourly pictures of crops.
    - A weather station (IoT-F3) to measure precipitation, air temperature (Ta), relative humidity (RH), radiation, and wind.
    - Three soil multi-depth probes (IoT-F4) for acquiring moisture (Ts) data and three respiration probes (IoT-F5) to measure CO_2 and H_2O.

Every one of these components or nodes exchanges messages with the Message Queuing Telemetry Transport (MQTT) protocol, carrying JavaScript Object Notation (JSON) serialized information from node sensors/cameras interpreted as the entity. While metering nodes (weather, soil probe, and respirometer) communicate by MQTT messages, camera nodes have to transmit images (maximum of 100 pictures/day for periodic snapshots of the area or alarms), and the use of FTP made a wide-band networking solution, such as WiFi, mandatory instead of narrowband solutions.

(e) Cloud Computing System

This study sets up a cloud-based data platform, which is an ecosystem that incorporates data acquired in the field. The data platform supports end-to-end data needs, such as ingestion, processing, and storage, to provide the following:

- A data lake repository for storing mission data to be downloaded in batches for post-mission analysis.
- A web interface for post-mission data analysis based on graphical dashboards, georeferenced visualizations, key performance indicators, and indices.
- A container framework for implementing "Decision Support System" functionalities that define missions to be sent to the robot. These functionalities (e.g., the mission planner) can be implemented and launched from the cloud platform.
- A soft real-time web interface for missions. The interface visualizes real-time robot activities and performances or sends high-level commands to the robot (e.g., start, stop, change mission).

These functionalities are ordered based on the strictness of real-time constraints.

The cloud-computing platform is based on the Hadoop stack and is powered by FIWARE. We adopted an open-source solution with well-known components that can be imported into different cloud service providers if no on-premises hardware is available. The core component of the platform is the (FIWARE) Orion Context Broker (OCB) from Telefonica [13], a publish/subscribe context broker that also provides an interface to query contextual information (e.g., obtain all images from the cameras in a specific farm), update context information (e.g., update the images), and be notified when the context is updated (e.g., when a new image is added into the platform). The images and raw data are stored in the HDFS (Hadoop distributed file system), while the NoSQL (not only structured query language) MongoDB database is used to collect the contextual data from FIWARE and further metadata necessary to manage the platform [14]. Additionally, we use Apache KAFKA, an open-source distributed event bus, to distribute context updates from FIWARE to all the modules/containers hosted on the cloud platform. The different cloud computing modules/containers used in this study are illustrated in Figure 5.

Central Manager

This central manager is an application that is divided into the following:

- Obstacle detection system. This module acquires visual information from the front of the robot (robot vision system) to detect obstacles based on machine vision techniques.
- Local mission planner and supervisor. The planner plans the motion of the robot near its surroundings. The local mission supervisor oversees the execution of the mission and reports malfunctions to the operator (see Section 2.1.5).
- Guidance system. This system is responsible for steering the mobile platform to follow the trajectory calculated by the planner. It is based on the GNSS if its signal is available. Otherwise, the system uses the information from the robot vision system to extract the crop row positions and follow them without harming the crop.
- Human–machine interface

A human–machine interface (HMI) is a device or program enabling a user to communicate with another device, system, or machine. In this study, a HMI using portable devices (android tablets) is addressed to allow farmers to perform the following:

- Supervise the mission.
- Monitor and control the progress of agricultural tasks.
- Identify and solve operational problems.
- Obtain real-time in-field access in an ergonomic, easy-to-use, and robust way.
- Maintain the real-time safety of the entire system.

To achieve these characteristics, a graphic device was integrated with the portable/remote controller of the mobile platform. This controller provides manual and remote vehicle control and integrates an emergency button.

2.1.6. Sequence of Actions

The relationships among these components and modules and the information flow are illustrated in Figures 2 and 3. The process is a repeated sequence of actions (A0 to A6), defined as follows:

A0 The system is installed in the field, The operator/farmer defines or selects a previously described mission using the HMI and starts the mission.
A1 The sensors of the perception module (M1) installed onboard the autonomous robot (M2) extract features from the crops, soil, and environment in the area of interest in front of the robot.
A2 The data acquired in action A1 are sent to the smart operation manager, determining the consequent instructions for the robots and the agricultural tool.
A3 The required robot motions and agricultural tool actions are sent to the robot controller, which generates the signal to move the robot to the desired positions.
A4 The robot controller forwards the commands sent by the smart navigation manager or generates the pertinent signals for the agricultural tool to carry out the treatment.
A5 The treatment is applied, and the procedure is repeated from action A1 to action A5 until field completion (A6).
A6 End of mission.

2.2. *Integration Methods*

Integrating all of the components defined in the previous section to configure an autonomous robot depends on the nature of the applications the robot is devoted to and the connections and communication among the different components that must be precisely defined. Thus, this section first describes the computing architecture of the controller, which integrates the different subsystems and modules. Second, the interfaces between subsystems are precisely defined. Finally, the operation procedure is defined.

2.2.1. Computing Architecture

A distributed architecture based on an open-source Robot Operating System (ROS) is proposed to integrate the system's main components onboard the mobile platform in this study. ROS is the operating system most widely accepted by software developers to create robotics applications. It consists of a set of software libraries and tools that include drivers and advanced algorithms to help developers build robot applications [15].

In this study, ROS, installed in the central controller, is used as a meta-operating system for the testing prototype. The necessary interfaces (bridges) are developed to establish communication with the autonomous vehicle, the perception system, and the laser-based weeding tool. Because of ROS versatility and its publisher/subscriber communication model, it is possible to adapt the messages to protocols commonly used in IoT, such as Message Queuing Telemetry Transport (MQTT).

ROS supports software developers in creating robotics functionalities to monitor and control robot components connected to a local network. However, this solution is not extendible to a wider network, such as the Internet. Fortunately, there exist some ROS modules that solve the problem. One is ROSLink, a protocol for extensions defining an asynchronous communication procedure between the users and the robots through the cloud [16]. ROSLink performance has been shown to be efficient and reliable, and it is widely accepted by the robotics software community [17]. Although ROSLink has been widely used to connect robotic systems with the cloud, it is oriented toward transmitting low-level messages. There is no convention to define standard data models that allow intelligent robotics systems to be scalable.

One alternative to a more internet-oriented communication framework is FIWARE, which offers interaction with the cloud using cloud services that provide well-known benefits, such as (a) cost and flexibility, (b) scalability, (c) mobility, and (d) disaster recovery [18]. FIWARE is an open software curated platform fostered by the European Commission and the European Information and Communication Technology (ICT) industry for the development and worldwide deployment of Future Internet applications. It attempts to provide a completely open, public, and free architecture and a collection of specifications that allows organizations (designers, service providers, businesses, etc.) to develop open and innovative applications and services on the Internet that fulfill their needs [19].

In this study, a cloud-based communication architecture has been implemented using FIWARE as the core, which allows messages between the edge and the cloud to be transferred and stored. The selection was made because this is an open-source platform that provides free development modules and has many enablers already developing and integrating solutions for smart agriculture.

In addition to FIWARE, we use KAFKA, a robust distributed framework for streaming data (see Section 2.1.5) that allows producers to send data and for consumers to subscribe to and process such updates. KAFKA enables the processing of streams of events/messages in a scalable and fault-tolerant manner, and decouples producers and consumers (i.e., a consumer can process data even after a producer has gone offline). For historic data, HDFS allows the download of batches of data at any time and replicates each data in three copies to prevent data loss.

The visual dashboard will also be available on the HMI for the field operations. Through the dashboard, the operator will also interact with the robot. FIWARE smart data models do not suffice to represent our application domain or to integrate the agricultural and robotic domains; therefore, we have extended the existing models and updated some existing entities. Since smart data models from FIWARE are overlapping and sometimes inconsistent, we had to envision a unified model to integrate and reconcile the data. To connect the robotic system with the cloud, specific data models were developed to represent the different robotic elements, following the guidelines of FIWARE and its intelligent data models [12].

The IoT devices deployed in the field must be able to establish connections through WiFi and LoRa technologies. WiFi is a family of wireless network protocols. These protocols are generally used for Internet access and communication in local area networks, allowing nearby electronic devices to exchange data using radio waves. LoRa technology is a wireless protocol designed for long-range connectivity and low-power communications and is primarily targeted for the Internet of Things (IoT) and M2M networks. LoRa tolerates noise, multipath signals, and the Doppler effect. The cost of achieving this is a very low bandwidth compared to other wireless technologies. This study uses a 4G LTE-M modem to connect to the Internet.

At a lower level of communication, CANbus or ISOBUS is generally used to control and monitor the autonomous vehicle. This study uses CANbus and its communication protocol CANopen. Autonomous vehicles and agricultural tools typically contain their own safety controllers. The first behaves as a master and, in the case of a risky situation, it commands the tool to stop.

The human–machine interface (HMI) will include a synchronous remote procedure call-style communication over the services protocol and asynchronous communications to ensure the robot's safety. In addition to these ROS-based protocols, the HMI has a safety control connected to the low-level safety system (by radiofrequency) for emergency stops and manual control.

Figure 6 illustrates the overall architecture, indicating the following:

- The modules (Mi), presented in the previous sections.
- The interconnection between modules, presented in the next section.
- The communication technologies and protocols to configure agricultural robotic systems that integrate IoT and cloud computing technologies.

The main characteristics of this architecture are summarized in Table 1.

(b) Field distribution

Figure 6. Experimental fields.

Table 1. Architecture components.

Architecture Component	Solutions/Comments
Operating system	ROS (Robot Operating System)
IoT–controller bridge	Hypertext Transfer Protocol (HTTP) to FIWARE Note: FIWARE is used as a communication protocol in the cloud; therefore, it is not necessary to use ROSLink.
ROS-based system for FIWARE tools	HTTP protocol to FIWARE Note: FIROS has several disadvantages when developing new data models to represent the robot, so a particular enabler will not be used to establish communication between the robot and the cloud.
Communication with IoT devices	WiFi, serial communication Note: Since a certain amount of data needs to be transmitted, WiFi would suffice.
The Internet	4G LTE-M modem
Devices onboard the mobile platform	CANopen, serial
Human–machine interface (HMI).	Synchronous remote procedure call-style communication over services protocol. Asynchronous communications to ensure the safety of the robot. Note: The HMI is used to provide access to SoM services through a web interface.

2.2.2. Interfaces between System Components

This architecture considers four main interfaces between systems and modules, as follows:

Smart Navigation Manager (M4)/Perception System (M1) interface

To receive the raw information from the perception system (sensors, cameras, etc.), the central manager uses direct connections via the transmission control protocol/Internet protocol (TCP/IP) for sensors and the universal serial bus (USB) for RGB and ToF cameras. All IoT devices use the available wireless communication technologies (WiFi and LoRa) to access the Internet and the cloud.

To guide the robot, the obstacle detection system obtains data from the guiding vision system (RGB and ToF cameras) through the Ethernet that communicates the central manager with the perception system. This communication is stated using the ROS manager and the perception–ROS bridge (see Figure 3).

Smart Navigation Manager (M4)/Agricultural Tool (M3) interface

These systems can communicate through ROS messaging protocols, where the publisher/subscriber pattern is preferred. This interface exchanges simple test messages to verify the communication interface.

It is worth mentioning that the perception system and the agricultural tool are connected directly in some specific applications. This solution decreases the latency of data communication but demands moving a portion of the decision algorithms from the smart navigation manager to the tool controller; therefore, the tool must exhibit computational features. This scheme is used in the weeding system to test the proposed architecture.

Smart Navigation Manager (M4)/Autonomous Robot (M2) interface

Initially, these systems communicate via CANbus with the CANopen protocol. The central manager uses this protocol to receive information on the status of the autonomous vehicle and basic information from the onboard sensors (GNSS, IMU, safety system, etc.). A CANbus–ROS bridge is used to adapt the communication protocols.

Autonomous Robot (M2)/Agricultural Tool (M3) interface

Usually, it is not necessary for the vehicle to directly communicate with the tool because the smart navigation manager coordinates them. However, as autonomous vehicles and agricultural tools usually have safety controllers, there is wired communication between the two safety controllers. In such a case, the autonomous vehicle safety controller works as a master and commands the tool safety controller to stop the tool if a dangerous situation appears.

Perception System (M1)/Agricultural Tool (M3)

This communication is required to inform the agricultural tools about the crop status. In weeding applications, the information is related to the positions of the weeds. In this specific application, the perception system (weed meristem detection module) sends the weed meristem positions to the laser scanner module of the agricultural tool. This communication is carried out using a conventional Ethernet connection. The metadata generated via the detection system are made available in the existing ROS network and sent to the smart navigation manager.

Smart Navigation Manager internal/cloud communications

The smart navigation manager is a distributed system that consists of three main modules:

- The central manager running on the central controller.
- The smart operation manager running on the cloud.
- The HMI running in a portable device.

The central manager and the smart operation manager communicate via NGSI v2, a FIWARE application programming interface, using a FIWARE–ROS bridge to adapt ROS protocols to NGSI v2 messages. In contrast, the HMI communicates with the central manager via WiFi and Internet, directly accessing the web services hosted in the cloud. The HMI exhibits a panic button connected via radiofrequency to the safety systems of the autonomous robot and the agricultural tool.

IoT system/Cloud

There is a direct link from the IoT system to the cloud using MQTT.

2.2.2.8. Operation Procedure

To use the proposed architecture and method, the user must follow the method below.

- Creating the map: The user creates the field map following the procedure described in the MapBuilder module (see Section 2.1.5).
- Creating the mission: The user creates the mission by selecting the mission's initial point (home garage) and destination field (study site).
- Sending the mission: The user selects the mission to be executed with the HMI (all defined missions are stored in the system) and sends it to the robot using the cloud services (see Section Smart Operation Manager (SoM)).
- Executing the mission: The mission is executed autonomously following the sequence of actions described in Section 2.1.6. The user does not need to act except for when alarms or collision situations are detected and warned of by the robot.
- Applying the treatment: When the robot reaches the crop field during the mission, it sends a command to activate the weeding tool, which works autonomously. The tool is deactivated when the robot performs the turns at the headland of the field and is started again when it re-enters. The implement was designed to work with its own sensory and control systems, only requiring the mobile platform for mobility and information when it must be activated/deactivated.
- Supervising the mission: When the robotic system reaches the crop field, it also sends a command to the IoT sensors, warning that the treatment is in progress. Throughout the operation, the mission supervisor module analyzes all the information collected by the cloud computing system, generated by both the robotic system and the IoT sensors. It evaluates if there is a possible deviation from the trajectory or risk of failure.
- Ending the mission: The mission ends when the robot reaches the last point in the field map computed by the MapBuilder. Optionally, the robot can stay in the field or return to the home garage. During the mission execution, the user can stop, resume, and abort the mission through the HMI.

3. Experimental Assessment

This section states the characteristics of the described autonomous robot with IoT and cloud computing connectivity. To achieve this purpose, the experimental field for this study is first described. Then, a test mission is defined to acquire data from the different subsystems. Finally, the system characteristics are analyzed and assessed.

The characteristics obtained are not compared with similar robotic systems due to the lack of such information in the literature. There are no published results in weeding applications; therefore, it is difficult to compare, and the indicators have been geared towards general cloud computing and mobile robotics characteristics. Therefore, cross-validation has been carried out, comparing the features of the autonomous robot with the general performance of the robot and cloud communication. Productivity, cost, and other indicators of the presented architecture are those of the general use of cloud computing.

3.1. Study Site

The system developed for this study was tested in an experimental field located in Madrid, Spain (40°18′45.166″, −3°28′51.096″). The climate of the study site is classified as a hot summer Mediterranean climate with an average annual temperature of 14.3 °C and precipitation of 473 mm.

The experimental field consisted of two areas of 60 × 20 m^2 that grew wheat (*Triticum aestivum* L.), with crop rows at a distance of 0.10 m, and maize (*Zea mays* L.), with crop rows at a distance of 0.50 m, respectively. Each area was divided into three sections of 20 × 20 m^2. The sections in one area were seeded in consecutive weeks, allowing us to conduct experiments in three-week windows. Figure 6 shows the experimental field and the distribution of the areas and sections.

3.2. Description of the Test Mission

Tests were conducted to assess the performance and quality of integrating new technologies in autonomous robots for agriculture. First, the testing prototype was integrated with the components introduced in Section 2; then, several IoT devices were disseminated throughout the field (RGB and multispectral cameras, weather stations, soil probes, etc.); finally, a mission was defined to acquire data in the study site to perform quantitative analyses. The mission consisted of covering sections of 20 × 20 m^2 with wheat and maize crops while the following occurred:

- Acquiring data from the IoT sensor network.
- Taking pictures of the crop.
- Acquiring data from the guidance system.
- Sending all the acquired information to the cloud.

The mission proposed by the planner is illustrated in Figure 7. The robot tracked the path autonomously, and the following procedures were carried out.

Figure 7. Robot's path from the home garage to the study site. The planner provides the mission for covering the study site.

Perception system procedure

- Guiding vision system: This experiment was conducted in the treatment stage, where the crop was detected to adjust the errors derived from planning and the lack of precision of the maps. YOLOv4 [20], a real-time object detector based on a one-stage object detection network, was the base model for detecting early-stage growth in maize [8], a wide-row crop. The model was trained using a dataset acquired in an agricultural season before these tests using the same camera system [21]. Moreover, in the case of wheat, which is a narrow-row crop, a different methodology was applied through the use of segmentation models, such as MobileNet, a convolutional neural network for mobile vision applications [22], trained using a dataset acquired in an agricultural season before these tests [23], with the same camera system. The detection of both crops was evaluated with regard to the GNSS positions collected manually for the different crop lines.

The maize and wheat datasets were built with 450 and 125 labeled images, respectively. Data augmentation techniques (rotating, blurring, image cropping, and brightness changes) were used to increase the size of the datasets. For both crops, 80% of the data was destined for training, 10% for validation, and 10% for testing.

- The AI vision system: This system uses data from the installed RGB cameras to enable robust automated plant detection and discrimination. For this purpose, the state-of-the-art object detection algorithm Yolov7 is used in combination with the Nvidia framework DeepStream. Tracking the detected plants is performed in parallel by a pretrained DeepSort algorithm [24]. The reliability of the object detection algorithm is evaluated using test datasets with the commonly used metrics "intersection over union" (IoU) and "mean average precision" (mAP). This system works cooperatively with laser scanners as a stand-alone system. The information is not stored in the cloud.

The dataset used for training weed/crop discrimination was generated in fields in several European countries. It contains 4000 images, 1000 of which are fully labeled. Distinctions are made according to the processing steps to be applied: weeds, grasses, and crops. In addition, the dataset was expanded to three times its original size through augmentation measures. As well as generating new training data, this enables robustness against changing environmental influences, such as changing color representation, motion blur, and camera distortion. The YoloV7 network achieved a mean average precision (mAP) of 0.891 after 300 epochs of training. The dataset was divided into 80%, 10%, and 10% for training, validation, and testing subsets, respectively.

Autonomous robot procedure

The navigation controller: Given a set of trajectories based on RTK-GNSS, the performance of the guidance controller was evaluated by measuring lateral and angular error through the incorporation of colored tapes on the ground and using the onboard RGB camera and ToF to extract the tape positions to compute the errors concerning the robot's pace.

Smart Navigation Manager procedure:

- Smart operation manager: The processing time, latency, success rate, response time, and response status based on requests of the mission planner, IoT sensors, and cloud computing services were evaluated using ROS functionalities that provide statistics related to the following:
    - The period of messages by all publishers.
    - The age of messages.
    - The number of dropped messages.
    - Traffic volume to be measured in real-time.
- Central manager: The evaluation is similar to that used for the navigation controller.
- Obstacle detection system: YOLOv4 and a model already developed based on the COCO database were introduced to detect common obstacles in agricultural environments and were also used for evaluation. YOLOv4 is a one-stage object detection model, and COCO (common objects in context) is a large-scale object detection, segmentation, and captioning dataset.

4. System Assessment and Discussion

The mission described in the previous section produced crop images, sensor data, and traffic information with the following characteristics:

- Crop images: During the robot's motion, images are acquired at a rate of 4 frames/s to guide the robot. The RGB images are 2048 × 1536 pixels with a weight of 2.2 MB (see Figures 8 and 9), and the ToF images feature 352 × 264 points (range of 300–5000 mm) (see Figure 10). The images are sent to the guiding and obstacle detection system through the Ethernet using ROS (perception–ROS bridge in the perception system and ROS manager in the central manager). A subset of these images is stored in the cloud for further analysis. Using a FIWARE–ROS bridge with the NGSI application programming interface, the system sends up to 4 frames/s.
- Sensor data: IoT devices send the acquired data using 2.4 GHz WiFi with the MQTT protocol and JSON format.
- Traffic information: The ROS functionalities mentioned above revealed that during a field experiment (10 min duration), the total number of delivered messages was 2,395,692, with a rate of only 0.63% dropped messages (messages that were dropped due to not having been processed before their respective timeout), with average traffic of 10 MB/s and maximum traffic of 160 MB at any instant of time. No critical messages (command messages) were lost, demonstrating robustness within the smart navigation manager. Regarding cloud traffic, during a period of time of approximately 3 h, the messages sent to the cloud were monitored, where the number of messages received by the cloud was measured; the delay time of the transmission of the messages between the robot (edge) and the OCB, and between the robot and the KAFKA bus (see Figure 3), were also measured. During this interval of time, around 4 missions were executed, and a total of 14,368 messages were sent to the cloud, mainly the robot status and the perception system data. An average delay of about 250 ms was calculated between the moment the message is sent from the robot and the moment it is received in the OCB (see Figure 11a). Moreover, the KAFKA overhead, i.e., the time it takes for a message received by the OCB to be forwarded to the KAFKA bus and eventually processed by a KAFKA consumer, was approximately 1.24 ms, demonstrating that the internal communications within the server and hosted cloud services are robust (see Figure 11b).

Figure 8. Example of a wheat image acquired with the guiding vision system and uploaded to the cloud.

Figure 9. Example of a maize image acquired with the guiding vision system and uploaded to the cloud.

The system has been tested in a field with two different crops. Data related to cloud communication and robot guidance algorithms have been collected. The communication performance is similar to that obtained using conventional mechanisms, so we benefit from using ROS and FIWARE without compromising performance.

Figure 10. Example of a ToF intensity image acquired with the guidance system and uploaded to the cloud.

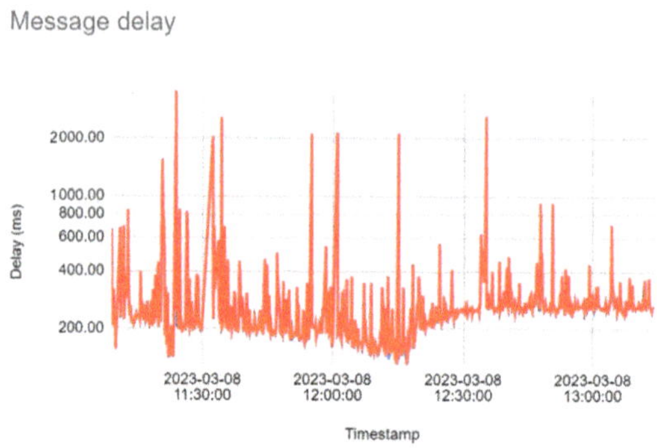

(a)

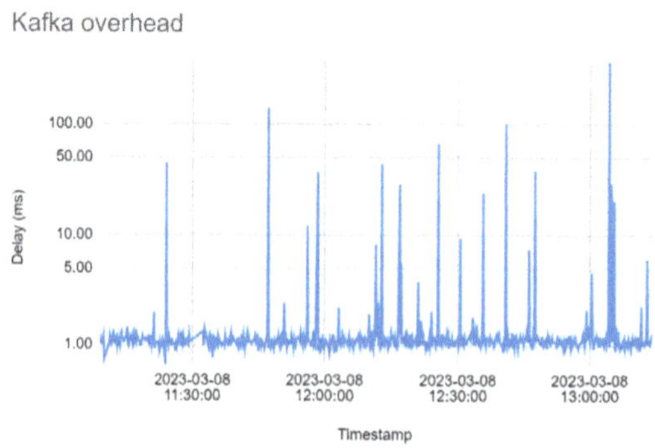

(b)

Figure 11. Example of a ToF intensity image acquired with the guidance system and uploaded to the cloud. (**a**) Message delay and (**b**) Kafka overhead.

5. Conclusions

An architecture is presented to configure autonomous robots for agriculture with access to cloud technologies. This structure takes advantage of new concepts and technologies, such as IoT and cloud computing, allowing big data, edge computing, and digital twins to be incorporated into modern agricultural robots.

The architecture is based on ROS, the most universally accepted collection of software libraries and tools for building robotic applications, and FIWARE, an open architecture that enables the creation of new applications and services on the Internet. ROS and FIWARE provide attractive advantages for developers and farmers. ROS and FIWARE offer powerful tools for developers to build control architectures for complex robots with cloud computing/IoT features, making development easier and leveraging open-source frameworks. ROS and FIWARE, as in the proposed integration, provide reusability, scalability, and maintenance using the appropriate hardware resources. In addition, integrating the robot controller into the Internet allows the exploitation of autonomous robot services for agriculture through the Internet.

On the other hand, the use of this type of architecture reveals to farmers the advantages of communicating autonomous robots with the cloud, providing them with leading benefits to storing data safely and efficiently, eliminating physical storage, and, thus, reducing the risk of data loss. Data stored in the cloud makes it easy to access data from anywhere and share it with other farmers or platforms. In addition, the services offered in the cloud are very flexible to contract the actual storage needed at all times, optimizing the farmer's resources. Finally, farmers can use the analysis tools available in the cloud to make their own decisions. In any case, working in the cloud requires an initial investment, which is usually recovered quickly.

The different components of the robot, particularized for a laser-based weeding robot, are described, and the general architecture is presented, indicating the specific interfaces. Based on these components, the article presents the action sequence of the robot and the operating procedure to illustrate how farmers can use the system and what benefits they can obtain.

Several experiments with two crops were conducted to evaluate the proposed integration based on the data communication characteristics, demonstrating the system's capabilities. The crop row detection system works correctly for both crops, tracking the rows with an accuracy of ± 0.02 m. The evaluation concluded that the system could send image frames to the cloud at 4 frames/s; messages between subsystems and modules can be passed with a 0.63% rejection rate. Regarding the traffic of the information exchanged, an average delay of 250 ms was detected in the messages between the robot and the OCB. In contrast, the OCB and the KAFKA bus measured an average message of 1.24 ms. This indicates the robustness of internal communications within the server and hosted cloud services. This performance is in the range obtained when a system communicates with the cloud using conventional methods, so ROS and FIWARE facilitate communication with the cloud without compromising performance.

Future work will focus on extending cloud computing architecture to integrate digital twins, orchestrate big data ensembles, and facilitate the work of robots with edge computing performance.

Author Contributions: Conceptualization, L.E., R.F., P.G.-d.-S., M.F., M.G., G.V., H.S., M.H. and M.W.; methodology, L.E. and R.F.; software, L.E., M.F., H.S. and M.W.; validation, L.E., M.F., G.V. and H.S.; investigation, L.E., R.F., P.G.-d.-S., M.F., M.G., G.V., H.S., M.H. and M.W.; writing—original draft preparation, P.G.-d.-S.; writing—review and editing, L.E., P.G.-d.-S. and R.F.; supervision, L.E. and P.G.-d.-S.; funding acquisition, P.G.-d.-S., G.V., M.G. and M.W. All authors have read and agreed to the published version of the manuscript.

Funding: This article is part of a project that has received funding from the European Union's Horizon 2020 research and innovation program under grant agreement No 101000256.

Institutional Review Board Statement: The study was conducted in accordance with the Declaration of Helsinki, and approved by the Ethics Committee of CSIC.

Data Availability Statement: Herrera-Diaz, J.; Emmi, L.A.; Gonzalez de Santos, P. Maize Dataset. 2022. Available online: http://doi.org/10.20350/digitalCSIC/14566 (accessed on 1 April 2023). Herrera-Diaz, J.; Emmi, L.; Gonzalez de Santos, P. Wheat Dataset. 2022. Available online: http://doi.org/10.20350/digitalCSIC/14567 (accessed on 1 April 2023).

Conflicts of Interest: The authors declare no conflict of interest.

References

1. Ghose, B. Food security and food self-sufficiency in China: From past to 2050. *Food Energy Secur.* **2014**, 3, 86–95. [CrossRef]
2. Zhang, N.; Wang, M.; Wang, N. Precision agriculture—A worldwide overview. *Comput. Electron. Agric.* **2002**, *36*, 113–132. [CrossRef]
3. Stentz, A.; Dima, C.; Wellington, C.; Herman, H.; Stager, D. A System for Semi-Autonomous Tractor Operations. *Auton. Robot.* **2002**, *13*, 87–104. [CrossRef]
4. Bergerman, M.; Maeta, S.M.; Zhang, J.; Freitas, G.M.; Hamner, B.; Singh, S.; Kantor, G. Robot Farmers: Autonomous Orchard Vehicles Help Tree Fruit Production. *IEEE Robot. Autom. Mag.* **2015**, *22*, 54–63. [CrossRef]
5. Gonzalez-De-Santos, P.; Ribeiro, A.; Fernandez-Quintanilla, C.; Lopez-Granados, F.; Brandstoetter, M.; Tomic, S.; Pedrazzi, S.; Peruzzi, A.; Pajares, G.; Kaplanis, G.; et al. Fleets of robots for environmentally-safe pest control in agriculture. *Precis. Agric.* **2017**, *18*, 574–614. [CrossRef]
6. Underwood, J.P.; Calleija, M.; Taylor, Z.; Hung, C.; Nieto JFitch, R.; Sukkarieh, S. Real-time target detection and steerable spray for vegetable crops. In Proceedings of the International Conference on Robotics and Automation: Robotics in Agriculture Workshop, Seattle, WA, USA, 9–11 May 2015.
7. Kongskilde. New Automated Agricultural Platform—Kongskilde Vibro Crop Robotti. 2017. Available online: http://conpleks.com/robotech/new-automated (accessed on 14 December 2022).

8. Emmi, L.; Herrera-Diaz, J.; Gonzalez-De-Santos, P. Toward Autonomous Mobile Robot Navigation in Early-Stage Crop Growth. In Proceedings of the 19th International Conference on Informatics in Control, Automation and Robotics (ICINCO 2022), Lisbon, Portugal, 14–16 July 2022; pp. 411–418. [CrossRef]
9. Bannerjee, G.; Sarkar, U.; Das, S.; Ghosh, I. Artificial Intelligence in Agriculture: A Literature Survey. *Int. J. Sci. Res. Comput. Sci. Appl. Manag. Stud.* **2018**, *7*, 3.
10. Osinga, S.A.; Paudel, D.; Mouzakitis, S.A.; Athanasiadis, I.N. Big data in agriculture: Between opportunity and solution. *Agric. Syst.* **2021**, *195*, 103298. [CrossRef]
11. Yang, D.; Li, D.; Sun, H. 2D Dubins Path in Environments with Obstacle. *Math. Probl. Eng.* **2013**, *2013*, 291372. [CrossRef]
12. Emmi, L.; Parra, R.; González-de-Santos, P. Digital representation of smart agricultural environments for robot navigation. In Proceedings of the 10th International Conference on ICT in Agriculture, Food & Environment (HAICTA 2022), Athens, Greece, 22–25 September 2022; pp. 1–6.
13. Orion Context Broker. Telefonica. Available online: https://github.com/telefonicaid/fiware-orion (accessed on 22 February 2023).
14. Francia, M.; Gallinucci, E.; Golfarelli, M.; Leoni, A.G.; Rizzi, S.; Santolini, N. Making data platforms smarter with MOSES. *Futur. Gener. Comput. Syst.* **2021**, *125*, 299–313. [CrossRef]
15. ROS—The Robot Operating System. 2023. Available online: https://www.ros.org/ (accessed on 24 April 2020).
16. ROSLink. 2023. Available online: https://github.com/aniskoubaa/roslink (accessed on 5 January 2023).
17. Koubaa, A.; Alajlan, M.; Qureshi, B. ROSLink: Bridging ROS with the Internet-of-Things for Cloud Robotics. In *Robot Operating System (ROS)*; Koubaa, A., Ed.; Studies in Computational Intelligence; Springer: Cham, Switzerland, 2017; Volume 707. [CrossRef]
18. Fiware Community Fiware: The Open Source Platform for Our Smart Digital Future. Available online: https://www.fiware.org/ (accessed on 5 January 2023).
19. López-Riquelme, J.; Pavón-Pulido, N.; Navarro-Hellín, H.; Soto-Valles, F.; Torres-Sánchez, R. A software architecture based on FIWARE cloud for Precision Agriculture. *Agric. Water Manag.* **2017**, *183*, 123–135. [CrossRef]
20. Bochkovskiy, A.; Wang, C.Y.; Liao, H.Y.M. YOLOv4: Optimal speed and accuracy of object detection. *arXiv* **2020**, arXiv:2004.10934.
21. Herrera-Diaz, J.; Emmi, L.A.; Gonzalez de Santos, P. Maize Dataset. 2022. Available online: https://digital.csic.es/handle/10261/264581 (accessed on 1 April 2023).
22. Howard, A.G.; Zhu, M.; Chen, B.; Kalenichenko, D.; Wang, W.; Weyand, T.; Andreetto, M.; Adam, H. MobileNets: Efficient Convolutional Neural Networks for Mobile Vision Applications. *arXiv* **2017**, arXiv:1704.04861.
23. Herrera-Diaz, J.; Emmi, L.; Gonzalez de Santos, P. Wheat Dataset. 2022. Available online: https://digital.csic.es/handle/10261/264622 (accessed on 1 April 2023).
24. Wojke, N.; Bewley, A.; Paulus, D. Simple Online and Realtime Tracking with a Deep Association Metric. In Proceedings of the IEEE International Conference on Image Processing (ICIP), Beijing, China, 17–20 September 2017.

agriculture

Article

Designing an Interactively Cognitive Humanoid Field-Phenotyping Robot for In-Field Rice Tiller Counting

Yixiang Huang [1,2], Pengcheng Xia [1,2], Liang Gong [1,2,*], Binhao Chen [1,2], Yanming Li [1,2,3] and Chengliang Liu [1,2]

1 School of Mechanical Engineering, Shanghai Jiao Tong University, Shanghai 200240, China
2 MoE Key Lab of Artificial Intelligence, AI Institute, Shanghai Jiao Tong University, Shanghai 200240, China
3 Key Laboratory of Intelligent agricultural technology (Yangtze River Delta), Ministry of Agriculture and Rural Affairs, Shanghai 200240, China
* Correspondence: gongliang_mi@sjtu.edu.cn

Abstract: Field phenotyping is a crucial process in crop breeding, and traditional manual phenotyping is labor-intensive and time-consuming. Therefore, many automatic high-throughput phenotyping platforms (HTPPs) have been studied. However, existing automatic phenotyping methods encounter occlusion problems in fields. This paper presents a new in-field interactive cognition phenotyping paradigm. An active interactive cognition method is proposed to remove occlusion and overlap for better detectable quasi-structured environment construction with a field phenotyping robot. First, a humanoid robot equipped with image acquiring sensory devices is designed to contain an intuitive remote control for field phenotyping manipulations. Second, a bio-inspired solution is introduced to allow the phenotyping robot to mimic the manual phenotyping operations. In this way, automatic high-throughput phenotyping of the full growth period is realized and a large volume of tiller counting data is availed. Third, an attentional residual network (AtResNet) is proposed for rice tiller number recognition. The in-field experiment shows that the proposed method achieves approximately 95% recognition accuracy with the interactive cognition phenotyping platform. This paper opens new possibilities to solve the common technical problems of occlusion and observation pose in field phenotyping.

Keywords: phenotyping; agricultural robot; tiller counting; deep learning; residual network

Citation: Huang, Y.; Xia, P.; Gong, L.; Chen, B.; Li, Y.; Liu, C. Designing an Interactively Cognitive Humanoid Field-Phenotyping Robot for In-Field Rice Tiller Counting. *Agriculture* **2022**, *12*, 1966. https://doi.org/10.3390/agriculture12111966

Academic Editors: Jin Yuan, Wei Ji and Qingchun Feng

Received: 1 October 2022
Accepted: 6 November 2022
Published: 21 November 2022

1. Introduction

The growing population places high demands on crop yields [1]. Crop breeding is a crucial technique to increase yields, disease resistance and other desirable properties by improving the genetic characteristics of crops [2]. Phenotyping is a process central to breeding, which refers to measuring the key parameters related to crop properties, such as plant height, leaf area, leaf angle, number of grains and number of tillers [3,4]. The phenotyping process is currently mainly performed by crop breeding experts, who measure these parameters with manual tools and their sufficient experience.

In order to acquire crop growth status at different growth stages, breeding experts need to perform in-field manual phenotyping for each crop at regular intervals. Undoubtedly, this work is labor-intensive and time-consuming. The traditional manual phenotyping method is highly experience-dependent and its efficiency and reliability are limited. As a result, the rate of plant genome research is restricted by the rate of phenotyping, which is defined as the "Phenotyping Bottleneck" [5].

To speed up the breeding process and relieve the bottleneck, studies on high-throughput phenotyping platforms (HTPPs) have been widely conducted [6]. Many advanced technologies have been applied for automatic phenotyping [7]. The Scanalyzer 3D High Throughput platform [8] developed by German research institute LemnaTec has high impact [9]. Plants are transported by conveyers through a sequence of imaging cabinets equipped with various sensors to acquire various phenotype data. This system is widely used in various phenotyping platforms, such as the Plant Accelerator of Australian Centre for Plant Functional Genomics (ACPFG) [10]. The Plant Accelerator, consisting of four greenhouses and two Scanalyzer 3D platforms, can accomplish high-throughput phenotyping, as well as watering and weighing the plants. Hartmann et al. [11] developed an open-source image analysis pipeline called HTPheno. It can acquire crop images using pipelines in greenhouses and measure various phenotypic parameters from the images. Liu et al. [12] presented a Digital Plant Phenotyping Platform for multiple trait measurement, such as leaf and tiller orientation. These HTPPs significantly increased the phenotyping efficiency compared with the traditional manual process. However, plants grown in greenhouses are not affected by soil condition, weather variation or many other natural factors, so phenotypes may differ from those grown naturally in fields. Moreover, to avoid the influence of leaf occlusion and overlap on measurement, plants are planted separately, which cannot simulate the plant interplay when planted closely in fields.

For the purpose of field high-throughput phenotyping, many field high-throughput phenotyping platforms have been developed to date. LemnaTec also developed a field HTPP named the Scanalyzer Field recently [13]. It is a fully automated gantry system with an extensive measurement platform equipped with cameras and sensors. It can measure up to 0.6 hectares of crops to acquire detailed phenotypic data. Researchers at the University of Arizona and the United States Department of Agriculture (USDA) [14] developed a field HTPP that included a sonar proximity sensor, sonar and GPS antenna and infrared radiometer (IRT) sensors. The system can measure canopy height, reflectance, and some other phenotypic parameters, but it can only acquire data overhead. The Robotanist developed by Mueller-Sim et al. [15] is a ground-based platform. It can autonomously navigate fields to measure stalk strength with a manipulator and collect phenotypic parameters with non-contact sensors. The platform developed by researchers at Iowa State University employs a stereo camera rig that consists of six stereo camera heads to accomplish high quality 3D reconstruction of sorghum plant architecture [16]. The system is carried by a self-navigate tractor equipped with RTK-GPS signals. Zhou et al. [17] introduced a rice panicle counting platform using images captured by an unmanned aerial vehicle based on deep learning algorithms.

Field HTPPs automatically conduct phenotyping in natural fields with high efficiency using automatic navigating and measurement systems. However, leaf occlusion and overlap in field environments severely restrict the measurement accuracy of some parameters. This has become a key challenge for automatic in-field phenotyping and restricts practical applications.

Tillers refer to the aboveground branches of gramineous plants, and the number of tillers is one of the most important parameters in ecology and breeding studies. The rice yield is usually dominated by primary tillers and some early secondary tillers [18]. As a result, tiller number is a key phenotypic trait for rice and the measurement and analysis of the tiller number are indispensable in phenotyping [19]. Rice tillers are currently manually counted using the separated shoots from a single plant by experts. The counting process is inefficient and labor-intensive. Automatic tiller counting methods have been studied in the past few years. For instance, Yang et al. [20] used an X-ray computed tomography (CT) system to measure rice tillers on a conveyer. In their work, a mean absolute error (MAE) of approximately 0.3 was reached. Huang et al. [21] proposed to measure rice tillers through magnetic resonance imaging (MRI). However, it is not suitable to perform in-field high-throughput measurements using these cumbersome and expensive systems. Scotford et al. [22] used spectral reflectance and ultrasonic sensing techniques to estimate tiller density and an accuracy of ± 125 tillers per m^2 was achieved. Deng et al. [23] presented a rice tiller counting platform based on in-field images captured by smartphones and they were measured after the rice plants were cut and the branches were removed. Yamagishi et al. [24] proposed to count rice tillers using proximal sensing data taken by an unmanned aerial vehicle. These methods provided some attempts for in-field tiller counting, but the key problem of occlusion and overlap was not addressed, restricting the recognition accuracy in practical applications.

To tackle the occlusion and overlap problem in in-field phenotyping in this paper, a novel phenotyping paradigm of interactive cognition is proposed. A detectable quasi-structured environment is actively constructed for in-field phenotyping; therefore, the cognition process can be accomplished smoothly. This method overcomes the problem of occlusion and overlap in traditional passive automatic phenotyping methods. Meanwhile, a field phenotyping robot is developed and a bio-inspired solution is adopted so that it mimics the manual operations of breeding experts in fields. In this way, the phenotyping operational schedules are regularized. Moreover, based on the interactive cognition phenotyping method, a rice tiller counting method based on attentional residual networks (AtResNet) is proposed using the structured light images captured by the robot. The main contributions of this paper are as follows:

(1) An interactive cognition methodology is proposed for full growth period in-field high-throughput phenotyping.
(2) To accomplish the interactive cognition-based field phenotyping, a humanoid robot is designed with human-in-the-loop interactive methodology.
(3) A high-accuracy rice tiller counting method based on the phenotyping platform is proposed.

The rest of this paper is organized as follows. Section 2 introduces the interactive cognition phenotyping method based on the humanoid robot. Section 3 presents the bio-inspired operational forms. Section 4 describes the rice tiller counting algorithm and Section 5 shows the experimental results. Section 6 concludes the paper.

2. Interactive Cognition Phenotyping Method

In many industrial applications, the machine vision techniques for object detection and measurement are mature. Industrial robots generally use non-interactive passive detection methods to achieve cognition of the surrounding environment. However, occlusion and overlap rarely exist in industrial scenes; in other words, the scenes are structured. Hence, non-interactive cognition methods can basically meet the cognition requirements. However, in fields, simple machine vision inspection methods are not compatible with complex unstructured agricultural scenes [25]. It is difficult to perform phenotyping for crops in occlusion scenes. To solve this problem, we propose a new phenotyping paradigm of interactive cognition. A phenotyping robot is introduced to interact with the surrounding plants. The robot mimics breeding experts' manual operations of removing occlusion and overlap, while performing phenotyping in fields. A detectable quasi-structured environment is constructed; therefore, full cognition of the crops can be achieved through machine vision-based detecting methods.

2.1. *Interactively Cognitive Humanoid Field Phenotyping Robot*

In order to interactive with crops and construct various detectable scenes for phenotyping, the robot needs to have high operational dexterity. To perform the phenotyping operations of experts, we used a bio-inspired design methodology to design the humanoid robot ontology. The robot ontology, as shown in Figure 1, is based on an open-source project named InMoov [26], and it has been redesigned to improve its adaptability to the agricultural working environment. Its shoulder and arm have five degrees of freedom, ensuring the completion of complex actions, such as those carried out by humans, and sufficient space for movement. The manipulator is a humanoid mechanical hand, inspired by an open-source project [27]. The mechanical hand has one degree of freedom. Five fingers can grip and stretch at the same time so that phenotyping actions, such as separating ears and handling stalks, can be performed.

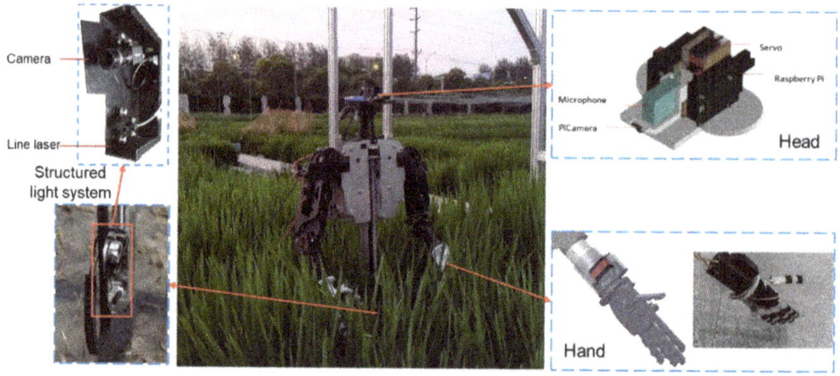

Figure 1. Interactively cognitive humanoid phenotyping robot.

The robot is placed on a field truss platform that can move along the tracks in field. The robot can move to a suitable position to interact with the plants under analysis. It can move along two mutually perpendicular horizontal tracks with a moving speed of 0.1 m/s to 0.3 m/s. It can descend 25 cm towards the ground and lift 75 cm above the ground.

The liftable line-structured light system equipped on the chest of the robot body is used for environmental detection and cognition. The system consists of a Basler acA2500-14gc color camera and line laser module that can scan up and down, driven by a stepper motor. The camera has a horizontal and vertical resolution of 2590 × 1942 px, a frame rate of 14 FPS, and a sensor area of 1/2.5 inch. The scanning speed is approximately 20 mm/s and the scanning stroke is 500 mm. The 3D reconstruction of plants and measurement of many phenotypic parameters can be realized using the structured light system.

An interactive system that consists of a raspberry Pi, a microphone and a PiCamera is mounted on the robot's head. The PiCamera can screen live video of the field and transmit video streams to the server built by the raspberry Pi. The video stream delay is about 0.5 s, and the resolution is 1280 × 960 with a 30-fps frame rate.

2.2. Interactive Cognition Phenotyping Process

When the robot moves to the front of the plant under analysis, it can actively interact with the plant to build a more detectable environment if there is evidence of occlusion and overlap. As shown in Figure 2, when the plant is sheltered by other plants, the robot arm can push aside the plants to remove the occlusion. Then, the plant can be detected by the vision system and full phenotypic data can be acquired. Similarly, when the back part of a plant is occluded by the front part, the same active interaction process can be used to build a phenotype detectable environment.

Figure 2. The robot removing occlusion.

The robot operates on a field truss platform and it can move along two mutually perpendicular horizontal tracks. A fixed position in the field can be taken as the origin of the absolute coordinate system, and the two moving directions are the X-axis and the Y-axis, respectively. We use the motor-driven signals of the servo motors as odometers. When the robot moves to a position to measure a specific plant, the moving distance along the two directions can be calculated by the pulse number of the motor-driven signals. Therefore, the geographic coordinates of the robot can be determined. The relative distance of the plant to the robot can be measured by the pre-calibrated structural light system and the geographic coordinate of the plant can be determined. In our experimental field spot, where the longest moving distance is 50 m, the measurement error of the robot geographic coordinates is approximately 2 cm. In the robot operating space, the structural light system measurement error is approximately 0.1 cm. In this manner, an electronic map of every plant in the field can be established. Phenotypic data of every plant measured by the robot platform can be recorded on the map. With the electronic map, the robot platform can measure the same plant at different growth stages, thus establishing a complete full growth cycle phenotype database to provide complete phenotypic data for crop breeding.

Despite the introduction of the robot technique and active interactive cognition method, the efficiency and accuracy of automatic phenotyping can still be considerably improved, which is required to release the "Phenotyping Bottleneck". In addition, with the use of electronic maps, automatic phenotyping of full growth cycles can be realized.

3. Bio-Inspired Operational Forms

In natural agricultural environments, it is extremely difficult for robots to perform fully autonomous measurements and cognition. To date, operation in these non-structured scenes cannot reach relatively high accuracy. As a result, phenotyping schedules and operation need to be formulated first. Due to the humanoid structure of the robot, a bio-inspired solution is proposed. By mimicking phenotyping operations of breeding experts, the phenotyping operational schedules are regularized.

The human–robot interactive technique (HRI) is used to regularize the phenotyping schedule. Breeding experts remotely control the robot platform to perform interactive phenotyping operations with the HRI system. The HRI framework is shown in Figure 3.

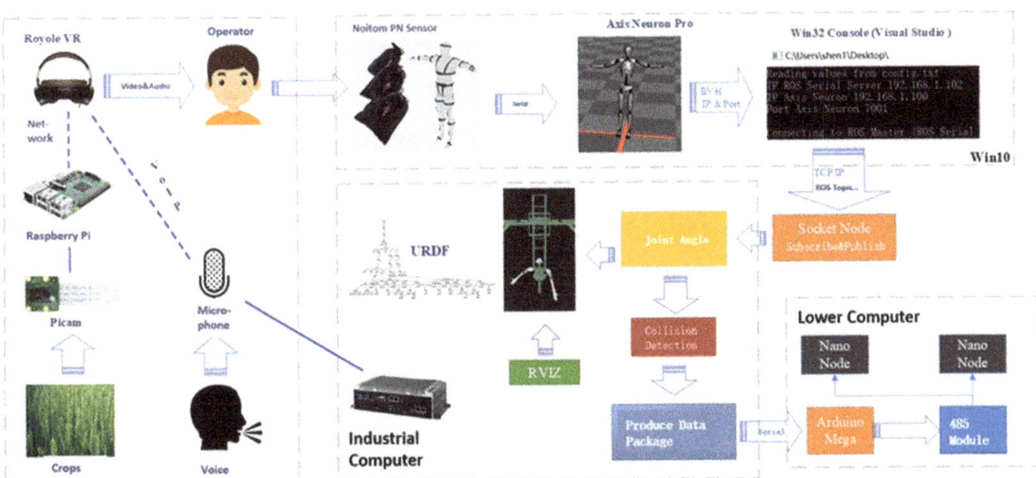

Figure 3. Human–robot interactive (HRI) framework for interactive phenotyping.

3.1. *Head-Mounted Interactive System*

A head-mounted interactive system is used to acquire the live scenes and voice, so that the operator can easily manipulate the robot to interact with the in-field environment. An approximately immersive operation experience can be obtained when breeding experts use this system.

The structure of the head-mounted interactive system is described in Section 2.1. The operator wears a Royole VR standalone headset to acquire the live video. In this way, the operator can remotely observe the scene of the robot in the current field of view in real time. The microphone mounted at the robot head can record the sound around the robot. The operator end and the robot end can communicate through the VR standalone headset and the robot computer. The raspberry Pi works as a server. In this way, the operator can hear the real-time voice that is "heard" by the robot to monitor the in-field situations better.

3.2. Motion Interactive System Based on Perception Neuron (PN) Sensor

In order to expediently control the complex movement of the multi-degree of freedom robot and improve control precision, a wearable sensor system is adopted to map the operator's movement to the robot's movement. Then, the robot can mimic the operator's phenotyping operations.

A perception neuron (PN) sensor system produced by Noitom Company® [28] is used. This sensor system includes thirty-two inertial measurement units, each of which has a three-axis gyroscope, a three-axis accelerometer, and a three-axis magnetometer.

A PN sensor can export a BioVision Hierarchy (BVH) file after acquiring human motion data. A BVH file is a universal human motion feature description format, which is often used in skeletal animation models [29]. The BVH file describes the human skeleton model in the joint diagram shown in Figure 4a. Each joint describes the motion information through three rotation parameters and a complete description of the human motion is achieved. After the BVH data collected by the PN sensor are transmitted to the robot controller through the TCP/IP protocol, the Euler angles in the BVH need to be converted into joint angles and sent to the lower computer.

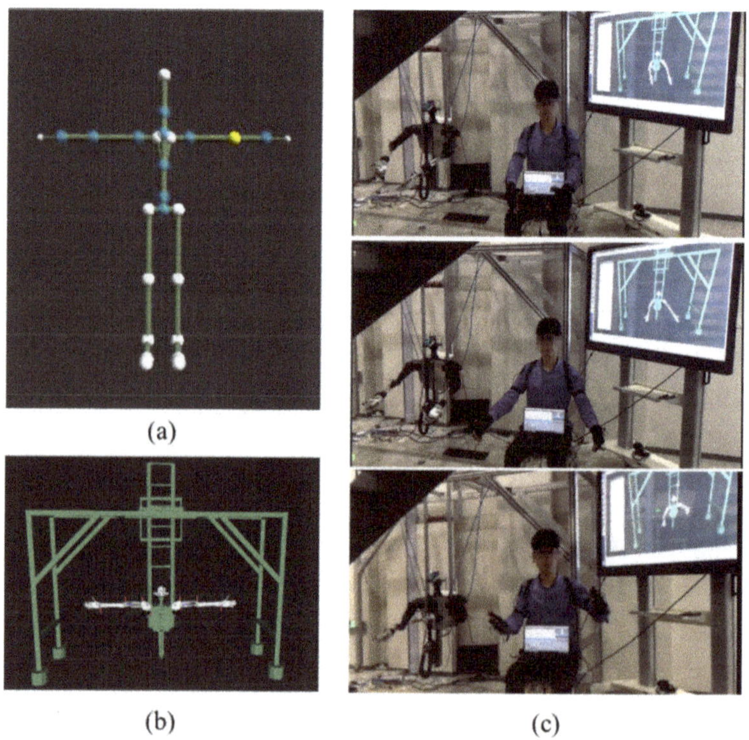

Figure 4. Motion interaction. (**a**) BVH joint diagram. (**b**) URDF visualization. (**c**) Motion interactive experiments.

However, the actual movement of the human body is physiologically constrained. Not every joint has three degrees of freedom, and some degrees of freedom are not independent of each other, so there is a large difference between the BVH model and human body. Therefore, the mapping of Euler angles to the joint angle of the robot requires a reasonable algorithm. For example, the human shoulder joint has three degrees of freedom, which is similar to the shoulder of the robot body, so the Euler angle of the shoulder joint motion can be directly mapped to the robot body through rotation matrix. Since the elbow joint of the robot body has only one bending degree of freedom and lacks a rotational one, the elbow bending angle can be obtained by calculating the angle between the direction vector of the large arm and forearm. The angle of rotation of the wrist joint is mapped by the angle of rotation of the human elbow. We denote vector $\vec{r_1}$ and $\vec{r_2}$ as the large arm and forearm, respectively, and $\vec{r_1}$ is the position direction of the X-axis. Therefore, the elbow bending angle can be calculated as

$$\theta = \pi - \left\langle \vec{r_1}, \vec{r_2} \right\rangle = \pi - arccos\left(\vec{r_1} \cdot \vec{r_2}\right). \tag{1}$$

We assume that the two rotation degrees of freedom are along the Y- and Z-axes, respectively. The PN sensor can acquire the Euler angles of ZYX axes of the human arm, i.e., α_z, β_y, γ_x. Since the rotation degree of freedom of the X-axis does not exist in the human arm, $\gamma_x \approx 0$. The rotation matrix of the elbow is formulated as

$$R = \begin{pmatrix} 1 & 0 & 0 \\ 0 & 1 & 0 \\ 0 & 0 & 1 \end{pmatrix} \begin{pmatrix} cos\beta_y & 0 & sin\beta_y \\ 0 & 1 & 0 \\ -sin\beta_y & 0 & cos\beta_y \end{pmatrix} \begin{pmatrix} cos\alpha_z & sin\alpha_z & 0 \\ -sin\alpha_z & cos\alpha_z & 0 \\ 0 & 0 & 1 \end{pmatrix}. \tag{2}$$

The direction vector of $\vec{r_1}$ is $\hat{r_1} = (1,0,0)^T$. Therefore, the direction vector of $\vec{r_1}$ is

$$\begin{aligned} \hat{r_2} = R\hat{r_1} &= \begin{pmatrix} 1 & 0 & 0 \\ 0 & 1 & 0 \\ 0 & 0 & 1 \end{pmatrix} \begin{pmatrix} cos\beta_y & 0 & sin\beta_y \\ 0 & 1 & 0 \\ -sin\beta_y & 0 & cos\beta_y \end{pmatrix} \begin{pmatrix} cos\alpha_z & sin\alpha_z & 0 \\ -sin\alpha_z & cos\alpha_z & 0 \\ 0 & 0 & 1 \end{pmatrix} \begin{pmatrix} 1 \\ 0 \\ 0 \end{pmatrix} \\ &= \begin{pmatrix} cos\beta_y cos\alpha_z \\ -sin\alpha_z \\ -sin\beta_y cos\alpha_z \end{pmatrix}. \end{aligned} \tag{3}$$

Finally, the elbow bending angle can be obtained by

$$\theta = \pi - arccos\left(\vec{r_1} \cdot \vec{r_2}\right) = \pi - arccos(\hat{r_1}, \hat{r_2}) = \pi - arccos\left(cos\beta_y cos\alpha_z\right). \tag{4}$$

The robot hand has only one degree of freedom. In order to map the human hand motion to the maximum extent, the hand degree of freedom selects the fold angle of the human middle finger. Because of the high degree of freedom of the human neck, the left and right rotational degrees of freedom of the robot are directly mapped by the left and right rotation angle of the human neck.

A Unified Robot Description Format (URDF) file is constructed in the robot operating system (ROS) that runs on the robot's industrial computer. It contains the joint relations of each mechanical parts of the robot and real-time simulation of the robot can be realized based on the URDF file, as shown in Figure 4b. ROS transmits the mapped joint angle data in real time through the serial port to the lower machine with a 10 Hz sampling frequency. Then, the lower machine drives the joint servos moving to the corresponding angle. Therefore, the operator's motion is mapped to the robot ontology. Some motion interactive experiments are shown as Figure 4c.

3.3. Bio-Inspired Operation

Through the head-mounted interactive system and the motion interaction system based on PN sensors, the operator can remotely control the robot in an immersive interactive way. The breeding expert must wear a full headset linked to the interactive system in the control room, so that it is possible to observe the real-time environment around the robot by the head-mounted interactive system through the movement of their head. The operator observes the plants that need to be measured and moves the robot to the appropriate position. The operator only needs to repeat the procedure and operations during the traditional manual phenotyping process, then the robot can be controlled to mimic his/her action to interact with the plant. The phenotype is then measured by the machine vision system. The naturally instructive paradigm is user-friendly and especially highly efficient with the first person view (FPV), which can accomplish efficient phenotyping operations [30]. The robot completely mimics the interactive operations of the breeding experts, so this interactive form has high efficiency and strong adaptability. With the help of the automated visual system, high-efficiency and high-precision phenotyping is achieved through the interactive cognition method.

Regularized phenotyping forms are formed through the bio-inspired operations based on the HRI technique. In the process of HRI, the typical operation schedules and actions of the breeding experts are recorded. In the long term, a large amount of data is recorded to form a manual teaching dataset. With a sufficiently large data set, the automation of interactive cognition can be continuously improved through continuous training using machine learning algorithms. We have conducted various studies on the human-in-the-loop imitation control method to improve robot adaptability to uncertain environments, although it is still challenging to realize entire task autonomy in a short period of time [31]. Eventually, fully automated bio-inspired phenotyping systems can be implemented to replace the traditional manual phenotyping pattern.

4. In-Field Rice Tiller Counting Method

4.1. Image Acquisition

When the occlusion is removed through the interactive method illustrated above, images of the rice plant can be captured by the camera for tiller counting. However, since the tillers have similar colors with the background, it is difficult to recognize each tiller from an RGB image without depth information. To provide depth information for the images captured by an RGB camera, we use a horizontal line laser to scan the tillers. While the structured light system scans up and down, multiple images that scan different heights of the plant can be recorded for further tiller number recognition.

To reduce the influence of natural light on the light spots of the laser, we capture images with a small aperture to reduce the amount of light. Under this circumstance, the laser light spots can still be clearly identified and the rest of the regions are relatively dark. The images are then transformed to grayscale images to reduce computation. These grayscale images are resized to 256×256 pixels through bilinear interpolation to further improve computation efficiency.

4.2. Rice Tiller Number Recognition Algorithm

After the images with laser light are obtained and preprocessed, a rice tiller counting algorithm is then used to obtain tiller numbers from the images. In practical applications, accurately counting the tiller number is difficult and unnecessary. In practice, the aim of gene-editing breeding is to promote effective tillering (tillers with panicles) to obtain high yields, while eliminating ineffective tillering (tillers without panicles) for reduced nutrition consumption [32]. Since the panicle numbers can be statistically estimated by drone detection, we aim to statistically estimate the total number of under-canopy tillers and then the number of effective tillers can be estimated. Therefore, we divide the tiller numbers into several grades and the task in this paper is to obtain the approximate ranges of tiller numbers.

In this paper, a deep learning method based on an attentional residual network (AtResNet) is proposed. Figure 5 illustrates the network structure. Resized grayscale images are directly input into the network, and they are processed through stacked layers. The backbone network is a deep convolutional neural network (CNN) with residual connections to ResNet [33] to prevent the overfitting problem. There are three convolutional blocks with similar structures, each of which firstly processes the input through a two-dimensional convolution operation as follows.

$$x_i^l = f_{conv}^l\left(x_i^{l-1};\theta^{c,l}\right) = x_i^{l-1} * w^{c,l} + b^{c,\,l}, \quad (5)$$

where x_i^{l-1} denotes the input of the convolutional layer and $\theta^{c,l} = \left\{w^{c,l}, b^{c,l}\right\}$ are the parameters of this layer. Then, a batch normalization (BN) [31] layer is introduced to speed up the network convergence, which is formulated for each mini-batch as follows.

$$\hat{x_i^l} = \frac{x_i^l - E\left[x_i^l\right]}{\sqrt{Var\left[x_i^l\right]}}, \quad (6)$$

$$y_i^l = \gamma^l \hat{x_i^l} + \beta^l, \quad (7)$$

where γ^l and β^l are learnable parameters. $E[\cdot]$ and $Var[\cdot]$ denote the mean and variance value, respectively. Then, a rectified linear unit (ReLU) layer is used with a rectified linear function, which is formulated as

$$ReLU(x) = \max(0,\ x). \quad (8)$$

Then, a max-pooling layer is adopted, which calculates the maximum values within the receptive field.

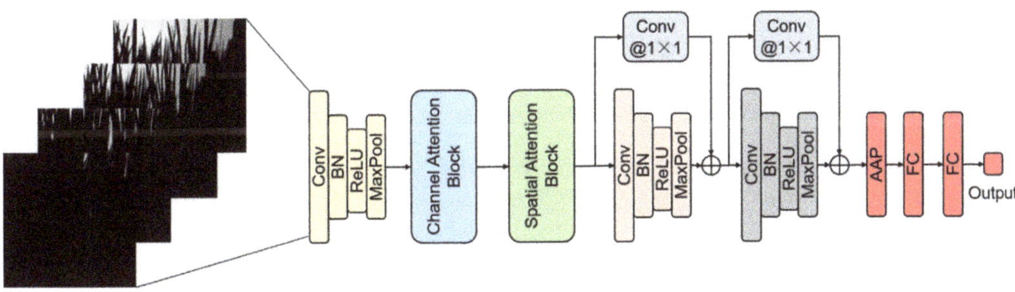

Figure 5. AtResNet model for rice tiller number recognition.

Residual connections are introduced to the second and last convolutional blocks to accelerate network training and prevent overfitting. A convolutional layer with a 1×1 kernel is used to perform identity mapping, which keeps the input and output size of the convolutional block the same. Then, the output of the l-th convolutional block can be calculated as follows.

$$x_i^l = \sigma\left[f_{CB}\left(x_i^{l-1};\theta^{CB}\right) + BN\left(f_{1\times1}\left(x_i^{l-1},\theta^{1\times1}\right)\right)\right], \quad (9)$$

where f_{CB} is the mapping function of the convolutional block, and $f_{1\times1}$ is the mapping function of the 1×1 convolutional layer in residual connections. σ denotes the ReLU function. The output of the last convolutional block is processed by an adaptive average pooling (AAP) layer and two fully connected (FC) layers and the final output is a vector whose length is the same as the tiller number grades.

Since these images are dark in most regions and the laser light spots only appear in some small areas, attention mechanisms [34] are introduced to help the model focus on the informative regions. Firstly, a channel attention block [35] is adopted to allocate different weights to different feature channels. The channel attention block firstly aggregates spatial information through adaptive average pooling and adaptive max pooling operations. Then, a shared convolutional network is used to generate attention maps for each aggregated feature vector. In addition, two maps are summed to obtain the final channel attention map. In short, these channel attention operations are summarized as follows.

$$A_c(x) = \sigma_s[f^c_{conv}(AvgPool(x)) + f^c_{conv}(MaxPool(x))], \tag{10}$$

where $x \in R^{W\times H\times C}$ represents the input features, and f^c_{conv} denotes the mapping function of the shared convolutional network, which consists of a 1×1 convolutional layer with C/r channels, a ReLU layer, and a 1×1 convolutional layer with C channels. σ_s denotes the sigmoid function. Finally, the calculated channel attention map $A_c(x)$ is applied to the input feature by element-wise multiplication, as follows:

$$x' = A_c(x) \otimes x. \tag{11}$$

Similarly, a spatial attention block [36] is adopted afterwards to obtain spatial attention maps to help the network to focus on informative spatial regions. Channel information is aggregated by average and maximum values. Two features are concatenated and then processed by a convolutional layer to produce the spatial attention map. The spatial attention operations can be summarized as follows.

$$A_s(x) = \sigma_s[f^s_{conv}([Avg(x);\ Max(x)])], \tag{12}$$

where f^s_{conv} denotes the mapping function of the convolutional layer. Finally, the calculated spatial attention map *A_s (x)* is applied to the input feature by element-wise multiplication, as follows:

$$x'' = A_s(x') \otimes x'. \tag{13}$$

The whole network outputs a vector $\hat{y}_i$, which represents the predicted probability of the i-th sample that belongs to each tiller number grade. $\hat{y}_i$ is obtained through a softmax function of the output y_{fc} of the last FC layer, as follows:

$$\hat{y_{i,j}} = \frac{e^{y_{fc,j}}}{\sum_{j=1}^{K} e^{y_{fc,j}}}, \tag{14}$$

where $y_{fc,j}$ and $\hat{y_{i,j}}$ denote the j-th element of y_{fc} and $\hat{y}_i$, respectively, and K is the number of all tiller number grades. The network is trained by minimizing the cross-entropy loss, which is defined as follows.

$$L = -\frac{1}{N}\sum_{i=1}^{N}\sum_{k=1}^{K} I(y_i = k)\log(\hat{y_{i,\ k}}), \tag{15}$$

where $I(\cdot)$ is the indicator function, y_i is the true tiller number grade label of the i-th sample and N is the sample number.

5. Experiment and Results

5.1. Data Description

Following the image acquisition procedure illustrated in Section 4.1, a set of images are obtained in fields using the structured light system. Then, these images are categorized into four classes according to the rice plant tiller number. In large-scale variant breeding, we found that the total tiller numbers of most variants are mainly between 21 and 25 [37]. We hoped to achieve relatively accurate tiller counting in this range. Therefore, we subdivide this range and the numbers fewer than 21 and more than 25 are divided roughly. Some image examples are shown as Figure 6. The details of these images are shown in Table 1. These images are transformed to grayscale images and resized to 256×256. Then, they are randomly split into a training set and a testing set with the ratio of 3:1.

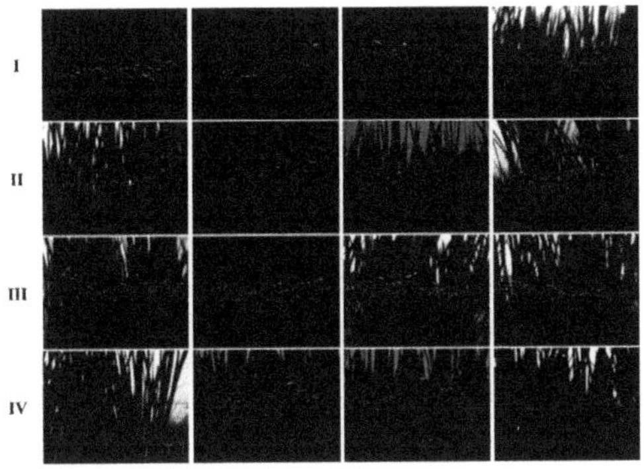

Figure 6. Some image examples of four tiller number grades (improved brightness).

Table 1. Dataset details.

Grade	Tiller Number	Image Number
I	<21	120
II	21~22	278
III	23~25	280
IV	>25	100

5.2. Experiment Setup

We use all the images in the training set to train the AtResNet and test the model using the testing set samples. The detailed parameter settings used in the experiment are listed in Table 2.

Table 2. Parameter details of AtResNet.

Layer	Parameter	Output Size
Conv1	Kernel size: 5 × 5 Stride: 2 × 2 Padding: 2 Kernel number: 16	128 × 128 × 16
Pool1	Kernel size: 2 × 2	64 × 64 × 16
Conv2	Kernel size: 3×3 Stride: 1 × 1 Padding: 1 Kernel number: 32	64 × 64 × 32
Pool2	Kernel size: 2 × 2	32 × 32 × 2
Conv3	Kernel size: 3 × 3 Stride: 1 × 1 Padding: 1 Kernel number: 64	32 × 32 × 64
Pool3	Kernel size: 2 × 2	16 × 16 × 64
Channel Attention	r: 16	-
Spatial Attention	Kernel size: 7 × 7 Stride: 1 × 1 Padding: 3 Kernel number: 1	-
AAP	Output size: 4 × 4	4 × 4 × 64
FC1	Unit number: 128	128 × 1
FC2	Unit number: 4	4 × 1

The network is implemented by PyTorch on an NVIDIA GTX 1660 GPU. It is trained by the Adam optimizer with a learning rate of 0.001 for 50 epochs. In each mini-batch, 64 samples are inputted into the system. A convolutional neural network (CNN) without residual connections and attention operations, and a ResNet without attention operations are also implemented for performance comparison. They share the same backbone structure and parameters with the AtResNet and all the experiments are repeated for 10 trials to reduce randomness.

5.3. Results

The experiment results of all the three methods are shown in Table 3. From the recognition results, we can observe that these deep learning-based methods achieved more than 93% tiller number recognition accuracy. This is satisfactory for practical applications. In addition, the proposed AtResNet outperforms the other two methods. We also illustrate the training and testing accuracy and loss values during the training process in Figure 7. We can observe that the AtResNet has lower accuracy and fewer loss fluctuations during model testing. It may be because the introduction of residual connections and attention operations helps the model to converge faster.

Table 3. Tiller number recognition accuracy (%) of three methods.

Method	Mean	Standard Deviation
CNN	93.49	1.64
ResNet	94.21	2.06
AtResNet	**94.72**	1.70

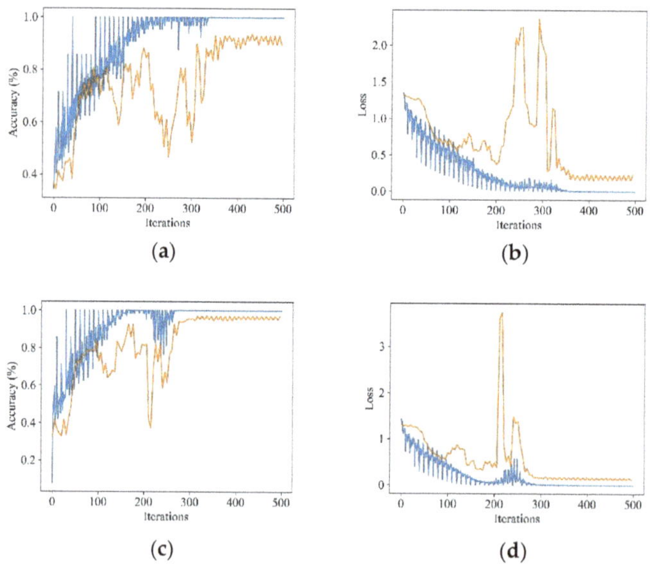

Figure 7. Training and testing accuracy and loss value curve. (**a**) CNN accuracy. (**b**) CNN loss. (**c**) AtResNet accuracy. (**d**) AtResNet loss. Blue line denotes training process and orange line denotes testing process.

To further explore the recognition results, we also analyze the confusion matrix of the results, as shown in Figure 8. It is observed that all the three methods can accurately recognize images with grade IV tiller numbers. For grade II and III, the AtResNet displays higher recognition accuracy compared with the other two methods. Figure 9 shows some examples of spatial attention maps. The different colors represent different relative attention values. We can observe that the laser spot regions have different attention values with other dark areas. So, the network can selectively focus on the informative regions.

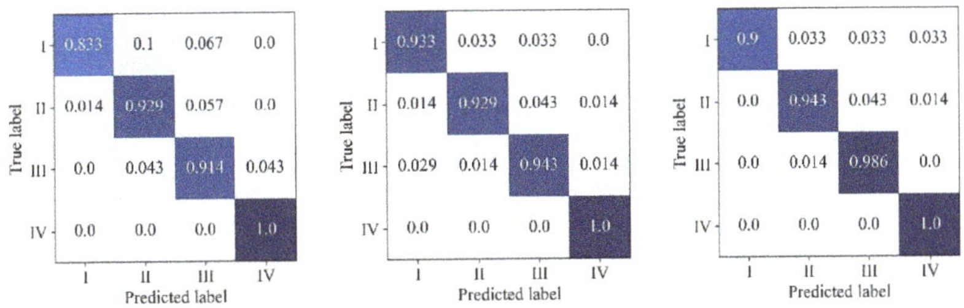

(**a**) CNN (**b**) ResNet (**c**) AtResNet

Figure 8. Confusion matrix of tiller number recognition results.

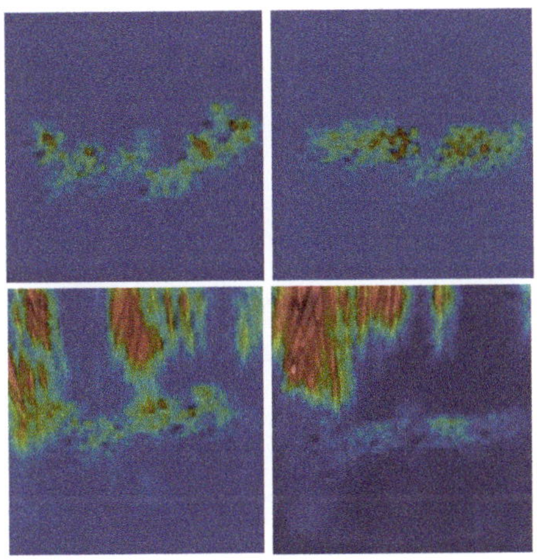

Figure 9. Examples of spatial attention maps in AtResNet.

6. Conclusions

This paper presents a new in-field phenotyping paradigm. An interactive cognition method is proposed to overcome the problem of occlusion and overlap in traditional passive automatic phenotyping methods. A bio-inspired solution is introduced so that the phenotyping robot can mimic the manual phenotyping operations. In this way, automatic high-throughput phenotyping of full growth cycles is realized. A tiller number recognition method (AtResNet) is proposed based on interactive cognition. In-field images are collected for the experiments. The experiment results show that the proposed method can achieve approximately 95% tiller number recognition accuracy and outperforms other deep learning-based methods. This paper provides a new solution to the occlusion and observation pose problems in field phenotyping. Although drone detection can estimate the panicle number in a more efficient way, the proposed method overcomes the difficulty of under-canopy tiller counting, which assists in effective and ineffective tillering counting. Compared with traditional manual breeding processes, the proposed in-field phenotyping paradigm offers a more efficient solution to repeating phenotyping across the full growth period. In future work, we will develop multiple phenotyping robots and explore the control scheme of switching between them to further improve in-field phenotyping efficiency. Moreover, the panicle counting method based on drone detection over the canopy will be studied to estimate effective tillering.

Author Contributions: Conceptualization, Y.H. and L.G.; methodology, P.X. and L.G.; software, Y.H. and P.X.; validation, Y.L. and B.C.; data curation, Y.H., P.X. and L.G.; writing—original draft preparation, Y.H., P.X. and L.G.; writing—review and editing, L.G. and C.L.; visualization, P.X.; project administration, C.L.; funding acquisition, L.G. All authors have read and agreed to the published version of the manuscript.

Funding: This research was funded by the National Key Research and Development Program under Grant No. 2019YFE0125200.

Institutional Review Board Statement: Not applicable.

Data Availability Statement: The data presented in this study are available from the corresponding authors upon reasonable request.

Conflicts of Interest: The authors declare no conflict of interest.

References

1. Godfray, H.C.J.; Beddington, J.R.; Crute, I.R.; Haddad, L.; Lawrence, D.; Muir, J.F.; Pretty, J.; Robinson, S.; Thomas, S.M.; Toulmin, C. Food Security: The Challenge of Feeding 9 Billion People. *Science* **2010**, *327*, 812–818. [CrossRef]
2. Crossa, J.; Fritsche-Neto, R.; Montesinos-Lopez, O.A.; Costa-Neto, G.; Dreisigacker, S.; Montesinos-Lopez, A.; Bentley, A.R. The Modern Plant Breeding Triangle: Optimizing the Use of Genomics, Phenomics, and Enviromics Data. *Front. Plant Sci.* **2021**, *12*, 651480. [CrossRef]
3. Hickey, L.T.; Hafeez, A.N.; Robinson, H.; Jackson, S.A.; Leal-Bertioli, S.C.M.; Tester, M.; Gao, C.; Godwin, I.D.; Hayes, B.J.; Wulff, B.B.H. Breeding crops to feed 10 billion. *Nat. Biotechnol.* **2019**, *37*, 744–754. [CrossRef]
4. Yang, Z.; Gao, S.; Xiao, F.; Li, G.; Ding, Y.; Guo, Q.; Paul, M.J.; Liu, Z. Leaf to panicle ratio (LPR): A new physiological trait indicative of source and sink relation in japonica rice based on deep learning. *Plant Methods* **2020**, *16*, 1–15. [CrossRef]
5. Furbank, R.T.; Tester, M. Phenomics—technologies to relieve the phenotyping bottleneck. *Trends Plant Sci.* **2011**, *16*, 635–644. [CrossRef]
6. Araus, J.L.; Cairns, J.E. Field high-throughput phenotyping: The new crop breeding frontier. *Trends Plant Sci.* **2014**, *19*, 52–61. [CrossRef]
7. Jin, X.; Zarco-Tejada, P.J.; Schmidhalter, U.; Reynolds, M.P.; Hawkesford, M.J.; Varshney, R.K.; Yang, T.; Nie, C.; Li, Z.; Ming, B.; et al. High-Throughput Estimation of Crop Traits: A Review of Ground and Aerial Phenotyping Platforms. *IEEE Geosci. Remote Sens. Mag.* **2020**, *9*, 200–231. [CrossRef]
8. Scanalyzer 3d, Lemnatec Gmbh. Available online: http://www.lemnatec.com/products/hardware-solutions/scanalyzer-3d/ (accessed on 30 September 2022).
9. Petrozza, A.; Santaniello, A.; Summerer, S.; Di Tommaso, G.; Di Tommaso, D.; Paparelli, E.; Piaggesi, A.; Perata, P.; Cellini, F. Physiological responses to Megafol®treatments in tomato plants under drought stress: A phenomic and molecular approach. *Sci. Hortic.* **2014**, *174*, 185–192. [CrossRef]
10. Appf. Plant Accelerator. Available online: https://www.plant-phenomics.org.au/ (accessed on 30 September 2022).
11. Hartmann, A.; Czauderna, T.; Hoffmann, R.; Stein, N.; Schreiber, F. HTPheno: An image analysis pipeline for high-throughput plant phenotyping. *BMC Bioinform.* **2011**, *12*, 148. [CrossRef]
12. Liu, S.; Martre, P.; Buis, S.; Abichou, M.; Andrieu, B.; Baret, F. Estimation of Plant and Canopy Architectural Traits Using the Digital Plant Phenotyping Platform. *Plant Physiol.* **2019**, *181*, 881–890. [CrossRef]
13. Scanalyzer Field, Lemnatec Gmbh. Available online: http://www.lemnatec.com/products/hardwaresolutions/scanalyzer-field/ (accessed on 30 September 2022).
14. Andrade-Sanchez, P.; Gore, M.A.; Heun, J.T.; Thorp, K.R.; Carmo-Silva, A.E.; French, A.N.; Salvucci, M.E.; White, J.W. Development and evaluation of a field-based high-throughput phenotyping platform. *Funct. Plant Biol.* **2014**, *41*, 68–79. [CrossRef]
15. Mueller-Sim, T.; Jenkins, M.; Abel, J.; Kantor, G. The Robotanist: A ground-based agricultural robot for high-throughput crop phenotyping. In Proceedings of the 2017 IEEE International Conference on Robotics and Automation (ICRA), Singapore, 29 May–3 June 2017; pp. 3634–3639. [CrossRef]
16. Bao, Y.; Nakami, A.D.; Tang, L. Development of a Field Robotic Phenotyping System for Sorghum Biomass Yield Component Traits Characterization. In Proceedings of the ASABE and CSBE/SCGAB Annual International Meeting, Montreal, QC, Canada, 13–16 July 2014; Available online: https://elibrary.asabe.org/abstract.asp?aid=44616&t=5 (accessed on 30 September 2022).
17. Zhou, C.; Ye, H.; Hu, J.; Shi, X.; Hua, S.; Yue, J.; Xu, Z.; Yang, G. Automated Counting of Rice Panicle by Applying Deep Learning Model to Images from Unmanned Aerial Vehicle Platform. *Sensors* **2019**, *19*, 3106. [CrossRef]
18. Xing, Y.; Zhang, Q. Genetic and Molecular Bases of Rice Yield. *Annu. Rev. Plant Biol.* **2010**, *61*, 421–442. [CrossRef]
19. Chen, J.; Gao, H.; Zheng, X.-M.; Jin, M.; Weng, J.-F.; Ma, J.; Ren, Y.; Zhou, K.; Wang, Q.; Wang, J.; et al. An evolutionarily conserved gene, *FUWA*, plays a role in determining panicle architecture, grain shape and grain weight in rice. *Plant J.* **2015**, *83*, 427–438. [CrossRef]
20. Yang, W.; Xu, X.; Duan, L.; Luo, Q.; Chen, S.; Zeng, S.; Liu, Q. High-throughput measurement of rice tillers using a conveyor equipped with x-ray computed tomography. *Rev. Sci. Instrum.* **2011**, *82*, 025102. [CrossRef]
21. Zhifeng, H.; Liang, G.; Chengliang, L.; Yixiang, H.; Qingliang, N. Measurement of Rice Tillers Based on Magnetic Resonance Imaging. *IFAC-Pap.* **2016**, *49*, 254–258. [CrossRef]

22. Scotford, I.; Miller, P. Estimating Tiller Density and Leaf Area Index of Winter Wheat using Spectral Reflectance and Ultrasonic Sensing Techniques. *Biosyst. Eng.* **2004**, *89*, 395–408. [CrossRef]
23. Deng, R.; Jiang, Y.; Tao, M.; Huang, X.; Bangura, K.; Liu, C.; Lin, J.; Qi, L. Deep learning-based automatic detection of productive tillers in rice. *Comput. Electron. Agric.* **2020**, *177*, 105703. [CrossRef]
24. Yamagishi, Y.; Kato, Y.; Ninomiya, S.; Guo, W. Image-Based Phenotyping for Non-Destructive in Situ Rice (Oryza Sativa L.) Tiller Counting Using Proximal Sensing. *Sensors* **2022**, *22*, 5547. [CrossRef]
25. Chen, Y.; Xiong, Y.; Zhang, B.; Zhou, J.; Zhang, Q. 3D point cloud semantic segmentation toward large-scale unstructured agricultural scene classification. *Comput. Electron. Agric.* **2021**, *190*, 106445. [CrossRef]
26. Langevin, G. Inmoov—Open Source 3d Printed Life Size Robot. Available online: http://inmoov.fr (accessed on 30 September 2022).
27. Kontoudis, G.P. Openbionics—Open-Source Robotic & Bionic Hands. Available online: http://www.openbionics.org (accessed on 30 September 2022).
28. Perception Neuron System. Available online: https://noitom.com/perception-neuron-series (accessed on 30 September 2022).
29. Ferenc, G.; Dimic, Z.; Lutovac, M.; Vidakovic, J.; Kvrgic, V. Open Architecture Platforms for the Control of Robotic Systems and a Proposed Reference Architecture Model. *Trans. Famena* **2013**, *37*, 89–100.
30. Gong, L.; Li, X.; Xu, W.; Chen, B.; Zhao, Z.; Huang, Y.; Liu, C. Naturally teaching a humanoid Tri-Co robot in a real-time scenario using first person view. *Sci. China Inf. Sci.* **2019**, *62*, 50205. [CrossRef]
31. Gong, L.; Chen, B.; Xu, W.; Liu, C.; Li, X.; Zhao, Z.; Zhao, L. Motion Similarity Evaluation between Human and a Tri-Co Robot during Real-Time Imitation with a Trajectory Dynamic Time Warping Model. *Sensors* **2022**, *22*, 1968. [CrossRef]
32. Ren, M.; Huang, M.; Qiu, H.; Chun, Y.; Li, L.; Kumar, A.; Fang, J.; Zhao, J.; He, H.; Li, X. Genome-Wide Association Study of the Genetic Basis of Effective Tiller Number in Rice. *Rice* **2021**, *14*, 1–13. [CrossRef]
33. He, K.M.; Zhang, X.Y.; Ren, S.Q.; Sun, J. Deep Residual Learning for Image Recognition. In Proceedings of the 2016 IEEE Conference on Computer Vision and Pattern Recognition (CVPR), Las Vegas, NV, USA, 27–30 June 2016; pp. 770–778.
34. Ioffe, S.; Szegedy, C. Batch Normalization: Accelerating Deep Network Training by Reducing Internal Covariate Shift. In Proceedings of the International Conference on Machine Learning, Lille, France, 6–11 July 2015; Volume 37, pp. 448–456.
35. Mnih, V.; Heess, N.; Graves, A.; Kavukcuoglu, K. Recurrent Models of Visual Attention. In Proceedings of the 27th International Conference on Neural Information Processing Systems, Montreal, QC, Canada, 8–13 December 2014.
36. Woo, S.H.; Park, J.; Lee, J.Y.; Kweon, I.S. CBAM: Convolutional Block Attention Module. In *Computer Vision—ECCV 2018*; PT VII; Springer: Berlin/Heidelberg, Germany, 2018; Volume 11211, pp. 3–19.
37. Yan, Y.; Wei, M.; Li, Y.; Tao, H.; Wu, H.; Chen, Z.; Li, C.; Xu, J.-H. MiR529a controls plant height, tiller number, panicle architecture and grain size by regulating SPL target genes in rice (Oryza sativa L.). *Plant Sci.* **2020**, *302*, 110728. [CrossRef]

agriculture

Article

A Real-Time Apple Targets Detection Method for Picking Robot Based on ShufflenetV2-YOLOX

Wei Ji *, Yu Pan, Bo Xu and Juncheng Wang

School of Electrical and Information Engineering, Jiangsu University, Zhenjiang 212013, China; ailqy369@163.com (Y.P.); xubo@ujs.edu.cn (B.X.); w18981051316@icloud.com (J.W.)
* Correspondence: jiwei@ujs.edu.cn

Abstract: In order to enable the picking robot to detect and locate apples quickly and accurately in the orchard natural environment, we propose an apple object detection method based on Shufflenetv2-YOLOX. This method takes YOLOX-Tiny as the baseline and uses the lightweight network Shufflenetv2 added with the convolutional block attention module (CBAM) as the backbone. An adaptive spatial feature fusion (ASFF) module is added to the PANet network to improve the detection accuracy, and only two extraction layers are used to simplify the network structure. The average precision (AP), precision, recall, and F1 of the trained network under the verification set are 96.76%, 95.62%, 93.75%, and 0.95, respectively, and the detection speed reaches 65 frames per second (FPS). The test results show that the AP value of Shufflenetv2-YOLOX is increased by 6.24% compared with YOLOX-Tiny, and the detection speed is increased by 18%. At the same time, it has a better detection effect and speed than the advanced lightweight networks YOLOv5-s, Efficientdet-d0, YOLOv4-Tiny, and Mobilenet-YOLOv4-Lite. Meanwhile, the half-precision floating-point (FP16) accuracy model on the embedded device Jetson Nano with TensorRT acceleration can reach 26.3 FPS. This method can provide an effective solution for the vision system of the apple picking robot.

Keywords: machine vision; picking robot; apple detection; YOLOX; ShufflenetV2

Citation: Ji, W.; Pan, Y.; Xu, B.; Wang, J. A Real-Time Apple Targets Detection Method for Picking Robot Based on ShufflenetV2-YOLOX. *Agriculture* **2022**, *12*, 856. https://doi.org/10.3390/agriculture12060856

Academic Editor: Surya Kant

Received: 21 April 2022
Accepted: 8 June 2022
Published: 13 June 2022

1. Introduction

China's apple planting area and output account for more than 50% of the world [1], but its picking is still dominated by manual picking, with high cost. Therefore, the apple picking robot is the development direction in the future. How to locate and detect apples quickly and accurately in the natural environment is the focus and difficulty of vision research of picking robots [2].

At present, the research on fruit detection at home and abroad is mainly divided into target detection based on the traditional algorithm and target detection based on the deep learning algorithm, and both have made some progress. Traditional algorithms require artificially designed features [3], and their accuracy and detection speed are not as good as those of deep learning algorithms. Currently, they are mostly used for image preprocessing. Xia [4] proposed a method for fruit segmentation based on the K-means clustering algorithm. The Canny edge detection operator was used to extract the fruit contour, the Y-node search algorithm was used for contour separation, and finally, the least squares method was used for contour reconstruction. Liu [5] used a simple linear iterative clustering algorithm to segment the apple image collected in the orchard into super-pixel blocks, and used the color features extracted by blocks to determine the target candidate region. Lv [6] calculated the distance of each fruit in the connected area by using the Euclidean distance method, extracted the effective peak from the smoothed curve by using the improved local extreme value method, and determined the shape of overlapping apples according to the number of peaks. Bochkovskiy [7] chose incandescent lighting to obtain images at night. In the image segmentation stage, the power transformation was used to improve the R-G color difference threshold segmentation method, and the genetic algorithm was introduced to optimize the solution of the maximum interclass variance. The accuracy was 94% and the detection speed was 2.21 FPS.

The detection algorithm based on deep learning has wider applicability than the traditional algorithm. When using a specific dataset, it can learn deeper features and obtain higher accuracy. It is easier to detect the target. In recent years, deep learning has been used in a wide range of industries. Some scholars have conducted in-depth research on apple target detection based on deep learning. Sa [8] basically achieved rapid detection and achieved an F1 score of 0.838 by using the improved Fast R-CNN training RGB color and near-infrared images to detect fruits. Zhao [9] used the improved YOLOv3 algorithm with 13 layers to prove that it is feasible to use the deep learning algorithm in the natural environment under the verification of different illumination directions, different growth stages of apples, and different picking times. Mazzia [10] achieved a detection speed of 30 FPS using a modified YOLOv3-Tiny network on a matched embedded device, the Jetson AGX Xaver. However, the Jetson AGX Xaver is very expensive and its AP is only 83.64%, which does not satisfy the need for detection accuracy. Yan [11] using the improved YOLOv5 can effectively identify grasping apples that are not obscured by leaves or only obscured by leaves, and nongrasping apples that are obscured by branches or other fruits. Wu [12] achieved 98.15% AP and 0.965F1 using an improved EfficientNet-YOLOV4 dataset augmented by foliage occlusion data. However, its model capacity is 158 M, and the real-time detection speed is only 2.95 FPS. Chu [13] designed a novel Mask-RCNN for apple detection. By adding a suppression branch to the standard Mask-RCNN to suppress nonapple features, its F1 index is 0.905, but the detection speed is only 4 FPS. The suppression branch of this method is designed according to color, which is only effective when the color difference between fruit and leaf is large. When the color difference is not large, due to light, disease, or debris, the detection effect may not be good.

Although the above studies have all achieved some results for apple recognition in different scenarios, they all have similar problems. That is, high detection speed and high detection accuracy cannot be satisfied simultaneously. At the same time, according to the current research literature, several directions have been little studied. First, most of the current research on apple recognition has focused on apples that are dense, overlapping, or obscured by foliage, with very little research on apples in the context of bagging. Secondly, there are few studies related to apple detection models running on edge devices to determine how the detection models will perform in practice. To solve the above problems, an apple detection algorithm based on YOLOX-Tiny is proposed in this paper. It can meet the needs of a picking robot working with high precision and in real time. Compared to similar studies, our main contributions are the following two.

(1) A novel lightweight apple detector was designed. The ShufflenetV2-YOLOX model was designed from a practical perspective based on the orchard environment and obtained excellent detection speed and detection accuracy.
(2) It was validated and deployed on the Jetson Nano, an edge device. It was validated that the model can meet the requirements for real-time and high-precision detection on an edge device, and can provide an effective solution for picking robots.

2. Materials and Methods

2.1. Apple Image Acquisition and Data Augmentation

This paper takes the Fuji apple, the largest main apple variety in China, as the research object, and collects apple images from the apple demonstration base in Feng County, Xuzhou City, Jiangsu Province, China. Considering the possible natural environment in the actual orchard picking, the images of unbagged apples, bagged apples, and apples under weak light at night are collected.

In the process of image acquisition, in order to ensure the clarity of the image and meet the working environment of the picking robot, we keep the distance between the camera and the fruit at 0.3 m–2 m. In the night apple image acquisition, a single LED lamp is used for illumination, and the brightness of the fruit area is changed by changing different illumination angles. A total of 1793 pictures are taken during the shooting, including apple images under different natural conditions such as forward light, backlight, side light, overlap, and occlusion, 577 apple images without bagging during the day, 567 apple images bagged during the day, and 649 apple images including bagging at night, as shown in Figure 1. Among them, the appearance of apples in the daytime will vary greatly due to the different angles and intensity of light. Bagging can not only prevent the fruit from being harmed by dust, pests, and pesticide residues, but also make the fruit surface smooth and beautiful, and increase the effective yield and income. However, due to a layer of plastic bags on the surface, the apple will be in an irregular state, and its surface and shape characteristics will be disturbed. This makes traditional image detection methods, such as texture, color difference, and Hough Circles transformation, unable to effectively detect apples [8]. At the same time, there are often water droplets in the plastic bag, which will bring greater difficulties to image detection. Because the image of apples at night is presented under the irradiation of a strong light source, there may be significant contrast on the same picture. For example, the surface of apples directly illuminated by the light source will be strong and bright, resulting in the lack of surface feature information, while those not directly directed will be relatively dark and difficult to detect. Therefore, apple images in the above cases will interfere with image detection to a certain extent [13].

(a) Unbagged apples **(b)** Bagged apples **(c)** Apples at night

Figure 1. Apple image in natural state.

The apple dataset images collected in this experiment are small in number and contain complexities such as bagging, nighttime, occlusion, and overlap. Deep learning has certain requirements on the size of the dataset. If the original dataset is relatively small, it cannot meet the training of the network model well, thus affecting the performance of the model. Image enhancement is the process of expanding the dataset by processing the original image, which can improve the performance of the model to a certain extent. Therefore, we use the imgaug algorithm for data enhancement, using mirror flip, changing brightness, flipping up and down, adding Gaussian noise, dropout, scaling, and other operations to mix and enhance the images with a 10-fold enhancement factor, while ensuring the morphological features are intact. Finally, 17,930 images are obtained, as shown in Figure 2. Although the augmented dataset is slightly different from the actual situation, the blurring is quite beneficial in improving the robustness of the model. The models trained with the data-enhanced dataset have higher accuracy compared to the unfuzzed dataset [14].

Figure 2. Image after data augmentation.

The annotation software used in this paper is LabelImg, and the annotation file format is "xml". To better compare different types of networks and training sets, the images are converted to Pascal VOC format. At the same time, the training set and verification set are generated according to the ratio of 9:1, and 30 apple images in the complex natural environment are selected as the test set to verify the detection effect of the model. All networks used in this paper are based on the pre-training of the ImageNet dataset, use migration learning to train 150 epochs on this dataset, and select the best one as the detection weight parameter to load into the network.

2.2. Design of Apple Object Detection Network

2.2.1. Baseline Selection

There are a number of deep learning methods available, and one of the most effective networks for target detection is the convolutional neural network. These are divided into one-stage networks and two-stage networks [15–19]. The one-stage network is superior in detection speed, and the accuracy rate is also high. The neural network used for target detection is divided into the one-stage network and two-stage network according to the detection stage. The one-stage network is better in detection speed and high in accuracy. The YOLO series is a representative one-stage network, and among them, YOLOX is the latest version [20], which is improved with YOLOv3 + Darknet53 as the baseline. YOLOX adopts understanding coupling, Mosaic and Mixup image enhancement technology, anchor-free, SimOTA, and other tricks, which is greatly improved compared with the previous YOLOv4 and YOLOv5.

YOLOX is divided into x, l, m, s, tiny, and nano models from large to small according to the proportion of network depth and network width. Different models of networks can be selected according to different use scenarios. Among them, YOLOX-Tiny is a lightweight network in the YOLOX series, and its detection accuracy and speed are better than YOLOv4-Tiny, which is suitable for deployment in apple picking robots. However, the detection accuracy and detection speed of YOLOX-Tiny still have room for improvement compared with advanced apple detection algorithms at home and abroad.

2.2.2. ShufflenetV2-YOLOX Network Design

To meet the needs of the apple picking robot, it is necessary to improve the accuracy and detection speed of the network based on YOLOX-Tiny. This paper proposes a ShufflenetV2-YOLOX network. Figure 3 shows its network structure. First, this method takes YOLOX-Tiny as the baseline and uses the lightweight network Shufflenetv2 added with CBAM as the backbone. At the same time, ASFF is added after the PANnet network to improve the accuracy of network detection. Deleting a feature extraction layer reduces the amount of parameter calculation of the whole network, improves the detection speed of the network, and makes it meet the needs of real-time and high precision on embedded devices. The head network adopts YOLOX's decoupled head. It is divided into two parts: object prediction and position regression, which are predicted separately and then integrated for prediction. The loss function of the detection frame position can choose to use the traditional Intersection over Union (IOU) loss and Generalized Intersection over Union (GIOU) loss [21,22], and both OBJ loss and CLS loss use the Binary Cross Entropy loss method. To deal with the complex situation in orchard apple target detection, we select the better GIOU loss as the IOU loss of the detection frame.

$$\mathrm{IOU} = \frac{\mathrm{S_{overlap}}}{\mathrm{S_{nuion}}} \tag{1}$$

$$\mathrm{GIOU} = \mathrm{IOU} - \frac{|\mathrm{A_C} - \mathrm{S_{nuion}}|}{\mathrm{A_C}} \tag{2}$$

where $\mathrm{S_{overlap}}$ is the area of intersection of the predicted bounding box and the true bounding box. $\mathrm{S_{union}}$ is the area of the union of the two bounding boxes [14]. $\mathrm{A_c}$ is the minimum enclosing rectangle that predicts the border and the true frame.

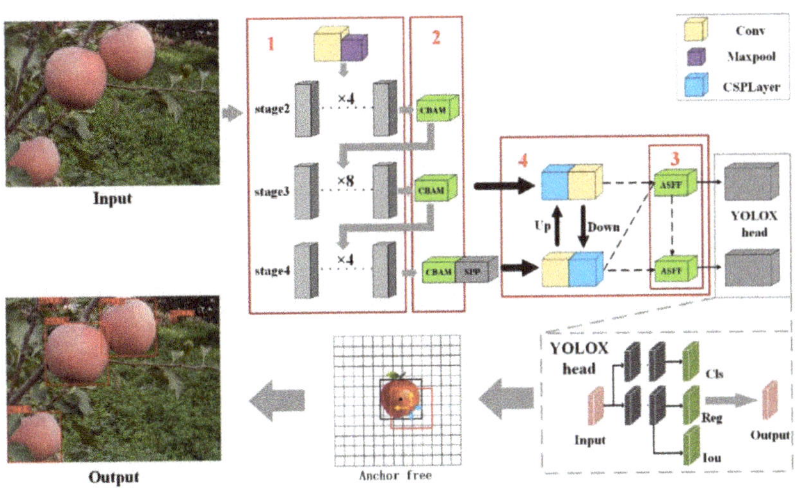

Figure 3. The network structure of ShufflenetV2-YOLOX. 1: Backbone Network Design; 2: Increase Attention Mechanism; 3: Add the ASFF Module; 4: Prune the Feature Layer.

Following the selection of the baseline model, the modeling phase is divided into four main stages. The first step is to replace the backbone network with ShufflenetV2. In the second step, the attention mechanism CBAM is added. The third step is to add the adaptive feature fusion mechanism ASFF. Finally, the feature extraction layer is reduced.

Backbone Network Design

YOLOX-Tiny is the lightweight network of YOLOX, which is achieved only by reducing the network width and depth. Compared with those specialized lightweight networks, it is not enough, so the first thing we need to do is to choose a lightweight network to replace YOLOX-Tiny backbone. ShuffleNetV2 is improved from ShuffleNet and has achieved excellent results in lightweight networks [23,24]. It inherits grouped convolution, depthwise separable convolution, and channel shuffle operations of ShuffleNet, and also improves the original unreasonable parts according to four efficient network pairs.

ShufflenetV2 is an image classification network in which the global average pooling and fully connected layers modules are added to achieve higher results in the ImageNet network competition and are useless for object detection networks. In order to replace the backbone of YOLOX-Tiny, we choose to keep only the network structure before stage4 in the ShufflenetV2 network, and then extract the output from each stage and connect it to PANet instead of CSPDarkNet. This can not only improve the running speed but also meet the design of the target detection network. The structure of ShufflenetV2 in YOLOX is shown in Figure 4.

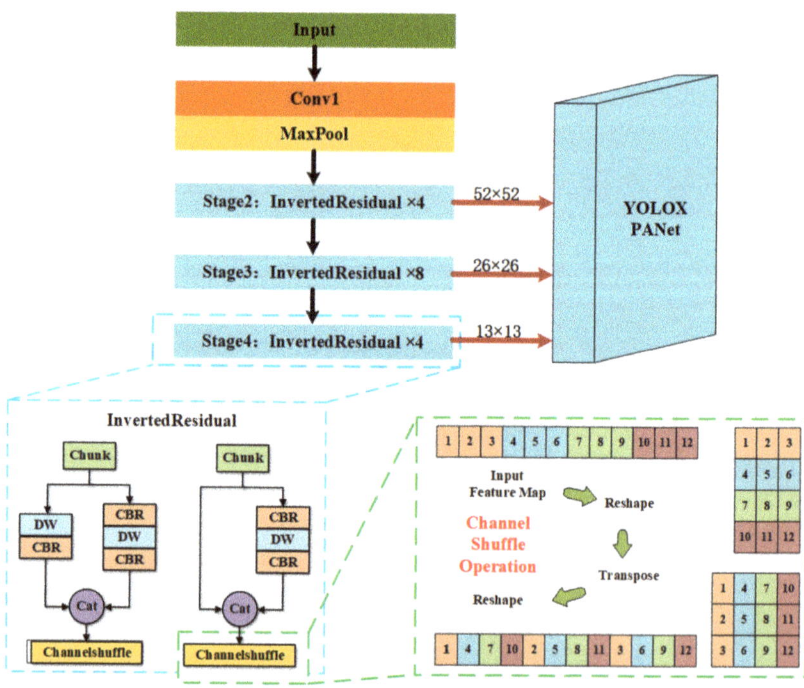

Figure 4. Shufflenetv2 network structure.

Increase Attention Mechanism

As the convolutional neural network (CNN) becomes deeper, the effective features become sparse. At this time, we need to introduce the "attention" mechanism. The attention mechanism can automatically learn and calculate the contribution of input data to output data so that it can ignore irrelevant noise information and focus on key information. CBAM is an attention mechanism module that combines space and channel [25]. Compared with the SE attention mechanism that only focuses on channels, it can achieve better results. CBAM consists of a Channel Attention Module and Spatial Attention Module, which carry out Attention on the channel and space, respectively, as shown in Figure 5. In this paper, the CBAM module is added to the stage of the ShufflenetV2 backbone network, which can strengthen the apple features learned by the network.

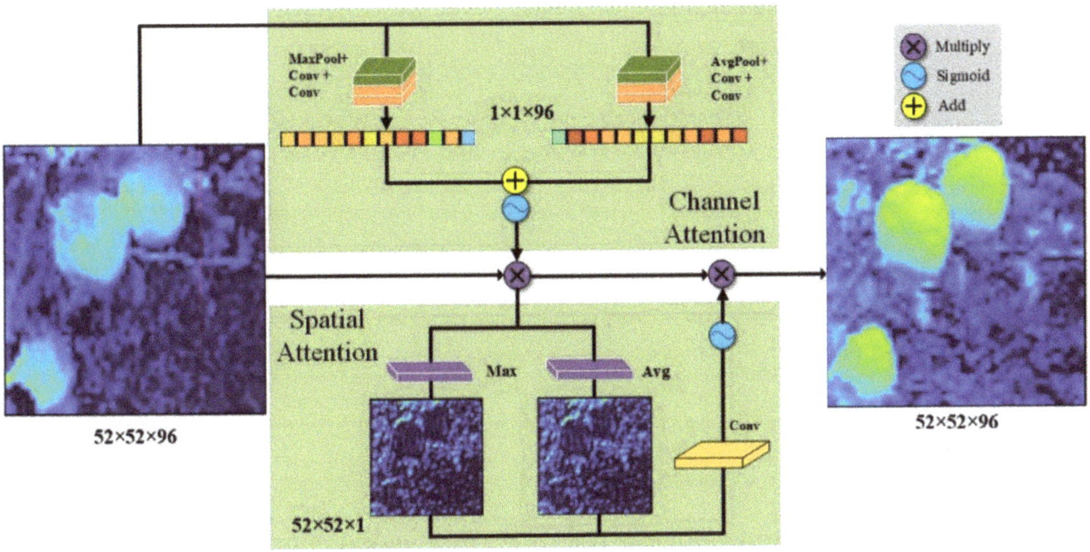

Figure 5. CBAM.

Add the ASFF Module

Feature pyramid can fuse features of different layers and detect images of different sizes, but the inconsistency between features of different scales is its main limitation. The ASFF module can make each feature layer focus on identifying objects that fit its grid size, spatially filter features on other layers, and retain only useful information for composition. This can solve the problem of indistinguishable fruits of different sizes clustered together in apple images [5]. Other layers in ASFF are adjusted to the same size as the current layer through convolution operations and fused to obtain adaptive weights. The adaptive weight is then combined with each layer to finally obtain a fusion module of the same size as the current layer. Its structure is shown in Figure 6. This paper adds an ASFF module after the PANet network to learn the relationship between different feature maps. This allows apples of different sizes to be predicted by the corresponding feature layers, improving the detection accuracy of the network.

Prune the Feature Layer

Adding modules can improve the detection accuracy of the network but also reduce the detection speed of the network. To improve the detection speed of the network to meet the real-time requirements, this paper chooses to delete one feature extraction layer in PANet and adjust the structure, and only uses two feature layers (TFL) to reduce the amount of calculation. Figure 7 shows the PANnet part of the YOLOX-tiny network. The black box shows the reduced network structure and the number of anchors. We only keep the 13 × 13 and 26 × 26 outputs, i.e., out2 and out3, respectively.

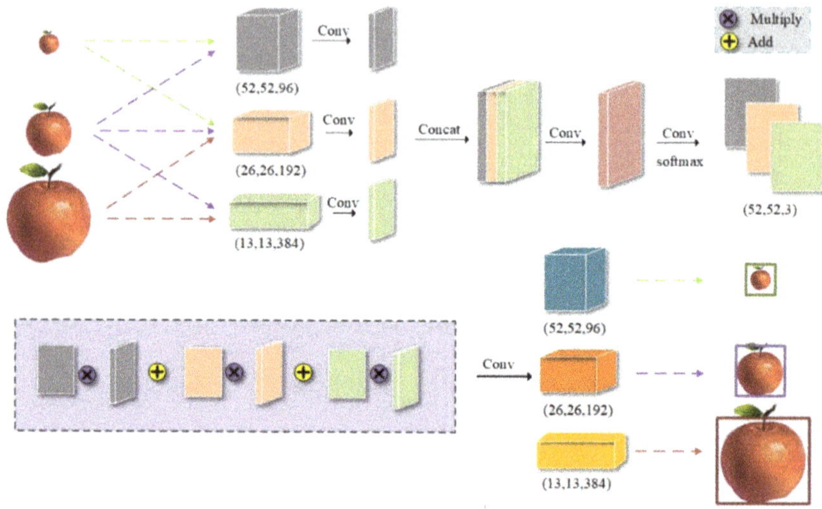

Figure 6. ASFF.

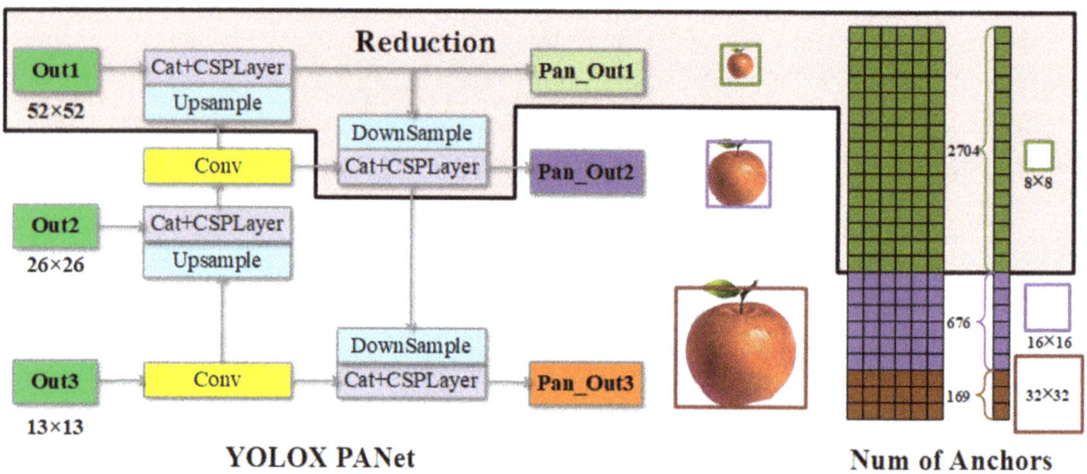

Figure 7. PANet with two feature layers.

Using only two feature layers can not only reduce the number of convolution kernels and computational complexity of PANet but also reduce the computing power required for prediction. In the case of 416 × 416 input size and num_class = 1, deleting a layer of feature layer Head will reduce the original 3549 anchors to 845 anchors. Although this will reduce the detection ability of small targets, it will not select long-distance small targets as objects during the operation of the apple picking robot, and the actual effect verification will not reduce the detection effect of the model too much.

3. Results and Discussion

The training equipment used in this paper is a PC device with the Windows 10 operating system. The system is equipped with an Intel e5-2683 processor, 64 GB of memory, four NVIDIA GTX1080ti graphics cards, and 11 GB of video memory. The algorithm programs used in this paper are written in the Python language on PyCharm, and CUDA and cuDNN are used for network training acceleration. The training epoch is set to 150, the batch size is set to 64, and the input image size is set to 416 × 416. Input size and detection speed are mutually exclusive quantities and a smaller image input size speeds up detection. Therefore, the input image size is set to 416 × 416 to improve the real-time performance of the model detection. For all network models to compare performance fairly, the same input size needs to be set in the comparison experiments. This has a significant impact on the performance of the network models. The test equipment uses the Windows 10 operating system, AMD Ryzen7 4800 h processor, 16 GB of memory, an NVIDIA GTX1650 graphics card, and 4 GB of graphics memory (Table 1).

Table 1. Test System Hardware.

Computer Configuration	Specific Parameters
Operating system	Windows 10
CPU	AMD Ryzen 7 4800H
Random Access Memory	16 GB
GPU	NVIDIA GTX1650

To verify the detection effect of the model on apples in the natural environment, this paper uses 30 complex orchard pictures as the test set, including 5 daytime unbagged apple pictures, 12 daytime bagged apple pictures, 10 nighttime unbagged apple pictures, and 3 nighttime bagged apple pictures. As nighttime and bagging are the focus and difficulty of the current research on picking robot vision, this paper chooses the nighttime and bagging test images to account for a higher proportion, which can better reflect the model's detection effect on apples in the natural environment.

In this paper, AP, Precision, Recall, Param, FPS, and F1 are selected as the comparison standards for detection effects to determine the pros and cons of the model. Param represents the number of parameters the network contains, and FPS represents the number of pictures the model can detect in one second. Taking the IOU threshold of 0.5 as the standard, the AP value is the area under the Precision—Recall (PR) curve formed by Precision and Recall. F1 score can be regarded as a weighted average of model accuracy and recall, which takes into account both the accuracy and recall of the model.

3.1. *ShufflenetV2-YOLOX Model Performance Verification*

To validate the effectiveness of the network improvement method, we chose to conduct ablation experiments to evaluate each step. AP, Param, and FPS were chosen as the evaluation metrics. The results of the ablation experiment are shown in Table 2.

Table 2. Ablation experiment.

YOLOX-Tiny	ShufflenetV2	CBAM	ASFF	TFL	AP	Param(M)	FPS
√					90.52%	5.03	55
√	√				91.69%	3.19	53
√	√	√			94.16%	3.61	52
√	√	√	√		97.29%	6.68	48
√	√	√	√	√	96.76%	5.40	65

It can be seen from the data in Table 2 that each step of improvement is an effective improvement, which effectively improves the detection speed or detection accuracy of the model. The AP value of the ShufflenetV2-YOLOX method is 96.76%, which is 6.24% higher than that of the original YOLOX-Tiny method. Although the Param is increased by 0.4 m, the detection speed is increased by 18% to 65 FPS. Both the CBAM module and ASFF module effectively improve the detection effect of the network, and the method of deleting the feature layer also improves the detection speed within the range of tolerable reductions in accuracy. Due to the use of depthwise separable convolution and channel shuffle operations in ShufflenetV2, when CSPDarknet is replaced, although the amount of network parameters is reduced, the detection speed is not improved.

Different deployment devices are suitable for different network structures. For example, a PC can use a CPU or GPU for inference. Depthwise separable convolutions are more suitable for running on CPUs, and normal convolutions are more suitable for running on GPUs. Due to the depthwise convolution and channel shuffle operations used in ShufflenetV2, inference on a GPU is not the best choice. Using ShufflenetV2 as the backbone network can achieve 15.6 FPS on the Ryzen7 4800 h(CPU), while YOLOX-Tiny can only achieve 11.5 FPS. In practice, we can choose different network structures based on different deployment devices.

3.2. Apple Detection Effect in Natural State

Apple recognition in complex environments has always been a research challenge. In this experiment, to verify the recognition effect of the trained model for different fruit states, apples without bags, apples with bags, and apples at night from the test set are detected. Figure 8 shows the apple detection results in a natural environment using the ShufflenetV2-YOLOX model. According to the detection results, the model proposed in this paper achieves good recognition results in various situations and meets the accuracy requirements of the apple picking robot.

Figure 8. Apple detection effect in natural environment based on ShufflenetV2-YOLOX network model.

For images of unbagged apples during the day, the model can detect most of the apples, with only a few overlapping and too distant apples having detection errors. Images of bagged apples are not only sticky, overlapping, and obscured, but also irregular in shape due to the film on the surface of the bagged fruit. There are gaps between the fruit and the film, which compromise the texture and color characteristics of the apple surface. As a result, bagged apples can be identified less accurately than nonbagged apples. Because of the low ambient light at night, apples close to a light source will have more distinctive features. As a result, apples close to the light source are easily detected, while apples away from the light source are difficult to detect. This is the biggest obstacle to nighttime image detection. In the future, the overall effect would be improved if more effort could be put into planning the lighting system to achieve more uniform illumination. Some obscured or small targets may not be detected, due to the limitation of the input image size of 416 × 416. All models have the same problem. Increasing the input size of the image can improve the detection of the model to some extent, but at the expense of detection speed. For example, ShufflenetV2-YOLOX has a detection speed of 65 FPS at an input size of 416 × 416 and 60 FPS at an input size of 640 × 640. Although this is a reduction of 5 frames, the detection is much better and many small targets can be detected. However, the small targets are apple targets that are further away from the picking robot. For the apple picking robot, the small targets are not its working targets and do not affect the actual results. In subsequent work, a threshold pattern can be used, ignoring apples with a detection area smaller than a certain percentage. A target that focuses on a larger proportion of the area is an apple with a shorter distance. This facilitates the work of the picking robot.

Table 3 shows the precision and recall of the model detection in the three cases, the number of apples in the pictures, and the number of apples detected. There were 31 images containing 372 apple targets, of which 345 were detected. Our model can effectively address the low recall of apple detection networks under bagged and nighttime conditions.

Table 3. Detection results in different scenarios.

	Number of Images	AP	Recall	Number of Apples	Number of Apples Detected
Total	31	96.76%	93.75%	372	345
Unbagged apples	11	97.29%	94.45%	134	125
Bagged apples	9	95.53%	93.15%	110	102
Apple at night	11	95.86%	93.45%	128	118

3.3. *Apple Detection Effects Contrast Experiment of Different Models*

To verify the superiority of the ShufflenetV2-YOLOX model proposed in this paper, it is compared with YOLOv5-s, YOLOv4-Tiny, Efficientdet-d0, Mobilenetv2-YOLOv4-lite, and YOLOX-Tiny [7,26,27]. Figure 9 shows the apple detection results of ShufflenetV2-YOLOX and other models in the natural environment. ShufflenetV2-YOLOX, YOLOv4-Tiny, YOLOX-Tiny, Mobilenetv2-YOLOv4-lite, and YOLOv5-s have an image input size of 416 × 416, and Efficientdet-d0 has better results because its network settings have a fixed input size of 512 × 512. To make each model have a clearer contrast effect, this paper selects the apples detected by all models as the total set and marks the detection effect diagram of each model. The white circle indicates the missed area, and the blue circle indicates the missed area. The more white and blue circles, the worse the effect of the model. As can be seen from Figure 9, apple targets during the day are bright in color and distinct in shape. Most models perform best on unbagged apples during the day. On the other hand, the plastic bags on the surface of the apples can blur their color and shape characteristics, resulting in the target and background being too close together. Bagged apples are therefore very susceptible to missed detection. At night, apples under strong light and low light are difficult to detect due to illumination problems. However, the ShufflenetV2-YOLOX model proposed in this paper has the least white and blue circles in the detection images, indicating that it has the highest recall rate. In particular, apple images in bagging and at night, although not all targets in the image are detected, have a significant advantage over other lightweight networks. This shows that the model can effectively solve the problem of low recall rate of the apple detection network under bagging and night conditions.

(**a**). ShufflenetV2-YOLOX

(**b**). YOLOX-Tiny

(**c**). YOLOv4-Tiny

(**d**).YOLOv5-s

(**e**). MobilenetV2-YOLOv4-lite

Figure 9. *Cont.*

(**f**). Efficientdet-d0

Figure 9. Comparison of ShufflenetV2-YOLOX with other advanced networks for apple detection effects.

Figure 10 shows a comparison of the PR curves of the different models for apple detection. Table 4 shows a comparison of AP, precision, recall, F1, parameters, and FPS for the different models. In terms of detection accuracy, YOLOv4-Tiny is a simplified lightweight network from YOLOv4 with an AP of 89.14%, which is close to the performance of YOLOX-Tiny. YOLOv5-s is currently one of the best detection results among lightweight networks, with a relatively high recall and detection accuracy. The AP and F1 reach 95.44% and 0.94, respectively. Mobilenet-YOLOv4-lite achieves an AP of 92.99%. It has the highest accuracy of the tested models with 95.96%, but it does not have a high recall of 83.59%, which does not meet the apple target detection requirements. The performance of Efficientdet-d0 is similar to that of Mobilenet-YOLOv4-lite. The ShufflenetV2-YOLOX model proposed in this paper has a high recognition accuracy with an AP of 96.76% and a detection accuracy of 95.62%. In particular, the recall rate is the highest score among all lightweight networks, reaching 93.75%. Compared to other models, our model can effectively detect bagged and nighttime apple targets from low-resolution images, which is responsible for its high recall rate.

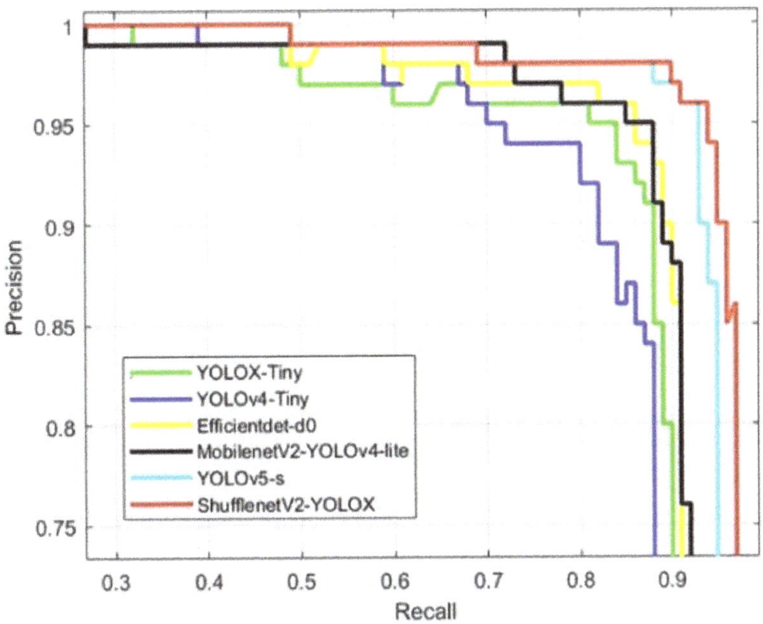

Figure 10. PR curve comparison of ShufflenetV2-YOLOX with other advanced networks.

Table 4. Comparison of ShufflenetV2-YOLOX with other lightweight networks.

Models	AP	Precision	Recall	F1	Param (M)	FPS
YOLOX-Tiny	90.52%	94.06%	74.22%	0.83	5.03	55
YOLOv4-Tiny	89.14%	89.64%	87.89%	0.89	5.77	54
Mobilenet-YOLOv4-lite	92.99%	95.96%	83.59%	0.89	10.30	22
YOLOv5-s	95.44%	94.82%	92.97%	0.94	7.20	18
Efficientdet-d0	92.89%	95.91%	86.42%	0.91	3.69	21
Ours	96.76%	95.62%	93.75%	0.95	5.40	65

In terms of detection speed, Yolov4-tiny and YOLOX-Tiny have an advantage in detection speed due to their lightweight network structure design, which can reach around 55 FPS. YOLOv5-s is a little slower at 18 FPS, and Efficientdet-d0 has fewer network parameters but is slow because it uses a lot of deeply divisible volume integrals. Although its floating-point operations per second (FLOPS) are small, it spends more time on memory access costs, so the speed is not ideal at 21 FPS. MobilenetV2-YOLOv4-lite uses MobilenetV2 to replace the YOLOv4 backbone, but the PANet is still large, and it uses deep detachable convolution instead of partial convolution, so the detection speed is not ideal, only 22 FPS. Our ShufflenetV2-YOLOX benefits from a lightweight backbone network with a low number of parameters. The anchor-free and two feature extraction layers can in turn reduce parameters and computations while satisfying the actual apple orchard detection. This results in a fast recognition speed of up to 65 FPS.

With higher detection accuracy and speed, ShufflenetV2-YOLOX enables real-time, accurate, and fast recognition of apples in natural environments, making it more suitable for deployment in apple picking robots.

3.4. Apple Detection Effect in Embedded Devices

Traditional deep learning algorithms use an Industrial Personal Computer (IPC) as the deployment device, which is not suitable for real-time apple detection in the field, due to its weight and power limitations. The edge device has powerful arithmetic power, small size, light weight and low power consumption. It can locally perform arithmetic processing on the collected data and is a good choice to replace IPCs, and NVIDIA Jetson Nano is the most cost-effective edge device available [10].

The apple picking experimental platform with Jetson Nano as the controller is shown in Figure 11. It mainly consists of a moving part, a gripper, a visual recognition system, and a robot arm. When the apple picking robot starts the picking task, it will first detect and select an apple through the visual recognition system. Then, it sends the apple's position information to the control system, and the robot arm is driven to approach the apple. The gripper will be driven to the designated position to grab the apple and use the cutter to cut off the stalk.

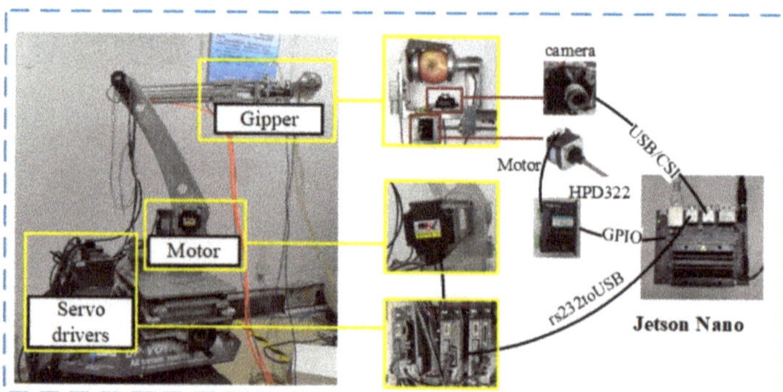

Figure 11. The apple picking experimental platform.

In this paper, we use Jetson Nano as an embedded deployment platform with software environment JetPack-4.5.1, TensorRT-7.1.3, and the image input size set to 416 × 416. The Pytorch model is first transformed into an ONNX model, and then TensorRT is used to quantify the accuracy of the parameters of the model and to merge the workflow so that it keeps the model running on the GPU as much as possible, thus allowing the model to run faster. We test the inference speed of the Pytorch Single-precision floating-point (FP32) model, ONNX INT64 model, TensorRT FP32 model, and TensorRT FP16 model on Jetson Nano. In Figure 12, the arrows refer to the increase or decrease in accuracy as a result of this operation compared to the previous phase.

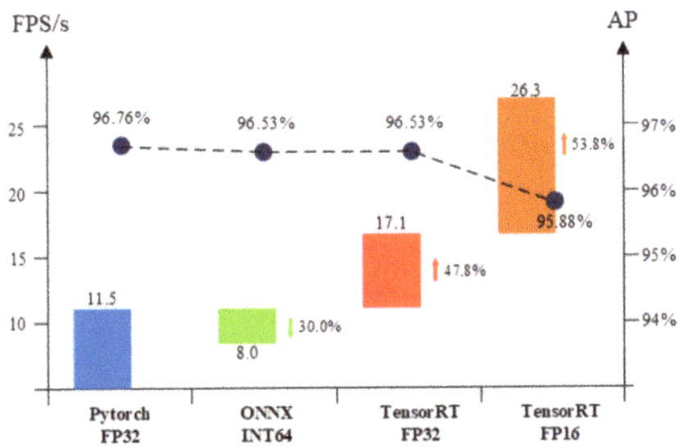

Figure 12. ShufflenetV2-YOLOX models for inference speed and AP accuracy on Jetson Nano.

On the Jetson Nano, the ShufflenetV2-YOLOX model with Pytorch FP32 can run at a speed of 11.5 FPS. The ONNX model, on the contrary, runs slower because of its parameter precision of double precision (INT64). As shown in Figure 11, we can see that TensorRT is very effective in accelerating the model. The TensorRT FP32 detects 47.8% faster with essentially no change in AP accuracy, reaching 17.1 FPS, while the TensorRT FP16 model detects 26.3 FPS with only a 0.88% loss in AP, a 53.8% improvement compared to the TensorRT FP32, and a 128.3% improvement compared to the original Pytorch FP32 model. ShufflenetV2-YOLOX is fully capable of meeting the real-time requirements of picking robots on embedded devices.

3.5. Comparison of ShufflenetV2-YOLOX with Existing Apple Target Recognition Methods

Table 5 gives the ShufflenetV2-YOLOX proposed in this paper as well as existing apple detection approaches. In the FPS column, PC and Edge indicate the speed at which the method runs in the computer and edge devices, respectively.

As can be seen from Table 5, the ShufflenetV2-YOLOX model proposed in this paper does not achieve the highest detection accuracy though, being 1.4 percentage points lower in AP compared to other methods mentioned in the literature. The possible reasons for this are considered: On the one hand, the dataset used in this thesis is complex, with three scenarios present. Each image contains an average of 12 apple targets, which raises the difficulty of apple detection. On the other hand, the network designed in this thesis is a light network, which focuses more on the operation speed of the network. Therefore, it is slightly lacking in detection accuracy. Compared with the methods in [12,13], the improved network in this thesis is more lightweight and improves the detection speed by 62 FPS and 61 FPS, respectively. The study in [10] can achieve a detection speed of 30 FPS on edge devices. However, the Jetson AGX Xaver it uses is eight times more expensive than the Jetson Nano used in this paper and is not cost-effective. Its AP is only 83.64%, well below our 96.76%.

Compared to the parameters in the literature, the ShufflenetV2-YOLOX model proposed in this paper has more outstanding advantages. Real-time detection can be achieved while ensuring detection accuracy.

Table 5. Comparison between ShufflenetV2-YOLOX and existing detection methods.

Methods	Data Sets	Detection Network	Input Size	AP	FPS	F1
Literature [10]	Dense apple	Improved YOLOv3-Tiny	1920 × 1080	83.64%	30 (Edge)	\
Literature [12]	Apple shaded by leaves	EfficientNet-YOLOV4	416 × 416	98.15%	2.95 (PC)	0.96
Literature [13]	Apple	Mask-RCNN	280 × 720	\	4 (PC)	0.905
Our	Unbagged apples, Bagged apples and Apple at night	ShufflenetV2-YOLOX	416 × 416	96.76%	65 (PC)/ 26.3 (Edge)	0.95

4. Conclusions

To solve the problems associated with apple object detection in natural environments, this paper presented ShufflenetV2-YOLOX, an improved apple object detection method based on YOLOX-Tiny. The method was trained using a dataset of apples under daytime, bagged, and nighttime conditions. By replacing the backbone network, adding an attention mechanism, adding adaptive feature fusion, and reducing the number of feature extraction layers, the detection speed and detection accuracy of the model were improved.

The AP, accuracy, recall, F1, and FPS of the trained model were 96.76%, 95.62%, 93.75%, 0.95, and 65 FPS, respectively. A 6.24% improvement in AP and 10 FPS improvement in detection speed were achieved compared to the original YOLOX-Tiny network work. In addition, compared to the advanced lightweight networks YOLOv5-s, Efficientdet-d0, YOLOv4-Tiny, and Mobilenet-YOLOv4-Lite, the AP increased by 1.32%, 3.87%, 7.62%, and 3.77%, respectively, and the detection speed increased by 47 FPS, 44 FPS, 11 FPS, and 43 FPS, respectively. This shows that the feature fusion mechanism and the attention mechanism can improve the accuracy of apple detection in natural environments at an additional cost. The application of anchorless detectors overcame the drawbacks of past Anchor-based detectors, which were computationally intensive and reduced the setting of hyperparameters and post-processing. At the same time, the application of a lightweight backbone network and the use of only two feature extraction layers reduced the size of the model and increased the detection speed. For some embedded devices with low computational power, such as the NVIDIA Jetson Nano, the detection speed could reach 11.5 FPS, while with TensorRT acceleration, the inference speed of the TensorRT FP16 model reached 26.3 FPS at the expense of only 0.88% AP.

In summary, it offers significant advantages over other current lightweight networks in terms of detection speed and detection accuracy, and significantly improves recall rates for night and bagged apples. It can meet the requirements of real-time and high-precision detection for embedded devices. The method can provide an effective solution for vision systems for apple-picking robots.

Author Contributions: Conceptualization, W.J. and Y.P.; methodology, W.J. and Y.P.; software, Y.P. and J.W.; validation, J.W.; formal analysis, Y.P.; investigation, B.X.; data curation, Y.P.; resources, B.X.; writing—original draft preparation, Y.P.; writing—review and editing, W.J. and J.W.; visualization, B.X.; supervision, W.J.; project administration, W.J.; funding acquisition, W.J. and B.X. All authors have read and agreed to the published version of the manuscript.

Funding: This research was funded by the National Natural Science Foundation of China (No. 61973141), the Jiangsu agriculture science and technology innovation fund (No. CX(20)3059), and A Project Funded by the Priority Academic Program Development of Jiangsu Higher Education Institutions (No. PAPD-2018-87).

Institutional Review Board Statement: Not applicable.

Data Availability Statement: Not applicable.

Conflicts of Interest: The authors declare no conflict of interest.

Abbreviations

ASFF	Adaptive Spatial Feature Fusion	AP	Average Precision
CBAM	Convolutional Block Attention Module	CNN	Convolutional Neural Network
FP32	Single-precision Floating-point	FP16	Half-precision Floating-point
FPS	Frames Per Second	FLOPS	Floating-point Operations Per Second
GIOU	Generalized Intersection over Union	IOU	Intersection over Union
INT64	Double Precision	IPC	Industrial Personal Computer
PR	Precision-Recall	TFL	Two Feature Layers

References

1. Wang, N.; Joost, W.; Zhang, F.S. Towards sustainable intensification of apple production in China-Yield gaps and nutrient use efficiency in apple farming systems. *J. Integr. Agric.* **2016**, *15*, 716–725. [CrossRef]
2. Ji, W.; Ding, Y.; Xu, B.; Chen, G.Y.; Zhao, D.A. Adaptive variable parameter impedance control for apple harvesting robot compliant picking. *Complexity* **2020**, *2020*, 4812657.
3. Sumit, S.; Harsh Stuart, K.; Utkal, M. Automated CNN Based Coral Reef Classification Using Image Augmentation and Deep Learning. *Int. J. Eng. Intell. Syst. Electr. Eng. Commun.* **2021**, *29*, 253–261.
4. Xia, X.; Zhou, G.M.; Qiu, Y.; Li, Z.; Wang, J. Detection of double overlapped fruits in natural scene for apple operation robot. *J. Agric. Sci. Technol.* **2018**, *20*, 63–73.
5. Liu, S.; Huang, D.; Wang, Y. Learning Spatial Fusion for Single-Shot Object Detection. *arXiv* **2019**, arXiv:1911.09516.
6. Lv, J.D.; Wang, F.; Xu, L.M.; Ma, Z.H.; Yang, B. A segmentation method of bagged green apple image. *Sci. Hortic.* **2019**, *246*, 411–417. [CrossRef]
7. Bochkovskiy, A.; Wang, C.Y.; Liao, H.M. Yolov4: Optimal speed and accuracy of object detection. *arXiv* **2020**, arXiv:2004.10934.
8. Sa, I.; Ge, Z.; Dayoub, F.; Upcroft, B.; Perez, T.; McCool, C. DeepFruits: A fruit detection system using deep neural networks. *Sensors* **2016**, *16*, 1222. [CrossRef] [PubMed]
9. Zhao, D.A.; Wu, R.D.; Liu, X.Y.; Zhao, Y.Y. Apple positioning based on YOLO deep convolutional neural network for picking robot in complex background. *Trans. Chin. Soc. Agric. Eng.* **2019**, *35*, 164–173.
10. Mazzia, V.; Khaliq, A.; Salvetti, F.; Chiaberge, M. Real-Time apple detection system using embedded systems with hardware accelerators: An edge AI application. *IEEE Access* **2020**, *8*, 9102–9114. [CrossRef]
11. Yan, B.; Fan, P.; Lei, X.; Liu, Z.; Yang, F. A real-time apple targets detection method for picking robot based on improved YOLOv5. *Remote Sens.* **2021**, *13*, 1619. [CrossRef]
12. Wu, L.; Ma, J.; Zhao, Y.; Liu, H. Apple detection in complex scene using the improved YOLOv4 model. *Agronomy* **2021**, *11*, 476. [CrossRef]
13. Chu, P.Y.; Li, Z.J.; Lammers, K.; Lu, R.F.; Liu, X.M. Deep learning-based apple detection using a suppression mask R-CNN. *Pattern Recognit. Lett.* **2021**, *147*, 206–211. [CrossRef]
14. Tian, Y.N.; Guo, D.Y.; Wang, Z.; Wang, H.; Li, E.; Liang, Z.Z. Apple detection during different growth stages in orchards using the improved YOLO-V3 model. *Comput. Electron. Agric.* **2019**, *157*, 417–426. [CrossRef]
15. Mosavi, A.; Ardabili, S.; Varkonyi-Koczy, A.R. List of Deep Learning Models. In *International Conference on Global Research and Education*; Springer: Cham, Switzerland, 2019; pp. 202–214.
16. Ji, W.; Gao, X.X.; Xu, B.; Pan, Y.; Zhang, Z.; Zhao, D.A. Apple target recognition method in complex environment based on improved YOLOv4. *J. Food Process Eng.* **2021**, *44*, e13866. [CrossRef]
17. Liu, W.; Anguelov, D.; Erhan, D.; Szegedy, C.; Reed, S.; Fu, C.Y.; Berg, A.C. SSD: Single shot multibox detector. In *European Conference on Computer Vision*; Springer: Cham, Switzerland, 2016; pp. 21–37.
18. Redmon, J.; Divvala, S.; Girshick, R.; Farhadi, A. You Only Look Once: Unified, Real-Time Object Detection. In Proceedings of the IEEE/CVF Conference on Computer Vision and Pattern Recognition, Las Vegas, NV, USA, 27–30 June 2016; pp. 779–788.
19. Ren, S.; He, K.; Girshick, R.; Sun, J. Faster R-CNN: Towards real-time object detection with region proposal networks. *IEEE Trans. Pattern Anal. Mach. Intell.* **2015**, *39*, 1137–1149. [CrossRef] [PubMed]
20. Ge, Z.; Liu, S.; Wang, F.; Li, Z.; Sun, J. YOLOX: Exceeding YOLO Series in 2021. *arXiv* **2021**, arXiv:abs/2107.08430.
21. Yu, J.H.; Jiang, Y.N.; Wang, Z.Y.; Cao, Z.M.; Huang, T. UnitBox: An Advanced Object Detection Network. In Proceedings of the 24th ACM international conference on Multimedia, Amsterdam, The Netherlands, 15–19 October 2016; pp. 516–520.
22. Rezatofighi, H.; Tsoi, N.; Gwak, J.; Sadeghian, A.; Reid, I.; Savarese, S. Generalized Intersection Over Union: A Metric and a Loss for Bounding Box Regression. In Proceedings of the 2019 IEEE/CVF Conference on Computer Vision and Pattern Recognition, Long Beach, CA, USA, 15–20 June 2019; pp. 658–666.
23. Ma, N.N.; Zhang, X.Y.; Zheng, H.T.; Sun, J. ShuffleNet V2: Practical Guidelines for Efficient CNN Architecture Design. In Proceedings of the 15th European Conference on Computer Vision, Munich, Germany, 8–14 September 2018; pp. 122–138.

24. Zhang, X.Y.; Zhou, X.Y.; Lin, M.X.; Sun, J. Shufflenet: An Extremely Efficient Convolutional Neural Network for Mobile Devices. In Proceedings of the IEEE/CVF Conference on Computer Vision and Pattern Recognition, Salt Lake City, UT, USA, 18–23 June 2018; pp. 6848–6856.
25. Woo, S.H.; Park, J.; Lee, J.Y.; Kweon, I.S. CBAM: Convolutional Block Attention Module. In Proceedings of the 15th European Conference on Computer Vision, Munich, Germany, 8–14 September 2018; pp. 3–19.
26. Tan, M.; Pang, R.; Le, Q.V. Efficientdet: Scalable and Efficient Object Detection. In Proceedings of the IEEE/CVF Conference on Computer Vision and Pattern Recognition, Seattle, WA, USA, 13–19 June 2020; pp. 10781–10790.
27. Sandler, M.; Howard, A.; Zhu, M.; Zhmoginov, A.; Chen, L.C. Mobilenetv2: Inverted Residuals and Linear Bottlenecks. In Proceedings of the IEEE/CVF Conference on Computer Vision and Pattern Recognition, Salt Lake City, UT, USA, 18–23 June 2018; pp. 4510–4520.

agriculture

MDPI

Article

Bale Collection Path Planning Using an Autonomous Vehicle with Neighborhood Collection Capabilities

Saira Latif [1,*], Torbjörn Lindbäck [1], Magnus Karlberg [1] and Johanna Wallsten [2]

1 Department of Engineering Sciences and Mathematics, Luleå University of Technology (LTU), SE-971 87 Luleå, Sweden
2 Department of Agricultural Research for Northern Sweden, Swedish University of Agricultural Sciences (SLU), SE-901 83 Umeå, Sweden
* Correspondence: saira.latif@ltu.se

Abstract: This research was mainly focused on the evaluation of path planning approaches as a prerequisite for the automation of bale collection operations. A comparison between a traditional bale collection path planning approach using traditional vehicles such as tractors, and loaders with an optimized path planning approach using a new autonomous articulated concept vehicle with neighborhood reach capabilities (AVN) was carried out. Furthermore, the effects of carrying capacity on reduction in the working distance of the bale collection operation was also studied. It was concluded that the optimized path planning approach using AVN with increased carrying capacity significantly reduced the working distance for the bale collection operation and can thus improve agricultural sustainability, particularly within forage handling.

Keywords: agriculture; path planning; neighborhood collection; autonomous vehicle; genetic algorithm; global optimization; bale collection problem; forage handling

Citation: Latif, S.; Lindbäck, T.; Karlberg, M.; Wallsten, J. Bale Collection Path Planning Using an Autonomous Vehicle with Neighborhood Collection Capabilities. *Agriculture* **2022**, *12*, 1977. https://doi.org/10.3390/agriculture12121977

Academic Editors: Jin Yuan, Wei Ji and Qingchun Feng

Received: 4 October 2022
Accepted: 14 November 2022
Published: 22 November 2022

1. Introduction

Up until present, the application of scientific and technological developments through increased mechanization and precision farming have provided several opportunities in agricultural production and within forage handling operations. Some promising engineering developments in the 20th century with regard to forage handling include forage harvesters, balers, and the automated wrapping equipment of balers using stretch films 25 μm thick to reduce the risks of dust, molds, spores, and mycotoxin respiratory allergenic disorders in livestock and humans. Baler machines have made it possible to trade silage (harvest and storage of moist grass using fermentation) in portable packages between farms, which typically weigh 600–800 kg freshly cut per bales and are more popular on smaller farms with limited labor and financial resources to construct silos [1,2].

Bales made up of hay or silage formed by hay are usually too heavy to be picked up by humans alone. Thus, they are picked up from fields using conventional utility vehicles such as tractors or loaders operated by a human. These kinds of operations are labor intensive and associated with health and accident risks [3]. There is also a potential to further improve the efficiency and environmental impact since most decisions are made by humans and thus limited to human capacities in terms of sensing, multitasking, planning, consequence analysis, etc.

Therefore, in this study, the possibility of using a new autonomous agricultural vehicle with the neighborhood pick-up capabilities concept (AVN) was investigated. The research focused on off-board path planning, which is a critical task within the complete automation process of the bale pick-up operation.

Research in the route or path planning of agricultural field tasks can be broadly categorized into two groups based on the similarity of operations: coverage path planning (CCP) and point-to-point path planning (P2P). It has been observed by [4] that agricultural operations that required coverage path planning have been slightly more investigated. Most solutions for the path planning of agricultural field operations are based on optimization methods utilizing heuristic approaches or metaheuristic approaches depending upon the size and context of the problem [5]. In situations where vehicle routes must be planned over large areas with high economical risk, methods such as metaheuristics perform an extensive search for a solution and should thus be preferred [6].

Route planning for agricultural field operations (AFOs) involving the use of vehicles is referred to as vehicle route planning (VRP), which is a well-studied problem in the field of operational planning. Recently, VRP solutions have been applied to the planning and execution of various agricultural field tasks by researchers for the scheduling of the transportation of livestock [7,8] mission planning for coverage operation such as grass mowing and seedling [9], biomass operation scheduling [10], farm-to-farm path determination for scheduling crop harvesting [11], and route planning for fertilizer application [12]. Recently, a decision tool to support farmers in the operational planning of field operations was proposed by [13] to assist in field partitioning, route generation, and evaluation.

Significant improvements have been shown for AFOs in research by the automation of the AFOs. A study [14] on field coverage operations for an autonomous tractor using a mission planner showed a 50% reduction in non-working distance. Coverage operations were then further studied for irregular shaped fields with obstacles [15,16]. In another implementation by [17], the optimal covering route and feasible positions for grain transfer between the combine harvesters and tractors were generated using VRP and the minimum cost network flow.

The application and comparability of metaheuristics for AFOs have been widely studied and is still ongoing. Recently, a hybrid genetic algorithm (GA) was tested by [18] for a capacitive vehicle route problem (CVRP) by utilizing Gillett and Miller, Downhill, and nearest neighbor heuristics to generate the initial population and refine solutions of GA. Experimental results showed that the hybrid approach generated good solutions for CVRP with low computational cost. In another research by [19] with regard to capacitated coverage path planning problem for arable field, two popular metaheuristics—simulated annealing optimization (SAO) and ant colony optimization (ACO) techniques—were evaluated and it was found that SAO performed better than ACO. Aside from AFOs, a multi-objective optimal solution to priority-based waste collection and transportation was proposed by [20] using particle swarm optimization, local search, and simulated annealing (SAO). The optimized solution resulted in a 42.3% reduction in the negative effects of greenhouse gas emissions compared to traditional waste management.

So far, few studies have investigated the bale management in fields. There exists few published studies on the sequence optimization of the bale collection operation using wagons or loaders. The intended bale field operation was described as a bale collection problem (BCP) and was solved as a traveling salesman problem using GA by [21]. While in another study on BCP in [22], a heuristic-based approach based on K-mean clustering and nearest neighbor techniques to optimize the bale collection route were tested in simulation. Comparative results from both studies showed significant improvement in the final generated route. However, no other research studies were found on the route optimization of bale collection and no single study was found on the bale collection on fields, especially with the prerequisite of neighborhood pick-up possibilities.

1.1. Objective

The objective of the research presented in this paper was to optimize the bale collection operation by means of travelled distance using notion of an autonomous articulated vehicle with neighborhood collection capability (AVN) and compare that with traditional collection methods.

1.2. Scope

The research focused on the development of a global route plan for bale collection operations in simulation for notion of using AVN. For a global route plan, a static and known environment was considered since bale positions and fields are static entities. Bale positions were assumed to been known from a previous baling operation.

The following additional general assumptions were made:

- Only bale collection operation was studied;
- A notion of new type of agricultural vehicle (AVN) was considered for the application;
- The AVN was considered to be a nonholonomic point like robot for the path generation;
- Kinetic constraints of the vehicle were excluded;
- Feasibility is measured only by total travelled distance.

2. Research Methodology

To investigate the effects of different bale collection strategies, a simulation approach was chosen. Path planning is typically performed in computer environments, which further makes feasibility evaluation easy compared to real life experimental strategies (i.e., to measure the feasibility on path suggestions on an actual field).

Two different approaches were studied and verified through the testing of situations with outcome pre-knowledge. The first approach imitates the bale collection strategy of farmers by always choosing the closest bales from the current position. The other approach instead uses a GA to optimize the collection order and position within a radius from which the AVN can reach. To investigate the differences in travelled distance (i.e., chosen feasibility) between a traditional and proposed collection approach, two different fields of the same size and with the same number of bales with a pre-determined distribution was studied. One was a simple rectangular field (field 1) and the other was a L-shaped field with more geometrical constraints (field 2). This enables investigations of possible dependencies on field complexity. With the fields selected, some simulation parameters could be set (e.g., grid size, inflation length, number of possible pick-up positions etc.) by conducting verifying tests to find a trade-off between the computational time and accuracy. Then, the experiments were designed by choosing which parameters to vary and thus which simulations to run. To enable comparison, the results from these simulations were then compiled into tables and some paths were also visualized, enabling the analysis of collection order as well as verification on the feasibility.

The traditional approach was generated by considering how humans would operate in a typical agricultural environment for bale collection operation. Generally, a human operator would pick-up the next visible bales closest to the present location. Such a heuristic approach could be programmed by using the nearest neighbor algorithm. Through this approach, two different cases were studied: one with a traditional pick-up vehicle which always has to go to the nearest bales and another with the AVN.

In addition, an optimization approach based on commonly used GA was further developed, thus enabling a comparison to the traditional approach. Here, two different strategies for initial population generation were used to show the effects on convergence.

Verification of the simulations were conducted by running a test simulation on configurations where the results were pre-known. In addition, the results from all simulations were analyzed manually to make sure that the paths were consistent.

2.1. Model Description

In this study, a notion of an AVN (see Figure 1) with a regular forwarder crane of 10 m long was used for the modeling. For comparison, traditional agricultural vehicles (e.g., tractors or loaders) were also modeled. These traditional vehicles are typically equipped with front loaders requiring additional traveling for the loading of each bale (i.e., they cannot load bales onto themselves). This effect is excluded in the traditional vehicle models in this study, leading to underestimation of the travelled distance.

Figure 1. Autonomous off-road vehicle platform—Autonomous articulated vehicle with neighborhood reach capability (AVN).

The problem formulation for the bale collection operation with a crane makes it somewhat unique. The AVN can collect bales at a radius R, which makes the situation a close enough traveling salesman problem (CETSP) [23]. The CETSP is a NP hard, combinatorial problem, and some recent solutions for CETSP have been proposed based on the discrete gravitation search algorithm and self-organizing maps [24,25]. However, the vehicle can have different carrying capacities, thus leading to a close enough traveling salesman problem with a capacity constraint. In this study, the collection sequence and collection positions minimizing the total travelled distance was searched for and thus the CETSP is defined as

$$min_L(\Sigma,\ BP, CP) \tag{1}$$

where Σ is the bale pick up sequence; CP is the desired collection position at radius R (specified by AVN reach radius) around the bale positions (BP); min_L is a function that calculates the minimum length tour at collection positions around each bale.

Agricultural Field Models

The agricultural fields were modeled in two steps. To represent "go" and "no go" areas (obstacles), binary occupancy maps (BOM) [26] were used and to find non-collision paths within the "go" areas, probability roadmaps (PRM) [27] were used.

In this research, to investigate the possible effects of field complexity, two different fields were studied. Field 1 (see Figure 2) is a rectangular field without any obstacle areas imitating a quite typical environment for bale collection operations. Field 2 (see Figure 3) on the other hand, is a representation of an irregular more complex agricultural field with obstacle or intrusion areas. For both fields, bales were positioned by calculating the distance, going in straight lines from one end to the other until the whole field was covered, and the harvesting vehicle had collected enough material to form a bale based on the average yield, etc. given in Table 1.

Figure 2a shows the BOM of field 1 where black dots indicates bales occupancy and Figure 2b shows the inflated BOM of field 1. To reduce the calculation intensity for simulations, only 10 discrete points on each bale collection radius were used, which are represented as black dots surrounding the inflated bales in Figure 2b.

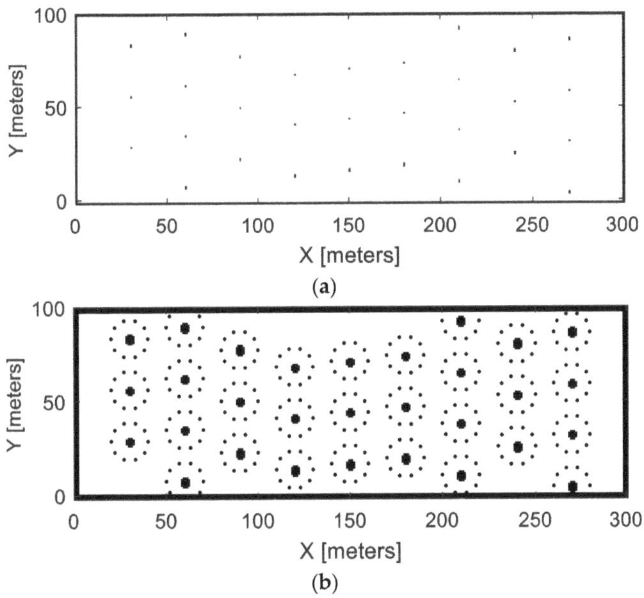

Figure 2. (**a**) Binary occupancy map of the bale position in field 1. (**b**) Inflated binary occupancy map of the bale position in field 1.

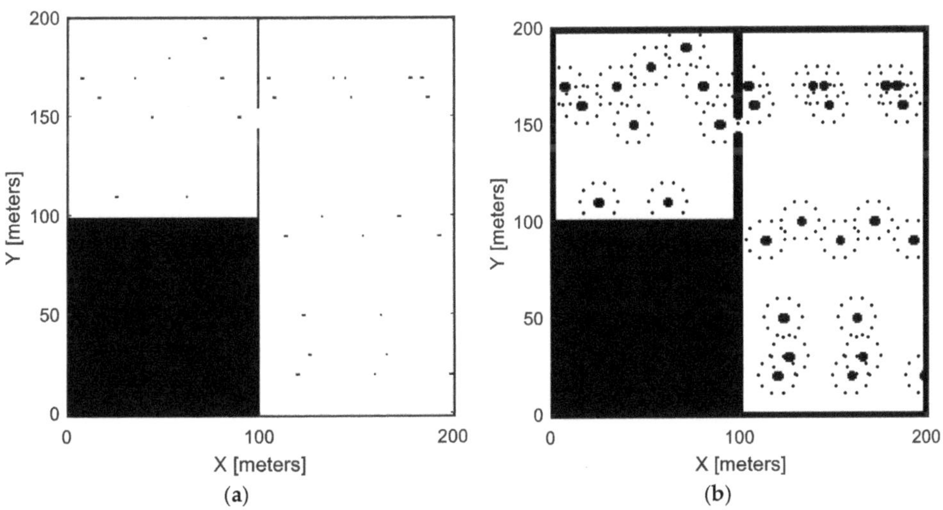

Figure 3. (**a**) Binary occupancy map (BOM) of the bale position in field 2. (**b**) Inflated binary occupancy map (BOM) of the bales and bale collection positions in field 2.

Table 1. Bale distribution parameters.

Bale Distribution Parameters	
Average grass yield	~7000 kg/ha
Average weight of bales	~700 kg
Harvester width	~3.0 m
Distance req to make one bale by harvester	~330 m

Figure 3a shows the BOM of field 2 where black dots indicates bales occupancy and Figure 3b shows the inflated BOM of field 2 including the discretized collection points at AVN's reach radius.

The distance traveled to release a bale can then be calculated through

$$d \times HW \times \check{Y}_{grass} = \widetilde{W}_{HB} \quad (2)$$

where d is the distance required to make one bale by harvester; HW is the harvester width; $\widetilde{W}_{HB}$ is the average weight of one bale; and $\check{Y}_{grass}$ is the average grass yield in a typical season. '×' represents multiplication operator. Based on the parameters in Table 1 and Equation (2), bales were released after a travelled distance of around 330 m (some minor adjustments were made if the release position coincided with the boundary of the field).

Binary Occupancy Maps for Field 1 and Field 2

A typical agricultural environment for the bale collection operation was modeled in 2D using binary occupancy maps. Bales are represented as occupied circle areas and once a bale is picked up, it is removed from the BOM. To take the collection vehicle size into consideration, the occupied areas were further inflated in the BOM. In Table 2, all BOM settings for both fields (simple and complex) are summarized.

Table 2. Binary occupancy map (BOM) setting for both fields.

Binary Occupancy Map Based Settings for Both Fields	
Total field area	3 hectares
Grid cell size	1 m
Grid resolution (cells/meter^2)	1 m
Inflation	1.3 m

Probabilistic Roadmaps

To further reduce the calculation intensity for the GA-simulations, static PRM was used (stationary nodes and connection lines) to generate the collision free paths. The same number of nodes and connection distance was used for both fields and the chosen PRM parameters are given in Table 3.

Table 3. Selected PRM settings for the simulation.

PRM Graph Parameters			
Number of nodes	1000 (Fixed position nodes + random nodes)	Fixed position nodes	Storage position, start position, end position and potential pickup points for each bale and/or each bales position
		Random nodes	Nodes besides fixed nodes are randomly generated once and remained fixed afterward for all cases
Connection distance	50		

The quality of the PRM depends on the number of nodes and connection distance and also impacts the calculation intensity. For this study, 1000 nodes and 50 m in connection distance was evaluated as a suitable trade-off.

The "bale storage position, pick-up positions (also bale positions for traditional pick-up vehicle), start- and end position of the vehicle" were pre-defined nodes and then another 1000 randomly generated nodes were added. PRMs for both fields were kept fixed, despite the changes in map (e.g., when bales are picked up) to speed up the computation. However, PRM connection lines did not cross the bale areas even after being removed.

Figure 4 shows the PRM for field 1 (a) and field 2 (b).

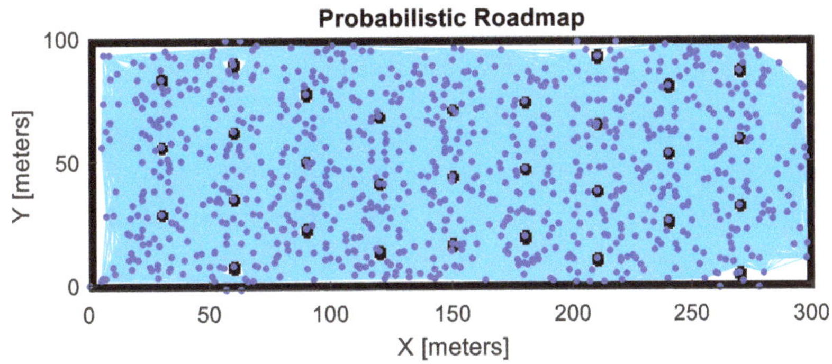

(a)

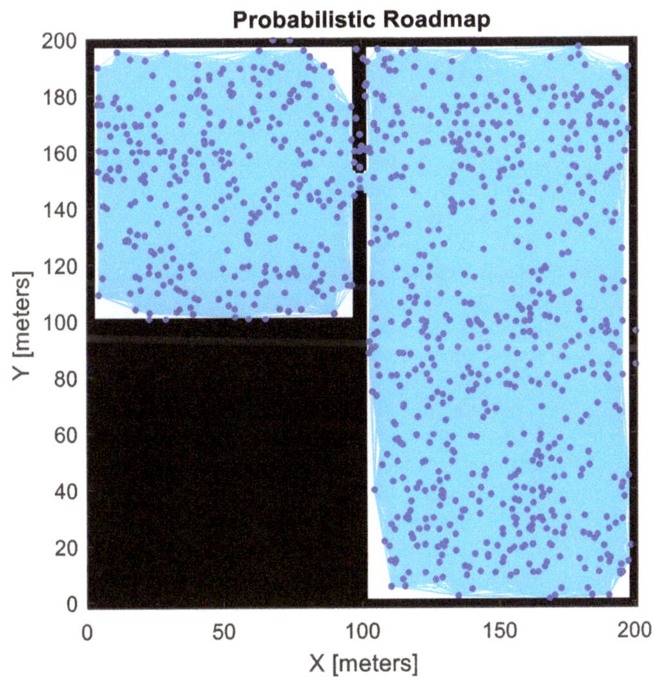

(b)

Figure 4. (**a**) Static PRM for field 1. (**b**) Static PRM for field 2.

2.2. Bales Collection Path Approaches

Two approaches to generate the bale collection paths were studied. The idea was to imitate the bale collection approach of a farmer and compare it to a bale collection approach based on optimization.

2.2.1. Nearest Neighbor Approach

One way of imitating how farmers collect bales, which was used for this study, is through the nearest neighbor approach. It was here assumed that a farmer will choose the nearest bales from its current position and then continue collecting one by one based on proximity. In the case of a traditional collection vehicle, the bale center is used as the collection position. On the other hand, for the AVN, the nearest bale is first derived and then the collection point around the bale that is closest to the Euclidian vector from the previous collection position to the current nearest bale center is derived. A straight-line path is used if no obstacles are intersected, otherwise a collision free path based on PRM is derived. This approach uses the MATLAB© built-in nearest neighbor search algorithm based on Euclidean distance between the set of points in free space. In case when there are obstacles in the space, it may result in false positive in comparison to the farmers' visual judgment in a real situation.

2.2.2. Optimization Approach

Optimization of the total distance travelled (fitness function) was carried out by use of a GA, which has good performance on finding the global optimum, has possibilities for parallelization, and can be applied to various types of problems. However, GA can become very calculation intensive and therefore, a lot of emphasis has been spent on simplifications, making each iteration as fast as possible.

Since the notion of an agricultural vehicle (see Figure 1) with neighborhood collection capability is used for this study, bales were collected not only in a certain order, but also from a point on a circle with a certain radius (corresponding to the crane length) surrounding the bales. Thus, a traveling solution is defined by a collection order and a set of points on the collection circumference (i.e., collection angles). Since the collection order is a permutation while collection angles are a set of constrained real numbers between 0 and 2π (not a permutation), it was decided to use two GAs. Hence, the first GA (GA1) was used to optimize the collection order represented as chromosome in the population of permutations of the bales' identities. For each collection order proposed by the first GA, a second GA (GA2) was then used to optimize the collection positions for each bale. To speed up the calculations, a discrete number of collection positions were defined from which GA2 had to choose. In this way, the number of possible combinations were significantly decreased, and integer representation was used for the chromosomes, which also contributes to computational efficiency. For both GAs, the built in "ga"-solver in MATLAB© was used. However, since GA1 is based on permutation chromosomes, custom functions for the initial population, crossovers, and mutations were developed (for GA2, default settings for these properties were used). To enable a comparison of the initial conditions, two different cases of population initialization were tested (i.e., randomized initialization and nearest neighbor initialization). Crossovers were conducted by flipping a random sized part of the chromosomes while the mutations were carried out by swapping two elements in the chromosome. After evaluating the performance by means of computational time and accuracy, the following settings were used for both GAs:

- Population size = 50
- Crossover fraction = 50%
- Function tolerance = 1×10^{-5}
- Elite count = 10
- Maximum nr. of stalling generations = 50
- Maximum nr. of generations = 100

For GA1, vectorization (i.e., working with the complete population for each iteration instead of sequentially working which each chromosome in sequence) and no parallelization was used, while the opposite was used for GA2, thus enabling GA2 to evaluate different sets of collection angles in parallel, which is possible since there exist no dependencies between those solutions.

At the lowest computational level (i.e., for a suggested collection order and set of collection angles), the total travelled distance can be calculated. Here, between two subsequent collection points, a straight line path was derived if no collision in the occupancy map occurred. Otherwise, the PRM was used to find the shortest collision free path (within the pre-generated PRM network). To further improve the computational efficiency, all simulated collection orders were stored together with the, for that order, optimized set of collection positions. For each new generation, this enabled an initial check of whether the suggested collection orders have already been optimized by means of collection angles or not. If not, a new optimization simulation is initiated, otherwise the already stored feasibility value is used.

A 20-core computer was used for the parallel computations, leading to a total simulation time for all set of parameters (field type, carrying capacities) of about 5 days.

3. Results

Simulations with the same set of parameters were carried out for both field 1 and field 2. The simulations included both the nearest neighbor and the optimization approaches. For the nearest neighbor, to enable a fair comparison, two different cases were studied. In the first case, notion of traditional vehicle without distance collection possibilities was modeled and referred to as the "benchmark". In the other case, the AVN notion was used and referred to as the "nearest neighbor with radius R" (referred as NNR). Additionally, the optimization approach was divided into two cases using the AVN notion. In the first case, random permutations of the pickup sequence were used for the initial population, which here is referred to as "random permutation initialization" (RPI). For the second case, the nearest neighbor collection sequence was included in the initial population, which is referred to as the "nearest neighbor permutation initialization" (NNPI). For each of these four cases, the three different carrying capacities 1, 10, and all bales were evaluated, leading to 12 different simulations for each field. The resulting paths for carrying capacity CC = 10 are shown in the main text while the paths for the remaining simulations can be found in Appendix A.

3.1. Nearest Neighbor Approach

Figure 5 shows the resulting paths for field 1 with CC = 10 of the benchmark-(U) and NNR case (L). Circle 'o' represents bales heuristically optimized pickup positions and dots '•' and '.' represents bales positions and discretized pickup position at reach radius respectively. By adding a reach radius, the traveled distance was reduced from 1750 m to 1590 m while the collection sequence remained.

Figure 6 shows the resulting paths for field 2 with CC = 10 of the benchmark-(L) and NNR case (R). By adding a reach radius, the traveled distance was reduced from 1470 m to 1300 m while the collection sequence remained.

3.2. Optimization Approach

Figure 7a shows the resulting paths for field 1 with CC = 10 of the RPI case where 'x' represents bales optimized pickup positions. Figure 7b shows the corresponding fitness convergence where black dots '·'represent the best fitness in each generation and marker '+' represents the average population fitness value in each generation. Figure 7c shows the resulting path of the NNPI case with the corresponding fitness convergence (d). By incorporating a nearest neighbor optimization as guess in the initial collection sequence population, the travelled distance was reduced from 1470 m to 1360 m.

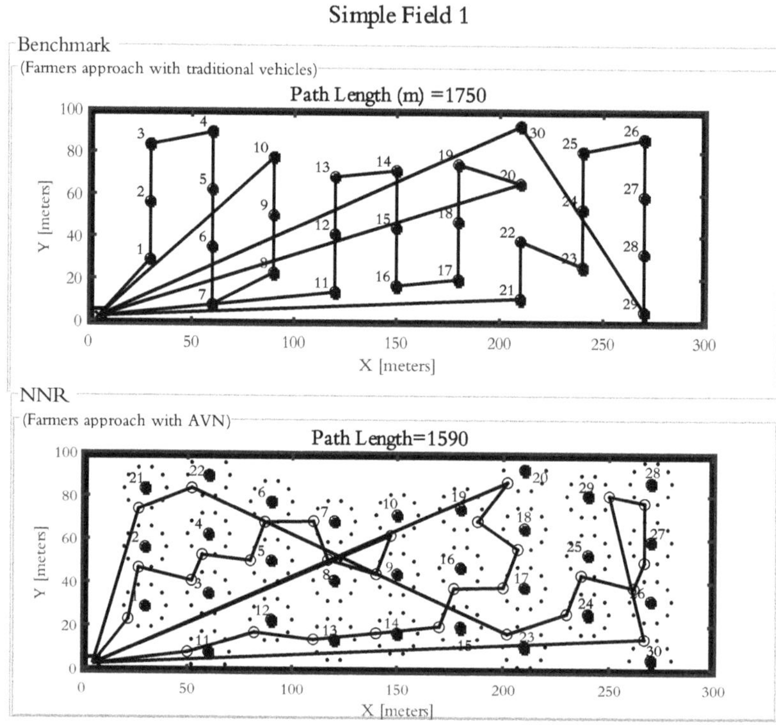

Figure 5. Resulting paths for field 1 with CC = 10 of benchmark-(U) and NNR (L).

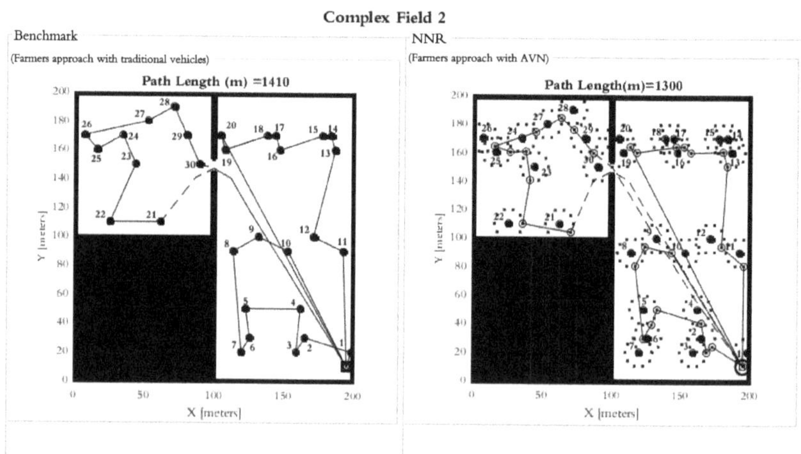

Figure 6. Resulting paths for field 2 with CC = 10 of benchmark-(L) and NNR (R).

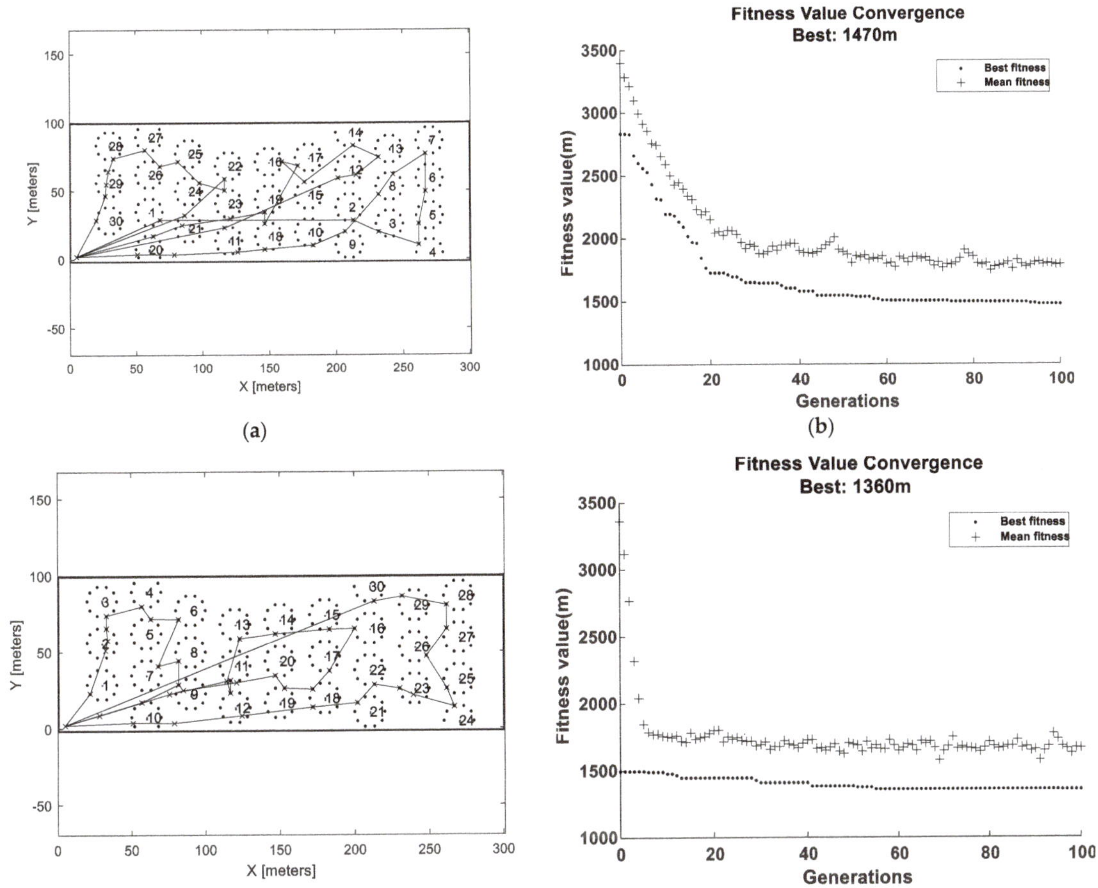

(a) (b) (c) (d)

Figure 7. Resulting paths for field 1 with CC = 10 of (**a**) RPI case (**b**) RPI convergence, (**c**) NNPI case and (**d**) NNPI convergence.

Figure 8a shows the resulting paths for field 2 with CC = 10 of the RPI case and the corresponding fitness convergence (b). Figure 8c shows the resulting path of the NNPI case with the corresponding fitness convergence (d). By incorporating a nearest neighbor in the initial collection sequence population, the travelled distance was reduced from 1490 m to 1230 m.

3.3. Results Compilation

Results of the travelled distance for all simulations are compiled in Tables 4 and 5 where the two path planning approaches and their respective subcases are arranged in columns from left to right for the three different carrying capacities given in rows. For the optimization approach, solutions for CC = 1 had weak dependency on the collection order. Some deviations compared to NNR might occur due to the fact that the discrete collection positions do not necessary coincide with a straight line from the storage location to the bales. Hence the NNR with CC = 1 is an approximation for the optimized approach. Table 4 shows the compiled results of the travelled distance for field 1.

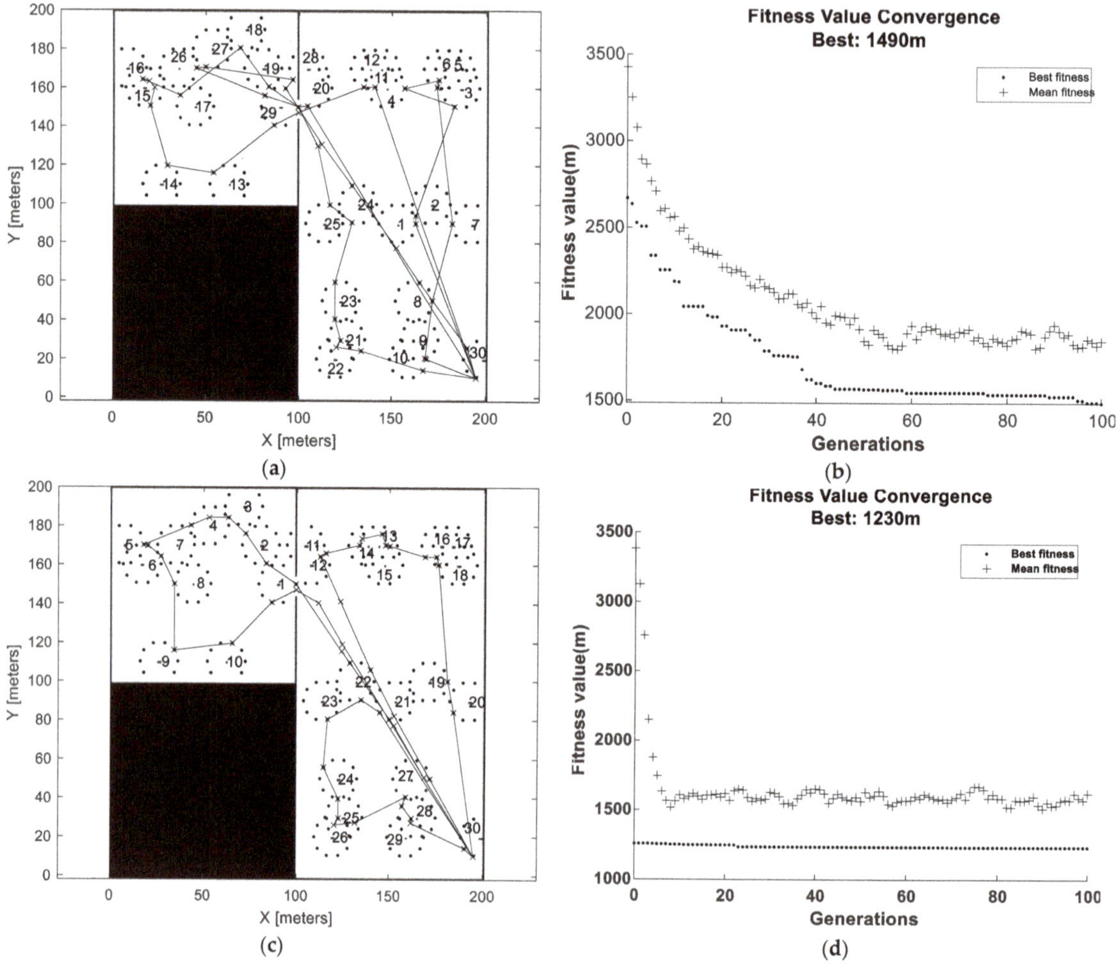

Figure 8. Resulting paths for field 2 with CC = 10 of (**a**) RPI case, (**b**) RPI convergence, (**c**) NNPI case, and (**d**) NNPI convergence.

Table 4. Compiled results for field 1.

		Path Distance (m)			
	Path Planning Approaches	Nearest Neighbor Approach		Optimization Approach	
		(Traditional vehicles)	(AVN notion)	(AVN notion)	
	Subcases	Benchmark	NNR	RPI	NNPI
Vehicles Carrying Capacity (CC)	CC = 1	9630	~9040	~9040	~9040
	CC = 10	1750	1550	1470	1360
	CC = all	1160	990	860	820

It can be observed in Table 4 that an increasing carrying capacity for all three cases resulted in a significant distance reduction. Percentage reduction in the travelled distance in field 2 for the three carrying capacities are shown in Figure 9.

Table 5. Result compilation for field 2.

		Path Distance (m)			
	Path Planning Approaches	Nearest Neighbor Approach		Optimization Approach	
		(Traditional vehicles)	(AVN notion)	(AVN notion)	
	Subcases	Benchmark	NNR	RPI	NNPI
Vehicles Carrying Capacity (CC)	CC = 1	8900	8380	~8380	~8380
	CC = 10	1470	1300	1490	1230
	CC = all	990	830	880	740

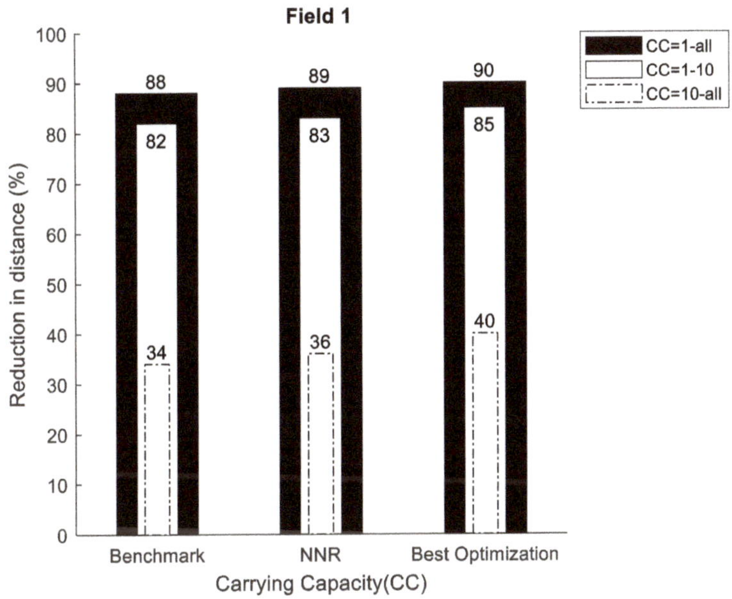

Figure 9. Travelled distance reduction for the three carrying capacities within each case for field 1.

Figure 10 shows a comparison of the path planning cases for two carrying capacities (CC = 1 will give approximately the same result for the different cases) by means of percentage reduction in the travelled distance. Black bars represent NNR over the benchmark, white bar with solid line borders NNPI over the benchmark and white bar with the dashed dotted border NNPI over NNR.

Table 5 shows the compiled results of travelled distance for field 2.

Percentage reduction in the travelled distance in field 2 for three carrying capacities are shown in Figure 11.

Figure 12 shows comparison path planning cases for two carrying capacities (CC = 1 will give approximately the same result for all cases) by means of a percentage reduction in the travelled distance. The black bar represents the NNR over benchmark, the white bar with solid line borders NNPI over benchmark, and the white bar with dashed dotted border is the NNPI over NNR.

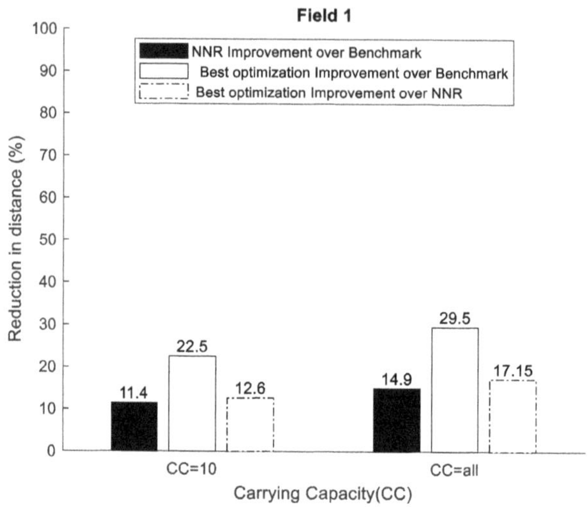

Figure 10. Comparison of the travelled distance reduction for two carrying capacities among each case for field 1.

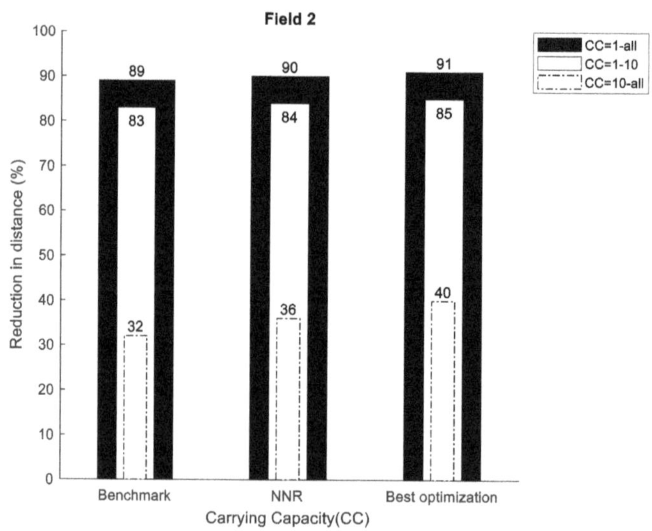

Figure 11. Travelled distance reduction for three carrying capacities within each case for Field 2.

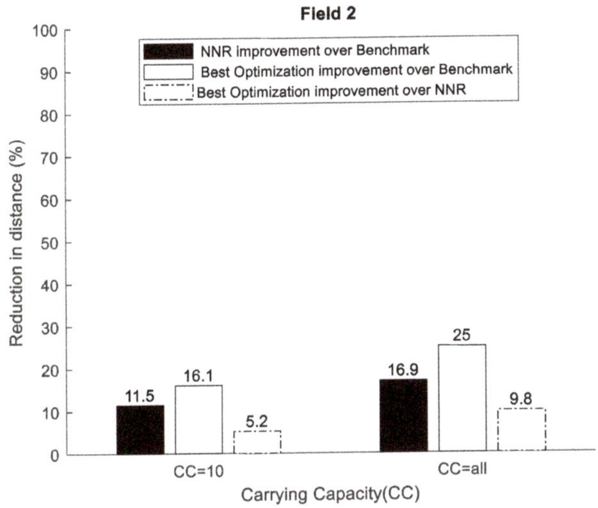

Figure 12. Comparison of the travelled distance reduction for two carrying capacities among each case for field 2.

4. Discussion

In order to simplify the computational intensity in optimizing the path planning task for the bale collection operations, there have been a number of approximations made in the modeling, as described in the scope and modeling part of the paper. This includes neglecting vehicle kinetics, considering bale collection only, keeping the PRM network static, discretization of the collection positions, etc. The GA is also significantly dependent on settings for the optimization algorithm, which effects both the accuracy and calculation time. Convergence to an optimal solution is, for instance, highly dependent on the size of the initial population and number of generations. Apart from CC = 1, the benchmark approach will always underestimate the travel distance since the loading stage is excluded from the distance calculation (i.e., relative improvements by the AVN will also be underestimated). Although these approximations will affect the output in an absolute manner, it is plausible that the relative behavior will remain, which was therefore focused on in making conclusions.

Taking the modeling limitations into consideration, some key insights were gained by analyzing the simulation results. It was found that adding carrying capacity significantly reduced the traveling distance for the bale collection operations. There was an exponential decaying trend in the distance reduction with respect to the carrying capacity. Hence, the bale collection procedure can be significantly improved, even with a small carrying capacity added. Comparing the benchmark with NNR showed that NNR reduced the travelled distance by about 10–20% (depending on field type and carrying capacity). Comparing the nearest neighbor strategy with optimization, the collection order may change for optimization (whether this is generally true or not cannot be concluded by the data presented in this paper). As would be expected, the simulations showed that the optimization approach reduced the travelled distance compared to the nearest neighbor approach. Compared to the benchmark, this reduction was about 20–30% for field 1 and 15–25% for field 2 and compared to NNR, this reduction was around 10–20% for field 1 and around 5–10% for field 2. Thus, the relative travelled distance reduction for the optimized solutions was slightly higher for the regular simple field (Field 1) compared to the complex field (Field 2). These travelled distance improvements can be compared to the similar studies by [21,22], which showed a 6.0 and 6.8% reduction for similar cases, respectively. It should be noted that the convergence to optimal solution strongly depended on the choice of initial population. The results indicate that the nearest neighbor initialization is a better choice than randomly permutated initialization independent of carrying capacities and field complexity (similar results for both fields).

5. Conclusions

It can be concluded that a vehicle with neighborhood collection capabilities and added carrying capacity can significantly reduce the travelled distance for bale collection operations (the benchmark model even gives an underestimation in this study). To generate short paths, the optimization approach is superior compared to the nearest neighbor approach and including the benchmark collection order in the initial population for the genetic algorithm improves the convergence compared to random initialization. Hence, implementing the optimization path planning approach, neighborhood collection capabilities, and adding a carrying capacity will have a significant effect on the farmers' economic and environmental sustainability. By reducing the working distance through optimized path planning implies less fuel consumption and more cost effectiveness. Although the primary focus in this study was on bale collection operation, it is plausible that the same approach is applicable in similar activities both within agriculture and beyond, for example, in forestry.

Author Contributions: Conceptualization, S.L., T.L. and M.K.; Methodology, S.L., T.L. and M.K.; Software, S.L.; Validation, S.L.; Formal analysis, S.L.; Investigation, S.L.; Resources, S.L., J.W.; Data curation, S.L.; Writing—original draft, S.L.; Writing—review & editing, S.L. and M.K.; Visualization, S.L.; Supervision, T.L., M.K. and J.W.; Project administration, M.K.; Funding acquisition, M.K. All authors have read and agreed to the published version of the manuscript.

Funding: This research was funded by The Royal Swedish Agricultural Academy (SLO-foundation) and the Swedish Governmental Agency for Innovation Systems (VINNOVA) as part of the project Automation for Autonomous Terrain Mobility (AUTO2).

Institutional Review Board Statement: Not applicable.

Data Availability Statement: Not applicable.

Conflicts of Interest: The authors declare no conflict of interest.

Appendix A

Figure A1 shows the resulting paths for field 1 with CC = 1 of the benchmark-(a) and NNR case (b). By adding a reach radius, the traveled distance was reduced from 9630 m to 9040 m while the collection sequence remained.

Figure A2 shows the resulting paths for field 1 with CC = all of the benchmark-(a) and NNR case (b). By adding a reach radius, the traveled distance was reduced from 1160 m to 990 m while the collection sequence remained.

Figure A3a shows the resulting paths for Field 1 with CC = all of the RPI case and the corresponding fitness convergence (b). Figure A3c shows the resulting path of the NNPI case with the corresponding fitness convergence (d). By incorporating a nearest neighbor in the initial collection sequence population, the travelled distance was reduced from 860 m to 820 m.

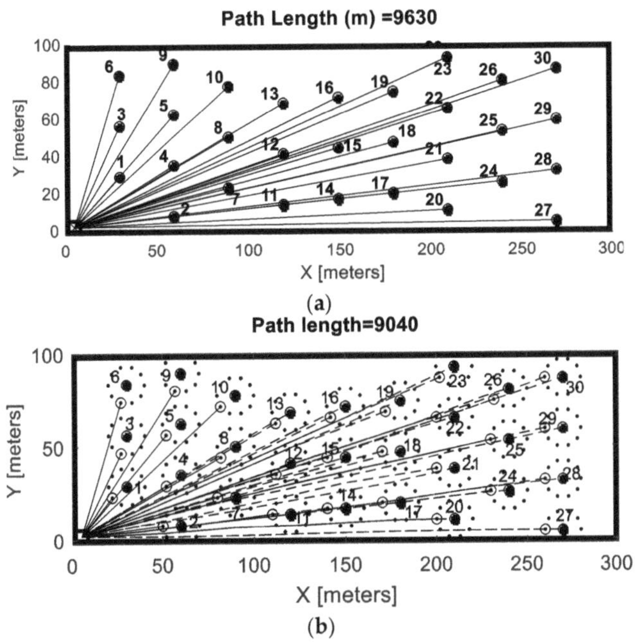

Figure A1. Resulting paths for field 1 with CC = 1 of (**a**) the benchmark and (**b**) NNR.

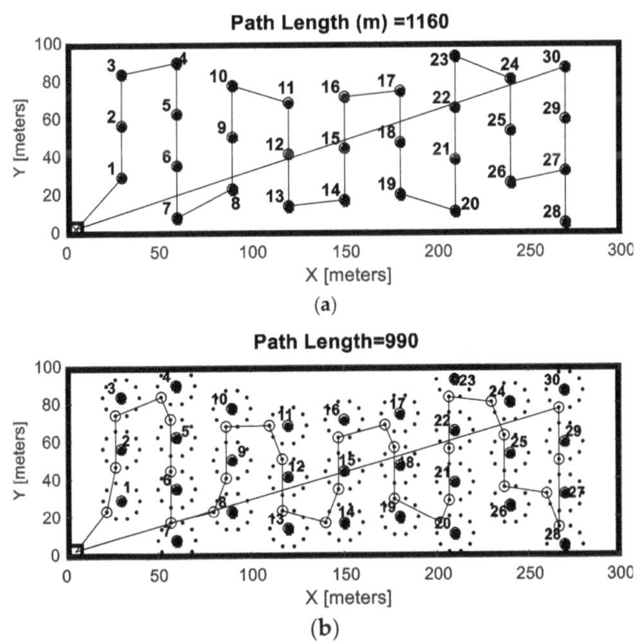

Figure A2. Resulting paths for field 1 with CC = all of (**a**) the benchmark and (**b**) NNR.

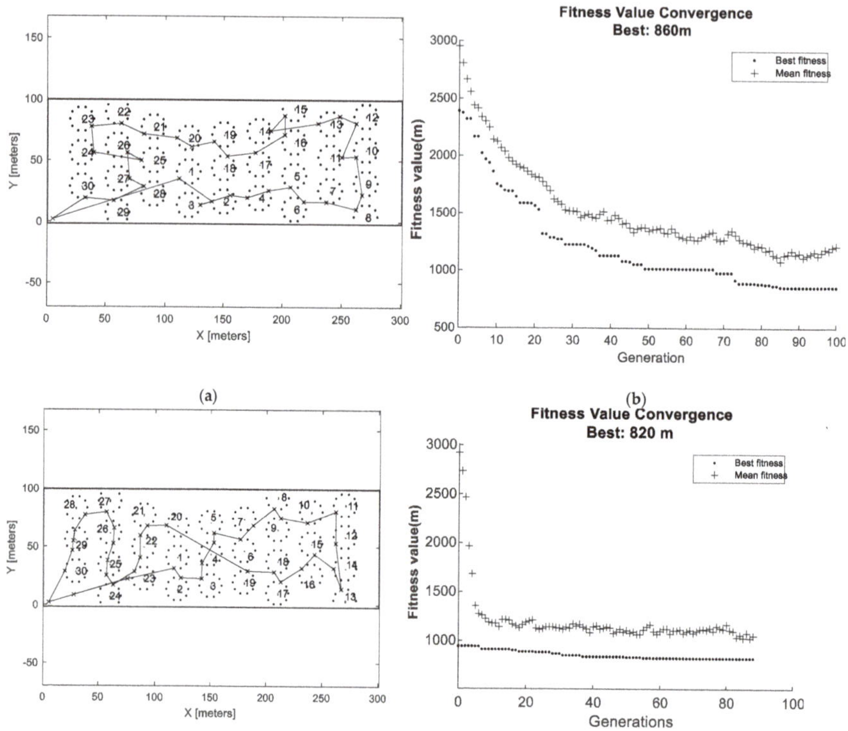

(a) (b) (c) (d)

Figure A3. Resulting paths for field 1 with CC = all of (**a**) the RPI case, (**b**) RPI convergence, (**c**) NNPI case, and (**d**) NNPI convergence.

Figure A4 shows the resulting paths for field 2 with CC = 1 of the benchmark-(a) and NNR case (b). By adding a reach radius, the traveled distance was reduced from 8900 m to 8380 m while the collection sequence remained.

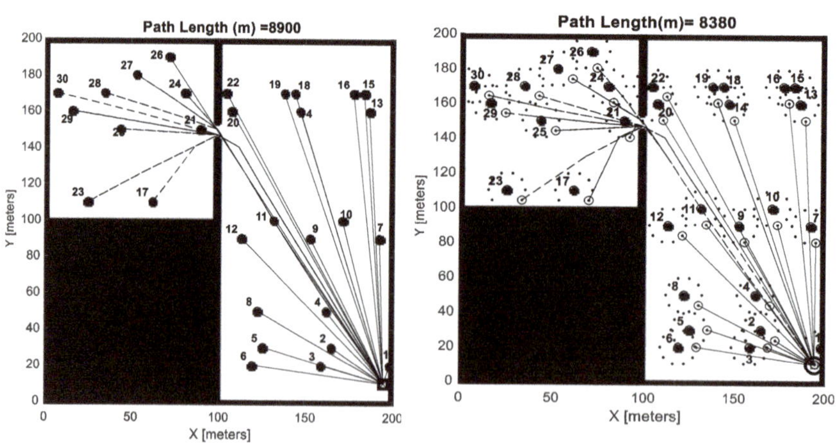

(a) (b)

Figure A4. Resulting paths for field 2 with CC = 1 of (**a**) the benchmark and (**b**) NNR.

Figure A5 shows the resulting paths for field 2 with CC = all of the benchmark-(a) and NNR case (b). By adding a reach radius, the traveled distance was reduced from 960 m to 830 m while the collection sequence remained.

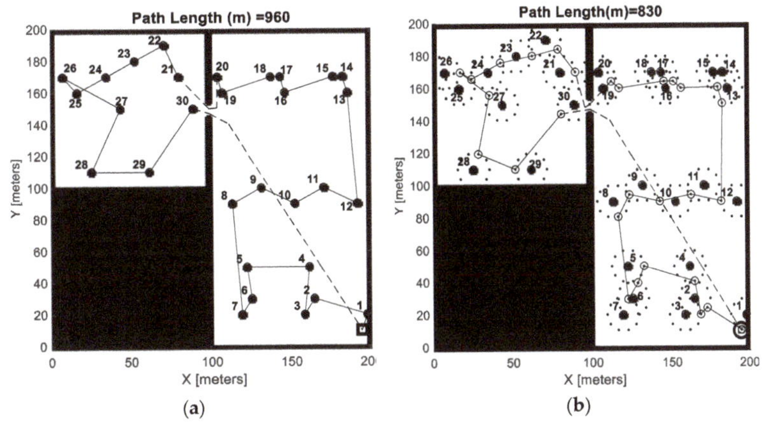

Figure A5. Resulting paths for field 2 with CC = all of (**a**) the benchmark and (**b**) NNR.

Figure A6a shows the resulting paths for field 2 with CC = all of the RPI case and the corresponding fitness convergence (b). Figure A6c shows the resulting path of the NNPI case with the corresponding fitness convergence (d). By incorporating a nearest neighbor in the initial collection sequence population, the travelled distance was reduced from 880 m to 740 m.

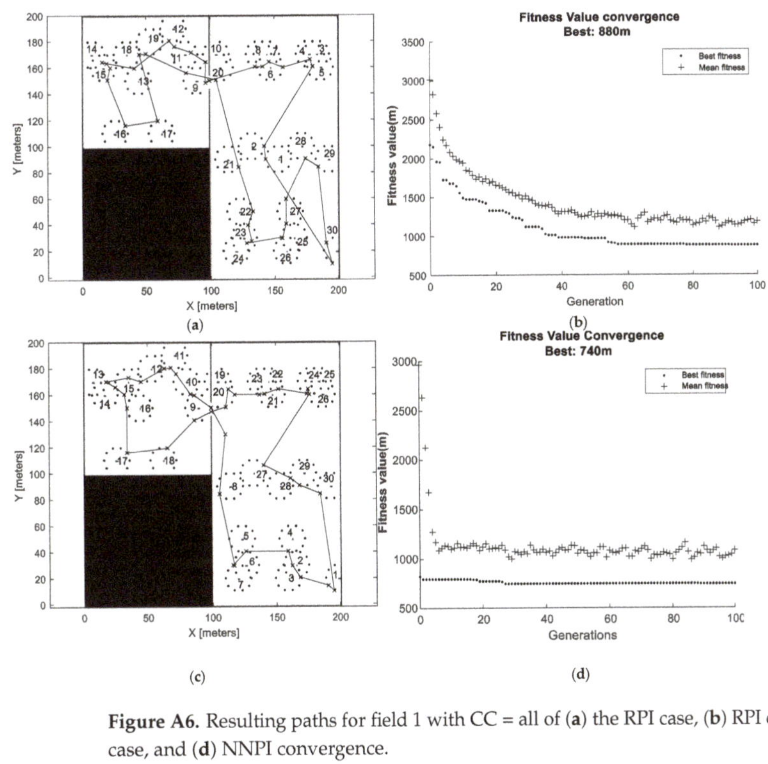

Figure A6. Resulting paths for field 1 with CC = all of (**a**) the RPI case, (**b**) RPI convergence, (**c**) NNPI case, and (**d**) NNPI convergence.

References

1. Müller, C.E. Silage and haylage for horses. *Grass Forage Sci.* **2018**, *73*, 815–827. [CrossRef]
2. Wilkinson, J.M.; Rinne, M. Highlights of progress in silage conservation and future perspectives. *Grass Forage Sci.* **2017**, *73*, 40–52. [CrossRef]
3. Pinzke, S.; Lavesson, L. Ergonomic conditions in manual harvesting in Swedish outdoor cultivation. *Ann. Agric. Environ. Med.* **2018**, *25*, 481–487. [CrossRef] [PubMed]
4. Santos, L.C.; Santos, F.N.; Pires, E.J.S.; Valente, A.; Costa, P.; Magalhaes, S. Path planning for ground robots in agriculture: A short review. In Proceedings of the 2020 IEEE International Conference on Autonomous Robot Systems and Competitions, ICARSC 2020, Ponta Delgada, Portugal, 15–17 April 2020. [CrossRef]
5. Filip, M.; Zoubek, T.; Bumbalek, R.; Cerny, P.; Batista, C.; Olsan, P.; Bartos, P.; Kriz, P.; Xiao, M.; Dolan, A.; et al. Advanced Computational Methods for Agriculture Machinery Movement Optimization with Applications in Sugarcane Production. *Agriculture* **2020**, *10*, 434. [CrossRef]
6. Cordeau, J.-F.; Gendreau, M.; Laporte, G.; Potvin, J.Y.; Semet, F. A guide to vehicle routing heuristics. *J. Oper. Res. Soc.* **2002**, *53*, 512–522. [CrossRef]
7. Oppen, J.; Løkketangen, A. A tabu search approach for the livestock collection problem. *Comput. Oper. Res.* **2008**, *35*, 3213–3229. [CrossRef]
8. Sigurd, M.; Pisinger, D.; Sig, M. Scheduling Transportation of Live Animals to Avoid the Spread of Diseases. *Transp. Sci.* **2004**, *38*, 197–209. [CrossRef]
9. Bochtis, D.; Sørensen, C. The vehicle routing problem in field logistics part I. *Biosyst. Eng.* **2009**, *104*, 447–457. [CrossRef]
10. Bochtis, D.; Dogoulis, P.; Busato, P.; Sørensen, C.; Berruto, R.; Gemtos, T. A flow-shop problem formulation of biomass handling operations scheduling. *Comput. Electron. Agric.* **2012**, *91*, 49–56. [CrossRef]
11. Basnet, C.B.; Foulds, L.R.; Wilson, J.M. Scheduling contractors' farm-to-farm crop harvesting operations. *Int. Trans. Oper. Res.* **2006**, *13*, 1–15. [CrossRef]
12. Utamima, A.; Reiners, T.; Ansaripoor, A.H. Automation in Agriculture: A Case Study of Route Planning Using an Evolutionary Lovebird Algorithm. In Proceedings of the ACM International Conference Proceeding Series, In Proceedings of the 12th International Conference on Computer and Automation Engineering, ICCAE, New York, NY, USA, 14–16 February 2020. [CrossRef]
13. Nilsson, R.S.; Zhou, K. Decision Support Tool for Operational Planning of Field Operations. *Agronomy* **2020**, *10*, 229. [CrossRef]
14. Bochtis, D.; Vougioukas, S. Minimising the non-working distance travelled by machines operating in a headland field pattern. *Biosyst. Eng.* **2008**, *101*, 1–12. [CrossRef]
15. Hameed, I.A.; Bochtis, D.; Sørensen, C.G. An Optimized Field Coverage Planning Approach for Navigation of Agricultural Robots in Fields Involving Obstacle Areas. *Int. J. Adv. Robot. Syst.* **2013**, *10*, 231. [CrossRef]
16. Zhou, K.; Jensen, A.L.; Sørensen, C.G.; Busato, P.; Bothtis, D. Agricultural operations planning in fields with multiple obstacle areas. *Comput. Electron. Agric.* **2014**, *109*, 12–22. [CrossRef]
17. Ali, O.; Verlinden, B.; Van Oudheusden, D. Infield logistics planning for crop-harvesting operations. *Eng. Optim.* **2009**, *41*, 183–197. [CrossRef]
18. De Araujo Lima, S.J.; De Araújo, S.A. An approach based on genetic algorithm and variable neighborhood search for solving the capacitated vehicle routing problem. In Proceedings of the 7th International Conference on Information Systems, Logistics and Supply Chain, ILS, Lyon, France, 8–11 July 2018.
19. Moghadam, E.K.; Vahdanjoo, M.; Jensen, A.L.; Sharifi, M.; Sørensen, C.A.G. An Arable Field for Benchmarking of Metaheuristic Algorithms for Capacitated Coverage Path Planning Problems. *Agronomy* **2020**, *10*, 1454. [CrossRef]
20. Wu, H.; Tao, F.; Yang, B. Optimization of Vehicle Routing for Waste Collection and Transportation. *Int. J. Environ. Res. Public Health* **2020**, *17*, 4963. [CrossRef]
21. Gracia, C.; Iglesias, B.D.; Barreiro, P. A hybrid genetic algorithm for route optimization in the bale collecting problem. *Span. J. Agric. Res.* **2013**, *11*, 603. [CrossRef]
22. Zamar, D.S.; Gopaluni, B.; Sokhansanj, S. A Constrained K-Means and Nearest Neighbor Approach for Route Optimization in the Bale Collection Problem. *IFAC-PapersOnLine* **2017**, *50*, 12125–12130. [CrossRef]
23. Gulczynski, D.J.; Heath, J.W.; Price, C.C. The close enough traveling salesman problem: A discussion of several heuristics. In *Perspectives in Operations Research*; Springer: New York, NY, USA, 2006. [CrossRef]
24. Antonescu, M.; Bira, C. Discrete gravitational search algorithm (DGSA) applied for the close-enough travelling salesman problem (TSP/CETSP). In Proceedings of the 2019 International Semiconductor Conference (CAS), Sinaia, Romania, 9–11 October 2019. [CrossRef]
25. Faigl, J. GSOA: Growing Self-Organizing Array—Unsupervised learning for the Close-Enough Traveling Salesman Problem and other routing problems. *Neurocomputing* **2018**, *312*, 120–134. [CrossRef]
26. Moravec, H.; Elfes, A. High resolution maps from wide angle sonar. In Proceedings of the 1985 IEEE International Conference on Robotics and Automation, St. Louis, MO, USA, 25–28 March 1985. [CrossRef]
27. Kavraki, L.; Svestka, P.; Latombe, J.-C.; Overmars, M. Probabilistic roadmaps for path planning in high-dimensional configuration spaces. *IEEE Trans. Robot. Autom.* **1996**, *12*, 566–580. [CrossRef]

agriculture

MDPI

Article

3D Locating System for Pests' Laser Control Based on Multi-Constraint Stereo Matching

Yajun Li [1,2], Qingchun Feng [2,*], Jiewen Lin [3], Zhengfang Hu [1], Xiangming Lei [1] and Yang Xiang [1,*]

1 College of Mechanical and Electrical Engineering, Hunan Agriculture University, Changsha 410128, China; lyj20210043@stu.hunau.edu.cn (Y.L.); victoria@stu.hunau.edu.cn (Z.H.); lxm@stu.hunau.edu.cn (X.L.)
2 Intelligent Equipment Research Center, Beijing Academy of Agriculture and Forestry Sciences, Beijing 100097, China
3 College of Engineering, China Agricultural University, Beijing 100083, China; b20203070536@cau.edu.cn
* Correspondence: fengqc@nercita.org.cn (Q.F.); xy@hunau.edu.cn (Y.X.)

Abstract: To achieve pest elimination on leaves with laser power, it is essential to locate the laser strike point on the pest accurately. In this paper, *Pieris rapae* (L.) (Lepidoptera: Pieridae), similar in color to the host plant, was taken as the object and the method for identifying and locating the target point was researched. A binocular camera unit with an optical filter of 850 nm wavelength was designed to capture the pest image. The segmentation of the pests' pixel area was performed based on Mask R-CNN. The laser strike points were located by extracting the skeleton through an improved ZS thinning algorithm. To obtain the 3D coordinates of the target point precisely, a multi-constrained matching method was adopted on the stereo rectification images and the subpixel target points in the images on the left and right were optimally matched through fitting the optimal parallax value. As the results of the field test showed, the average precision of the ResNet50-based Mask R-CNN was 94.24%. The maximum errors in the *X*-axis, the *Y*-axis, and the *Z*-axis were 0.98, 0.68, and 1.16 mm, respectively, when the working depth ranged between 400 and 600 mm. The research was supposed to provide technical support for robotic pest control in vegetables.

Keywords: robotic pest control; Mask R-CNN; skeleton extraction; binocular vision; stereo matching

Citation: Li, Y.; Feng, Q.; Lin, J.; Hu, Z.; Lei, X.; Xiang, Y. 3D Locating System for Pests' Laser Control Based on Multi-Constraint Stereo Matching. *Agriculture* **2022**, *12*, 766. https://doi.org/10.3390/agriculture12060766

Academic Editor: Andrea Colantoni

Received: 12 April 2022
Accepted: 25 May 2022
Published: 27 May 2022

1. Introduction

Physical pest control with laser power is widely considered as effective in reducing the pollution to the environment and even the damage to human health from the chemical pesticide [1,2]. Since 1980, many researchers have explored the outcome of pest elimination with lasers [3–5]. It has been demonstrated in these studies that laser power can cause damage to the exoskeleton and underlying tissues of pests, disrupt the anabolism of tissue cells, and ultimately kill pests [6,7]. Li et al. [5] found that the 24 h mortality rate of the fourth larval instar of *Pieris rapae* (L.) (Lepidoptera: Pieridae) reached 100% under the optimal working parameter combination of laser power of 7.5 W, an irradiation area of 6.189 mm^2, the laser opening time of 1.177 s, and the irradiation position in the middle of the abdomen. Therefore, to make laser pest control technology applicable in engineering settings, a pest control device is required to accurately focus the laser on the middle of the pest's abdomen to ensure that the laser kills the pests precisely under intense energy.

In this respect, machine vision technology can be applied to identify the pests present in the field [8,9]. However, most pests have a protective color for defense. In particular, the image background is complex and pest image features are less than prominent due to the intensive planting of crops [10]. Moreover, prior research on pest identification has mainly focused on the classification and counting of the pest species, with little attention paid to the 3D location of pests. Therefore, deep learning technology and binocular vision are integrated in this study to accurately identify and locate the laser strike point on the pest, thus providing technical support for robotic pest control in vegetables.

The mask regional convolutional neural network (Mask R-CNN) model first proposed by He et al. [11] can be used for instance segmentation and detection of pest images and achieves multiple research results in pest detection tasks [12,13]. Wang et al. [14] constructed a *Drosophila* instance segmentation model for automatically detecting and segmenting *Drosophila* wing, chest, and abdomen images, with an average precision of 94%. The instance segmentation can obtain target contour information without image morphological processing and is more suitable for accurate pest identification in laser pest control tasks. However, the above methods are used to segment RGB images of pests in specific environments, such as laboratory environments [15] and yellow sticky traps [16]. Existing algorithms still accurately segment pest targets with protective color characteristics in field environments.

As an extension of computer vision technology, near-infrared (NIR) imaging technology is used in insect species identification [17] and plant disease monitoring [18] widely. Sankaran et al. [19], based on visible-near infrared and thermal imaging technology, quickly identified citrus greening with an average precision of 87%. Luo et al. [20] used NIR imaging technology to track and monitor the structure and physiological phenology of Mediterranean tree-grass ecosystems under seasonal drought. Our team [21] proposed a monocular camera unit with an 850 nm optical bandpass filter to capture the image for identifying the pests, and the NIR image was confirmed to highlight the gray difference between the larvae of *P. rapae* and the vegetable leaves (Figure 1).

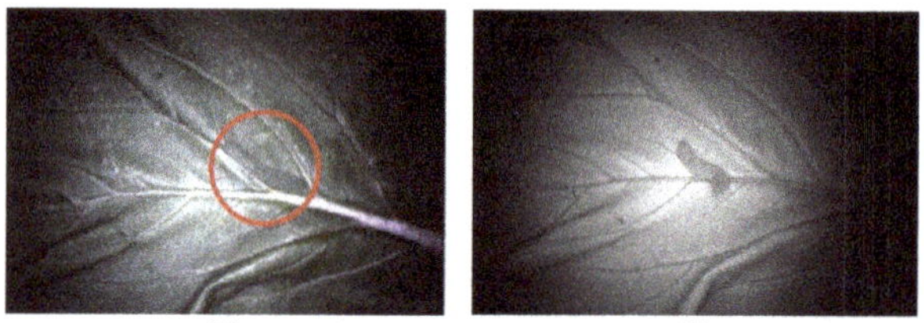
(a) (b)

Figure 1. Comparison of near-infrared imaging effects of *Pieris rapae* on cabbage leaves. (**a**) The original image. (**b**) Near-infrared image. In the process of image acquisition, *P. rapae* and cabbage leaves were placed in a black box and an 850 nm infrared filter was installed on the camera to collect near-infrared images with an 850 nm ring light source. The original image is not equipped with a filter but is equipped with a white ring light with the same power as the 850 nm.

After identifying and segmenting pests in the field, the laser strike point is located in three dimensions based on binocular stereo vision. Stereo matching is an important factor affecting the location accuracy of binocular vision. Based on the constraint range and search strategy, the matching algorithm can be divided into local [22,23], global [24,25], and semi-global [26,27] stereo matching. However, the smaller larvae of *P. rapae* remain. With the 4th and 5th instar larvae of *P. rapae* as an example, their average widths reach 1.564 mm and 2.738 mm, respectively [28]. The above stereo matching of the global parallax map for the small target pests will result in low matching efficiency and poor location accuracy. Therefore, on the basis of the determined operation range, the candidate matching region was narrowed by the multi-constrained method to improve the efficiency and location accuracy of the stereo matching.

In this study, we designed a 3D locating system for pests' laser control to eliminate the above problems of inconspicuous pest image features, unclear location of strike points, and inefficient matching algorithms. A binocular camera unit with an optical filter of 850 nm wavelength was designed to capture the pest image. The ResNet50-based Mask R-CNN extracted the bounding box and the segmentation mask of the *P. rapae* pixel area, and the laser strike point was located in the middle of the pest abdomen, which was extracted through an improved ZS thinning algorithm with smoothing iterations. Furthermore, a multi-constrained matching method was adopted on the stereo rectification images. The subpixel target points in the images on the left and right were optimally matched by fitting the optimal parallax value with the most similar feature between the template area among the two images. The 3D coordinates of each laser strike point were located according to its pixel coordinates in the two images. Finally, the recognition and localization performance of the system for targets at different locations was evaluated by implementing it on a field test platform. The research results can provide theoretical reference for the automatic laser strike of the pest control robot.

2. Materials and Methods

2.1. Binocular NIR Vision Unit

The 3D locating system was composed of a binocular vision system, a light source module, and host computer software, as shown in Figure 2a. In this system, the binocular vision system was composed of two gigabit industrial cameras produced by Hangzhou Haikang Robot Technology Co., Ltd. (Zhejiang, China). The camera model was MV-CA060-10GC, which is equipped with the lens model MVL-HF0628M-6MPE and a near-infrared filter of 850 nm. The resolution of each camera is 3072 (H) × 2048 (V), the focal length is 6 mm, and the frame rate is 15 fps. The two cameras were installed on the camera frame in parallel, and the baseline length was 50 mm. In addition, the system was illuminated by an 850 nm diffuse light bar, which can emit light evenly without shadows. The image processing platform adopted a Lenovo notebook ThinkPad P1, 24 GB RAM, Inter-Core i7-8750H@2.20 GHz, Windows 10, 64-bit system. The software system was mainly based on the OpenCV visual library and the TensorFlow deep learning framework.

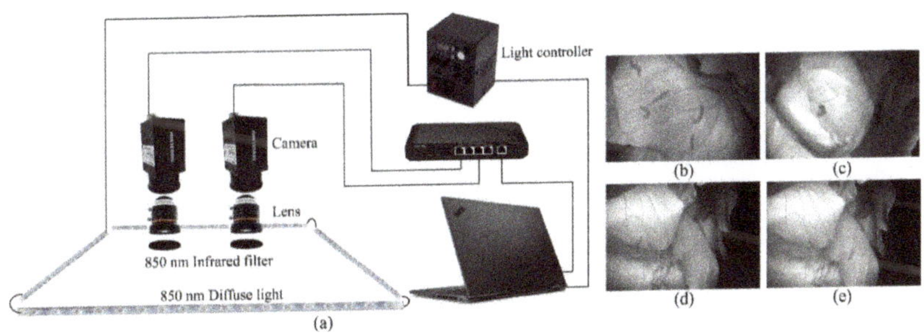

Figure 2. The binocular NIR vision unit and the example of collected images. (**a**) The visual system composition. (**b–e**) Regular cabbage as background and *P. rapae* larvae in different positions and postures taken from the collected NIR images.

Before image acquisition, a chessboard calibration board with a square size of 30 mm × 30 mm was used to perform stereo correction on the binocular camera [29]. In the process of image acquisition, the acquisition device was placed immediately above the cabbage leaves under natural illumination to collect images of *P. rapae* in the field. The collected images are shown in Figure 2b–e.

2.2. System Architecture

The flow of the field pest 3D locating system proposed in the study is shown in Figure 3, which mainly includes three parts: (1) pest identification and instance segmentation of the Mask R-CNN, (2) locating the laser strike point by extracting the skeleton of the pest, and (3) the 3D localization of laser strike point involved matching template preprocessing, multi-constraint narrowing of the matching region, subpixel stereo matching, and 3D coordinate extraction.

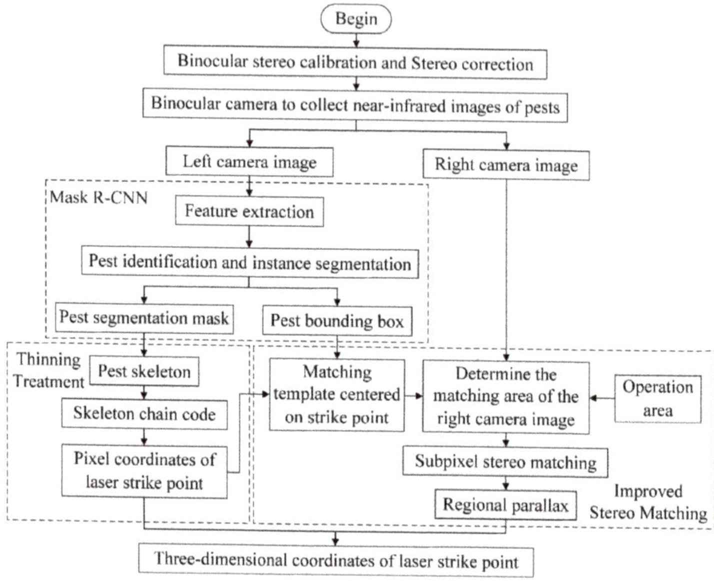

Figure 3. Structure diagram of the 3D locating system for field pests.

2.2.1. Instance Segmentation of *Pieris rapae* Image Area Based on Mask R-CNN

(1) Mask R-CNN Model

The accuracy of pest contour segmentation directly affects the accuracy of the laser strike point and stereo matching parallax. Based on the self-built NIR field *P. rapae* image dataset, this paper selected ResNet50-based Mask R-CNN [11] to identify and segment the pests' image area. The model structure is shown in Figure 4, which mainly includes the following steps:

1. The feature extraction network ResNet50 [30] extracted multi-scale information from the input image and generated a series of feature maps.
2. According to the mapping relationship between the feature map and the input image, the region proposal network (RPN) used the sliding window of the convolution layer to scan the anchor box in the feature map and generated a series of regions of interest (RoI) through classification and regression.
3. The RoI Align determined the eigenvalue of each point in the RoI and then performed pooling and other operations to match and align the target candidate region obtained by the RPN network with the feature map.
4. The feature maps output by RoI Align were input to the fully connected (FC) layers and the fully convolutional network (FCN). The former identified *P. rapae* and located the respective bounding boxes, and the latter segmented the pixel area of the larvae.

(2) Dataset augmentation and labeling

In total, 1000 images of *P. rapae* larvae in different poses were collected in the Brassica oleracea field. The sample numbers were expanded to 2000 by rotation, magnification, and horizontal and vertical mirroring, which improves the robustness of the recognition model [31]. Among them, each image contains at least one *P. rapae* larvae. We then marked the outline of *P. rapae* with the help of the open-source tool LabelMe. This tool can pick *P. rapae* masks from images and output a dataset in COCO format. Finally, the dataset was divided into a training set and a validation set according to the ratio of 8:2 for model training.

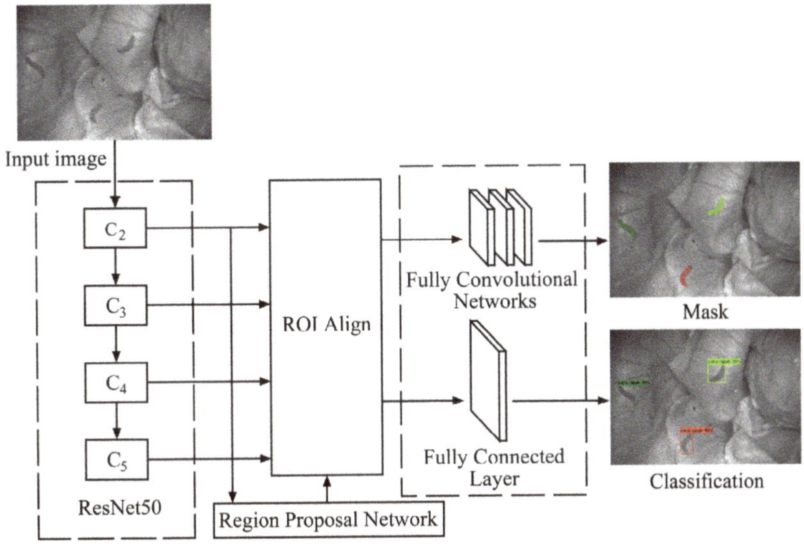

Figure 4. Overall Mask R-CNN with the ResNet50 model structure.

(3) Transfer training

The model training was completed using a PC with the following hardware environment: 32 GB RAM, Inter-Xeon E5-2623 v3*2@3.00 GHz, and NVIDIA GeForce RTX2080. The software system uses the TensorFlow deep learning framework under Windows 10 and 64-bit operating systems for coding and training and was configured with Python3.6, Anaconda 5.3.1, and CUDA10.0 compilation environments.

The training method adopted the transfer training method. The Mask R-CNN was initialized with the feature extraction network weights of the pre-trained model, while the object classification, bounding box regression, and FCN parameters were randomly initialized. During training, the initial learning rate was 0.001, the momentum parameter was 0.9, and the batch size was set to 1. In the RPN structure, the anchor point sizes were 32, 64, 128, 256, and 512. The anchor point frame ratio was 0.5:1:2.

The model object detection and region segmentation results are shown in Figure 5. The high-quality segmentation mask distinguishes pests from the background, which can be used to calculate the location of the laser strike point directly.

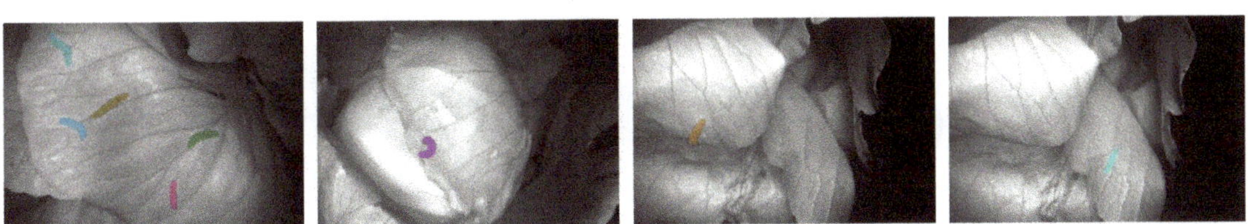

(a) (b) (c) (d)

Figure 5. Visualization results of the ResNet50-based Mask R-CNN. (**a–d**) *P. rapae* larvae in different positions and postures taken from the collected NIR images. (**a**) Multiple pests, (**b**) curled pests, (**c**) occlusion state, and (**d**) dorsal position of the leaf.

2.2.2. Pest Skeleton Extraction and Strike Point Location

(1) Laser strike point

Laser pest control requires focusing the laser on the middle of the pest abdomen to ensure that the laser kills the pests with intense energy. The body of *P. rapae* larvae is tubular and segmented, as shown in Figure 6. The middle part of the abdomen irradiation position was between the 8th and 9th segments, near the midpoint of the skeleton [5,32]. Therefore, this paper set the laser strike point as the midpoint of the skeleton of the pest image area. The improved ZS thinning algorithm was used to extract pest skeletons. Then, pest skeleton chain code was established to extract the skeleton midpoint coordinates to determine the final strike point.

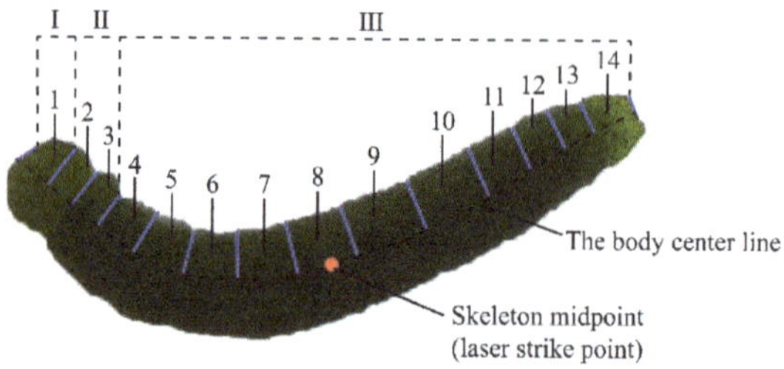

Figure 6. Characteristics of the *Pieris rapae* larvae and locating the laser strike point. The body of *P. rapae* larvae can be divided into the head (I), the thorax (II), and the abdomen (III). The numbers 1–14 denote the different segments of the larvae, separated by blue lines.

(2) Pest skeleton extraction based on improved ZS thinning algorithm

The skeleton consists of a single pixel, which provides an orientation for extracting the laser strike point coordinates. However, due to the different positions and postures of pests in the field and the sensitivity of the traditional skeleton extraction algorithm to the boundary, the extracted pest skeletons display the phenomenon of a non-single-pixel width and end branches, as shown in Figure 7.

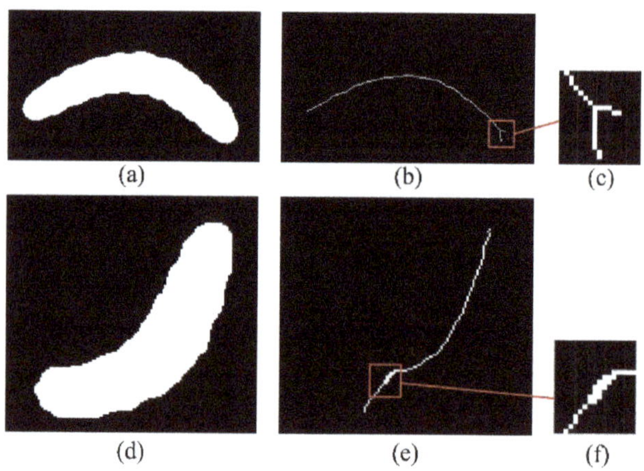

Figure 7. Visualization of the ZS thinning algorithm. (**a**,**d**) The segmentation mask of the Mask R-CNN. (**b**,**e**) The pest skeleton images. (**c**,**f**) The local details of the pest skeleton.

To solve these problems in the above-mentioned thinning process, this paper introduced an improved ZS thinning algorithm [33] with smoothing iterations to extract pest skeletons. The whole skeleton process was divided into three iterative processes: smooth iteration, global iteration, and two-stage scanning.

In the smooth iteration, the candidate deletion points were extracted based on the refinement constraints of the traditional ZS algorithm. Then, the smooth pixel points in the candidate deletion points were preserved in the smooth iteration process, which suppress the branching at the end of the pest skeleton, as shown in Figure 8. Among them, the definition of smooth pixel points satisfies Equation (1):

$$5 \leq N_b(P_0) \leq 6 \quad (1)$$

where $N_b(P_0)$ denotes the number of pixels with value 1 in the neighborhood of the scanning point P_0.

Figure 8. Example of the smooth pixel point determination. The numbers 1–6 denote the candidate deletion points extracted by the ZS thinning algorithm, where 2, 4, and 6 denote the smooth pixel points.

In smoothing iteration and global iteration, the reserved template under 24 neighborhood subdomains was added. The candidate deletion points that meet the retention template were reserved, which avoided the problem of topological structure deletion. Figure 9a–i shows the pixel set of the retention templates. The 24 neighborhood pixels were divided into 4 × 4 subdomains in 4 different directions for generating specific structures in different directions. Figure 9a–h was used to maintain diagonal lines of two-pixel widths, and Figure 9i was used to maintain the 2 × 2 square structure.

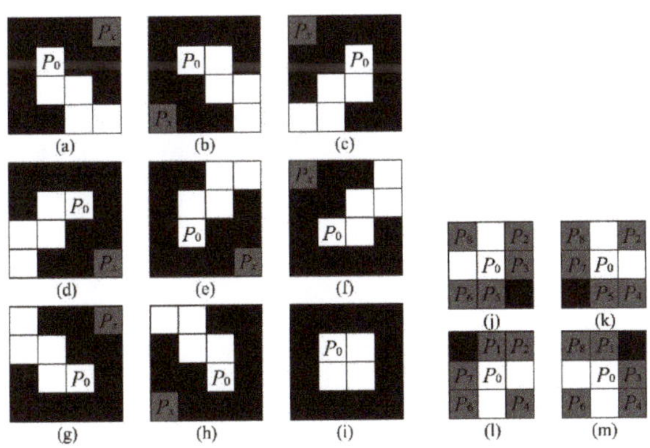

Figure 9. The retention templates and the deletion templates. **(a–i)** The retention templates in different directions. **(j–m)** The deletion templates in different directions. The pixels of scanning points are marked as P_0, and pixel sets P_x of 8 neighborhoods and 24 neighborhoods are constructed, where x = 1, 2, . . . 24. The pixel P_x in the gray square can be either 1 or 0.

In the two-stage scanning, the deletion templates under 8 neighborhoods were used to eliminate the pixels with non-single-pixel widths that form an included angle of 90. The definition of the deletion templates satisfied Figure 9j–m.

Based on the improved ZS thinning algorithm, the pest skeletons in Figure 7a,d were extracted again. The visualization is shown in Figure 10.

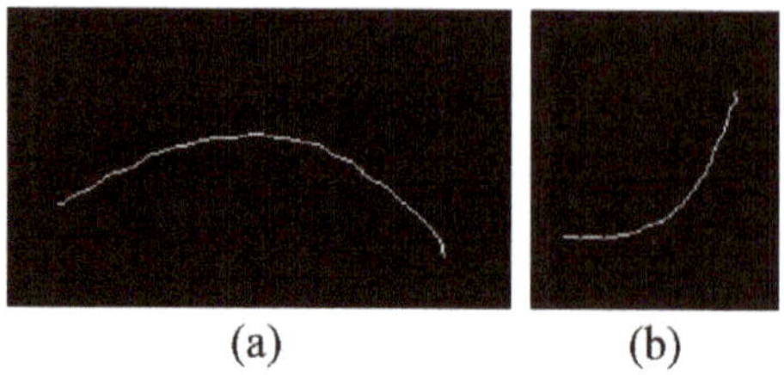

Figure 10. Visualization of the improved ZS thinning algorithm. (a,b) The pest skeleton images extracted from Figure 7a,d.

(3) Strike point location

After extracting the skeleton of pests with a single-pixel width, the system used Freeman chain code notation [34] to extract the linked list. Then, the skeleton pixel length was calculated by combining the chain code and the midpoint position coordinate was located according to the pointer. The visualization results of different processing stages are shown in Figure 11.

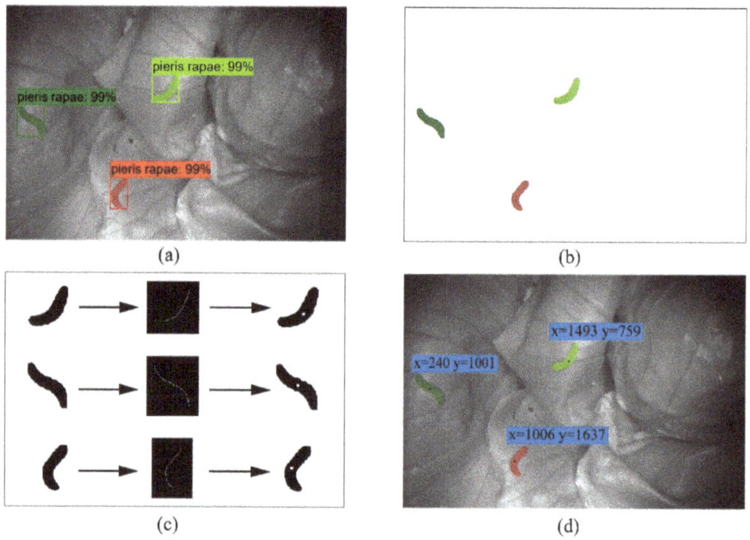

Figure 11. Visualization of the pest skeleton extraction and laser strike point location for different stages: (a) The identification and segmentation result of an NIR *P. rapae* image, (b) extracted segmentation mask image, (c) thinning treatment, and (d) coordinates of laser strike points.

2.2.3. The Multi-Constrained Stereo Matching Method

In this study, we only need to calculate the 3D spatial coordinates of the laser strike point and, thus, a multi-constraint stereo matching algorithm was proposed. As shown in Figure 12, the algorithm constructs two constraints in the matching process.

(1) The first construct: Row Constraint

After the binocular camera (Figure 12a) completed the camera calibration and stereo correction, the same pest satisfied the constraint of peer-to-peer sequential consistency in the stereo rectification images [35]. Therefore, using the pest segmentation mask in the image on the left as the template, template matching was performed on the same row in the image on the right according to the row constraint.

Assuming that the coordinate of the laser strike point in the image on the left was $p_1(x_1, y_1)$, the range of the coordinate $p_2(x_2, y_2)$ of the center point of the matching box in the image on the right can be limited to $y_2 = y_1$, as shown in Figure 12b.

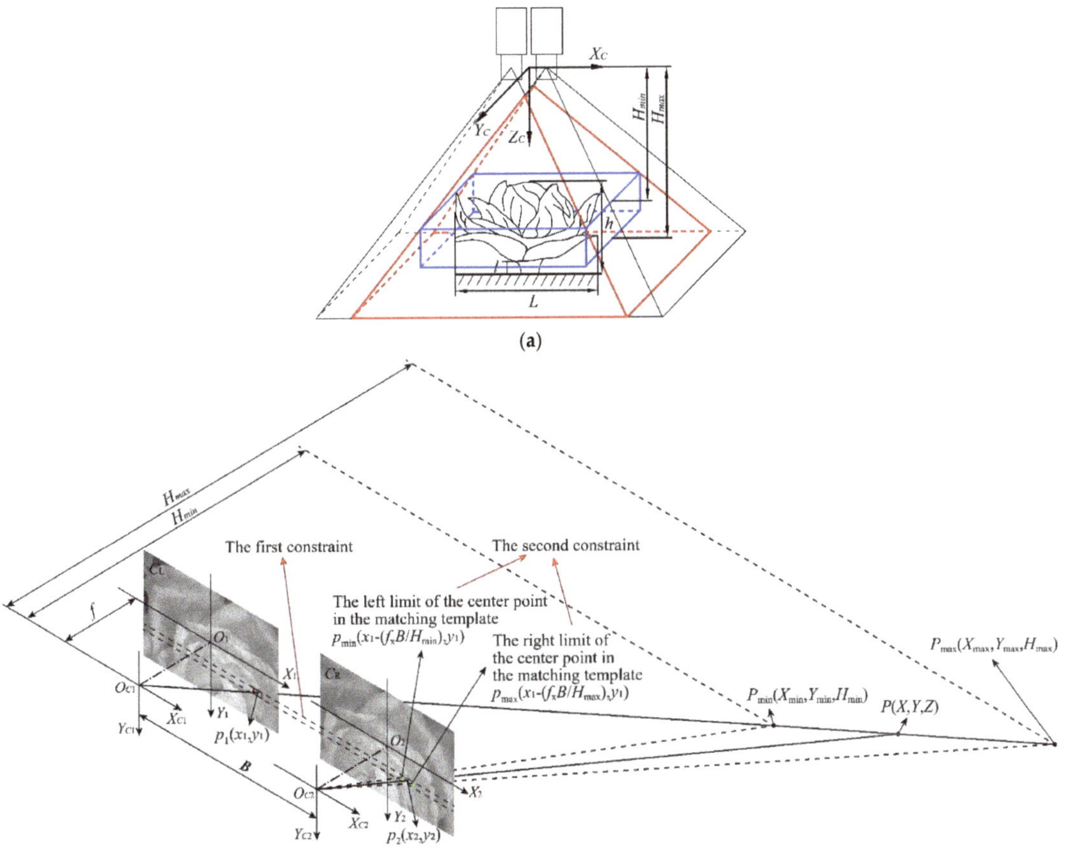

Figure 12. Search range of the multi-constraint stereo matching method. (**a**) The binocular vision locating system. The red frame is the binocular public area, the blue frame is the operation area for locating pests, and the depth range is $H_{min} \sim H_{max}$. L is the leaf spreading degree; h is the plant height; and l is the bottom leaf height of the cabbage. f_x is the camera fixed parameter. (**b**) The spatial geometric diagram. O_{C1} and O_{C2} are the optical centers of the cameras on the left and right, C_L and C_R are the imaging planes of the binocular cameras, and the image coordinate systems are $X_1O_1Y_1$ and $X_2O_2Y_2$, respectively. $p_1(x_1, y_1)$ is the laser strike point in the image on the left; $p_2(x_2, y_2)$ is the center point of the best matching box in the image on the right; $P(X, Y, Z)$ are the target pests.

(2) The second construct: Column Constraint

For the laser pest control robot to effectively identify field pests and facilitate the trajectory planning of its striking equipment, the working area was regarded as a cuboid (Figure 12a). According to the principle of triangulation [35], the coordinate of the target point in the world coordinate system can be calculated by Equation (2):

$$Z = \frac{fB}{(x_1 - x_2)\mu_x} = \frac{f_x B}{x_1 - x_2} \tag{2}$$

where B is the baseline distance of the binocular cameras, f is the focal length of the cameras, μ_x is the physical size of each pixel in the X-axis direction of the imaging plane, and f_x is the fixed parameter of the camera, which is determined during camera calibration.

In Equation (2), if the depth range of the operation area, the coordinate $p_1(x_1, y_1)$ of the target in the image on the left, and the camera fixed parameter f_x were known, the range of the X-axis of the target in the image on the right can be limited. The specific equation of x_2 was as follows.

$$x_1 - \frac{f_x B}{H_{min}} \le x_2 \le x_1 - \frac{f_x B}{H_{max}} \tag{3}$$

where H_{min} and H_{max} are the value ranges of the Z-axis of the system operation area in the world coordinate system (Figure 12).

Based on the multiple constraints above, the matching range of the template on the polar line of the target image on the right can be further restricted.

In the matching process, the normalized cross-correlation coefficient with linear illumination invariance was selected to measure the match similarity [36]:

$$R(x,y,d) = \frac{\sum_{i=1}^{n}\sum_{j=1}^{m}\left[T(x+i,y+j) - \overline{T}(x,y)\right]\left[I(x+i-d,y+j) - \overline{I}(x-d,y)\right]}{\sqrt{\sum_{i=1}^{n}\sum_{j=1}^{m}\left[T(x+i,y+j) - \overline{T}(x,y)\right]^2}\sqrt{\sum_{i=1}^{n}\sum_{j=1}^{m}\left[I(x+i-d,y+j) - \overline{I}(x-d,y)\right]^2}}, d \in \left[\frac{f_x B}{H_{max}}, \frac{f_x B}{H_{min}}\right] \tag{4}$$

where $R(x,y,d)$ is the normalized correlation quantity when the midpoint (x,y) is located in parallax d in the matching area of the camera image on the right. Here, n is the width of the template window; m is the height of the template window; $T(x+i, y+j)$ is the pixel value of the template window point $(x+i, y+j)$; and $\overline{T}(x,y)$ is the average pixel value of the template window. $I(x+i-d, y+j)$ is the pixel value of the matching area point $(x+i-d, y+j)$; and $\overline{I}(x-d,y)$ is the average pixel value of a template window with a side length of $m \times n$ defined by the point $(x-d, y)$ as the center.

After obtaining the parallax d_0 with the maximum similarity (Equation (4)), the algorithm extracted the matching similarity $R(x,y,d)$ of the adjacent parallaxes $(d_0-2, d_0-1, d_0+1, d_0+2)$ with phase-pixel-level accuracy and constructed a parallax-similarity (d-R) pointset, as shown in Figure 13. Then, the quadratic, cubic, and quartic polynomial fitting curves were performed on the pointset to obtain the polynomial curve with the highest fitting degree (R^2). The abscissa of the crest (Figure 13, Point S) at the best fitting curve was the parallax under subpixel accuracy. Finally, the 3D coordinates of each pest in the world coordinate system were calculated by the subpixel parallax.

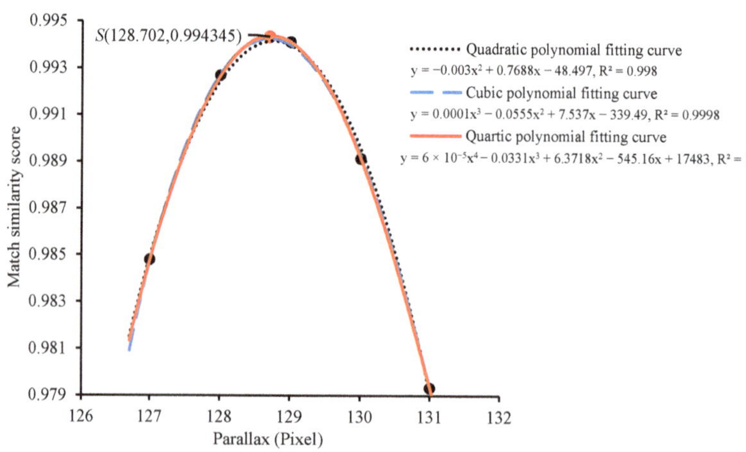

Figure 13. Polynomial fitting curves of disparity and similarity.

3. Test and Results

3.1. Experiments

To evaluate the recognition and localization accuracy of the laser strike point, combined with the characteristics of the actual operating conditions of the cabbage greenhouse, we further collected the *P. rapae* images at different positions in the vegetable field to construct a test set (Experiment 1: n = 70, Experiment 2: n = 30). The system automatically outputs and saves the identification and segmentation results of the *P. rapae* pixel area and records the 3D coordinates of the laser strike point.

The experiment was conducted in the cabbage field (28.18 N, 113.07 E) of Hunan Agricultural University in Changsha, Hunan Province, as shown in Figure 14. According to the leaf spreading degree (350 $\pm$ 46.6 mm), plant height (300 $\pm$ 25.6 mm), and bottom leaf height (32 $\pm$ 6.7 mm) of the field cabbage, the distance between the origin of the binocular camera and the effective operation area of the laser was set to 400–600 mm. The length of the working area along the X_C-axis was 400 mm and the Y_C-axis was 260 mm.

Figure 14. Accuracy test platform site. Key: 1. visual processing platform; 2. binocular camera with an 850 nm filter; 3. linear displacement sensor; 4. fixed support frame; 5. digital display for displacement sensor; 6. cabbage.

3.1.1. Experiment 1: Accuracy Evaluation of Pest Identification and Instance Segmentation Network

Combined with the test sample images (n = 70) of different scenarios, the number of *P. rapae* that were manually labeled and automatically identified by the model were recorded. Three indicators, precision value (Equation (5)), recall value (Equation (6)), and F_1-measure (Equation (7)), were used to evaluate the recognition performance of the Mask R-CNN model on the target.

$$P = \frac{TP}{TP + FP} \quad (5)$$

$$R = \frac{TP}{TP + FN} \quad (6)$$

$$F_1 = \frac{2 \times PR}{P + R} \quad (7)$$

where TP is a correctly predicted positive sample, FP is an incorrectly predicted negative sample, and FN is an incorrectly predicted positive sample.

3.1.2. Experiment 2: Performance Evaluation of the 3D Locating System

The image coordinate deviation and the actual depth deviation between the auto-location results of the laser strike point and the manual annotation results were used to evaluate the performance of the 3D locating system.

Given that the absolute deviation of coordinates represents different physical distances in images of different scales, it is impossible to characterize the true locating error quantitatively. In experiment 2, we collected 30 pairs of binocular images of the same *P. rapae* at different locations in the vegetable field. Therefore, it is assumed that the physical diameter of the *P. rapae* body width in the area of the laser strike point was constant and d represented the pixel width of *P. rapae* body in images of different scales (Figure 6). The X-axis, Y-axis location error of the world coordinate system was represented by the ratio of the pixel deviations (e_x,e_y) and d of the system output and the manually marked point on the x coordinate, y coordinate of the image.

In experiment 2, a linear displacement sensor (provided by Shenzhen Howell Technology Co., Ltd. (Shenzhen, China), KPM18-255) was used to measure the vertical distance from the pest surface to the camera plane. The sensor position accuracy was 0.05 mm. The displacement sensor is installed in a base with a magnet. The base can be adsorbed on the top plate in such a way that the displacement sensor is always perpendicular to the imaging plane and can move horizontally in the plane of the top plate, as shown in Figure 15.

Figure 15. Accuracy testing experiment equipment. Key: 1. digital display for displacement sensor; 2. 850 nm diffuse light bar; 3. binocular camera with an 850 nm filter; 4. base with a magnet; 5. linear displacement sensor; 6. *Pieris rapae*.

3.2. Validity Results of Mask R-CNN

The model training (Section 2.2.1) results showed that the average precision (AP), $AP^{0.50}$, and $AP^{0.75}$ of the ResNet50-based Mask R-CNN model constructed on the self-built NIR field *P. rapae* image dataset reached 94.24%, 98.74%, and 96.79%, respectively.

Manual detection was performed on 70 images in the test set. The target distribution of the test set was actually 158 *P. rapae* larvae, and each image contains at least one.

Then, the test set images were input into the above models. The object detection results of the larvae in the image samples of the test set by the model are shown in Table 1. The values of precision, recall, and F_1 were 96.65%, 97.47%, and 96.55%, respectively, showing the effectiveness of the proposed model.

Table 1. Identification results for the *P. rapae* larvae in the test set.

Number [1]				Precision (%)	Recall (%)	F_1 (%)
N	*TP*	*FP*	*FN*			
158	154	3	4	95.65	97.47	96.55

[1] *N* is the total number of larvae in the test set. *TP*, *FP*, and *FN* are the quantities of correctly predicted positive samples, incorrectly predicted negative samples, and incorrectly predicted positive samples, respectively.

3.3. 3D Localization Results of Field Pests

The binocular stereo vision system completed the camera calibration and stereo correction, and the results are shown in Table 2. The reprojection error was 0.36 pixels, and the calibration results meet the test requirements [37].

Table 2. The internal and external parameters of the binocular stereo vision system.

Parameters	Left Camera	Right Camera
Focus/mm	6	
Cell size/µm	2.4 (Sx) × 2.4 (Sy)	
Center column (Cx)/pixel	1589.60	1609.84
Center row (Cy)/pixel	1034.15	1051.87
2nd order radial distortion (K1)/1/pixel^2	−0.087540	−0.086044
4th order radial distortion (K2)/1/pixel^4	0.162294	0.155954
6th order radial distortion (K3)/1/pixel^6	0.000185	0.000337
2nd order tangential distortion (P1)/1/pixel^2	0.000210	−0.000308
2nd order tangential distortion (P2)/1/pixel^2	−0.065631	−0.056233
Image size/pixel	3072(H) × 2048(V)	
Baseline distance/mm	49.50	
Reprojection error/pixel	0.36	

3.3.1. *X*-Axis and *Y*-Axis Location Error

In this paper, the ratio of the image positioning deviation of the laser strike point of different scales to the pixel width of the *P. rapae* body was used as the *X*-axis and *Y*-axis location error, and the results are shown in Figure 16. In the sample images of the whole test set (N = 30), all larvae were correctly recognized and segmented and the average image location errors in the *x* coordinate and the *y* coordinate of the laser strike point were 0.09 and 0.07, respectively. The maximum errors in different scenarios were 0.23 and 0.16.

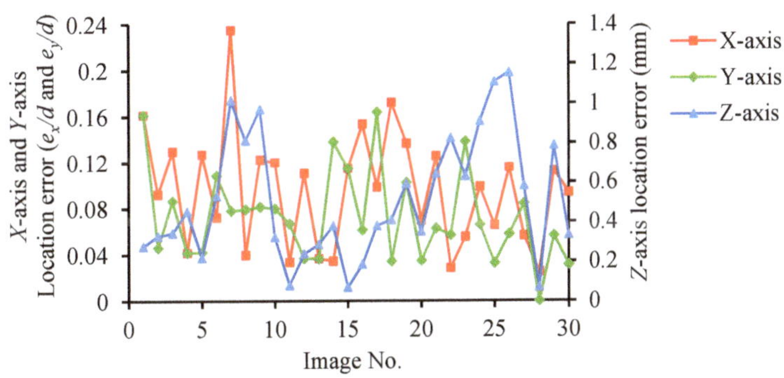

Figure 16. The location error of the laser strike point on the *X*-axis, the *Y*-axis, and the *Z*-axis. *d* denotes the pixel width of the *P. rapae* body in images of different scales. The location error is represented by the ratio of the x coordinate, y coordinate deviation (e_x,e_y) and *d*.

In the experiment, the same *P. rapae* larvae were used in different locations of the vegetable field and the larval body width was 4.16 mm (Manual measurement). Therefore, the average absolute error of the *X*-axis of the laser strike point was 0.40 mm and the maximum error was 0.98 mm. The average absolute error of the *Y*-axis was 0.30 mm, and the maximum error was 0.68 mm.

Considering the distance between the real and the located point, the average absolute error of the total location error in the *X*–*Y* plane was 0.53 mm, and the maximum error was 1.03 mm. All the located point were within the effective strike range in the middle of the pest abdomen (Figure 6).

3.3.2. *Z*-Axis Location Error

Analysis of Figure 16 shows the visual location error in the depth direction of the system when the working depth was between 400 and 600 mm. The average absolute error was 0.51 mm, and the maximum error value was 1.15 mm. The root mean square error and the mean absolute percentage error of the system were 0.58 mm and 0.10%, respectively, which shows that there is a strong explicit correlation between the estimated depth and the actual depth of the system.

4. Discussion

An automatic laser strike point localization system was established in this study based on the multi-constraint stereo matching method, which provided a basis for pests' laser control. Three aspects of the proposed model will be discussed in this section, i.e., the effects of the segmentation model, the effect of the location method, and the effect of the stereo matching method. Further improvements for the 3D locating system will also be pointed out in this section.

4.1. Analyses of Instance Segmentation Result

Experiment 1 showed that the segmentation results (AP, $AP^{0.50}$, and $AP^{0.75}$) of the ResNet50-based Mask R-CNN model were higher than 94% on the self-built NIR image dataset of *P. rapae*. The good segmentation performance of the network proves that the application of near-infrared imaging technology is feasible for pest identification, with protective color characteristics in multi-interference scenes.

In the sample images of the whole test set, the number of correctly predicted, incorrectly predicted, and unrecognized *P. rapae* were 154, 3, and 4, respectively. Among them, the number of incorrectly predicted and unrecognized *P. rapae* in a single *P. rapae* image was 0. The main causes of errors are: (1) When two or more *P. rapae* larvae overlap each other, the larvae bodies are blocked. This situation increases the difficulty of identification, resulting in multiple pests being identified as a whole or a single pest being only partially segmented (Figure 17a). (2) In the near-infrared image, the soil color is close to that of the cabbage bugs. When a leaf has a hole to expose the soil and the shape is a long strip, the model will misjudge it as a *P. rapae* larva (Figure 17b). Furthermore, the complicated network structure also makes the training time of Mask R-CNN longer. The detection time for a single image in the segmentation network was 460 ms.

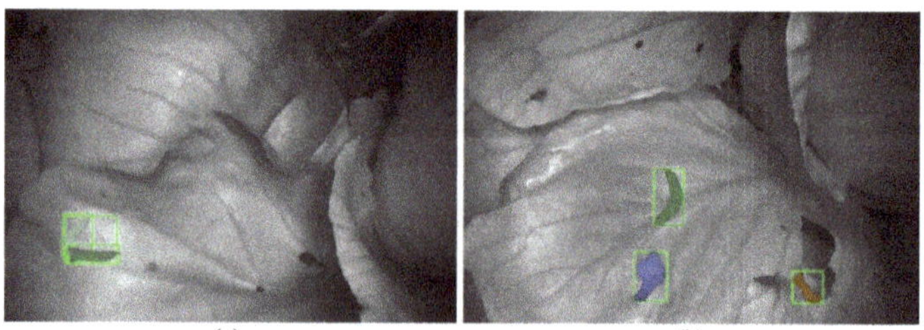

(a) (b)

Figure 17. False identification results. (**a**) Two *P. rapae* larvae overlap each other and (**b**) leaf holes mistakenly identified as *P. rapae*.

4.2. Analyses of Location Result

According to the segmentation mask in the bounding box, the laser strike point was located as the midpoint of the skeleton of pest image area, which was extracted through an improved ZS thinning algorithm. This method solves the problem of pest contour extraction based on deep learning, which greatly improves the robustness and efficiency of the algorithm.

However, this method cannot accurately locate the laser strike point in some special cases. The main causes of errors are: (1) When the *P. rapae* is partially occluded by leaves or the inclination angle is large, the method of locating the laser strike point through the midpoint of the skeleton is inaccurate because only a part of the pest skeleton is extracted (Figure 18a). (2) If the *P. rapae* larvae curl up in a ring, the pest segmentation mask is a circle. The laser strike points finally obtained by the above location method is near the center of the circle and is not within the effective strike range (Figure 18b).

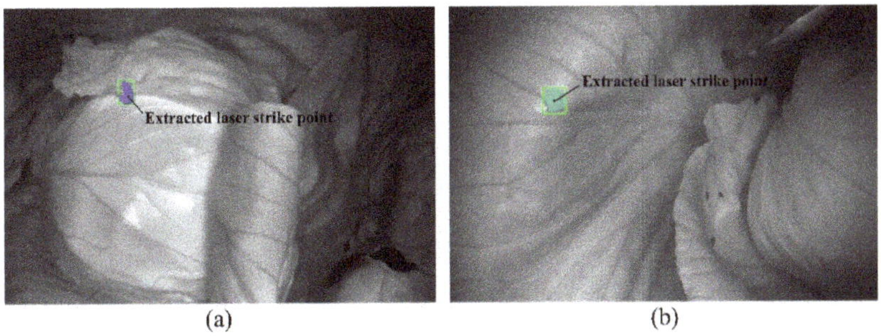

Figure 18. Incorrect location results in special cases. (**a**) The body of the *P. rapae* shaded by leaves. (**b**) The larvae curl up in a ring.

Fortunately, the above situation is not common. Fieldwork indicates that the *P. rapae* larvae are mostly found on the leaf surface in the morning, sunset, and night and are mainly located on the petioles, leaf veins, and undeveloped new leaves of the outer leaves. Especially at sunrise and at night, the *P. rapae* larvae can be clearly seen from the top of the plant when illuminated with light. The larvae curl up only when hit by external stimuli and usually become strip shaped. In general, the location method is suitable in most cases. However, the method still needs to be further improved to adapt to complex working conditions.

4.3. Analyses of the Multi-Constraint Stereo Matching Result

Experiment 2 showed that the average location errors on the *X*-axis, the *Y*-axis, and the *Z*-axis of the laser strike point were 0.40, 0.30, and 0.51 mm, respectively, and the maximum errors were 0.98, 0.68, and 1.16 mm. The system has high location accuracy on the *X*-axis and the *Y*-axis. Considering the distance between the real and the located point, the average absolute error of the total location error in the world coordinate system was 0.77 mm. The maximum error was 1.45 mm.

With the fourth and fifth instar larvae of *P. rapae* as an example, their average widths reach 1.564 mm and 2.738 mm, respectively [28]. Considering that the laser strikes vertically downward and the irradiation area is 6.189 mm^2 (diameter 2.8 mm) [5], the effective stroke of the laser end effector is increased by a maximum of 1.45 mm for accommodating the location error of the laser strike point. The extra travel poses less technical risk to the design and motion control of the laser strike device. The results satisfy the localization requirements of lasers to strike *P. rapae* larvae accurately.

The reasons for the errors are as follows: As the depth increases, the proportion of the pest area in the whole image is smaller, which results in pest segmentation and location errors. There are errors in internal and external parameters, which lead to an increase in the system error. Moreover, manual measurement error of the displacement sensor can also result in errors.

Overall, the average time of the entire pest localization process, including field pest identification, contour segmentation, and 3D coordinate position, was 0.607 s. Because the matching area was reduced, the stereo matching algorithm proposed in the study takes only 24.2% of the total time, approximately 0.147 s, which shows that the matching algorithm can quickly and accurately locate the three-dimensional coordinates of pests in the field after obtaining the pest segmentation results.

4.4. Discussion about Further Improvement Aspects

The data for this experiment were mainly collected at a depth of 400–600 mm above the ground. In the follow-up research, the relationship between the spatial resolution of the image and the laser strike point location accuracy of the proposed system can be further analyzed to obtain the best spatial solution. In this experiment, all images were collected from directly above. However, this will result in a lack of image information for pests that may be occluded by leaves or have a larger body inclination. This is somewhat detrimental to understanding the overall situation of pest infestation. In future research, the data of pests located on leaves should be collected from multiple angles to generate well-established and accurate 3D location information of pests.

5. Conclusions

A novel 3D locating system based on binocular vision was proposed for laser pest control, combining a Mask R-CNN, pest skeleton extraction, and multi-constraint stereo matching. The ResNet50-based Mask R-CNN model was trained and validated with a self-built NIR field *P. rapae* image dataset collected in a real-world agriculture scene. The AP, recall, and F_1 values were 94.24%, 97.47%, and 96.55% of the Mask R-CNN, respectively, showing the adaptability of the proposed model.

Furthermore, when the working depth varied between 400 and 600 mm, the average location errors were 0.40 mm, 0.30 mm, and 0.51 mm and the maximum errors were 0.98, 0.68, and 1.16 mm for the 3D system in the X-axis, Y-axis, and Z-axis direction. The conclusions of this study provide a design basis for the follow-up research and development of the laser pest control execution system.

Since the laser strike point extraction in this paper was limited to the processing of two-dimensional image features, there is still room for improvement in object point localization methods and accuracy evaluation experiments. In the future, the depth camera can be further used to obtain the overall 3D pose information of the pests to improve the target localization accuracy.

Author Contributions: Conceptualization, Y.L., Y.X. and Q.F.; methodology, Y.L., Y.X. and Q.F.; software, Y.L.; validation, Y.L., J.L., X.L. and Z.H.; formal analysis, Y.L.; investigation, Y.L., J.L., X.L. and Z.H.; resources, Y.X. and Q.F.; data curation, Y.L.; writing—original draft preparation, Y.L.; writing—review and editing, Y.X. and Q.F.; visualization, Y.L.; supervision, Y.X. and X.L.; project administration, Y.X.; funding acquisition, Y.X. and Q.F. All authors have read and agreed to the published version of the manuscript.

Funding: This research was funded by the National Key Research and Development Plan Project (2019YFE0125200), the Natural Science Foundation of Hunan Province of China (2021JJ30363), the BAAFS Innovation Capacity Building Project (KJCX20210414), and the Science and Technology General Project of Beijing Municipal Education Commission (KM202112448001).

Institutional Review Board Statement: Not applicable.

Informed Consent Statement: Not applicable.

Data Availability Statement: All data are presented in this article in the form of figures and tables.

Conflicts of Interest: The authors declare no conflict of interest.

References

1. Sumesh, N.; Chang, C.; Hsu, F.; Su, C.; Chen, S. Rapid laser pest control system with 3D small object detection. In Proceedings of the International Society for Optical Engineering (SPIE 11299)—AI and Optical Data Sciences, San Francisco, CA, USA, 25 March 2020. [CrossRef]
2. Kaierle, S.; Marx, C.; Rath, T.; Hustedt, M. Find and Irradiate—Lasers Used for Weed Control: Chemical free elimination of unwanted plants. *Laser Tech. J.* **2013**, *10*, 44–47. [CrossRef]
3. Enserink, M. Can the war on locusts be won? *Science* **2004**, *306*, 1880. [CrossRef] [PubMed]
4. Keller, M.D.; Leahy, D.J.; Norton, B.J.; Mullen, E.R.; Marvit, M.; Makagon, A. Laser induced mortality of Anopheles stephensi mosquitoes. *Sci. Rep.* **2016**, *6*, 20936. [CrossRef] [PubMed]
5. Li, Y.; Xiang, Y.; Yang, Z.; Han, X.; Lin, J.; Hu, Z. A Laser irradiation method for controlling Pieris rapae larvae. *Appl. Sci.* **2021**, *11*, 9533. [CrossRef]
6. Faruki, S.I.; Das, D.R.; Khan, A.R.; Khatun, M. Effects of ultraviolet (254 nm) irradiation on egg hatching and adult emergence of the flour beetles, *Tribolium castaneum*, *T. confusum* and the almond moth, *Cadra cautella*. *J. Insect Sci.* **2007**, *7*, 36. [CrossRef]

7. Sorungbe, A.A.; Badmus, H.A.; Sulaimon, A.M. Effect of ultraviolet irradiation on egg hatching of tropical warehouse moth (Ephestia Cautella), development of its adult and mortality. *Int. J. Res. Pharma. Bio. Sci.* **2016**, *3*, 23–27.
8. Cheng, X.; Zhang, Y.; Chen, Y.; Wu, Y.; Yue, Y. Pest identification via deep residual learning in complex background. *Comput. Electron. Agric.* **2017**, *141*, 351–356. [CrossRef]
9. Ren, F.; Liu, W.; Wu, G. Feature reuse residual networks for insect pest recognition. *IEEE Access* **2019**, *7*, 122758–122768. [CrossRef]
10. Ngugi, L.C.; Abelwahab, M.; Abo-Zahhad, M. Recent advances in image processing techniques for automated leaf pest and disease recognition—A review. *Inf. Process. Agric.* **2020**, *8*, 27–51. [CrossRef]
11. He, K.; Gkioxari, G.; Dollár, P.; Girshick, R. Mask r-cnn. In Proceedings of the IEEE International Conference on Computer Vision, Venice, Italy, 22–29 October 2017. [CrossRef]
12. Jiao, L.; Dong, S.; Zhang, S.; Xie, C.; Wang, H. AF-RCNN: An anchor-free convolutional neural network for multi-categories agricultural pest detection. *Comput. Electron. Agric.* **2020**, *174*, 105522. [CrossRef]
13. Tassis, L.M.; de Souza, J.E.T.; Krohling, R.A. A deep learning approach combining instance and semantic segmentation to identify diseases and pests of coffee leaves from in-field images. *Comput. Electron. Agric.* **2021**, *186*, 106191. [CrossRef]
14. Wang, J.; Chen, Y.; Hou, X.; Wang, Y.; Zhou, L.; Chen, X. An intelligent identification system combining image and DNA sequence methods for fruit flies with economic importance (Diptera: Tephritidae). *Pest Manag. Sci.* **2021**, *77*, 3382–3396. [CrossRef]
15. Minakshi, M.; Bharti, P.; Bhuiyan, T.; Kariev, S.; Chellappan, S. A framework based on deep neural networks to extract anatomy of mosquitoes from images. *Sci. Rep.* **2020**, *10*, 13059. [CrossRef]
16. Zhong, Y.; Gao, J.; Lei, Q.; Zhou, Y. A Vision-Based Counting and Recognition System for Flying Insects in Intelligent Agriculture. *Sensors* **2018**, *18*, 1489. [CrossRef]
17. Zhang, Y.; Yu, G.; Han, L.; Guo, T. Identification of four moth larvae based on near-infrared spectroscopy technology. *Spectrosc. Lett.* **2015**, *48*, 1–6. [CrossRef]
18. Türker-Kaya, S.; Huck, C.W. A review of mid-infrared and near-infrared imaging: Principles, concepts and applications in plant tissue analysis. *Molecules* **2017**, *22*, 168. [CrossRef]
19. Sankaran, S.; Maja, J.M.; Buchanon, S.; Ehsani, R. Huanglongbing (Citrus Greening) Detection Using Visible, Near Infrared and Thermal Imaging Techniques. *Sensors* **2013**, *13*, 2117–2130. [CrossRef]
20. Luo, Y.; El-Madany, T.S.; Filippa, G.; Ma, X.; Ahrens, B.; Carrara, A.; Gonzalez-Cascon, R.; Cremonese, E.; Galvagno, M.; Hammer, T.W.; et al. Using Near-Infrared-Enabled Digital Repeat Photography to Track Structural and Physiological Phenology in Mediterranean Tree–Grass Ecosystems. *Remote Sens.* **2018**, *10*, 1293. [CrossRef]
21. Hu, Z.; Xiang, Y.; Li, Y.; Long, Z.; Liu, A.; Dai, X.; Lei, X.; Tang, Z. Research on Identification Technology of Field Pests with Protective Color Characteristics. *Appl. Sci.* **2022**, *12*, 3810. [CrossRef]
22. Chen, Y.; Wang, J.; Zeng, Z.; Zou, X.; Chen, M. Research on vision pre-positioning for litchi picking robot under large field of view. *Trans. Chin. Soc. Agric. Eng.* **2019**, *35*, 48–54. [CrossRef]
23. Jiao, J.; Wang, R.; Wang, W.; Dong, S.; Wang, Z.; Gao, W. Local stereo matching with improved matching cost and disparity refinement. *IEEE MultiMedia* **2014**, *21*, 16–27. [CrossRef]
24. Zhang, F.; Prisacariu, V.; Yang, R.; Torr, P.H. Ga-net: Guided aggregation net for end-to-end stereo matching. In Proceedings of the IEEE/CVF Conference on Computer Vision and Pattern Recognition, Long Beach, CA, USA, 16–20 June 2019. Available online: https://arxiv.org/pdf/1904.06587v1.pdf (accessed on 11 April 2022).
25. Liu, J.G.; Ji, G.; Yan, F.W.; Shen, J.H.; Sun, Y.F. Stereo matching network based on disparity optimization. *Comput. Eng.* **2022**, *48*, 220–228. [CrossRef]
26. Scharstein, D.; Taniai, T.; Sinha, S.N. Semi-global stereo matching with surface orientation priors. In Proceedings of the 2017 International Conference on 3D Vision (3DV), Qingdao, China, 10–12 October 2017. [CrossRef]
27. Lu, Z.; Wang, J.; Li, Z.; Chen, S.; Wu, F. A resource-efficient pipelined architecture for real-time semi-global stereo matching. *IEEE T. Circ. Syst. Vid.* **2021**, *32*, 660–673. [CrossRef]
28. Chen, Y.N.; Ma, J. Study on the larva age markers of three important vegetable pests. *J. Changjiang Veg.* **1994**, *2*, 17–18.
29. Yu, X.; Fan, Z.; Wan, H.; He, Y.; Du, J.; Li, N.; Yuan, Z.; Xiao, G. Positioning, Navigation, and Book Accessing/Returning in an Autonomous Library Robot using Integrated Binocular Vision and QR Code Identification Systems. *Sensors* **2019**, *19*, 783. [CrossRef]
30. He, K.; Zhang, X.; Ren, S.; Sun, J. Deep residual learning for image recognition. In Proceedings of the IEEE Conference on Computer Vision and Pattern Recognition, Las Vegas, NV, USA, 26 June–1 July 2016.
31. Yu, X.G.; Wu, X.M.; Luo, C.B.; Ren, P. Deep learning in remote sensing scene classification: A data augmentation enhanced convolutional neural network framework. *GISci. Remote Sens.* **2017**, *54*, 741–758. [CrossRef]
32. Powell, J.A. Lepidoptera. In *Encyclopedia of Insects*, 2nd ed.; Resh, V.H., Cardé, R.T., Eds.; Academic Press: New York, NY, USA, 2009; pp. 559–586.
33. Yuan, L.Y.; Zhou, H.; Han, D.; Xu, G.D. Improved skeleton extraction algorithm with smoothing iterations. *Comput. Eng. Appl.* **2020**, *56*, 188–193. [CrossRef]
34. Žalik, B.; Mongus, D.; Lukač, N.; Žalik, K.R. Efficient chain code compression with interpolative coding. *Inform. Sci.* **2018**, *439*, 39–49. [CrossRef]
35. Kaehler, A.; Bradski, G. *Learning OpenCV 3*; OReilly Media, Inc.: Sebastopol, CA, USA, 2016; pp. 350–353.

36. Luo, L.; Zou, X.; Ye, M.; Yang, Z.; Zhang, C.; Zhu, N.; Wang, C. Calculation and localization of bounding volume of grape for undamaged fruit picking based on binocular stereo vision. *Trans. Chin. Soc. Agric. Eng.* **2016**, *32*, 41–47. [CrossRef]
37. Zhang, L.; Li, C.; Fan, Y.; Zhang, X.; Zhao, J. Physician-Friendly Tool Center Point Calibration Method for Robot-Assisted Puncture Surgery. *Sensors* **2021**, *21*, 366. [CrossRef] [PubMed]

agriculture MDPI

Article

3D Positioning Method for Pineapple Eyes Based on Multiangle Image Stereo-Matching

Anwen Liu [1], Yang Xiang [1,*], Yajun Li [1,2], Zhengfang Hu [1], Xiufeng Dai [1], Xiangming Lei [1] and Zhenhui Tang [1]

1 College of Mechanical and Electrical Engineering, Hunan Agriculture University, Changsha 410128, China
2 Intelligent Equipment Research Center, Beijing Academy of Agriculture and Forestry Sciences, Beijing 100097, China
* Correspondence: xy@hunau.edu.cn; Tel.: +86-13319585960

Abstract: Currently, pineapple processing is a primarily manual task, with high labor costs and low operational efficiency. The ability to precisely detect and locate pineapple eyes is critical to achieving automated pineapple eye removal. In this paper, machine vision and automatic control technology are used to build a pineapple eye recognition and positioning test platform, using the YOLOv5l target detection algorithm to quickly identify pineapple eye images. A 3D localization algorithm based on multiangle image matching is used to obtain the 3D position information of pineapple eyes, and the CNC precision motion system is used to pierce the probe into each pineapple eye to verify the effect of the recognition and positioning algorithm. The recognition experimental results demonstrate that the mAP reached 98%, and the average time required to detect one pineapple eye image was 0.015 s. According to the probe test results, the average deviation between the actual center of the pineapple eye and the penetration position of the probe was 1.01 mm, the maximum was 2.17 mm, and the root mean square value was 1.09 mm, which meets the positioning accuracy requirements in actual pineapple eye-removal operations.

Keywords: pineapple eye; three-dimensional; YOLOv5; stereo-matching

Citation: Liu, A.; Xiang, Y.; Li, Y.; Hu, Z.; Dai, X.; Lei, X.; Tang, Z. 3D Positioning Method for Pineapple Eyes Based on Multiangle Image Stereo-Matching. *Agriculture* **2022**, *12*, 2039. https://doi.org/10.3390/agriculture12122039

Academic Editor: Yanbo Huang

Received: 13 October 2022
Accepted: 26 November 2022
Published: 28 November 2022

1. Introduction

Pineapple is a fruit with a high added economic value. In 2018, China's yearly pineapple production was approximately 1.64 million tons [1]. Approximately 30% of pineapples are utilized for production and processing [2]. The processing of pineapple is complicated, especially because even after the pineapple has been skinned, there are still many eyes on its surface that need to be removed. Currently, the main way to remove pineapple eyes is to do so manually with special tools, which is labor intensive and has high labor costs and low production efficiency. The key to automatically removing pineapple eyes is to rapidly and accurately identify and locate pineapple eyes.

Machine vision technology is frequently utilized in fruit recognition and quality inspection because of its noncontact nature, high speed, and high precision [3]. In traditional machine vision technology, targets are primarily recognized based on characteristics such as color, shape, and texture. Li et al. [4] proposed a field recognition system for pineapple based on monocular vision through threshold segmentation, morphological processing, and other operations to recognize pineapples and obtain pineapple center point information. Lin et al. [5] presented a segmentation method using texture and color features, and Leung-Malik textures and HSV color features were fused to realize the detection and recognition of citrus fruit. Lv et al. [6] proposed a method to deepen the fruit region and improve the edge definition in images by using a histogram equalization algorithm. Then, the R-B color difference image based on histogram equalization was obtained, and green apple recognition was realized. Kurtulmus et al. [7] used circular Gabor texture analysis for green citrus object recognition. When the fruit surface is uneven in color, shadowed, or obscured due to environmental factors such as light, the recognition quality of traditional machine vision technology is significantly reduced [8].

By applying machine learning technology to fruit image analysis, a better applica-tion effect and higher efficiency can be obtained [9]. Li Han et al. [10] used a naive Bayes classifier to classify fruit and nonfruit regions, and the interference caused by the color similarity of green tomatoes and green foliage backgrounds was eliminated to improve the fruit recognition accuracy. Wang et al. [11] proposed a litchi recognition algorithm based on K-means clustering, which can better resist the influence of illumination changes and can maintain high accuracy for recognition under occlusion and different lighting conditions. Zhao et al. [12] extracted the Haar-like features of grayscale images and used the AdaBoost classifier for classification and recognition. In the actual environment, the detection accuracy rate of ripe tomatoes reached 96%, and the classifier structure was simple.

In recent years, object detection based on deep learning has shown great advantages in the field of fruit image recognition [13,14]. The convolutional neural network, with its fast detection speed and excellent ability to extract target features, not only reduces the workload but also improves the recognition speed and accuracy [15]. Zhang Xing et al. [16] proposed a study on pineapple picking and recognition under a complex field environment based on the improved YOLOv3. The multiscale fusion training network was used to detect single-category pineapple, and a detection and recognition rate of approximately 95% was achieved using this method. Tian et al. [17] proposed an improved YOLOv3 model to identify apples at different growth stages in orchards. The model was used with the DenseNet method to process low-resolution feature layers; this method effectively enhances feature propagation, promotes feature reuse, improves network performance, and has good recognition performance under apple overlap and occlusion conditions. Yu et al. [18] proposed a mask R-CNN algorithm to identify 100 wild strawberry images. The results demonstrated that the average recognition accuracy was 95.78% and the recall rate was 95.41%. Zhang et al. [19] proposed a real-time detection method for grape clusters based on the YOLOv5s deep learning algorithm. By training and adjusting the parameters of the YOLOv5s model on the data set, the fast and accurate detection of grape clusters was finally realized. The test results showed that the precision, recall, and mAP of the grape cluster detection network were 99.40%, 99.40%, and 99.40%, respectively.

Studies related to fruit positioning, which are widely used, have mainly focused on the three-dimensional positioning of fruit for robot picking, and methods include binocular stereo vision, structured light stereo vision, and monocular stereo vision. In binocular stereo vision, not only can the image information of different angles of the target be obtained, but the three-dimensional position information of the target through stereo matching can also be obtained [20]. Therefore, this is a widely used approach in fruit and vegetable recognition [21], positioning [22], and acquisition of phenotypic parameters [23]. Luo et al. [24] proposed a method for solving and positioning enclosure based on binocular stereo vision. When the depth distance was within 1000 mm, the positioning error was less than 5 mm. However, the calibration process of the binocular camera is complex, and the calculational burden of the algorithm was relatively large [25]. Stereovision, which is based on structured light, is a combination of structured light technology and binocular stereo vision technology. Through structured light matching, the corresponding pixels of the left and right cameras are subjected to stereo matching, parallax calculation, and recovery of the three-dimensional data of the scene. Zhang et al. [26] used a machine vision system based on a near-infrared array structure and three-dimensional reconstruction technology to realize the recognition and positioning of apple stems and calyxes. However, structured light stereo vision is easily affected by illumination [27]. Monocular stereo vision positioning can be divided into monocular camera positioning of one, two, or more images. The positioning of a single image mainly relies on the mapping relationship between the known spatial information of the characteristic light points, lines, or other image features to obtain the position coordinate information [28]. Generally, images from different perspectives are obtained using the positioning method by changing the position of the camera, and the matching relationship of image feature points is used to obtain the relative positional relationship between the cameras during multiple shots to realize the positioning of the target. Zhao et al. [29] used a monocular color camera to build a vision system to locate the picking point of litchi clusters and realize the three-dimensional positioning of litchi clusters.

To date, there have been no research reports on pineapple eye machine-vision recognition or positioning. Based on the analysis of the existing research in the field of fruit recognition and positioning, deep learning technology based on convolutional neural networks is proposed in this paper to carry out research on pineapple eye recognition. On this basis, combined with the entire circumference-image-acquisition-of-pineapple method, the three-dimensional localization of pineapple eyes is realized by applying the stereo-matching method of monocular and multiangle images.

2. Materials and Methods

2.1. Structure and Working Principle of the Test Platform

The structure of the pineapple eye recognition and positioning test platform is shown in Figure 1. The notebook is an HP-Shadow Elf equipped with an Intel i7-10750H CPU@2.60 GHz processor, 16 GB RAM, and an NVIDIA GeForce GTX1650Ti graphics card. The 64-bit Windows 10 operating system is installed, and the software development environment is Visual Studio2017 + OpenCV4.0.0. The color camera is an Imaging Source DFK41BU02 with a resolution of 1280(H) $\times$ 960(V), a frame rate of 15 fps, and an 8.5 mm Computar lens. A CR-9600-R ring light source is installed directly under the camera lens. The Mitsubishi FX3U-32MT PLC controller is used as the control core, and the PLC is connected to the notebook through the serial communication port. The motion platform is composed of a clamping cylinder, servo motor, linear slide, probe cylinder, and probe. The peeled pineapple is clamped using the clamping cylinder and rotated at a precise angle by the servo motor to acquire the entire circumference of the pineapple image. In this paper, a probe is used to evaluate the accuracy of the identification and positioning algorithm. The probe is installed on the probe cylinder and can be inserted into the pineapple through the telescopic movement of the probe cylinder. The probe cylinder, which can accurately move, is installed and positioned in the direction parallel to the pineapple axis.

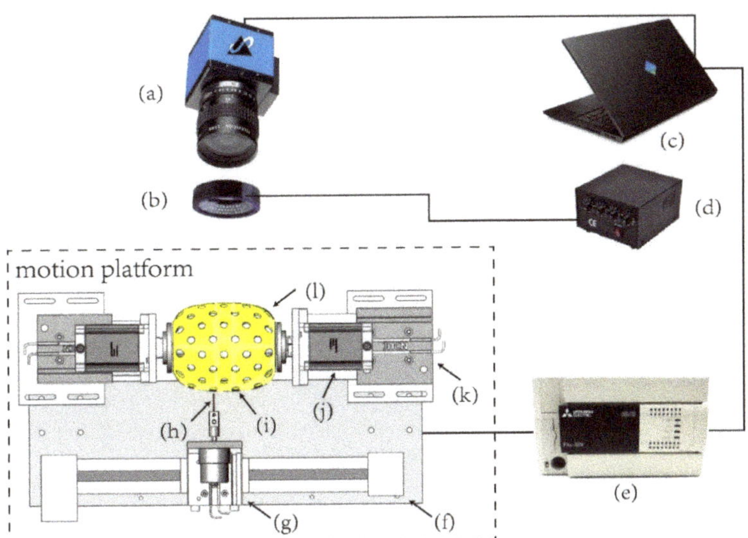

Figure 1. Structure of the test platform. (**a**) color camera, (**b**) ring light source, (**c**) notebook, (**d**) light source controller, (**e**) PLC controller, (**f**) linear slide, (**g**) probe cylinder, (**h**) probe, (**i**) pineapple eye, (**j**) servo motor, (**k**) clamping cylinder, and (**l**) pineapple.

2.2. *Image Acquisition of Pineapple Eyes*

Goodfarmer Philippine pineapples, which were manually peeled and placed on the test platform for image acquisition, were used for the experiments. Before image acquisition, the dot calibration plate was used to reduce the lens distortion and perspective distortion caused by the tilt of the camera [30]. To obtain the images of all pineapple eyes and provide a sufficient number of images for multiangle image stereo matching, images of pineapples were collected in 60° intervals, and 6 images were collected for each pineapple. Figure 2 shows images of the same pineapple collected from different angles. From this figure, there are obvious differences in the shape and size of pineapple eyes.

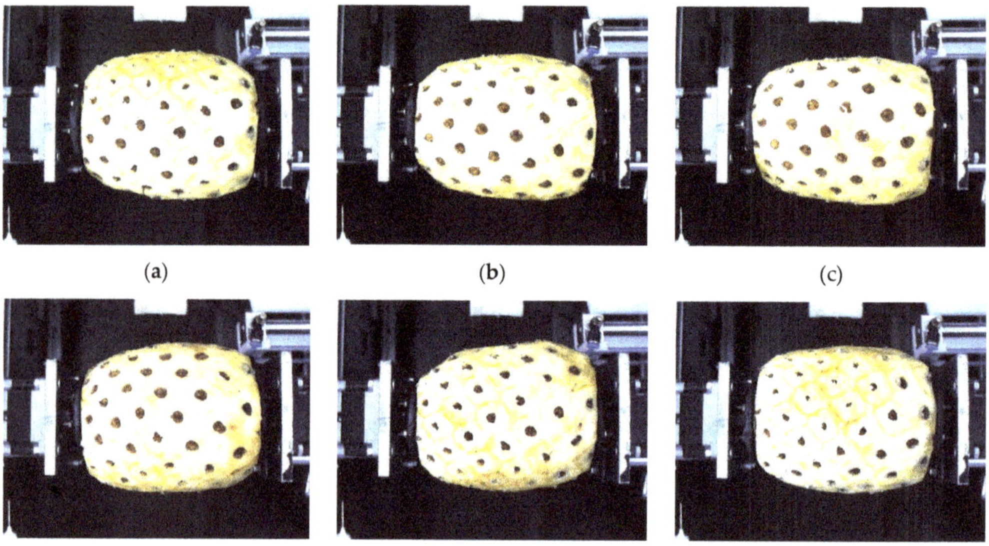

(a) (b) (c)

(d) (e) (f)

Figure 2. Images of the same pineapple at different angles. (**a**) 0 degrees; (**b**) 60 degrees; (**c**) 120 degrees; (**d**) 180 degrees; (**e**) 240 degrees; (**f**) 300 degrees.

2.3. *Pineapple Eye Recognition Algorithm Based on YOLOv5*

In this paper, YOLOv5 is selected as the target detection network for pineapple eye recognition. Among the commonly used object detection networks, strong detection performance is achieved with the YOLOv5 network [31], which uses mosaic data enhancement, adaptive anchor frame calculation, and adaptive image scaling at the input end. In the backbone network, the features of the target adopted through Focus and CSPNet (cross-stage partial network) can be quickly extracted. In the neck network, FPN (feature pyramid network) and PANet are used for multiscale fusion of the extracted features. GIoU (generalized intersection over union) loss is used as the loss function of the target detection frame in the output end. NMS (nonmaximum suppression) is introduced to filter out the overlapping candidate frames and obtain the best prediction output. These improvements ensure the detection accuracy and speed of small targets and have the advantages of a shallow structure, small weight file, and relatively low requirements for the configuration of the mounted equipment.

There are 4 versions of YOLOv5 [32]: YOLOv5s, YOLOv5m, YOLOv5l, and YOLOv5x. The width and depth of the YOLOv5s model are the initial values. This model is small and fast and is suitable for the detection of small and simple datasets. The YOLOv5m and YOLOv5x models have the deepest depths and are suitable for detection on large and complex datasets. As the depth of the network increases, the detection accuracy is improved, while the detection speed is reduced. In YOLOv5, the learning ability of the neural network improves, the amount of calculation is reduced, and high detection accuracy is maintained. To maximize the detection speed while maintaining sufficient detection accuracy, YOLOv5l is used in this paper as the pineapple eye detection model. The structure of YOLOv5l is shown in Figure 3.

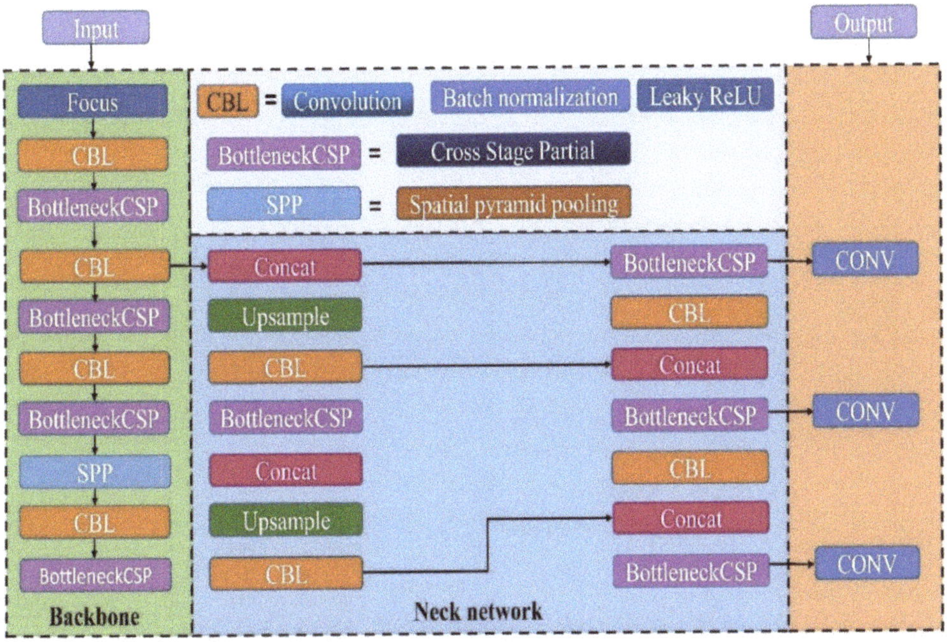

Figure 3. YOLOv5l model structure.

To construct the experimental dataset, 240 pineapple images were obtained from 40 pineapples. Then, the image was processed with data enhancements, such as rotation and horizontal and vertical mirror images, to improve the robustness of the recognition mode, and 600 pineapple images were finally obtained, with a total of approximately 18,000 pineapple eyes. The pineapple eye images were manually labeled one by one by labeling software. Pineapple eyes in the image were marked with a rectangular box and then named P. The labeling information was stored in the PASCALVOC (Pattern Analysis, Statical Modeling and Computational Learning, Visual Object Classes) format [33], in which the coordinates, labels, and serial numbers of each box are contained. The pineapple eye image, labeled data, and other files were saved according to the PASCALVOC dataset directory structure to build the pineapple eye dataset.

The 600 pineapple eye images enhanced by the dataset were divided into a training set, validation set, and test set at an 8:1:1 ratio. Because the size of the pineapple eye target is small, to improve the detection accuracy, the input size is 640 × 640 pixels, 32 images were taken as a batch, and the weight parameters were updated once for each batch of images trained.

YOLOv5 incorporates the current mainstream detection approach FPN (feature pyramid network) [34] and inherits the grid generation idea of the YOLO algorithm. The 640 × 640 feature plot is divided into grid areas of equal size S × S cells (usually 80 × 80, 40 × 40, or 20 × 20). After maximum suppression, the output end of the network outputs the prediction information of all grid information. The prediction information of each grid includes the classification probability and confidence of the target as well as the center coordinates and length and width of the box surrounding the detection target. The classification probability represents the classification information of the predicted target in the grid region, and the confidence represents the probability of the detection target in the grid region. The central coordinates and length-and-width information of the box represent the specific size and position of the target predicted by the grid.

2.4. Three-Dimensional Positioning Algorithm for Pineapple Eyes

In this paper, images of pineapples are collected using 30° intervals; obviously, the same pineapple eye appears on multiple consecutive images. By analyzing these images and matching the same pineapple eyes in different images, the parallax information of the pineapple eye can be obtained. The depth information of the pineapple eye can be obtained through triangulation. In this paper, two images with an angle difference of 90° are used as a group for stereo matching analysis to obtain the three-dimensional position information of all pineapple eyes. Considering the high similarity of pineapple images from different angles, the traditional stereo-vision-matching algorithm is not expected to perform well. In addition, a large amount of calculation is required in this algorithm, which also has low efficiency. Therefore, this algorithm is not suitable for the needs of actual production. Figure 4a,b show the comparison of the γ degree and γ + 90-degree pineapple eye images. Here, one pineapple eye appears in both images.

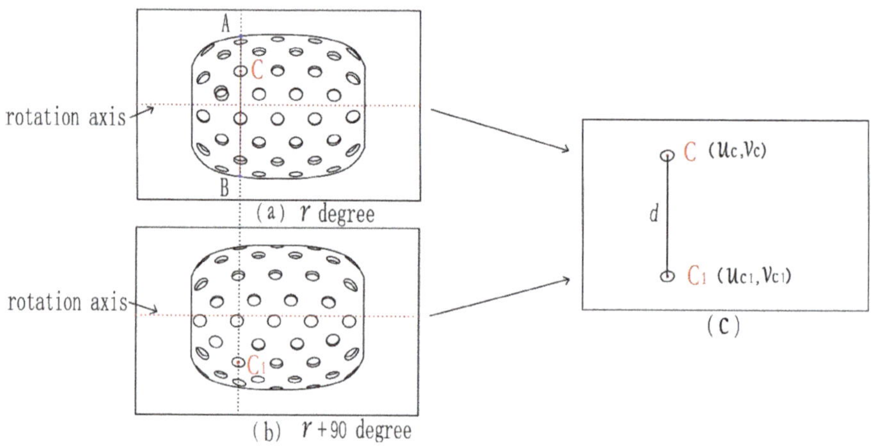

Figure 4. Epipolar constrained stereo matching. (**a**) γ degree; (**b**) γ + 90 degree; (**c**) calculate d of distance schematic diagram.

The central coordinates (u_c, v_c) and (u_{c1}, v_{c1}) are used to describe the position of the pineapple eye in the two images. Therefore, the position of the pineapple eye in the two images must satisfy the following two constraints: (1) the center of the pineapple eye is located on the same vertical line in the two images, that is, $v_{c1} = v_c$. (2) The row coordinates of the center of the pineapple eye on the two images can be predicted by the displacement of the center of the pineapple eye after the pineapple is rotated by 90°, namely, $u_{c1} = u_c + d$.

In order to obtain the value d in Figure 4c, Figure 5 is used to describe the solution process in detail, f is the focal length, and S is the distance between point O and point p, the optical center of the camera; R is the radius on the contour of the pineapple cross-section through C of the pineapple eye. We can obtain Formula (1).

$$\begin{cases} \eta = arc\tan\left(\frac{l_0 d_x}{f}\right) \\ R = S\sin\eta \\ \frac{l_1 d_x}{R\sin r} = \frac{f}{S - R\cos r} \\ d = d_1 + d_2 = R\sin r + R\cos r \end{cases} \quad (1)$$

where d_x represents the physical size of a pixel on the u-axis, which is 0.00465 mm in this paper, η is the angle between the Op and the Ap, and r is the OG and the OC_1.

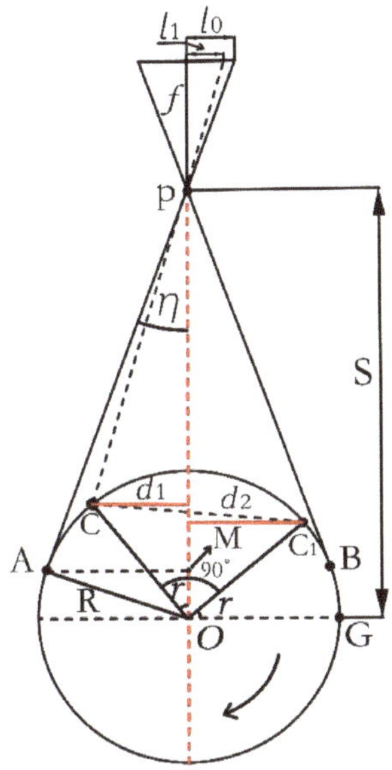

Figure 5. Schematic diagram of the 90° rotation distance of the pineapple eye center point.

Since the contour of the pineapple cross section through C of the pineapple eye is not an ideal circle, and due to system errors such as installation and imaging, u_{c1} and v_{c1} cannot fully meet the above constraints; therefore, a certain tolerance Δ is added when finding a matching pineapple eye in the γ + 90 degree image. In other words, we search for the target pineapple eye within the rectangular box ($u_{c1} - \Delta$,$v_{c1} - \Delta$,$u_{c1} + \Delta$,$v_{c1} + \Delta$). To ensure that there is only one pineapple eye in the rectangular box, Δ is set to a third of the minimum distance between the two pineapple eyes in the image. Obviously, according to the above constraints, the pineapple eye below the rotation axis in Figure 4a is not found in Figure 4b, so there is no need to perform a matching operation.

In this paper, a 3D localization algorithm for pineapple eyes based on monocular multiangle image matching is proposed. After obtaining the image coordinates of the same pineapple eye in two images with a difference of 90°, the depth of the pineapple eye is calculated by triangulation. The information is then used to calculate the three-dimensional position information of the pineapple eye. The camera coordinate system $O_c\_X_cY_cZ_c$ is established with the camera optical center as the origin, as shown in Figure 6. The center point C of any pineapple eye is selected as the measurement object. (u_c, v_c) represents the pixel value of pineapple eye center C under the imaging plane, O_1 is the intersection of the imaging plane of the pineapple eye center point C and the camera optical axis, and the pixel value is (u_0, v_0).

The circle in Figure 7 is the cross-sectional profile of the pineapple through point C. ψ is the angle between the line segment OC and the optical axis of the camera, which satisfies the formula $\psi = \arctan\left(\frac{h_1}{h_2}\right)$. The pineapple is rotated clockwise in the direction indicated by the arrow in the figure. p refers to the optical center of the camera. The distance between point C and point p of the camera optical center is W, the distance between point O and point p of the camera optical center is S, and l_1 is the number of pixels in the axial direction of the pineapple eye imaging plane. When the pineapple rotates clockwise by 90°, which is equivalent to a 90° counterclockwise rotation of the camera, as shown in the dotted line in Figure 8, l_2 is the number of pixels in the axial direction of the pineapple eye imaging plane after rotation. The following formula can be obtained from Figure 7:

$$\begin{cases} \alpha = \arctan\left(\frac{l_1 \times d_x}{f}\right) \\ \beta = \arctan\left(\frac{l_2 \times d_x}{f}\right) \end{cases} \quad (2)$$

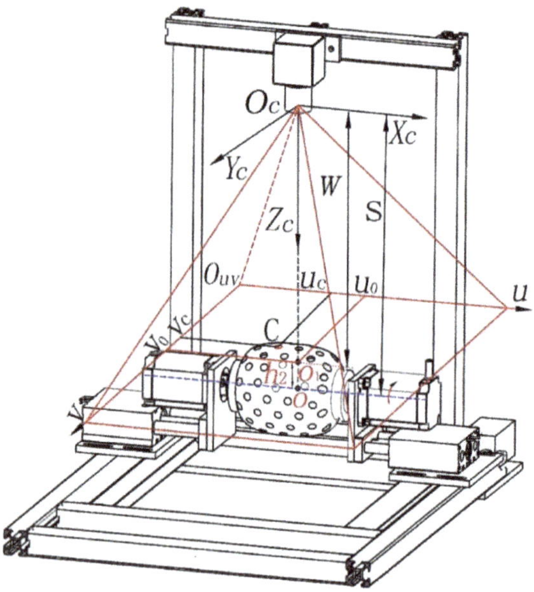

Figure 6. Camera coordinate system for the pineapple eye.

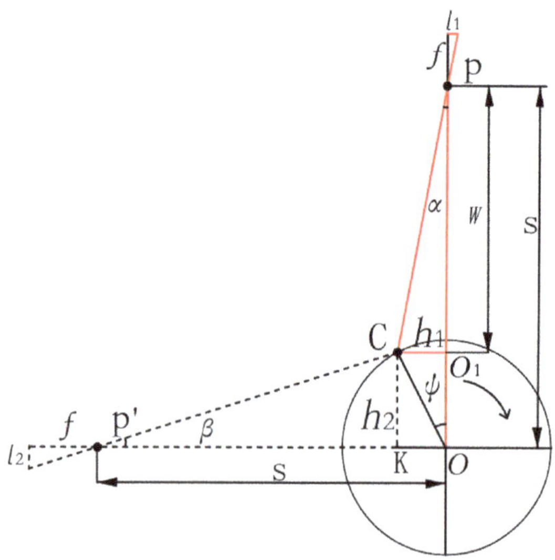

Figure 7. Schematic diagram of the pineapple eye depth information calculation.

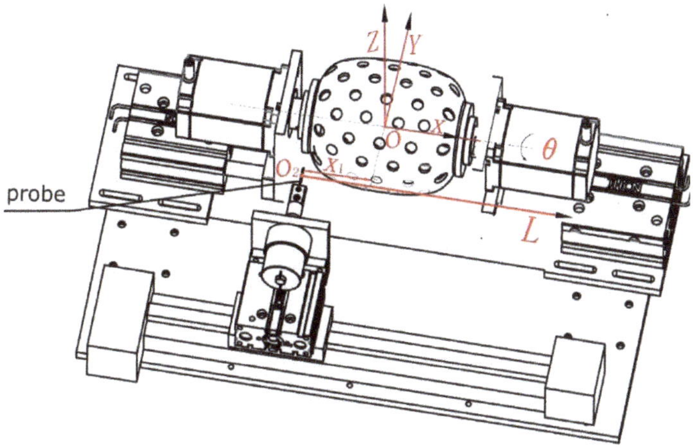

Figure 8. Three-dimensional positioning schematic diagram.

In this formula, d_x represents the physical size of a pixel on the u-axis, which is 0.00465 mm in this paper. α and β can be solved by using Formula (2), and h_1 and h_2 in Figure 7 can be simultaneously solved according to the following equations.

$$\begin{cases} h_1 = \frac{S(1-tan(\beta))tan(\alpha)}{1-tan(\alpha)tan(\beta)} \\ h_2 = \frac{S(1-tan(\alpha))tan(\beta)}{1-tan(\alpha)tan(\beta)} \end{cases} \quad (3)$$

In Formula (3), h_1 is the distance between point C and point O_1, mm. h_2 is the distance between point C and point K, mm. Figure 7 shows that the center point C of the pineapple eye is imaged at time t, and the object distance of the imaging plane is $W = S - h_2$. Then, the number of pixels of pineapple eye point C on the imaging plane and in the camera coordinate system are determined using the following equation:

$$\begin{cases} \frac{d_x(u_c-u_0)}{f} = \frac{X_c}{W} \\ \frac{d_x(v_c-v_0)}{f} = \frac{Y_c}{W} \end{cases} \quad (4)$$

In other words, at time t, the center point C of the pineapple eye fulfills the matrix in the camera coordinate system, with the camera optical center serving as the origin:

$$\begin{bmatrix} X_c \\ Y_c \\ Z_c \end{bmatrix} = W \begin{bmatrix} u_c & -u_0 & 0 \\ v_c & -v_0 & 0 \\ 0 & 0 & 1 \end{bmatrix} \begin{bmatrix} \frac{d_x}{f} \\ \frac{d_x}{f} \\ 1 \end{bmatrix} \quad (5)$$

To facilitate subsequent experiments and the operation of removing pineapple eyes in practical engineering applications, the pineapple three-dimensional space coordinate with O as the center is established O_XYZ. The geometric vector approach is used to translate the camera coordinates into the 3D space coordinates of the pineapple eye, as shown in Figure 8.

$$\begin{cases} X = -X_c \\ Y = Y_c \times \cos(-\alpha) + (S - Z_c) \times \sin(-\alpha) \\ Z = S - ((S - Z_c) \times \cos(\alpha) + Y_c \times \sin(\alpha)) \end{cases} \quad (6)$$

Furthermore, the three-dimensional coordinates of the pineapple eye (X, Y, Z) are converted to the probe, which can be used for eye removal after changing to the eye-removal tool. Position L and pineapple rotation angle θ are represented by the space coordinates

(L, θ), and O_2 is the starting point of the probe. As shown in Figure 9, the corresponding conversion formula is as follows:

$$\begin{cases} L = X_1 - X \\ \theta = \frac{Y}{S-Z} \times 180/PI \end{cases} \tag{7}$$

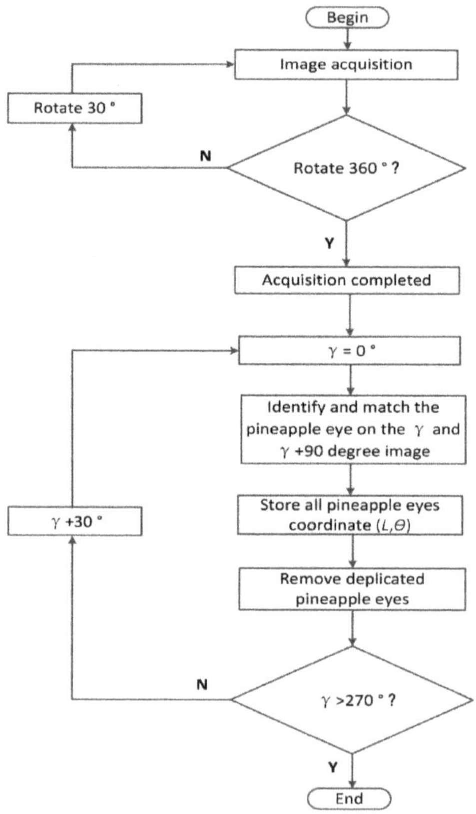

Figure 9. Flow diagram of the 3D positioning method for pineapple eyes.

In Equation (7), X_1 is the distance from the optical center of the horizontal axis camera to the starting point of the probe. Because the pineapple is rotating during the image acquisition process, all the calculated coordinates of the pineapple eyes are the result of the calculation of the current pineapple angle conditions. To obtain the coordinates of all pineapple eyes for the whole pineapple in the same coordinate space, we should reverse rotate the coordinates of all pineapple eyes to the 0° position. Therefore, Formula (7) should be modified to the following:

$$\begin{cases} L = X_1 - X \\ \theta = \frac{Y}{S-Z} \times \frac{180}{PI} - \gamma \end{cases} \tag{8}$$

The position information of all pineapple eyes can be obtained after image stereo matching and pineapple eye position computation. To ensure that the position information of each pineapple eye is calculated, the image acquisition angle interval is set to 30 degrees, which leads to the same pineapple eye being calculated in multiple sets of images. This results in more calculated pineapple eyes than the actual number of pineapple eyes. To avoid the same pineapple eye being repeatedly calculated, a successful match of a pineapple eye in the image is marked. When using the image and the next picture, the marked pineapple eye does not participate in the matching calculation.

2.5. Flow Diagram of 3D Positioning Algorithm

The flow diagram of the 3D positioning algorithm for pineapple eyes based on multiangle image stereo matching in the study is shown in Figure 9. It mainly includes all pineapple eye image acquisition to identify and match the pineapple eye on the γ and γ + 90-degree image. When matching images on the γ degree and γ + 90 degree, all the pineapple eye coordinates (L, θ) are stored in a list. When matching the next set of images (γ + 30 degree and γ + 120 degree), some pineapple eyes which are duplicated with the previous set of images will inevitably be obtained. Because the pineapple eye coordinate (L, θ) is a global coordinate, the coordinates (L, θ) are approximate. By comparing the newly obtained pineapple eye coordinates with the pineapple eye coordinates stored in the list, it is easy to find and eliminate duplicate pineapple eyes. In this paper, the Euclidean distance judgment is used as the judgment basis; when the distance between the two pineapple eyes is less than 1 mm, the two pineapple eyes are considered to be duplicate pineapple eyes.

2.6. Probe Positioning Test

In this paper, a probe test method is proposed for evaluating the positioning accuracy of the positioning system. The probe mounted on the linear slide, as illustrated in Figure 10, may be accurately moved and positioned in the direction of the pineapple axis. At the same time, the servo drive motor rotates the pineapple at a precise angle. Therefore, according to the coordinates (L, θ) of any pineapple eye, the probe can be moved to the position of the pineapple eye and inserted into the pineapple eye through the extension action of the probe cylinder. The deviation *er* (error) between the actual center of the pineapple eye and the probe penetration position can be calculated to evaluate the positioning accuracy of the pineapple eye:

$$er = \sqrt{(W_2/2 - W_1 - 0.99)^2 + (H_2/2 - H_1 - 0.99)^2} \tag{9}$$

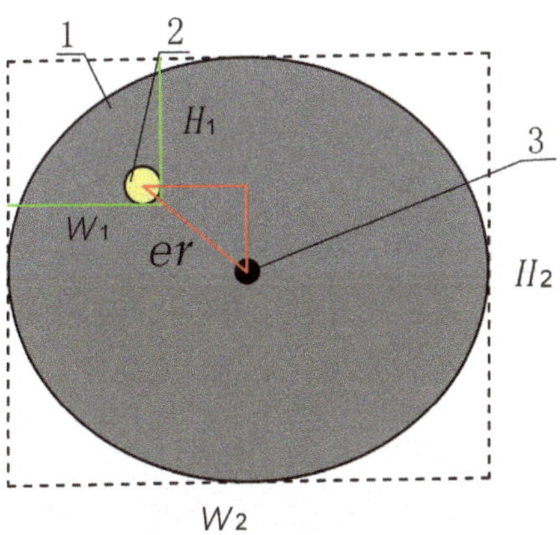

Figure 10. Measurement principle of the probe position error. 1. pineapple eye, 2. probe, and 3. pineapple eye center point.

In Equation (9), *er* is the error, and W_1 is the distance between the left edge of the pineapple eye and the right edge of the probe, in mm. W_2 is the maximum length of the pineapple eye in the horizontal direction, in mm. H_1 is the distance between the upper edge of the pineapple eye and the lower edge of the probe, in mm. H_2 is the maximum length of the pineapple eye in the vertical direction, in mm. The probe radius is 0.99 mm.

Using five Goodfarmer Philippine pineapples, the diameter of the pineapple eye was 9–12 mm (manual measurement) after manual peeling. The positioning test is carried out on the built-in test platform. When the probe reaches each pineapple eye position, a Vernier caliper is used to successively measure the distances W_1, W_2, H_1, and H_2, as shown in Figure 11.

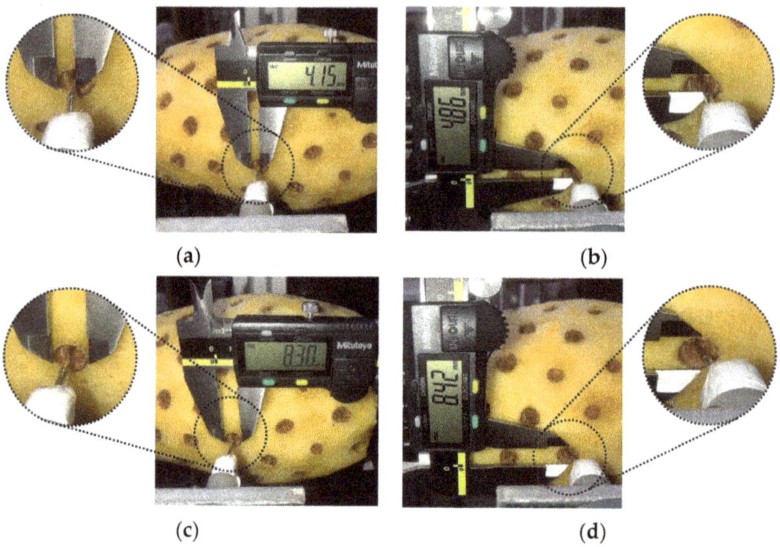

Figure 11. Measuring the pineapple eye error with a Vernier caliper. (**a**) W_1 measurement; (**b**) H_1 measurement; (**c**) W_2 measurement; (**d**) H_2 measurement.

3. Results and Discussion

3.1. YOLOv5 Model Performance Evaluation

To evaluate the detection effect of the pineapple eye recognition model, the model recognition accuracy and detection efficiency are mainly measured from four parameters: recall (*R*), precision (*P*), average accuracy (*AP*), and detection time of a single pineapple eye.

$$\begin{cases} P = \frac{TP}{TP+FP} \\ R = \frac{TP}{TP+FN} \\ AP = \int_0^1 PdR \end{cases} \tag{10}$$

The *AP* value in Formula (10) is the area between the *P*–*R* curve and the coordinate axis, *TP* represents the number of positive samples (pineapple eyes) correctly predicted as positive samples, *TN* denotes the number of negative samples correctly predicted as negative samples, *FP* indicates the number of negative samples predicted as positive samples, and *FN* suggests the number of positive samples predicted as negative samples.

The curve of network model training is shown in Figure 12. Figure 12a shows the loss function curve of training, with a minimum value of 0.01689. Figure 12b shows the accuracy *P* (precision) curve, and the maximum accuracy is 97.8%. Figure 12c shows the recall rate *R* (recall) curve, and the maximum recall rate is 97.5%. Figure 12d shows the mean average precision curve when the IOU threshold is set to 0.5.

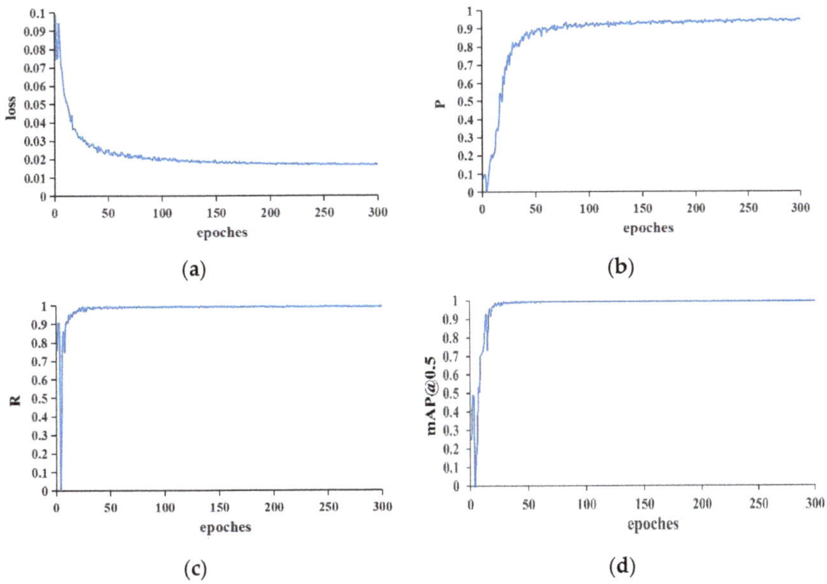

Figure 12. Model training results. (**a**) Value of loss varies with the number of iterations; (**b**) P vary with the number of iterations; (**c**) R vary with the number of iterations; (**d**) mAP@0.5 vary with the number of iterations.

The *P–R* curve is a graph that depicts the relationship between precision and recall. The abscissa represents *R*, while the ordinate represents *P*. The region contained in the *P–R* curve and the coordinate axis is *AP*. The larger the area between the curve and the coordinate axis is, the better the model recognition effect. Figure 13 shows the *P–R* curve with a threshold of 0.5 generated in the training process. Since there is only one recognition target in this paper, the *AP* is equal to the mAP (mean Average Precision). The mAP is 99.2%.

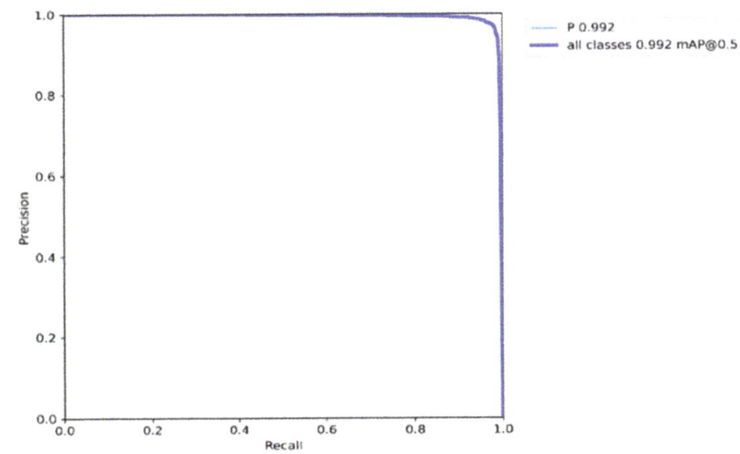

Figure 13. *P–R* curve.

To further verify the YOLOv5l model performance for pineapple eyes, the YOLOv5l network was compared with YOLOv5s, YOLOv5m, and YOLOv5x on 60 images in the test set; the target distribution of the test set was actually 1806 pineapple eyes. Then, the test set images were input into the above models, respectively. The target recognition results of the pineapple eyes in the image samples of the test set by the model are shown in Table 1. The YOLOv5 (l, s, m, and x) values of mAP at a confidence of 0.5 were 98%, 97.6%, 97.8%, and 98%, respectively, showing the effectiveness of the proposed model. Additionally, the average times required to detect one pineapple eye image were 0.015 s, 0.012 s, 0.019 s, and 0.024 s, respectively. Figure 14 shows the YOLOv5l detection effect diagram with a confidence level greater than 0.5.

Table 1. Identification results for the pineapple eyes in test set.

Models	Precision (%)	Recall (%)	mAP (%)	Average Time(s)
YOLOv5l	98.0	96.6	98.0	0.015
YOLOv5s	98.3	96.2	97.6	0.012
YOLOv5m	97.9	96.3	97.8	0.019
YOLOv5x	98.1	96.5	98.0	0.024

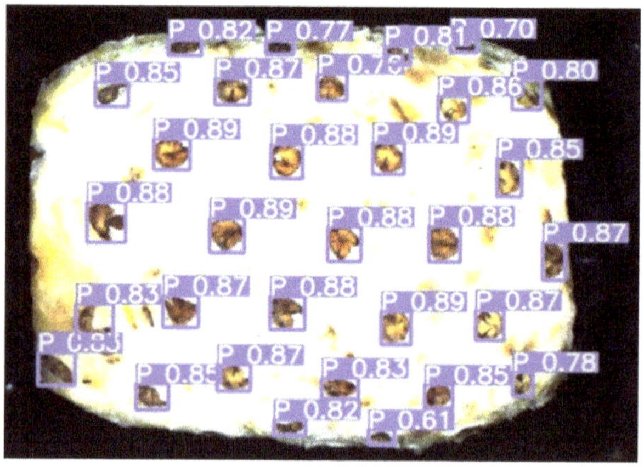

Figure 14. YOLOv5l detection effect diagram.

Average time is the time to detect one pineapple eye image.

In order to further analyze the accuracy of the YOLOv5l model in pineapple eye image detection, the training results of YOLOv5l and the target detection model Mask R-CNN were compared with a threshold of 0.5, as shown in Table 2. As can be seen from Table 2, the mAP and detection speed of YOLOv5l are significantly higher than Mask R-CNN.

Table 2. Comparison models of YOLOv5l and Mask R-CNN.

Models	mAP (%)	Average Time (s)
YOLOv5l	99.2	0.015
Mask R-CNN	97.5	0.021

3.2. Result of Probe Positioning Test

The probe positioning test result, as shown in Figure 15, reveals that of the five Goodfarmer Philippine pineapples after manual peeling (460 pineapple eyes in total, 444 pineapple eyes were successfully recognized), the deviation between the actual center of the pineapple eye and the probe puncture position was 1.01 mm, and the maximum was 2.17 mm, with a root mean square value of 1.09 mm.

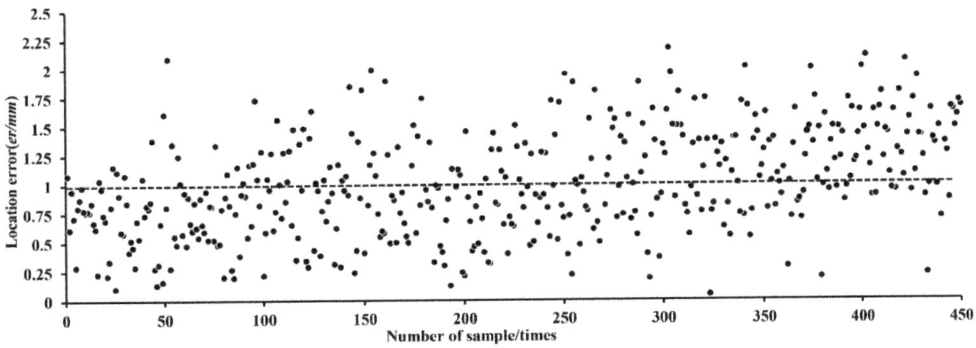

Figure 15. Probe positioning test.

3.3. Discussion

The YOLOv5 model has high detection accuracy on the self-built pineapple eye dataset. In the sample images of the whole test set, the accuracy, recall, and *AP* of the model are higher than 96%, indicating that the YOLOv5 recognition algorithm is feasible. The reason why a few pineapple eyes could not be successfully identified is that the pineapple eyes on both sides of the image are prone to distortion. This situation increases the recognition difficulty, resulting in some pineapple eye recognition errors. Therefore, further research on the optimization methods of models and parameters is needed to improve detection accuracy.

The localization experiment demonstrates that collecting images of the entire pineapple circumference at even intervals and employing multiangle image matching with high positioning precision may effectively accomplish three-dimensional localization of the pineapple eye. Simultaneously, pineapple eye coordinates have been converted into a form that can be directly applied by the actuator, which provides a good foundation for the further development of pineapple eye-removal equipment for practical operations.

4. Conclusions

A pineapple eye recognition algorithm was presented based on deep learning. YOLOv5 was used as the target detection network for pineapple eye recognition. The 600 pineapple eye images enhanced by the dataset are divided into a training set, validation set, and test set with an 8:1:1 ratio. The values in the final model validation of precision, recall, and mAP (mean average precision) were 97.8%, 97.5%, and 99.2%, respectively. The YOLOv5l network was compared with YOLOv5s, YOLOv5m, and YOLOv5x on 60 images in the test set. The YOLOv5 (l, s, m, and x) values of mAP were 98%, 97.6%, 97.8%, and 98%, showing the effectiveness of the proposed model. Additionally, the average times required to detect one pineapple eye image were 0.015 s, 0.012 s, 0.019 s, and 0.024 s. The detection results of YOLOv5l and Mask R-CNN were further compared, and the results showed that YOLOv5l was significantly higher than that of Mask R-CNN in both the mAP and detection speed.

A pineapple eye location algorithm based on monocular multiangle image stereo matching was proposed. Two images with different angles of 90° were selected as a group for stereo-matching analysis to obtain the three-dimensional position information of all pineapple eyes, establish a camera three-dimensional coordinate system with the camera optical center as the origin, and obtain the three-dimensional space coordinates (X, Y, Z) of the all pineapple eye through the geometric vector method. To facilitate subsequent experiments and the operation of removing pineapple eyes in practical engineering applications, in this paper, the three-dimensional space coordinate (X, Y, Z) of the pineapple eye was transformed into the space coordinate (L, θ) with the probe (or eye-removal tool) position L and the rotation angle θ of the pineapple as the reference. The probe test results showed that the average deviation between the actual center of the pineapple eye and the puncture position of the probe was

1.01 mm, the maximum was 2.17 mm, the root mean square value was 1.09 mm, and the positioning accuracy met the needs of the automated eye-removal operations.

The pineapple eye recognition and positioning algorithm proposed in this paper provides an important theoretical basis for the development of automatic pineapple-eye-removal equipment. The practical application performance of the algorithm needs to be verified and improved in the actual eye-removal operation. At the same time, only one variety of pineapple was tested, and the peeling operation was performed manually. The applicability of the algorithm to different varieties of pineapples and machine-peeled pineapples also needs to be further verified.

Author Contributions: Conceptualization, A.L., Y.X. and Y.L.; methodology, A.L., Y.X. and Y.L.; software, Y.L.; validation, A.L., Y.L., Z.H. and X.D.; formal analysis, A.L.; investigation, A.L., Z.H., X.L. and Z.T.; resources, Y.X. and Y.L.; data curation, A.L.; writing—original draft preparation, A.L.; writing—review and editing, Y.X. and Y.L.; visualization, A.L.; supervision, X.L.; project administration, A.L.; funding acquisition, Y.X. All authors have read and agreed to the published version of the manuscript.

Funding: This research was funded by the Natural Science Foundation of Hunan Province of China, grant number 2021JJ30363.

Institutional Review Board Statement: Not applicable.

Data Availability Statement: All data are presented in this article in the form of figures and tables.

Acknowledgments: We gratefully acknowledge Mingliag Wu, Ying Xiong and anonymous referees for thoughtful review of this research as well as the assistance of Yanfei Li with statistical analyses.

Conflicts of Interest: The authors declare no conflict of interest.

References

1. Jin, Y. Pineapple Market and Industry Investigation and Analysis Report. *Agric. Prod. Mark.* **2021**, *8*, 46–47.
2. Gong, Y. Research on Strategies for Optimization and Upgrading of Pineapple Industry in Zhanjiang. Master's Thesis, Guangdong Ocean University, Zhanjiang, China, 2020.
3. Jia, W.; Zhang, Z.; Shao, W.; Hou, S.; Ji, Z.; Liu, G.; Yin, X. FoveaMask: A Fast and Accurate Deep Learning Model for Green Fruit Instance Segmentation. *Comput. Electron. Agric.* **2021**, *191*, 106488. [CrossRef]
4. Li, B.; Ning, W.; Wang, M.; Li, L. In-Field Pineapple Recognition Based on Monocular Vision. *Trans. Chin. Soc. Agric. Eng.* **2010**, *26*, 345–349. [CrossRef]
5. Lin, G.; Zou, X. Citrus Segmentation for Automatic Harvester Combined with AdaBoost Classifier and Leung-Malik Filter Bank. *IFAC-Pap.* **2018**, *51*, 379–383. [CrossRef]
6. Lv, J.; Wang, F.; Xu, L.; Ma, Z.; Yang, B. A Segmentation Method of Bagged Green Apple Image. *Sci. Hortic.* **2019**, *246*, 411–417. [CrossRef]
7. Kurtulmus, F.; Lee, W.S.; Vardar, A. Green Citrus Detection Using "Eigenfruit", Color and Circular Gabor Texture Features under Natural Outdoor Conditions. *Comput. Electron. Agric.* **2011**, *78*, 140–149. [CrossRef]
8. Wang, D.; He, D. Fusion of Mask RCNN and Attention Mechanism for Instance Segmentation of Apples under Complex Background. *Comput. Electron. Agric.* **2022**, *196*, 106864. [CrossRef]
9. Kasinathan, T.; Singaraju, D.; Uyyala, S.R. Insect Classification and Detection in Field Crops Using Modern Machine Learning Techniques. *Inf. Process. Agric.* **2021**, *8*, 12. [CrossRef]
10. Li, H.; Zhang, M.; Gao, Y. Green ripe tomato detection method based on machine vision in greenhouse. *Trans. Chin. Soc. Agric. Eng.* **2017**, *33*, 328–334+388.
11. Wang, C.; Zou, X.; Tang, Y.; Luo, L.; Feng, W. Localisation of Litchi in an Unstructured Environment Using Binocular Stereo Vision. *Biosyst. Eng.* **2016**, *145*, 39–51. [CrossRef]
12. Zhao, Y.; Gong, L.; Zhou, B.; Huang, Y.; Liu, C. Detecting Tomatoes in Greenhouse Scenes by Combining AdaBoost Classifier and Colour Analysis. *Biosyst. Eng.* **2016**, *148*, 127–137. [CrossRef]
13. Altaheri, H.; Alsulaiman, M.; Muhammad, G. Date Fruit Classification for Robotic Harvesting in a Natural Environment Using Deep Learning. *IEEE Access* **2019**, *7*, 117115–117133. [CrossRef]
14. Koirala, A.; Walsh, K.B.; Wang, Z.; McCarthy, C. Deep Learning—Method Overview and Review of Use for Fruit Detection and Yield Estimation. *Comput. Electron. Agric.* **2019**, *162*, 219–234. [CrossRef]
15. Lv, J.; Xu, H.; Han, Y.; Lu, W.; Xu, L.; Rong, H.; Yang, B.; Zou, L.; Ma, Z. A Visual Identification Method for the Apple Growth Forms in the Orchard. *Comput. Electron. Agric.* **2022**, *197*, 106954. [CrossRef]

16. Zhang, X.; Gao, Q.; Pan, D. Picking recognition research of pineapple in complex field environment based on improved YOLOv3. *J. Chin. Agric. Mech.* **2021**, *42*, 201–206. [CrossRef]
17. Tian, Y.; Yang, G.; Wang, Z.; Wang, H.; Li, E.; Liang, Z. Apple Detection during Different Growth Stages in Orchards Using the Improved YOLO-V3 Model. *Comput. Electron. Agric.* **2019**, *157*, 417–426. [CrossRef]
18. Yu, Y.; Zhang, K.; Yang, L.; Zhang, D. Fruit Detection for Strawberry Harvesting Robot in Non-Structural Environment Based on Mask-RCNN. *Comput. Electron. Agric.* **2019**, *163*, 104846. [CrossRef]
19. Zhang, C.; Ding, H.; Shi, Q.; Wang, Y. Grape Cluster Real-Time Detection in Complex Natural Scenes Based on YOLOv5s Deep Learning Network. *Agriculture* **2022**, *12*, 1242. [CrossRef]
20. Ji, W.; Meng, X.; Qian, Z.; Xu, B.; Zhao, D. Branch Localization Method Based on the Skeleton Feature Extraction and Stereo Matching for Apple Harvesting Robot. *Int. J. Adv. Robot. Syst.* **2017**, *14*, 1729881417705276. [CrossRef]
21. Rong, X.; Jiang, H.; Ying, Y. Recognition of Clustered Tomatoes Based on Binocular Stereo Vision. *Comput. Electron. Agric.* **2014**, *106*, 75–90. [CrossRef]
22. Wang, C.; Tang, Y.; Zou, X.; Luo, L.; Chen, X. Recognition and Matching of Clustered Mature Litchi Fruits Using Binocular Charge-Coupled Device (CCD) Color Cameras. *Sensors* **2017**, *17*, 2564. [CrossRef]
23. Ge, L.; Yang, Z.; Sun, Z.; Zhang, G.; Zhang, M.; Zhang, K.; Zhang, C.; Tan, Y.; Li, W. A Method for Broccoli Seedling Recognition in Natural Environment Based on Binocular Stereo Vision and Gaussian Mixture Model. *Sensors* **2019**, *19*, 1132. [CrossRef]
24. Luo, W.; Schwing, A.G.; Urtasun, R. Efficient Deep Learning for Stereo Matching. In Proceedings of the IEEE Conference on Computer Vision & Pattern Recognition, Las Vegas, NV, USA, 27–30 June 2016; pp. 5695–5703.
25. Guo, H.; Liu, Y.F.; Wang, Y.; Shen, X. Calibration of Binocular Vision Measurement of Large Gear Workpiece Welding. *J. Donghua Univ. Sci.* **2013**, *4*, 455–459.
26. Zhang, B.; Huang, W.; Wang, C.; Gong, L.; Zhao, C.; Liu, C.; Huang, D. Computer Vision Recognition of Stem and Calyx in Apples Using Near-Infrared Linear-Array Structured Light and 3D Reconstruction. *Biosyst. Eng.* **2015**, *139*, 25–34. [CrossRef]
27. Hongsheng, S.; Zhenwei, W.; Hong, C. Three-Dimensional Reconstruction of Complex Spatial Surface Based on Line Structured Light. In Proceedings of the IECON 2021—47th Annual Conference of the IEEE Industrial Electronics Society, Toronto, ON, Canada, 13–16 October 2021; pp. 1–5.
28. Chen, C.; Tian, Y.; Lin, L.; Chen, S.; Li, H.; Wang, Y.; Su, K. Obtaining World Coordinate Information of UAV in GNSS Denied Environments. *Sensors* **2020**, *20*, 2241. [CrossRef]
29. Zhao, D.-A.; Lv, J.; Ji, W.; Zhang, Y.; Chen, Y. Design and Control of an Apple Harvesting Robot. *Biosyst. Eng.* **2011**, *110*, 112–122. [CrossRef]
30. Tsai, R. A Versatile Camera Calibration Technique for High-Accuracy 3D Machine Vision Metrology Using off-the-Shelf TV Cameras and Lenses. *IEEE J. Robot. Autom.* **1987**, *3*, 323–344. [CrossRef]
31. Olenskyj, A.G.; Sams, B.S.; Fei, Z.; Singh, V.; Raja, P.V.; Bornhorst, G.M.; Earles, J.M. End-to-End Deep Learning for Directly Estimating Grape Yield from Ground-Based Imagery. *Comput. Electron. Agric.* **2022**, *198*, 107081. [CrossRef]
32. Wu, T.-H.; Wang, T.-W.; Liu, Y.-Q. Real-Time Vehicle and Distance Detection Based on Improved Yolo v5 Network. In Proceedings of the 2021 3rd World Symposium on Artificial Intelligence (WSAI), Guangzhou, China, 18–20 June 2021; pp. 24–28.
33. Zhou, X.; Wei, G.; Fu, W.L.; Du, F. Application of Deep Learning in Object Detection. In Proceedings of the 2017 IEEE/ACIS 16th International Conference on Computer and Information Science (ICIS), Wuhan, China, 24–26 May 2017.
34. Xu, X.; Zhang, X.; Zhang, T. Lite-YOLOv5: A Lightweight Deep Learning Detector for On-Board Ship Detection in Large-Scene Sentinel-1 SAR Images. *Remote Sens.* **2022**, *14*, 1018. [CrossRef]

agriculture

Article

Motion Planning of the Citrus-Picking Manipulator Based on the TO-RRT Algorithm

Cheng Liu [1], Qingchun Feng [2,*], Zuoliang Tang [1], Xiangyu Wang [3], Jinping Geng [1] and Lijia Xu [1,*]

1 College of Mechanical and Electrical Engineering, Sichuan Agricultural University, Ya'an 625014, China; 2020317017@stu.sicau.edu.cn (C.L.); zuoliang_tang@stu.sicau.edu.cn (Z.T.); 2020217010@stu.sicau.edu.cn (J.G.)
2 Intelligent Equipment Research Center, Beijing Academy of Agriculture and Forestry Sciences, Beijing 100097, China
3 Institute of System Science and Technology, School of Electrical Engineering, Southwest Jiaotong University, Chengdu 611756, China; wxy1998@my.swjtu.edu.cn
* Correspondence: fengqc@nercita.org.cn (Q.F.); xulijia@sicau.edu.cn (L.X.)

Abstract: The working environment of a picking robot is complex, and the motion-planning algorithm of the picking manipulator will directly affect the obstacle avoidance effect and picking efficiency of the manipulator. In this study, a time-optimal rapidly-exploring random tree (TO-RRT) algorithm is proposed. First, this algorithm controls the target offset probability of the random tree through the potential field and introduces a node-first search strategy to make the random tree quickly escape from the repulsive potential field. Second, an attractive step size and a "step-size dichotomy" are proposed to improve the directional search ability of the random tree outside the repulsive potential field and solve the problem of an excessively large step size in extreme cases. Finally, a regression superposition algorithm is used to enhance the ability of the random tree to explore unknown space in the repulsive potential field. In this paper, independent experiments were carried out in MATLAB, MoveIt!, and real environments. The path-planning speed was increased by 99.73%, the path length was decreased by 17.88%, and the number of collision detections was reduced by 99.08%. The TO-RRT algorithm can be used to provide key technical support for the subsequent design of picking robots.

Keywords: picking manipulator; motion planning; TO-RRT; step-size dichotomy; regression superposition

Citation: Liu, C.; Feng, Q.; Tang, Z.; Wang, X.; Geng, J.; Xu, L. Motion Planning of the Citrus-Picking Manipulator Based on the TO-RRT Algorithm. *Agriculture* **2022**, *12*, 581. https://doi.org/10.3390/agriculture12050581

Academic Editor: Wen-Hao Su

Received: 16 March 2022
Accepted: 19 April 2022
Published: 21 April 2022

1. Introduction

Citrus is one of the most economically important crops in the world, and it is also the most cultivated fruit in southwestern China. Currently, citrus fruits are mainly picked manually, which is time-consuming, laborious, and expensive. According to a survey, the labor used in citrus picking operations accounts for $33 \sim 50\%$ of the whole production process. With the sharp decline in the number of rural employees in China, the development of the citrus industry has been severely restricted. To improve the efficiency of picking and enhance the competitiveness of China's citrus industry, both the research and development of citrus-picking robots have become research hotspots at home and abroad, and the path planning of the picking manipulator is one of the most difficult technologies.

In recent years, a series of path-planning methods have been proposed. The artificial potential field (APF) can be used to prevent the manipulator from colliding with obstacles when approaching the target. However, the APF easily falls into a local minimum, and it easily falls into oscillation in a complex environment [1]. Compared to the APF, the rapidly-exploring random tree (RRT) is more adaptable, faster, and more variable, but it is difficult to find the best path when using this approach [2]. Bidirectional RRT and RRT-connect algorithms are used to generate two random trees at the initial node and the target node, respectively, which improves the search speed compared with the RRT algorithm, but the path is still not optimal [3,4]. The RRT-star (RRT*) algorithm is used to make the path gradually converge with the optimum in the search process by reselecting the parent node and rerouting, but its running time is longer than that of the RRT algorithm [5]. Mohammed et al. [6] defined a straight line connecting the initial node and the target node so that the generation probability of the random tree node was normally distributed with the distance from the straight line, preventing excessive searching and avoiding falling into local extreme values. However, the searching ability in a complex environment still needs to be improved. Akgun et al. [7] combined the bidirectional RRT and RRT* algorithms to optimize the search time. Jeong et al. [8] proposed an RRT*-Quick method, which caused the nodes to tend to share the same parent node in a circular (or spherical) neighborhood. While the path generated using this method was smoother than the path generated by the RRT* algorithm, the search time increased slightly. Jeong et al. [9] introduced an informed-RRT algorithm into the RRT*-quick method to limit the sampling space of the random tree and solve the problem of increased search time caused by expanding the search domain in the process of improving the quality of the solution. When the tree nodes reached the maximum, the RRT* Fixed Nodes (RRT*FN) algorithm was used to remove a weak node and add a high-performance node so that the generated tree node was much smaller than the one in the RRT* algorithm. However, this method had little performance gap with the RRT* algorithm before the tree nodes reached the maximum number of nodes [10].

The RRT* algorithm has a strong ability to optimize the path cost, but its search efficiency is low. Cao et al. [11] introduced the target gravity to the RRT algorithm, and the attraction generated by the random node and the attraction generated by the target node were used to jointly guide the generation of new nodes in the random tree. This method improved the search speed of the random tree, but it could not escape the obstacle area quickly when blocked by obstacles. Wang et al. [12] changed the sampling area and assigned node state values so that the random tree could only be expanded through boundary nodes to reduce the generation of invalid nodes, but many redundant nodes were generated near obstacles. Zhang et al. [13] screened new nodes based on a biased-RRT algorithm. If the distance between the new node and the parent node was greater than the distance between the new node and any other nearby node, the new node was discarded. This method can be used to prevent excessive searching of the space and reduce the total number of nodes. Gong et al. [14] made the search direction of the random tree always point to the target node and performed local path planning near the obstacles. Although this method could reduce excessive searching of the space, its escape speed was slow when the random tree was blocked by many large obstacles. Li et al. [15] put forward an adaptive RRT-connect (ARRT-connect), which allowed the random tree to still have good performance in a narrow environment, and path planning could still be completed in a short time. Gao et al. [16] proposed a planning method based on an independent potential field that made the manipulator explore the gradient direction when it was far away from the target and avoided obstacles through the random search. Wang et al. [17] selected tree nodes according to the geometric structure and position of obstacles so that a path with a lower cost could be quickly obtained, but the effect of avoiding obstacles with irregular shapes was poor.

In this paper, based on a citrus tree environment, taking the shortest time as the optimization goal, and taking the Franka manipulator as the experimental platform, the RRT algorithm is improved in multiple dimensions. Its main contributions are as follows:

1. On the basis of the biased-RRT, the potential field function and the adaptive probability threshold are introduced, so that the random tree has corresponding growth strategies in different potential fields. The above strategies improve the directional search ability of random trees in the repulsive potential field and enhance the escape ability of random trees in the repulsive potential field;
2. To solve the problem of "falling into a trap" in the repulsive potential field of random trees, a node-first search strategy is proposed, which makes the selection of extended nodes of random trees more purposeful;

3. Proper step size is crucial to improve search ability. Using an attractive step size is helpful to reduce the number of collision detections and computational complexities outside the repulsive potential field. "Step-size dichotomy" solves the problem of random trees colliding with obstacles many times due to too large of step size in the repulsive potential field;
4. By introducing a regression superposition algorithm, the random tree can avoid over-searching space in the repulsive potential field and enhance the escape ability.

The rest of this article is organized as follows: The basic principles of the RRT algorithm, as well as some improvement methods and the design process of the TO-RRT algorithm, including the adaptive probability threshold, the node-first search strategy, an attractive step size, "step-size dichotomy", and a regression superposition algorithm are introduced in Section 2. In Section 3, the performance of various algorithms in MATLAB, MoveIt!, and the real environment are compared. The main contributions of the article and future work are discussed in Section 4. The full text is summarized in Section 5.

2. Materials and Methods

2.1. RRT Algorithm

The RRT algorithm, which is a spatial search algorithm based on random sampling, aims to generate a collision-free random tree connecting the first and the last positions [18,19].

Each time the *Tree* grows, a random node q_{rand} is generated in the space. Then, the tree node q_{near} closest to q_{rand} is found in the tree, and a new tree node q_{new} is found in the direction of $q_{near} \rightarrow q_{rand}$ with a fixed step λ and is connected to q_{near} as $\overline{q_{near}q_{new}}$. If neither q_{new} nor $\overline{q_{near}q_{new}}$ collide with obstacles, q_{new} and $\overline{q_{near}q_{new}}$ are added to the random tree. After several expansions, if the distance between q_{new} and q_{goal} is less than the given threshold, the *Tree* finds a path connecting q_{init} to q_{goal}, as shown in Algorithm 1.

Algorithm 1. RRT Algorithm.

$Tree \leftarrow q_{init}$
for $i = 1$ to n do
$q_{rand} \leftarrow RandomSample(C_{free})$;
$q_{near} \leftarrow NearestPoint(Tree, q_{rand})$;
$q_{new} \leftarrow Extend(q_{near}, q_{rand}, \lambda)$;
if $CollisionFree(q_{near}, q_{new})$ then
$AddNewPoint(Tree, q_{new})$;
end if
if $Distance(q_{new}, q_{goal}) < \rho_{min}$ then
return *Tree*
end if
end for

2.2. Some Improvement Methods

The RRT algorithm can be used to effectively explore high-dimensional space, but the path cost is high, and the algorithm takes a long time to reach completion. The biased-RRT algorithm can be used to effectively solve the shortcomings of the RRT algorithm [20–23], as shown in Algorithm 2.

Algorithm 2. Biased-RRT Algorithm.

$Tree \leftarrow q_{init}$
for $i = 1$ to n do
if $RandomNumber <= m$ then
$q_{rand} \leftarrow RandomSample(C_{free})$;
else
$q_{rand} \leftarrow q_{goal}$;
end condition
$q_{near} \leftarrow NearestPoint(Tree, q_{rand})$;
$q_{new} \leftarrow Extend(q_{near}, q_{rand}, \lambda)$;
if $CollisionFree(q_{near}, q_{new})$ then
$AddNewPoint(Tree, q_{new})$;
end if
if $Distance(q_{new}, q_{goal}) < \rho_{min}$ then
return *Tree*
end if
end for

In Algorithm 2, q_{rand} is determined by the size between the random number, $RandomNumber$, and the probability threshold m. If $RandomNumber <= m$, q_{rand} takes any point in the space; otherwise, the target node is taken as the sampling point.

The biased-RRT algorithm is used to guide the growth of the random tree, increase the effectiveness of sampling points, and shorten the time of path planning. However, when obstacles obstruct the growth of random trees, the biased-RRT algorithm cannot escape the obstacles quickly. Therefore, some scholars have put forward corresponding solutions, as shown in Table 1.

Table 1. Comparison of RRT improvement methods.

RRT Type	Algorithm Name	Solutions
Biased-RRT	NC-RRT [12]	The random tree search is guided by gradually changing the sampling area, and it is expanded through the boundary nodes as much as possible through the node control mechanism.
Biased-RRT	RRT-BCR [13]	A regression mechanism is introduced to prevent excessive searching, and an adaptive expansion mechanism is introduced to avoid the repeated search of expansion nodes.
RRT*	MOD-RRT* [24]	An initial path planner and a path replanner are proposed. When encountering obstacles, the path replanner selects alternative paths to avoid collision.
P-RRT	PBG-RRT [25]	By giving weights to the goal and random points, the random tree deviates from obstacles.
RRT*	HSRRT* [26]	The random tree is guided to deviate from an obstacle through the APF, and the heuristic sampling scheme of Gaussian function is used to generate sampling points near the obstacle to improve the search efficiency.

Note: NC-RRT, Node Control-RRT; RRT-BCR, Biased-RRT algorithm with boundary expansion mechanism and regression mechanism; MOD-RRT*, multi-objective RRT*; PBG-RRT, rapidly exploring random tree based on heuristic probability bias-goal; HSRRT*, heuristically sampling-based rapidly exploring random tree.

2.3. *TO-RRT Algorithm*

2.3.1. Adaptive Probability Threshold

At present, some improved RRT algorithms have been used to add potential field functions to the target node q_{goal}, random nodes q_{rand}, and obstacles. The random tree changes its growth direction under the action of a combined potential field, which makes it expand to the target when avoiding obstacles [27–33]. This kind of algorithm improves the search efficiency of the random tree, but each expansion of the random tree requires several vector operations of the potential field force, which occupies a large amount of the system memory. In addition, if the repulsive potential field of the obstacles is considered, the algorithm may fall into a local minimum problem, as is common in APF, resulting in q_{near} being unable to generate a new node q_{new}, as shown in Figure 1.

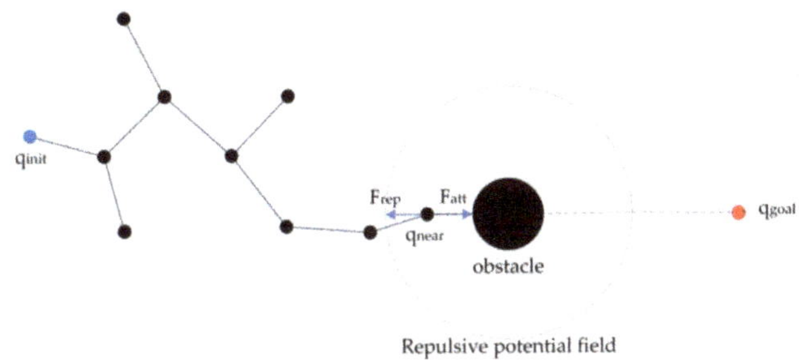

Figure 1. The algorithm falls into a local minimum.

Considering the complexity and uncertainty of the above algorithm in different environments, an attractive potential field and a repulsive potential field for the target node and obstacle, respectively, based on the biased-RRT algorithm were established. Therefore, the probability threshold changed according to the different types of potential fields.

Similar to the APF algorithm, the range of the attractive potential field was the whole operating space, while the range of the repulsive potential field was limited to a certain distance around the obstacle. In the range of the repulsive potential field, if the random tree tended to search for the target node q_{goal}, the random tree had a strong ability to grow biased. At this time, if the obstacle blocked q_{goal}, multiple failed growth near obstacles could occur for the random tree, so it tended to search randomly within the repulsive potential field. When the random tree left the range of the repulsive potential field, it continued to tend to search for q_{goal}, as shown in Algorithm 3.

Algorithm 3. Probability Threshold under the Control of Potential Field.

```
1: if RandomNumber <= threshold(q_new, obstacle) then
2: q_rand ← RandomSample(C_free);
3: else
4: q_rand ← q_goal;
5: end if
6: return q_rand
```

The growths of the random tree under both the control of the constant probability threshold and the adaptive probability threshold are shown in Figure 2a,b, respectively. Figure 2a shows that, if the random tree maintained a constant probability threshold during the search process, the obstacle did not affect the goal of random tree expansion. If the adaptive probability threshold was adopted, the random tree chose a better growth direction according to the location tendency of the new node. It was learned through many experiments that the probability threshold outside the scope of the repulsive potential field was 0.3, and the probability threshold inside the range of the repulsive field was 0.7.

2.3.2. Node-First Search Strategy

According to the biased-RRT algorithm, when $RandomNumber > m$, q_{rand} takes the coordinate value of q_{goal} and then selects the q_{near} closest to q_{rand} in the random tree as the parent node of q_{new}. If the random tree only expands to the target in each search round without considering the random search, then the new node in this search round will become the parent node of the new node in the next search round, and the random tree is a straight line segment connecting q_{init} and q_{goal}.

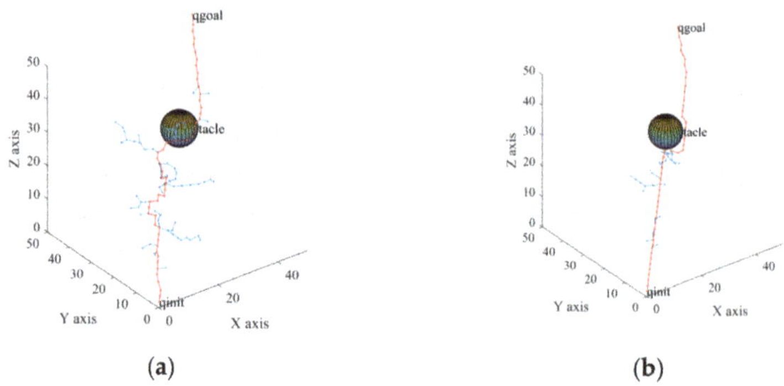

Figure 2. Threshold comparison. (**a**) Constant probability threshold; (**b**) Adaptive probability threshold.

As in Section 2.3.1, when the end node of the random tree expands to the range of the repulsive potential field, the random tree tends to select any node in the space as q_{rand} in the next search selection, so the probability of random expansion of the end node is small. If the next round of search satisfies $RandomNumber > threshold$, since the end node of the random tree is closest to q_{goal}, q_{goal} will be expanded, causing the newly generated path to collide with the obstacle. To summarize, when $RandomNumber > threshold$, the end node collides with the obstacle; when $RandomNumber <= threshold$, any node in the tree will be selected for expansion, which is no different from the traditional RRT algorithm. This phenomenon is called "falling into a trap", as shown in Figure 3.

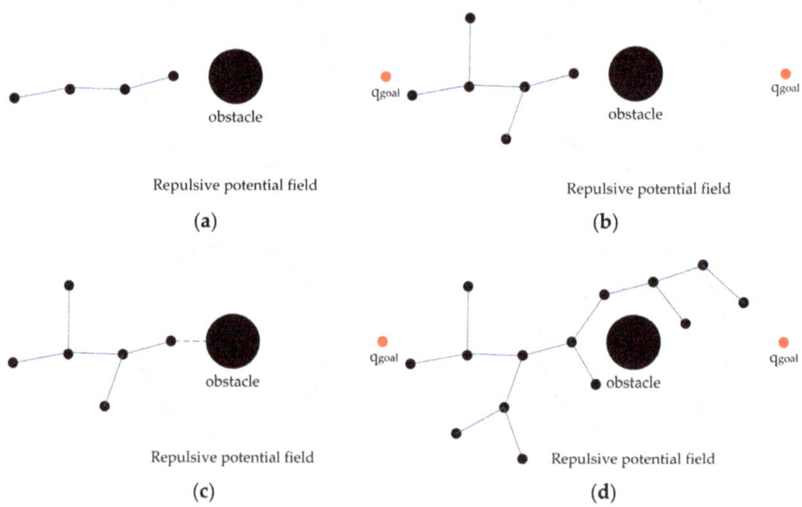

Figure 3. The random tree falls into a trap in the repulsive potential field. (**a**) The random tree entering obstacle potential field; (**b**) The random tree begins to expand randomly; (**c**) The random trees collide with obstacles; (**d**) The random tree is expanded several times.

For this reason, a node-first search strategy was proposed in this paper, as shown in Algorithm 4. When the q_{new} of the random tree grew into the range of the repulsive potential field of obstacles, a virtual spherical surface with a radius r and center q_{new} was generated. If $RandomNumber <= threshold$ was satisfied in the next round of search, point q_{rand} on the virtual spherical surface was preferentially selected, and q_{new} was used as the parent node of the next round of search to generate a new node q_{new2}. If q_{new2} and the line segment $\overline{q_{new}q_{new2}}$ did not collide with obstacles, the path and q_{new2} were kept. A new search round continued until the end node of the random tree was separated from the obstacles, as shown in Figure 4.

Algorithm 4. Node-First Search Algorithm.

```
if RandomNumber <= threshold(q_new, obstacle) then
if Distance(q_new, obstacle) < R_rpf then
q_rand ← sphere(q_new, r_virtual);
else
q_rand ← RandomSample(C_free);
end if
else
q_rand ← q_goal;
end if
return q_rand
```

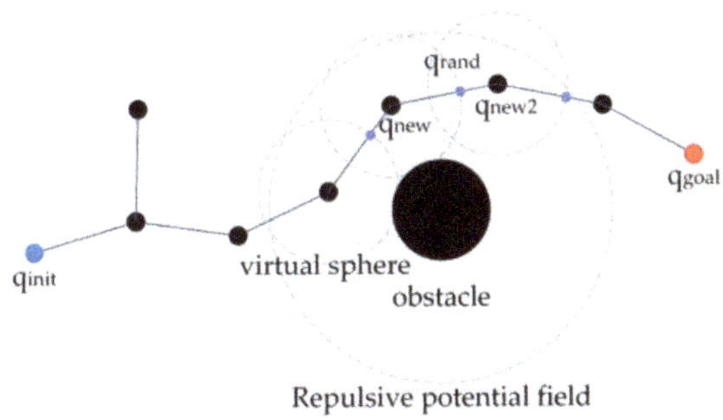

Figure 4. Schematic diagram of the node-first search strategy.

2.3.3. Attractive Step Size and Step-Size Dichotomy

From the above description, the node-first search strategy was used to prevent the random tree "falling into a trap" within the range of the repulsive potential field. Since the random tree has a certain probability of random search outside the range of the repulsive potential field of obstacles, more iterations will be generated. An appropriate step size can effectively reduce the iterations of the random tree. In the case that the length of the path is determined, a small step size will cause more collision detections and distance calculations, and a large step size will often make the random tree collide with obstacles. Therefore, the step size should be expanded as much as possible on the premise of reducing the number of collisions [34,35].

According to the APF algorithm, the attractive force of q_{goal} acts on the whole operating space and is proportional to the distance between the end joints of the manipulator, which is beneficial to control the growth step of the RRT. If obstacles are not considered, the random tree should increase the step size when it is far away from q_{goal} to quickly expand to q_{goal}. When the random tree is closer to q_{goal}, if it continues to maintain a large step size, a large number of redundant nodes will be generated at q_{goal}, as shown in Figure 5a. Therefore, the random tree should gradually approach q_{goal} with small step sizes, as shown in Figure 5b.

For this reason, an attractive step size was proposed, which was defined as:

$$attStepsize = k \times Distance(q_{near}, q_{goal}) \tag{1}$$

where $attStepsize$ represents the attractive step size, and k is the attractive parameter.

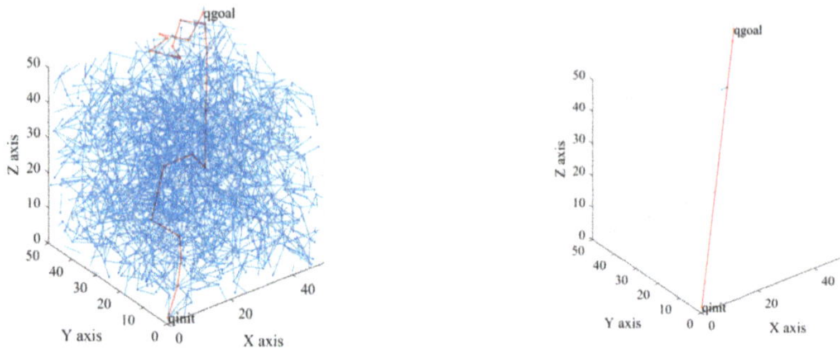

(a) **(b)**

Figure 5. Fixed step size and attractive step size. (**a**) Fixed step size; (**b**) Attractive step size.

If obstacles are considered, the step size of the random tree in the random search is fixed, and an attractive step size is taken when growing toward q_{goal}. This method ensures that the random tree grows toward q_{goal} as soon as possible outside the range of the repulsive potential field and avoids collisions with obstacles due to excessive steps within the range of the repulsive potential field.

The parameters of the potential field function of the manipulator are different in different operating spaces. For example, when the attractive parameter k is too large, $attStepsize$ will increase accordingly. If $Distance(q_{near}, obstacle) < attStepsize$, q_{new} will collide with obstacles. In addition, the end nodes tend to grow toward q_{goal} outside the range of the repulsive potential field. Therefore, the random tree still has a high probability of colliding with obstacles in the next round of search.

For this reason, a "step-size dichotomy" was introduced to solve the problem of excessive step size. When q_{near} grew toward q_{goal} and there were obstacles between them, the distance d_{nob} between q_{near} and the obstacles was calculated. If $d_{nob} <= attStepsize$, the $attStepsize$ was shortened to the original value of 2^{-1}, and the sizes of $attStepsize$ and d_{nob} were compared again until $d_{nob} > attStepsize$; see Algorithm 5.

Algorithm 5. Step-size Dichotomy.

if $Collision(q_{near}, q_{goal})$ then
while $adpStepsize > Distance(q_{near}, obstacle)$ do
$adpStepsize = adpStepsize/2$;
end while
else
$attStepsize = k \times Distance(q_{near}, q_{goal})$;
end if
return $attStepsize$

2.3.4. Regression Superposition Algorithm

From Section 2.3.3, if the random tree grows within the range of the obstacle repulsive potential field, a large number of redundant nodes will be generated on the surface of the obstacle due to the high probability of the random search, as shown in Figure 6a. As a result, a regression superposition algorithm is proposed in this section to adaptively select extended nodes and change the step size of the random search, as shown in Figure 6b.

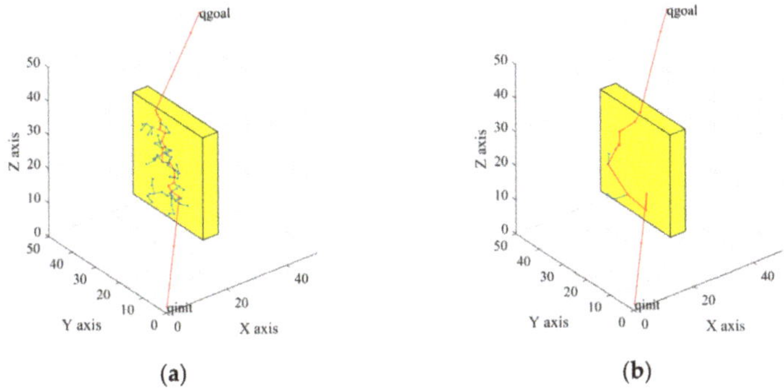

Figure 6. Differences before and after improvement. (a) The random trees generated a large number of nodes on the surface of obstacles; (b) The random tree had fewer nodes on the obstacle surface.

The regression superposition algorithm consists of a regression algorithm [36] and a step-size superposition algorithm. In the regression algorithm, if the distance between q_{new} and q_{near} was larger than the distance between q_{new} and any node q_i in the random tree except q_{near}, it was considered to meet the regression conditions:

$$\begin{cases} Distance(q_{near}, q_{new}) > Distance(q_{near}, q_i) \\ q_i \in Tree \end{cases} \quad (2)$$

If Formula (2) was satisfied, q_{new} was regarded as a regression node. The regression node would not become the tree node of the random tree, but it was removed until a new node that did not meet the regression condition was found, as shown in Figure 7a.

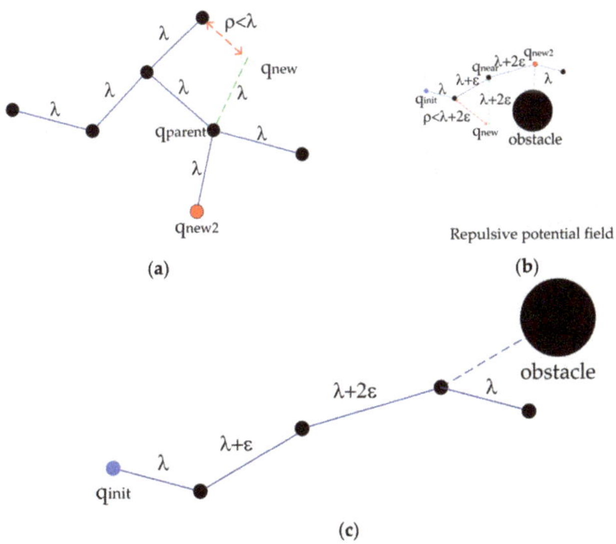

Figure 7. Regression superposition algorithm. (a) Regression algorithm; (b) Regression superposition algorithm; (c) Step-size superposition algorithm. Note: λ, the initial step size; ε, the step size of superposition; ρ, the distance between q_{new} and any node.

To further reduce the number of tree nodes, the step-size superposition algorithm was incorporated based on a regression algorithm. When the random tree was searched randomly, the initial step size was set to λ, and the step size was increased by ε after each round of the random search until the extended branches of the random tree collided with obstacles. Then, the search step size of the next round was returned to the initial step size λ, and the step size was superimposed again until the random tree searched toward q_{goal}, as shown in Figure 7c.

The random tree used an attractive step size when searching toward q_{goal} to reduce the generation of redundant nodes. During the random search of the random tree, the regression superposition algorithm was used to enhance the ability of the random tree to search the unknown space, as shown in Figure 7b.

The TO-RRT algorithm was used to dynamically adjust the growth direction of the random tree by the probability threshold controlled by the potential field and to define two different growth methods according to the different growth directions. Therefore, the random tree could quickly grow to the target outside the range of the repulsive potential field and quickly determine the escape path within the range of the repulsive potential field. The algorithm flow chart is shown in Figure 8.

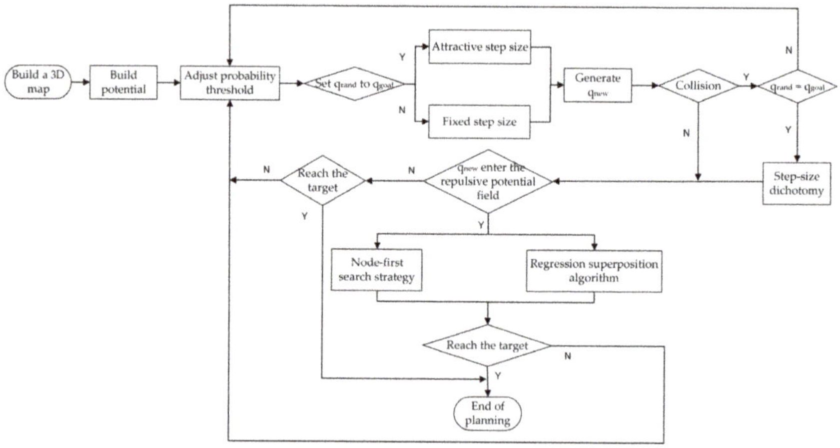

Figure 8. Flowchart of the TO-RRT algorithm. Note: A, yes; N, no.

3. Results

3.1. Comparative Experiment of Path Planning in a Complex Environment

To verify the speed, stability, and low path cost of the TO-RRT algorithm, the RRT algorithm, the biased-RRT algorithm with a target offset probability of 50%, the TO-RRT algorithm, the RRT-BCR algorithm, and the NC-RRT algorithm are compared in this section using complex environments (i.e., a multi-sphere environment, a multi-rectangle environment, a single-channel environment, and a multi-channel environment).

In the simulation experiment, the initial step size was 2, the maximum number of failed growth times was 100,000, the map size was 50 × 50 × 50, the starting point was (1, 1, 1), and the target point was (49, 49, 49). The blank area in the map represented the obstacle-free area, other colors represented the obstacle area, the blue path represented the random tree, the black path represented the collision-free path from the starting point to the target point, and the red path represented the path optimized by the greedy algorithm.

Figure 9a,e,i,m,q show that, although the RRT algorithm can be used to find a collision-free path from the initial point to the target point, the whole space was searched, so that the highest amount path nodes were generated. Compared with the RRT algorithm, the biased-RRT algorithm did not search too much invalid space, so there were fewer path nodes. When using the RRT-BCR algorithm and the NC-RRT algorithm, the sizes of the random trees were reduced through a regression mechanism and an adaptive sampling area, respectively. The TO-RRT algorithm was used to greatly reduce the number of nodes in the space, and its complexity was the lowest. Figure 9b,f,j,n,r show that the RRT algorithm still searched the whole space. Although the biased-RRT algorithm generated fewer nodes than the RRT algorithm, the search tree generated a large number of nodes on the surface of obstacles, which increased the number of iterations. The NC-RRT algorithm made the random tree tend to expand through boundary nodes through the node control mechanism, so it had fewer redundant nodes. It can be seen from Figure 9c,d,g,h,k,l,o,p,s,t that the RRT algorithm and the biased-RRT algorithm could not quickly find the "escape channel". Although the RRT-BCR algorithm limited the expansion of nodes that were prone to collision, it increased the expansion times of other nodes. Due to the regression superposition algorithm and node-first search strategy introduced into the TO-RRT algorithm, the random tree could quickly search the nearby area to find the "escape channel" in the repulsive potential field.

There are certain errors and contingencies in a single experiment. To better reflect a real situation, 10 simulation experiments were carried out in the same environment as described above, shown in Figure 10.

Figure 10 shows that the TO-RRT algorithm maintained strong stability in 10 experiments and did not traverse the whole space due to being blocked by obstacles, while the RRT algorithm and the biased-RRT algorithm both generated a large number of nodes in the space. In addition, the RRT-BCR algorithm had fewer path nodes than the biased-RRT algorithm, and in the NC-RRT algorithm, there was little difference in the path in each search. The comparison of the running times of the three algorithms in different environments is shown in Figure 11. Figure 11 shows that the RRT algorithm had the longest running time and poor running-time stability, especially in a single-channel environment, with the longest running time at 45.6057 s and the shortest running time at 1.2880 s. Compared with the RRT algorithm, the biased-RRT algorithm had a much shorter running time and strong running-time stability, but the search time in a complex environment was longer. The longest running times of the TO-RRT algorithm in the four environments were 0.0225 s, 0.0420 s, 0.0618 s, and 0.0443 s, and the shortest running times were 0.0056 s, 0.0134 s, 0.0101 s, and 0.0115 s. The difference between the longest search time and the shortest search time in a single environment did not exceed 0.06 s, which not only indicated a short search time but also a strong and stable running time. The NC-RRT algorithm performed poorly in a multi-rectangle environment, with a difference of 4.44 times between the longest running time and the shortest running time, while the RRT-BCR algorithm was only 3.82 times.

Table 2 shows the average values of each index of the 3 algorithms over 10 experiments (biased-RRT represents the biased-RRT algorithm with a target offset probability of 50%). In the multi-sphere environment, the TO-RRT algorithm had a running time that was 99.74% less than the RRT algorithm, which was mainly because the number of collision detections and the number of failed node growths of the former were reduced by 99.39% and 97.17%, respectively, compared with the latter. In addition, compared with the RRT algorithm, the number of path nodes in the TO-RRT algorithm was reduced by 82.92%, which shortened the length of its search path by 18.99%. When the random tree encountered a large area of obstacles, the TO-RRT algorithm was used to reflect the advantages in the search time more than the RRT algorithm. For example, the number of tree nodes and the number of failed growths of nodes of the RRT algorithm in the multi-rectangle environment reached 17,358.3 and 3144.8, respectively, resulting in a running time of 7.8822 s, while the running time of the TO-RRT algorithm was only 0.0213 s. In addition, the RRT-BCR algorithm performed better than the NC-RRT algorithm in a multi-rectangle environment, and its running time was shortened by 29.14% compared with the NC-RRT algorithm because the RRT-BCR algorithm removed nodes that collided many times when facing obstacles with large occlusion areas. The biased-RRT algorithm produced too much failure growth when encountering obstacles with large areas. For example, in a multi-channel environment, the node failure growth rate of the biased-RRT algorithm was 62.54%, while the RRT algorithm and TO-RRT algorithm had node failure growth rates of only 36.40% and 15.82%, respectively. Therefore, the biased-RRT algorithm was not ideal in a complex environment. Since the NC-RRT algorithm always took the area between the configuration point and the target as the sampling radius and tended to use boundary nodes for expansion, it could not produce valid nodes when the obstacle was between the configuration point and the target. For example, in multi-channel and multi-rectangle environments, the collision detection times of the NC-RRT algorithm were 21,487 times and 55,077 times. In summary, compared with the other algorithms, the TO-RRT algorithm had significant advantages in searching speed and the number of nodes in the random tree.

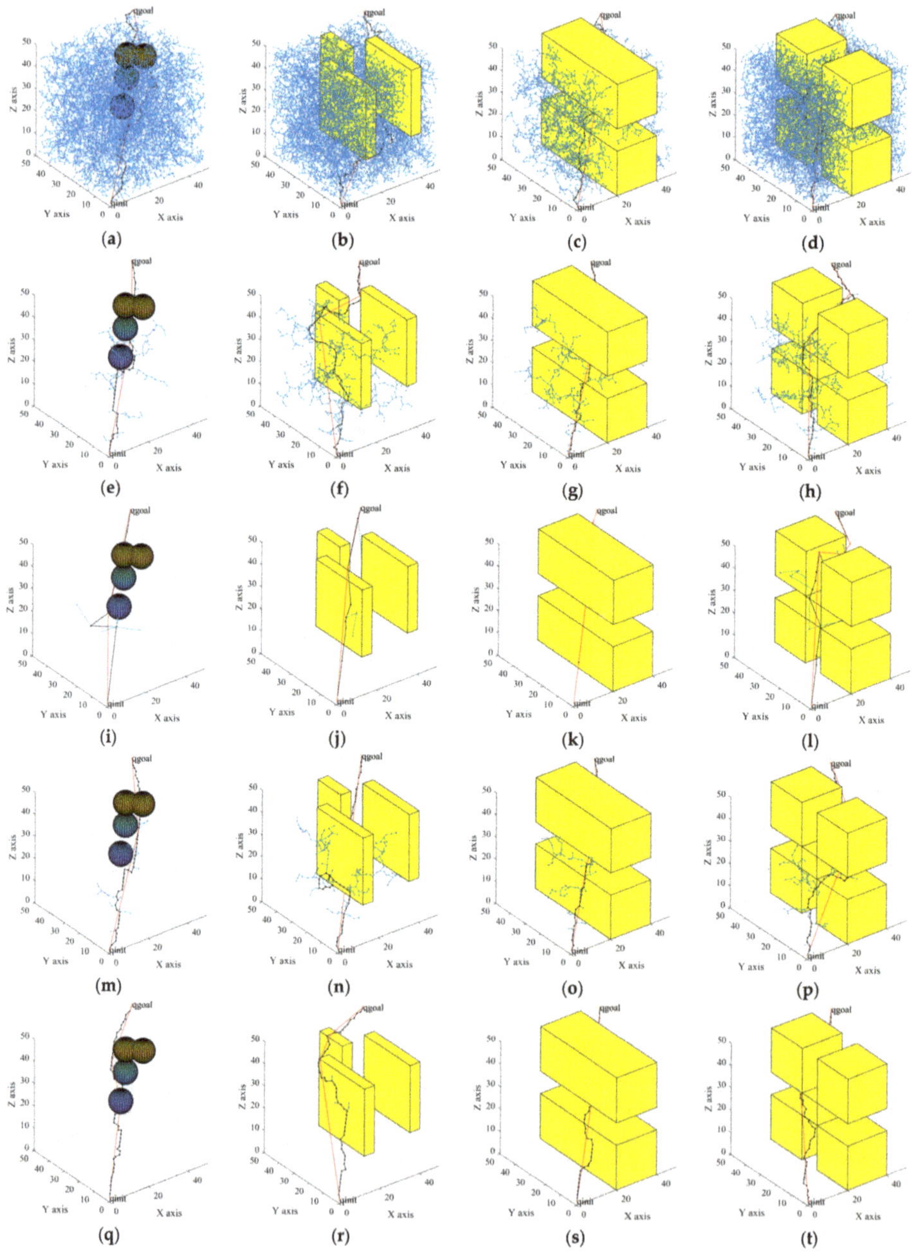

Figure 9. The performances in different environments of: the RRT algorithm (**a–d**); the biased-RRT algorithm with a target offset probability of 50% (**e–h**); the TO-RRT algorithm (**i–l**); the RRT-BCR algorithm (**m–p**); and the NC-RRT algorithm (**q–t**).

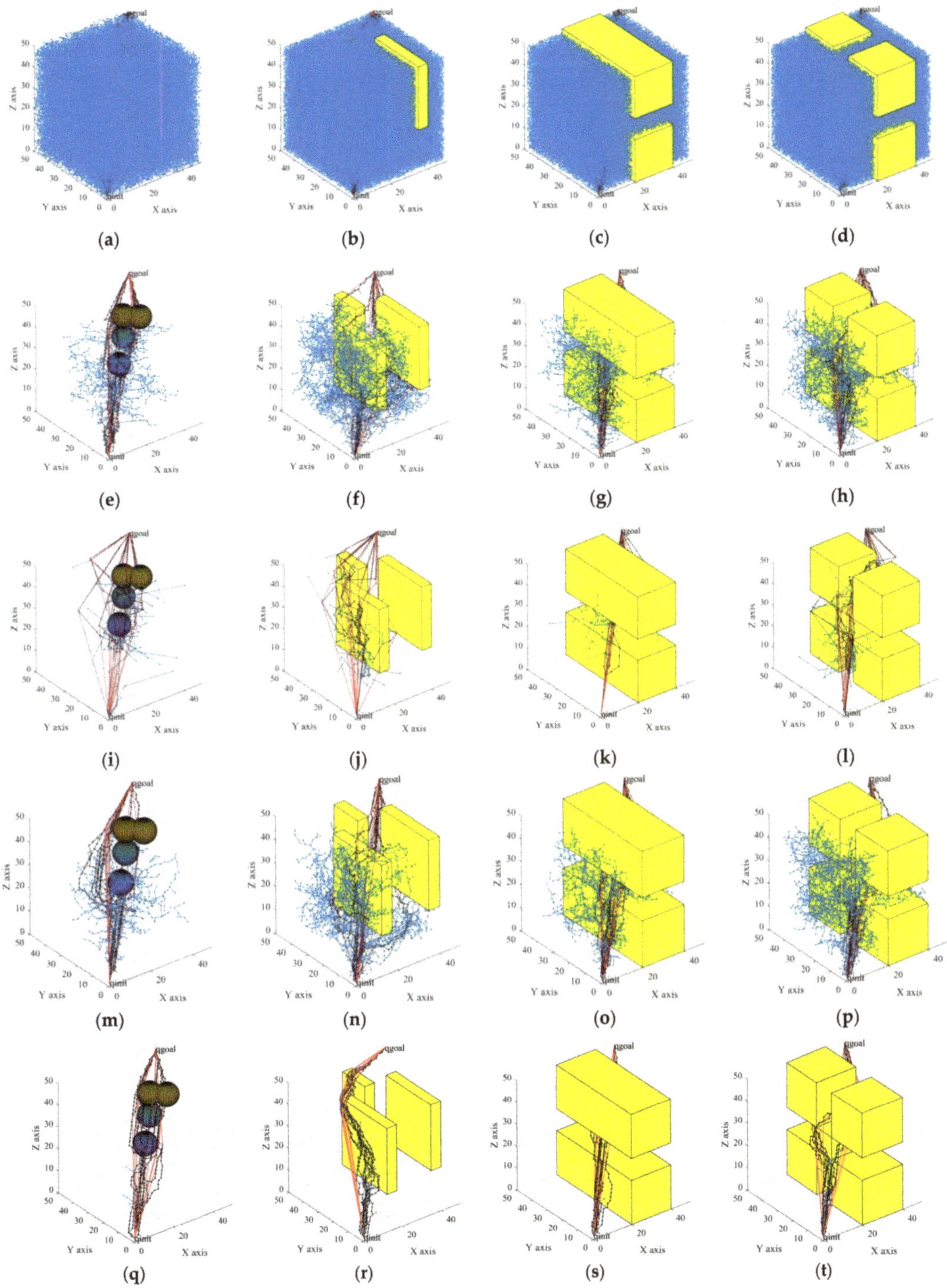

Figure 10. Ten experiments each of: the RRT algorithm (**a–d**); the biased-RRT algorithm with a target offset probability of 50% (**e–h**); the TO-RRT algorithm (**i–l**); the RRT-BCR algorithm (**m–p**); and the NC-RRT algorithm (**q–t**).

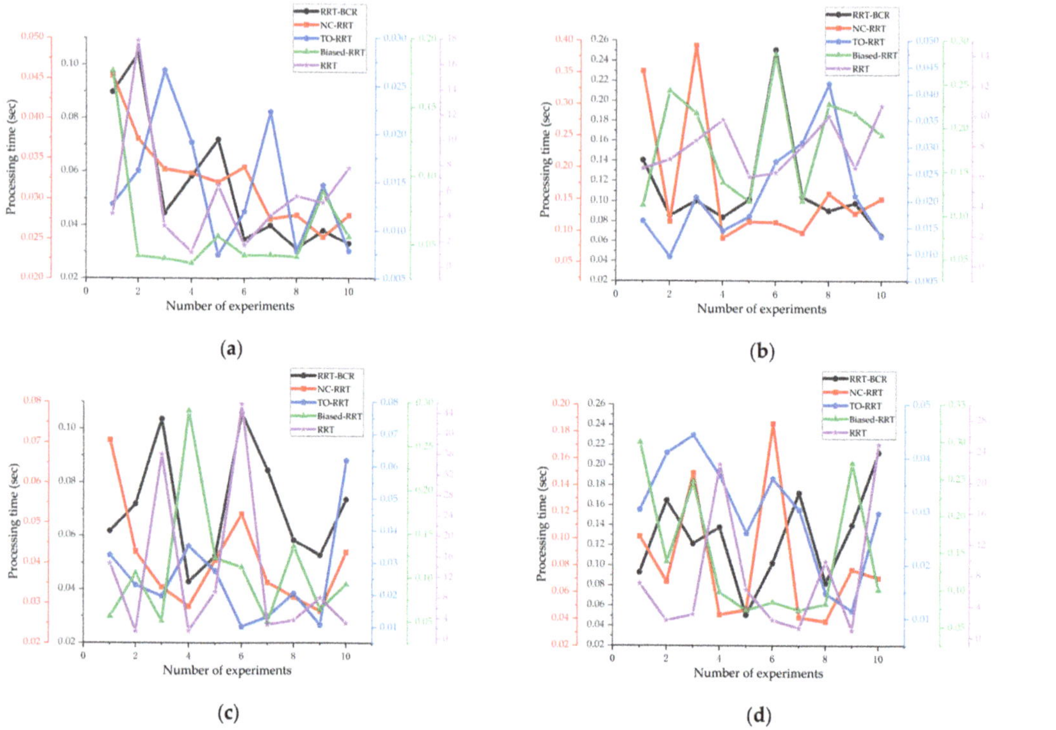

Figure 11. The running times of the RRT algorithm, the biased-RRT algorithm with a target offset probability of 50%, the TO-RRT algorithm, the RRT-BCR algorithm, and the NC-RRT algorithm. (**a**) Multi-sphere environment; (**b**) Multi-rectangle environment; (**c**) Single-channel environment; (**d**) Multi-channel environment.

Table 2. Experimental results of each algorithm in different environments.

	Algorithm Type	Running Time (s)	Path Length (cm)	Tree Nodes (Number)	Path Nodes (Number)	Collision Detection (Number)	Failed Node Growth (Number)	Node Failure Growth Rate (%)
Multi-sphere	RRT	5.6342	124.6008	10,454.3	60.9	10,693.7	229.4	2.15
	Biased-RRT	0.0617	100.1367	140.1	54	228	87.9	38.55
	TO-RRT	0.0147	100.9338	22.9	10.4	65.5	6.5	9.92
	RRT-BCR	0.0545	101.9241	113.4	54.3	123.2	9.8	7.95
	NC-RRT	0.0324	94.3765	50.6	50.2	183.7	133.1	78.46
Multi-rectangle	RRT	7.8822	140.9832	14,213.5	68.7	17,358.3	3144.8	18.12
	Biased-RRT	0.1860	125.8082	414.3	62.9	1033.9	619.6	59.93
	TO-RRT	0.0213	110.1866	32.7	13.2	135.5	17.3	12.77
	RRT-BCR	0.1121	121.8465	243.8	60.4	294.4	50.6	17.19
	NC-RRT	0.1709	107.2454	55.8	53.6	55,077	54,519	99.99
Single-channel	RRT	12.4436	131.1145	8333.9	64.2	13,560.3	5226.4	38.54
	Biased-RRT	0.1074	108.6431	242.2	55.5	607.3	365.1	60.12
	TO-RRT	0.0254	107.4978	32.8	12.7	130.7	20	15.30
	RRT-BCR	0.0707	109.4179	159.3	55.8	203.8	44.5	21.83
	NC-RRT	0.0406	96.7172	49.8	49.8	659	609.2	92.44

Table 2. *Cont.*

	Algorithm Type	Running Time (s)	Path Length (cm)	Tree Nodes (Number)	Path Nodes (Number)	Collision Detection (Number)	Failed Node Growth (Number)	Node Failure Growth Rate (%)
Multi-channel	RRT	8.0047	134.4688	11,702.5	64.7	18,399.5	6697	36.40
	Biased-RRT	0.1461	114.4721	322.9	56.9	861.9	539	62.54
	TO-RRT	0.0301	117.5516	51.8	16.1	222.5	35.2	15.82
	RRT-BCR	0.1276	120.4389	278.8	61.8	369.3	90.5	24.51
	NC-RRT	0.0821	102.7622	55.3	52.8	21,487	20,934	97.43
Average index	RRT	8.4912	132.7918	11,176.05	64.625	15,002.95	3824.4	23.8025
	Biased-RRT	0.1253	112.2650	279.875	57.325	682.775	402.9	55.285
	TO-RRT	0.0229	109.0425	35.05	13.1	138.55	19.75	13.4525
	RRT-BCR	0.0912	113.4070	198.825	58.075	247.675	48.85	17.87
	NC-RRT	0.0815	100.2753	52.875	51.6	19,351.675	19,061.48	92.08

Note: RRT, rapidly-exploring random tree; Biased-RRT, rapidly-exploring random tree with target Bias; TO-RRT, time-optimal rapidly-exploring random tree; RRT-BCR, Biased-RRT with boundary expansion mechanism and regression mechanism; NC-RRT, Node Control-RRT.

3.2. Obstacle Avoidance Test Based on the Robotics Toolbox

To verify the feasibility of the TO-RRT algorithm on the manipulator, Robotics Toolbox 10.2 in MATLAB was used to model the Franka manipulator. Franka is a 7-DOF robot with high precision and fast response. Its payload is 3 kg, and the maximum contact area is 855 mm. The Franka manipulator can realize two-way communication between itself and the workstation through the Franka Control Interface (FCI) and an Ethernet connection. Therefore, complete real-time control can be achieved with a sampling frequency of 1 kHz. In terms of picking performance, Franka's pose repeatability is within 0.1 mm. Even at the highest speed of 2 m/s, the path deviation can be ignored, which provides good working conditions for fruit picking. The physical object of the Franka manipulator and its D-H parameters are shown in Figure 12a and Table 3, respectively.

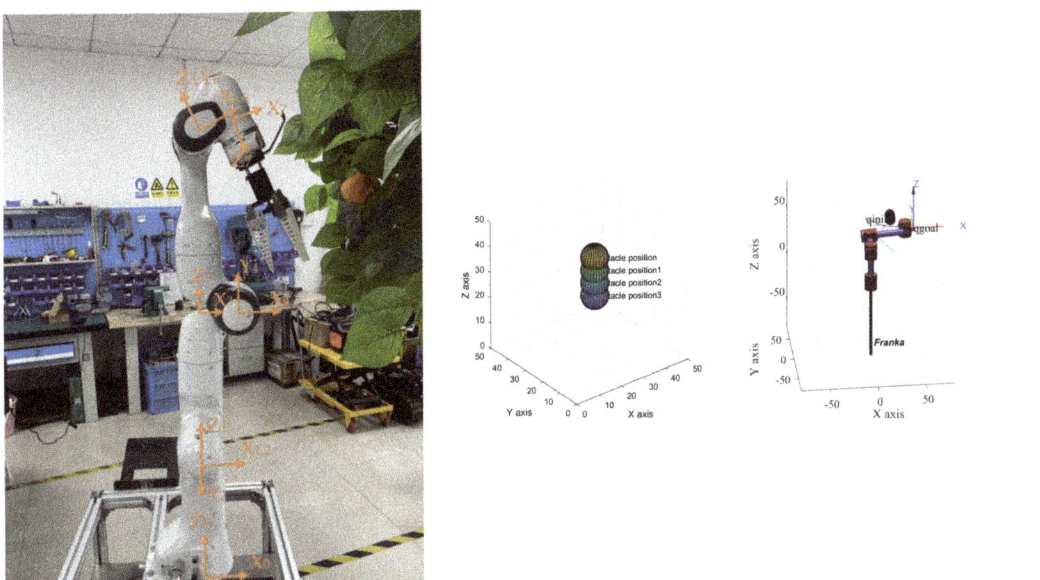

(a) **(b)** **(c)**

Figure 12. Materials and results of simulation experiments based on using Robotics Toolbox. (**a**) The physical object of the Franka manipulator; (**b**) Trunk model; (**c**) The Franka manipulator avoids obstacles.

Table 3. D-H parameters.

Link i	Link Offset a_i(m)	Link Length d_i(m)	Link Twist α_i(rad)	Link Twist θ_i(rad)
1	0	0.333	$\frac{\pi}{2}$	θ_1
2	0	0	0	θ_2
3	0	0.316	0	θ_3
4	0.0825	0	$\frac{\pi}{2}$	θ_4
5	−0.0825	0.384	$-\frac{\pi}{2}$	θ_5
6	0	0	0	θ_6
7	0.088	0	$\frac{\pi}{2}$	θ_7

To simplify the trunk and improve the operation speed of the TO-RRT algorithm, the trunk was regarded as a combination of spheres [29], as shown in Figure 12b and Table 4. To judge whether the manipulator collided with obstacles, the shortest distance $d_{collision}$ from the center of the sphere to the origin of the coordinate system of adjacent links of the manipulator was used. The three-dimensional coordinates of each joint of the manipulator were obtained through a forward kinematics solution, and if the manipulator did not collide with the tree trunk, the following conditions must be met:

$$d_{collision} > R + r \tag{3}$$

Table 4. Obstacle parameters.

Number	Obstacle Coordinates (cm)	Obstacle Radius (cm)
1	(25,55,48)	5
2	(25,53,47)	5
3	(25,51,46)	5
4	(25,49,45)	5

In the formula, $R = 5$ cm is the radius of the obstacle ball, and $r = 3$ cm is the radius of the cylinder.

Figure 12c shows the Franka manipulator using the TO-RRT algorithm to plan its path, and the minimum-snap trajectory optimization algorithm was used to smooth the trajectory of the manipulator [37,38]. Figure 13 shows the shortest distance.

3.3. Comparative Experiments in a Virtual Picking Environment

The motion-planning experiment of the Franka manipulator was initially realized through Robotics toolbox, which proved that the TO-RRT algorithm was feasible in the motion of the manipulator. MoveIt! was used in this section to build a virtual picking environment and to conduct comparative experiments on different algorithms in this environment. The experimental parameters are shown in Table 5.

During the experiment, the maximum search time was 10 min, the maximum number of failed searches was 10,000, and the search domain was $\{x, y, z| -1 < x < 1, -1 < y < 1, -1 < z < 1\}$(m). Due to the large number of sampling points generated, the global search time of the RRT algorithm was 243.322451 s. Compared with the RRT algorithm, the search time of the biased-RRT algorithm was only 3.720342 s. However, affected by the nature of obstacles and the probability threshold, the collision-free path generated by the biased-RRT algorithm was less smooth. In contrast, since the NC-RRT algorithm controlled the sampling interval, its trajectory was the smoothest among all the algorithms. Compared with the previous algorithms, the TO-RRT search time and path length were only 0.074915 s and 0.63548128 m, respectively, due to the generation of smaller random trees. The simulation results are shown in Table 6 and Figure 14.

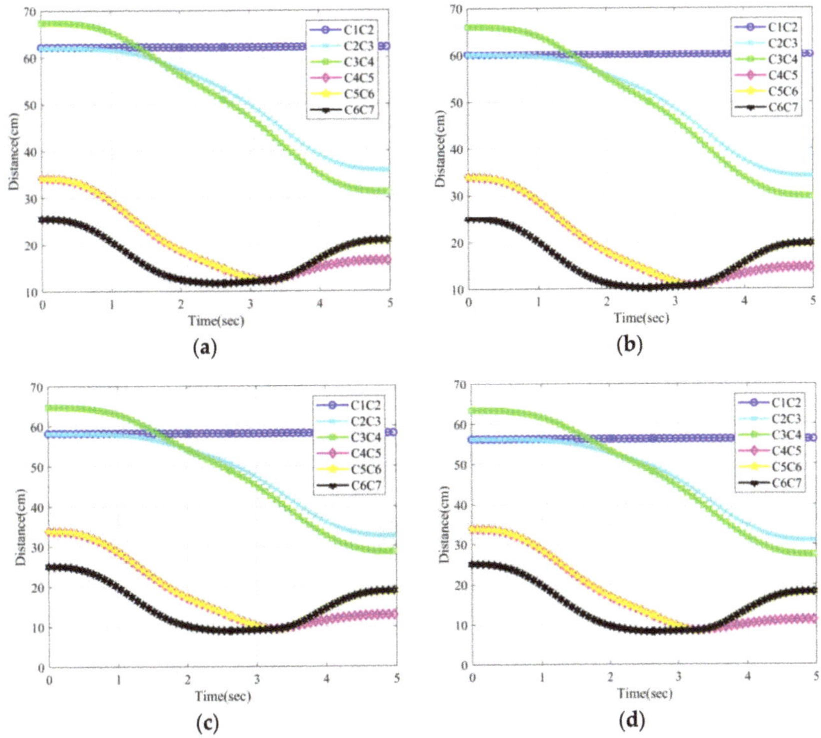

Figure 13. The shortest distance from the center of the sphere to the origin of the coordinate system of adjacent links of the manipulator. (**a**) Obstacle with coordinates (25,55,48); (**b**) Obstacle with coordinates (25,53,47); (**c**) Obstacle with coordinates (25,51,46); (**d**) Obstacle with coordinates (25,49,45). C_iC_{i+1} represents the distance between the line segment between coordinate system i and coordinate system $i + 1$ and the center of the sphere.

Table 5. Experimental parameters.

	Initial Pose	Pose of Citrus 1	Pose of Citrus 2
Position	(0.3595, 0, 0.643499)	(0.106155, 0.227978, 0.744871)	(−0.234434, 0.360095, 0.737649)
Orientation	(−0.65328, −0.270598, 0.653283, 0.270599)	(−0.636052, 0.309414, 0.231336, 0.66797)	(−0.771505, 0.309187, 0.226895, 0.507644)

Table 6. Experimental data using MoveIt!.

	RRT	Biased-RRT	TO-RRT	RRT-BCR	NC-RRT
Global planning time(s)	243.322451	3.720342	0.074915	1.222014	0.181070
Global waypoints(number)	41	29	7	20	15
Path length at obstacle avoidance(m)	1.89919096	1.46801193	0.63548128	0.592291	0.53239712

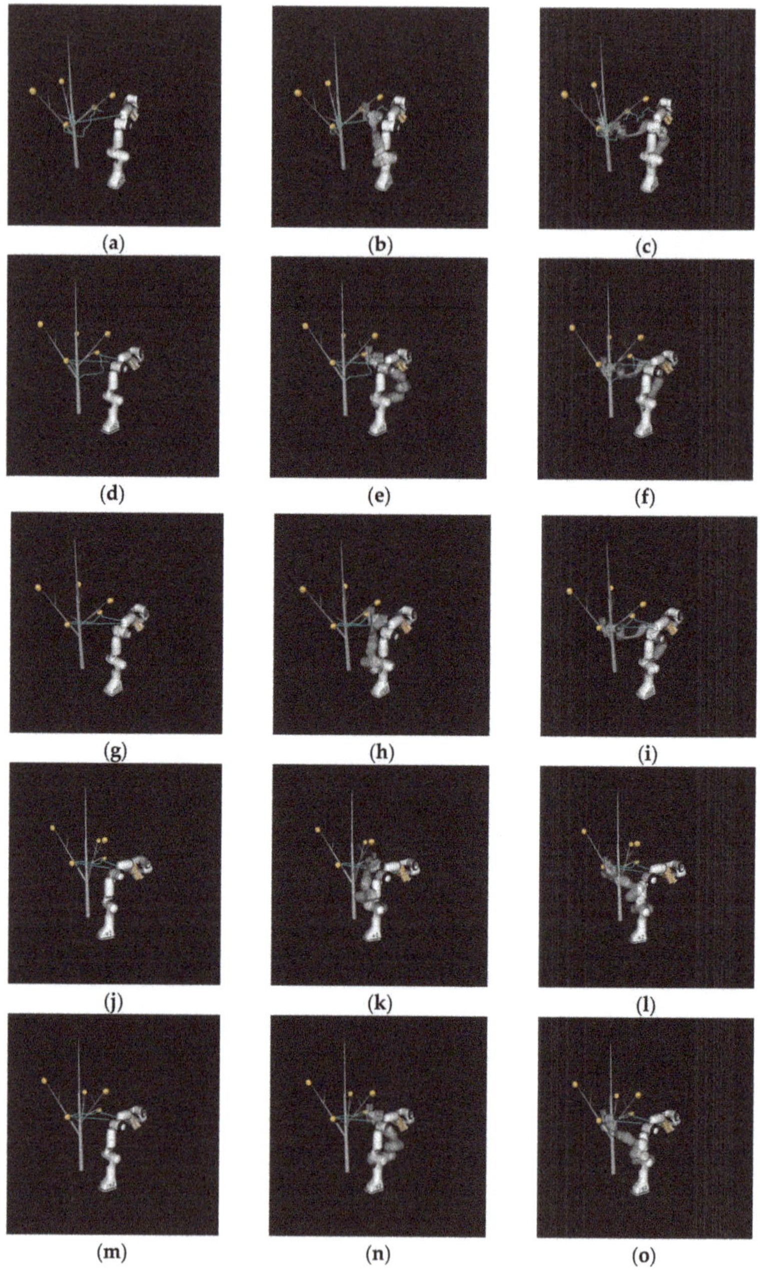

Figure 14. The use of MoveIt! with: the RRT algorithm (**a–c**); the biased-RRT algorithm with a target offset probability of 50% (**d–f**); the TO-RRT algorithm (**g–i**); the RRT-BCR algorithm (**j–l**); and the NC-RRT algorithm (**m–o**).

3.4. Contrastive Experiments in Real Environments

To test the performance of TO-RRT in actual picking, the Franka manipulator was taken as the moving object, the citrus as the operation object, and the tree trunk as the obstacle avoidance object to construct a multi-objective citrus-picking environment. The environmental parameters are shown in Tables 7 and 8. First, the manipulator adjusted its pose to the initial state, and its joint angle was $(0, -\frac{\pi}{4}, 0, -\frac{\pi}{2}, 0, \frac{\pi}{3}, 0)$. Second, the three-dimensional coordinates of the citrus, the parameter information of obstacles, and the picking pose of the manipulator were transmitted to the planning thread, and the continuous and collision-free trajectory was obtained through inverse kinematics. Finally, MoveIt! published the trajectory through moveit_commander to move_group and transmitted the control signal to the robot controllers to complete the picking action. The control block diagram is shown in Figure 15. The experimental results showed that the TO-RRT algorithm could be used to effectively reduce the nodes, shorten the planning time, and reduce the movement time of the manipulator, as shown in Figure 16 and Table 9.

Table 7. Obstacle information.

Number	Obstacle Coordinates (m)	Obstacle Radius (cm)
1	(0.369822, −0.153781, 1.04791)	1
2	(0.426765, −0.149826, 1.00189)	1
3	(0.45418, −0.186812, 0.947317)	1
4	(0.330284, −0.344084, 1.01095)	1.5
5	(0.384351, −0.371103, 0.94411)	1.5
6	(0.48388, −0.335959, 0.897789)	1.5

Table 8. Target information.

	Coordinates (m)
Base coordinates	(0,0,0)
Citrus 1 coordinates	(0.208763, −0.432806, 0.764728)
Citrus 2 coordinates	(0.423718, 0.0602042, 0.994)

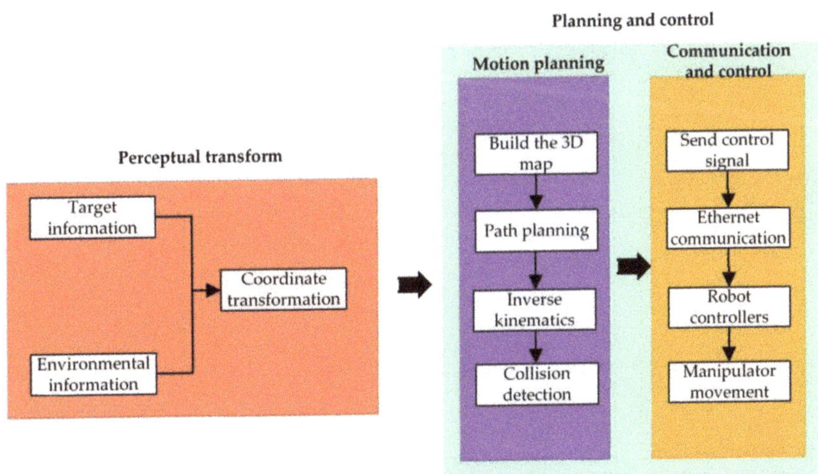

Figure 15. Control block diagram.

(**a**) (**b**) (**c**) (**d**)

Figure 16. The manipulator reached Citrus 1 and Citrus 2 and avoided the branches. (**a**) Initial state of manipulator; (**b**) The manipulator reaches the first citrus; (**c**) Obstacle avoidance of the manipulator; (**d**) The manipulator reaches the second citrus.

Table 9. Comparison of the planning time and movement time.

Algorithm Type	Planning Time(s)	Movement Time(s)
RRT	53.873985	84.3975
Biased-RRT	0.0883	18.0498
TO-RRT	0.0508	17.3703
RRT-BCR	0.0771	17.9238
NC-RRT	0.0649	17.7131

4. Discussion

4.1. Analysis

From Figure 10a–d, since the RRT algorithm did not consider the effect of target offset probability, the entire workspace was searched in all environments. The above problems led to the huge scale of the random tree and caused more collision detection times. Therefore, the path length and movement time of the manipulator were the longest among all the algorithms, as shown in Tables 6 and 9. From Table 2, the biased-RRT algorithm avoided redundant searching through heuristic guidance, effectively reducing the number of tree nodes and collision detection times. From the average index in Table 2, since the RRT-BCR algorithm removed nodes that collided multiple times, its node failure growth rate was very low. However, this approach took a considerable amount of computation time, only 0.0112 s less than the biased-RRT algorithm, as shown in Table 9. From the average index in Table 2, the path length of the NC-RRT algorithm was the shortest, and the running time was second only to the TO-RRT algorithm. As can be seen from the multi-rectangle environment in Table 2, the NC-RRT algorithm had to continuously expand its sampling space when facing obstacles with large occlusion areas, resulting in 55,077 collision detections (which was the highest among all the algorithms). From Table 2, the TO-RRT algorithm reduced the numbers of path nodes and collision detections through an attractive step size, reduced the number of node failure growth through the node-first search strategy, and, finally, enhanced the escape ability through the regression superposition algorithm. However, the TO-RRT algorithm produced larger steps near obstacles, which led to a slightly longer path length than the other improved algorithms, as shown in Table 6.

4.2. Future Work

Industry 5.0 is a new generation of the industrial revolution representing "personalization", in which personalized products and services are created for humans by using the creativity of human experts to interact with efficient, intelligent, and precise machines. The key technologies of Industry 5.0, such as human–computer interaction, collaborative robots, and edge computing (EC), can provide ideas and technical support for Agriculture 5.0 [39].

As the number of China's aging population increases by the year, the number of rural employees has dropped sharply, and original agricultural production methods can no longer meet the development needs of the current citrus industry. Through the high integration of artificial intelligence and mechanical equipment, the transformation and upgrade of the production mode of China's agricultural industry can be realized. The improved method proposed in this paper can be used in the fields for picking robots and pruning robots and for the path planning of orchard patrol robots [40–42]. By analyzing the characteristics of a citrus tree environment, the work presented in this paper aimed to optimize the time required and improve it on the basis of a traditional algorithm to greatly shorten the planning time of the manipulator and reduce the movement time of the manipulator to a certain extent. However, the detection of obstacles is an objective challenge faced by this method.

In recent years, path planning through deep reinforcement learning (DRL) has become a research hotspot. A robot senses environmental information through sensors and trains the samples in the process of continuous interaction with the environment to complete an efficient, accurate, and low-environment-dependence path-planning method. The fusion of deep reinforcement learning and traditional path-planning algorithms has gradually become a research trend. For example, LM-RRT determines the selection probability of extension and connection trees based on reinforcement learning and guides the trees to pass through narrow channels quickly [43]. Based on this, the research on improving the TO-RRT algorithm by reinforcement learning will be discussed in the next stage.

5. Conclusions

A time-optimal RRT algorithm based on the characteristics of the complex environment of citrus trees was proposed in this paper. The constructed algorithm had an attractive potential field and a repulsive potential field for the target node and obstacle, respectively. In addition, dynamic adjustment of the probability threshold under the action of the superimposed potential field was achieved, and a node-first search strategy was used to solve the "falling into a trap" problem. In addition, an attractive step size and a "step-size dichotomy" were introduced in this algorithm so that the random tree could expand the step size as much as possible on the premise of reducing the number of collisions. Finally, a regression superposition algorithm was used to improve the search efficiency of the random tree in the range of the obstacle repulsive potential field. The TO-RRT algorithm was simulated in complex environments, and the motion-planning of the Franka manipulator was carried out using Robotics Toolbox and MoveIt! It can be seen from the simulation results that the TO-RRT algorithm had fewer tree nodes, collision detection times, and failed growth times, so this algorithm had a shorter planning time than the RRT algorithm, the biased-RRT algorithm, the RRT-BCR algorithm, and the NC-RRT algorithm, especially when the random tree faced a large obstacle area. To obtain the performance of the algorithm in real work, we built a real picking environment indoors. Through the performance evaluation of various indicators of the different algorithms, it was proved that the TO-RRT algorithm still had a good performance in movement time.

Author Contributions: Conceptualization, C.L., L.X. and Q.F.; methodology, C.L., L.X. and Q.F.; software, C.L., Z.T. and X.W.; writing—original draft preparation, C.L. and L.X.; writing—review and editing, C.L., L.X. and Q.F.; visualization, C.L., Z.T., J.G. and X.W.; supervision, L.X. and Q.F.; funding acquisition, L.X. and Q.F. All authors have read and agreed to the published version of the manuscript.

Funding: This research was funded by the Beijing Science and Technology Plan Project: Z201100008020009; Key R&D project of Science and Technology Department of Sichuan Province: 2020YFN0025; Key Projects of Innovation and Entrepreneurship of Sichuan Science and Technology Department: 2021JDRC0091; Chengdu Technology Innovation R&D Project: 2021-YF05-01744-SN.

Institutional Review Board Statement: Not applicable.

Informed Consent Statement: Not applicable.

Data Availability Statement: Data are contained within the article.

Conflicts of Interest: The authors declare no conflict of interest.

References

1. Khatib, O. Real-Time Obstacle Avoidance for Manipulators and Mobile Robots. In Proceedings of the 1985 IEEE International Conference on Robotics and Automation, St. Louis, MO, USA, 25–28 March 1985; pp. 396–404. [CrossRef]
2. LaValle, S.M. Rapidly-Exploring Random Trees: A New Tool for Path Planning. 1998. Available online: http://citeseerx.ist.psu.edu/viewdoc/download?doi=10.1.1.35.1853&rep=rep1&type=pdf (accessed on 10 March 2022).
3. LaValle, S.M.; Kuffner, J.J. Randomized Kinodynamic Planning. *Int. J. Robot. Res.* **2001**, *20*, 378–400. [CrossRef]
4. Kuffner, J.J.; LaValle, S.M. RRT-connect: An efficient approach to single-query path planning. Proceedings 2000 ICRA. Millennium Conference. In Proceedings of the IEEE International Conference on Robotics and Automation. Symposia Proceedings (Cat. No.00CH37065), San Francisco, CA, USA, 24–28 April 2000; pp. 995–1001. [CrossRef]
5. Karaman, S.; Frazzoli, E. Incremental sampling-based algorithms for optimal motion planning. In Proceedings of the Robotics Science and Systems 2010, Zaragoza, Spain, 27 June 2010; Volume 104. Available online: http://www.roboticsproceedings.org/rss06/p34.pdf (accessed on 15 March 2022). [CrossRef]
6. Mohammed, H.; Romdhane, L.; Jaradat, M.A. RRT* N: An efficient approach to path planning in 3D for Static and Dynamic Environments. *Adv. Robot.* **2021**, *35*, 168–180. [CrossRef]
7. Akgun, B.; Stilman, M. Sampling heuristics for optimal motion planning in high dimensions. In Proceedings of the 2011 IEEE/RSJ International Conference on Intelligent Robots and Systems, San Francisco, CA, USA, 25–30 September 2011; pp. 2640–2645. [CrossRef]
8. Jeong, I.B.; Lee, S.J.; Kim, J.H. RRT*-quick: A motion planning algorithm with faster convergence rate. In *Robot Intelligence Technology and Applications 3*; Intelligent Systems and Computing: Cham, Switzerland, 2015; pp. 67–76. [CrossRef]
9. Jeong, I.B.; Lee, S.J.; Kim, J.H. Quick-RRT*: Triangular inequality-based implementation of RRT* with improved initial solution and convergence rate. *Expert Syst. Appl.* **2019**, *123*, 82–90. [CrossRef]
10. Adiyatov, O.; Varol, H.A. Rapidly-exploring random tree based memory efficient motion planning. In Proceedings of the 2013 IEEE International Conference on Mechatronics and Automation (ICMA), Takamatsu, Japan, 4–7 August 2013; pp. 354–359. [CrossRef]
11. Cao, X.; Zou, X.; Jia, C.; Chen, M.; Zeng, Z. RRT-based path planning for an intelligent litchi-picking manipulator. *Comput. Electron. Agric.* **2019**, *156*, 105–118. [CrossRef]
12. Wang, X.; Luo, X.; Han, B.; Chen, Y.; Liang, G.; Zheng, K. Collision-free path planning method for robots based on an improved rapidly-exploring random tree algorithm. *Appl. Sci.* **2020**, *10*, 1381. [CrossRef]
13. Zhang, H.; Wang, Y.; Zheng, J.; Yu, J. Path planning of industrial robot based on improved RRT algorithm in complex environments. *IEEE Access* **2018**, *6*, 53296–53306. [CrossRef]
14. Gong, H.; Yin, C.; Zhang, F.; Hou, Z.; Zhang, R. A path planning algorithm for unmanned vehicles based on target-oriented rapidly-exploring random tree. In Proceedings of the 2017 11th Asian Control Conference (ASCC), Gold Coast, QLD, Australia, 17–20 December 2017; pp. 760–765. [CrossRef]
15. Li, B.; Chen, B. An Adaptive Rapidly-Exploring Random Tree. *IEEE/CAA J. Autom. Sin.* **2021**, *9*, 283–294. [CrossRef]
16. Gao, X.; Wu, H.; Zhai, L.; Sun, H.; Jia, Q.; Wang, Y.; Wu, L. A rapidly exploring random tree optimization algorithm for space robotic manipulators guided by obstacle avoidance independent potential field. *Int. J. Adv. Robot. Syst.* **2018**, *15*, 1729881418782240. [CrossRef]
17. Wang, J.; Li, X.; Meng, M.Q.H. An improved RRT algorithm incorporating obstacle boundary information. In Proceedings of the 2016 IEEE International Conference on Robotics and Biomimetics (ROBIO), Qingdao, China, 3–7 December 2016; pp. 625–630. [CrossRef]
18. Veras, L.G.; Medeiros, F.; Guimaraes, L. Systematic literature review of sampling process in rapidly-exploring random trees. *IEEE Access* **2019**, *7*, 50933–50953. [CrossRef]
19. Zu, W.; Fan, G.; Gao, Y.; Ma, H.; Zhang, H.; Zeng, H. Multi-UAVs Cooperative Path Planning Method based on Improved RRT Algorithm. In Proceedings of the 2018 IEEE International Conference on Mechatronics and Automation (ICMA), Changchun, China, 5–8 August 2018; pp. 1563–1567. [CrossRef]
20. Li, H.; Liang, Y.; Wang, M.; Dan, T. Design and implementation of improved RRT algorithm for collision free motion planning of high-dimensional robot in complex environment. In Proceedings of the 2012 2nd International Conference on Computer Science and Network Technology, Changchun, China, 29–31 December 2012; pp. 1391–1397. [CrossRef]
21. Kang, G.; Kim, Y.B.; You, W.S.; Lee, Y.H.; Oh, H.S.; Moon, H.; Choi, H.R. Sampling-based path planning with goal oriented sampling. In Proceedings of the 2016 IEEE International Conference on Advanced Intelligent Mechatronics (AIM), Banff, AB, Canada, 12–15 July 2016; pp. 1285–1290. [CrossRef]
22. Ahmadyan, S.N.; Kumar, J.A.; Vasudevan, S. Goal-oriented stimulus generation for analog circuits. In Proceedings of the 49th Annual Design Automation Conference, New York, NY, USA, 3 June 2012; pp. 1018–1023. [CrossRef]
23. Wang, J.; Wu, S.; Li, H.; Zou, J. Path planning combining improved rapidly-exploring random trees with dynamic window approach in ROS. In Proceedings of the 2018 13th IEEE Conference on Industrial Electronics and Applications (ICIEA), Wuhan, China, 31 May–2 June 2018; pp. 1296–1301. [CrossRef]

24. Qi, J.; Yang, H.; Sun, H. MOD-RRT*: A sampling-based algorithm for robot path planning in dynamic environment. *IEEE Trans. Ind. Electron.* **2020**, *68*, 7244–7251. [CrossRef]
25. Yuan, C.; Zhang, W.; Liu, G.; Pan, X.; Liu, X. A heuristic rapidly-exploring random trees method for manipulator motion planning. *IEEE Access* **2019**, *8*, 900–910. [CrossRef]
26. Wen, N.; Zhang, R.; Liu, G.; Wu, J.; Qu, X. Online planning low-cost paths for unmanned surface vehicles based on the artificial vector field and environmental heuristics. *Int. J. Adv. Robot. Syst.* **2020**, *17*, 1729881420969076. [CrossRef]
27. Qureshi, A.H.; Ayaz, Y. Potential functions based sampling heuristic for optimal path planning. *Auton. Robot.* **2016**, *40*, 1079–1093. [CrossRef]
28. Zaid, T.; Qureshi, A.H.; Yasar, A.; Raheel, N. Potentially guided bidirectionalized rrt* for fast optimal path planning in cluttered environments. *Robot. Auton. Syst.* **2018**, *108*, 13–27. [CrossRef]
29. Yang, Y.; Liu, J.; Zheng, Y.; Huang, Q. Obstacle Avoidance Path Planning of Manipulator of Forestry Felling & Cultivation Machine. *Sci. Silvae Sin.* **2021**, *57*, 179–192. [CrossRef]
30. Liu, Y.; Zhao, H.; Liu, X.; Xu, Y. An Improved RRT Industrial Robot Path Avoidance Planning Algorithm. *Inf. Control* **2021**, *50*, 235–246. [CrossRef]
31. Liu, C.; Han, J.; An, K. Dynamic Path Planning Based on an Improved RRT Algorithm for RoboCup Robot. *Robot* **2017**, *39*, 8–15. [CrossRef]
32. Li, Y.; Xu, D. Cooperative Path Planning of Dual-arm Robot Based on Attractive Force Self-adaptive Step Size RRT. *Robot* **2020**, *42*, 606–616. [CrossRef]
33. Ruan, X.; Zhou, J.; Zhang, J.; Zhu, X. Robot goal guide RRT path planning based on sub-target search. *Control Decis.* **2020**, *35*, 2543–2548. [CrossRef]
34. Lin, N.; Zhang, Y. An adaptive RRT based on dynamic step for UAVs route planning. In Proceedings of the 2014 5th IEEE International Conference on Software Engineering and Service Science (ICSESS), Beijing, China, 27–29 June 2014; pp. 1111–1114. [CrossRef]
35. Wang, C.; Meng, Q.H. Variant step size RRT: An efficient path planner for UAV in complex environments. In Proceedings of the 2016 IEEE International Conference on Real-Time Computing and Robotics (RCAR), Angkor Wat, Cambodia, 6–10 June 2016; pp. 555–560. [CrossRef]
36. Li, Y.; Wang, S.; Jiang, L.; Meng, J.; Xie, Y. Motion Planning of Mobile Manipulator Based on RRT with Sparse Nodes. *China Mech. Eng.* **2020**, *32*, 9.
37. Mellinger, D.; Kumar, V. Minimum snap trajectory generation and control for quadrotors. In Proceedings of the 2011 IEEE International Conference on Robotics and Automation, Shanghai, China, 9–13 May 2011; pp. 2520–2525. [CrossRef]
38. Richter, C.A.; Bry, A.P.; Roy, N. Polynomial Trajectory Planning for Aggressive Quadrotor Flight in Dense Indoor Environments. In *Robotics Research*; Springer Tracts in Advanced Robotics; Springer: Cham, Switzerland, 23 April 2016; Volume 114, pp. 649–666. [CrossRef]
39. Reddy, P.; Pham, Q.V.; Prabadevi, B.; Deepa, N.; Dev, K.; Gadekallu, T.; Ruby, R.; Liyanage, M. Industry 5.0: A survey on enabling technologies and potential applications. *J. Ind. Inf. Integr.* **2021**, *26*, 100257. [CrossRef]
40. Gadekallu, T.R.; Rajput, D.S.; Reddy, M.P.K.; Lakshmanna, K.; Bhattacharya, S.; Singh, S.; Jolfaei, A.; Alazab, M. A novel PCA–whale optimization-based deep neural network model for classification of tomato plant diseases using GPU. *J. Real-Time Image Process.* **2021**, *18*, 1383–1396. [CrossRef]
41. Li, T.; Feng, Q.; Qiu, Q.; Xie, F.; Zhao, C. Occluded Apple Fruit Detection and Localization with a Frustum-Based Point-Cloud Processing Approach for Robotic Harvesting. *Remote Sensing.* **2022**, *14*, 482. [CrossRef]
42. Li, T.; Qiu, Q.; Zhao, C. Task planning of multi-arm harvesting robots for high-density dwarf orchards. *Trans. Chin. Soc. Agric. Eng. (Trans. CSAE)* **2021**, *37*, 1–10. [CrossRef]
43. Wang, W.; Zuo, L.; Xu, X. A learning-based multi-RRT approach for robot path planning in narrow passages. *J. Intell. Robot. Syst.* **2018**, *90*, 81–100. [CrossRef]

agriculture

Article

Contour Resampling-Based Garlic Clove Bud Orientation Recognition for High-Speed Precision Seeding

Jian Liu [1,2], Jin Yuan [1,3,*], Jiyuan Cui [1], Yunru Liu [1] and Xuemei Liu [1,3]

1 College of Mechanical & Electronic Engineering, Shandong Agricultural University, Tai'an 271018, China
2 College of Artificial Intelligence, Shandong University of Engineering and Vocational Technology, Jinan 250200, China
3 Shandong Agricultural Equipment Intelligent Engineering Laboratory, Tai'an 271018, China
* Correspondence: jyuan@sdau.edu.cn; Tel.: +86-151-6538-7381

Abstract: Achieving fast and accurate recognition of garlic clove bud orientation is necessary for high-speed garlic seed righting operation and precision sowing. However, disturbances from actual field sowing conditions, such as garlic skin, vibration, and rapid movement of garlic seeds, can affect the accuracy of recognition. Meanwhile, garlic precision planters need to realize a recognition algorithm with low-delay calculation under the condition of limited computing power, which is a challenge for embedded computing platforms. Existing solutions suffer from low recognition rate and high algorithm complexity. Therefore, a high-speed method for recognizing garlic clove bud direction based on deep learning is proposed, which uses an auxiliary device to obtain the garlic clove contours as the basis for bud orientation classification. First, hybrid garlic breeds with the largest variation in shape were selected randomly and used as research materials, and a binary image dataset of garlic seed contours was created through image sampling and various data enhancement methods to ensure the generalization of the model that had been trained on the data. Second, three lightweight deep-learning classifiers, transfer learning based on MobileNetV3, a naive convolutional neural network model, and a contour resampling-based fully connected network, were utilized to realize accurate and high-speed orientation recognition of garlic clove buds. Third, after the optimization of the model's structure and hyper-parameters, recognition models suitable for different levels of embedded hardware performance were trained and tested on the low-cost embedded platform. The experimental results showed that the MobileNetV3 model based on transfer learning, the naive convolutional neural network model, and the fully connected model achieved accuracy of 98.71, 98.21, and 98.16%, respectively. The recognition speed of the three including auxiliary programs was 19.35, 97.39, and 151.40 FPS, respectively. Theoretically, the processing speed of 151 seeds per second achieves a 1.3 hm^2/h planting speed with single-row operation, which outperforms state-of-the-art methods in garlic-clove-bud-orientation recognition and could meet the needs of high-speed precise seeding.

Keywords: garlic seeding; orientation recognition; garlic clove righting; deep learning; fully connected neural network

Citation: Liu, J.; Yuan, J.; Cui, J.; Liu, Y.; Liu, X. Contour Resampling-Based Garlic Clove Bud Orientation Recognition for High-Speed Precision Seeding. *Agriculture* **2022**, *12*, 1334. https://doi.org/10.3390/agriculture12091334

Academic Editor: Jin He

Received: 5 August 2022
Accepted: 24 August 2022
Published: 29 August 2022

1. Introduction

Garlic is a globally cultivated crop due to its rich nutritional and medicinal value. According to 2022 statistical data from the FAO, the garlic planting area in China in 2020 was about 830,000 hectares, and garlic production reached 20 million tons, the largest in the world. However, the current mechanized planting of garlic is not efficient, and the sowing period of garlic is very short, so high-speed, high-efficiency, and accurate planters are urgently needed.

Many studies have shown that the orientation of garlic cloves buds during garlic sowing into the soil significantly affects the time and consistency of seedling emergence, garlic yield, and garlic bulb quality [1,2]. One study showed that when the garlic clove buds were facing upward and the inclination angle was within ±45°, all the indexes of garlic plants performed well. When the garlic clove buds were placed horizontally, the performance of each index was slightly inferior to that of the garlic clove buds facing upward. When the garlic clove buds were facing downward and the inclination angle was within ±45°, the performance of each index was the worst, making them prone to garlic seed necrosis, uneven seedling emergence time, disordered and weak growth, and other problems [3]. Therefore, the precise sowing of garlic first needs to meet the agronomic requirements of garlic planting with clove buds being placed upright.

Cangshan and *Jinxiang* garlic are the most widely cultivated garlic breeds in China. At present, existing garlic planters mostly adopt a righting mechanism to adjust the garlic clove bud direction. The garlic cloves of *Cangshan* are neat and uniform, and their weight, geometric shape, and the center of gravity are consistent, which could be utilized by a mechanical mechanism to achieve garlic bud upright sowing into soil [4,5]. *Jinxiang* garlic, the most commonly planted variety, is a hybrid breed with variable sizes of cloves, irregular geometric shape, and unstable center of gravity, and the mechanical righting method often has a poor effect [6]. The righting of hybrid garlic seeds remains an open problem, and beyond that, high-speed precision sowing requires shorter cycling time for righting seeds.

The correct recognition of garlic clove bud orientation is the foundation of garlic clove righting operation, and computer vision is the only feasible way to judge the clove bud orientation of hybrid breed garlic. In the early stage, some studies tried to use artificial feature engineering to solve the orientation recognition of garlic seeds, such as the density of edges [7], the position of centroid [8], the curvature of contour [9], etc. These methods are effective for garlic cloves with a standard shape, but poor for garlic cloves with residual garlic husks and abnormal spikes, while commercial garlic seeds often have residual husks and irregular geometric shapes, so the robustness of the artificial features engineering algorithm is not ideal, and the actual use is very poor.

At present, as automatic feature-learning methods, deep-learning methods perform well and have been widely used in the agricultural field [10], including in the orientation recognition of garlic clove buds [11]. However, some methods can only identify the position of qualified garlic clove buds, lack a description of unqualified positions, and cannot provide position information to support the righting operation of the garlic planter.

The above-mentioned studies are limited to the scope of algorithms and theory, while some other studies are focused on practical application, including the integration of algorithms in embedded hardware that can be equipped with garlic planters [12]. Li et al. designed an automatic righting device for garlic clove buds based on the Jetson Nano processor. The success rate of garlic clove bud righting of the device reached 96.25%, and when the number of parallel sowing rows was 12, its sowing efficiency was 0.099–0.132 hm^2/h [13]. The righting method of Li et al. requires a Jetson Nano processor in each righting channel to achieve the planting efficiency of 0.099–0.132 hm^2/h. However, the hardware cost of Jetson Nano is relatively high (US $99), so this design may not be conducive to commercial application.

So far, no research has tried to realize fast and accurate recognition of garlic seed orientation that can meet the needs of high-speed and accurate sowing of garlic with a low-cost embedded processor, and no research has attempted to solve the problem that the abnormal shape of garlic seeds, such as garlic skin residue, etc., affects orientation recognition. The above two research gaps hinder the practical application and large-scale promotion of machine-vision-based garlic seed orientation identification methods. Therefore, this paper proposes a robust, lightweight, and high-performance garlic bud orientation recognition method based on deep learning to achieve high-speed and accurate orientation recognition based on a single low-cost embedded processor.

Disturbances from actual field sowing conditions, such as garlic skin, vibration, and rapid movement of garlic seeds, can affect the accuracy of recognition. Meanwhile, garlic precision planters are in need of a recognition algorithm with a low delay calculation under the condition of limited computing power, which is a challenge for embedded computing platforms. In order to solve these problems, this study carried out the following work:

- Building of a special data set for model training, including shape anomalies such as garlic residue and motion blur, to ensure the generalization ability of the model to real scenes.
- Use of multiple deep-learning and feature-compression methods to realize garlic bud direction recognition and optimization of the model by tune operations.
- Performance tests of the models on low-cost embedded boards, selection of the optimal model, comparison with other methods to verify the superiority of this method.

The main contributions of this paper are as follows: an efficient method for obtaining a contour map is proposed, and a data enhancement method is proposed on this basis; quick-recognition models of lightweight CNN MobileNetV3 and naive CNN based on the contour map are proposed for high-speed recognition of garlic seed orientation; a high-speed contour orientation recognition method based on highly compressed contour features is proposed that realizes ultra-high-speed recognition on low-cost embedded platform.

Finally, a recognition speed of 151.40 FPS was achieved on the OrangePi 3 LTS, which can support sowing operations at a speed of 1.3 hm^2/h, which is superior to the state-of-the-art method of garlic orientation recognition.

2. Materials and Methods

2.1. Garlic Clove Data Collection

In the field of deep learning, especially in image recognition, the collection of complete datasets that cover all application conditions is critical. The operator can judge the direction of a garlic clove bud mainly based on an outline of visual information. Based on this, the binary contour image of garlic seeds is used as the basis for judging the orientation of garlic cloves. Along with the support of a specific device, it is very easy to obtain an outline of garlic cloves. This paper used a strong light source as the background, obtained the shadow image and binary image of the garlic seed, and then applied the *findcontours* function of the graphics library OpenCV. This design has the following advantages: first, the binary contour image eliminates the imaging differences between different image sensors. Second, using a single-channel image as the input of the CNN model helps to reduce the amount of computation. Third, many traditional methods [7,9] also use contour images as input data, and using binary contours as model input is conducive to algorithm integration between different devices.

2.1.1. Garlic Clove Image Sample Collection

Deep learning requires that the training data and test data meet the conditions of being independent and identically distributed to ensure the generalization ability of the model. In the practical application stage of the model, the input data of the model must be independent and identically distributed with the training set in order to make the model work effectively. Therefore, considering the practical application of a deep learning model in a garlic planter, the training samples should cover wider morphological diversity distribution to enhance the robustness of model. When selecting training samples, one should not only ensure the garlic clove sizes, weights, and appearance, but also consider the influence of garlic seed production technology and other factors on garlic seed morphology, such as skin residue.

In this paper, the binary contour image was used as the model input, and the morphological features such as color and texture were discarded in the process of extracting the contours, while some edge features were preserved. Commercial garlic seeds are prone to having residual garlic skins and abnormal spikes. These garlic skin residues have a great impact on the extraction of contour images, and sometimes the extracted contour images may seriously deviate from the standard shape of garlic seeds. Therefore, the selection of training samples should also consider the situation of carried garlic skins while seeding. Because the individual shape of garlic cloves of hybrid garlic breed is the most diverse, we randomly selected *Jinxiang* garlic and divided it into cloves, retained all the garlic cloves without screening, and obtained a total of 735 garlic cloves for image samples. When dividing garlic cloves, about $\frac{1}{2}$ of the skin residue of the garlic cloves was retained to ensure consistency with real sowing of garlic seeds, as shown in Figure 1a.

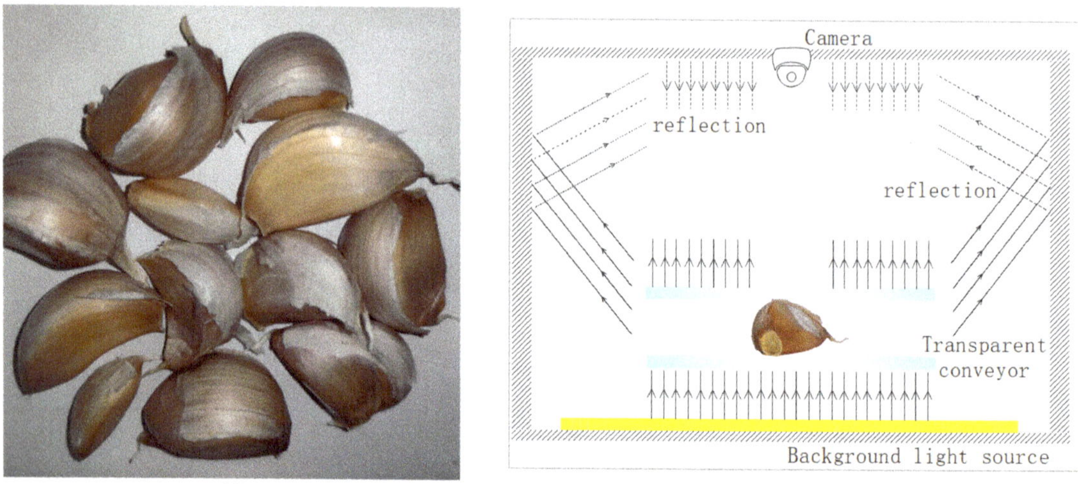

(a) (b)

Figure 1. (**a**) Garlic clove samples and (**b**) image acquisition device.

Sample Acquisition Device and Image Preprocessing

In order to directly obtain the contour images of the garlic cloves, a garlic seed shooting device was designed that uses a transparent clamping belt to clamp and transmit the garlic cloves and adopts the method of back illumination of area light source. The area light source is placed below the transparent clamping belt, and the image sensor is placed above the transparent clamping belt. The clamping transmission module is wrapped by an opaque shell to avoid the influence of external light on image acquisition. The light emitted by the area light source passes through a transparent clamping tape to form a clear garlic clove shadow image on the vision sensor, as shown in Figure 1b. The image collected under the ideal state is shown in Figure 2a. However, because the reflection in the shell cannot be completely eliminated, some reflected light will still be cast on the upper surface of the garlic clove, and the continuous transmission of the garlic clove will bring the dust that adhered to the garlic clove into the shell, reducing the contrast between the shadow area of the garlic clove and the background, as shown in Figure 2b.

The above situation increases the difficulty of binarization of shadow image. Because the shadow of the garlic seed image is too dark, the binarization performance to achieve contour is poor. Manually adjusting the binarization threshold can alleviate the problem of misclassifying the area around the shadow image, but the shadow of the garlic clove will be lost and cannot be applied automatically, as shown in Figure 2c,d. An extremely low-computation pixel compensation method is proposed to solve this problem. The control system records an image of the empty conveyor belt without cloves, and then calculates the pixel difference matrix between this image and a pure white image and saves it as a pixel compensation matrix. When intercepting the garlic clove shadow image frame, the intercepted image frame is added to the pixel difference matrix, and then the Otsu binarization method [14] is used to obtain a high-quality binarized image, as shown in Figure 3. The calculation rules are shown in Equations (1) and (2), where O represents the image with no load when the device is initialized; C stands for the pixel compensation matrix; X represents the image frame collected in real time; X' represents the image frame after compensation; and m and n represent the number of rows and columns of the pixel matrix, respectively.

$$C = 255 - O = \left[255 - o_{ij}\right]_{m \cdot n} \tag{1}$$

$$X' = X + C = \left[\max\left(x_{ij} + c_{ij},\ 255\right)\right]_{m \cdot n} \tag{2}$$

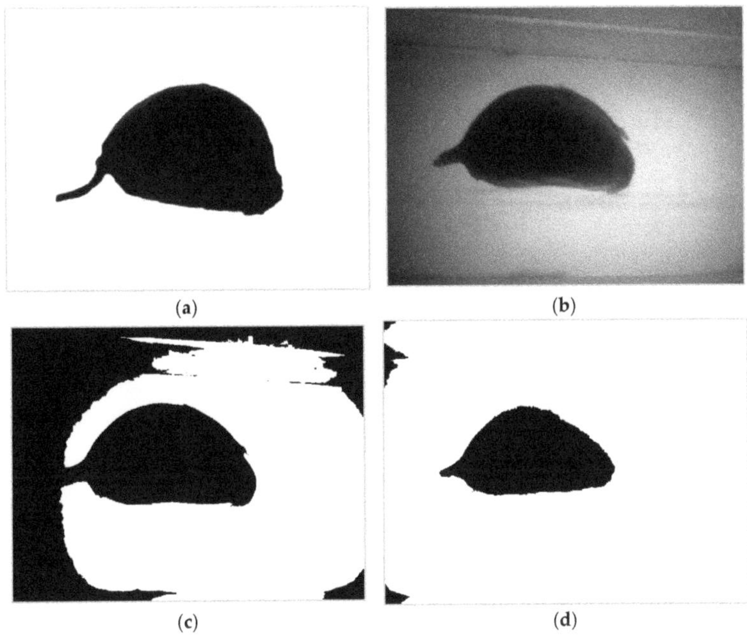

Figure 2. The lighting environment in the device affects shadow imaging. (**a**) Idealized shadow image, (**b**) Actual shadow image, (**c**) Otsu-based binary image, (**d**) Fixed-threshold-based binary image.

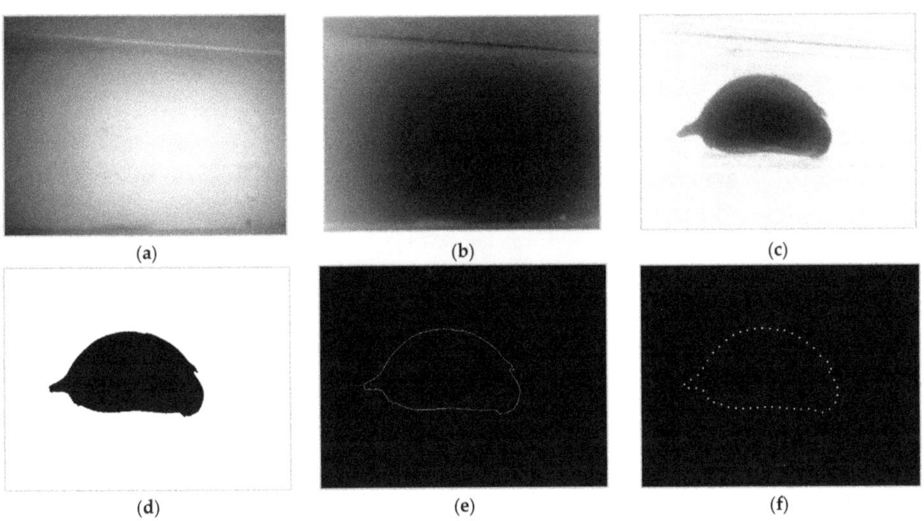

Figure 3. Contour extraction after pixel compensation. (**a**) Background of conveyor, no clove, (**b**) Pixel compensation matrix, (**c**) Compensated garlic clove shadow, (**d**) Binary shadow image, (**e**) Outline of garlic seed, (**f**) Contour sampling points.

Dataset Acquisition Method

During the sample image acquisition, the mechanical device introduced in the previous section was used to transmit garlic seeds, the vision sensor on it was used to record a video, and then the image frames were extracted from the acquired video. A total of 1470 original image samples were obtained. Among them, 1172 images were randomly selected as the training set, and the remaining 298 images were used as the validation set. Since the length–width ratio of most image sensors is 4:3, when applied to the seeder, the long side of the picture was parallel to the travel direction of garlic seeds to obtain a larger observation field of garlic seeds. In order to meet this demand, the image samples used for model training were processed with the same length–width ratio and were finally saved with an image size of 640 × 480 by cutting or expanding the image boundary (Figure 4). Image rotation does not change the shape of garlic cloves. In this study, the original image samples were all adjusted to the upward state of garlic clove buds through image rotation operation. In the data enhancement stage, image samples with other orientations were generated through image rotation.

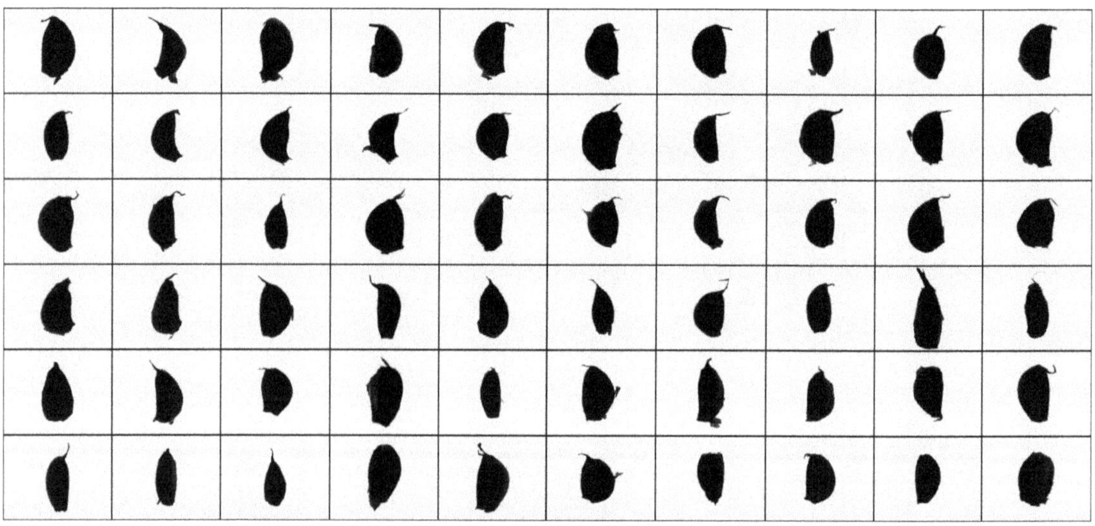

Figure 4. Part of the original sample.

2.1.2. Data Enhancement for Datasets

Because the original images were adjusted to the garlic clove bud upward state, the key task in data enhancement was to generate image samples with left, bottom, and right orientation. In addition, some image transformations need to be performed on the image samples to make the images of the dataset more diverse to ensure the generalization ability of the model. In order to make the training data and the validation data conform to the conditions of independent and identical distribution, the same data augmentation operation was performed on the training samples and the validation samples, and the samples in the training set and validation set were always isolated during this process. The image enhancement methods include horizontal flipping, stretching, shearing, translation, rotation, and motion blur. All these methods except motion blur can be realized by two-dimensional geometric transformation, which can be completed by multiplying the pixel matrix of the image by a homogeneous transformation matrix. The mathematical expression of this process is shown in Equation (3). In order to enhance the generalization of the sample to the image acquisition environment in the garlic seeder, these methods need to follow a certain logical order.

$$X' = M_f \cdot M_s \cdot M_d \cdot M_t \cdot M_r \cdot X \tag{3}$$

$$M_f = \begin{bmatrix} -1 & 0 & w \\ 0 & 1 & 0 \\ 0 & 0 & 1 \end{bmatrix}; M_s = \begin{bmatrix} s_x & 0 & 0 \\ 0 & s_y & 0 \\ 0 & 0 & 1 \end{bmatrix}; M_d = \begin{bmatrix} 1 & d_x & 0 \\ d_y & 1 & 0 \\ 0 & 0 & 1 \end{bmatrix}; M_t = \begin{bmatrix} 1 & 0 & t_x \\ 0 & 1 & t_y \\ 0 & 0 & 1 \end{bmatrix}; M_r = \begin{bmatrix} \cos\theta & -\sin\theta & 0 \\ \sin\theta & \cos\theta & 0 \\ 0 & 0 & 1 \end{bmatrix}$$

where X represents the original image, and X' represents the image after transformation; M_f, M_s, M_d, M_t, and M_r represent the transformation matrix of horizontal inversion, stretching, shearing, translation, and rotation, respectively; w represents the width of the image; s_x and s_y represent the stretching ratio in two directions; d_x and d_y represent the shearing amplitude in two directions; t_x and t_y represent the translation distance in two directions; θ represents the rotation angle of the image.

Morphological Diversity

First, a flip operation on the image was performed. Due to the irregular shape of garlic cloves, they usually show different external contours when the two sides of their abdomen are facing vertically downward. Therefore, the diversity of the dataset can be increased through the horizontal flipping operation (Figure 5). After this operation, the sample size doubled to 2940.

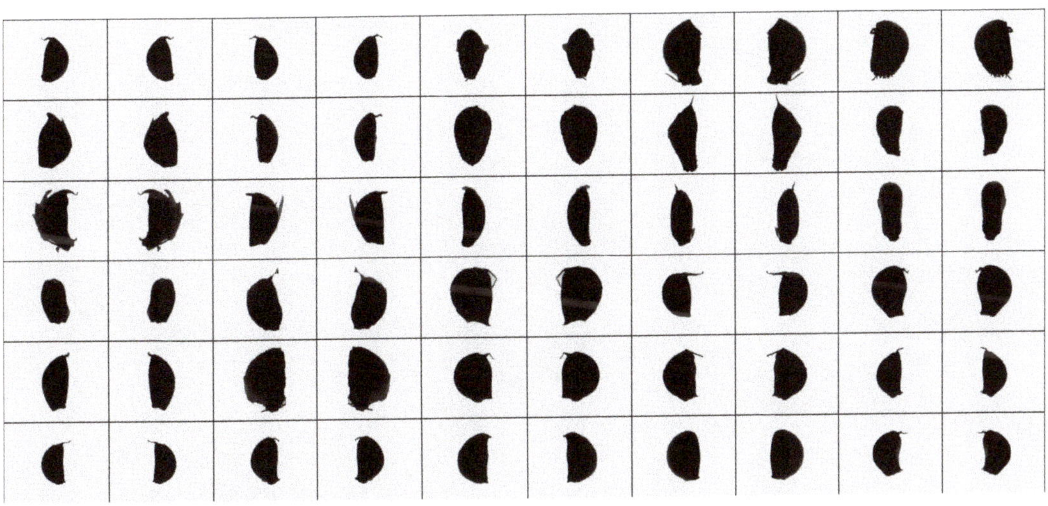

Figure 5. Horizontal reversal amplification sample.

Then, stretch, shear, and translation were performed. These three operations can effectively increase the morphological diversity of the image and are still effective after image rotation. The stretching operation range was a random value in the range of 0–20%. The strength of the shear operation was a random value in the range of 0–10. The range of translation operation amplitude was a random value in the range of 0–10%. Through the overlapping operation of the above three transformations, the image samples were amplified to 29,400. The amplified samples are shown in Figure 6.

Image Rotation and Class Generation

When the plane is divided into four equal regions, upper, left, lower and right, the range of each region is 90°. In order to ensure the generalization ability of the model for irregular orientation, before generating image categories, a random small-amplitude rotation operation was performed on the image samples. The rotation amplitude in the ideal state should be ± 45°, but the original image was manually rotated and righted, and there might be subtle deflection that is not easy to detect. Furthermore, because the data enhancement operation includes shear transformation of random amplitude, a rotation transformation in the range of ±30° was performed on each image, acting directly on the original image, without generating new image samples. The transformed samples are shown in Figure 7.

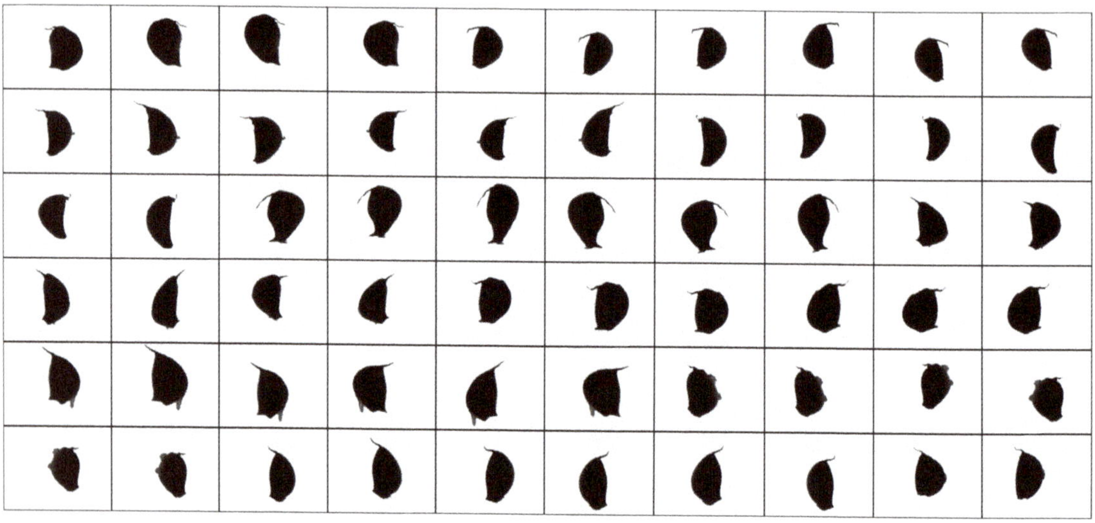

Figure 6. Stretching, shearing, and translation.

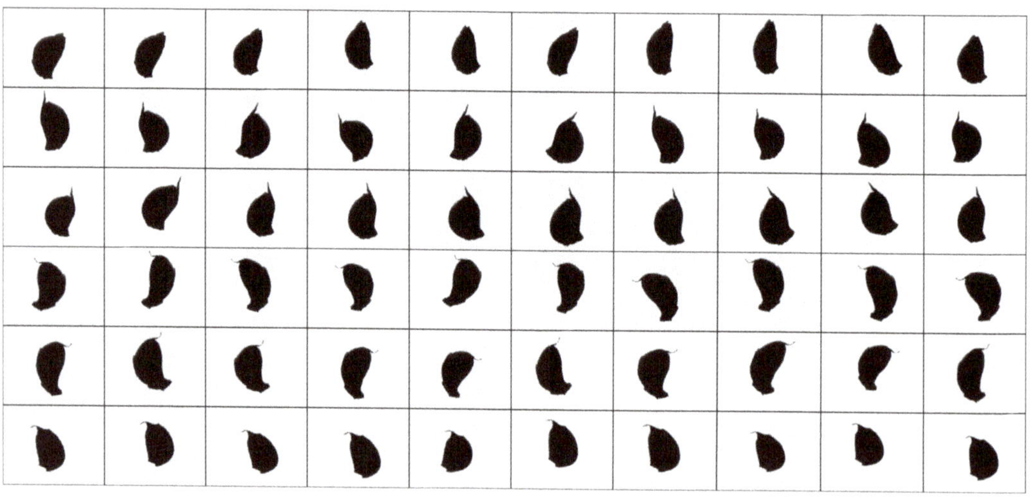

Figure 7. Image samples after rotation in the range of ± 30°.

After completing the above operations, each original image was rotated by 90, 180, and 270° counterclockwise to obtain the standard left, lower, and right images. At this time, the sample size expanded to 117,600. The samples of each orientation class are shown in Figure 8.

Motion Blur and Contour Extraction

Seeding speed is an important performance index of garlic seeders. In order to achieve high-speed seeding, garlic seed images should be collected in motion, which may lead to motion blur in the collected images. Because of the influence of uncertain motion blur on contour extraction, the data enhancement operation should also include motion blur with a certain probability and range. In this study, image samples were randomly selected with a probability of 50%, and motion blur with random amplitude was applied in the direction parallel to the long edge of the image (Figure 9).

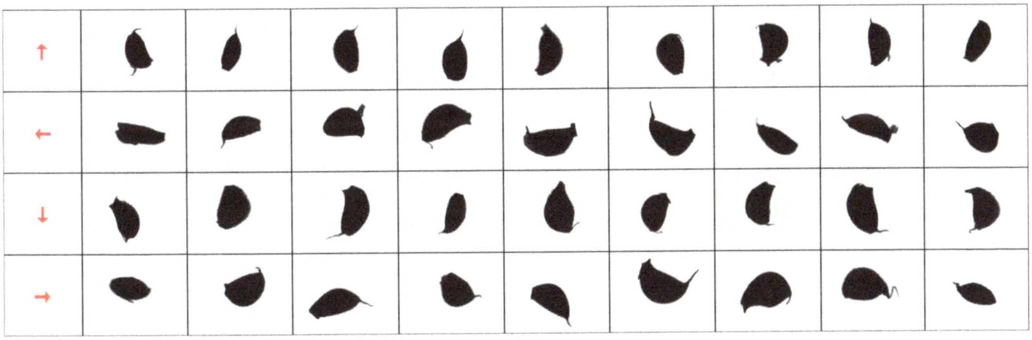

Figure 8. Samples of each orientation class.

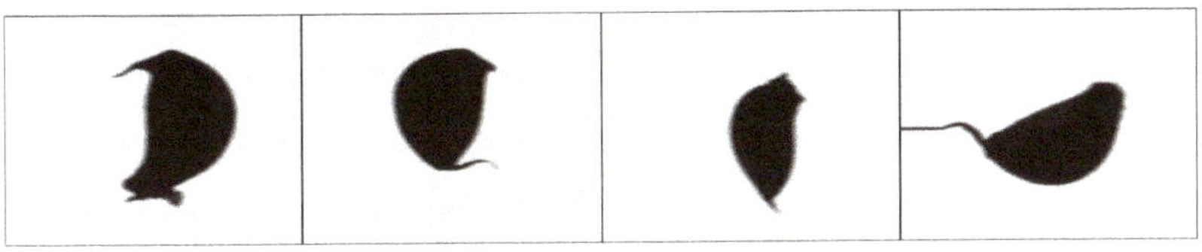

Figure 9. Motion blur in different direction classes.

After all the data enhancement operations were completed, the contour of the image samples were extracted one by one, finally forming the garlic seed contour dataset for model training (Figure 10).

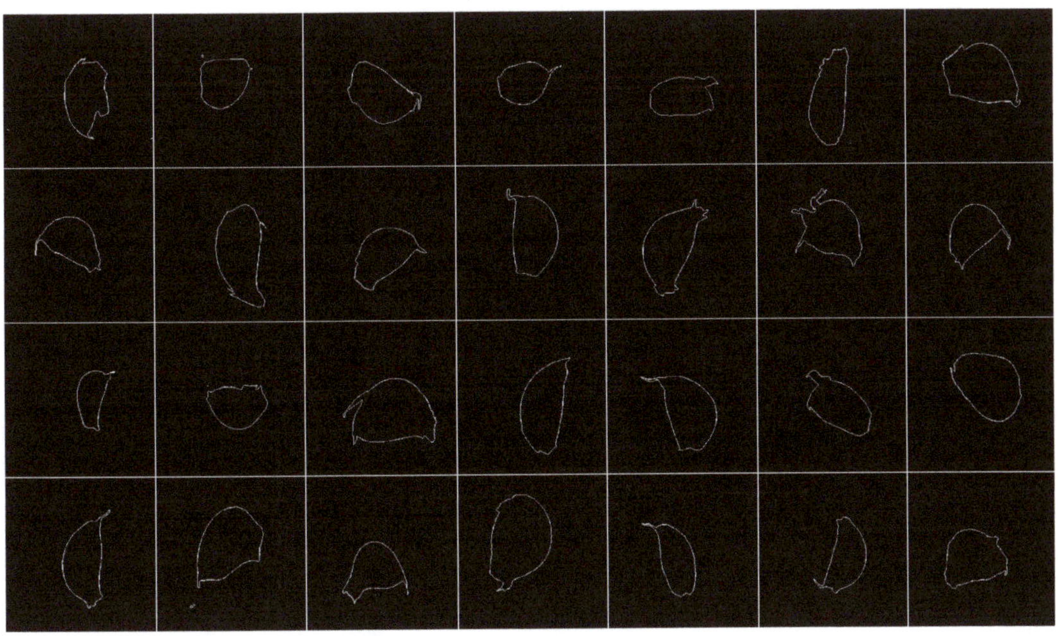

Figure 10. Final generated outer contour image samples.

Logical Sequence of Data Enhancement Operations

If the rotation operation used to generate the orientation classification is performed before the zoom, stretch, shear, and translation operations, image samples with greater morphological differences can be generated, theoretically promote the generalization ability of the deep-learning model during training. However, it was found in experiments that datasets with the same orientation classification morphology have better performance. A possible reason is that when the four classifications contain samples with the same morphology, the deep learning model can suppress the influence of morphology on classification and pay more attention to the high-level semantic feature of "orientation".

Because motion blur is directional, motion blur transformation should be carried out after rotation transformation, and motion blur may affect the edge contour, so contour extraction should be carried out after motion blur operation.

Storage of Dataset

The lightweight CNN model discussed has low computation complexity. A large batch size can be used in training on PC, and the training/inferring time of each batch is very short. In the initial practice, it was found that the transmission speed of training samples was often lower than the processing speed of the model; therefore, the way the dataset is stored has been improved. In the experimental environment of this paper, when using TFRecord format defined in TensorFlow [15] to store datasets, the input speed of samples could reach more than 10 times that of batch reading image files, which is faster than the inference speed of all deep learning models introduced in this paper. Therefore, this format is used as one of the storage schemes for datasets and testing of several CNN models.

The fully connected model proposed in Section 2.2.3 takes the pixel coordinates of garlic clove contours as the input. When the dataset composed of image samples is converted into the form of pixel coordinate array, the volume of the dataset is further reduced, and the whole dataset can be loaded into memory during training. The format DataFrame of Pandas [16] is used to store an array of contour point coordinates for all samples, which is changed to H5 format for loading on each training task. For the above two dataset formats, the shuffle operation was implemented for each training epoch to obtain better training results.

2.2. Lightweight Recognition Model of Garlic Clove Bud Orientation

For the garlic seed orientation recognition method studied in this paper, its accuracy is the first important criterion. Secondly, it is of practical significance to improve the running speed of the recognition model under the premise of ensuring accuracy. At the same time, the hardware cost of the algorithm application is also one of the factors considered in this paper, which is a necessary condition to ensure the generalizability of the application. Low hardware cost means low computing performance, so the complexity of the recognition model needs to be greatly reduced, which should be key for input features and lightweight models. Therefore, the main contribution of this paper is to propose a deep-learning model, that is, to improve the recognition rate and running speed of the model, give priority to ensuring the accuracy of the algorithm, and try to lighten the model on this basis to adapt to low-cost embedded platforms. The application of a convolutional network and a fully connected neural network in garlic-clove orientation recognition was attempted in this study. The convolutional network included MobileNetV3 [17] with relatively complex structure and the naive CNN model, composed of convolutional-pooling stacking only. These directly used garlic contour image samples as input, and automatically completed feature extraction through image convolutional operations. The fully connected model used the contour point coordinate set sampled from the image samples as the input, and the contour point sampling operation can be regarded as a feature extraction method.

2.2.1. Transfer Learning Based on MobileNetV3

MobileNetV3 is an excellent lightweight deep-learning model with two versions, large and small, which can be used as solutions for different levels of hardware performance. The TensorFlow framework comes with MobileNetV3 implementation code based on Keras API and provides six groups of pre-training parameters for training from an ImageNet dataset, which correspond to three forms of large and small models: standard width, 0.75 width, and standard width minimal mode. Based on this, transfer learning was tested.

The input size of the model is directly related to the amount of calculation required. In terms of ensuring the recognition performance of the model, the smaller the input size, the better. It was found in the experiment that when the image sample size was scaled to 120 × 160, the recognition performance of the model did not decrease significantly, so the input size of the MobileNetV3 model was modified to (120, 160, 1). The orientation of garlic clove buds was divided into four categories; the output size of the corresponding model is a 4-dimensional vector. Because the input and output of the model are redefined, only the weight values of the intermediate layers that are consistent with the original model parameter structure were loaded when loading the pre-training weights, and the intermediate layers with different structures were equivalent to training from zero.

Twelve model structures, including six with pre-training weights, were trained. The training results (Table 1) show that the transfer learning is effective. Using the pre-trained weights on the ImageNet dataset to perform transfer learning on the garlic seed outline, the image dataset could obtain a higher accuracy than starting training from zero. Overall, the large model performed better than the small model. The performance of the minimalistic mode was lower than that of the non-minimalistic mode, but this gap was not noticeable when pre-trained weights were not used. By comparing the number of parameters and calculation of different models, it can be seen that reducing the width factor mainly reduces the amount of calculation required for the model, which can improve the running speed of the model, while the minimalistic mode mainly reduces the number of parameters of the model, which can reduce the memory consumption of the model. The accuracy rate of all the model forms can reach above 0.96, and they all have certain application value. The structure of MobileNetV3-Large is shown in Figure 11.

Table 1. Overview of the performance of the transfer learning model.

Model Form	Parameter Quantity	Calculated Quantity (FLOPs)	Accuracy of Training from Zero	Accuracy of Transfer Learning
Large 1.0	4.04 M	0.178 G	0.97468	0.98352
Large 0.75	2.61 M	0.126 G	0.97423	0.98037
Large 0.5	1.36 M	0.0592 G	0.97111	-
Large 1.0 minimalistic	2.55 M	0.168 G	0.97224	0.97808
Large 0.75 minimalistic	1.71 M	0.119 G	0.96860	-
Large 0.5 minimalistic	0.97 M	0.0544 G	0.96541	-
Small 1.0	1.46 M	0.0453 G	0.96981	0.97434
Small 0.75	0.98 M	0.0339 G	0.96723	0.97528
Small 0.5	0.55 M	0.0177 G	0.96381	-
Small 1.0 minimalistic	0.99 M	0.0401 G	0.96733	0.96848
Small 0.75 minimalistic	0.69 M	0.0298 G	0.96194	-
Small 0.5 minimalistic	0.41 M	0.0149 G	0.96035	-

Note: - indicates that the test could not be performed due to a lack of pre-trained weights.

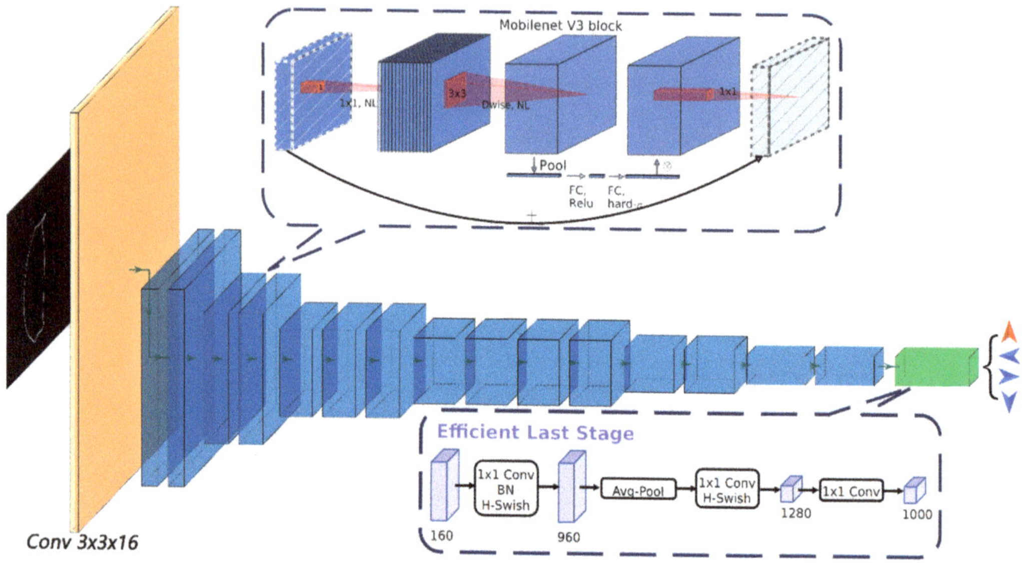

Figure 11. Structure of MobileNetV3.

2.2.2. Naive CNN Model

Compared with classic CNN models such as AlexNet [18] and VGG NETS [19], MobileNetV3 has a relatively complex model structure, and includes some advanced designs, such as depthwise separable convolution [20], inverse residual block structure [21], squeeze-and-excitation block [22] and h-swish [17] activation function. Along with their support, MobileNetV3 shows excellent classification performance for some large-scale natural image datasets. The garlic contour image is very different from natural images. As a binary image, its content density and information density are very low. In order to explore which designs of MobileNetV3 are most helpful to the classification task of garlic contour images, some experiments were done. By training a modified model that separately applies the squeeze-and-excitation module, h-swish activation function, and 5×5 convolution kernel, this study found that the 5×5 convolution kernel had the greatest impact on model performance among the three, while the squeeze-and-excitation and the h-swish activation function had little effect on model performance. After determining the importance of convolutional kernel size, a series of naive CNN models with a structure similar to VGG NETS were constructed, which were compared with MobileNetV3 to analyze the impact of inverse residual block structure on the performance of the model and further verify the importance of convolutional kernel size.

In order to make the training results of the models more comparable, the same training conditions as MobileNetV3 transfer learning were used to train these models. The performance achieved after full convergence is shown in Table 2. Because these models have a simple structure, compared with the MobileNetV3 model, the parameters and calculation of the naive CNN model with similar performance are greatly reduced. This seems to indicate that the model with a simple structure is more suitable for solving the direction judgment problem of garlic clove contour images, but the naive CNN model does not match the performance achieved by MobileNetV3-Large transfer learning using the same training strategy.

Table 2. List of Naive CNN models.

Out Stride	**1**	**2**	**3**	**4**	**5**	**6**	**7**
1	Input and rescaling						
2	Conv3 × 4 Conv3 × 4 Max Pool	Conv3 × 8 Conv3 × 8 Max Pool	Conv5 × 8 Conv5 × 8 Max Pool	Conv3 × 8 Conv3 × 8 Conv3 × 8 Max Pool	Conv3 × 8 Conv3 × 8 Conv3 × 8 (Stride = 2)	Conv5 × 8 Max Pool	Conv5 × 16 Max Pool
4	Conv3 × 8 Conv3 × 8 Max Pool	Conv3 × 16 Conv3 × 16 Max Pool	Conv5 × 16 Conv5 × 16 Max Pool	Conv3 × 16 Conv3 × 16 Conv3 × 16 Max Pool	Conv3 × 16 Conv3 × 16 Conv3 × 16 (Stride = 2)	Conv5 × 16 Max Pool	Conv5 × 32 Max Pool
8	Conv3 × 16 Conv3 × 16 Max Pool	Conv3 × 32 Conv3 × 32 Max Pool	Conv5 × 32 Conv5 × 32 Max Pool	Conv3 × 32 Conv3 × 32 Conv3 × 32 Max Pool	Conv3 × 32 Conv3 × 32 Conv3 × 32 (Stride = 2)	Conv5 × 32 Conv5 × 32 Max Pool	Conv5 × 64 Conv5 × 64 Max Pool
16	Conv3 × 32 Conv3 × 32 Max Pool	Conv3 × 64 Conv3 × 64 Max Pool	Conv5 × 64 Conv5 × 64 Max Pool	Conv3 × 64 Conv3 × 64 Conv3 × 64 Max Pool	Conv3 × 64 Conv3 × 64 Conv3 × 64 (Stride = 2)	Conv5 × 64 Conv5 × 64 Max Pool	Conv5 × 128 Conv5 × 128 Max Pool
32	Conv3 × 64 Conv3 × 64 Max Pool	Conv3 × 128 Conv3 × 128 Max Pool	Conv5 × 128 Conv5 × 128 Max Pool	Conv3 × 128 Conv3 × 128 Conv3 × 128 Max Pool	Conv3 × 128 Conv3 × 128 Conv3 × 128 (Stride = 2)	Conv5 × 128 Conv5 × 128 Conv5 × 128 Max Pool	Conv5 × 256 Conv5 × 256 Conv5 × 256 Max Pool
*	Conv3 × 128 Conv3 × 128 GAP	Conv3 × 256 Conv3 × 256 GAP	Conv5 × 256 Conv5 × 256 GAP	Conv3 × 256 Conv3 × 256 Conv3 × 256 GAP	Conv3 × 256 Conv3 × 256 Conv3 × 256 GAP	Conv5 × 256 Conv5 × 256 Conv5 × 256 GAP	Conv5 × 512 Conv5 × 512 Conv5 × 512 GAP
*	Efficient last stage (from MobileNetV3)						
Accuracy	0.95448	0.96892	0.97079	0.97249	0.96569	0.97617	0.97844
Params	39.18 K	143.8 K	155.9 K	238.2 K	238.2 K	248.3 K	921.8 K
FLOPs	9.94 M	29.4 M	42.0 M	49.9 M	36.3 M	29.9 M	97.2 M

Note: * Indicates that after Global Average Pooling, the size of the feature map will already be 1. The calculation of Out Stride no longer makes sense.

The performance of the naive CNN model provides some guidance for the optimization of the model. Comparing model 1 and model 2 in Table 2, there is a large gap in the performance of the model when the number of channels of each convolutional layer is doubled. It can be seen that ensuring the width of the model is one of the key factors to improve the performance of the model, but the cost of doubling the width is high, and the number of parameters and calculations is doubled. Comparing model 2 and model 3, it is further verified that the convolutional kernel of 5 × 5 is more efficient than the convolutional kernel of 3 × 3, and because of the depthwise separable convolution, the increase in the number of parameters and calculation is not large. Comparing model 4 and model 5, the max-pooling operation is more reliable than the down-sampling method using a convolutional layer with a step size of two as a characteristic graph. Comparing model 3 and model 6, the position of the lower sampling layer in the model will also affect the performance of the model. In general, the low layer of the model (close to the input layer) does not need to stack too many convolutional layers, while the high layer of the model (close to the output layer) needs to stack more convolutional layers.

2.2.3. Contour-Resampling-Based Fully Connected Network

Because the information density of the binarized contour image samples is extremely low, using CNN to solve the classification problem of such images seems to be a waste of performance, so another solution was tried.

The contour image samples in the dataset can be represented as a coordinate set of contour pixel points that only contains elements twice the number of contour pixel points (horizontal and vertical coordinate values). The number of contour pixels in some 480 × 640 size contour maps is counted, and the number of contour pixels is within 800, while the total number of pixels in the overall contour map is up to 307,200. Therefore, the set of contour point coordinates of each image sample is used as the model input, and building a fully connected neural network can also solve the orientation recognition problem of garlic clove contour images. Although this method needs to increase the steps of extracting contour points from the collected image, the increased amount of calculation is very small. Along with the help of OpenCV, the extraction process of contour points is also very easy to implement.

Uniform Input Size

It is very difficult to realize the variable length input of a neural network. Because the number of contour pixels contained in each contour image sample is different, it is necessary to unify the number of contour pixels of the sample first. Hence, the equidistant sampling method is used to sample a fixed number of point coordinates from the contour of each image, and then draw a polygon with these sampled points and observe its ability to reconstruct the original sample through artificial vision (Figure 12). It was found that when 50 contour points were sampled, the polygon formed was very close to the shape of the original sample. Through the above sampling method, 50, 100, and 200 contour point sets of all contour image samples were collected and combined with the orientation classification of the samples as the training dataset of the fully connected model. A fully connected model with three Hidden layers and 512, 256, and 128 neurons was used for testing. It was found that there was no significant difference in the recognition rate of the model when 200, 100, or 50 sampling points were used, but using fewer sampling points could effectively reduce the number of parameters and calculations of the model, so 50 points is preferable.

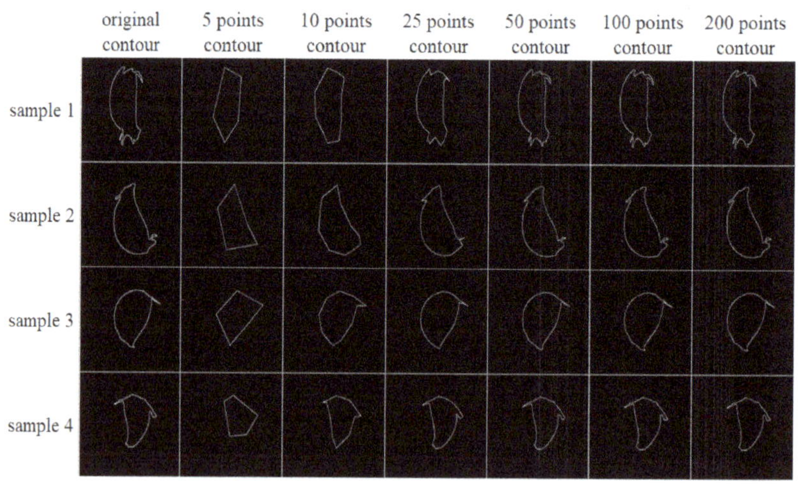

Figure 12. The ability of contour points with different sampling rates to restore the original contour.

The matrix shape of the contour point set obtained by the findcontours function of OpenCV is $[n, 2]$, where n is the number of contour points, and the matrix shape of the contour point set after sampling is $[m, 2]$, where m is the number of sampling points, and the dimension with length of 2 contains the horizontal and vertical coordinates of each contour point. For the fully connected model discussed in this section, the input of each layer of the model should be a one-dimensional vector, so it is necessary to flatten the contour point set. There are two ways to flatten: the first from the point dimension, the other from the coordinate dimension. The first flattening method was chosen (the value of the data_format parameter corresponding to the Keras Flatten layer is "channels_last").

Structure of Fully Connected Model

Through the testing of several fully connected models defined by Keras API, it was found that when the number of Hidden layers of the model was less than three, increasing the number of fully connected layers was effective. When the number of layers exceeded three, increasing the number of fully connected layers could not significantly improve the recognition rate of the model. Using more neurons can improve the performance of the model, but increasing the number of neurons will greatly increase the number parameters of the model, resulting in the model becoming bloated. The preliminary test results of typical models are shown in Table 3. The accuracy of the model with 4096, 2048, and 1024 neurons in the Hidden layer reached 0.97893. The accuracy of the model with 1024, 512, and 256 neurons in the Hidden layer was 0.97465. The accuracy of the model with 512, 256, and 128 neurons in the Hidden layer was 0.97241.

Table 3. Overview of fully connected models.

No.	**1**	**2**	**3**	**4**	**5**	**6**	**7**	**8**
				Input and Flatten				
Number of Neurons in Each Layer	4096 2048 1024	2048 1024 512	1024 512 256	1024 512 256 128 64 32	512 256 128	512 256 128 64 32	1024 ... ... ×10	512 ... ... ×10
Accuracy	0.97893	0.97619	0.97465	0.97341	0.97241	0.97249	0.97469	0.97253
Params	12.16 M	3.460 M	1.075 K	1.118 M	220.4 K	230.7 K	9.902 M	2.581 M
FLOPs	24.3 M	6.92 M	2.15 M	2.23 M	0.444 M	0.465 M	19.8 M	5.18 M

The structure of the fully connected model is shown in Figure 13. Each fully connected layer includes a batch normalization [23] layer. It is particularly noteworthy that adding batch normalization layers after the flat layer can greatly improve the convergence speed of the model.

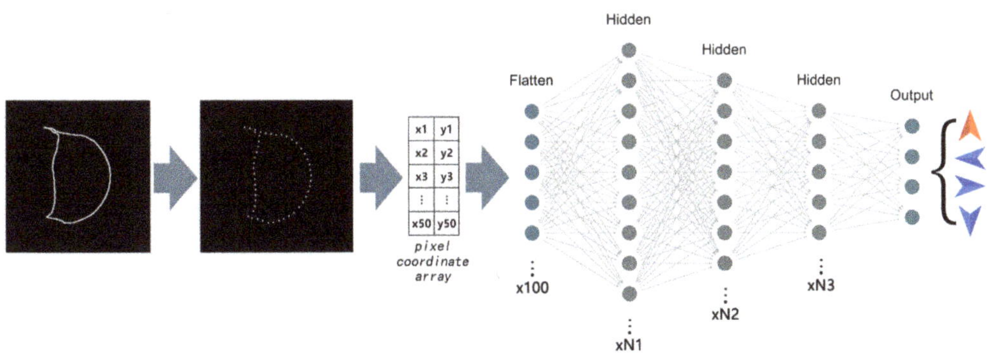

Figure 13. Fully connected model. Note: both the Flatten and Hidden layers are connected to the Batch Normalization layer, but only the Batch Normalization layer of the Hidden layer applies the activation function. N1, N2 and N3 are the number of undetermined Hidden layer neurons.

2.2.4. Model Optimization

In order to obtain faster computing speed and higher accuracy, the three deep learning models have been greatly optimized, and the optimization directions include model lightweighting and model training tuning.

Implementation of Lightweight Convolutional Model

It was found that when the size of the input image of MobileNetV3 was reduced to 60×80, the recognition rate of the model decreased significantly, while in the relevant test of the naive CNN model, the input of 60×80 did not greatly reduce performance of the model.

The stride of the first standard convolution layer of the MobileNetV3 model is 2. When it was modified to 1, the performance of small-sized input of 60×80 was improved. Removing the 1×1 standard convolutional layer before the Global Average Pooling [24] layer did not reduce the recognition rate of the model, but it could reduce the number of parameters and computation of the model and improve the convergence speed of the model.

For the naive CNN model, the actual receptive field of the 3×3 convolutional kernel when using the input size of 60×80 was larger than the actual receptive field of the 5×5 convolution kernel when using the input size of 120×160. Since the edge of the garlic contour image is a background does not contain anything, the convolutional layers in the first two groups of the convolutional-pooling modules could be modified to valid padding. Due to the above adjustments and halved input size, two convolutional-pooling modules can be removed to achieve the same feature map size, which can greatly reduce the number of parameters and computations in the model. Since the input of the model is only a single-channel image, the use of standard convolution at the input of the model only increases the number of calculations and parameters by very little compared with the depthwise separable convolution. This change has been tested to slightly improve the performance of the model. The naive CNN model structure obtained after the abovementioned optimization procedure is shown in Figure 14.

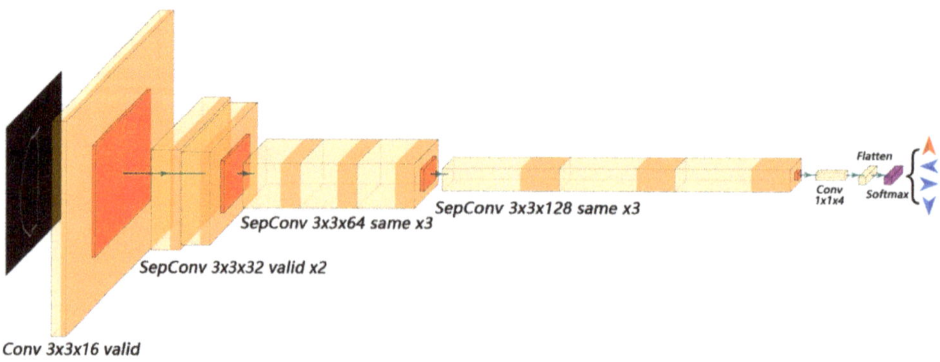

Figure 14. Naive CNN model after lightweighting.

Model Training and Tuning

Three optimizers: Adam [25], Nadam [26], and SGD were tested in model training. The final convergence results of Adam's optimizer in multiple training tests of the same model were unstable. Nadam and SGD were more stable than Adam, but Nadam had the greatest computational complexity of the three and the slowest performance. SGD is theoretically less efficient than Adam and Nadam, but the fully connected model proposed in Section 2.2.3 could converge stably when the learning rate of SGD was set to 1.0 or even higher. In this way, both the convergence speed of the model and convergence stability could be guaranteed. In addition, the computational complexity of SGD was the lowest of the three, and the computational speed was the fastest.

Five activation functions comprising Tanh, Relu6 [27], Gelu [28], Swish [29], and h-swish [17] were tested, and the convergence curve for the fully connected model is shown in Figure 15. When using the SGD optimizer, the effect of swish and h-swish was better (1000-epochs validation set accuracy is 0.97611 and 0.97586, respectively), and because the computational complexity of Hard-Swish was lower than that of Swish, and the model using h-swish was more reliable in weight value quantization, the fully connected model uses the h-swish activation function.

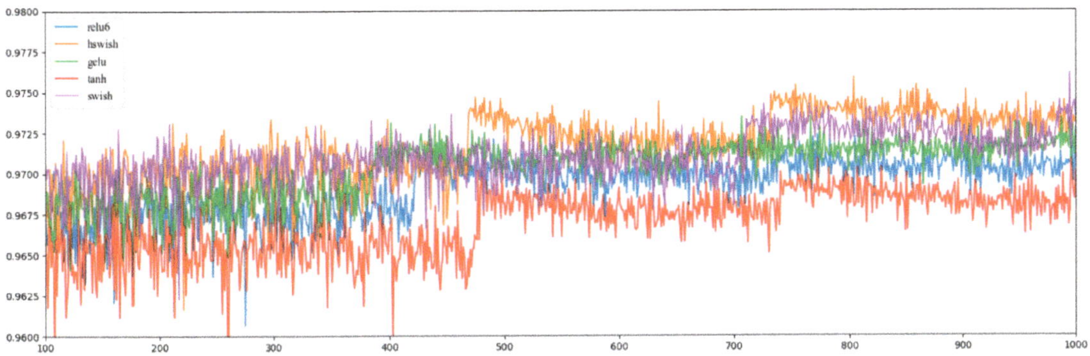

Figure 15. Convergence curves of different activation functions. Note: the above convergence curves were all measured on the fully connected model proposed in Section 2.2.3, and the range shown in the figure is 100 to 1000 epochs.

For the garlic seed contour dataset, the training loss values of the three types of models tested are close to 0 in the later stage of the training process, the training set accuracy can be close to 100%, and the validation set accuracy is different. This is an overfitting phenomenon, and the loss flooding method [30] has a significant effect on it. The idea of this method is to keep the training loss value always above a certain threshold delta, so that the model can continue to learn and possibly converge to a better performing state. In the optimization process of the model, there may be a large number of local optimum points. The random walk strategy of the loss flooding method requires the optimizer to have a large enough optimization stride to ensure that the model escapes the local optimum point. When the model is optimized to a good state range, the weight value needs to be saved in time to prevent missing the state. In the later stages of the finite number of training iterations, the probability of the random walk method obtaining a better state becomes very low, but continuing to train the model with a smaller learning rate can often make the model's performance improve again in the short term, so the learning rate decay method combined with the loss flooding method is very effective. In order to ensure that the model is in an ideal state when the learning rate decay is triggered, a program is written to dynamically load the weights saved during the last state boost each time the learning rate decays.

The LSR [31] method was also used. When the LSR method was applied alone, the model was trained with a label_smoothing parameter of 0.2, and the obtained validation set accuracy was comparable to the loss flooding method with a delta of 0.1. When the loss flooding method was combined with the LSR method, the delta and label_smoothing parameters were set to 0.7 and 0.2, and the accuracy of the validation set obtained was slightly improved, but the accidental components were not excluded.

Based on the above methods, L1/L2 regularization and Dropout [32] regularization were further tried. L1/L2 regularization is effective for convolutional models, but not for fully connected models. Dropout regularization looks simple and crude, but it significantly improves the performance of the fully connected model.

2.3. Application Method in Embedded System

The application of deep learning models in seeders requires some additional support programs and control programs. First, the deep-learning model will give a direction judgment for any input image, including the image when no garlic seeds pass by. Therefore, in order to avoid meaningless direction judgment and device linkage, for each frame of a collected image, a judgment should be made on whether it contains garlic seeds. Secondly, since the deep learning models constructed in this paper are all based on the contour image of garlic seeds or their sampling point sets as the basis for classification, an additional program is required to extract binarized contour image or resampling the contour points. A flow chart of the complete orientation judgment process is shown in Figure 16.

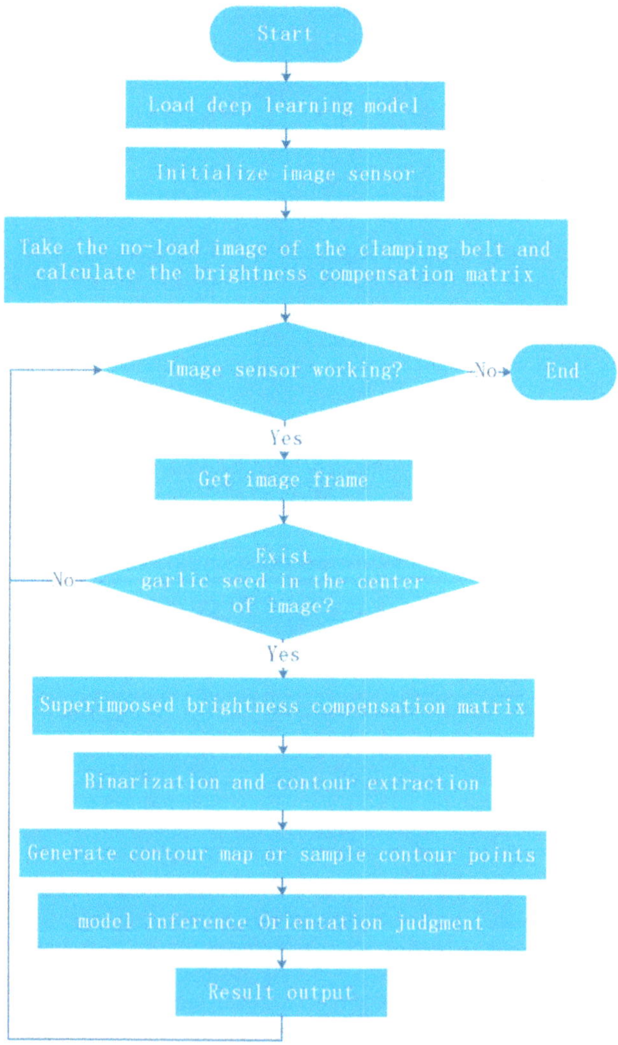

Figure 16. Flow chart of orientation judgment procedure.

Under backlight illumination, it can be judged whether a garlic seed is passing by monitoring the change of the average pixel value of the central area of the camera's field of view. Figure 17 shows the relationship between the average pixel value and the movement of a garlic seed in the camera's field of view. When a garlic seed passes through the camera's field of view, it is captured with multiple frames of images, then compared to the image frames of single garlic seed, and an image with the lowest average pixel value in the central area is obtained, which is the optimal image frame. This process is shown in Figure 18.

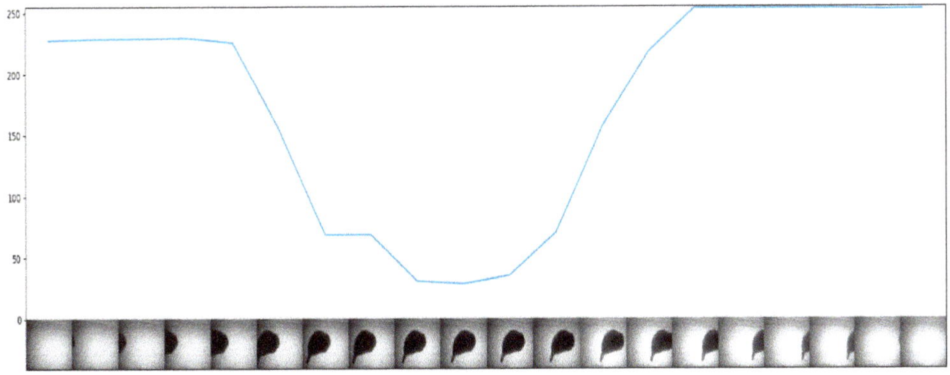

Figure 17. Relationship between the pixel mean value of the central area of the field of view and the position of garlic seeds.

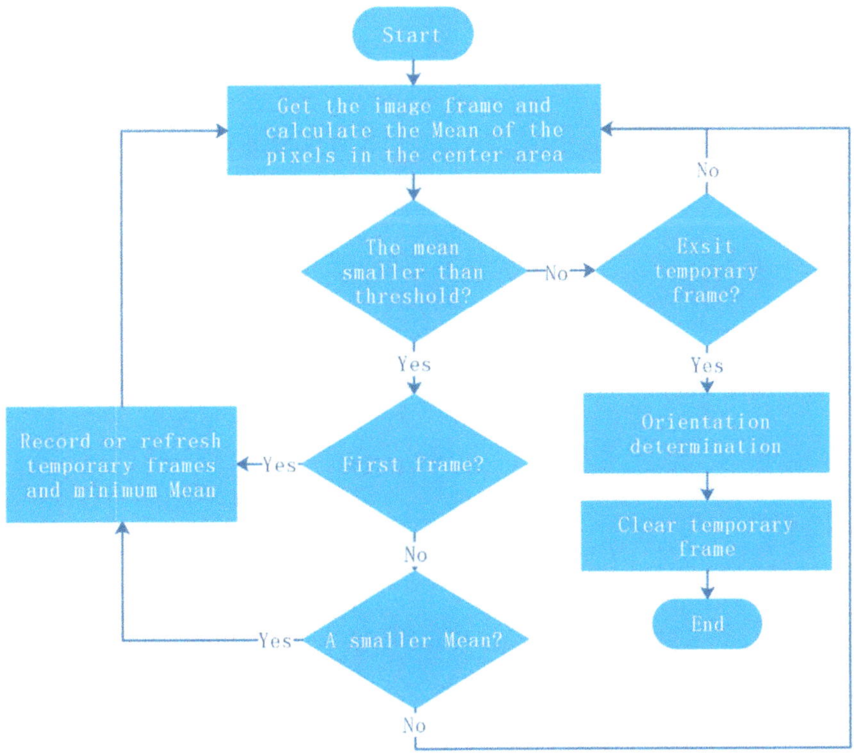

Figure 18. Frame retrieval flow chart.

After obtaining the optimal image frame, brightness compensation is performed, binarization of the image is completed, and the contours from the binarized image are extracted. The output of the contour extraction algorithm is a set of contour points. For the CNN model that takes the contour image as input, the set of contour points needs to be drawn as a contour image. For the fully connected model that takes the set of contour points as input, it is necessary to reduce the number of contour point coordinates to a number suitable for the input of the model through sampling.

3. Results and Discussion

3.1. Model Test and Result

After a series of optimization operations, some typical models were retrained, and their performances are shown in Table 4. Of these models, the transfer learning model based on MobileNetV3-Large has the highest recognition rate of 98.71% on the validation set. However, compared to other models in Table 4, MobileNetV3-Large is too bloated. The recognition rate of the standard-width MobileNetV3-Small model is second only to the MobileNetV3-Large model, but its parameters and computation are still too large. The naive CNN model in Figure 14 performs better than the MobileNetV3-Small with reduced width factor, and its performance is close to that of the standard-width MobileNetV3-Small, but it has the lowest number of parameters among all the models in the table. The fully connected model with 512, 256, and 128 neurons in the Hidden layer achieves almost the same accuracy as the naive CNN model with extremely low computational cost and parameter cost. It has the fastest speed and the most cost-effective application.

Table 4. Performance of the optimized model.

Model	Parameter Quantity	Calculated Quantity (FLOPs)	Accuracy of Training from Zero (%)	Accuracy of Transfer Learning (%)	F1 Score (Macro)
MobileNetV3-Large 1.0	4.04 M	0.178 G	*	98.71	0.98717
Small 1.0	894.1 K	41.8 M	*	98.42	0.98412
Small 0.5 minimalistic	122.5 K	13.7 M	97.67	*	*
Small 0.25 minimalistic	46.9 K	7.48 M	97.58	*	*
Small 0.1 minimalistic	23.2 K	5.44 M	97.34	*	*
Naive CNN model	61.9 K	17.9 M	98.21	*	0.98216
Fully Connected Model	220.4 K	0.444 M	98.16	*	0.98157

The * in Table 4 means that there is no test.

The last column of Table 4 shows the macro F1 score of the models. The F1 score of the four models are almost equal to the accuracy rate, which indicates that the recognition rate of the models for the four orientation categories are very balanced. The ROC curve and AUC value of the models also support this point. The ROC curves of the four orientations are almost identical with only a small gap, and they cover each other in the graph and are difficult to distinguish, as shown in Figure 19. Meanwhile, the macro average AUC and the AUC of each classification are close to 1, which indicates that the recognition effect of the models for each orientation classification are very good.

Based on the program flow introduced in Section 2.3, representative experimental models were selected and converted to TFlite format for speed testing on OrangePi 3 LTS. The test results are shown in Table 5. For the three CNN models in the table, due to their own calculation being more complicated, adding the complete process has little impact on its speed. The fully connected model itself has simple calculation and fast inference speed, but the ability of the support program to provide input data for the model is limited, and it finally reached a speed of about 151.40, which is still more than 50% faster than the fastest CNN model.

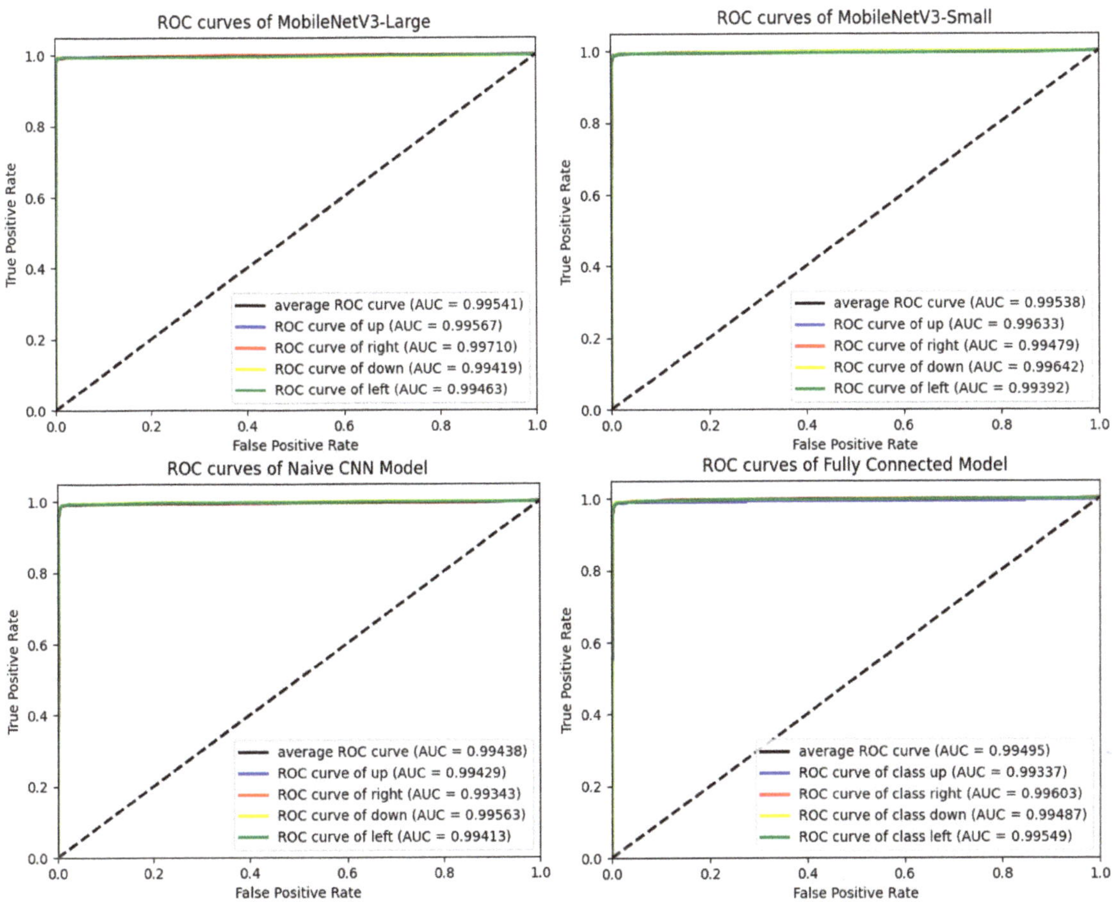

Figure 19. ROC and AUC for models with recognition rates over 98%.

Table 5. Inference speed of the model on OrangePi 3 LTS.

Model	MobileNetV3 Large	MobileNetV3 Small	Naive CNN Model	Fully Connected Network
Model inference speed (FPS)	23.86	76.20	136.30	929.75
Complete process speed (FPS)	19.35	59.80	97.39	151.40

3.2. Discussion

3.2.1. Reliability Verification Experiment

In order to verify the validity of the data and model, a program was written to rotate the image samples 90° counterclockwise before extracting the contours and then extract the contour to identify its orientation, as shown in Figure 20. Since the original orientation categories up, left, bottom, and right correspond to labels 0, 1, 2, and 3 respectively, the category labels output by the model after the rotation should be 1, 2, 3, and 0. In the test, all the tested models can achieve almost the same recognition rate as before the sample rotation, as shown in Table 6.

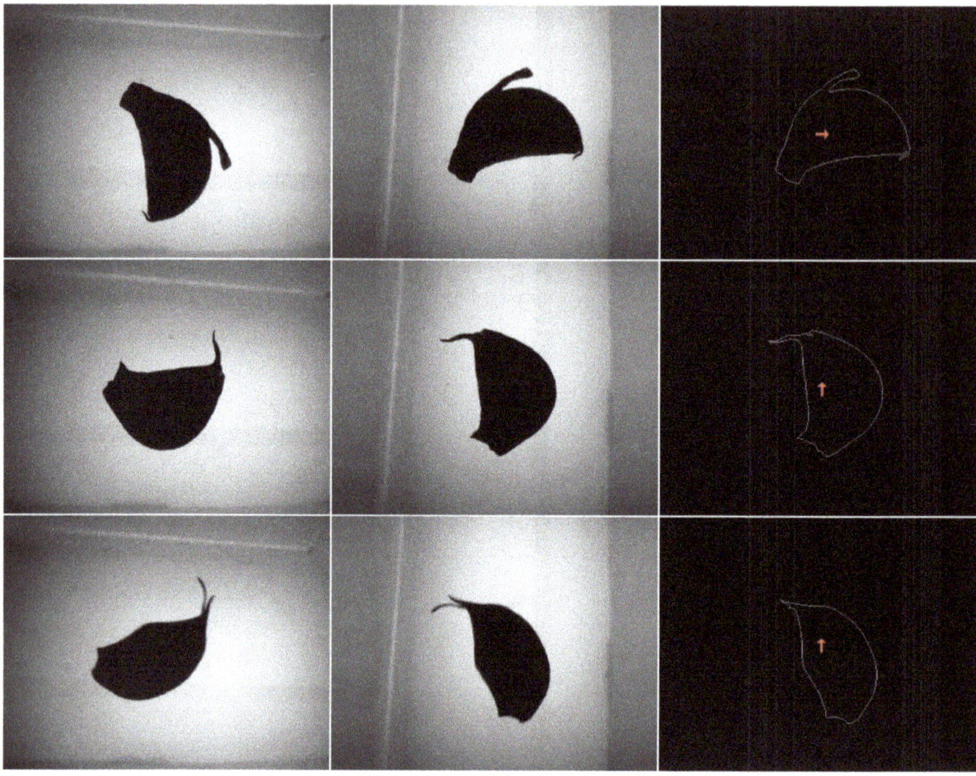

Figure 20. Sample rotation test of the model.

Table 6. Orientation recognition rate of the model to the rotated sample.

Model	MobileNetV3 Large	MobileNetV3 Small	Naive CNN Model	Fully Connected Network
Accuracy (%)	98.64	98.42	98.04	98.02

3.2.2. Comparison with Statistical Learning

The difficulty in applying statistical learning methods to image recognition is how to extract image features. The method of sampling fixed coordinate points at equal intervals of contour lines introduced in Section Uniform Input Size greatly reduces the feature dimension of the image, which can be regarded as a kind of feature extraction method. Based on this, several statistical learning algorithms such as KNN, SVM, and lightGBM [33] were fitted and tested using a dataset of 50 contour sampling points, but none of them matched the classification performance of the neural network algorithm.

In the test of the KNN algorithm, the NCA [34] algorithm is used to reduce the dimension of the data samples of 50 sampling points to generate a vector of a specific dimension and use it as the input of the KNN algorithm. In the parameter adjustment test, when the dimension of the model input vector, that is, the NCA output vector, was reduced to 25, and the number of adjacent elements of the KNN model is 16, the recognition rate of the validation set of the KNN model reaches a peak at 93.70%.

In the test of the SVM algorithm, the performance of the RBF kernel function was significantly higher than that of linear and poly kernel functions. The randomized search CV method was used to select C and gamma parameters. When C is 98.21 and gamma is

0.0044, the accuracy of the validation set of the SVM model reaches 94.06% of the optimal figure in the experiment.

In the test of the lightGBM algorithm, PCA algorithm was used to reduce the dimension of data samples at 50 sampling points. When the dimension of data samples was reduced to 25, num_leaves and max_depth parameter of the lightGBM algorithm were 127 and 8, respectively, and the recognition rate of the validation set of the lightGBM model could reach an optimal 96.56% in the experiment.

If PCA is not used, the lightGBM model can only achieve a recognition rate of less than 92% of the validation set, which indicates that the processing of the PCA algorithm not only reduces the dimension of the sample vector but also improves the ability of the data to represent the original sample. After follow-up tests, the improvement of model accuracy by PCA preprocessing is limited to gbdt-based algorithms such as XGBoost [35] and lightGBM and cannot greatly improve the recognition rate of validation sets of KNN, SVM, and fully connected neural networks. Using PCA to convert the coordinate data of 50 sampling points into a 25-dimensional vector can reduce the complexity of the model. For the fully connected model in Table 4, after modifying the model input to a 25-dimensional vector, the number of parameters was reduced to 183.6 K. The amount of computation was reduced to 0.36 M, but the recognition rate on the validation set dropped to 97.97%.

As a comparison, the accuracy and running speeds of KNN, SVM, lightGBM, and the fully connected model on the embedded platform are shown in Table 7. Obviously, the speed of the fully connected model is better than that of the statistical learning model.

Table 7. Performance comparison of statistical learning models and fully connected model.

Model	KNN	SVM	lightGBM	Fully Connected Model
Accuracy (%)	93.70	94.06	96.56	98.16
Model inference speed (FPS)	37.34	82.63	118.30	929.75

3.2.3. Comparison with Methods in Other Literature

Table 8 lists the garlic orientation recognition methods and their recognition rates described in the literature in recent years. It can be seen that the recognition rate of the method proposed in this paper is higher than other methods. Since all these studies use private datasets, this horizontal comparison is only for reference. However, because the samples contained in the dataset constructed in this paper uniquely retain the common morphological abnormalities and motion blur phenomena in the real scene, the reliability of the recognition rate achieved by the model in this study is at least not lower than that of other studies.

Table 8. Comparison of recognition rate of methods in related literatures.

Article	Ref. [7]	Ref. [11]	Ref. [12]	Ref. [13]	This Article
Accuracy (%)	92.67	97.50	>90.56 *	97.25	>98

Note: * Ref. [12] only published that the success rate of garlic seeds righting is 90.56. It can be inferred that the recognition rate must be greater than this value.

The generally high recognition rates of the models proposed in this paper indicate that the dataset enhancement method and the contour-image-based garlic-clove-bud orientation recognition models adopted in this paper are effective. The form of binarized contour image unifies the pixel value distribution of contour points, so that the information of image samples can be completely expressed by the coordinate set of contour points. The feature extraction method of contour point equidistant sampling further reduces the dimension of the input data, so that the extremely lightweight fully connected neural network can also complete the orientation classification task of garlic seeds with high accuracy and speed.

3.2.4. Application Prospect

The operating speed of the garlic planter can be calculated by Equation (4), where η represents the sowing efficiency (hm^2/h), w represents the plant spacing (m), h represents the row spacing (m) and v represents the sowing speed (pieces/second).

$$\eta = 0.36 \cdot v \cdot w \cdot h \tag{4}$$

The garlic sowing efficiency of the existing garlic seed adjustment method is in the range of 0.05–0.2 hm^2/h [36,37]. According to the planting standards of 0.2 m row spacing and 0.12 m plant spacing, the four orientation recognition models in Table 5 can reach sowing speeds of 0.16, 0.51, 0.84, and 1.30 hm^2/h, respectively. The above speed is the ideal single-row seeding speed. It can also be used in multi-row seeders in the form of controlling multiple rows through a single board. It only needs a single embedded board with the same performance as the OrangePi 3 LTS. If there are performance bottlenecks in the other devices that make up the garlic planter, the hardware configuration can be further reduced, thereby reducing the manufacturing cost of the planter.

4. Conclusions

To meet the need of high-speed garlic seed righting operations and low-cost onboard embedded computing platforms, the contour-based multiple lightweight deep-learning models including transfer learning based on MobileNetV3, naive CNN model, and a contour resampling-based fully connected neural network are proposed for garlic-clove-bud orientation recognition and tested by the image garlic seed samples with the same conditions as a field planter, and the best model was selected for parameter optimization. All of the models' recognition rate of garlic clove bud orientation exceeded 98%. The MobileNetV3 model based on transfer learning, the naive CNN model, and the fully connected model achieved accuracy of 98.71, 98.21, and 98.16%, respectively, all far exceeding statistical learning methods. The parameters of the three are 4.04 M, 61.9 K, and 220.4 K, respectively. The calculation amount of the three is 0.178 G, 17.9 M, and 0.44 M FLOPs, respectively. The recognition speed of the three including auxiliary programs is 19.35, 97.39, and 151.40 FPS, respectively.

Experimental results showed that the contour-image-based garlic-clove-bud orientation recognition method is effective. The form of binarized contour image unifies the pixel value distribution of contour points, so that the information of garlic clove samples can be completely expressed by the coordinate set of contour points. Resampling of contour points further compresses sample features and simplifies the structure of deep-learning models. Ideally, a fully connected neural network based on contour resampling could support a seeding rate of 1.3 hm^2/h. Therefore, the garlic-clove-bud orientation recognition based on deep learning proposed by this paper can meet the needs of high-speed and accurate sowing of garlic.

The main goals of this research for the future are to complete the integration of garlic species orientation recognition algorithm and orientation device, verify the effect of system integration, and continuously improve the device; collect more garlic seed contour image samples to join the dataset and train the model to continuously enhance its generalization ability; and try to generalize the orientation recognition algorithm proposed in this paper to other problems in the agricultural field.

Author Contributions: J.L.: writing—original draft, writing—review and editing, conceptualization, methodology, investigation, data curation, formal analysis, software, and validation. J.Y.: writing—original draft, writing—review and editing, funding acquisition, software and production of related equipment. J.C.: investigation, device design, and production of related equipment. Y.L.: investigation, data curation, validation, and production of related equipment. X.L.: writing—review and editing, funding acquisition, project administration. All authors have read and agreed to the published version of the manuscript.

Funding: This research was funded by cotton industry technology system and industrial innovation team projects in Shandong Province (SDAIT-03-09) and supported by the National Natural Science Foundation of China (51675317).

Institutional Review Board Statement: Not applicable.

Informed Consent Statement: Not applicable.

Data Availability Statement: Anyone can access the data by sending an email to jyuan@sdau.edu.cn.

Conflicts of Interest: The authors declare no conflict of interest.

References

1. Ahn, Y.; Choi, G.; Choi, H.; Yoon, M.; Suh, H. Effect of planting depth and angle of non-cloved bulb garlics on the garlic growth and yield. *Korean J. Hortic. Sci. Technol.* **2007**, *25*, 82–88.
2. Jin, C.; Yuan, W.; Wu, C.; Zhang, M. Experimental study on effects of the bulbil direction on garlic growth. *Trans. CSAE* **2008**, *24*, 155–158.
3. Liu, J. Effects of Different Garlic Varieties and Bulbil Seeding Directions on Growth Characteristic and Quality. Master's Thesis, Shandong Agricultural University, Tai'an, China, 2018.
4. Huang, H.S. Numerical Simulation and Experimental Study on the Adjusting Device and Inserting Device of Double Duckbill Garlic Planter. Master's Thesis, Shandong Agricultural University, Tai'an, China, 2019.
5. Zhang, X.; Yi, S.; Tao, G.; Zhang, D.; Chong, J. Design and Experimental Study of Spoon-Clamping Type Garlic Precision Seeding Device. *Wirel. Commun. Mob. Comput.* **2022**, *2022*, 5222651. [CrossRef]
6. Tian, L. Design and Experiment of Garlic Seed Orientation Device Based on Image Recognition Technology. Master's Thesis, Shandong Agricultural University, Tai'an, China, 2020.
7. Wu, X.; Hu, W. Research on Garlic Clove Orientation Recognition Based on Observation Window. *Meas. Control. Technol.* **2016**, *35*, 35–39.
8. Zhao, L.Q.; Song, W.; Yin, Y.Y.; Zhou, C. System of Garlic Shape Recognition Based on Vision Technology. *J. Agric. Mech. Res.* **2014**, *36*, 206–208, 212.
9. Guo, Y.F.; Lu, B.Y. Research on the Identification of the Roots of Garlic Based on SUSAN Algorithm. *J. Yangling Vocat. Tech. Coll.* **2011**, *10*, 1–4.
10. Kamilaris, A.; Prenafeta-Boldú, F.X. Deep learning in agriculture: A survey. *Comput. Electron. Agric.* **2018**, *147*, 70–90.
11. Fang, C.; Sun, F.Z.; Ren, C.G. Identifying bulbil direction of garlic based on deep learning. *Appl. Res. Comput.* **2019**, *36*, 598–600.
12. Hou, J.L.; Tian, L.; Li, T.H.; Niu, Z.R.; Li, Y.H. Design and experiment of test bench for garlic bulbil adjustment and seeding based on bilateral image identification. *Trans. CSAE* **2020**, *36*, 50–58.
13. Li, Y.H.; Liu, Q.C.; Li, T.H.; Wu, Y.Q.; Niu, Z.R.; Hou, J.L. Design and experiments of garlic bulbil orientation adjustment device using Jetson Nano processor. *Trans. CSAE* **2021**, *37*, 35–42.
14. Otsu, N. A threshold selection method from gray-level histograms. *IEEE Trans. Syst. Man Cybern.* **1979**, *9*, 62–66.
15. Abadi, M.; Agarwal, A.; Barham, P.; Brevdo, E.; Chen, Z.; Citro, C.; Corrado, G.S.; Davis, A.; Dean, J.; Devin, M. Tensorflow: Large-scale machine learning on heterogeneous distributed systems. *arXiv* **2016**, arXiv:1603.04467.
16. McKinney, W. Data structures for statistical computing in python. In Proceedings of the 9th Python in Science Conference, Austin, TX, USA, 28 June–3 July 2010; pp. 51–56.
17. Howard, A.; Sandler, M.; Chen, B.; Wang, W.; Chen, L.-C.; Tan, M.; Chu, G.; Vasudevan, V.; Zhu, Y.; Pang, R. Searching for MobileNetV3. In Proceedings of the 2019 IEEE/CVF International Conference on Computer Vision (ICCV), Seoul, Korea, 27 October–2 November 2019; pp. 1314–1324.
18. Krizhevsky, A.; Sutskever, I.; Hinton, G.E. Imagenet classification with deep convolutional neural networks. *Adv. Neural Inf. Processing Syst.* **2012**, *25*, 1097–1105. [CrossRef]
19. Simonyan, K.; Zisserman, A. Very deep convolutional networks for large-scale image recognition. *arXiv* **2014**, arXiv:1409.1556.
20. Sifre, L.; Mallat, S. Rigid-motion scattering for texture classification. *Comput. Sci.* **2014**, *3559*, 501–515.
21. Sandler, M.; Howard, A.; Zhu, M.; Zhmoginov, A.; Chen, L.-C. Mobilenetv2: Inverted residuals and linear bottlenecks. In Proceedings of the IEEE Conference on Computer Vision and Pattern Recognition, Salt Lake City, UT, USA, 18–23 June 2018; pp. 4510–4520.
22. Hu, J.; Shen, L.; Sun, G. Squeeze-and-excitation networks. In Proceedings of the IEEE Conference on Computer Vision and Pattern Recognition, Salt Lake City, UT, USA, 18–23 June 2018; pp. 7132–7141.
23. Ioffe, S.; Szegedy, C. Batch normalization: Accelerating deep network training by reducing internal covariate shift. In Proceedings of the International Conference on Machine Learning, Lille, France, 7–9 July 2015; pp. 448–456.
24. Lin, M.; Chen, Q.; Yan, S. Network in network. *arXiv* **2013**, arXiv:1312.4400.
25. Kingma, D.P.; Ba, J. Adam: A method for stochastic optimization. *arXiv* **2014**, arXiv:1412.6980.
26. Timothy, D. Incorporating nesterov momentum into adam. *Nat. Hazards* **2016**, *3*, 437–453.
27. Howard, A.G.; Zhu, M.; Chen, B.; Kalenichenko, D.; Wang, W.; Weyand, T.; Andreetto, M.; Adam, H. MobileNets: Efficient convolutional neural networks for mobile vision applications. *arXiv* **2017**, arXiv:1704.04861.

28. Hendrycks, D.; Gimpel, K. Gaussian error linear units (gelus). *arXiv* **2016**, arXiv:1606.08415.
29. Ramachandran, P.; Zoph, B.; Le, Q.V. Searching for activation functions. *arXiv* **2017**, arXiv:1710.05941.
30. Ishida, T.; Yamane, I.; Sakai, T.; Niu, G.; Sugiyama, M. Do we need zero training loss after achieving zero training error? *arXiv* **2020**, arXiv:2002.08709.
31. Szegedy, C.; Vanhoucke, V.; Ioffe, S.; Shlens, J.; Wojna, Z. Rethinking the inception architecture for computer vision. In Proceedings of the IEEE Conference on Computer Vision and Pattern Recognition, Las Vegas, NV, USA, 27–30 July 2016; pp. 2818–2826.
32. Srivastava, N.; Hinton, G.; Krizhevsky, A.; Sutskever, I.; Salakhutdinov, R. Dropout: A simple way to prevent neural networks from overfitting. *J. Mach. Learn. Res.* **2014**, *15*, 1929–1958.
33. Ke, G.; Meng, Q.; Finley, T.; Wang, T.; Chen, W.; Ma, W.; Ye, Q.; Liu, T.-Y. Lightgbm: A highly efficient gradient boosting decision tree. In Proceedings of the Advances in Neural Information Processing Systems 30 (NIPS 2017), NeurIPS, Long Beach, CA, USA, 4–9 December 2017; Volume 30.
34. Goldberger, J.; Roweis, S.; Hinton, G.; Salakhutdinov, R. Neighbourhood components analysis. In Proceedings of the 17th International Conference on Neural Information Processing Systems; 2004; pp. 513–520. Available online: https://dl.acm.org/doi/10.5555/2976040.2976105 (accessed on 4 August 2022).
35. Chen, T.; Guestrin, C. Xgboost: A scalable tree boosting system. In Proceedings of the 22nd Acm Sigkdd International Conference on Knowledge Discovery and Data Mining, San Francisco, CA, USA, 13–17 August 2016; pp. 785–794.
36. Geng, A.J.; Li, X.Y.; Hou, J.L.; Zhang, Z.L.; Zhang, J.; Chong, J. Design and experiment of automatic directing garlic planter. *Trans. Chin. Soc. Agric. Eng.* **2018**, *34*, 17–25.
37. Geng, A.; Li, X.; Hou, J.; Zhang, Z.; Zhang, J.; Chong, J. Optimal design of a general type-2 fuzzy classifier for the pulse level and its hardware implementation. *Eng. Appl. Artif. Intell.* **2021**, *97*, 104069.

agriculture

MDPI

Article

Research on the Adaptability of High-Performance Film for Full Recycling to the Curl-Up Film Collecting Method

Jie Liu [1], Xuanfeng Liu [2], Yongxin Jiang [2], Xin Zhou [2], Li Zhang [2] and Xuenong Wang [1,*]

1 College of Mechanical and Electrical Engineering, Xinjiang Agricultural University, Urumqi 830052, China; xcx@xjau.edu.cn
2 Institute of Agricultural Mechanization, Xinjiang Academy of Agricultural Sciences, Urumqi 830091, China; njsyyx@xaas.ac.cn (X.L.); jyx@xaas.ac.cn (Y.J.); 171847298@masu.edu.cn (X.Z.); zhangli@xaas.ac.cn (L.Z.)
* Correspondence: xjwxn@xaas.ac.cn

Abstract: Given the problem of the low tensile performance of the plastic film used in China, which brings about difficulties in curl-up film collecting, in this study, a contrast test was carried out on the tensile property of high-performance film for full recycling and the ordinary polyethylene film (PE film) that is used extensively in China. Test results showed that, within the service period, the elongation at break and tensile yield stress of the high-performance film were higher than those of ordinary polyethylene film, and, within the film-laying period of 0~30 days, the reduction scale of the elongation at break and tensile yield stress was higher than that within the film-laying period of 30~180 days. In this study, in order to obtain the lowest tensile performance of the film by curl-up film collecting, the operation principles of the curl-up film collectors were analyzed. The test on the force of curling up the film in the process of overcoming the force between the film and soil was analyzed. Test and analysis results showed that, for different sampling positions, film pick-up angles, and film types, the tensile stress on the film while pulling it up was within a range of 15.97~21.86 MPa. In order to verify the curling up effect of differently structured film collectors on different types of film with different thicknesses, a field test on film curl-up collecting was designed. A contrast test was carried out on two types of curl-up film collectors, 1JRM-2000 and 11SM-1.2, and the test results showed that the film recycling rate and working performance on the film laid in the same year by the film collector with a fixed film pick-up angle were higher than those for varying film pick-up angles. The curl-up film collector fixed with an automatic film-guiding mechanism is not affected by the velocity difference between the linear velocity of the film curl-up mechanism and the advancing velocity of the machine. The film recycling rate and working performance on the film laid in the same year by the 11SM-1.2 curl-up film collector can meet the operational requirements for collecting high-performance film with thicknesses of 0.008 mm and 0.01 mm. This research can provide a reference for simplifying the structure of residual plastic film collectors, increasing the film recycling rate, and reducing the cost.

Keywords: high-performance film for full recycling; film recycling; field experiment; film recycling rate

Citation: Liu, J.; Liu, X.; Jiang, Y.; Zhou, X.; Zhang, L.; Wang, X. Research on the Adaptability of High-Performance Film for Full Recycling to the Curl-Up Film Collecting Method. *Agriculture* **2022**, *12*, 1051. https://doi.org/10.3390/agriculture12071051

Academic Editors: Jin Yuan, Wei Ji and Qingchun Feng

Received: 18 June 2022
Accepted: 16 July 2022
Published: 19 July 2022

1. Introduction

Film mulching technology has the advantages of increasing temperature and moisture; preventing plant diseases, insects and weeds; and promoting crop growth [1]. In 2019, the amount of plastic film used in China reached 1.379×10^6 t, and the area covered by plastic film reached 1.76281×10^7 hm^2 [2], which ranked first in the world. However, the farmland residual film recycling technology in China started relatively late, and the long-term, large-scale use of ultra-thin and low-strength plastic film has caused a series of problems, such as soil compaction, a decreased seedling rate, and crop yield reduction [3].

At present, manual recycling is mainly adopted in the treatment of non-point source pollution of farmland residual film, mechanical recycling, and the use of degradable plastic film. Manual recycling of residual film is time-consuming, laborious, and costly, and it is difficult to motivate farmers [4]. Residual film recycling is not required after laying degradable residual film, since the film can decompose through natural degradation [5], however, degradable plastic film is not yet mature in cost control and production technology; the high cost of use and the unpredictable degradation effect during use make it difficult to implement large-scale promotion and use [6,7]. Mechanical recycling of residual film is currently the most widely used method for its high operating efficiency and low operating cost. The mulch film widely used in China has two levels of thickness, 0.008 mm and 0.01 mm, and its tensile property is lower than the mulch film with a thickness of 0.025 mm or more, which is commonly used abroad. Residual plastic film collectors used abroad are mostly curl-up residual film recycling machines with a simple mechanical structure demanding a good tensile performance of plastic film [8], while development of the residual plastic film collectors used in China is restricted by the poor tensile properties of plastic film. According to the planting mode of crops, a variety of film collectors with different mechanical structures has been developed, mainly including drum type, spring tooth type, and tooth chain type [9], which are not only complex in structure but also have a lower film recycling rate than those developed in foreign countries. Marí et al. [10] studied the application of biodegradable plastic mulch films (BDMs) in strawberry planting, and the research results showed that BDMs are a viable alternative to PE mulch. However, Anunciado [6] pointed out in the study of BDMs that the extent of change to the physicochemical properties of BDMs, due to agricultural weathering, is greatly affected by the polymeric composition and is greater in warmer climates. Steinmetz [11] studied BDMs and mentioned that the high use cost restricted the popularization of BDM. Therefore, due to the high cost of agricultural weathering, the technology of BDMs cannot effectively solve the problem of non-point source pollution of residue film in fields. Zhang et al. [12] performed parameter optimization on the Arc-Shaped Nail-Tooth Roller-Type Recovery Machine for Sowing Layer Residual Film, and the field test results showed that this machine type could achieve a normal residual film collection rate of 66.8% on common polyethylene mulching film. Zhou et al. [13] developed a kind of film collector with a film-removing plate, and this device can achieve a film collection rate of 86.93% on common polyethylene mulching film in ideal conditions. However, in the process of collecting the polyethylene mulching film, there are still residue films uncollected in the field, thus, the film-collecting effect was not satisfactory. Qu et al. [14] replaced the traditional rheological processing of drag and shear on high polymer materials with plasticizing transport based on volume elongational rheology, which reduced the macromolecular chain breakage of high polymer materials and greatly improved the mechanical properties of film molded by processing extreme rheological plastics, such as polyethylene. Based on the complex blow-molding technology, through dynamic distribution, the film can be overlaid for 3–5 layers, and the macromolecules are oriented in different directions between the layers to achieve an interweaving effect; thus, the tensile performance of the film is greatly enhanced, and the "high-performance film for full recycling to the curl-up film recycling method" (which can be called "high-performance film") was developed [15,16]. Since the tensile performance of the high-performance film is better than that of common polyethylene film, laying the high-performance film for full recycling can greatly improve the film collecting rate, and the production cost of the high-performance film is very low compared with BDMs; therefore, this technology has become an effective means to solve non-point source pollution of residue films in agricultural fields.

A contrast test on the tensile property of high-performance film and ordinary polyethylene film under different test factors was carried out, and the variation rules of the tensile properties of both films during the film-laying period of 0–180 days, as well as the minimum tensile level for the 180-day film-laying period, were obtained. Moreover, the operation principles of the curl-up residual plastic film collector were analyzed, and the curl-up collecting of the film for the 180-day film-laying period was carried out. Through an analysis on overcoming the force between the soil and the film during curl-up collecting of the film, the tensile stresses on the film while the curl-up film collector pulled it up under different test factors were obtained. The field test on the curl-up collecting of film was carried out. By comparing the film recycling rate on the film laid in the same year and the working performance of the two residual plastic film collectors of different structures, the proper structure adaptable to the curl-up collecting of high-performance film was obtained. This research can provide theoretical support for simplifying the structure of residual plastic film collectors, enhancing the film recycling rate, and reducing the cost of film recycling.

2. Contrast Test on the Tensile Properties of High-Performance Film and Ordinary Polyethylene Film

In order to obtain the variation law of the tensile properties of the high-performance film and the ordinary polyethylene film laid in a cotton field in Xinjiang within their service period and the minimum tensile level at the end of the service period, the film-laying period, the film thickness, sampling direction, and sampling position were used as test factors; the elongation at break and tensile yield stress were used as test indexes to carry out the contrast test on the two types of films.

2.1. Basic Information of the Test Field

Maigaiti County is located in the southwestern part of Xinjiang Uygur Autonomous Region, which includes the western part of the Tarim Basin, the eastern part of the Kashgar region, the southwestern edge of the Taklimakan Desert, the northern foot of the Karakoram Mountains, the lower reaches of the Yarkant River, and the lower reaches of the Tiznafu River (77°28′–79°05′ east longitude, 38°25′–39°22′ north latitude). This county has a temperate continental dry climate with sufficient sunshine, a large temperature difference between day and night, very little precipitation, hot summers and cold winters, and a windy and sandy spring. The average annual sunshine is 2836.5 h, the annual average temperature is 11.8 °C, and the annual average precipitation is 56.5 mm.

2.2. Test Materials and Field Management

Considering local production conditions, the high-performance film and ordinary polyethylene film with thicknesses of 0.008 mm and 0.01 mm were laid on the cotton test field in Maigaiti county on 30 April 2021. The film-laying site is shown in Figure 1. The planting mode of one film, which covered three pipes and six rows with 660 mm + 100 mm of machine-harvested cotton was adopted in the test field. The plant spacing was 12.5 cm, and routine management of the field was adopted for water–fertilizer management. The high-performance film was manufactured by Guangdong Siico Technology Co., Ltd., (Guangdong, China); the ordinary polyethylene film is manufactured by Xingnong Industry and Trade Co., Ltd. in Bayingolin Mongol Autonomous Prefecture, in Xinjiang province, China. The film-laying situation in the test field is shown in Figure 1.

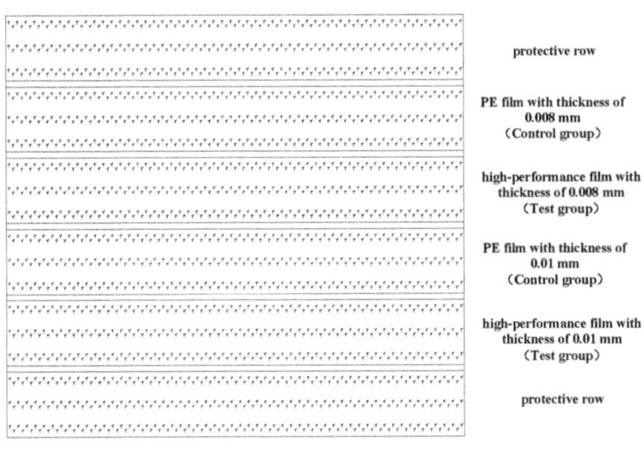

Figure 1. Diagram of plastic film laying in test field.

2.3. Test Design

2.3.1. Test Factors and Levels

Both high-performance film and ordinary polyethylene film are made from high-molecular compounds [17]. Therefore, at the same sampling spot, their tensile performance is affected mainly by natural erosion, material aging, material thickness, and material anisotropy [18]. Thus, the film-laying period, sampling position, film thickness, and sampling direction were chosen as the test factors in the test on the film tensile property.

Material anisotropy determines that different tensile properties are obtained by testing the film from different directions. Hence, the direction along the film-laying direction was defined as the horizontal direction, while the perpendicular direction of the film-laying direction was defined as the vertical direction. The degree of sunniness and the natural erosion effect on the different positions of the film (near and far away from the plants) may vary due to degree of shading of the cotton plants on the film, thus, the sample-taking positions on the film were divided into near-end positions and far-end positions.

2.3.2. Test Indexes

According to the requirements of GB/T 1040.3-2006 Plastics—Determination of Tensile Properties, the elongation at break of the film and the tensile yield stress were taken as the test indexes, and the calculation method is as follows:

$$\varepsilon_t = \frac{L - L_0}{L_0} \times 100\% \quad (1)$$

where L is the distance between the marked lines when the sample is torn off, mm; L_0 is the distance between the original graticule lines, mm.

$$\sigma_t = \frac{F_b}{bd} \quad (2)$$

where F_b is the breaking load of the sample, N; b is the sample width, mm; and d is the sample thickness, mm.

2.3.3. Determination of Test Parameters

The strain data sample frequency is obtained based on test speed, the ratio of the distance between the original graticule lines of the standard sample and the original clamp distance, and the minimum resolution of the obtained strain signal of the accurate data, and its calculation method is as follows:

$$f_{\min} = \frac{vL_0}{60L_c r} \quad (3)$$

where $f_{\min}$ is the sampling frequency of minimum strain data, Hz; v is the test speed, mm/min; L_c is original clamp distance, mm; and r is the minimum resolution of the obtained strain signal of the accurate data, mm.

According to the recommended test speed and the original clamp distance of the standard samples in GB/T 1040.1-2018, v = 10 mm/min, L_c = 115 mm, the CMT-6103 electronic universal testing machine, which is controlled by a microcomputer, obtained the minimum resolution of the obtained strain signal of the accurate data, which was 0.008 mm. After calculation, the sampling frequency of the minimum strain data was obtained, and $f_{\min}$ = 9.06 Hz.

The load data sampling frequency is based on the test speed, strain range, minimum resolution of the obtained strain signal of accurate data, and the initial clamp distance, in which the elastic modulus, test speed, and clamp distance determine the load growth rate. The ratio between the load growth rate and the minimum resolution of the obtained strain signal of accurate data determines the load data sampling frequency of the test machine. The calculation method is as follows:

$$f_{force} = \frac{\dot{F}}{r} = \frac{v}{\Delta\varepsilon \times 60 \times L_c \times 5 \times 10^{-3}} \tag{4}$$

where $\dot{F}$ is the load growth rate, %, and $\Delta\varepsilon$ is the strain range of the samples. $\Delta\varepsilon = 3 \times 10^{-2}$ was selected according to standard requirements, and the sampling frequency of the load data was calculated to be 9.66 Hz.

In this test, an extensometer is used as the strain indicating device, and it should be a Level 1 extensometer as required by GB/T 12160-2019, that is, the relative error of the gauge length is ±1%, the percent of reading is 0.5%, the absolute value is 1 μm, the relative error is ±1%, and the absolute error is ±3 μm.

In order to avoid the toe at the initial stage in the stress–strain curve, in measuring the related stress, the prestress on the sample before the test should satisfy Equation (5) as follows:

$$0 < \sigma_0 \leq \sigma^*/100 \tag{5}$$

where σ_0 is the prestress at the beginning of the test, MPa; σ^* is the tensile yield stress of the material, MPa. In order to make the prestress at the beginning of the test adapt to the two types of film, σ^* should be less than the lower value of the tensile yield stress of the two types of film; thus, $\sigma_0 = 0.09$ Mpa was selected [19].

2.3.4. Sample Collection

The service period of the film laid on the cotton field of south Xinjiang in China is about 180 d. In order to reflect the tensile property variation process of the two types of film during their service periods, film samples were collected every 30 d from the film-laying date to carry out the tensile property test; the samples were collected seven times. Each time, the sampling objects included two sets of high-performance film and ordinary polyethylene film of 0.008 mm and 0.01 mm in thickness, with a width of slightly more than 300 mm and a length of slightly more than 660 mm. After sample collection, the film samples were rinsed to remove the impurities for airing. On each selected sample film, eight standard tensile pieces were cut down by a cutter and used as test material, as shown in Figure 2. The size of the standard tensile film pieces is shown in Figure 3. During each instance of sample collection, the intact film sample pieces were obtained on dry, hard, flat land, and the sampling positions were marked on the film.

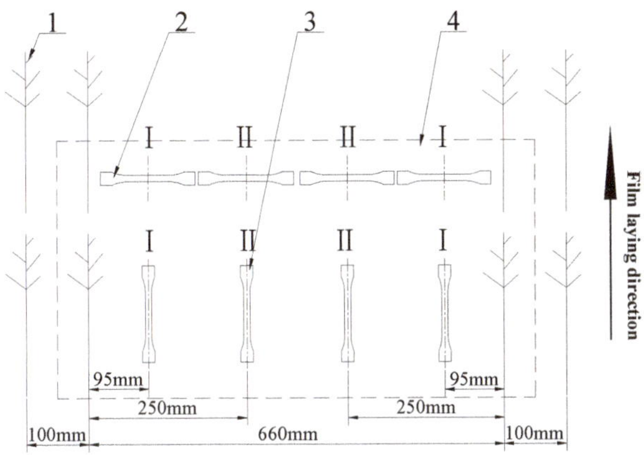

Figure 2. Schematic diagram of sampling location: 1—cotton plant, 2—vertical film sampling, 3—horizontal film sampling, 4—film sample piece, I—near-end position, II—far-end position.

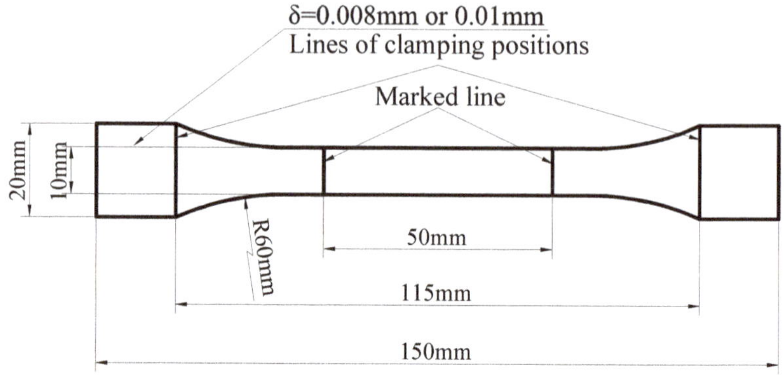

Figure 3. Standard tensile sample of film.

2.3.5. Test Scheme

Before the test, a low-power magnifying glass was used to check the test samples; the sample pieces with unsmooth and frayed edges or damages were eliminated to avoid test errors caused by stress concentration on the damaged parts of the sample pieces in the test. The CMT-6103 electronic universal testing machine controlled by a microcomputer was used to carry out a test on the film tensile property. According to Equations (1) and (2), the elongation at break and tensile yield stress of the film were calculated. The test was repeated four times, and test results were averaged. The test process is shown in Figure 4. Figure 4a shows the state of the sample after prestressing, and Figure 4b–d show the tensile process of the sample after loading.

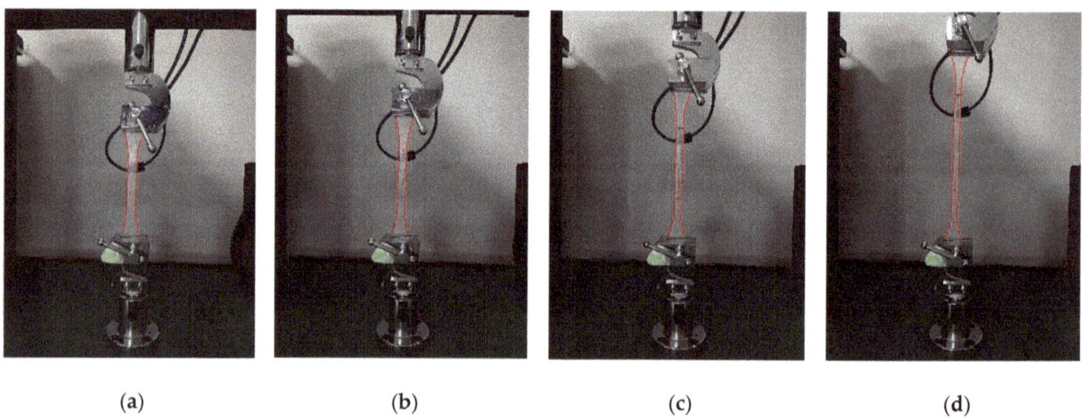

Figure 4. Process of the tensile test of film. (**a**) shows the state of the sample after prestressing, (**b–d**) show the tensile process of the sample after loading.

3. Test on Curl-Up Force in Film Collecting

The curl-up residual plastic film collector is generally composed of the film pick-up mechanism, film-guiding mechanism, film-curling mechanism, impurity separation mechanism, and film-unloading mechanism [20]. During operation, the film pick-up mechanism loosens the soil on the film surface on both sides of the film and separates the film from the soil [21]. Then, the film-guiding mechanism transmits the film to the impurity separation mechanism to the film-curling mechanism. The impurity separation mechanism separates the soil, roots, and stems from the film through vibration or sweeping. The film-curling mechanism curls up the film to a suitable size, and, finally, the film-unloading device unloads the residue film package after curling up.

In the test on the curl-up force during film collecting, by simulating the process of overcoming the force from the soil to the film during curl-up collecting of the residue film, the tensile stresses on the film while the curl-up film collector pulls up the film under different test factors were obtained. In collecting film, the film pick-up mechanism separates the film from the soil and forms a film pick-up angle α; the curl-up force F is formed in curl-up collecting film. The force between the film and soil under the effect of the curl-up force is shown in Figure 5. Since the soil on the film's surface at the slope has the tendency to move downwards, there is a friction f_2 from the film against the soil on the film at the slope. At the same time, the film is uncovered by the film pick-up mechanism along the film pick-up angle α. The cohesion force between the film and soil prevents the film from moving and forms a downward force F_a along the film pick-up angle α.

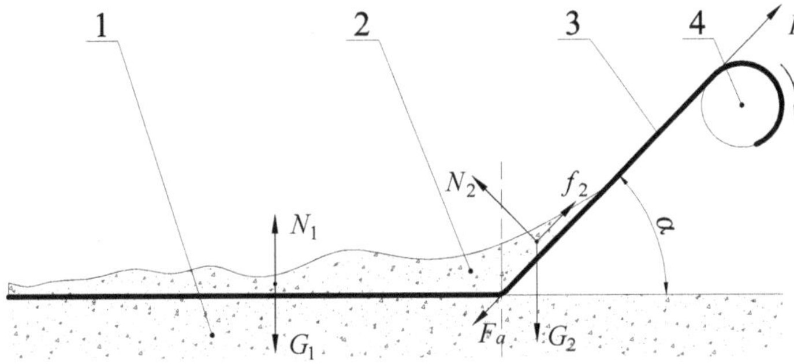

Figure 5. Diagram of force between plastic film and soil under the action of curl-up force: 1—soil under the film, 2—soil on the film, 3—film, 4—film-curling mechanism.

In Figure 5, N_1 is the support force from the soil and film on the flat ground to the soil on the film; G_1 is the gravity of the soil on the film; N_2 is the support force from the film at the slope to the soil on the film; and G_2 is the gravity of soil on the film at the slope. Then, the mechanics equilibrium equation during operation of the curl-up residual plastic film collector is established as follows:

$$\begin{cases} F = F_a + G_2 \sin\alpha - f_2 \\ N_2 - G_2 \cos\alpha = 0 \\ N_1 - G_1 = 0 \end{cases} \tag{6}$$

In order to prevent the film from being torn down due to the speed difference between the linear velocity of the film-curling mechanism and the advancing speed of the machine, the linear velocity of the curling speed should be equal to the advancing speed of the machine, and the speed should be uniform, so as to avoid tearing down the film with the rigid impact from an abrupt change in the film collecting speed. The test on the curl-up force in film collecting was carried out. By measuring the curl-up force F, the tensile stresses on film during the curl-up collecting process under different factor levels were obtained.

3.1. Test Conditions

The field test was carried out at the field research and development base of the Northwest Oasis Agricultural Environment Key Laboratory of the Ministry of Agriculture, Tuobuliqi Town, Korla City, Bayingolin Mongolian Autonomous Prefecture, Xinjiang Uygur Autonomous Region in early November 2021. The planting mode (660 mm (wide row) + 100 mm (narrow row)) with protective rows on both sides was adopted, and the film thicknesses were 0.008 mm and 0.01 mm for both the high-performance film and ordinary polyethylene film. The ground was relatively flat, and the drip irrigation belt had been recycled. Using the TZS-1 soil moisture tester, the moisture content of the surface soil was

16.2%. Before the test, the height of the stubbles in the test field was controlled within 120 mm. The test field is shown in Figure 6.

Figure 6. Test field on curl-up force in film collecting: 1—HP-50 type Digital Display Pull and Push Strength Calculator, 2—film.

3.2. Test Method and Design

3.2.1. Test Factors and Levels

It can be known based on Equation (6) that the value of the curl-up force F is related to cohesion between the soil under the film and the film F_a, the film pick-up angle, the gravity of the soil on the film at the slope in the film pick-up G_2, and the friction f_2 between the film and the soil on the film. Since the moisture content of the soil under the film of different types is different, the higher the moisture content under the film, the higher the cohesion Fa of the soil under the film to the film. The mass of soil on the film is related to the film-laying position. Since cotton plants can shield sandstorms, with the passage of time, the mass of soil near the middle part of the field is lower, and the friction of the film to the soil on the film at the slope is related to the friction coefficient between the soil friction and soil, as well as the mass of soil on the film. Therefore, the sampling position, film pick-up angle, and the types and positions of the laid film were used as test factors. For each planting line of 100 m, the front point of each line was defined as position 1, and 25 m from position 1 along the film-laying direction was defined as position 2; 50 m from position 1 along the film-laying direction was defined as sampling position 3. According to the film pick-up angle of the 1JRM-2000 curl-up film collector, the standard range of the film pick-up angle was determined to be 30–75°. The table of test factor levels in the test on the curl-up force during film collecting is shown in Table 1.

Table 1. Test factor levels.

Levels	Sampling Position	Film Pick-Up Angle	Type of Film
1	Position 1	30°	High-performance film
2	Position 2	45°	Ordinary polyethylene film
3	Position 3	60°	
4		75°	

3.2.2. Test Method

The tensile stress on the film was selected as the test index, which is calculated by Equation (7):

$$\sigma = \frac{F}{bd} \tag{7}$$

where σ is the tensile stress on the film, MPa.

In the test, the process of generating the curl-up force on the film with the curl-up film collector was simulated. Figure 7 shows the diagram of the operation process of the 1JRM-2000 curl-up film collector.

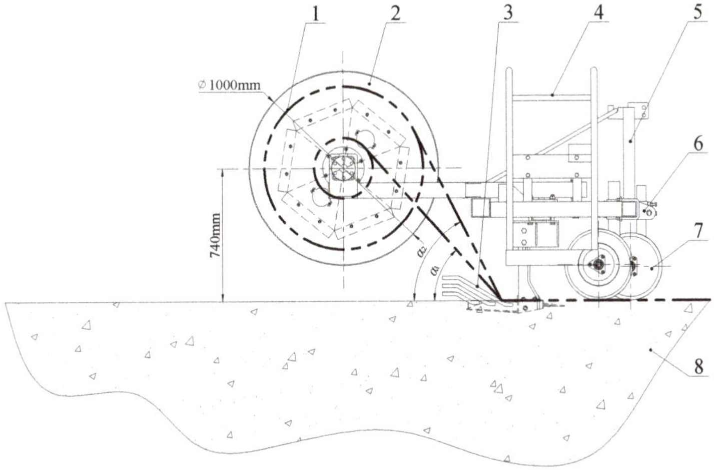

Figure 7. Operation process of the 1JRM-2000 curl-up film collector: 1—film, 2—film-curling mechanism, 3—film pick-up mechanism, 4—operation platform, 5—body frame, 6—traction mechanism, 7—deep limiter, 8—soil.

During operation, the variation range of the film pick-up angle is α_1-α_2. According to Figure 8, during the operation process of the curl-up film collector, the collected residue film would continually wrap around the film-curling device, increasing the film pick-up angle with the increase in the diameter of the residue film wrapping around the film-curling device. The HP-50 digital display pull- and push-strength calculator was adopted to measure the curl-up force. During the force measurement, one end of the film was connected with the pull and push strength calculator, and the other end was at different angles with the ground to simulate the changing process of film pick-up angle during the curl-up collecting of film. The value of the film pick-up angle is controlled by the digital display angle ruler. When the film is initially pulled up, the soil on the film accumulates, and the film is subject to greater soil gravity. When the film is pulled up higher, the accumulation speed of the soil is similar to that of soil falling down from the film. At this time, the soil gravity is in dynamic equilibrium, and the curl-up force becomes stable. The digital display pull- and push-strength calculator was used to record the maximum value of the curl-up force in pulling up the film, and the obtained curl-up force was substituted into Equation (7) to calculate the tensile stress of the film.

(a) (b)

Figure 8. Test field of curl-up collecting of film. (**a**) 11SM-1.2 curl-up film collector; (**b**) 1JRM-2000 curl-up film collector.

3.3. Results and Analysis

3.3.1. Results and Analysis of Contrast Test on the Tensile Properties of High-Performance Film and Ordinary Polyethylene Film

Table 2 shows the contrast test results of the tensile properties of the high-performance film and the ordinary polyethylene film laid in the Xinjiang cotton fields with a service period of 0–180 days.

Table 2 shows that the elongation at break and tensile yield stress of the high-performance film before and during use were higher than those of the ordinary polyethylene film; the elongation at break and tensile yield stress of the film with a thickness of 0.01 mm were higher than those of the film with a thickness of 0.008 mm. The tensile property of the film at a near-end position was higher than that of the film at a far-end position. When the sampling direction was horizontal, the elongation at break and tensile yield stress of the ordinary polyethylene film were higher than those when the film was collected vertically. For the high-performance film, and the elongation at break collected horizontally was higher than that collected vertically; its tensile yield stress was lower than that collected vertically. This is due to the different anisotropy of the high-performance film from the ordinary polyethylene film caused by the orientation of the macromolecules between the layers of the high-performance film. With the increase in the film-laying period, both the elongation at break and tensile yield stress of the high-performance film and ordinary polyethylene film decreased. The variation in the scales of the decrease in the elongation at break and tensile yield stress of the film is shown in Table 3. During the film-laying period of 0~30 days, the scales of the decrease in the elongation at break and tensile yield stress were higher than those during the film-laying period of 30~180 days. When the film-laying period was 120 days and 180 days, the scale of decrease in the elongation at break of the ordinary polyethylene film with a thickness of 0.01 mm collected horizontally at a far-end position and the high-performance film with a thickness of 0.008 mm collected horizontally at a near-end position were negative. This is caused by difference in the thickness of the film and different sampling positions, since the thickness error of film is +0.003~−0.002 mm. Each instance of sampling is located at that of the previous instance; thus, it may have little effect on the scale of decrease in the elongation at break of the film, which shows that there was little variation in the tensile property of the film when the film-laying periods were 90~120 days and 150~180 days.

Table 2. Results of film tensile test during service period of 0–180 d.

Samples	0.008 mm Horizontal				0.008 mm Vertical				0.01 mm Horizontal				0.01 mm Vertical			
	Ordinary Polyethylene Film		High-Performance Film		Ordinary Polyethylene Film		High-Performance Film		Ordinary Polyethylene Film		High-Performance Film		Ordinary Polyethylene Film		High-Performance Film	
Days of Film laying/Day	Elongation at Break/%	Tensile Yield Stress/MPa	Elongation at Break/%	Tensile Yield Stress/MPa	Elongation at Break/%	Tensile Yield Stress/MPa	Elongation at Break/%	Tensile Yield Stress/MPa	Elongation at Break/%	Tensile Yield Stress/MPa	Elongation at Break/%	Tensile Yield Stress/MPa	Elongation at Break/%	Tensile Yield Stress/MPa	Elongation at Break/%	Tensile Yield Stress/MPa
0	286.715	22.37	504.052	29.814	270.146	19.738	400.837	39.313	331.216	26.773	823.628	34.55	297.147	24.957	554.794	41.35

Samples	0.008 mm Horizontal near-end position				0.008 mm Horizontal far-end position				0.008 mm Vertical near-end position				0.008 mm Vertical far-end position			
	Ordinary polyethylene film		High-performance film		Ordinary polyethylene film		High-performance film		Ordinary polyethylene film		High-performance film		Ordinary polyethylene film		High-performance film	
Days of film laying/day	Elongation at break/%	Tensile yield stress/MPa	Elongation at break/%	Tensile yield stress/MPa	Elongation at break/%	Tensile yield stress/MPa	Elongation at break/%	Tensile yield stress/MPa	Elongation at break/%	Tensile yield stress/MPa	Elongation at break/%	Tensile yield stress/MPa	Elongation at break/%	Tensile yield stress/MPa	Elongation at break/%	Tensile yield stress/MPa
30	246.164	19.16	481.141	25.738	217.43	18.671	467.345	24.681	223.514	17.211	371.159	33.746	211.214	15.71	366.024	32.12
60	231.313	17.13	461.457	24.363	186.431	16.241	440.531	22.1	209.131	15.678	356.166	30.925	193.864	14.133	342.711	28.167
90	226.173	16.734	455.63	24.025	177.214	15.716	433.131	21.437	197.216	15.13	347.01	30.173	179.317	13.326	331.463	26.857
120	213.163	16.037	450.57	23.71	163.517	14.971	423.53	20.973	188.314	14.316	339.751	29.613	165.214	12.1	322.3	26.382
150	201.21	15.616	447.214	23.338	149.73	14.1	417.502	20.313	178.132	13.18	329.383	29.088	147.612	10.937	313.918	25.626
180	191.147	15.183	450.307	22.961	136.246	13.17	412.137	19.782	170.214	12.873	321.088	28.617	131.371	9.864	305.866	24.791

Samples	0.01 mm Horizontal near-end position				0.01 mm Horizontal far-end position				0.01 mm Vertical near-end position				0.01 mm Vertical far-end position			
	Ordinary polyethylene film		High-performance film		Ordinary polyethylene film		High-performance film		Ordinary polyethylene film		High-performance film		Ordinary polyethylene film		High-performance film	
Days of film laying/day	Elongation at break/%	Tensile yield stress/MPa	Elongation at break/%	Tensile yield stress/MPa	Elongation at break/%	Tensile yield stress/MPa	Elongation at break/%	Tensile yield stress/MPa	Elongation at break/%	Tensile yield stress/MPa	Elongation at break/%	Tensile yield stress/MPa	Elongation at break/%	Tensile yield stress/MPa	Elongation at break/%	Tensile yield stress/MPa
30	289.416	23.114	781.976	29.875	263.41	22.371	740.591	29.617	260.179	21.011	513.261	36.351	243.633	20.843	463.48	35.56
60	266.147	20.416	751.88	27.2	239.172	19.997	702.467	27.542	237.638	19.371	487.531	34.172	216.137	18.361	415.167	32.3
90	251.147	19.824	746.02	26.613	220.491	18.625	681.016	25.971	218.697	18.019	469.837	33.713	197.083	16.91	406.014	31.163
120	243.732	18.863	739.515	26.17	223.863	17.173	663.389	24.663	208.136	17.164	460.173	33.164	181.691	16.052	398.845	29.622
150	237.281	18.014	728.165	25.841	221.316	16.538	644.807	23.546	201.066	16.62	447.136	32.721	174.25	15.166	390.649	28.65
180	226.391	17.631	717.243	25.17	204.851	16.031	628.324	22.927	196.213	16.033	438.217	32.467	166.261	14.527	383.439	28.038

Table 3. Variation of scale of decrease in film tensile properties with film-laying period.

Samples	**0.008 mm Horizontal Near-End Position**				**0.008 mm Horizontal Far-end Position**				**0.008 mm Vertical Near-end Position**				**0.008 mm Vertical Far-End Position**			
	Ordinary Polyethylene Film		**High-Performance Film**		**Ordinary Polyethylene Film**		**High-Performance Film**		**Ordinary Polyethylene Film**		**High-Performance Film**		**Ordinary Polyethylene Film**		**High-Performance Film**	
Days of Film Laying/Day	**Scale of Decrease in Elongation at Break/%**	**Scale of Decrease in Tensile Yield Stress/%**	**Scale of Decrease in Elongation at Break/%**	**Scale of Decrease in Tensile Yield Stress/%**	**Scale of Decrease in Elongation at Break/%**	**Scale of Decrease in Tensile Yield Stress/%**	**Scale of Decrease in Elongation at Break/%**	**Scale of Decrease in Tensile Yield Stress/%**	**Scale of Decrease in Elongation at Break/%**	**Scale of Decrease in Tensile Yield Stress/%**	**Scale of Decrease in Elongation at Break/%**	**Scale of Decrease in Tensile Yield Stress/%**	**Scale of Decrease in Elongation at Break/%**	**Scale of Decrease in Tensile Yield Stress/%**	**Scale of Decrease in Elongation at Break/%**	**Scale of Decrease in Tensile Yield Stress/%**
30	14.14331	14.34958	4.54536	13.67143	24.16511	16.53554	7.28238	17.21674	17.26178	12.80272	7.40401	14.16071	21.81487	20.40734	8.68508	18.29675
60	6.03297	10.59499	4.09111	5.3423	14.257	13.01484	5.73752	10.45744	6.43494	8.90709	4.03951	8.35951	8.21442	10.03819	6.36925	12.30697
90	2.2221	2.31173	1.26274	1.38735	4.94392	3.23256	1.67979	3	5.69739	3.49534	2.57071	2.43169	7.50371	5.71004	3.28207	4.65083
120	5.75223	4.16517	1.11055	1.31113	7.72907	4.74039	2.21665	2.16448	4.51383	5.38004	2.09187	1.85596	7.86484	9.20006	2.76441	1.76863
150	5.60745	2.62518	0.74483	1.56896	8.43154	5.81791	1.42328	3.1469	5.40693	7.93518	3.05165	1.77287	10.65406	9.61157	2.60068	2.86559
180	5.00124	2.7728	−0.69162	1.61539	9.00554	6.59574	1.28502	2.61409	4.44502	2.32929	2.51834	1.61922	11.00249	9.81073	2.565	3.25841

Samples	0.01 mm Horizontal near-end position				0.01 mm Horizontal far-end position				0.01 mm Vertical near-end position				0.01 mm Vertical far-end position			
	Ordinary polyethylene film		High-performance film		Ordinary polyethylene film		High-performance film		Ordinary polyethylene film		High-performance film		Ordinary polyethylene film		High-performance film	
Days of film laying/day	Scale of decrease in elongation at break/%	Scale of decrease in tensile yield stress/%	Scale of decrease in elongation at break/%	Scale of decrease in tensile yield stress/%	Scale of decrease in elongation at break/%	Scale of decrease in tensile yield stress/%	Scale of decrease in elongation at break/%	Scale of decrease in tensile yield stress/%	Scale of decrease in elongation at break/%	Scale of decrease in tensile yield stress/%	Scale of decrease in elongation at break/%	Scale of decrease in tensile yield stress/%	Scale of decrease in elongation at break/%	Scale of decrease in tensile yield stress/%	Scale of decrease in elongation at break/%	Scale of decrease in tensile yield stress/%
30	12.62016	13.66675	5.05714	13.53111	20.47184	16.44194	10.08186	14.27786	12.44098	15.8112	7.4862	13.75203	18.00927	16.48435	16.45908	14.00242
60	8.03998	11.67258	3.84871	8.95397	9.20162	10.61195	5.14778	7.00611	8.66365	7.80544	5.01304	5.99433	11.28583	11.90807	10.42397	9.1676
90	5.63598	2.89969	0.77938	2.15809	7.8107	6.86103	3.05367	5.70402	7.97053	6.97951	3.62931	1.3432	8.8157	7.90262	2.20465	3.52941
120	2.95245	4.84766	0.87196	1.6646	−1.52931	7.79597	2.58834	5.03639	4.82906	4.74499	2.05688	1.62845	7.80991	5.07392	1.7657	4.94223
150	2.64676	4.50087	1.53479	1.25716	4.13535	3.69766	2.80107	4.52905	3.39682	3.16942	2.83306	1.33579	4.09541	5.51956	2.05493	3.27481
180	4.5895	2.12612	1.49993	2.59665	7.43959	3.06567	2.55627	2.6289	2.41364	3.53189	1.9947	0.77626	4.58479	4.21337	1.84565	2.16405

3.3.2. Results and Analysis of Test on Curl-Up Force in Film Collecting

The software Allpairs was used to generate a hybrid orthogonal table for the test [22], and the test results are shown in Table 4.

Table 4. Test plans and results.

Test No.	Sampling Position	Film Pick-Up Angle/°	Type of Film	The Tensile Stress on the Film/MPa
1	Position 1	30	High-performance film	21.86
2	Position 2	30	Ordinary polyethylene film	19.125
3	Position 1	45	Ordinary polyethylene film	19.364
4	Position 2	45	High-performance film	19.83
5	Position 3	60	High-performance film	16.427
6	Position 1	60	Ordinary polyethylene film	18.217
7	Position 3	75	Ordinary polyethylene film	15.97
8	Position 1	75	High-performance film	17.039
9	Position 3	30	High-performance film	17.513
10	Position 3	45	Ordinary polyethylene film	16.824
11	Position 2	60	High-performance film	17.726
12	Position 2	75	Ordinary polyethylene film	16.013
$(k_1)_1$	19.12	19.499	18.399	
$(k_1)_2$	18.174	18.673	17.586	
$(k_1)_3$	16.684	17.457		
$(k_1)_4$		16.341		
R_1	2.436	3.158	0.813	

According to the analysis of the results in Table 4, it can be obtained that under different test factors, the required film tensile stress for the operation of the curl-up film collector was 15.97–21.86 MPa. By comparing the value with the results of the film tensile property test, the minimum tensile yield stress of the high-performance film with a thickness of 0.01 mm was higher than the required minimum film tensile stress during normal operation of the curl-up film collector. The results of the range analysis showed that the influence order of the test factors on the film tensile stress was *Film Pick-up Angle* > *Sampling Position* > *Type of Film*; the film tensile stress achieved the maximum value when position 1 was chosen as the sampling position, the film pick-up angle was 30°, and the film type was high-performance film.

In order to find out the significance level of the test factors on the test indexes, a variance analysis was made on the above test results, and the analysis results are shown in Table 5.

Table 5. Variance analysis.

Indexes	Sources of Variance	Sum of Squares	Degree of Freedom	Mean Square	*F* Value	Significance
The tensile stress *Y* on the film/MPa	*Sampling Position*	12.07	2	6.035	6.771	**
	Film Pick-up Angle	16.07	3	5.357	6.01	**
	Type of Film	0.81	1	0.81	0.909	
	Residual error	4.457	5	0.891		
	Sum	33.407	11			

Note: ** means the effect is very significant.

According to the analysis results in Table 4, the required film tensile stress for the operation of the curl-up film collector under different test factors was 15.97~21.86 MPa. By comparing this range with the results of the film tensile property test, only the minimum tensile yield stress of the high-performance film with a thickness of 0.01 mm was higher than the minimum film tensile stress required in normal operation of the curl-up film collector. The range analysis results showed that the influence order of the test factors on the film tensile stress was Film Pick-up Angle > *Sampling Position* > *Type of Film*; the film tensile stress achieved the maximum value when position 1 was used as the sampling position, the film pick-up angle was 30°, and the film type was high-performance film.

In order to verify the significance level of each test factor on the test indexes, a variance analysis was made on the above test results, and the analysis results are shown in Table 5.

It can be observed from Table 5 that the sampling position and film pick-up angle had significant influence on the film tensile stress, while the type of film had an insignificant influence on the film tensile stress. During the service period of the film, due to various reasons, such as the wind-blown sand, the soil on the film accumulates. Since the cotton plants can stop the sand, the soil accumulated around the center of each row along the film-laying direction decreases; the longer the service period of the film, the more obvious this tendency becomes. Therefore, when the sampling position was the front point of each row, the soil quantity on the film was highest; thus, the curl-up force required to pull up the film is very high. With the shift of the sampling position to the center of each row and, therefore, with less soil on the film, the curl-up force required to pull up the film reduces. According to Equation (7), the film tensile stress is directly proportional to the curl-up force; thus, the sampling position had a significant influence on the film tensile stress. The angle between the direction of the curl-up force and the ground is equal to the film pick-up angle. The larger the film pick-up angle, the larger the valid component force to pull up the film would become, and the smaller the curl-up force is required. Thus, the film pick-up angle had a significant influence on the film tensile stress. Although the type of film has influence on the soil's moisture content under the film, it has small influence on the cohesion of the soil under the film and the gravity of the soil on the film; thus, the type of film has an insignificant influence on the film tensile stress.

4. Field Test on Curl-Up Collecting of Film

By considering the test results of the contrast test on the tensile property under different test factors and the field test on film curl-up collecting between high-performance film and ordinary polyethylene film, the high-performance film with a thickness of 0.01 mm satisfied the requirements for the tensile stress of film in curl-up collecting. Since the value of the film pick-up angle is inversely proportional to the required curl-up force during curl-up collecting of the film, the film pick-up angle of the machine was set to 45°–75° for film collecting. In order to verify the effect of the curl-up collecting of the film for film collectors with different structures on different types of film with different thicknesses, a test on field film curl-up collecting was designed, and the test site is shown in Figure 8.

4.1. Test Method and Design

The 1JRM-2000 curl-up film collector and the 11SM-1.2 curl-up film collector were used for a contrast test in the field research and development base of the Key Laboratory of Northwest Oasis Agricultural Environment of Ministry of Agriculture, in Tuobuliqi Town, Korla City, Bayingolin Mongolian Autonomous Prefecture of Xinjiang Uygur Autonomous Region, China, during March of 2022. According to the standard GB/T25412-2021, the film recycling rate of the device on the film laid in the same year and the working performance of the device were used as test indexes. The structures of the two types of collectors are shown in Figure 9.

According to Figure 9, when the 11SM-1.2 curl-up film collector was working, the eight groups of film pick-up mechanisms at the front and the two groups of side-film shovels separated the film and soil; the film-guiding and impurity separation mechanisms separated impurities from the film and sent the film to the film-curling mechanism. The film-curling mechanism rotated and winded the film on it. While unloading the film, the hydrocylinder was manually controlled, and the film unloading mechanism unloaded the film package. During the working process, the film pick-up angle remained unchanged and was determined by the angle of the film pick-up mechanism. If the film pick-up angle is too large, the soil penetration angle of the film pick-up mechanism is too large, and the soil produces high resistance against the film pick-up mechanism. If the film pick-up angle is too small, it produces high film tensile stress and tears off the film. Thus, the film pick-up angle was determined to be 45°. During operation of the 1JRM-2000 curl-up film collector, the soil-loosening shovel on the deep limiter in the front of the film collector first loosens the soil around the side film. The film-cutting mechanism cuts the soil from the center along the film-laying direction, and then the film pick-up mechanism in the middle of the machine separates the cut film from the soil. With the forward movement of the machine, by manually controlling the hydrocylinder, the film-unloading mechanism opens, and, during the working process of the machine, the film pick-up angle increases with the increase in the diameter of the film package.

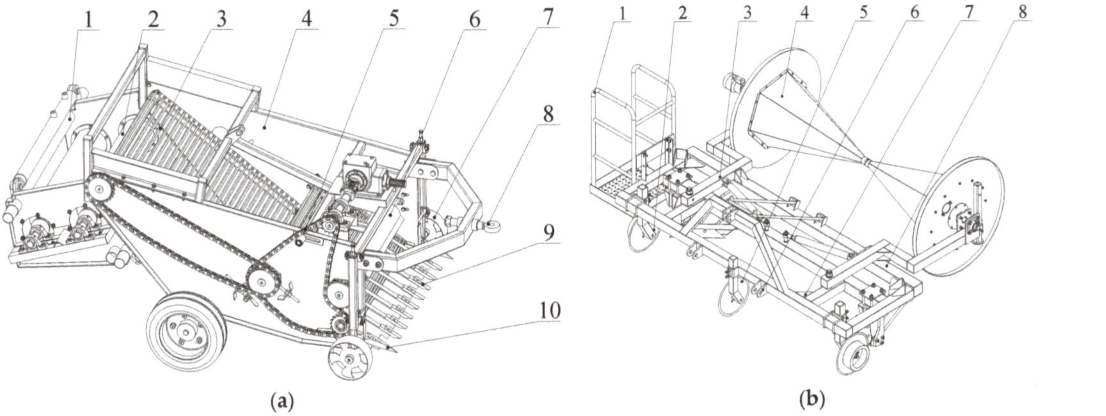

Figure 9. Schematic diagram of structure of two types of curl-up film collectors. (**a**) 11SM-1.2 curl-up film collector: 1—film-unloading mechanism, 2—film-curling mechanism, 3—film-guiding and impurity separation mechanism, 4—body frame, 5—drive system, 6—film-pulling mechanism, 7—depth wheel components, 8—traction mechanism, 9—film pick-up mechanism, 10—side-film shovel; (**b**) 1JRM-2000 curl-up film collector: 1—operation platform, 2—depth limiter, 3—film-unloading mechanism, 3—film-curling mechanism, 4—film-cutting mechanism, 5—traction mechanism, 6—film pick-up mechanism, 8—body frame.

According to the requirements of "five-point random sampling" [23], a measurement area of 200 m × 3.8 m was selected, and test points were chosen within this area. From the four corners of the measurement area along the diagonal lines, four measurement points were randomly determined within the range of one-quarter to one-eighth of the diagonal length, in addition to the intersection of the diagonal lines as the five pre-operation measuring points. Then, five points were selected as post-operation measuring points in the same area near to, but not overlapping, the five pre-operation measuring points. The measuring points cover a length of 5 m and the width of the film, which is 1.25 m. After controlling the stubble height of the cotton plants within 120 mm, the two film collectors started working simultaneously from the start of each row on the same type of film for an operation length of 200 m. The test was repeated three times, and the test results were averaged. Before the machine reached the operation position, a length of 50 was set as the accelerating region to let the machine adjust to a suitable speed. Timing started when the machine entered the operation position, and timing stopped after the machine completed an operation length of 200 m. After operation, residue film pieces were taken from the five pre-operation and post-operation measurement points in the two measurement areas. The residual film taken from each measuring point was washed, dried, and weighed, and the average value was calculated. The film recycling rate on the film laid in the same year can be calculated according to Equation (8):

$$J = (1 - \frac{W}{W_0}) \times 100\% \tag{8}$$

where J is the film recycling rate of the film laid in the same year by the machine, %; W is the mass of residue film laid in the same year in the field after machine operation, g; and W_0 is the mass of residue film laid in the same year before machine operation, g.

4.2. Test Results and Analysis

Test results are listed in Table 6. The results of the test come from "five-point sampling", which is suitable for the survey objects with relatively uniform population distribution and good representativeness. In order to exclude errors caused by accidental factors, three replicate groups were arranged for each sampling, and the final results were averaged.

Table 6. Results of field test on curl-up collecting of film.

Test No.	Type of Collector	Type of Film	Thickness of Film/mm	Film Recycling Rate of the Film Laid in the Same Year/%	Working Performance/km·h
1	1JRM-2000	High-performance film	0.01	81.16	6.15
2	1JRM-2000	High-performance film	0.008	73.26	4.79
3	1JRM-2000	Ordinary polyethylene film	0.01	57.31	3.76
4	1JRM-2000	Ordinary polyethylene film	0.008	55.43	3.13
5	11SM-1.2	High-performance film	0.01	96.11	9.37
6	11SM-1.2	High-performance film	0.008	85.45	8.24
7	11SM-1.2	Ordinary polyethylene film	0.01	78.52	7.88
8	11SM-1.2	Ordinary polyethylene film	0.008	72.49	7.64

Table 6 shows that, during the curl-up film collecting of the 1JRM-2000 curl-up film collector on film with different thicknesses, the film recycling rate of the film laid in the same year and the working performance were lower than that of the 11SM-1.2 curl-up film collector. During the working process of the 1JRM-2000 curl-up film collector, with the increase in the film pick-up angle, the curl-up force changes, and the film is easily broken down during film pick-up. In order to collect the film more easily, the 1JRM-2000 film collector used soil-loosening shovels to loosen the soil around the side film to reduce the force on film. After the soil was loosened, some side film still adhered to the soil and could not be collected, making the film recycling rate of this device lower than that of the 11SM-1.2 curl-up film collector. When the type of film to be collected was high-performance film, since the mechanical properties of the high-performance film were higher than those of the ordinary polyethylene film, the film-cutting mechanism could not effectively cut off the high-performance film, thereby preventing the machine from improving the working performance. When the type of film to be collected was high-performance film, since the mechanical properties of the high-performance film were higher than those of the ordinary polyethylene film, the film-cutting mechanism could not effectively cut it off, which shows the low working performance of the machine. When the type of film to be collected was ordinary polyethylene film, whose minimum tensile yield stress should be lower than the required film tensile stress for the normal operation of the curl-up film collector, the force direction on the film kept changing during operation, and the film was easily broken. In this case, it was necessary to pull the broken film manually to the film-curling mechanism, and thus the working performance of the machine was greatly affected. Since the film pick-up angle of the 11SM-1.2 curl-up film collector is a fixed value, during collecting of the film, the curl-up force is only determined by factors such as the soil quantity on the film. When there is little change in the curl-up force, the film is not broken, and, moreover, with the assistance of the film-guiding mechanism, in the case of film breakage during curl-up collecting, the film-guiding mechanism can transmit the newly separated film from the soil to the film-curling mechanism without manual operation. It can be obtained from the results of the field test on the curl-up collecting of the film that the 11SM-1.2 curl-up film collector achieved film recycling rates of 85.45% and 96.11% on the high-performance film with thicknesses of 0.008~0.01 mm laid in the same year; the 1JRM-2000 curl-up film collector achieved the film recycling rate of 81.16% on the high-performance film with a thickness of 0.01 mm laid in the same year, which could satisfy the requirements of GB/T25412-2021 and achieved working performances of 8.24 km/h, 9.37 km/h, and 6.15 km/h and satisfied the requirements for agricultural production.

Due to the long-term use of ultra-thin and low-strength plastic films in China, the residual film collectors developed in China are mainly aimed at collecting low-tensile strength plastic films. The current related researches includes: The Agricultural Mechanization Research Institute of Xinjiang Academy of Agricultural Sciences [24] has developed a 4JSM-2.1A arc-reciprocating residual film collector; Jiangsu University [25] has developed a combined residual film reclaimer with upper conveyor chain; and China Agricultural University [26] has developed a collecting and separating device for strip plastic film baler. The residual film collected by this device is fragmented, and the film collection mechanism also collects some impurities into the film collecting box during the recycling process, so the collected film can only be reused through granulation, and it is difficult to completely remove impurities, such as the straw, soil and other impurities mixed in the residual film fragments. The cost of using residual film for granulation remains high, and many downstream enterprises of residual film recycling should only rely on government subsidies to support them. It can be concluded in this study that the tensile strength and weather resistance of the high-performance film for full recycling are better than those of the ordinary polyethylene film, and the residual film can be recycled by means of pick-up recycling. The collection of low-tensile strength plastic film and the collected plastic film with high integrity have relatively few impurities, which greatly reduces the cost of collecting residual film for downstream enterprises.

5. Conclusions

(1) A contrast test was carried out on the tensile properties of high-performance film and ordinary polyethylene film, and the test results showed that the elongation at break and the yield stress of the high-performance film before and during the operation were higher than those of the ordinary polyethylene film. The tensile property at a near-end position of the cotton plants was higher than that for a far-end position. When the sampling direction was horizontal, the elongation at break and the tensile yield stress of the ordinary polyethylene film were higher than those when the sampling direction was vertical, and the elongation at break of the high-performance film was higher than that when the sampling direction was vertical, its tensile yield stress was lower than that when the sampling direction was vertical. With the increase in the film laying period, the elongation at break and tensile yield stress had downward tendencies, and, within 0–30 days, the scales of decrease in the elongation at break and tensile yield stress were higher than those during 30–180 days.

(2) Test results showed that the range in tensile stress on the film was 15.97~21.86 MPa when the film is pulled up from different sampling positions, at different film pick-up angles, and with different types of film. The minimum tensile yield stress of the high-performance film with a thickness of 0.01 mm was higher than the maximum film tensile stress required for pulling up the film by the curl-up film collector. The influence order of the test factors on the film tensile stress was film pick-up angle > sampling position > type of film. After a variance analysis on the test data, the results showed that the sampling position and film pick-up angle had significant influences on the tensile stress of the film, while the type of film had an insignificant influence.

(3) Test results showed that during operation of the film collectors, the 11SM-1.2 curl-up film collector with a fixed film pick-up angle achieved a higher film recycling rate on the film laid in the same year and a higher working performance in collecting film of different types and with different thicknesses than the 1JRM-2000 curl-up film collector. The 11SM-1.2 curl-up film collector achieved a film recycling rate of 85.45% and 96.11% on the high-performance film with thicknesses of 0.008 mm and 0.01 mm. The 1JRM-2000 curl-up film collector achieved a film recycling rate of 81.16% on the high-performance film with a thickness of 0.01 mm laid in the same year, which satisfied the requirements of

GB/T25412-2021. Its working performances were 8.24 km/h, 9.37 km/h, and 6.15 km/h, respectively, which could satisfy the demand in production.

(4) In real production, the linear velocity of the film-curling mechanism and the advancing speed of the machine cannot be equally consistent; therefore, the monitoring-feedback–control system is generally adopted to realize a dynamic equilibrium between the linear velocity of the film-curling mechanism and the advancing speed of the machine, thus enhancing the complexity of the machine. If the difference between the linear velocity of the film-curling mechanism and the advancing speed of the machine is too large, the film is easily torn off. Since the automatic film-guiding mechanism can automatically supply film, the working performance of the 11SM-1.2 curl-up film collector is not affected by the difference between the linear velocity of the film-curling mechanism and the advancing speed of the machine.

(5) In the future, we can optimize the curl-up collecting method of film collectors from the perspective of a simulation analysis, and subsequent tests should consider test indexes, such as the number of instances of film breakage and the impurity rate of the film, to find out the optimal mechanical structure and working parameters, and to make preparations for secondary or multiple utilizations of the collected film.

Author Contributions: Conceptualization, methodology, investigation, data curation, formal analysis, and writing—original draft preparation: J.L.; software, visualization, and validation: X.L., L.Z. and X.Z.; writing—review and editing, X.W.; project administration, funding acquisition, resources, and supervision: Y.J. All authors have read and agreed to the published version of the manuscript.

Funding: This research was funded by the National Natural Science Foundation of China, grant number 51965059; Xinjiang Academy of Agricultural Sciences' Key Cultivation Project of Scientific and Technological Innovation, grant number xjkcpy—2021003; The Special Project of Basic Scientific Activities for Non-profit Institutes supported by the government of Xinjiang Uyghur Autonomous Region, grant number KY2022018.

Institutional Review Board Statement: Not applicable.

Informed Consent Statement: Not applicable.

Data Availability Statement: Not applicable.

Conflicts of Interest: The authors declare no conflict of interest.

References

1. Abduwaiti, A.; Liu, X.; Yan, C.; Xue, Y.; Jin, T.; Wu, H.; He, P.; Bao, Z.; Liu, Q. Testing Biodegradable Films as Alternatives to Plastic-Film Mulching for Enhancing the Yield and Economic Benefits of Processed Tomato in Xinjiang Region. *Sustainability* **2021**, *13*, 3093. [CrossRef]
2. Zhongbing, W.; Jun, W. *China Rural Statistical Yearbook 2020*; Aihua, L., Zhicai, Y., Eds.; China Statistics Press: Beijing, China, 2020; Volume 3, p. 42.
3. Can, H.; Xufeng, W.; Xiuying, T.; Yan, Z.; Rongchang, Y. Current situation and control strategies of residual film pollution in Xinjiang. *Trans. Chin. Soc. Agric. Eng.* **2019**, *35*, 233–234. [CrossRef]
4. Zhang, D.; Liu, H.; Hu, W.; Qin, X.; Ma, X.; Yan, C.; Wang, H. The status and distribution characteristics of residual mulching film in Xinjiang, China. *J. Integr. Agric.* **2016**, *15*, 2639–2646. [CrossRef]
5. Scarascia-Mugnozza, G.; Schettini, E.; Vox, G.; Malinconico, M.; Immirzi, B.; Pagliara, S. Mechanical properties decay and morphological behaviour of biodegradable films for agricultural mulching in real scale experiment. *Polym. Degrad. Stab.* **2006**, *91*, 2801–2808. [CrossRef]
6. Anunciado, M.B.; Hayes, D.G.; Wadsworth, L.C.; English, M.E.; Schaeffer, S.M.; Sintim, H.Y.; Flury, M. Impact of Agricultural Weathering on Physicochemical Properties of Biodegradable Plastic Mulch Films: Comparison of Two Diverse Climates Over Four Successive Years. *J. Polym. Environ.* **2020**, *29*, 1–16. [CrossRef]
7. Willett, J.L. Mechanical properties of LDPE/granular starch composites. *J. Appl. Polym. Sci.* **1994**, *54*, 1685–1695. [CrossRef]
8. Zhao, Y.; Chen, X.; Wen, H.; Zheng, X.; Niu, Q.; Kang, J. Research status and prospect of control technology for residual plastic film pollution in farmland. *Trans. Chin. Soc. Agric. Mach.* **2017**, *48*, 1–14. [CrossRef]
9. Rongqing, L.; Xuegeng, C.; Bingcheng, Z.; Hewei, M.; Peng, J.; Xiangbin, P.; Za, K.; Weimin, L. Problems and countermeasures of cotton residue film recovery and resource reuse in xinjiang. *Trans. Chin. Soc. Agric. Eng.* **2019**, *35*, 1–13. [CrossRef]
10. Marí, A.; Pardo, G.; Cirujeda, A.; Martínez, Y. Economic Evaluation of Biodegradable Plastic Films and Paper Mulches Used in Open-Air Grown Pepper (*Capsicum annum* L.) Crop. *Agronomy* **2019**, *9*, 36. [CrossRef]

11. Steinmetz, Z.; Wollmann, C.; Schaefer, M.; Buchmann, C.; David, J.; Troger, J.; Munoz, K.; Fror, O.; Schaumann, G.E. Plastic mulching in agriculture. Trading short-term agronomic benefits for long-term soil degradation? *Sci. Total Env.* **2016**, *550*, 690–705. [CrossRef]
12. Zhiyuan, Z.; Jingbin, L.; Xianfei, W.; Yongman, W.; Shuaikang, X.; Zipeng, S. Parameters Optimization and Test of an Arc-Shaped Nail-Tooth Roller-Type Recovery Machine for Sowing Layer Residual Film. *Agriculture* **2022**, *12*, 660. [CrossRef]
13. Zhou, X.; Zhang, X.; Shi, Z.; Zhang, C.; Bai, S.; Huang, S. Design and test of film lifting separation conveying device of stripper-plate-type residual film recovery machine. *J. Shihezi Univ. (Nat. Sci.)* **2022**, 1–9. [CrossRef]
14. Jinping, Q. Independent innovation achievements have been implemented to solve the problem of residual membrane pollution. *Sci. Technol. Rev.* **2021**, *39*, 1.
15. Fan, W.; Zongli, M.; Bin, D.; Wen, L.; Duopeng, W.; Shangtao, C.; Yanfang, W. Study on Tensile Rheological Properties of Polyethylene Melt. *China Plast.* **2021**, *35*, 13–18. [CrossRef]
16. Tingting, Q.; Zhaoxia, H.; Heping, X.; Jinping, Q. Progress on Volume Elongational Rheology Processing Technology of Polymer. *Plastics* **2020**, *49*, 85–89 + 94.
17. Li, L. Residue status and pollution control of farmland mulch film. *Farm Staff* **2021**, *39*, 1.
18. Jiaxi, Z.; Xuenong, W.; Li, Z.C.Y.; Yongxin, J.; Haichun, Z.; Xuanfeng, L.; Yuanyuan, Q.; Xiangjin, W.; Shanshan, Z. Effects of mechanical tensile properties of plastic film on plastic recycling method. *Trans. Chin. Soc. Agric. Eng.* **2015**, *31*, 41–47. [CrossRef]
19. Jingang, W.; Lingyan, G.; Rui, G.; Ling, L. Prediction model of tensile yield stress-molecular structure of polyethylene. *J. Synth. Resin. Plast.* **2021**, *38*, 7–10. [CrossRef]
20. Liwen, S. Design and Experimental Study of Residual Film Rolling Machine. Master's Thesis, Northwest A&F University, Xianyang, China, 2019.
21. Qiufang, Z.; Dongxing, Z. Design of Film Removing Wheel for a Polythene Film Collector during Seedling Period. *J. Agric. Mech. Res.* **2007**, *9*, 84–86. [CrossRef]
22. Yan, W.; Silian, S. *Experimental Design and MATLAB Data Analysis*; Lei, S., Congmian, Z., Eds.; Tsinghua University Press: Beijing, China, 2012; Volume 6, pp. 153–155.
23. Junyi, G. *Experimental Statistical Methods*, 4th ed.; Jiaju, P., Fanling, K., Eds.; China Agricultural Press: Beijing, China, 2016; Volume 16, pp. 318–320.
24. Xuenong, W.; Jie, L.; Xuanfeng, L.; Yongxin, J.; Haichun, Z.; Jiaxi, Z.; Li, Z. Design and test of a coreless-roll packing device for residual film. *Trans. Chin. Soc. Agric. Eng.* **2022**, *38*, 28–35.
25. Lingyang, Z.; Aiping, S.; Qian, S.; Juxiang, D.; Jian, L.; Dongxu, L. Parameter optimization and experiment of the combined residual film reclaimer with upper conveyor chain. *J. Agric. Mech. Res.* **2022**, *45*, 164–170. [CrossRef]
26. Qi, N.; Chao, J.; Yan, Z.; Xuegeng, C.; Xuan, Z.; Hongwen, L. Design and Experiment on Collecting and Separating Device for Strip Plastic Film Baler. *Trans. Chin. Soc. Agric. Mach.* **2017**, *48*, 101–107.

agriculture

MDPI

Article

Dynamic Fresh Weight Prediction of Substrate-Cultivated Lettuce Grown in a Solar Greenhouse Based on Phenotypic and Environmental Data

Lin Liu [1,2], Jin Yuan [1,3,*], Liang Gong [4], Xing Wang [1] and Xuemei Liu [1,2,*]

[1] College of Mechanical & Electronic Engineering, Shandong Agricultural University, Tai'an 271018, China
[2] Shandong Provincial Key Laboratory of Horticultural Machinery and Equipment, Tai'an 271018, China
[3] Shandong Agricultural Equipment Intelligent Engineering Laboratory, Tai'an 271018, China
[4] School of Mechanical Engineering, Shanghai Jiao Tong University, Shanghai 200240, China
* Correspondence: jyuan@sdau.edu.cn (J.Y.); lxmywj@126.com (X.L.); Tel.: +86-151-6538-7381 (J.Y.)

Abstract: The fresh weight of vegetables is an important index for the accurate evaluation of growth processes, which are affected by factors such as temperature and radiation fluctuation, especially in a passive solar greenhouse. Predicting dynamic growth indexed by fresh weight in a solar greenhouse remains a challenge. A novel method for predicting the dynamic growth of leafy vegetables based on the in situ sensing of phenotypic and environmental data of batches is proposed herein, enabling prediction of the dynamic fresh weight of substrate-cultivated lettuce grown in a solar greenhouse under normal water and fertilizer conditions. Firstly, multibatch lettuce cultivation experiments were carried out and batch datasets constructed by collecting growth environmental data and lettuce canopy images in real time. Secondly, the cumulative environmental factors and instantaneous fresh weights of the lettuce batches were calculated. The optimum response time in days was then explored through the most significant correlations between cumulative environmental factors and fresh weight growth. Finally, a dynamic fresh weight prediction model was established using a naive Bayesian network, based on cumulative environmental factors, instantaneous fresh weight, and the fresh weight increments of batches. The results showed that the computing time setpoint of cumulative environmental factors and instantaneous fresh weight of lettuce was 8:00 AM and the optimum response time was 12 days, and the average R^2 values among samples from three batches reached 95.95%. The mean relative error (MRE) of fresh weight prediction 4 days into the future based on data from the current batch was not more than 9.57%. Upon introducing another batch of data, the prediction 7 days into the future dropped below 8.53% MRE; upon introducing another two batches, the prediction 9 days into the future dropped below 9.68% MRE. The accuracy was improved by the introduction of additional data batches, proving the model's feasibility. The proposed dynamic fresh weight growth prediction model can support the automatic management of substrate-cultivated leafy vegetables in a solar greenhouse.

Keywords: fresh weight prediction; growth model; naive Bayesian network; solar greenhouse; substrate-cultivated lettuce

Citation: Liu, L.; Yuan, J.; Gong, L.; Wang, X.; Liu, X. Dynamic Fresh Weight Prediction of Substrate-Cultivated Lettuce Grown in a Solar Greenhouse Based on Phenotypic and Environmental Data. *Agriculture* **2022**, *12*, 1959. https://doi.org/10.3390/agriculture12111959

Academic Editor: Maciej Zaborowicz

Received: 12 October 2022
Accepted: 17 November 2022
Published: 20 November 2022

1. Introduction

In line with an annual increase in greenhouse planting area in recent years [1,2], the solar greenhouse, a relatively low-cost, environmentally controllable, and highly productive option for farmers, has become the predominant facility type used to provide year-round vegetable production in northern China [3,4]. A solar greenhouse has a large roof area along the south side which is passively heated by sunlight during the daytime [5]. Meanwhile, a thermal blanket is rolled over the greenhouse at night to hold heat inside the structure, and a northern brick wall preserves heat inside the structure [6]. Compared with Venlo greenhouses [7], passive solar greenhouses generally provide only basic environmental control with low-cost equipment [8]. In addition, the use of advanced automated fertigators to ensure sufficient water and fertilizer absorption of vegetables in solar greenhouses has become popular owing to significant labor savings [9]. In production in a Venlo greenhouse, environmental control technology [10,11] is used to regulate indoor environmental parameters such as light, temperature, humidity, and carbon dioxide, making the vegetable growth environment close to optimal. Solar greenhouses have the disadvantages of large temperature fluctuation ranges and frequently weak solar radiation [12], which is not conducive to crop growth, making the crop growth models established in Venlo greenhouse systems unsuitable for application in solar greenhouses.

In a suitable environment, vegetable growth adheres to certain inherent laws throughout the plant's life cycle [13]. Scholars have studied many crop growth models [14–17] with the aim of guiding future crop production in greenhouse systems through regulation of the environment, water, and fertilizer. In one greenhouse crop growth system, a machine learning method based on the expectation maximum algorithm was applied to link environmental parameters with crop growth [18]. Based on only a small number of samples, future crop growth could be predicted several days in advance. Thus, the feasibility of using environmental parameters to predict vegetable growth in greenhouse systems has been verified. However, in the above-mentioned model, leaf area index, evapotranspiration, and dry weight were taken as crop growth indicators, and the leaf area index and dry weight were obtained by destructive methods at intervals of one week. For one thing, indicators obtained using destructive methods cannot provide growth indicators over the whole life cycle sequence of a specific plant, and indicators for shorter time intervals were not obtained. For another, many vegetables needed to be planted in order to assess indicators using destructive sampling during the vegetable growth period, and the process was inefficient and cumbersome. Moreover, the indicators used to measure vegetable growth could not directly reflect the current vegetable yield (i.e., fresh weight).

The fresh weight of vegetables is an important index for accurate evaluation of the growth process, so it is of great significance to apply the fresh weight index to the prediction of crop growth. Compared with hydroponic vegetables, the online, nondestructive monitoring of the fresh weight of substrate-cultured vegetables during the growth process is a challenge. In view of the importance of fresh weight, Yanes et al. [19] proposed a deep learning image segmentation method to obtain information from canopy images for the estimation of fresh weight of hydroponic lettuce, and a regression model relating lettuce size and fresh weight was established. Jung et al. [20] established a model of the relationship between the projected area of lettuce canopy and fresh weight in an environmentally controllable, water-based lettuce cultivation system based on the morphological analysis machine vision method. Jiang et al. [21] developed a fresh weight estimation system for hydroponic lettuce based on online image processing, which realized high-precision estimation of the fresh weight of lettuce and allowed environmental control for high-quality production. In hydroponic vegetable production systems, the plants can be removed from the nutrient solution temporarily and directly weighed without hindering their continuous growth. This is convenient for nondestructive calibration of fresh weight and makes it easy to realize nondestructive, high-precision fresh weight estimation. In substrate culture systems, the plants can be taken out of the substrate and directly weighed to accurately obtain the fresh weight. However, plants weighed in this way will not continue to grow [22,23], and the subsequent fresh weight growth cannot be obtained. It is difficult to achieve nondestructive estimation of the fresh weight of substrate-cultivated vegetables. In order to solve this problem, Liu et al. [24] proposed a fresh weight estimation method based on the fusion of phenotypic characteristics and environmental parameters, which was used to realize nondestructive estimation of the individual and population fresh weights of substrate-cultured lettuce in a solar greenhouse.

However, accurate prediction of dynamic fresh weight growth based on in situ sensing in solar greenhouse systems is still a challenge. Fresh weight growth of vegetables is affected by many complex environmental factors [25]. Large indoor temperature fluctuations and frequently weak solar radiation in solar greenhouse systems lead to differences in the fresh weight growth of different batches. There is a complex and uncertain relationship between vegetable fresh weight growth and environmental factors. Therefore, in contrast to the static modeling of fresh weight under hydroponic conditions [19], a novel prediction method for the dynamic growth of leafy vegetables based on phenotypic and environmental data of batches is proposed herein, which is able to predict the dynamic fresh weight of substrate-cultivated lettuce in a solar greenhouse system under normal water and fertilizer conditions.

The main contributions of this paper are as follows:

(1) Multibatch substrate-cultivated lettuce cultivation experiments were carried out, with the growth environment and lettuce canopy images monitored in real time. A dataset was built using phenotypic and environmental data of batches.

(2) Computation of the cumulative environmental factors and instantaneous fresh weight of batches of lettuce was achieved. The optimum response time was explored via the most significant correlations between cumulative environmental factors and fresh weight growth.

(3) A dynamic fresh weight prediction model was established using a naive Bayesian network, based on cumulative environmental factors, instantaneous fresh weight, and fresh weight increments of batches, which can be used to predict the dynamic fresh weight of substrate-cultured lettuce in a solar greenhouse system.

2. Materials and Methods

2.1. Experimental Design

The experimental site was Solar Greenhouse No. 6 in Shandong Agricultural University Science and Technology Innovation Park, located in Tai'an City, Shandong Province, China (36.16° N, 117.16° E). The greenhouse has a span of about 8 m, a height of about 4 m, and a length of about 50 m from east to west. The experimental material was Italian lettuce, which was produced by Hebei Maohua Seed Industry Limited Company. The main characteristics of this lettuce are a semi-erect form, plant height of about 26 cm, development of about 28 cm, and nearly round leaves. The color is emerald green, and the loose leaves do not form a ball. In order to improve the accuracy of the dynamic fresh weight prediction model, multiple batches of planting experiments were carried out. The same variety of lettuce was used for the multiple batches of planting experiments. When the lettuce seedlings in a batch had grown to five leaves and a heart, the batch was transplanted into a planting tank filled with substrate.

The aboveground growth environment of the lettuce was the closed microclimate environment of the passive solar greenhouse. Due to the structural characteristics of a passive solar greenhouse, only simple environmental regulation could be achieved during the lettuce growth process, barring the introduction of heating, fans, supplementary lights, etc. For example, in the morning, the thermal blanket was opened to allow storage of heat from the sunlight. At noon, the vent was opened to allow natural ventilation for dehumidification, cooling, and air exchange. In the evening, the thermal blanket was closed for insulation, so as to ensure a normal indoor lettuce growth environment and prevent frostbite of the lettuce plants. The underground growth environment of the lettuce plants was the substrate. The substrate had the characteristics of good ventilation and a good drainage effect, but the water retention effect was relatively poor. Therefore, Yamazaki formula nutrient solution at a 100% concentration was used for irrigation via the water and fertilizer application system in the greenhouse (Figure 1), ensuring normal water and fertilizer conditions throughout the lettuce cultivation experiment.

Figure 1. Lettuce cultivation experiment.

2.2. Acquisition of Environmental Data and Lettuce Images in the Solar Greenhouse

An environmental monitoring and image acquisition platform (Figure 1) was used to record the temperature, humidity, photosynthetically active radiation, carbon dioxide concentration, and lettuce canopy images in the solar greenhouse during the lettuce cultivation experiment. The platform was mainly composed of a support mechanism, guide rail slide, hanger, cross bar, sensor, and controller. The support mechanism was used to support the guide rail slide so that the guide rail slide could move horizontally in the north–south direction at a certain height from the ground. The guide rail slide was fixed at the upper end of the support mechanism and the cross bar equipped with the sensor was connected through the hanger, so that the sensor could move in the north–south direction synchronously with the cross bar. The height of the cross bar could be adjusted according to the current situation, and the cross bar and the guide rail slide were kept vertical in the horizontal direction. The guide rail slide was controlled by the controller and the cross bar equipped with sensors was moved to complete the environmental monitoring and image acquisition tasks in the upper part of the planting area.

2.3. Calculation of Environmental Factors and Instantaneous Fresh Weight

2.3.1. Calculation of Cumulative Radiant Heat Product

The effects of temperature and radiation on the fresh weight of lettuce can be measured by the cumulative radiant heat product. The specific calculation formula is as follows [26]:

$$R_{TE} = \begin{cases} 0 & (T < T_b) \\ \frac{T-T_b}{T_{ob}-T_b} & (T_b \le T < T_{ob}) \\ 1 & (T_{ob} \le T \le T_{ou}) \\ \frac{T_m-T}{T_m-T_{ou}} & (T_{ou} < T \le T_m) \\ 0 & (T > T_m) \end{cases} \quad (1)$$

$$T_{EP} = \sum R_{TEP} \quad (2)$$

$$R_{TEP} = \sum_{i=1}^{24} \left(R_{TEi} \times P_{ARi} \times 3600/10^6 \right) \quad (3)$$

where T_b is the lower limit of growth temperature (°C), T_m is the upper limit of growth temperature (°C), T_{ob} is the lower limit of optimum growth temperature (°C), T_{ou} is the upper limit of optimum growth temperature (°C), T is the ambient temperature (°C), R_{TE} is the relative thermal effect, R_{TEP} is the daily cumulative radiant heat product ($MJ \cdot m^{-2} \cdot d^{-1}$), R_{TEi} is the relative thermal effect in the i-th hour, P_{Ari} is the average photosynthetically active radiation in the i-th hour ($MJ \cdot m^{-2} \cdot d^{-1}$), and T_{EP} is the cumulative radiation heat product ($MJ \cdot m^{-2}$).

2.3.2. Calculation of Crop Evapotranspiration

If reference evapotranspiration is used to replace crop evapotranspiration, there will be a large error. Therefore, in order to improve the accuracy of calculation of crop evapotranspiration, the crop coefficient was used to correct the reference evapotranspiration. The specific calculation formula is as follows [27]:

$$ET_c = ET_0 \cdot K_c \tag{4}$$

$$ET_{0i} = \frac{0.408\Delta(Rn - G) + \gamma\frac{1713}{T+273}(e_s - e_a)}{\Delta + 1.64\gamma} \tag{5}$$

$$\Delta = \frac{2505 \cdot \exp\left(\frac{17.27T}{T+237.3}\right)}{(T+237.3)^2} \tag{6}$$

$$e_s = \frac{e_s(T_{\max}) + e_s(T_{\min})}{2} \tag{7}$$

$$e_s(T_{\max/\min}) = 0.6108 \cdot \exp\left(\frac{17.27T_{\max/\min}}{T_{\max/\min} + 237.3}\right) \tag{8}$$

$$e_a = \frac{e_s(T_{\min})\frac{RH_{\max}}{100} + e_s(T_{\max})\frac{RH_{\min}}{100}}{2} \tag{9}$$

$$R = K \cdot P_{AR} \tag{10}$$

$$R_n = a \cdot R + b \tag{11}$$

where ET_0 is the reference evapotranspiration under full irrigation ($cm \cdot d^{-1}$), Δ is the slope of the saturated vapor pressure curve ($kPa \cdot °C^{-1}$), R_n is the net radiation of the crop canopy ($MJ \cdot m^{-2} \cdot d^{-1}$), G is the soil heat flux density ($MJ \cdot m^{-2} \cdot d^{-1}$), γ is the dry and wet table constant ($kPa \cdot °C^{-1}$), T is the daily average temperature at the height of 1.5 to 2.5 m above the surface (°C), $T_{max/min}$ is the daily maximum or minimum air temperature at the height of 1.5 to 2.5 m above the surface (°C), e_s is the average saturated vapor pressure at the height of 1.5 to 2.5 m above the surface (kPa), e_a is the average actual vapor pressure at the height of 1.5 to 2.5 m above the surface (kPa), $RH_{max/min}$ is the daily maximum or minimum relative humidity at the height of 1.5 to 2.5 m above the surface (%), ET_{ci} is the evapotranspiration of crops on the i-th day under full irrigation (cm/d), K_c is the crop coefficient, R is the total solar radiation ($MJ \cdot m^{-2} \cdot d^{-1}$), P_{AR} is the photosynthetically active radiation ($MJ \cdot m^{-2} \cdot d^{-1}$), K is the conversion coefficient between photosynthetically active radiation and total solar radiation, and a and b are the conversion coefficients between net radiation and total radiation.

If $G = 0$, $\gamma = 0.067$, $K_c = 0.7$, 1.00 or 0.95 [28], $K = 80/39$ [29], a = 0.8277, and b = 0.2909 [30], then ET_c can be calculated using Formulas (4)–(11) and the indoor temperature, humidity, and photosynthetically active radiation.

2.3.3. Calculation of Instantaneous Fresh Weight and Fresh Weight Increment

Based on the previous research results of this research group [24], the online, nondestructive calculation of the fresh weight of substrate-cultivated lettuce grown in a solar greenhouse was realized by combining the data of phenotypic characteristics and environmental characteristics. Firstly, the collected lettuce canopy images were used to extract phenotypic characteristics such as shape, color, and texture. Then, using the online monitoring values of temperature and photosynthetically active radiation, cumulative radiant heat product was calculated as an environmental factor. Finally, the above factors were introduced into the model for fresh weight estimation of substrate-cultivated lettuce grown in a solar greenhouse, and the instantaneous fresh weight of the lettuce was obtained (Figure 2). The fresh weight increment was obtained by subtracting the instantaneous fresh weight at one time point from another.

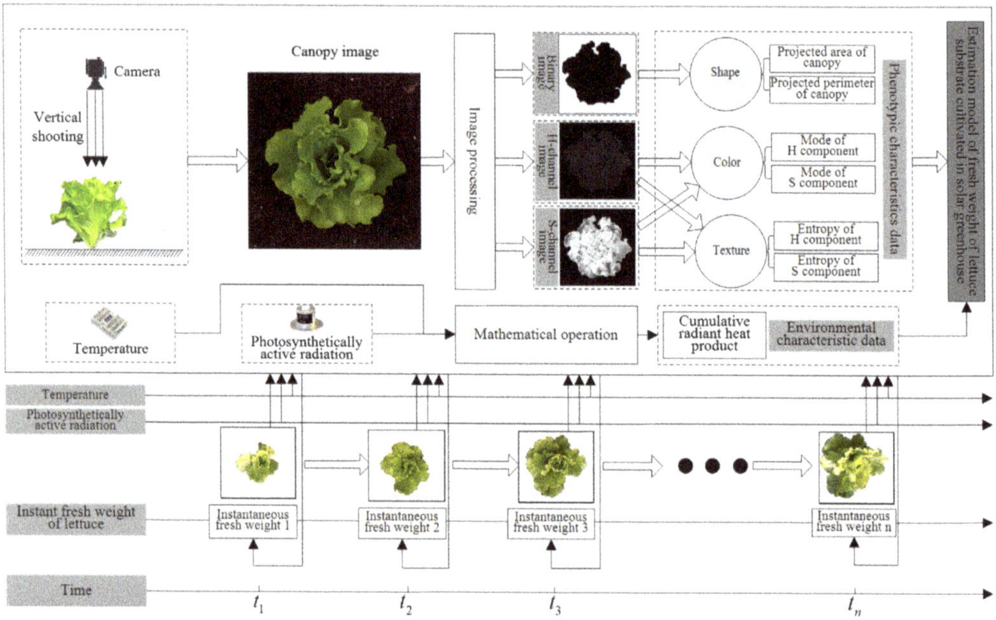

Figure 2. Schematic diagram of calculation of instantaneous fresh weight.

2.4. Exploration of Optimum Response Time in Days

In order to study the optimum response time of the most significant correlations between cumulative environmental factors and fresh weight growth, a naive Bayesian network [31–33] was used to establish the relationship model. There were $n - k$ elements in the dataset, including cumulative environmental factors, instantaneous fresh weight, and fresh weight increments of the previous k days, and the dataset was divided into a training set and a test set. The training set was introduced into the naive Bayesian network for model training, and the test set was used for model testing.

The determination coefficient of the model was calculated by referring to Formula (12) using predicted values and measured values, and was used to examine the degree of correlation between predicted values and measured values of the samples in the dataset. The normal value range is from 0 to 1, and the closer it is to 1, the better the model fits the data. The calculation formula is as follows:

$$R^2 = 1 - \frac{\sum_{i=1}^{n}(y\_test_i - y\_pre_i)^2}{\sum_{i=1}^{n}(y\_test_i - y\_mean)^2} \tag{12}$$

where y_test_i is the measured value of the i-th sample in the dataset (g), y_pre_i is the predicted value of the i-th sample in the dataset (g), and y_mean is the average of the measured values of all samples in the dataset (g).

The coefficient of determination was used as the evaluation index of the model. The larger the coefficient of determination, the more significant the relationship between cumulative environmental factors and fresh weight growth.

The solution process with the most significant response between cumulative environmental factors and fresh weight growth in the previous k days is shown in Figure 3. The figures on the y axis represent the environmental parameters (temperature, humidity, photosynthetically active radiation, and carbon dioxide concentration) or the instantaneous fresh weight of lettuce at a certain time. Firstly, instantaneous fresh weight on day 1, cumulative environmental factors (cumulative radiant heat product, crop evapotranspiration, and average carbon dioxide concentration), and fresh weight increment from day 1 to day $k + 1$ were taken as the first element group in constructing the dataset. The instantaneous fresh weight on day 2, cumulative environmental factors, and fresh weight increment from day 2 to day $k + 2$ were used as the second element group in constructing the dataset. Correspondingly, instantaneous fresh weight on day $n - k$, cumulative environmental factors, and fresh weight increment from day $n - k$ to day n were taken as the last element group in constructing the dataset, which had a total of $n - k$ element groups. The dataset was then divided into a training set and a test set, and the training set was substituted into the naive Bayesian network for model training. Finally, the test set was substituted into the above model and the determination coefficient was calculated, which was used as the evaluation index for the significance of the response between cumulative environmental factors and fresh weight growth in the previous k days.

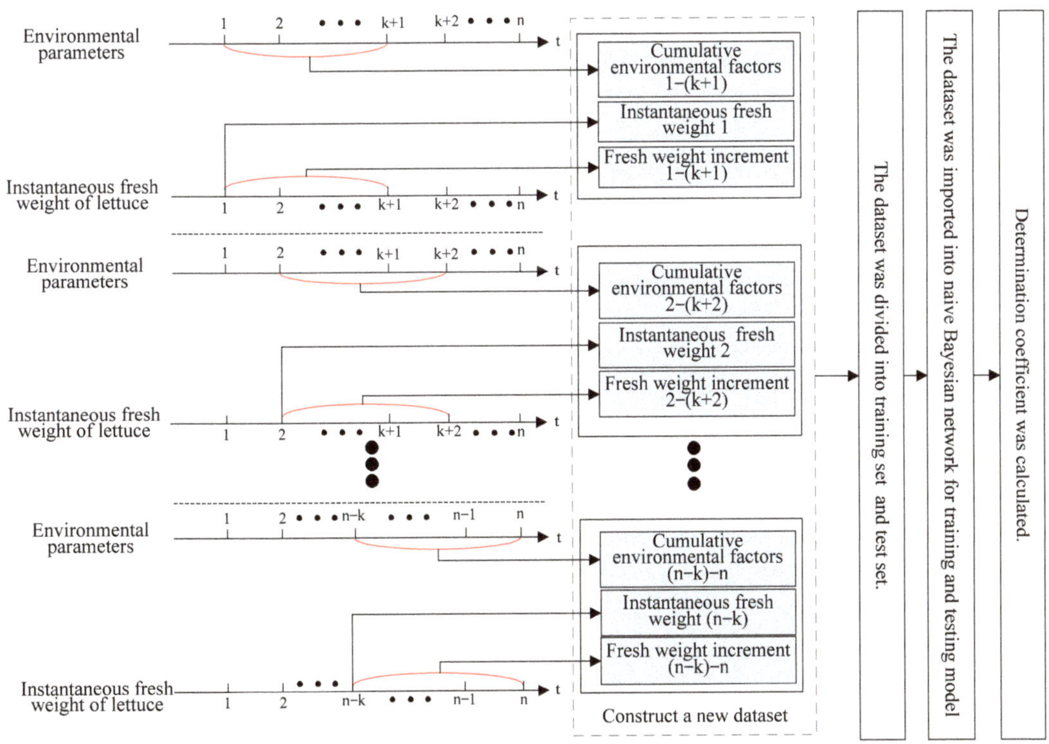

Figure 3. Schematic diagram of the solution process for the most significant response between cumulative environmental factors and fresh weight growth in the previous k days.

2.5. Establishment of Dynamic Fresh Weight Growth Prediction Model

Using the above methods, it was easy to obtain the optimum response time of the most significant correlations between cumulative environmental factors and the fresh weight growth of substrate-cultivated lettuce grown in a solar greenhouse. Thus, a dynamic fresh weight prediction model was constructed, using the collected data to predict the dynamic fresh weight growth of lettuce.

2.5.1. Predicting the Fresh Weight on the Next Day

Firstly, a dataset labeled 1 is constructed using instantaneous fresh weight, cumulative environmental factors, and fresh weight increment in the previous k days from day 1 to day n_0, with a total of $n_0 - k$ elements. The dataset labeled 1 is imported into the naive Bayesian network for training and testing of the model. Then, instantaneous fresh weight on day $n_0 - k + 1$ and cumulative environmental factors from day $n_0 - k + 1$ to day $n_0 + 1$ are taken as the inputs of the above model, and the fresh weight increment from day $n_0 - k + 1$ to day $n_0 + 1$ is derived by substituting the above model. Finally, the instantaneous fresh weight on day $n_0 + 1$ is calculated and the relative error is calculated. The specific calculation formula is as follows:

$$m'_{n0} = m_k + \Delta m'_{n0-k} \tag{13}$$

$$RE = \frac{|m'_{n0-} \ m_{n0}|}{m_{n0}} \tag{14}$$

$$MRE = \frac{1}{n}\sum\nolimits_{i=1}^{n} RE_i \tag{15}$$

$$\sigma = \sqrt{\frac{1}{n-1}\sum\nolimits_{i=1}^{n}(RE_i - MRE)^2} \tag{16}$$

where $\Delta m'_{n0-k}$ is the predicted value of fresh weight increment from day n_0 to day k (g), m_k is the measured value of instantaneous fresh weight on day k (g), m'_{n0} is the predicted value of instantaneous fresh weight on day n_0 (g), m_{n0} is the measured value of instantaneous fresh weight on day n_0 (g), RE is the relative error between the predicted value and measured value of instantaneous fresh weight (%), MRE is the mean relative error (%), and σ is the standard deviation of relative error (%).

2.5.2. Predicting the Fresh Weight in the Next 2 Days

① Using the method of predicting the fresh weight on the next day, the fresh weight increment from day $n_0 - k - 1$ to day $n_0 + 1$ can be obtained.

② The instantaneous fresh weight on day $n_0 - k + 1$, cumulative environmental factors, and predicted fresh weight increment from day $n_0 - k + 1$ to day $n_0 + 1$ are taken as the last element group to construct a new dataset labeled 2, with a total of $n_0 - k + 1$ elements. The dataset labeled 2 is imported into the naive Bayesian network for training and testing of the model.

③ The cumulative environmental factors from day $n_0 - k + 2$ to day $n_0 + 2$ and the instantaneous fresh weight on day $n_0 - k + 2$ are taken as the inputs of the above model, and the fresh weight increment from day $n_0 - k + 2$ to day $n_0 + 2$ is derived by substituting them into the above model.

④ With reference to Equations (13) and (14), the instantaneous fresh weight on day $n_0 + 2$ and the relative error are calculated.

2.5.3. Predicting the Fresh Weight in the Next m_0 Days

Schematic diagram of the solution process for predicting fresh weight in the next m_0 days based on the phenotypic and environmental data from the previous k days is shown as Figure 4.

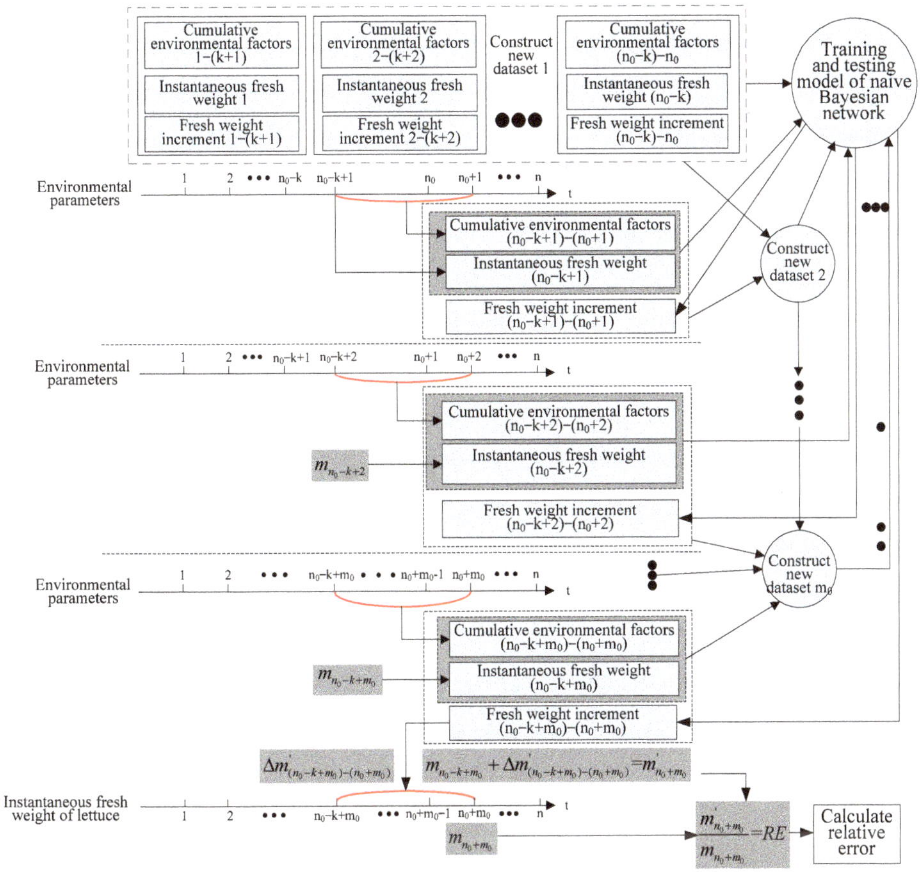

Figure 4. Schematic diagram of the solution process for predicting fresh weight in the next m_0 days based on the phenotypic and environmental data from the previous k days.

① According to the method of predicting the fresh weight in the next 2 days, the fresh weight increment from day $n_0 - k - 2$ to day $n_0 + 2$ can be obtained.

② By analogy, the instantaneous fresh weight on day $n_0 - k + m_0 - 1$, cumulative environmental factors, and predicted fresh weight increment from day $n_0 - k + m_0 - 1$ to day $n_0 + m_0 - 1$ are taken as the last element to construct a new dataset labeled m_0, which has a total of $n_0 - k + m_0 - 1$ elements. The dataset labeled m_0 is imported into the naive Bayesian network for training and testing of the model.

③ The cumulative environmental factors from day $n_0 - k + m_0$ to day $n_0 + m_0$ and the instantaneous fresh weight on day $n_0 + m_0$ are taken as the inputs of the above model, and the fresh weight increment from day $n_0 - k + m_0$ to day $n_0 + m_0$ is derived by substituting them into the above model.

④ With reference to Equations (13) and (14), the instantaneous fresh weight on day $n_0 + m_0$ and the relative error are calculated.

Therefore, through the above methods, the future fresh weight can be predicted using phenotypic and environmental data. For example, if instantaneous fresh weight on the next day is predicted, the cumulative environmental factors from day $n_0 - k + 1$ to day $n_0 + 1$ will be used, which from day $n_0 - k + 1$ to day n_0 are real and known. However, the cumulative environmental factors from day n_0 to day $n_0 + 1$ have not occurred and are unknown. Even if there is an error in estimating the environmental factors from day n_0 to day $n_0 + 1$, the impact on the accuracy of the cumulative environmental factors from day $n_0 - k + 1$ to $n_0 + 1$ is only $1/k$. The overall error generated in fresh weight dynamic prediction is not too large.

3. Results and Discussion

3.1. Fresh Weight Growth Curve of Lettuce

It can be seen from Figure 5 that on both sunny and cloudy days, the changes in fresh weight at nighttime are not obvious, while the changes in fresh weight during the daytime are relatively obvious. The fresh weight tends to decrease in the morning when the sun suddenly becomes stronger. The fresh weight then rises slowly and gradually recovers. When the sun is shining brightly at noon, the fresh weight tends to decrease again. The fresh weight recovers slowly in the afternoon, and it tends to remain stable. The main reason is that the transpiration during the daytime is obviously higher than that at nighttime [34], and the lettuce water content changes faster under the high temperatures, strong light, and low humidity of the daytime. Transpiration is an important indicator for measuring plant water content [35], and its strength is closely related to the degree of water loss in plants [36]. Moreover, water absorption through roots is the main way that water content is maintained in plants [37]. When the water lost by transpiration is higher than that absorbed by roots, the fresh weight of lettuce shows a downward trend. With the decrease of water content in a lettuce plant, a larger pull force is created, forcing the root to absorb more water to maintain normal metabolism and to supplement the water lost through transpiration. When the rate of water absorption by the roots increases slowly, approaching and exceeding the rate of water loss by transpiration, the fresh weight decreases slowly, stops gradually, and begins to increase. Finally, the fresh weight approaches the previous fresh weight range. During the processes of losing water through transpiration and absorption of water through the roots, and with the increase of photosynthesis of the lettuce leaves, the content of organic matter produced by photosynthesis gradually increases in the plant, making the lettuce larger in volume and allowing more water to be stored in the plant. The fresh weight of lettuce will then increase.

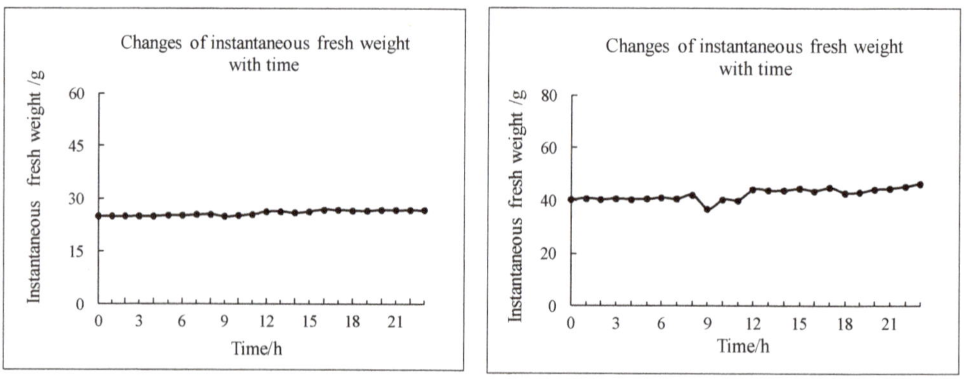

(a) (b)

Figure 5. Fresh weight growth curves in different weather conditions. (**a**) On a cloudy day, (**b**) on a sunny day.

By comparing the change of fresh weight on a sunny day with that on a cloudy day, it was found that the variation in fresh weight growth on the sunny day was higher than that on the cloudy day. This was mainly due to the higher temperatures, stronger illumination, and lower humidity on the sunny day than on the cloudy day, meaning that the volumes of water lost through transpiration and absorbed by the roots were greater and the variations of fresh weight were stronger. There is no sunlight at nighttime and there is little change in temperature and humidity. The water lost by transpiration and the water absorbed by roots is relatively stable. At the same time, compared with a cloudy day, lettuce has a higher level of photosynthesis and accumulates more organic matter on a sunny day, which enables lettuce to absorb more water, increasing its fresh weight.

In order to accurately construct the relationship between the environmental factors and fresh weight growth, the calculation time of cumulative environmental factors and instantaneous fresh weight of lettuce was set at 8:00 AM.

3.2. Optimum Response Time

It can be seen from Figure 6 that the response relationship between cumulative environmental factors and fresh weight growth over different cumulative days was different during the growth process among different samples in the same batch. With the increase in the number of cumulative days, the predicted determination coefficient showed a trend of increase at first. There was an individual decline in this process, but it did not affect the trend of increase. When the cumulative time reached 12 days, the determination coefficients for samples 1, 2, and 3 reached maximum values of 97.02%, 95.64%, and 97.06%, followed by a trend of decrease. In this process, there was an individual increase, but it did not affect the decreasing trend. The optimum response time of the most significant correlation between cumulative environmental factors and fresh weight growth among the different samples in the same batch was 12 days.

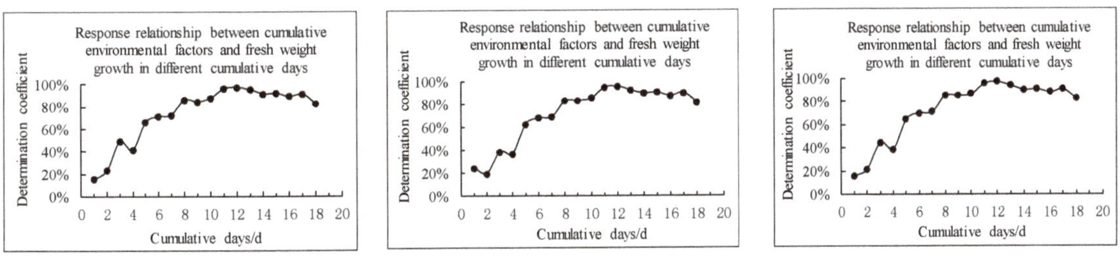

(a) (b) (c)

Figure 6. The response relationship between cumulative environmental factors and fresh weight growth among the different samples in the same batch. (**a**) Sample 1, (**b**) sample 2, (**c**) sample 3.

It can be seen from Figure 7 that the response relationship between cumulative environmental factors and fresh weight growth in different cumulative days was different during the growth process among the different samples in different batches. With the increase of cumulative days, the determination coefficient showed a trend of gradual increase at first. In this process, there was a decline in some cases, but it did not affect the trend of increase. When the determination coefficient reached the maximum value, it began to decrease. In this process, there was an increase in some cases, but it did not affect the decreasing trend. In the samples from the first batch, the coefficient of determination reached a maximum value of 97.57% for 13 cumulative days. The determination coefficient for 12 cumulative days was 97.29%, which was very close to the maximum value of the determination coefficient, and only 0.28% lower. In the samples of the second batch, the determination coefficient reached a maximum value of 94.14% for 13 cumulative days. The determination coefficient for 12 cumulative days was 93.47%, which was very close to the maximum value of the determination coefficient, and only 0.67% lower. In the samples of the third batch, the determination coefficient reached a maximum value of 97.72% for 10 cumulative days. The determination coefficient for 11 cumulative days was 97.39%, which was very close to the maximum value, and only 0.33% lower. The determination coefficient for 12 cumulative days was 97.09%, which was very close to the maximum value, and only 0.63% lower.

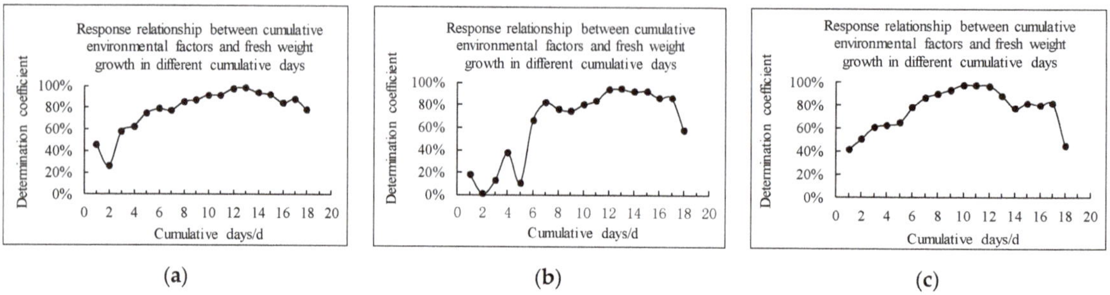

Figure 7. The response relationship between cumulative environmental factors and fresh weight growth among the samples in different batches. (**a**) Samples from the first batch, (**b**) samples from the second batch, (**c**) samples from the third batch.

It can be seen in Table 1 that with the increasing number of cumulative days, the average value of the determination coefficient in the three batches of samples showed a trend of first increasing and then decreasing. When the cumulative time reached 12 days, the average of the determination coefficient reached its maximum value of 95.95%, indicating that the optimum response time of the most significant correlation between cumulative environmental factors and fresh weight growth among different samples in the different batches was 12 days.

Table 1. Numerical distribution table of adjacent regions with maximum values of the coefficient of determination in different batches.

Cumulative Days	First Batch Samples	Second Batch Samples	Third Batch Samples	Average
10	0.9117	0.7976	0.9772	0.8955
11	0.9122	0.8339	0.9739	0.9067
12	0.9729	0.9347	0.9709	0.9595
13	0.9757	0.9414	0.8866	0.9346

3.3. *Using Batch Data to Predict the Dynamic Fresh Weight of Lettuce*

It is obvious from Figure 8 that the fresh weight on the next day can be predicted by using only the data from the current batch (MRE_1 = 6.25%, σ_1 = 7.05%). The relative error (Figure 9) of predicting fresh weight using only the data from the current batch fluctuated greatly at first, and there was one point with a relative error of 40.9%. Subsequently, the relative error fluctuation began to stabilize. This is mainly because the number of elements constructed from the data of the current batch was relatively small at the initial stage, and the accuracy of the model trained by the naive Bayesian network was relatively low. With the increase of the number of elements in the dataset, the accuracy of the model trained by the naive Bayesian network gradually improved, and the relative error started to decrease.

It can be seen from Table 2 that only the data from the current batch were used to predict fresh weight, and the relative error gradually increased with the increasing number of future days (MRE: 6.25% < 6.50% < 7.88%, σ: 7.05% < 6.76% < 11.17%). The data from the current batch with the introduction of another batch were used to predict fresh weight, and the relative error had a tendency to increase with the increasing number of future days (MRE: 4.86% < 5.57% < 6.50%, σ: 5.77% < 6.04%, 5.77% < 5.78%). The data from the current batch with the introduction of another two batches were used to predict fresh weight, and the relative error gradually increased with the increasing number of future days (MRE: 4.35% < 5.40% < 5.29%, σ: 4.87% < 5.38% < 6.11%).

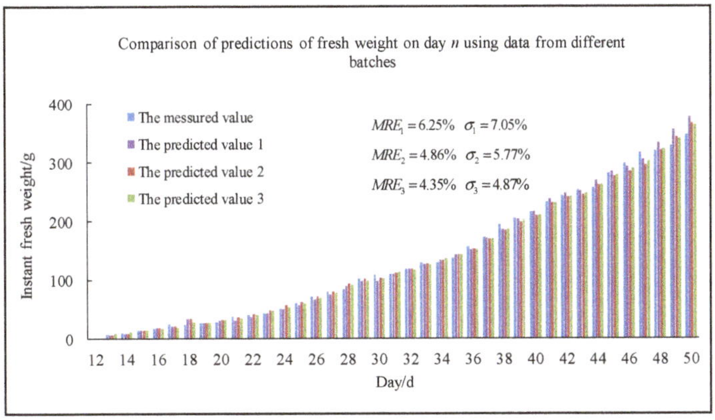

Figure 8. Prediction of fresh weight on the next day. Note: Predicted value 1 is the value of fresh weight on the next day predicted using the data of the current batch. Predicted value 2 is the value of fresh weight on the next day predicted by introducing another batch. Predicted value 3 is the value of fresh weight on the next day predicted by introducing the data from another 2 batches.

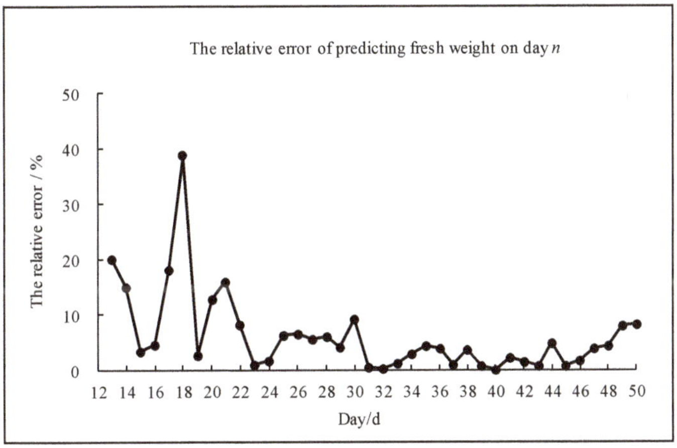

Figure 9. The relative error of predicting fresh weight using only the data from the current batch.

Table 2. Prediction of fresh weight error over the next 3 days.

Error / Batches	Day 1 in the Future		Day 2 in the Future		Day 3 in the Future	
	MRE	σ	MRE	σ	MRE	σ
Current batch	6.25%	7.05%	6.50%	6.76%	7.88%	11.17%
Introducing another batch	4.86%	5.77%	5.57%	6.04%	6.50%	5.78%
Introducing another 2 batches	4.35%	4.87%	5.40%	5.38%	5.29%	6.11%

As shown in Figure 10, the data from the current batch were used to predict the fresh weight in the future. With the increasing number of future days, the MRE of fresh weight prediction gradually increased. In other words, the accuracy of predicting fresh weight in the future gradually decreased, and the MRE of fresh weight prediction over 4 days based on data from the current batch was not more than 9.57%.

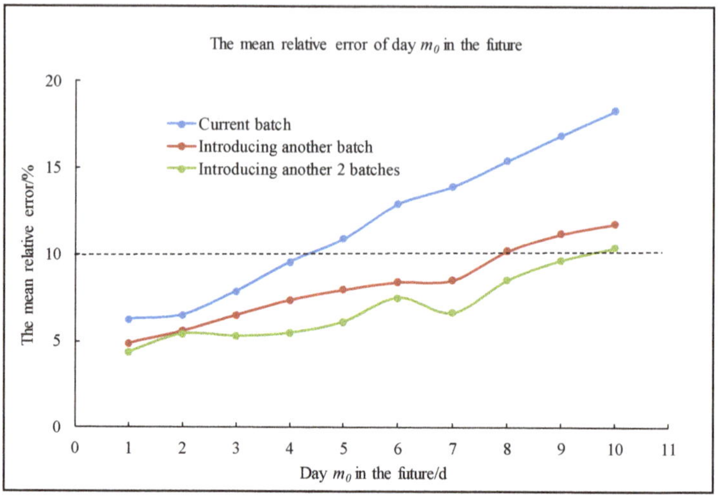

Figure 10. Comparison chart of the mean relative error of predicted future fresh weight.

Upon introducing another batch of data, the MRE of fresh weight prediction gradually increased with the increasing number of future days. However, it was lower than that of the fresh weight predicted using only the data from the current batch, and the MRE of fresh weight prediction in the next 7 days based on the introduction of another batch of data was not more than 8.53%, indicating that the accuracy of predicting fresh weight was improved by introducing another batch.

After introducing the data from another two batches, the MRE of fresh weight prediction tended to increase with increasing number of future days. However, it was lower than that for the fresh weight predicted using the data with only one additional batch, and the MRE of fresh weight prediction over 9 days based on the introduction of data from another two batches was not more than 9.68%, indicating that the accuracy of the fresh weight prediction could be further improved by introducing more batches.

4. Conclusions and Future Work

A dynamic fresh weight growth prediction model based on phenotypic and environmental batch data was proposed, and was used to predict the dynamic fresh weight growth of substrate-cultivated lettuce in a solar greenhouse under normal water and fertilizer conditions. The computation of cumulative environmental factors and instantaneous fresh weight of batches of lettuce was achieved. The optimum response days were explored through the most significant correlations between cumulative environmental factors and fresh weight growth. A dynamic fresh weight prediction model was established using a naive Bayesian network based on cumulative environmental factors, instantaneous fresh weight, and fresh weight increments of batches. Experimental results showed that the calculation time setpoint of cumulative environmental factors and instantaneous fresh weight of lettuce was 8:00 AM and the optimum response time was 12 days. The MRE of fresh weight prediction over 4 days based on data from the current batch was not more than 9.57%; upon introducing another batch of data, the prediction over 7 days dropped to not more than 8.53% MRE; upon introducing another two batches, the prediction over 9 days dropped to not more than 9.68% MRE, proving the model's feasibility.

In future work, the proposed dynamic growth prediction model of fresh weight will be integrated with an automatic management system and sensing data to support an autonomous fertigation strategy for substrate-cultivated leafy vegetables in a solar greenhouse system, playing an important role in promoting the automatic cultivation and management of vegetables in agricultural applications.

Author Contributions: L.L.: writing—original draft, conceptualization, methodology, investigation, data curation, software, and validation. J.Y.: writing—review and editing, conceptualization, and funding acquisition. X.W.: investigation. X.L.: writing—review and editing, formal analysis, funding acquisition, and project administration. L.G.: writing—review and editing, funding acquisition, project administration. All authors have read and agreed to the published version of the manuscript.

Funding: This research was supported by the National Natural Science Foundation of China (52275262), and the Shanghai Science and Technology Innovation Action Plan Project (21N21900100).

Institutional Review Board Statement: Not applicable.

Informed Consent Statement: Not applicable.

Data Availability Statement: Anyone can access the data by sending an email to jyuan@sdau.edu.cn.

Conflicts of Interest: The authors declare no conflict of interest.

References

1. Kalkhajeh, Y.K.; Huang, B.; Hu, W.; Ma, C.; Gao, H.; Thompson, M.L.; Hansen, H.C.B. Environmental soil quality and vegetable safety under current greenhouse vegetable production management in China. *Agric. Ecosyst. Environ.* **2021**, *307*, 107230. [CrossRef]
2. Sun, Z.; You, L.; Muller, D. Synthesis of agricultural land system change in China over the past 40 years. *J. Land Use Sci.* **2019**, *13*, 473–479. [CrossRef]
3. Chen, C.; Yu, N.; Yang, F.; Mahkamov, K.; Han, F.; Li, Y.; Ling, H. Theoretical and experimental study on selection of physical dimensions of passive solar greenhouses for enhanced energy performance. *Sol. Energy* **2019**, *191*, 46–56. [CrossRef]
4. Wang, H.Y.; Zu, G.; Yang, F.J.; Li, D.Z.; Tian, L.M. Design of multi-functional solar greenhouses in high latitude areas. *Trans. CSAE* **2020**, *36*, 170–178.
5. Bao, E.C.; Cao, Y.F.; Zou, Z.R.; Shen, T.T.; Zhang, Y. Research progress of thermal storage technology in energy-saving solar greenhouse. *Trans. CSAE* **2018**, *34*, 1–14.
6. Nemali, K. History of Controlled Environment Horticulture: Greenhouses. *HortScience* **2022**, *57*, 239–246. [CrossRef]
7. Berkers, E.; Geels, F.W. System innovation through stepwise reconfiguration: The case of technological transitions in Dutch greenhouse horticulture (1930–1980). *Technol. Anal. Strateg. Manag.* **2011**, *23*, 227–247. [CrossRef]
8. Wang, T.; Wu, G.; Chen, J.; Cui, P.; Chen, Z.; Yan, Y.; Zhang, Y.; Li, M.; Niu, D.; Li, B.; et al. Integration of solar technology to modern greenhouse in China: Current status, challenges and prospect. *Renew. Sustain. Energy Rev.* **2017**, *70*, 1178–1188. [CrossRef]
9. Koukounaras, A. Advanced Greenhouse Horticulture: New Technologies and Cultivation Practices. *Horticulturae* **2021**, *7*, 1. [CrossRef]
10. Lin, D.; Zhang, L.; Xia, X. Model predictive control of a Venlo-type greenhouse system considering electrical energy, water and carbon dioxide consumption. *Appl. Energy* **2021**, *298*, 117163. [CrossRef]
11. Yu, G.; Zhang, S.; Li, S.; Zhang, M.; Benli, H.; Wang, Y. Numerical Investigation for Effects of Natural Light and Ventilation on 3D Tomato Body Heat Distribution in a Venlo Greenhouse. *Inf. Process. Agric.* **2022**, *in press*. [CrossRef]
12. Liu, W. Engineering Technology Development of Temperature and Lighting Control Promote Productivity of Chinese Solar Greenhouse. *Agric. Eng.* **2015**, *5*, 15–17+42.
13. Rajasekar, M.; Arumugam, T.; Kumar, S.R. Influence of weather and growing environment on vegetable growth and yield. *J. Hort.* **2013**, *5*, 160–167.
14. Hang, T.; Lu, N.; Takagaki, M.; Mao, H. Leaf area model based on thermal effectiveness and photosynthetically active radiation in lettuce grown in mini-plant factories under different light cycles. *Sci. Hortic.* **2019**, *252*, 113–120. [CrossRef]
15. Zhang, L.; Xu, Z.; Xu, D.; Ma, J.; Chen, Y.; Fu, Z. Growth monitoring of greenhouse lettuce based on a convolutional neural network. *Hortic. Res.* **2020**, *7*, 124. [CrossRef] [PubMed]
16. Chang, C.-L.; Chung, S.-C.; Fu, W.-L.; Huang, C.-C. Artificial intelligence approaches to predict growth, harvest day, and quality of lettuce (Lactuca sativa L.) in a IoT-enabled greenhouse system. *Biosyst. Eng.* **2021**, *212*, 77–105. [CrossRef]
17. Graamans, L.; Baeza, E.; Dobbelstena, A.; Tsafaras, I.; Stanghellini, C. Plant factories versus greenhouses: Comparison of resource use efficiency. *Agric. Syst.* **2018**, *160*, 31–43. [CrossRef]
18. Kocian, A.; Massa, D.; Cannazzaro, S.; Incrocci, L.; Di Lonardo, S.; Milazzo, P.; Chessa, S. Dynamic Bayesian network for crop growth prediction in greenhouses. *Comput. Electron. Agr.* **2020**, *169*, 105167. [CrossRef]
19. Yanes, A.R.; Martinez, P.; Ahmad, R. Real-time growth rate and fresh weight estimation for little gem romaine lettuce in aquaponic grow beds. *J. Comput. Electron. Agric.* **2020**, *179*, 105827. [CrossRef]
20. Jung, D.-H.; Park, S.H.; Han, H.; Kim, H.-J. Image processing methods for measurement of lettuce fresh weight. *J. Biosyst. Eng.* **2015**, *40*, 89–93. [CrossRef]
21. Jiang, J.-S.; Kim, H.-J.; Cho, W.-J. On-the-go image processing system for spatial mapping of lettuce fresh weight in plant factory. *IFAC-PapersOnLine* **2018**, *51*, 130–134. [CrossRef]

22. Liu, Z.G.; Wang, J.Z.; Xu, Y.F.; Li, P.P. Lettuce yield and root distribution in substrates under drip irrigation and micro-sprinkler irrigation. *Trans. Chin. Soc. Agric. Mach.* **2014**, *45*, 156–160.
23. Chen, X.L.; Yang, Q.C.; Ma, T.G.; Xue, X.Z.; Qiao, X.J.; Guo, W.Z. Effects of red and blue LED irradiation in different alternating frequencies on growth and quality of lettuce. *Trans. Chin. Soc. Agric. Mach.* **2017**, *48*, 257–262.
24. Liu, L.; Yuan, J.; Zhang, Y.; Liu, X.M. Non-destructive Estimation Method of Fresh Weight of Substrate Cultured Lettuce in Solar Greenhouse. *Trans. Chin. Soc. Agric. Mach.* **2021**, *52*, 230–240.
25. Bian, Z.H.; Yang, Q.C.; Liu, W.K. Effects of light quality on the accumulation of phytochemicals in vegetables produced in controlled environments: A review. *J. Sci. Food Agric.* **2015**, *95*, 869–877. [CrossRef]
26. Zhou, T.M.; Zhen, W.; Wang, Y.C.; Su, X.J.; Qin, C.X.; Huo, H.Q.; Jiang, F.L. Modelling seedling development using thermal effectiveness and photosynthetically active radiation. *J. Integr. Agric.* **2019**, *18*, 2521–2533. [CrossRef]
27. Gu, Z.; Yuan, S.Q.; Qi, Z.M.; Wang, X.K.; Cai, B. Real-time precise irrigation scheduling and control system in solar greenhouse based on ET and water balance. *Trans. Chin. Soc. Agric. Mach.* **2018**, *34*, 101–108.
28. Allen, R.G.; Pereira, L.; Raes, D.; Smith, M. Crop Evapotranspiration-Guidelines for Computing Crop Water Requirements. In *FAO Irrigation and Drainage Paper 56*; FAO: Rome, Italy, 1998.
29. Mao, J.X. Analysis on conversion between solar radiance and illuminance. *J. Henan Agric. Sci.* **1995**, *240*, 11–12.
30. Li, X.; Luo, X.L.; Wang, R.; Li, T.L.; Xu, H.; Zhao, X.Y. Analysis on Solar Radiation's Changing Laws and Utilization in Solar Greenhouse. *Chin. Agric. Sci. Bull.* **2012**, *28*, 104–108.
31. Langarizadeh, M.; Moghbeli, F. Applying naive bayesian networks to disease prediction: A systematic review. *Acta Inform. Med.* **2016**, *24*, 364. [CrossRef]
32. Li, L.X.; Rahman, S.S.A. Students' learning style detection using tree augmented naive Bayes. R. *Soc. Open Sci.* **2018**, *5*, 172108. [CrossRef] [PubMed]
33. Harzevili, N.S.; Alizadeh, S.H. Mixture of latent multinomial naive Bayes classifier. *Appl. Soft Comput. J.* **2018**, *69*, 516–527. [CrossRef]
34. Resco de Dios, V.; Roy, J.; Ferrio, J.P.; Alday, J.G.; Landais, D.; Milcu, A.; Gessler, A. Processes driving nocturnal transpiration and implications for estimating land evapotranspiration. *Sci. Rep.* **2015**, *5*, 10975. [CrossRef] [PubMed]
35. Maes, W.H.; Steppe, K. Estimating evapotranspiration and drought stress with ground-based thermal remote sensing in agriculture: A review. *J. Exp. Bot.* **2012**, *63*, 4671–4712. [CrossRef] [PubMed]
36. Adeyemi, O.; Grove, I.; Peets, S.; Domun, Y.; Norton, T. Dynamic modelling of lettuce transpiration for water status monitoring. *Comput. Electron. Agric.* **2018**, *155*, 50–57. [CrossRef]
37. Jackson, R.B.; Sperry, J.S.; Dawson, T.E. Root water uptake and transport: Using physiological processes in global predictions. *Trends Plant Sci.* **2000**, *5*, 482–488. [CrossRef]

agriculture

MDPI

Article

Design and Test of Duckbill Welding Robot for Cotton Seeder

Yu Ren [1,2], Wensong Guo [1,2,*], Xufeng Wang [1,2], Can Hu [1,2], Long Wang [1,2], Xiaowei He [1,2] and Jianfei Xing [1,2]

1 College of Mechanical and Electrical Engineering, Tarim University, Alar 843300, China
2 Modern Agricultural Engineering Key Laboratory at Universities of Education Department of Xinjiang Uygur Autonomous Region, Tarim University, Alar 843300, China
* Correspondence: 120120004@taru.edu.cn

Abstract: To improve the automation, welding efficiency, and welding quality of duckbill welding of the cotton seeder, this study designed a cotton seeder duckbill welding robot. According to the characteristics of the duckbill weldment and welding requirements, the overall structure of the welding robot was determined, including the girdle feeding mechanism, static duckbill feeding mechanism, hinge feeding mechanism, welding fixture, welding actuator, and control system. To realize the continuous automatic feeding, positioning, fixing, welding, and unloading of the workpiece in the duckbill welding, the feeding mechanism adopts the method of cooperative cooperation of inductive proximity switch, electromagnet, and cylinder. The main body of the welding fixture adopts the pneumatic clamping method; the welding actuator adopts the synchronous belt module electric drive so that the welding torch can move in a straight line along the X axis and the Z axis. The welding process of the duckbill was simulated by Simufact Welding software, and the deformation and stress changes of the weldment were compared and analyzed when the single-sided single welding, the bilateral symmetrical double welding torch, two welding forms, and two welding process parameters were used to determine the welding process parameters of the welding robot. The prototype was made and the welding test was carried out. The test results show that the duckbill welding robot of the cotton seeder has stable feeding, solid clamping, accurate positioning, and high welding efficiency. According to the national standard, the appearance of the duckbill weld is inspected. The surface of the duckbill weld and the heat-affected zone has no cracks, incomplete fusion, slag inclusion, crater, and porosity. The forming quality of the welded parts is good. The design of the duckbill welding robot for cotton seeder is helpful in solving the problems of cumbersome positioning and clamping and low efficiency in manual and semi-automatic duckbill welding robots, which provides a strong guarantee for the large-scale and standardized welding production of the dibbler duckbill.

Keywords: cotton seeder; duckbill; Simufact Welding; welding robot; automated welding

Citation: Ren, Y.; Guo, W.; Wang, X.; Hu, C.; Wang, L.; He, X.; Xing, J. Design and Test of Duckbill Welding Robot for Cotton Seeder. *Agriculture* **2023**, *13*, 31. https://doi.org/10.3390/agriculture13010031

Academic Editors: Jin Yuan, Wei Ji and Qingchun Feng

Received: 21 November 2022
Revised: 9 December 2022
Accepted: 20 December 2022
Published: 22 December 2022

1. Introduction

The plastic mulching technique is one of the most widely used and effective technical measures to improve soil water storage capacity and plant water use efficiency [1,2]. At present, cotton sowing in Xinjiang is based on the method of sowing on film, which is carried out on the soil covered with the film [3,4]. The duckbilled dibbler is used for sowing on film in Xinjiang. The duckbill of the dibbler will cut the film at the sowing position and form holes in the soil. The welding quality of the duckbill of the dibbler is the key to affecting the quality of the hole and the speed of operation [5].

The number of welded duckbills in Xinjiang is about millions every year. Before 2017, the welding method of cotton planter duckbill was manual welding. Welding workers manually position, fix, and weld the three parts of the duckbill, static duckbill, hinge, and girdle. Manual welding has the following problems: unstable welding quality, low efficiency, high labor intensity, and low degree of automation. In 2017, we developed a semi-automatic duckbill welding robot. This semi-automatic welding robot needs to be loaded, positioned, fixed, and unloaded manually, and the welding operation is completed by the robot. The welding robot improves the welding efficiency of the duckbill. The purpose of this study is to further improve duckbill welding automation, welding efficiency, and welding quality, as well as reduce the labor intensity of welding duckbills.

There are many ways to improve welding efficiency and welding quality. For example, on the one hand, some research can be conducted on harmonic drive transmission [6–8]. This method can improve the welding quality by improving the positioning accuracy of the welding robot manipulator. On the other hand, according to the characteristics of the welding object and the welding requirements, a special welding robot can be designed to replace manual welding [9–13]. For example, Süleyman ERSÖZ et al. [14] proposed a robot system that can automatically complete measurement and welding operations for products that are difficult to manually complete standard measurement or welding operations. Namkug Ku et al. [15] designed a self-driving mobile welding robot for double-hull structures in shipbuilding. Stephen Mulligan et al. [16] developed and demonstrated an autonomous, mobile welding robot capable of fabricating large-scale customized structures. Jiang Yi et al. [17] designed a series-parallel-series hybrid structure mobile welding robot for welding corrugated plates of liquefied natural gas (LNG) membrane tanks. At present, there is little research on the application of special welding robots in the field of duckbill welding of the cotton planters.

Different from the traditional plane welding operation, the weld of the duckbill part is a fillet weld. Its processing technology is complex, the welding workload is large, and the weld is prone to defects [18]. The traditional welding process relies on experience to determine the amount of deformation, the reasonable welding method, and welding process parameters, which rely on experience and cannot fully and quantitatively grasp the law of welding deformation. With the development of finite element technology, the welding simulation is fully applied to the actual production and used to guide the process design and gradually became an effective means to provide technical support for the control of process measures in the welding robot manufacturing process [19–22].

To design a duckbill welding robot to improve the welding quality, stability, and welding efficiency of the duckbill parts of the cotton seeder, this study first analyzes the characteristics of the duckbill parts and then uses Simufact Welding software to simulate and analyze the duckbill welding process, which effectively provides technical support for the welding deformation control process measures of the duckbill welding robot in the manufacturing process. Finally, a cotton seeder duckbill welding robot is designed, and the reliability of the welding robot is verified by the welding test. The research results provide ideas for further improving the quick automatic feeding, clamping, positioning, and welding of the duckbill of the dibbler, as well as provide basic and technical support for the automatic welding of the duckbill of the dibbler.

2. Welding Object Characteristics

2.1. Assembly Structure of Duckbill Welding Parts

The assembly relationship diagram of duckbill welding parts is shown in Figure 1. The dibbler is one of the key components of the cotton mulching seeder, which is used to complete the seeding process. Sowing quality has a significant impact on crop growth and yield [23,24]. The duckbill part cuts the film at the seeding position during the seeding operation and forms holes in the soil. The duckbill is welded by three parts: the static duckbill, hinge, and girdle. The welding quality has a very important influence on the hole-forming effect of the dibbler and the seed falling position.

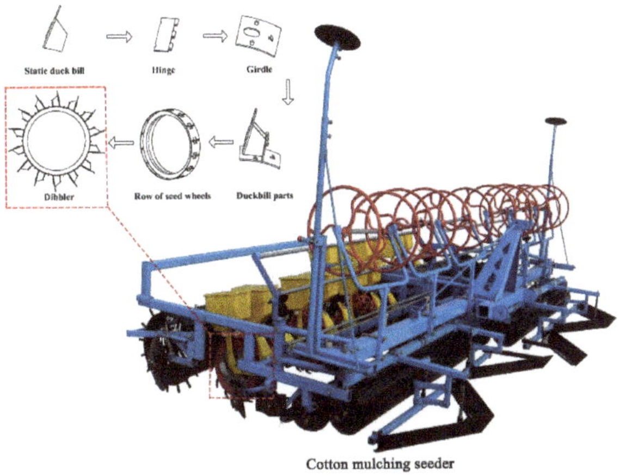

Figure 1. The assembly relationship diagram of duckbill welding parts.

2.2. *Material Properties of Duckbill Welding Parts*

As shown in Figure 2, the duckbill of a cotton seeder is composed of a girdle, static duckbill, and hinge, and its structural parameters are shown in Table 1. The material of duckbill parts is Q235, which is an ordinary carbon structural steel. The chemical composition and mechanical properties are shown in Table 2. Q235 has low carbon and alloy element content and excellent welding performance. Generally, special process measures, such as preheating and post-weld heat treatment, are not required during welding. However, when the incorrect welding form is adopted, the appearance of the weld will also appear poor, forming cracks.

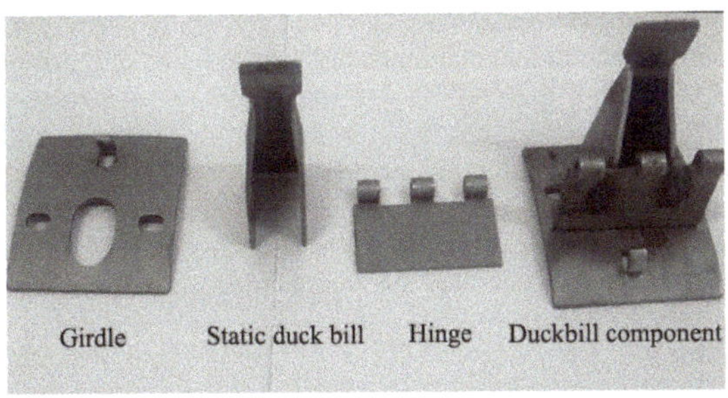

Figure 2. Physical drawing of the duckbill welding parts of the cotton seeder.

Table 1. Structural parameters of duckbill welding parts.

Parts	Length (mm)	Width (mm)	Thickness (mm)	Height (mm)	Mass (g)
Girdle	78.12	68.30	2.11	5.90	73.046
Static duck bill	34.09	27.52	2.57	74.97	77.747
Hinge	69.02	36.11	2.08	7.97	39.281

Table 2. Material properties of Q235.

C (Mass Fraction)/%	Mn	Si	S	P
0.14~0.19	0.30~0.65	0.30	≤0.050	≤0.045
Tensile strength (MPa)		Yield point (MPa)		Elongation (%)
375~500		235		26

2.3. Weld and Welding Requirements Analysis

As shown in Figure 3, the weld of the duckbill welding part is two fillet welds, which are: weld 1 formed by the static duckbill and the hinge and girdle, and weld 2 formed on another back symmetrical surface.

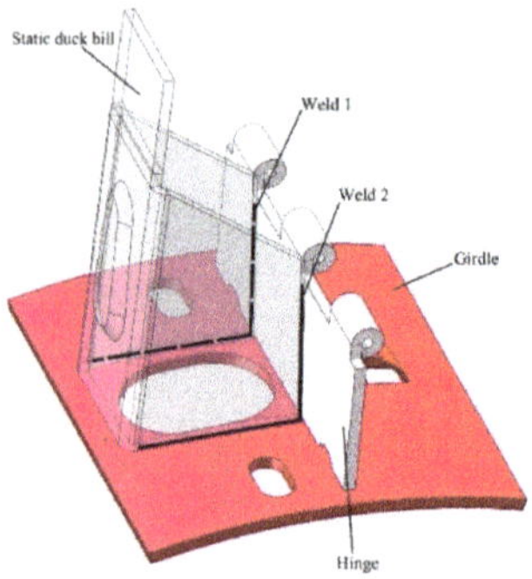

Figure 3. The weld diagram of duckbill welding parts.

An inappropriate welding process will increase the deformation of duckbill welding, resulting in the following problems: (1) the girdle and seeding wheel being difficult to assemble; (2) the following performance being affected; and (3) the quality of the hole being unstable. Duckbill parts in the process of sowing operation need to film soil punching and work under great pressure. The duckbill parts of the welding quality requirements are very high, including the ability to weld duckbill weld surface without cracks, crater shrinkage, and welding tumor defects.

3. Simulation and Analysis of the Welding Process

Welding deformation is the most important factor affecting welding quality. Welding deformation will lead to a manufacturing delay, economic cost, and reduced productivity. Excessive deformation may seriously damage manufacturing in extreme cases, leading to failure [25]. At the same time, high welding residual stresses in the weld can adversely affect the safety and performance of welded components [26,27]. In this study, Simufact Welding software is used to simulate the welding process of duckbill welding parts, and the influence of deformation and the stress of weldments under a single-sided single welding torch and bilateral symmetrical double welding torch, two welding forms, and two welding process parameters, is analyzed.

3.1. Heat Source Model

In welding simulation, a reasonable heat source model is very important for the accurate calculation of post-weld deformation and welding stress [28]. To realize the simulation calculation, the commonly used heat source models are the classical Gaussian distribution heat source model and the double ellipsoid heat source model [29,30]. The Gaussian model can obtain better calculation accuracy for planar high-energy beam welds in simulation calculations. The double ellipsoid heat source model is more close to the actual welding situation of a fillet weld, so this study chooses the double ellipsoid heat source model for calculation.

The heat flux density expression of the front part of the double ellipsoid heat source is:

$$q_f(x,y,z) = \frac{6\sqrt{3}f_t q_0}{abc_f\pi\sqrt{\pi}}\exp(-\frac{3x^2}{c_f^2} - \frac{3y^2}{a^2} - \frac{3z^2}{b^2}). \tag{1}$$

The heat flux distribution expression of the second half of the double ellipsoid heat source is:

$$q_b(x,y,z) = \frac{6\sqrt{3}f_t q_0}{abc_b\pi\sqrt{\pi}}\exp(-\frac{3x^2}{c_b^2} - \frac{3y^2}{a^2} - \frac{3z^2}{b^2}). \quad (2)$$

In the formula: a, b, c_f, and c_b are oval shape parameters of the heat source; q_0 is the heat input power, and $q_0 = \eta$UI; and f_f, f_b are the heat flux distribution coefficients of the ellipsoid before and after the heat source, $f_f + f_b = 2$.

3.2. Establishment of Welding Model

The solid model of duckbill welded parts was established by SolidWorks, and then the model was imported into Hypermesh for hexahedral meshing. The number of finite element mesh nodes was 37,394, and the number of finite elements was 27,997. The divided model was imported into Simufact Welding for assembly and configuration, as shown in Figure 4. In this study, the weldment material is Q235, and the energy input per unit length of the weld (line energy) is calculated according to Equation (3).

$$Q = \eta\frac{IU}{v} \quad (3)$$

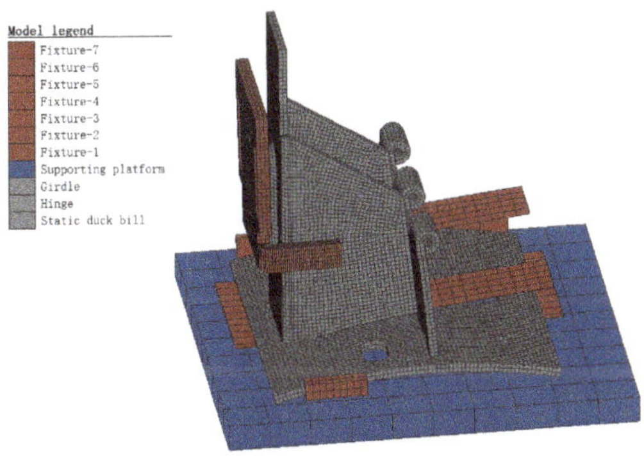

Figure 4. Meshing model of duckbill welding parts.

In the formula: Q is the line energy; I is the welding current; U is the welding voltage; v is the welding speed; and η is the welding thermal efficiency. As the weld of duckbill weldment is fillet weld, the welding heat is relatively concentrated. In this study, the welding thermal efficiency is taken as 0.8 in the simulation process [31].

3.3. Welding Simulation Results and Analysis

3.3.1. Effect of the Unilateral Single Welding Torch and Bilateral Symmetrical Double Welding Torch on Welding Deformation and Stress

Figure 5 shows the deformation of the duckbill welding parts under the single welding torch and the bilateral symmetrical double welding torch. By comparing and analyzing their total displacement cloud diagrams, the following conclusions were obtained: The area of deformation was larger under the condition of the single welding torch. This is because the two sides of the workpiece are uniformly heated and uniformly contracted at the same time by using the bilateral symmetrical double welding torch to reduce the distribution of welding deformation. The maximum displacement difference under the two conditions is 0.09 mm.

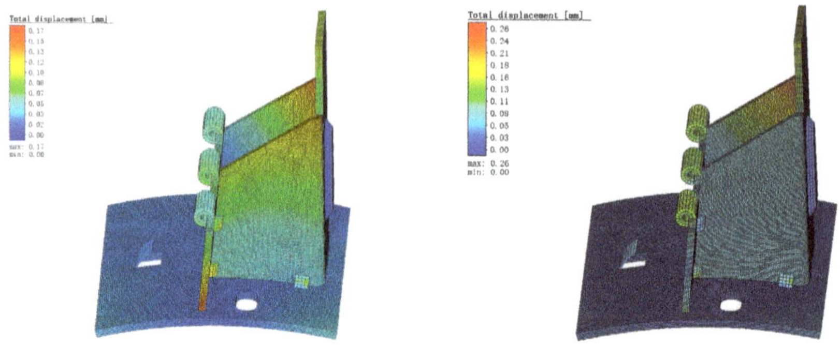

(**a**) (**b**)

Figure 5. Total displacement diagram of the single welding torch and bilateral symmetrical double welding torch. (**a**) Single welding torch; (**b**) bilateral symmetrical double welding torch.

Figure 6 shows the equivalent stress diagram under the condition of the single welding torch and the bilateral symmetrical double welding torch. It can be seen from the figure that under the two conditions, the equivalent stress decreases rapidly from the center of the weld generation area, and then tends to be gentle until it is close to zero. A large stress is generated in the weld zone, which is one of the main reasons for the deformation of the static duckbill. After welding, the weldment is cooling, and the volume shrinkage around the weld is caused by the decrease in temperature. However, the weldment is constrained to prevent its shrinkage, so large tensile stress is generated in the weld area. Under both conditions, the maximum stress difference produced by the duckbill component is 7.28 MPa, but welding a duckbill component with a single torch takes more time than with a bilateral symmetrical double torch. Therefore, this study finally chose the welding method of the bilateral symmetrical double welding torch.

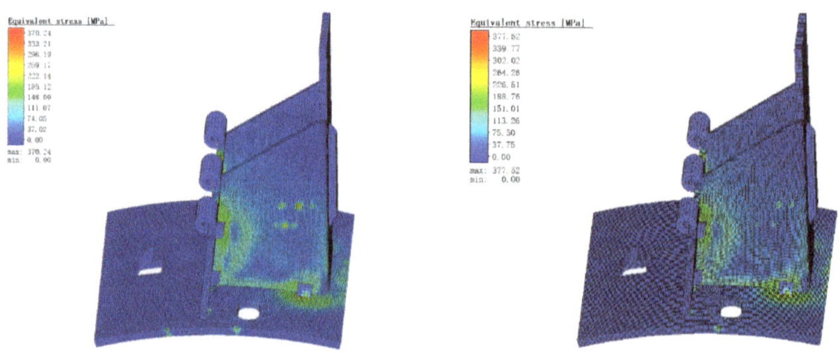

(**a**) (**b**)

Figure 6. Equivalent stress diagram under the condition of the single welding torch and bilateral symmetrical double welding torch. (**a**) Single welding torch; (**b**) bilateral symmetrical double welding torch.

3.3.2. Effect of Welding Form on Welding Deformation and Stress

Figure 7, respectively, shows the use of continuous welding and spot welding under the two forms of total displacement cloud. From Figure 7, it can be seen that the displacement areas of the two were mainly distributed at the top of the static duckbill, and the deformation of the rest was relatively small. This is because the deformation of the fixed part is smaller than that of the free part. The position and deformation of the fixed part will be greatly limited under the action of the clamping device, so the thermal deformation is reduced during the welding cycle. The maximum displacement of continuous welding is 0.98 mm, and that of spot welding is 0.26 mm. This is because in the weld, continuous welding, compared to spot welding, outputs greater thermal energy.

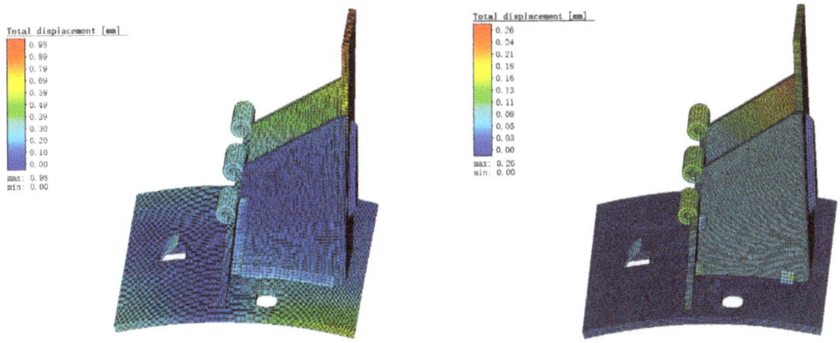

(**a**) (**b**)

Figure 7. Total displacement diagram under continuous welding and spot welding conditions. (**a**) Continuous welding; (**b**) spot welding.

Figure 8 is the equivalent stress diagram of continuous welding and spot welding. It can be seen from Figure 8 that the stress distribution of spot welding is smaller than that of continuous welding, and the difference in their maximum stress value is 121.89 MPa. Their stress distribution is similar, the stress distribution appears to diffuse from the weld to the distance and then weaken, but it is obvious that the stress distribution of continuous welding is wider and wider. This study finally chose the welding form of spot welding.

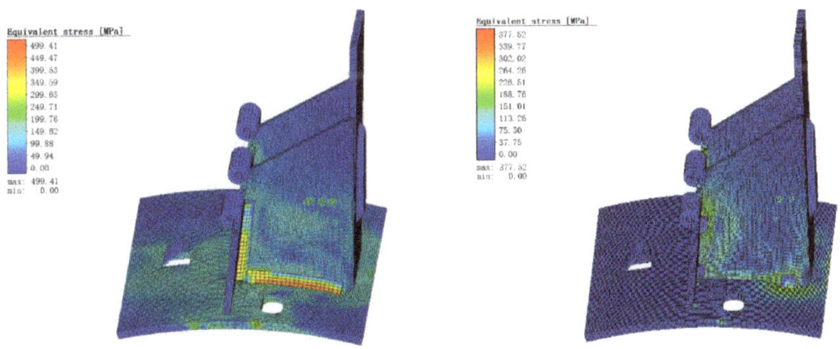

(**a**) (**b**)

Figure 8. Equivalent stress diagram under continuous welding and spot welding. (**a**) Continuous welding; (**b**) spot welding.

3.3.3. Effect of Welding Process Parameters on Welding Deformation and Stress

Figure 9 is the total displacement diagram of the duckbill welded parts when the welding speed is 4 mm/s and 10 mm/s. It can be seen from the figure that the total displacement difference between the two welding speeds is 0.13 mm, but at the welding speed of 4 mm/s, the deformation area is relatively larger. This is because the deposition amount of the wire metal on the unit-length weld is inversely proportional to the welding speed, and the melting width is inversely proportional to the square of the welding speed. Therefore, when the welding speed increases, the energy decreases, the penetration depth and width decrease, and the deformation area is relatively reduced.

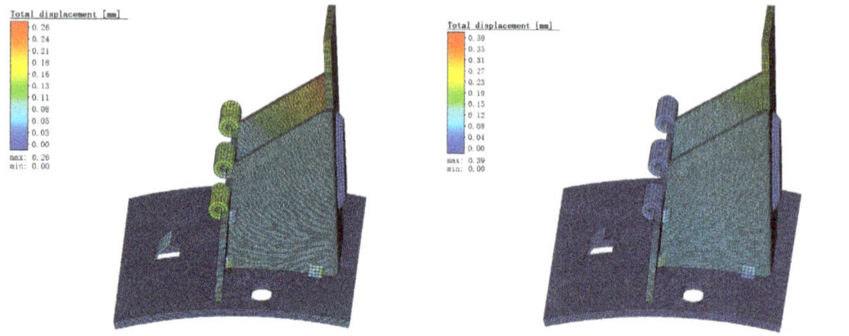

(**a**) (**b**)

Figure 9. Total displacement diagram at the welding speed of 4 mm/s and 10 mm/s; (**a**) 4 mm/s; (**b**) 10 mm/s.

Figure 10 is the equivalent stress diagram under the two welding speeds of 4mm/s and 10 mm/s. As can be seen from the figure: 4 mm/s welding speed under the maximum equivalent stress is larger and the equivalent stress of a wider range of areas. Welding speed is directly related to the size of the welding productivity, and to obtain the maximum welding speed, should be on the premise of quality assurance as far as possible, according to the specific circumstances of the appropriate adjustment of welding speed, to ensure that the weld height and width are the same. In this study, the welding speed is finally selected as 10 mm/s.

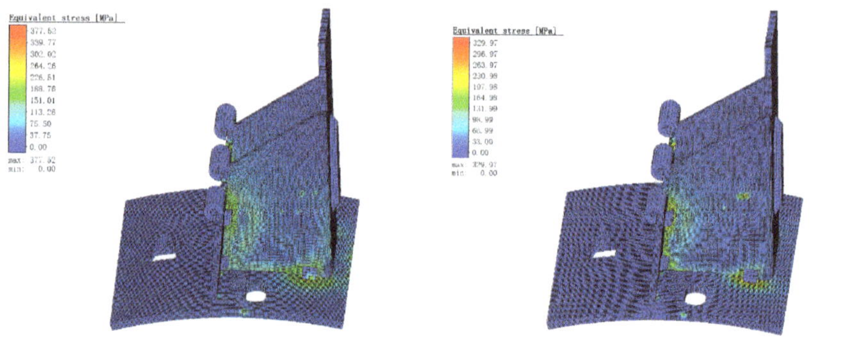

(**a**) (**b**)

Figure 10. Equivalent stress diagram at the welding speed of 4 mm/s and 10 mm/s; (**a**) 4 mm/s; (**b**) 10 mm/s.

4. Design of Duckbill Welding Robot for Cotton Seeder

4.1. Structure Composition and Working Principle

The duckbill welding robot of the cotton planter is mainly composed of a girdle feeding mechanism, static duckbill feeding mechanism, hinge feeding mechanism, support table, welding fixture, welding actuator, and control system, as shown in Figure 11.

Working process: Firstly, the girdle feeding mechanism completes the girdle feeding, and then the hinge and the static duckbill feeding structure completes the feeding work in turn. After the three welding parts of the girdle, the hinge, and the static duckbill are all loaded, the workpiece enters the position to be welded, the clamping cylinder works to clamp the workpiece, and the welding actuator moves and performs welding. After the welding is completed, the welding platform is opened, and the weldment falls to the ground.

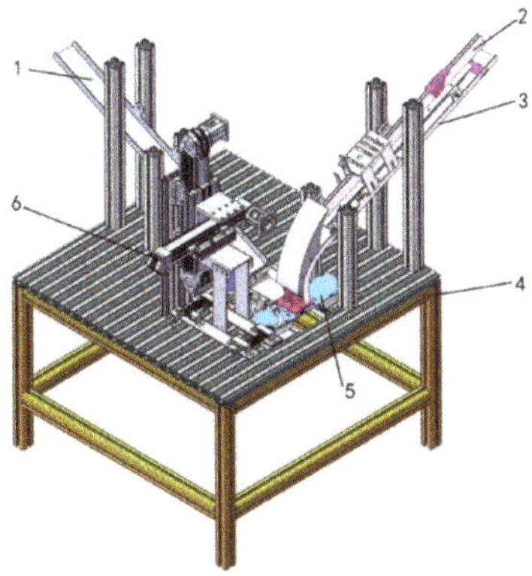

Figure 11. Structure diagram of duckbill welding robot for cotton seeder. 1. Girdle feeding mechanism; 2. static duckbill feeding mechanism; 3. hinge feeding mechanism; 4. support platform; 5. welding fixture; 6. welding actuator.

4.2. Design of Girdle Feeding Mechanism

According to the analysis of the assembly requirements of the duckbill parts, the feeding mechanism needs to meet the following requirements: (1) the hinge and the girdle should be vertical; (2) the static duckbill and the hinge are symmetrically distributed in the transverse center when they are matched with the girdle; (3) the static duckbill should avoid shielding girdle under the mouth. According to the above assembly requirements and the structural parameters of duckbill welding parts, the feeding structure is designed. The feeding mechanism realizes the sequential feeding action of welded parts through the cooperation of an inductive proximity switch, electromagnet, and cylinder.

The structure size of the girdle feeding mechanism is 800 mm × 68 mm × 22 mm. It adopts a modular design and is installed on the support platform through the aluminum profile pillar. The working process is as follows: When the inductive proximity switch detects that there is a girdle in the storage chute, the electromagnet is energized and absorbs the second girdle, and the cylinder shrinks. The first girdle falls freely to the girdle waiting area due to gravity, and finally, the mini cylinder pushes the girdle into the welding area. The girdle feeding mechanism is shown in Figure 12.

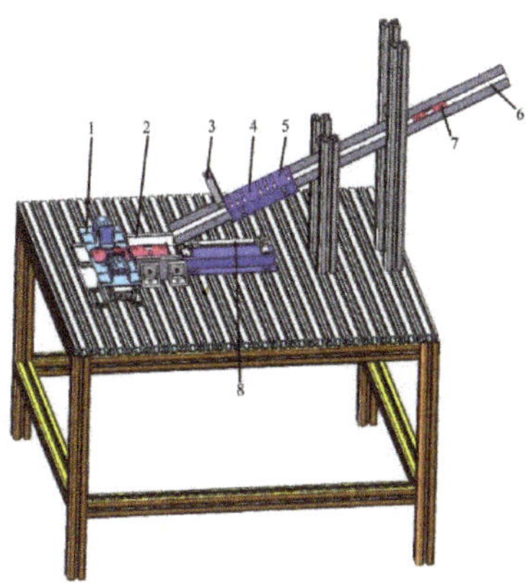

Figure 12. Structure diagram of the girdle feeding mechanism. 1. Welding area; 2. girdle blanking waiting for area; 3. cylinder; 4. inductive proximity switch; 5. electromagnet; 6. storage chute; 7. girdle; 8. mini cylinder.

4.3. Design of Static Duckbill and Hinge Feeding Mechanism

To save space, the static duckbill feeding mechanism and the hinge feeding mechanism adopt an integrated design, and the assembly relationship of the parts is shown in Figure 13. The static duckbill and the hinge feeding mechanism are equipped with fixed plates to fix inductive proximity switches, electromagnets, and cylinders.

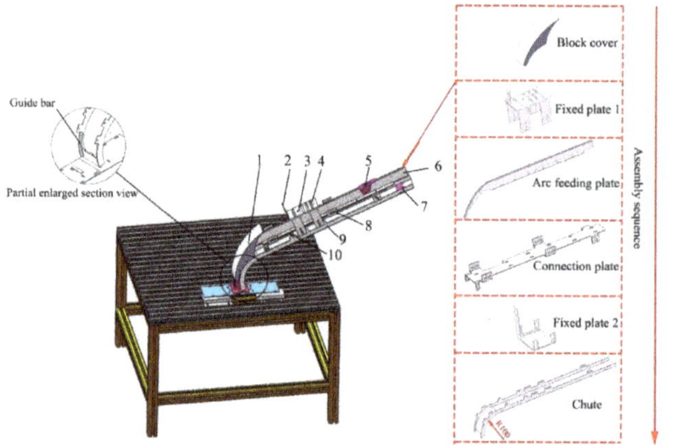

Figure 13. Structure diagram of static duckbill and hinge feeding mechanism. 1. Block cover; 2. cylinder; 3. inductive proximity switch; 4. electromagnet; 5. static duckbill; 6. arc feeding plate; 7. girdle; 8. connecting plate; 9. storage chute; 10. fixed plate 2.

The width of the storage chute of the hinge feeding mechanism is bent according to the dimensions of the hinge, and the bending angle is 90°. To ensure that the hinge is perpendicular to the girdle during blanking, the lower end of the storage chute adopts a circular arc design, and its arc inner diameter is 100 mm. To prevent the hinge from sliding out of the arc guide rail when feeding, the guide bars are symmetrically distributed on both sides to guide and limit displacement. The verticality of the hinge is ensured by limiting the outer side of the guide bar and the arc guide rail. The guide bar is shown in the partially enlarged section view in Figure 13.

The main component of the static duckbill feeding mechanism is an arc feeding plate, and the arc feeding plate is connected with the hinge storage chute through a connecting plate. When the static duckbill is feeding, the contact with the feeding plate is strip contact, and the contact area is small, which greatly reduces the friction when sliding. When sliding, the static duckbill slides along the outer edge of the arc feeding plate. To prevent it from sliding out directly at the outer arc position, a block cover is placed at the lug of the hinge storage chute. The feeding accuracy of the static duckbill will directly affect the welding quality. Therefore, there are multiple through holes on the arc feeding plate and the connection plate, respectively, and the porous coordination ensures structural stability. The static duckbill and hinge feeding mechanism structure diagram is shown in Figure 13.

4.4. Design of Welding Fixture

As shown in Figure 14, the welding fixture is mainly composed of three parts: girdle clamping mechanism, hinge clamping mechanism, and static duckbill clamping mechanism. The girdle clamping device is positioned by a limit block and clamped by a girdle pusher. The girdle first slides down from the girdle storage chute to the girdle waiting area, and the girdle push plate sticks out. According to the four-point positioning principle, the transverse and longitudinal positioning and clamping of the girdle are completed.

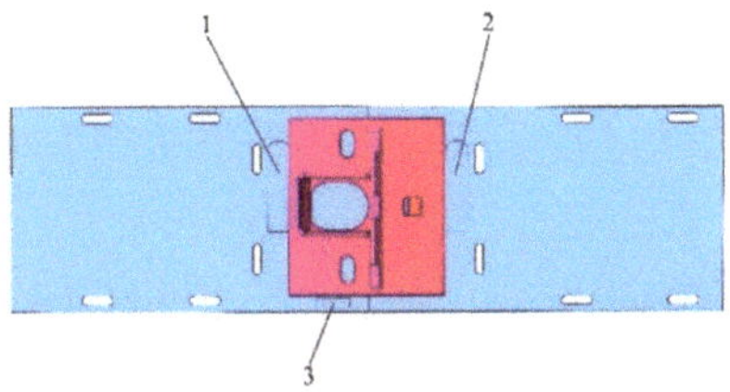

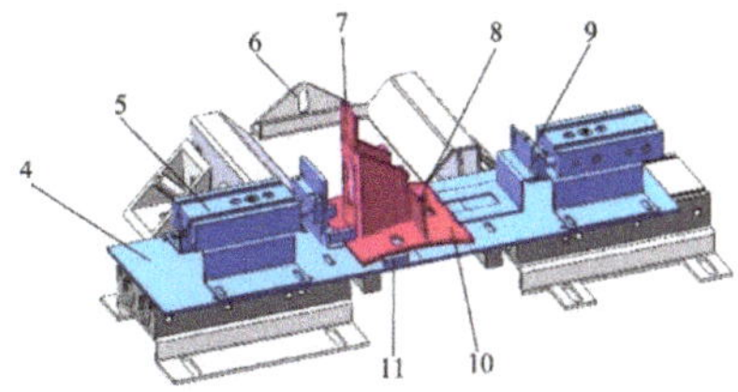

Figure 14. Structure diagram of welding fixture. 1. Limit block; 2. limit block; 3. limit block; 4. welding workbench; 5. static duckbill clamping device; 6. girdle push plate; 7. static duckbill; 8. hinge; 9. hinge clamping device; 10. girdle; 11. limit block.

The hinge clamping mechanism is composed of a guide bar and a hinge push plate. The guide bar is close to the side wall of the storage chute, symmetrically distributed on both sides, and plays a guiding and limiting role to the hinge. The guide bar is shown in the partially enlarged section of Figure 13. After the hinge is loaded onto the welding platform, the hinge is pushed out to complete the positioning of the hinge.

The static duckbill clamping mechanism is mainly composed of a cylinder and clamp push plate. The arc feeding plate supports and guides the static duckbill. After the static duckbill slides down to the welding workbench, the clamp push plate is pushed out to complete the horizontal and vertical positioning of the static duckbill.

The bottom of the welding workbench is composed of two welding bottom plates and two cylinders. The welding workbench can open and close under the action of the cylinder.

4.5. Welding Actuator

The schematic diagram of the welding actuator is shown in Figure 15. The welding actuator can move back and forth in a straight line along the X axis and Z axis. The stroke in the X axis direction is 100~150 mm, and the stroke in the Z axis direction is 150~200 mm. The double welding torch is symmetrically distributed on the welding torch bracket of the X axis linear slider. The movement of the X axis and Z axis is completed by the stepper motor electric drive synchronous belt module, and the movement speed is controlled by Siemens S7-1200PLC and the stepper motor driver. The Z axis selection has a brake stepper motor, which is locked when power fails, to prevent sliding.

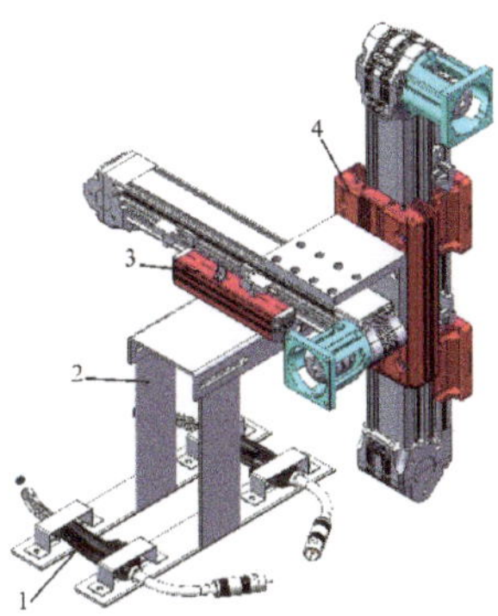

Figure 15. Welding actuator structure diagram. 1. Welding torch; 2. welding torch bracket; 3. X axis linear slider; 4. Z axis linear slider.

4.6. Control System Design

The cotton seeder duckbill welding robot controller is the Siemens S7-1200PLC. The communication between PLC and human–machine interaction (HMI) is Ethernet. PLC realizes manual and automatic control of the girdle, static duckbill, and hinge feeding operation. It also controls welding parameters, fixtures, welding actuators, and welding platforms. HMI enables manual and automatic program switching of welding robots and monitors the working conditions of welding robots to ensure the safe and smooth operation of welding operations. The control system flow chart is shown in Figure 16.

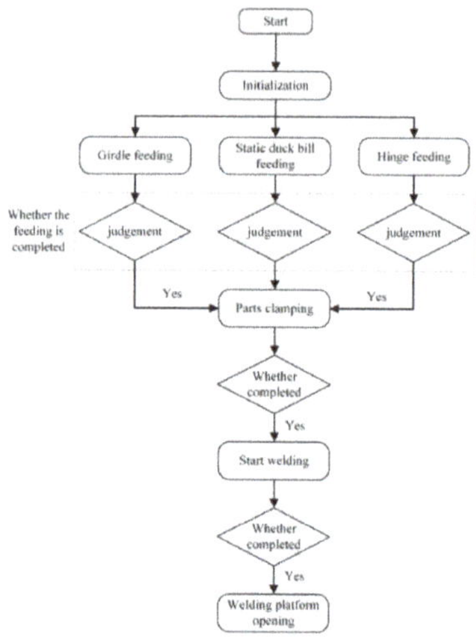

Figure 16. Control system flow chart.

5. Results and Discussion

5.1. Cotton Seeder Duckbill Welding Robot Test Results and Analysis

The welding wire used in the welding test is a 1.2 mm diameter solid wire (JQ·MG50-6; Tianjin Golden Bridge Welding Materials Group Co., Ltd., Tianjin, China), the protective gas is a mixture of CO_2 and argon gas, and the cotton planter duckbill welding robot was tested. The welding process parameters used in the test are shown in Table 3. The cotton seeder duckbill welding robot is shown in Figure 17.

Table 3. Welding process parameters.

Welding Current (A)	Welding Voltage (V)	Welding Speed (mm s)
38	26	10

Figure 17. Cotton seeder duckbill welding robot.

The cotton seeder duckbill welding robot factory test photo is shown in Figure 18. The Human Machine Interface (HMI) of the cotton seeder duckbill welding robot is shown in Figure 19. According to the national standard DL/T 868-2004 welding procedure qualification procedure [32], the appearance of the weld after duck beak welding is analyzed. It can be seen from Figure 20 that there are no defects such as unmelted, porosity, and undercutting on the weld surface, and the welding quality is good. After testing, the welding efficiency of the cotton seeder duckbill welding robot is 6–7 times faster than that of the manual, and 600–800 duckbills can be welded per hour. The weld is well-formed. The welding pass rate is 85%, which can meet the needs of practical engineering. The development of the cotton seeder duckbill welding robot will greatly improve the welding efficiency of the duckbill parts and promote the large-scale and standardized production of the duckbill of the cotton seeder. The forming of welding parts is shown in Figure 20. The cotton seeder duckbill welding robot performance comparison is shown in Table 4.

(**a**) (**b**)

Figure 18. Factory test of cotton seeder duckbill welding robot. (**a**) Welding test site; (**b**) welding test in progress.

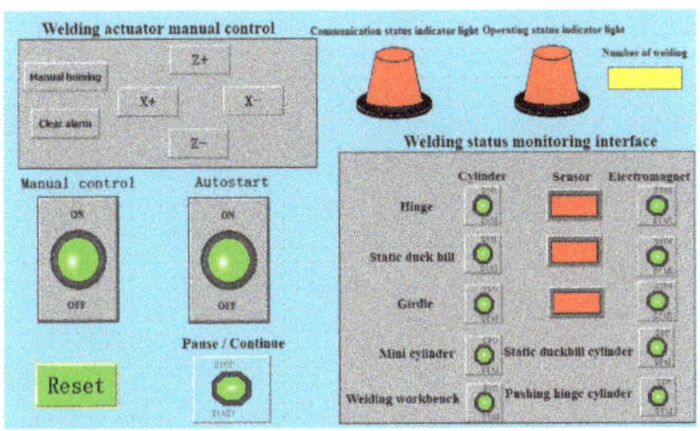

Figure 19. HMI.

Figure 20. Welding forming of duckbill parts.

Table 4. The cotton seeder duckbill welding robot performance comparison.

Welding Method / Assessment Indicators	Welding Duckbill Efficiency (Piece/h)	Welding Qualification Rate
Manual welding	100–130	99%
Semi-automatic duckbill welding equipment	200–300	100%
Cotton seeder duckbill welding robot	600–800	85%

5.2. Discussion

In this paper, a duckbill welding robot for cotton seeder is designed, including the mechanical structure and control system of the welding robot. The efficiency of a cotton seeder duckbill welding robot was greatly improved compared with manual work and semi-automatic welding robots, but there is still unqualified welding in the duckbill welding test. The main reason for this phenomenon is that there are some errors in the manufacturing and assembly of the parts of the duckbill welding robot for the cotton seeder. Mechanical vibration will occur during the operation, which will affect the accuracy of welding parts and the accuracy of welding gun welding. In the follow-up study, improving the welding robot parts manufacturing and assembly accuracy, and further optimizing the structure, will improve the welding robot welding qualification rate.

6. Conclusions

In this study, the characteristics of the duckbill parts were analyzed first, and then the welding process of the duckbill parts was simulated by Simufact Welding software. The whole process of welding was observed intuitively. At the same time, the deformation and stress changes of the weldment were compared and analyzed when the unilateral single welding torch and the bilateral symmetrical double welding torch, two welding forms, and two welding process parameters, were used. On this basis, a kind of cotton seeder duckbill welding robot was designed, and the welding test was carried out. The results show that the cotton seeder duckbill welding robot has high welding efficiency and good forming quality of welded parts. The design of the cotton seeder duckbill welding robot greatly improves the welding efficiency of the duckbill, which helps to solve the problems of low welding efficiency and unstable welding quality in manual welding and semi-automatic welding robots, and provides a strong guarantee for large-scale and standardized welding production of the duckbill.

Author Contributions: Writing—original draft preparation, Y.R.; writing—review and editing, W.G.; resources, X.W.; data curation, C.H.; visualization, L.W.; supervision, X.H. and J.X.; project administration, W.G. All authors have read and agreed to the published version of the manuscript.

Funding: This work was supported by the Bingtuan Science and Technology Program, grant number 2020CB034, and the Innovation research team project of Tarim University, grant number TDZKCX202103.

Institutional Review Board Statement: Not applicable.

Informed Consent Statement: Not applicable.

Data Availability Statement: The data presented in this study are available on request from the corresponding author.

Acknowledgments: Thanks for being grateful for the support of the Bingtuan Science and Technology Program (Grant No. 2020CB034), the Innovation research team project of Tarim University (Grant TDZKCX202103). The authors are grateful to anonymous reviewers for their comments.

Conflicts of Interest: The authors declare no conflict of interest.

References

1. Xie, Z.K.; Wang, Y.J.; Li, F.M. Effect of plastic mulching on soil water use and spring wheat yield in arid region of northwest China. *Agric. Water Manag.* **2005**, *75*, 71–83. [CrossRef]
2. Han, J.; Jia, Z.K.; Han, Q.F.; Zhang, J. Application of Mulching Materials of Rainfall Harvesting System for Improving Soil Water and Corn Growth in Northwest of China. *J. Integr. Agric.* **2013**, *12*, 1712–1721. [CrossRef]
3. Zhang, X.J.; Zhang, H.T.; Shi, Z.L.; Jin, W.; Cheng, Y.; Yu, Y.L. Design and experiments of seed pickup status monitoring system for cotton precision dibblers. *Trans. Chin. Soc. Agric. Eng.* **2022**, *38*, 9–19. [CrossRef]
4. Wang, Y.; Wei, M.; Dong, W.; Li, W.; He, J.; Han, C.; Jiang, Z. Design and Parameter Optimization of a Soil Mulching Device for an Ultra-Wide Film Seeder Based on the Discrete Element Method. *Processes* **2022**, *10*, 2115. [CrossRef]
5. Xiang, Y.; Kang, J.M.; Zhang, C.Y.; Peng, Q.J.; Zhang, N.N.; Wang, X.Y. Analysis and Optimization Test of the Peanut Seeding Process with an Air-Suction Roller Dibbler. *Agriculture* **2022**, *12*, 1942. [CrossRef]
6. Hrcek, S.; Brumercik, F.; Smetanka, L.; Lukac, M.; Patin, B.; Glowacz, A. Global Sensitivity Analysis of Chosen Harmonic Drive Parameters Affecting Its Lost Motion. *Materials* **2021**, *14*, 5057. [CrossRef]
7. Majchrák, M.; Kohár, R.; Hrček, S.; Brumerčík, F. The Comparison of the Amount of Backlash of a Harmonic Gear System. *Teh. Vjesn.* **2021**, *28*, 771–778. [CrossRef]
8. Rheaume, F.-E.; Champliaud, H.; Liu, Z. Understanding and modelling the torsional stiffness of harmonic drives through finite-element method. *Proc. Inst. Mech. Eng. Part C J. Mech. Eng. Sci.* **2009**, *223*, 515–524. [CrossRef]
9. Subham, A.; Santonab, C.; Kanika, P.; Shankar, C. A Rough Multi-Attributive Border Approximation Area Comparison Approach for Arc Welding Robot Selection. *Jordan J. Mech. Ind. Eng.* **2021**, *15*, 169–180.
10. Stephen, C.; Joseph, O.; Stephen, Z. Zero Moment Control for Lead-Through Teach Programming and Process Monitoring of a Collaborative Welding Robot. *J. Mech. Robot.* **2021**, *13*, 031016. [CrossRef]
11. Shah, H.; Sulaiman, M.; Shukor, A.; Jamaluddin, M. A Review Paper on Vision Based Identification, Detection and Tracking of Weld Seams Path in Welding Robot Environment. *Mod. Appl. Sci.* **2016**, *10*, 83. [CrossRef]
12. Wang, X.W.; Xue, L.K.; Yan, Y.X.; Gu, X.S. Welding Robot Collision-Free Path Optimization. *Appl. Sci.* **2017**, *7*, 89. [CrossRef]
13. Yin, T.; Wang, J.P.; Zhao, H.; Zhou, L.; Xue, Z.H.; Wang, H.H. Research on Filling Strategy of Pipeline Multi-Layer Welding for Compound Narrow Gap Groove. *Materials* **2022**, *15*, 5967. [CrossRef] [PubMed]
14. ERSÖZ, S.; TÜRKER, A.; AKTEPE, A.; ATABAŞ, i.; KOKOÇ, M. Design of a Tracking Welding Robot Automation System for Manufacturing of Steam and Heating Boilers. *Acad. Platf. J. Eng. Sci.* **2018**, *6*, 19–24. [CrossRef]
15. Ku, N.; Cha, H.; Lee, K.; Kim, J.; Kim, T.; Ha, S.; Lee, D. Development of a mobile welding robot for double-hull structures in shipbuilding. *J. Mar. Sci. Technol.* **2010**, *15*, 374–385. [CrossRef]
16. Mulligan, S.; Melton, J.; Lylynoja, A.; Herman, K. Autonomous welding of large steel fabrications. *Ind. Robot Int. J.* **2005**, *32*, 346–349. [CrossRef]
17. Jiang, Y.; Han, Q.Q.; Dai, Z.E.; Zhou, C.; Yu, J.F.; Hua, C.J. Structural design and kinematic analysis of a welding robot for liquefied natural gas membrane tank automatic welding. *Int. J. Adv. Manuf. Technol.* **2022**, *122*, 461–474. [CrossRef]
18. Yue, J.F.; Dong, X.T.; Guo, R.; Liu, W.J.; Li, L.Y. Numerical simulation of equivalent heat source temperature field of asymmetrical fillet root welds. *Int. J. Heat Mass Transf.* **2019**, *130*, 42–49. [CrossRef]
19. Zhao, H.Q.; Zhang, Q.Y.; Niu, Y.X.; Du, S.Z.; Lu, J.H.; Zhang, H.F.; Wang, J.C. Influence of triangle reinforcement plate stiffeners on welding distortion mitigation of fillet welded structure for lightweight fabrication. *Ocean Eng.* **2020**, *213*, 107650. [CrossRef]
20. Maarten, R.; Richard, P.; Henk, S.; Johan, M. A combined experimental and numerical examination of welding residual stresses. *J. Mater. Process. Technol.* **2017**, *261*, 98–106. [CrossRef]
21. Ahn, J.; He, E.; Chen, L.; Pirling, T.; Dear, J.P.; Davies, C.M. Determination of residual stresses in fibre laser welded AA2024-T3 T-joints by numerical simulation and neutron diffraction. *Mater. Sci. Eng. A* **2018**, *712*, 685–703. [CrossRef]
22. Li, Z.; Peng, W.X.; Chen, Y.Z.; Liu, W.H.; Zhang, H.Q. Simulation and experimental analysis of Al/Ti plate magnetic pulse welding based on multi-seams coil. *J. Manuf. Process.* **2022**, *83*, 290–299. [CrossRef]
23. Lu, B.; Ni, X.D.; Li, S.F.; Li, K.Z.; Qi, Q.Z. Simulation and Experimental Study of a Split High-Speed Precision Seeding System. *Agriculture* **2022**, *12*, 1037. [CrossRef]
24. Bai, S.H.; Yuan, Y.W.; Niu, K.; Shi, Z.L.; Zhou, L.M.; Zhao, B.; Wei, L.G.; Liu, L.J.; Zheng, Y.K.; An, S.; et al. Design and Experiment of a Sowing Quality Monitoring System of Cotton Precision Hill-Drop Planters. *Agriculture* **2022**, *12*, 1117. [CrossRef]
25. Mehran, J.; Antti, A.; Joseph, A.; Timo, B. Numerical and experimental investigations on the welding residual stresses and distortions of the short fillet welds in high strength steel plates. *Eng. Struct.* **2022**, *260*, 114269. [CrossRef]
26. Go, B.S.; Oh, K.H.; Kwon, S.; Bang, H.S. Reduction Characteristics of Welding Deformation According to Cooling Distance in Heat Sink Welding. *Int. J. Precis. Eng. Manuf.* **2022**, *23*, 1229–1236. [CrossRef]
27. Ikushima, K.; Maeda, S.; Shibahara, M. Ultra Large-Scale Nonlinear FE Analysis of Welding Mechanics. *Int. Conf. Comput. Exp. Eng. Sci.* **2019**, *22*, 175. [CrossRef]
28. Farias, R.M.; Teixeira, P.R.F.; Vilarinho, L.O. Variable profile heat source models for numerical simulations of arc welding processes. *Int. J. Therm. Sci.* **2022**, *179*, 107593. [CrossRef]
29. Farias, R.M.; Teixeira, P.R.F.; Vilarinho, L.O. An efficient computational approach for heat source optimization in numerical simulations of arc welding processes. *J. Constr. Steel Res.* **2021**, *176*, 106382. [CrossRef]

30. Goldak, J.; Chakravarti, A.; Bibby, M. A new finite element model for welding heat sources. *Metall. Mater. Trans. B* **1984**, *15*, 299–305. [CrossRef]
31. Yan, X.H.; Chen, G.Q.; Xu, H.L.; Yan, Z.Y.; Li, P. Welding Quality Control of Control Rod of Lawn Mower Cutter Plate Based on Simufact. *J. Netshape Form. Eng.* **2022**, *14*, 46–50.
32. *DL/T 868-2004*; The Code of Welding Procedure Qualification [S]. Power Standards: China, 2004.

agriculture

MDPI

Article

Wavelet Scattering Convolution Network-Based Detection Algorithm on Nondestructive Microcrack Electrical Signals of Eggs

Chenbo Shi [1], Yanhong Cheng [1], Chun Zhang [1], Jin Yuan [2], Yuxin Wang [1], Xin Jiang [1] and Changsheng Zhu [1,*]

[1] College of Intelligent Equipment, Shandong University of Science and Technology, Tai'an 271019, China; shichenbo@gmail.com (C.S.); 202183230018@sdust.edu.cn (Y.C.)
[2] College of Mechanical and Electronic Engineering, Shandong Agricultural University, Tai'an 271018, China; jyuan@sdau.edu.cn
* Correspondence: cs.zhu@sdust.edu.cn

Abstract: The detection of poultry egg microcracks based on electrical characteristic models is a new and effective method. However, due to the disorder, mutation, nonlinear, time discontinuity, and other factors of the current data, detection algorithms such as support-vector machines (SVM) and random forest (RF) under traditional statistical characteristics cannot identify subtle defects. The detection system voltage is set to 1500 V in the existing method, and higher voltages may cause damage to the hatched eggs; therefore, how to reduce the voltage is also a focus of research. In this paper, to address the problem of the low signal-to-noise ratio of microcracks in current signals, a wavelet scattering transform capable of extracting translation-invariant and small deformation-stable features is proposed to extract multi-scale high-frequency feature vectors. In view of the time series and low feature scale of current signals, various convolutional networks, such as a one-dimensional convolutional neural network (1DCNN), long short-term memory (LSTM), bi-directional long short-term memory (Bi-LSTM), and gated recurrent unit (GRU) are adopted. The detection algorithm of the wavelet scattering convolutional network is implemented for electrical sensing signals. The experimental results show that compared with previous works, the accuracy, precision, recall, F1-score, and Matthews correlation coefficient of the proposed wavelet scattering convolutional network on microcrack datasets smaller than 3 μm at a voltage of 1000 V are 99.4393%, 99.2523%, 99.6226%, 99.4357%, and 98.8819%, respectively, with an average increase of 2.0561%. In addition, the promotability and validity of the proposed detection algorithm were verified on a class-imbalanced dataset and a duck egg dataset. Based on the good results of the above experiments, further experiments were conducted with different voltages. The new feature extraction and detection method reduces the sensing voltage from 1500 V to 500 V, which allows for achieving higher detection accuracy with a lower signal-to-noise ratio, significantly reducing the risk of high voltage damage to hatching eggs and meeting the requirements for crack detection.

Keywords: nondestructive detection; poultry eggs; wavelet scattering convolutional network; microcurrent signal analysis; egg's electrical characteristic model

Citation: Shi, C.; Cheng, Y.; Zhang, C.; Yuan, J.; Wang, Y.; Jiang, X.; Zhu, C. Wavelet Scattering Convolution Network-Based Detection Algorithm on Nondestructive Microcrack Electrical Signals of Eggs. *Agriculture* **2023**, *13*, 730. https://doi.org/10.3390/agriculture13030730

Academic Editor: Quan-Sheng Chen

Received: 28 February 2023
Revised: 13 March 2023
Accepted: 20 March 2023
Published: 22 March 2023

1. Introduction

Eggs are one of the best sources of nutrition for human beings and are known as the best protein food. To ensure the quality of eggs, cracked eggs produced in packaging, processing, transportation, and other links should be eliminated. Cracks in eggs will not only lead to a reduction of freshness but also breed bacteria and do harm to health [1,2]. The quality and safety of eggs is a matter of concern to the public, so the detection of egg cracks is a necessary operation before the eggs enter the market for consumption. Manual detection is high-cost, low-efficiency, and depends on detection experience. After a long period of operation, the probability of missing cracks and false detection increases due to visual fatigue. This can also cause some damage to the operators' vision [3]. With the continuous development of industrialization and informatization, people are trying to find an automatic detection method to reduce the labor intensity of this task while simultaneously improving efficiency and accuracy.

Numerous techniques for egg crack detection have been put forth by researchers and academics, of which acoustic [4–7] and visual [8–11] methods are the most well-known and established. Crack detection based on acoustic characteristics is realized by collecting knock response signals with sound sensors after mechanical excitation on several parts of the egg surface, analyzing and processing the signal differences between the intact area and the cracked area of the egg surface, and finally extracting characteristic parameters. However, its accuracy easily suffers from interference by eggshell thickness, external noise, and the knock location. The detection of cracks in poultry eggs based on visual characteristics involves taking pictures of the surface of poultry eggs with an industrial camera and transmitting the images to supporting software. By analyzing the images, the software can identify the characteristics of cracked and intact eggs and then construct a model for discrimination. Nevertheless, many simple factors must be considered in practice, such as the camera's resolution, the recognition algorithm, and the shape of the egg crack. Additionally, the online detection speed and accuracy for microcracks cannot meet today's requirements. Hence, significant limitations remain on industrial sites.

The detection method based on high-voltage discharge has a good ability to detect leakage. Hu [12] proposed a "conductivity method" for ampoule bottle detection, applying a 3000 V∼5000 V direct current(DC) on the tested drug bottle and judging the leakage of the test sample by setting the threshold value of the measured current. Yan et al. [13] used 20 kHz of high voltage to carry out their experiment and finally verified the feasibility of high-voltage discharge for the detection of small leaks of large infusion plastic bottles. The micro-current high-voltage discharge method is similar to the traditional high-voltage discharge method. It requires that the package itself is not conductive and the contents are conductive, creating a difference between the electrical parameters of a defect-free package and a defective package. The micro-current high-voltage discharge method can also be used to detect tiny cracks or damages to sealed packaging in the food industry. With the further deepening of the basic theoretical research on the dielectric properties of eggs, egg crack detection based on electrical properties will become our research direction.

Recent research shows that detecting cracked eggs using the response model of egg electrical characteristics can effectively compensate for the shortcomings of the above methods [14]. Figure 1 illustrates how the electrical characteristic model can be used to detect egg cracks. The capacitors include the two electrodes and the egg in the center, and the eggshell is the insulator. Due to its low resistance value, the egg liquid is approximately the conductor. The current value detected is either a capacitance jump, a breakdown, or a combination of both. When the egg is intact, the whole system is in equilibrium, and the current value will remain stable . When the egg has a crack, the electrical characteristic model will change when the electrode passes through the crack, resulting in a transient current. Therefore, if the voltage of the two electrodes is kept constant and only the breakdown of the air medium is allowed, it is possible to identify cracked eggs based on the change in the current signal. Traditional machine learning methods have been used to classify the current signals, and the accuracy of detecting eggs with microcracks less than 3 μm compared to intact eggs was found to be 99%. Relevant experiments also prove that the proposed method has a certain universality and versatility. However, it has three shortcomings. In terms of the experimental device, the electrode position is fixed, which can easily cause the egg size to not fit with the upper electrode, thus causing errors in detection. In signal analysis, the extracted features are inadequate to allow capturing the details and changes of cracked eggs in the current signal, so further study of the feature extraction method of the current signal is necessary. A further area of research is how to reduce the 1500 V voltage used in the experiments because it is too high and might harm the unhatched eggs.

Therefore, this research aimed to propose an accurate and effective nondestructive method based on the electrical characteristics of eggs to detect cracks in poultry eggs in the industrial field on the basis of existing experiments [14]. The main objectives of the study are as follows:

- To modify the experimental device so that the electrode fits the egg to avoid missed or wrong detection due to data acquisition;
- To adjust the analysis algorithm, to analyze the characteristics of the microcurrent signal, and to design a feature extraction and classification algorithm that is effective for the current signal;
- To investigate how to reduce the voltage while maintaining detection accuracy and reducing the damage caused by high voltage to unhatched eggs.

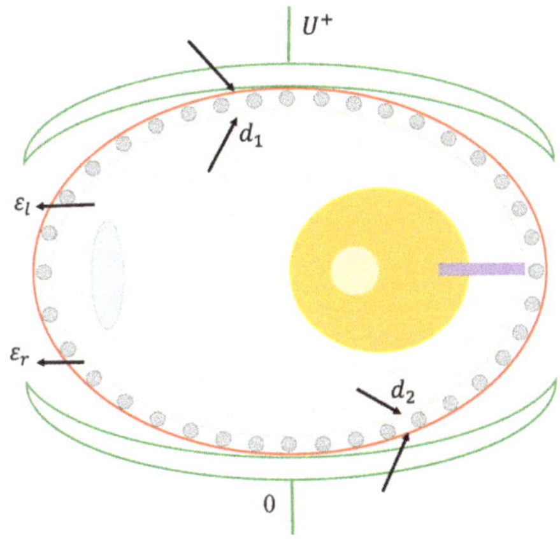

Figure 1. Schematic diagram of capacitor system composed of electrode and egg body in poultry egg crack detection method based on egg electrical characteristic model [14]. d_1 and d_2 represent the thickness of the upper and lower layers of the eggshell; ε_l represents the dielectric constant of egg liquid; and ε_r represents the relative dielectric constant, which represents the dielectric or polarization physical parameter of the dielectric material.

Employing wavelet scattering transform, Jinghui Li et al. [15] stratified and dispersed the information of a heart sound signal to each layer so that the information would not be lost and maintained the stability of signal features, which effectively expressed the corresponding characteristic information of the signal. Then they obtained the characteristic matrix of the signal , which was input into a double support-vector machine for classification, yielding excellent results. Ahmad A. Al-Taee et al. [16] used deep wavelet scattering transform as a feature extraction method for electromyographic signals and compared it with other wavelet methods, significantly improving the performance of electromyographic pattern recognition. Liu Hui et al. [17] adopted a wavelet scattering network to extract the features of the fault zero sequence current signal. Academics have found that wavelet scattering transform can extract both simple and complex characteristics from low-frequency and high-frequency signals, respectively, and thus can reduce intra-class differences to the greatest extent while preserving inter-class differentiation. The fault feature vector is input into bi-directional LSTM, which has the advantages of high identification accuracy and short training time compared with the fault identification methods in other literature. In addition, wavelet scattering convolutional neural networks have been widely used in sound, image , and time series fields due to their superior performance and perfect mathematical properties, and some achievements have been made [18–21]. To our knowledge, no researchers have applied appropriate research methods and ideas to detect microcurrent signals in eggs.

In recent years, deep learning has been rapidly developed. Because of its advantages of automatic feature extraction, ability to process unstructured data, and high accuracy, it has been widely used and studied in industrial defect detection. Yu et al. [22] combined convolutional neural networks(CNN), migration learning, and decision-level images to propose a vision-based automatic recognition method for the identification of the surface condition of concrete structures and designed an improved Dempster–Shafer (D-S) algorithm, which greatly improved the recognition accuracy of concrete structures. Yu et al. [23] used the improved bird swarm algorithm to optimize a 2DCNN, which performed better than other machine learning models in the evaluation of the torsional capacity of reinforced concrete (RC) beams. Deep learning has also shown strong performance in the processing of one-dimensional(1D) data. Hu et al. [24] designed a fault diagnosis method based on a one-dimensional convolutional neural network (1DCNN) and L2-support-vector machine(L2-SVM) for unbalanced data, which, compared with other intelligent methods, significantly improved the recognition accuracy and diagnostic performance of the model compared with other intelligent methods in processing unbalanced data. Le et al. [25] developed a 1DCNN for automated fault diagnosis, which can autonomously learn damage-sensitive features without pre-processing and can accurately diagnose potential faults that damage the smart chain. Relevant research [26–28] also shows that deep learning exhibits good performance in processing time series classifications.

Therefore, this study focuses on applying wavelet scattering on poultry eggs based on real-time electrical characteristics. The main innovations and contributions of this research can be summarized as follows:

- In order to solve the problem of missing detection caused by the lack of fit between the egg and the electrode due to the fixed position of the upper electrode, we improved the micro-crack detection system based on discharge analysis . We added a size identification device so that the upper electrode position can be automatically moved and fitted according to the size of the egg.
- Given the problem that the extracted features are inadequate to capture the details and changes in a cracked egg in the current signal, we applied the wavelet scattering transform to obtain the features of the micro-current signal. Using deep learning methods, we classified the features extracted and attained satisfactory results with local translation invariance and elastic deformation stability.
- The proposed and existing methods' effects were compared and analyzed. The class-imbalanced dataset and duck egg dataset were used to verify the versatility and effectiveness of the proposed method. Finally, the experiments with different voltage effects were conducted to verify that the proposed method can obtain higher accuracy on data with a low signal-to-noise ratio while reducing the sensing voltage and further reducing the risk of damage to unhatched eggs caused by high voltages.

The remaining chapters are organized as follows. Section 2 introduces the deployment of experimental equipment and the method of microcurrent signal analysis. In Section 3, we compare and analyze the experimental results to verify the effectiveness of the proposed method. Finally, the whole paper is summarized and the future research on egg crack detection is prospected in Section 4.

2. Materials and Methods

2.1. Egg Electrical Characteristic Crack Detection System

In this study, a crack detection system for poultry eggs based on real-time electrical characteristics was developed. As shown in Figure 2a, the system is mainly composed of four parts: a data acquisition platform, a high-voltage power supply, a controller, and a computer processing system. The data acquisition platform includes a visual detection part and a discharge detection part. The visual detection part is used to detect the size of eggs and control the automatic electrode adjustment mechanism. The discharge detection part comprises a rotating mechanism, a discharge electrode, an automatic electrode regulating mechanism, etc. The rotating mechanism ensures that eggs of different sizes can rotate flexibly and uniformly. The automatic electrode regulating mechanism adjusts up and down according to the data from the visual detection part to provide stable and reliable surface contact, as shown in Figure 2b. Through preliminary experiments and debugging, conductive silica gel was used as the electrode material for crack detection under a voltage of 1000 V, average humidity of 40% , and an average temperature of 18.5 °C. In this experiment, the current value is strictly controlled, and the automatic protection mechanism is triggered when the current exceeds 1 mA so that no damage will be caused to the eggs. In this study, the microcurrent signal obtained from an egg under the electrical characteristic crack detection system was used to evaluate and detect egg cracks. The detection and collection frequency of the detection system was set at 100 Hz, the collection time was 4.5 s, and the data length was 450 points. The data of egg rotation were collected 2.5 times in a single time, and one datum was collected for each egg.

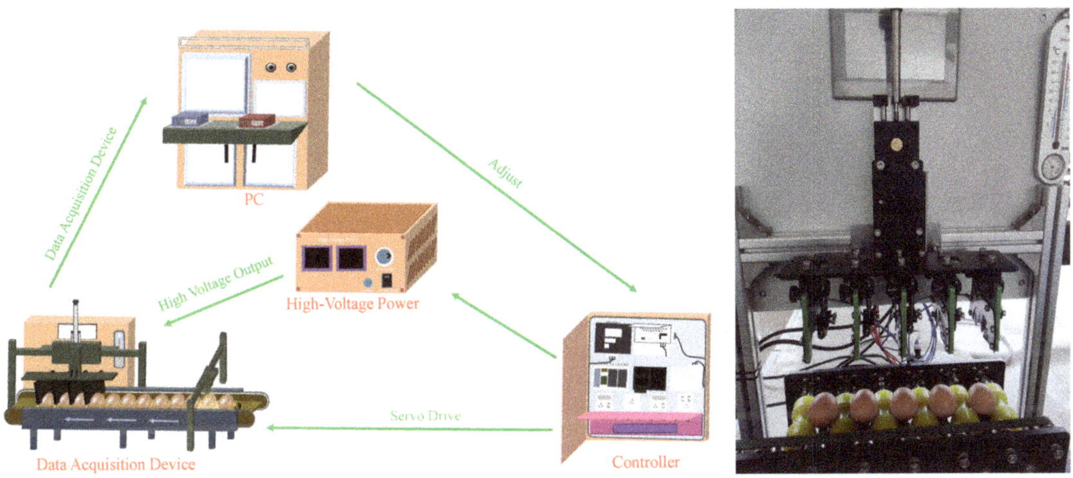

(**a**) (**b**)

Figure 2. Discharge principle and experimental setup. (**a**) Schematic diagram of poultry egg crack detection system based on real-time electrical characteristics. (**b**) The nondestructive detecting device based on electrical characteristics of eggs used in this paper.

After the eggs are subjected to the application of a DC voltage, the data acquisition card acquires the micro-current signal from 450 sample points on the eggs. Three representative egg samples are selected for analysis, as shown in Figure 3a–c, which are examples of cracked egg samples, minimally cracked egg samples, and intact egg samples, respectively. Figure 3d shows a micro-current detection signal of the cracked egg sample in Figure 3a. When the electrode is swept to the crack location, a strong current response is generated, and the sample shows a sudden rise in a steady microcurrent signal. Figure 3f shows the microcurrent detection signal of the intact egg sample, which does not produce a strong current reaction under DC voltage. There is no special change in the whole rotation process, and the sample is very stable. The existing classification algorithm based on the electrical characteristic signals of eggs can have a good classification effect on the above samples. However, for some extremely tiny cracks, such as thsoe in Figure 3c, a microcurrent signal is used, as shown in Figure 3e. The strong current response is small, and the performance is similar to that of intact egg samples. Currently, the existing classification algorithms based on the electrical signal characteristics of eggs struggle to distinguish those extremely small cracks.

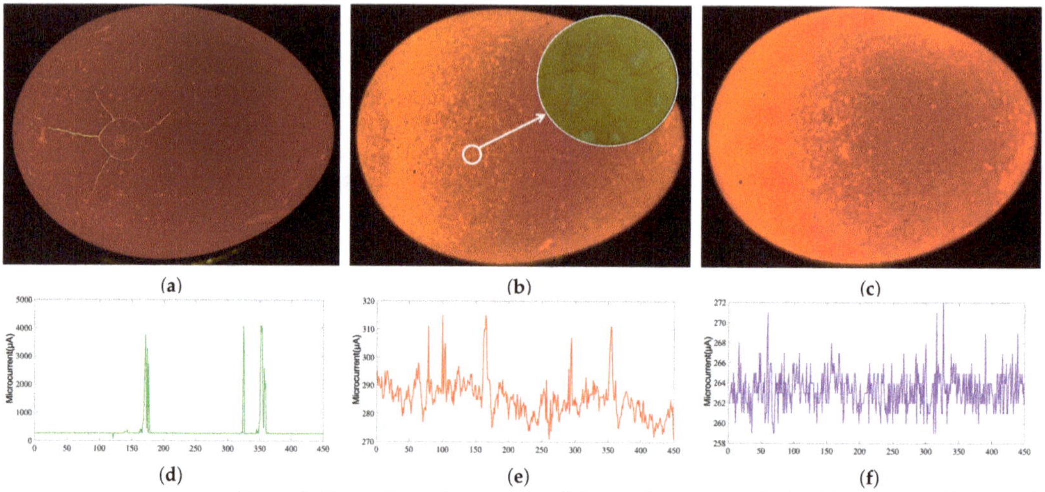

Figure 3. Comparison of current signal changes between cracked and intact eggs, where the abscissa is the sampling point, and the ordinate is the microcurrent. (**a**) A cracked egg sample. (**b**) An indistinguishable cracked egg sample. (**c**) An intact egg sample. (**d**) Micro-current signal of sample (**a**). (**e**) Micro-current signal of sample (**b**). (**f**) Micro-current signal of sample (**c**).

In order to verify the accuracy of the results, each experiment was executed multiple times. In each experiment, the training set and test set used by different methods remained the same. We randomly selected 70% of the sample data as the training set and the remaining 30% as the testing set for scattering feature extraction and then input these data to the deep learning network for classification for different experiments.

2.2. Detection Framework

The framework of the current crack detection algorithm based on the egg electrical characteristics model is shown in part A of Figure 4. The time, frequency , and wavelet domain features were manually extracted, and the traditional machine learning algorithm was used for classification [14]. The features extracted by such methods are fixed, making it challenging to capture multi-scale features and, therefore, easy to overlook feature variations in detail. At the same time, traditional machine learning algorithms have low compatibility with feature dimensions and high dependence on features, leading to limitations in classification. Considering this situation, this paper investigates the feature extraction and classification of current signals separately, using wavelet scattering transform in conjunction with deep learning methods for experiments.Part Proposed of Figure 4 illustrates the algorithm framework based on the wavelet scattering convolution network used in this paper. In the feature extraction part, we used the wavelet scattering transform to extract the features of the current signal, which has the properties and advantages of translation invariance and local deformation stability and does not require a manual design to extract features from the signal. It is immune to translational and slight deformation information and discards the parts of the signal sensitive to translation and deformation while obtaining high-frequency information. The classification part was analyzed using deep learning methods, which have the advantages of high learning ability, portability, and compatibility with feature dimensions compared to traditional methods. In this paper, we tried different deep-learning methods to study the algorithms that can be used for the classification of microcurrent signals.

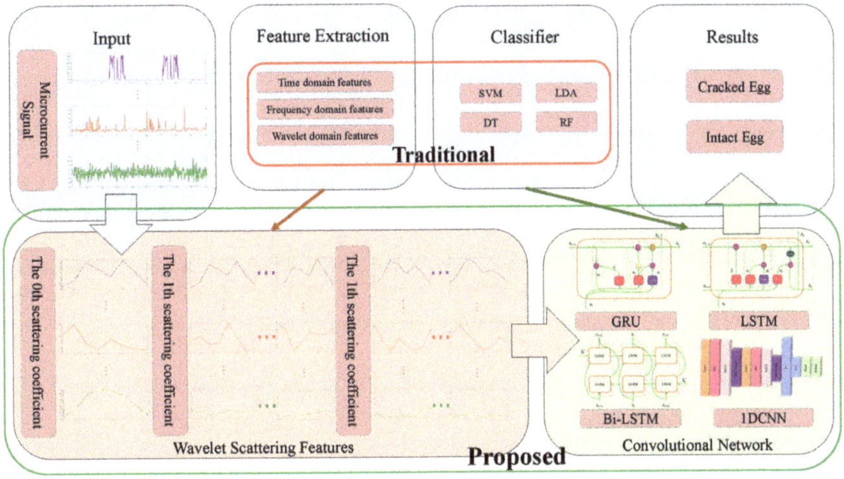

Figure 4. Research ideas for a non-destructive detection algorithm based on electrical characteristics of microcracks.

2.2.1. Wavelet Scattering Transform-Based Feature Extraction

Wavelet scattering transform (WST) [29–31] is an improved time-frequency analysis method based on wavelet transform, which solves the shortcomings of wavelet transform changing with time and has the advantages of translation invariance and local deformation stability of signal analysis and feature extraction. The wavelet scattering network is a framework for automatically extracting input signal features, requiring convolutional, nonlinear, and average cascade operations. After constructing the wavelet scattering network, the scattering coefficient matrix of training data can be obtained. The framework is shown in Figure 5a. In the wavelet scattering transform network, the convolution task is performed by the wavelet, the modular operator is used for nonlinearization, and the filtering function of the wavelet low-pass filter completes the pooling task.

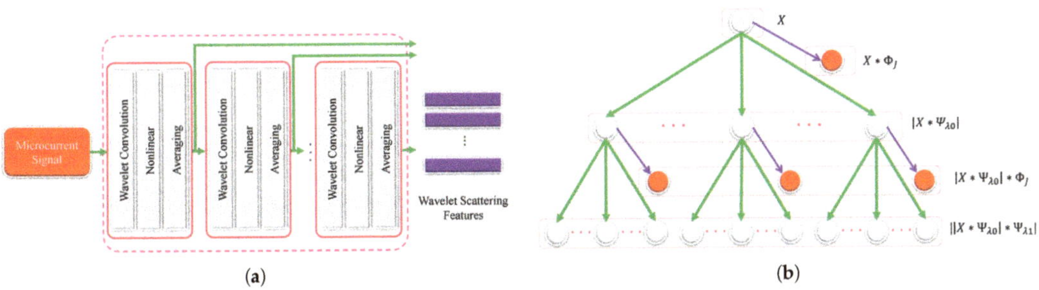

Figure 5. Principle of wavelet scattering transform. (**a**) Framework of wavelet scattering network. (**b**) Structure of wavelet scattering transform.

Wavelet scattering takes the modulus of the wavelet transform, mainly because the mean value of the wavelet coefficient is 0. In contrast, the mean value of linear transformation will not change, so the non-0 wavelet coefficient can be obtained through nonlinear transformation to carry specific information. After removing the complex phase of all wavelet scattering coefficients, operator $|W|$ can be obtained. By convolving with input signal X, the operator of the wavelet modulus transformation can be obtained:

$$|W|X = \{S(X), U(X)\} \tag{1}$$

$$S(X) = X \times \phi_J \tag{2}$$

$$U(X) = |X \times \psi_\lambda| \tag{3}$$

Equation (1) contains two parts: $S(X)$ (see Equation (2), where ϕ_J is the low-pass filter) is the invariant part, namely the scattering coefficient, which is the low-pass filtering and averaging of the input signal. It represents the translation invariance of the wavelet scattering transform, with the purpose of extracting the low-frequency information of the input signal and removing the high-frequency information. $U(X)$ (see Equation (3), where ψ_λ is high frequency wavelet) is the covariant part, representing the high frequency information on scale λ and acting as the exchange modulus of the nonlinear wavelet change. It is also the input to the next transformation and represents the stability of the operator, with the purpose of recovering the high-frequency information lost by the operation of the invariant part.

Therefore, the 0th-order wavelet scattering transform information is:

$$|W_0|X = \{S_0(X), U_0(X)\} \tag{4}$$

$$S_0(X) = X \times \phi_J \tag{5}$$

$$U_0(X) = |X \times \psi_{\lambda 0}| \tag{6}$$

By taking the 0th-order high-frequency $U_0(X)$ as the input X_1 of the first-order scattering transform and calculating with the new wavelet modulus operator, the information of the 1st-order wavelet scattering transform is obtained.

$$|W_1|X_1 = \{S_1(X_1), U_1(X_1)\} \tag{7}$$

$$S_1(X_1) = X_1 \times \phi_J = |X \times \psi_{\lambda 0}| * \phi_J \tag{8}$$

$$U_1(X_1) = |X_1 \times \psi_{\lambda 1}| = ||X \times \psi_{\lambda 0}| \times \psi_{\lambda 1}| \tag{9}$$

Similarly, by iterating the above process, the $g-1$-order high-frequency $U_{g-1}(X_{g-1})$ can be used as the input X_g of the first-order scattering transform to operate with the new wavelet modulus operator for the higher g-order scattering transform:

$$|W_g|X_g = \{S_g(X_g), U_g(X_g)\} \tag{10}$$

The structure of the scattering transformation is shown in Figure 5b. Input current signals are iterated successively in layers, forming a series of scattering coefficients from order 0 to order g:

$$S_G(X) = \{S_0(X), S_1(X_1), \cdots, S_g(X_g)\} \tag{11}$$

The scattering coefficient is the extracted micro-current wavelet scattering feature. After collecting the microcurrent signal of the egg, the wavelet scattering network is constructed to extract the wavelet scattering characteristics of the microcurrent signal, and a wavelet time scattering network with two cascaded filter banks is constructed. The first filter bank has eight wavelets per octave, the second filter bank has one wavelet per octave, the wavelet filter in the two filter banks is shown in Figure 6a,b, and the Littlewood–Paley sums of the filter banks are shown in Figure 6c. Figure 6d shows the Gabor small wave used for its low-pass filter to prove the invariance scale, obtain the inverse Fourier transform of the scale function, and center it in 0 s.

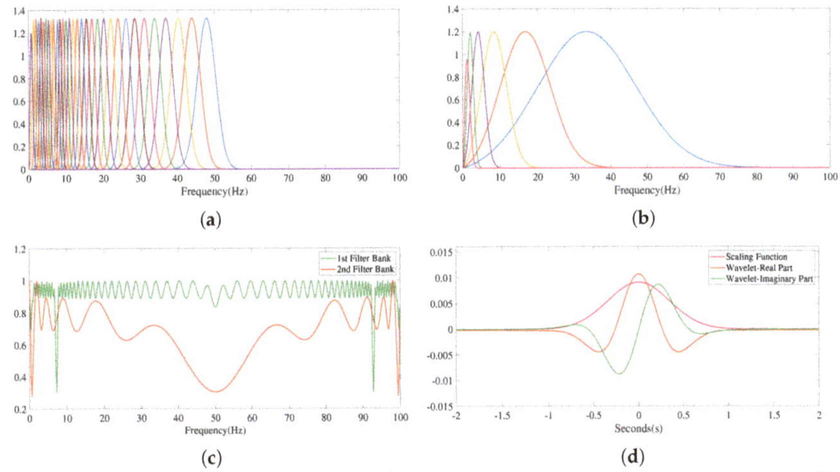

Figure 6. The principle of wavelet scattering characteristic parameters used in this paper. (**a**) The first filter bank with 8 wavelets per octave. (**b**) The second filter bank with 1 wavelet per octave. (**c**) The Littlewood −Paley sums of the filter banks. (**d**) The real and imaginary parts of the coarsest-scale wavelet from the first filter bank.

2.2.2. Classification Modeling Methods

Compared with traditional machine learning methods, deeper networks can accommodate richer semantic information and are more effective in processing temporal order, feature dimensionality, and scale. LSTM can effectively capture the association between long sequences, which is very effective in processing temporal information and can also alleviate the phenomena of gradient disappearance or explosion . Bi-LSTM is a combination of forward LSTM and backward LSTM, which can effectively use the input forward and backward feature information. GRU can effectively capture the association of long sequences, and the number of parameters used is small, which can also reduce the risk of overfitting. 1DCNN features scale shifts and invariance, region awareness, comprehensive feature learning, and fault tolerance, which can compensate for the shortcomings of other shallow classification algorithms with inadequate feature learning and poor generalization ability.

The data used in this paper are characterized by disorder, mutation, nonlinearity, and time discontinuity. Therefore, in this study, LSTM, Bi-LSTM, GRU, and 1DCNN were used to classify the features extracted from wavelet scattering. The details of these methods are as follows.

Long short-term memory (LSTM) [32] is a particular recurrent neural network (RNN), which is an improved neural network used to solve the problems of gradient disappearance and gradient explosion inherent to traditional RNNs in long sequence samples. The neural unit of LSTM contains three different gate structures: forgetting gates, input gates, and output gates to control information. Its appearance successfully solves the problem that the original RNN cannot store information for a long time and can effectively reduce the feature input dimension of the signal and enhance the temporal modeling capability.

Bi-directional long short-term memory (Bi-LSTM) [33] is a variant of LSTM, which refers to the combination of forward LSTM and reverse LSTM. The forward and reverse layers are simultaneously connected to the input and output layers through different weights. There is input each time, while the hidden layer has two node vectors, which can be transmitted from forward to backward or backward to forward. The state transmission is bidirectional. It can learn not only the rule of forward data but also the rule of backward data. Applying LSTM twice can improve the long-term dependence on learning and thus improve the model's accuracy.

A gated recurrent unit (GRU) [34] is a simple variant of LSTM, and its working principle is similar to LSTM. It has two gates: update doors and reset doors. GRU aims to reduce parameters to optimize the cell structure and improve operational efficiency, making converging during training more manageable. Its flexible and straightforward structure guarantees accuracy while significantly reducing the time required for classification.

A convolutional neural network (CNN) [35] is a typical representative of deep learning, consisting of an input layer, convolutional layer, pooling layer, fully connected layer, and output layer. Similar in structure to CNN, a one-dimensional convolutional neural network (1DCNN) [36] refers to a CNN whose input is a one-dimensional vector; similarly, its convolution kernel is one-dimensional, and it carries out a one-dimensional convolution operation. It is a deep learning algorithm suitable for processing one-dimensional data. Compared with 2DCNN, the convolution kernel of 1DCNN only needs to be scanned in one direction. The convolutional layer performs convolution operations on the one-dimensional input signal and the one-dimensional convolution kernel and then extracts the features through the activation function. The pooling layer scales and maps the input data through the pooling kernel and extracts features while reducing the dimension of the data. This paper uses max pooling, which selects the maximum value of each small region as the output. Pooling operations can reduce the size of spatial information, improve operational efficiency, and reduce the risk of overfitting.

As we all know, the selection of hyperparameters affects the performance of deep learning to a large extent. The improper setting of hyperparameters may lead to under-fitting and over-fitting of the network model during training, so that the samples outside the training samples cannot be accurately classified. In this paper, the method of grid search [37] is used for hyperparameter selection first, and the optimal hyperparameter method is determined by traversing the given parameter combination to determine the optimal value performance. Finally, we used the methods of LSTM, Bi-LSTM, GRU, and 1DCNN to classify the extracted feature data. The maximum number of rounds of training was 150. The Adam optimizer was applied. The learning rate was 0.01. The learning rate decline factor was set to 0.2. The number of iterations when the learning rate decreases was 5. The fully connected layer had two layers, and the number of corresponding classifications in the second layer was finally classified using the probability of the Softmax layer. The prediction label of the test sample was the output.

2.3. Performance Indicators

We divided all samples in the classification model into positive and negative samples. In this paper, cracked eggs were labeled as positive samples, and intact eggs were marked as negative samples. The prediction category determined by the model was obtained by predicting the sample data output in the model. In order to better evaluate the classification performance of the proposed detection model, five evaluation indicators, accuracy (ACC), precision (P), recall (R), F1-score ($F1$), and Matthews correlation coefficient (MCC), were used to comprehensively evaluate the algorithm models regarding their classification ability.

We define the false-positive (FP) rate as the number of negative samples that the model incorrectly predicts as positive samples. We define the true-positive (TP) rate as the number of positive samples correctly predicted by the model as positive samples. We define the false-negative (FN) rate as the number of positive samples that the model incorrectly predicts as negative samples and the true-negative (TN) rate as the number of negative samples correctly predicted by the model as negative samples.

Accuracy (ACC), precision (P), recall (R), and F1-Score ($F1$) are common performance indicators used to evaluate the predictive ability of classification models, and their calculation formulas are as follows:

$$ACC = \frac{TP + TN}{TP + FP + TN + FN} \times 100\% \tag{12}$$

$$P = \frac{TP}{TP + FP} \tag{13}$$

$$R = \frac{TP}{TP + FN} \tag{14}$$

$$F1 = 2 \times \frac{P \times R}{P + R} \tag{15}$$

The Matthews correlation coefficient (*MCC*) [38] comprehensively considers *TP*, *TN*, *FP*, and *FN*. It is considered to be a better measure of the classifier's performance. The value range of the *MCC* is $[-1, 1]$. A value of 1 means that the prediction is entirely consistent with reality, a value of 0 means that the predicted result is not as good as the result of random prediction, and a value of -1 means that the predicted result is inconsistent with the actual result. *MCC* is defined as follows:

$$MCC = \frac{TP \times TN - FP \times FN}{\sqrt{(FP + TP)(TP + FN)(FN + TN)(TN + FP)}} \tag{16}$$

We used *ACC*, *P*, *R*, *F*1, and *MCC* as the evaluation indicators of the proposed method. In addition, the training time and preference time of the model were considered as a metric for performance evaluation, as they are of great significance to the real-time detection of cracked eggs.

2.4. Experimental Environment

All experimental calculations in this study were performed using MATLAB R2022a software, and the experimental computer processor was an 11th Gen Intel (R) Core (TM) i5-11400H @ 2.70GHz 2.69 GHz, Windows 10 (64-bit) Professional version.

3. Results and Discussion

3.1. Experimental Data

The experimental data came from fresh eggs purchased at the farmers' market near the laboratory . The eggs were cleaned and transported to the laboratory. The mass of each egg was between 43.2 g∼62.3 g. The intact samples were observed under 10× magnification, and a total of 400 eggs were observed. To quickly obtain a sufficient number of egg microcrack samples, we selected 220 cracks of different types and positions by exerting external forces on different positions of eggs through the egg crack collision machine . The width of the artificial microcracks was generally less than 3 microns, which is usually not easy to observe with the human eye. The samples that could not be subjected to a discharge test due to excessive force or improper operation during the production process were rejected . Finally, a total of 356 egg samples that met the requirements were selected for experimentation, as shown in Table 1.

Table 1. The number of egg samples used by the electrical characteristic crack detection system to obtain a microcurrent signal.

Label	Count	Percent
Cracked	187	52.5281
Intact	169	47.4719

3.2. Wavelet Scattering Transform Features

Based on the settings described above, a wavelet scattering network was constructed to extract the wavelet scattering characteristics of the microcurrent signal. After the signal input network, the wavelet scattering transformation was carried out layer by layer, and the 0th scattering output was the convolution coefficient of the original signal and scale function. Figure 7 shows the partial scattering results obtained from the cracked egg sample in

Figure 3e and the intact egg sample in Figure 3f and the scattering coefficient of 8 time windows, respectively.

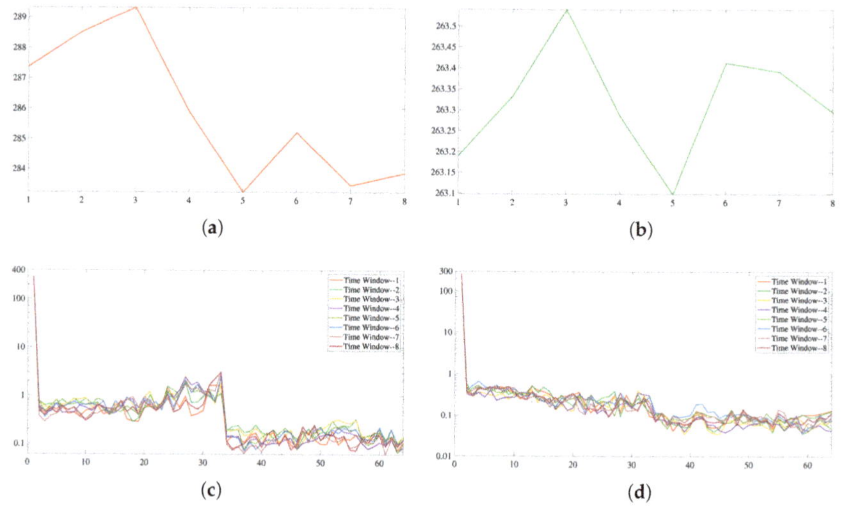

Figure 7. Comparison of wavelet scattering characteristics of cracked eggs and intact eggs. (**a**) The 0th scattering output of Figure 3e of cracked egg sample. (**b**) The 0th scattering output of Figure 3f of intact egg sample. (**c**) Scattering coefficients for 8 time windows of Figure 3e of cracked egg sample. (**d**) Scattering coefficients for 8 time windows of Figure 3f of intact egg sample.

The scattering feature of the cracked egg signal maintains the undulation information at the crack, and the scattering feature of the intact egg signal clarifies the small deformation in the original signal process and maintains the elastic deformation stability of the signal. The feature matrix extracted by the wavelet scattering network maintains the stability of the signal feature while ensuring that the information is not lost so that the cracked egg and intact egg signal features extracted by wavelet scattering transformation have an apparent distinction. The first and second order output a matrix of wavelet scattering coefficients, the dimensions of which represent the scattering path and wavelet scale, respectively. A current signal with a data size of 450×1 is input to the wavelet scattering network to extract the wavelet scattering feature with a data size of 64×8, where each row and column correspond to one scattering path and one time window, respectively.

3.3. Comparison of Experimental Results and Analysis

3.3.1. Experimental Results

Based on the optimal hyperparameters, we constructed and trained the wavelet scattering convolutional network to classify the microcurrent signals. In addition, we also used the previously studied methods to classify the microcurrent signals. The resulting comparison of the final classification is shown in Table 2. By comparing the five models, it can be found that WST+1DCNN gets higher results on index *ACC* (99.4393%), *F*1 (99.4357%) and *MCC* (98.8819%). WST+GRU and WST+1DCNN get higher results on *R* (99.6226%) index, WST+Bi-LSTM obtains higher results on index *P* (99.6154%). In general, the detection effects of WST+LSTM, WST+Bi-LSTM, WST+GRU, and WST+1DCNN based on the wavelet scattering convolutional network are superior to existing methods [14]. The algorithm based on WST+1DCNN has the best recognition effect and can effectively and accurately detect cracked eggs. The accuracy is 2.0561% higher than the accuracy of the existing method [14]. In terms of the real-time implementation of the method, the training time of the previous method is faster, being almost half of that of the method studied in this paper. Considering that all models can be pre-trained, the research in this paper is acceptable in terms of training time. We found that all five methods were able to keep the reasoning time within 0.01 s, with the previous research method processing the fastest at 0.0009 s. In contrast, previous studies used machine learning to extract manual features, while this study uses deep learning to implement, which requires the integration of multiple convolutional blocks, so the network architecture is deep and the time is relatively long. However, the results obtained in this study have met the requirements of industrial real-time detection of egg cracks, and the processing speed is faster than other microcrack detection techniques; see Table 3 for details. In conclusion, the method proposed in this paper can detect cracked eggs effectively and accurately and is acceptable in practical application.

Table 2. Experimental results of electrical signal classification algorithm based on wavelet scattering transformation feature.

Methods	*ACC*	*P*	*R*	*F1*	*MCC*	Training Time (s)	Inference Time (s)
Existing Method [14]	97.3832%	97.3577%	98.0984%	97.7210%	95.5056%	**6.7906**	**0.0009**
WST+LSTM	97.9439%	98.4761%	97.3585%	97.9119%	95.8971%	11.4690	0.0028
WST+Bi-LSTM	98.5047%	**99.6154%**	97.3585%	98.4725%	97.0355%	13.3880	0.0029
WST+GRU	99.2523%	98.8819%	**99.6226%**	99.2488%	98.5115%	12.0574	0.0036
WST+1DCNN	**99.4393%**	99.2523%	**99.6226%**	**99.4357%**	**98.8819%**	13.2714	0.0033

Table 3. Comparison of inference time between the proposed method and other crack detection techniques.

Author	Detection Mode	Inference Time (s)
Sun et al. [39]	Vision-based	1.65 ± 0.50
Wang et al. [40]	Acoustic-based	0.2
Purahong et al. [41]	Vision-based	0.08
Sun et al. [42]	Vision-based	0.98 ± 0.06
Sun et al. [43]	Acoustic-based	0.281
Proposed Method	Electrical-based	0.0033

To further verify the validity of the method proposed in this paper, we re-experimented on the dataset used in the existing method [14] and conducted a comparative study. A total of 770 egg signals were collected, including 367 intact egg signals and 403 cracked egg signals. The final results are shown in Table 4. Compared with the existing egg microcrack classification algorithm based on the electrical characteristics model, our proposed method has improved the accuracy rate by 0.3478% in the dataset . Considering the results above, the results obtained in this study on the detection of microcracks based on the electrical characteristics of eggs are better than those obtained by the preliminary experimental methods. It mainly considers feature extraction and classification algorithm. In the aspect of feature extraction, the existing method extracted the time domain feature, frequency domain feature and wavelet feature of the micro-current signal. A specific function calculates the features extracted by the manual design-based feature extraction method, so the extracted features will ignore the changes in detail, resulting in some of the distinguishing representative features being ignored, such as the cracked egg feature shown in Figure 8a, where the crack changes are subtle. The features extracted by traditional manual design methods focus on the general information of the signal and therefore struggle to capture the variations in detail. The wavelet scattering transform used in this paper extracts invariant and small deformation-stable features to extract multi-scale high-frequency feature vectors. The features extracted from the sample in Figure 8a using the wavelet scattering transform are shown in Figure 8b, which is significantly different from the regular intact egg signal features shown in Figure 8c. It can maintain the undulation information at the crack so that the features at the crack can have a clear representation in the whole feature matrix for classification and differentiation. In the aspect of classification algorithm, the machine learning method was used for classification in the previous experiment. Compared with the machine learning algorithm, the deep learning algorithm used in this study can effectively capture the correlation between long sequences and has a good effect in the processing of time sequences, feature dimensions, and scales and can carry out the comprehensive learning of features. Therefore, the method proposed in this paper can improve the performance of the model according to the electrical characteristics of eggs and can be used online in the detection of microcracked eggs in industrial applications.

Table 4. Comparison of the results of the dataset used in the existing method [14].

	Existing Method [14]	Proposed Method
ACC	99.0435%	99.3913%
P	99.2331%	99.6911%
R	99.0769%	99.2308%
*F*1	99.1529%	99.4601%
MCC	98.0598%	98.7649%

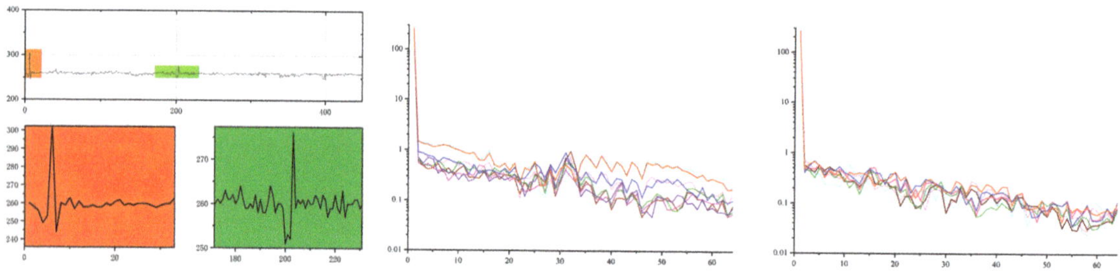

(a) (b) (c)

Figure 8. Comparison of sample features extracted using wavelet scattering transform. (**a**) Variation in detail of hard-to-capture microcrack sample signals. (**b**) Extraction of 8 time windows of feature information for such cracked eggs using WST, where eight different colors lines in the figure represent the information of the eight time windows. (**c**) Extraction of 8 time windows of feature information for intact eggs using WST, where eight different colors lines in the figure represent the information of the eight time windows.

3.3.2. Ablation Study

To evaluate the superior performance of the microcrack nondestructive detection algorithm for egg electrical characteristics based on wavelet scattering convolution network proposed in this paper, the extracted current original signal and the characteristics processed by wavelet scattering transformation were inputted to four classification methods for comparison, including LSTM, Bi-LSTM, GRU, and 1DCNN. The results are shown in Figure 9a. For the four classification methods, the accuracy of the wavelet scattering feature extraction was 2.243%, 1.8692%, 3.3644%, and 2.9907% higher than the accuracy of directly feeding microcurrent signals into the deep learning network. The average statistical error of accuracy of LSTM, Bi-LSTM, GRU, and 1DCNN is 0.9622%, 0.6339%, 0.9157%, and 0.5449% respectively. The results indicate that under the same conditions, the features extracted by wavelet scattering transformation were more recognizable and more capable of distinguishing cracked eggs from intact eggs.

The wavelet scattering network finally constructed in this paper has two cascaded filter banks. The first filter bank has eight wavelets per octave, and the second has one wavelet per octave. As for the selection of filter banks, the experiment proves that the scattering coefficient energy converges rapidly with the deepening of the network, and the energy after the two-layer network structure is about 1% [30], so two cascades of filter banks are constructed in this paper. For the number of wavelets per octave in the filter bank, different experiments as shown in Figure 9b have been made, proving that the combination of (8, 1) has the best result.

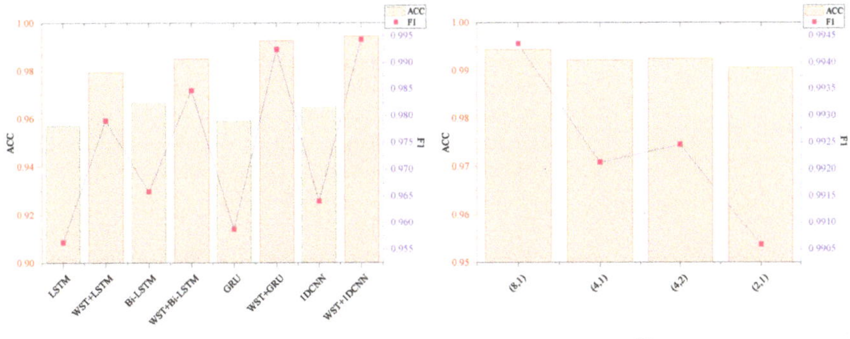

Figure 9. Comparison of ablation study results. (**a**) Results comparison using wavelet scattering transform versus without wavelet scattering transform methods. (**b**) Comparison of the results of the combination of the number of wavelets per octave of the filter bank.

3.3.3. More Results on Imbalanced and Duck Egg Datasets

It is worth noting that the data in a real industrial scenario are unpredictable.To verify the performance and versatility of the proposed method, the class-imbalanced dataset of eggs collected in this experiment and the duck egg dataset experiment were set up. The distribution of cracks in the duck egg dataset is shown in Figure 10. The experimental results for the class-imbalanced and duck egg datasets are shown in Table 5. The class-imbalanced dataset removed some collected data to simulate an imbalanced state. The dataset has 200 entries, including 169 intact and 31 cracked egg signal data. The MCC (98.0788%) was obtained under the condition of fewer cracked eggs, which is slightly lower than the previous experiments. It is still in the high-accuracy range, fully demonstrating its stability under data imbalance. A total of 267 fresh duck eggs were purchased from the advanced breeding duck incubation base. Signal data of 130 intact and 137 cracked eggs were obtained in this experiment, and an accuracy of 99.6169% was finally obtained. This experiment shows that the method proposed in this paper has good universality and extensibility.

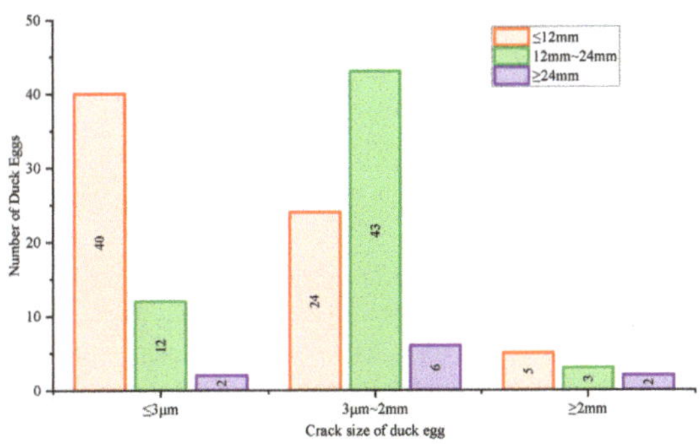

Figure 10. Crack size distribution of cracked duck eggs.

3.4. Discussion

This paper used the micro-current high-voltage discharge method to detect microcracks in eggs, but there is a small amount of literature and patents that investigate this technique. The most studied application for this technique is the case of plastic container leaks. Regarding the range of voltages, some studies [12] suggest that the voltage applied is typically 3000 V∼5000 V DC, but no relevant theoretical description is given. The electrode may cause certain damage to the detected object when it is in direct contact with the detected object. In the previous experiment, 1500 V was used, which may cause damage to eggs that are being incubated. The voltage in this study was set to 1000 V. The results obtained from the above four experiments proved the effectiveness of the proposed method.Especially for the detection of eggs to be hatched, the lower the voltage used, the safer the eggs will be, and the less they will be damaged. In this study, twenty eggs (ten intact and ten microcracked) were selected to conduct classification experiments on the signals extracted from poultry eggs with different voltages. The extracted signals were tested by the existing method [14] and the method proposed in this paper. The results are shown in Figures 11 and 12.

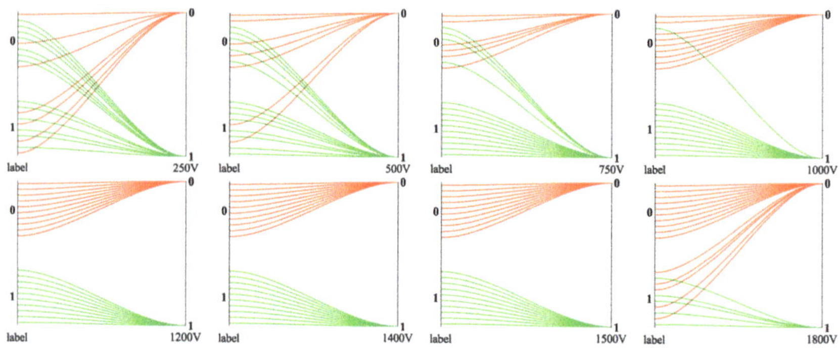

Figure 11. Crack detection results of eggs with different voltages by existing method [14]. Where, 0 in the figure represents cracked eggs, and 1 in the figure represents intact eggs; The green line indicates that the tested sample is classified as intact eggs, and the orange line indicates that the tested sample is classified as cracked eggs.

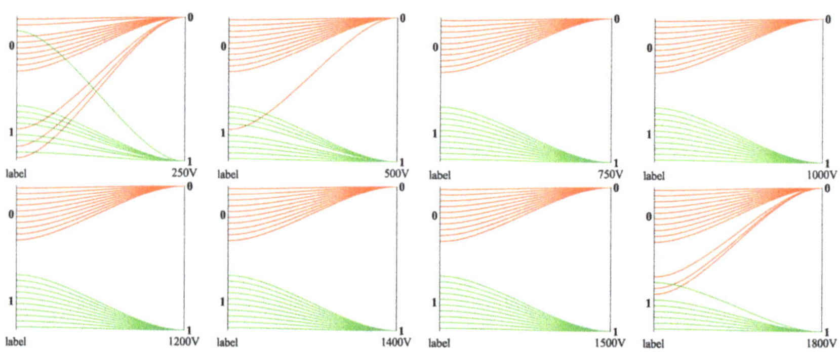

Figure 12. Crack detection results of eggs with different voltages by the method proposed in this paper. Where, 0 in the figure represents cracked eggs, and 1 in the figure represents intact eggs; The green line indicates that the tested sample is classified as intact eggs, and the orange line indicates that the tested sample is classified as cracked eggs.

Table 5. Method performance and versatility experiments.

	Imbalanced Dataset	**Duck Egg Dataset**
ACC	99.4444%	99.6169%
P	97.9798%	99.4949%
R	98.8889%	99.7416%
*F*1	98.3570%	99.6139%
MCC	98.0788%	99.2423%

The selected eggs were tested using 250 V, 500 V, 750 V, 1000 V, 1200 V, 1400 V, 1500 V, and 1800 V. From Figures 11 and 12, it can be concluded that the existing method and the method based on the wavelet scattering convolutional network proposed in this paper can be used to distinguish the microcurrent signals obtained from cracked eggs and intact eggs in the voltage between 1000 V and 1500 V. Both methods have detection errors when the voltage is higher than 1500 V. This is because when the voltage is increased, the current signal generated by the system fluctuates strongly, resulting in the voltage breakdown of the intact egg, which will reflect on the current and lead to false detection. The effect of the existing method on the classification of current signals below 1000 V is not obvious because of the weakening of the current response at the crack for voltages below 1000 V. The method proposed in this paper has the characteristics of translation invariance and elastic deformation stability and still maintains a good effect on the classification of current signals between 500 V and 1000 V. When the voltage is less than 500 V, the system cannot generate enough feedback signals, so it is difficult to detect extremely small cracks. The egg crack detection method based on electrical characteristics may cause damage to incubating eggs under high-voltage conditions, while the wavelet scattering convolutional network method proposed in this paper can adapt the sensing voltage from 1500 V to 500 V and can obtain higher detection accuracy on the data with a low signal-to-noise ratio, thus greatly reducing the risk of high-voltage damage to incubating eggs.

4. Conclusions

This paper constructs a crack detection system for the electrical characteristics of eggs . With the help of a size recognition device, which automatically adapts the upper electrode position to the size of the egg, the system gives a more detailed and consistent view of the egg's surface, resulting in a more representative collected signal. Given the limitations of disorder, mutation, nonlinearity, and time discontinuity of microcurrent signals, the reliability of signal features dominates the performance of the ultimate classification model. This paper suggests an electrical-based nondestructive detection model for microcracks in poultry eggs, which employs wavelet scattering transform to extract features. Wavelet scattering transform can effectively avoid the loss of valid information and produce a signal representation insensitive to small changes in the input signal. This paper discusses the feature extraction mechanism of wavelet scattering by visualizing the output results of the scattering feature process. Finally, the study feeds the acquired feature vector into the deep learning network for classification. The following are the conclusions that can be drawn from our experiments. The microcurrent signal has unpredictable and sudden transient characteristics. The wavelet scattering transform utilized to extract signal features and develop the corresponding matrix shows a distinguished capacity to collect signals with apparent differentiation and ensure satisfactory results. In this paper, we implement this feature extraction approach combined with appropriate classifiers to discuss the classification of egg microcurrent signals. The results show that WST+1DCNN has the best performance, and the average *ACC*, *P*, *R*, *F*1, and *MCC* obtained are 99.4393%, 99.2523%, 99.6226%, 99.4357%, and 98.8819%, respectively. In addition, we set the eggs' class-imbalanced dataset and the duck egg dataset to verify the performance and universality. Finally, we conduct experiments on egg detection at different voltages. The novel feature extraction and detection method proposed in this paper can reduce the sensing voltage from 1500 V to 500 V and obtain higher detection accuracy on lower signal-to-noise data, dramatically reducing the risk of damage to hatching eggs from high-voltage electricity. In the future, the main direction of our research is how to achieve higher precision in egg crack detection under lower-voltage conditions, which specifically includes the following several aspects. The first is how to improve the shape of the brush so that it can cover a larger area of the eggshell during rotation and reduce the amount of missed area. An increase in the contact region means that we can realize the distinction at lower voltage, as it can also obtain enough current accumulation values in the crack regions. Secondly, at the algorithm level, we hope to extract more abundant and high-dimensional current features in the crack region and improve the existing algorithm to make it more representative. Finally, multi-sensor fusion is also one of our directions. We speculate that an algorithm based on current features combined with image features or acoustic features can have higher accuracy than an algorithm based on single-current features.

Author Contributions: Conceptualization, C.S. and C.Z. (Changsheng Zhu); methodology, C.S., C.Z. (Changsheng Zhu) and Y.C.; software, Y.C., Y.W. and C.Z. (Changsheng Zhu); validation, C.S., C.Z. (Changsheng Zhu), C.Z. (Chun Zhang) , J.Y. and Y.C.; formal analysis, C.S., C.Z. (Chun Zhang), J.Y., Y.C., Y.W. and X.J.; investigation, C.S., Y.C., X.J. and Y.W.; resources, C.S. and J.Y.; data curation, Y.C., Y.W. and X.J.; writing—original draft preparation, C.S., C.Z. (Changsheng Zhu), Y.C. and Y.W.; writing—review and editing, C.S., C.Z. (Chun Zhang), C.Z. (Changsheng Zhu), Y.C., Y.W., X.J. and J.Y.; visualization, Y.C., Y.W. and X.J.; supervision, C.S., C.Z. (Chun Zhang), C.Z. (Changsheng Zhu) and J.Y. All authors have read and agreed to the published version of the manuscript.

Funding: This research was funded by the Tai'an Science and Technology Innovation Development Plan (No. 2021GX050 and No. 2020GX055), by the National Natural Science Foundation of China (52275262).

Institutional Review Board Statement: Not applicable.

Data Availability Statement: The data presented in this study are available on demand from the corresponding author at (cs.zhu@sdust.edu.cn).

Conflicts of Interest: The authors declare no conflict of interest.

References

1. Reu, K.D.; Grijspeerdt, K.; Messens, W.; Heyndrickx, M.; Uyttendaele, M.; Debevere, J.; Herman, L. Eggshell factors influencing eggshell penetration and whole egg contamination by different bacteria, including Salmonella enteritidis. *Int. J. Food Microbiol.* **2006**, *112*, 253–260. [CrossRef] [PubMed]
2. Munoz, A.; Dominguez-Gasca, N.; Jimenez-Lopez, C.; Rodriguez-Navarro, A.B. Importance of eggshell cuticle composition and maturity for avoiding trans-shell *Salmonella* contamination in chicken eggs. *Food Control* **2015**, *55*, 31–38. [CrossRef]
3. Mansoory, M.S.; Ashtiyani, M.; Sarabadani, H. Automatic Crack Detection in Eggshell Based on SUSAN Edge Detector Using Fuzzy Thresholding. *Mod. Appl. Sci.* **2011**, *5*, 117. [CrossRef]
4. Wang, S.C.; Ren, Y.L.; Chen, H.; Xiong, L.R.; Wen, Y.X. Detection of cracked-shell eggs using acoustic signal and fuzzy recognition. *Trans. CSAE* **2004**. *20*, 130–133.
5. Lin, H.; wen Zhao, J.; sheng Chen, Q.; rong Cai, J.; Zhou, P. Eggshell crack detection based on acoustic response and support vector data description algorithm. *Eur. Food Res. Technol.* **2009**, *230*, 95–100. [CrossRef]
6. Chen, C.; Qiang, M.; Cai, J.; Zhang, S.; Sun, L. Design of On-line Detection System for Poultry Eggshell Crack. *Agric. Eng.* **2020**, *10*, 22–27.
7. Kertész, I.; Zsom-Muha, V.; András, R.; Horváth, F.; Németh, C.; Felföldi, J. Development Of A Novel Acoustic Spectroscopy Method For Detection Of Eggshell Cracks. *Molecules* **2021**, *26*, 4693. [CrossRef]
8. Deng, X.; Wang, Q.; Chen, H.; Xie, H. Eggshell crack detection using a wavelet-based support vector machine. *Comput. Electron. Agric.* **2010**, *70*, 135–143. [CrossRef]
9. Bao, G.; Jia, M.; Xun, Y.; Cai, S.; Yang, Q. Cracked egg recognition based on machine vision. *Comput. Electron. Agric.* **2019**, *158*, 159–166.
10. Turkoglu, M. Defective Egg Detection Based On Deep Features And Bidirectional Long-Short-Term-Memory. *Comput. Electron. Agric.* **2021**, *185*, 106152. [CrossRef]
11. Zhao, Z.X.; Luo, Y.F.; Huang, X.B.; Yuan, K.; Huang, Y.; Cao, Y.Y. Research on Online Detection of Broken Eggs Based on Machine Vision and YOLOv4. *Mod. Agric. Equip.* **2022**, *43*, 8–16.
12. Hu, X. An Antomatically-Inspecting Method for Ampoule Sealing Quality. *J. Zhuzhou Inst. Technol.* **1997**, *11*. Available online: https://www.cnki.com.cn/Article/CJFDTotal-ZZGX199704001.htm (accessed on 19 March 2023).
13. Yan, Z.; Ren, D.; Chen, R.; Gao, Y.; Fu, X. Research on Large Infusion Plastic Bottle Leakage Detection Technology Based on High Voltage Discharge. *Machinery* **2018**, *45*, 19–21,72.
14. Shi, C.; Wang, Y.; Zhang, C.; Yuan, J.; Cheng, Y.; Jia, B.; Zhu, C. Nondestructive Detection of Microcracks in Poultry Eggs Based on the Electrical Characteristics Model. *Agriculture* **2022**, *12*, 1137. [CrossRef]
15. Li, J.; Ke, L.; Du, Q.; Ding, X.; Chen, X.; Wang, D. Heart Sound Signal Classification Algorithm: A Combination Of Wavelet Scattering Transform And Twin Support Vector Machine. *IEEE Access* **2019**, *7*, 179339–179348. [CrossRef]
16. Al-Taee, A.A.; Khushaba, R.N.; Zia, T.; Al-Jumaily, A. Feature Extraction Using Wavelet Scattering Transform Coefficients for EMG Pattern Classification. In *Lecture Notes in Computer Science, Proceedings of the 34th Australasian Joint Conference on Artificial Intelligence, Sydney, NSW, Australia, 2–4 February 2022*; Springer: Cham, Switzerland, 2022; Volume 13151, pp. 181–189.
17. Hui, L.; Yongkang, L.; Fang, G.; Wenyang, M.; Wenhao, C. Transmission line fault diagnosis based on wavelet scattering with BiLSTM. *Foreign Electron. Meas. Technol.* **2021**, *40*, 165–172.

18. Wen, J.-H.; Tan, L.-H. Image Retrieval Based on Wavelet Scattering Network. *J. Guangdong Univ. Technol.* **2019**, *36*, 63–67,80.
19. Fan, X.; Cheng, J.; Wang, Y.; Li, S.; Yan, B.; Zhang, Q. Automatic Events Recognition in Low SNR Microseismic Signals of Coal Mine Based on Wavelet Scattering Transform and SVM. *Energies* **2022**, *15*, 2326. [CrossRef]
20. Buriro, A.B.; Ahmed, B.; Baloch, G.; Ahmed, J.; Shoorangiz, R.; Weddell, S.J.; Jones, R.D. Classification of alcoholic EEG signals using wavelet scattering transform-based features. *Comput. Biol. Med.* **2021**, *139*, 104969. [CrossRef]
21. Liu, Z.; Yao, G.; Zhang, Q.; Zhang, J.; Zeng, X. Wavelet Scattering Transform For Ecg Beat Classification. *Comput. Math. Methods Med.* **2020**, *2020*, 3215681. [CrossRef]
22. Yu, Y.; Samali, B.; Rashidi, M.; Mohammadi, M.; Nguyen, T.N.; Zhang, G. Vision-based concrete crack detection using a hybrid framework considering noise effect. *J. Build. Eng.* **2022**, *61*, 105246. [CrossRef]
23. Yu, Y.; Liang, S.; Samali, B.; Nguyen, T.N.; Zhai, C.; Li, J.; Xie, X. Torsional capacity evaluation of RC beams using an improved bird swarm algorithm optimised 2D convolutional neural network. *Eng. Struct.* **2022**, *273*, 115066. . [CrossRef]
24. Hu, B.; Liu, J.; Zhao, R.; Xu, Y.; Huo, T. A New Fault Diagnosis Method for Unbalanced Data Based on 1DCNN and L2-SVM. *Appl. Sci.* **2022**, *12*, 9880. [CrossRef]
25. Le, B.T.; Le, T.C.; Luu, T.H.T.; Ho, D.D.; Huynh, T.C. Fault Assessment in Piezoelectric-Based Smart Strand Using 1D Convolutional Neural Network. *Buildings* **2022**, *12*, 1916. [CrossRef]
26. Ismail Fawaz, H.; Forestier, G.; Weber, J.; Idoumghar, L.; Muller, P.A. Deep Learning for Time Series Classification: A Review. *Data Min. Knowl. Discov.* **2019**, *33*, 917–963. [CrossRef]
27. Zhao, B.; Lu, H.; Chen, S.; Liu, J.; Wu, D. Convolutional neural networks for time series classification. *J. Syst. Eng. Electron.* **2017**, *28*, 162–169. [CrossRef]
28. Karim, F.; Majumdar, S.; Darabi, H.; Chen, S. LSTM Fully Convolutional Networks for Time Series Classification. *IEEE Access* **2018**, *6*, 1662–1669. [CrossRef]
29. Oyallon, E.; Mallat, S.; Sifre, L. Generic Deep Networks with Wavelet Scattering. *arXiv* **2013**, arXiv:1312.5940.
30. Bruna, J.; Mallat, S. Invariant Scattering Convolution Networks. *IEEE Trans. Pattern Anal. Mach. Intell.* **2013**, *35*, 1872–1886. [CrossRef]
31. Wiatowski, T.; Bölcskei, H. A Mathematical Theory of Deep Convolutional Neural Networks for Feature Extraction. *IEEE Trans. Inf.* **2018**, *64*, 1845–1866. [CrossRef]
32. Hochreiter, S.; Schmidhuber, J. Long Short-Term Memory. *Neural Comput.* **1997**, *9*, 1735–1780. [CrossRef]
33. Siami-Namini, S.; Tavakoli, N.; Namin, A.S. The Performance Of Lstm And Bilstm In Forecasting Time Series. In Proceedings of the 2019 IEEE International Conference on Big Data (Big Data), Los Angeles, CA, USA, 9–12 December 2019; pp. 3285–3292.
34. Lv, W.; Lv, Y.; Guo, J.; Ma, J. A Lane-Changing Decision-Making Model of Bus Entering considering Bus Priority Based on GRU Neural Network. *Comput. Intell. Neurosci.* **2022**, *2022*, 4558946. [CrossRef]
35. Gu, J.; Wang, Z.; Kuen, J.; Ma, L.; Shahroudy, A.; Shuai, B.; Liu, T.; Wang, X.; Wang, G.; Cai, J.; et al. Recent advances in convolutional neural networks. *Pattern Recognit.* **2018**, *77*, 354–377. [CrossRef]
36. Eren, L.; Ince, T.; Kiranyaz, S. A Generic Intelligent Bearing Fault Diagnosis System Using Compact Adaptive 1D CNN Classifier. *J. Signal Process. Syst.* **2019**, *91*, 179–189. [CrossRef]
37. Priyadarshini, I.; Cotton, C. A novel LSTM–CNN–grid search-based deep neural network for sentiment analysis. *J. Supercomput.* **2021**, *77*, 13911–13932. [CrossRef]
38. Jurman, G.; Riccadonna, S.; Furlanello, C. A Comparison Of Mcc And Cen Error Measures In Multi-Class Prediction. *PLoS ONE* **2012**, *7*, e41882. [CrossRef]
39. Sun, K.; Zhang, W.; Pan, L.; Tu, K. Recognition of a Cracked Hen Egg Image Using a Sequenced Wave Signal Extraction and Identification Algorithm. *Food Anal. Methods* **2018**, *11*, 1223–1233. [CrossRef]
40. Wang, H.; Mao, J.; Zhang, J.; Jiang, H.; Wang, J. Acoustic feature extraction and optimization of crack detection for eggshell. *J. Food Eng.* **2016**, *171*, 240–247. . [CrossRef]
41. Purahong, B.; Chaowalittawin, V.; Krungseanmuang, W.; Sathaporn, P.; Anuwongpinit, T.; Lasakul, A. Crack Detection of Eggshell using Image Processing and Computer Vision. *J. Phys. Conf. Ser.* **2022**, *2261*, 012021. [CrossRef]
42. Sun, K.; Ma, L.; Pan, L.; Tu, K. Sequenced wave signal extraction and classification algorithm for duck egg crack on-line detection. *Comput. Electron. Agric.* **2017**, *142*, 429–439. [CrossRef]
43. Sun, L.; Feng, S.; Chen, C.; Liu, X.; Cai, J. Identification of eggshell crack for hen egg and duck egg using correlation analysis based on acoustic resonance method. *J. Food Process Eng.* **2020**, *43*, e13430. [CrossRef]

agriculture

Article

Nondestructive Detection of Microcracks in Poultry Eggs Based on the Electrical Characteristics Model

Chenbo Shi [1], Yuxin Wang [1], Chun Zhang [1], Jin Yuan [2], Yanhong Cheng [1], Baodun Jia [1] and Changsheng Zhu [1,*]

[1] College of Intelligent Equipment, Shandong University of Science and Technology, Tai'an 271019, China; shichenbo@gmail.com (C.S.); 202083230017@sdust.edu.cn (Y.W.); 13811989049@163.com (C.Z.); 202183230018@sdust.edu.cn (Y.C.); 202183230035@sdust.edu.cn (B.J.)
[2] College of Mechanical and Electronic Engineering, Shandong Agricultural University, Tai'an 271018, China; jyuan@sdau.edu.cn
* Correspondence: cs.zhu@sdust.edu.cn

Abstract: The eggshell is the major source of protection for the inside of poultry eggs from microbial contamination. Timely detection of cracked eggs is the key to improving the edible rate of fresh eggs, hatching rate of breeding eggs and the quality of egg products. Different from traditional detection based on acoustics and vision, this paper proposes a nondestructive method of detection for eggshell cracks based on the egg electrical characteristics model, which combines static and dynamic electrical characteristics and designs a multi-layer flexible electrode that can closely fit the eggshell surface and a rotating mechanism that takes into account different sizes of eggs. The current signals of intact eggs and cracked eggs were collected under 1500 V of DC voltage, and their time domain features (TFs), frequency domain features (FFs) and wavelet features (WFs) were extracted. Machine learning algorithms such as support vector machine (SVM), linear discriminant analysis (LDA), decision tree (DT) and random forest (RF) were used for classification. The relationship between various features and classification algorithms was studied, and the effectiveness of the proposed method was verified. Finally, the method is proven to be universal and generalizable through an experiment on duck eggshell microcrack detection. The experimental results show that the proposed method can realize the detection of eggshell microcracks of less than 3 μm well, and the random forest model combining the three features mentioned above is proven to be the best, with a detection accuracy of cracked eggs and intact eggs over 99%. This nondestructive method can be employed online for egg microcrack inspection in industrial applications.

Keywords: electrical characteristics; poultry eggs; nondestructive detection; cracked eggs; machine learning

Citation: Shi, C.; Wang, Y.; Zhang, C.; Yuan, J.; Cheng, Y.; Jia, B.; Zhu, C. Nondestructive Detection of Microcracks in Poultry Eggs Based on the Electrical Characteristics Model. *Agriculture* **2022**, *12*, 1137. https://doi.org/10.3390/agriculture12081137

Academic Editor: Maciej Zaborowicz

Received: 28 June 2022
Accepted: 29 July 2022
Published: 31 July 2022

1. Introduction

As one of the main sources of protein nutrition in human daily life, the importance of poultry eggs and related products is self-evident. From 2010 to 2030, global egg production will be increased by 35%, with Asia contributing 64.8% to the total global growth [1]. In the preliminary processing of eggs, such as cleaning, testing, transportation, and other procedures, eggshell damage may occur, and bacteria and other microorganisms may enter the eggs from the cracks and cause spoilage before infecting the surrounding good eggs. This may not only lead to a shortened shelf life and lowered value but also make the food unsafe to eat, causing economic losses to enterprises in the end. Traditional eggshell crack detection mainly depends on artificial light or the sound produced upon impact to the eggs to identify cracked ones, which not only sets high requirements for the experience and physical condition of workers, but the efficiency and reliability cannot meet the growing market demand. Therefore, it is of great significance for consumers, enterprises, and agricultural modernization to study a highly reliable, non-destructive, and automatic cracked egg removal system [2].

In recent years, researchers both nationally and internationally have been trying to find an automatic online approach to detecting eggshell cracks so as to lower labor intensity and improve the efficiency and accuracy of detection. The related research mainly focuses on acoustic analysis and machine vision. Acoustic analysis has been proven to be an effective method for detecting cracked eggs [3]. Li Sun et al. built an equivalent mechanical model based on an automatic excitation device and analyzed the transient impact. They used cross-correlation analysis and Bayesian classification to detect eggshell cracks, which could reach a detection level of 97% [4]. P. Coucke et al. proposed exciting the eggs with a small impact hammer and extracting the spectral features of acoustic signals as the input vector of the classification algorithm, whose detection accuracy could reach 90% [5]. A non-destructive detection technique proposed by Cho et al. based on acoustic impulse response employed eight frequency domain indexes, such as the average area of the power spectrum, as input vectors, where multivariate discriminant analysis and multivariate regression analysis were used to establish a classification model. Its detection accuracy could reach 95% [6]. Deng et al. proposed a crack detection method based on a continuous wavelet transform and support vector machine (SVM). They integrated four wavelet features such as the first resonance scale and achieved a detection accuracy of 98.9% [7]. Sun et al., based on acoustic resonance, analyzed the difference in the frequency response signals between intact eggs and cracked eggs, extracted five excitation resonance frequency characteristics, such as the spectral peak, as input vectors, and achieved a detection accuracy of 96.11% [8]. Lai et al. measured the acoustic signals of intact duck eggs and cracked duck eggs, and they selected the five most significant frequency features as the input vectors in logistic regression analysis. The overall detection accuracy could reach 87.6% [9]. Wang et al. developed an excitation device driven by solenoids, producing sound signals by striking the eggs. The time domain and frequency domain features of 12 kinds of sound signals were extracted while using a neural network with feature dimension reduction as the classifier. The detection accuracy was as high as 99.2% in the training set, but it was slightly lower—about 95%—in the test set [10]. The detection approach for cracked eggs based on acoustic signals is very effective, and its accuracy can basically meet the requirements for application. However, this approach is susceptible to such factors as an uneven eggshell thickness, surrounding noise, the egg shape, crack position, striking angle, and so on. Moreover, a lack of stability or knocking with too much strength may cause secondary damage to the eggs [11,12]. In addition, the detection of egg cracks based on machine vision has also made great progress [13,14]. Elster et al. first applied machine vision technology to eggshell crack detection and could find the cracked eggs among the samples with an accuracy of 95.6%. However, it took 25.3 s to identify a cracked egg, which was relatively slow [15]. The cracked egg detection system designed by Gooddrum et al. was able to adjust the speed of the rotating device according to the size of the egg. Three images were taken for each egg at an interval of 120° near the equator, with a recognition accuracy of up to 90% [16]. Li et al. proposed a vacuum pressure chamber-based detection system which imposed a vacuum pressure of 18 kPa to enlarge the microcrack and reached a detection accuracy of 100% under a situation without stains [17]. However, at present, this method is only in the laboratory stage, so it cannot be used in actual industrial production lines. Wang et al. proposed a method for detecting cracks in eggs using multi-information fusion of a natural light image and polarization image with an accuracy of 94% [18]. Bao et al. aimed at the phenomenon of dark spots on the surface of eggshells under backlight conditions, adopted a negative LOG operator for image enhancement, set a threshold to eliminate black spots, and finally detected cracks through the LFI index, with the detection accuracy being up to 92.5% [19]. Muammer proposed a machine vision detection system based on deep learning, in which six images of egg surfaces were captured in the process of the continuous rotation of an egg, and the depth features were extracted by a pretrained residual network in parallel and then input into the BiLSTM network to carry out the detection of cracked eggs. The accuracy rate was up to 99.17% under experimental conditions [20]. Most of the above research methods were implemented in a laboratory environment, and in industrial production, the identification accuracy will be greatly affected due to the difference in the size, shape, depth, and other features of eggshell cracks, as well as such factors as the shooting angle and bright spots and stains on the shell. Therefore, it is still an urgent problem to study a new method that can meet the requirements of automatic detection of eggshell cracks in industrial production.

High-voltage leak detection(HVLD) is a type of common and mature non-destructive defect detection technology which is mainly used in the pharmaceutical industry and the food industry, with the advantages of high speed and high precision. In this method, electrodes are usually linked at both ends of a container with a non-flammable conducting liquid, and a high voltage is applied. Equivalent capacitance is generated between the liquid and the electrode due to the bottle wall. When the container is intact, there is a tiny current in the circuit, but if the container has cracks and leaks, the capacitance disappears and causes a discharge between the electrodes. Therefore, the discharge analysis technology can be used to detect the tiny cracks and damages of the container with high precision. Yoon S. Song et al. studied the important role the key variables play in the detection of tiny pinhole leakages in flexible bags and semi-rigid cups using high-voltage technology. They applied 0.25–10 kV to the packaging, and the results showed that the HVLD technology could detect defects of less than 10 μm [21]. Moll et al. studied and verified the use of high voltage to detect the defects of blow-fill-seal containers with an accuracy of 100% [22]. Sun Jun et al. studied a non-destructive method to identify egg varieties based on their dielectric properties. They used parallel plates to measure the dielectric properties of eggs at 10∼200 kHz and established a fast identification and classification model of egg varieties by using the SVM algorithm, which met the requirements of classification well [23]. However, as far as we know, no researchers have applied the relevant research methods and ideas to the detection of cracks in poultry eggs.

Therefore, this research is aimed at proposing a novel and convenient nondestructive method to detect cracks in poultry eggs in the industrial field. The main objectives of the study are as follows:

- Focused on the analysis of the physical properties of the eggshell, study the electric field characteristics inside and outside the eggs under the action of electrodes and establish the dynamic capacitance model and electrical breakdown model of an egg innovatively;
- Design a microcrack detection system based on discharge analysis, with which microcracks on an eggshell can be detected by analyzing the weak current changes in the circuit;
- Analyze and compare the crack detection algorithms and feature selection of eggs;
- Verify the universality and generalization of the proposed method.

2. Electrical Characteristics of Poultry Eggs

2.1. Physical Characteristics of Poultry Eggs

A complete poultry egg is composed of an eggshell, egg membrane, egg white, yolk, air chamber, etc. as shown in Figure 1a. The main component of the eggshell is calcium carbonate, which accounts for about 11% of the volume of the whole egg. They are hard and play an important role in protecting the egg white and yolk, exchanging gases with the outside world, and providing minerals for embryonic development. Its structure is shown in Figure 1b [24]. The common eggshell includes three layers with slightly different microstructures, and its radial cross section is shown in Figure 1c [25]. The outermost part of the shell is dense, meticulous, and has a certain strength, and thus it is called the cuticle. The middle layer is spongy and densely covered with many small holes, while the innermost layer, called the papillary layer, is pyramidal, and the spaces between the layers can hold air. On the surface of the eggshell lie pores of about 30 microns in diameter. These are called stomata, through which gas exchange and water evaporation occur.

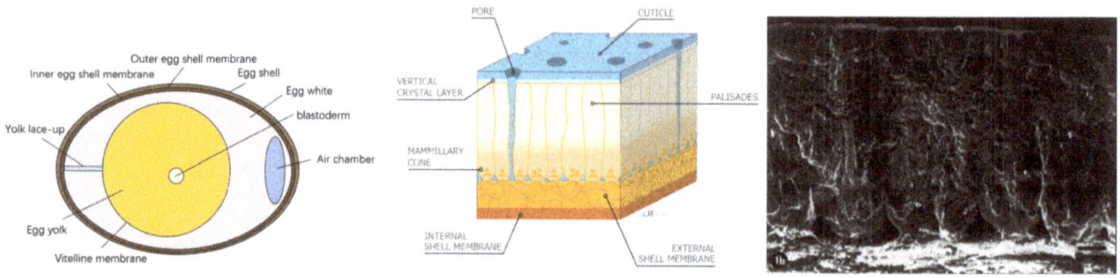

Figure 1. Schematic diagram of egg and eggshell structure. (**a**) Structure of the egg. (**b**) Microstructure of the eggshell. (**c**) Radial cross section of the eggshell.

2.2. Model of Electrical Characteristics of Poultry Eggs

A dielectric in the electric field produces an equivalent bound charge on the atomic scale under the electric field force, and this phenomenon is called dielectric polarization. For an eggshell, when the electric field intensity exceeds a certain value, the bound charge is forced to flow, causing dielectric breakdown and losing its insulation. Therefore, it is very important for the detection of the eggshell cracks to calculate the electrostatic fields of eggs and analyze the current change in the circuit. For this reason, we designed a dynamic detection method for cracks. The microcurrent will be generated at the crack of an eggshell when the egg rotates dynamically in the detection device, which is jointly generated by two models that will be discussed below: one is the electrical breakdown, and the other is capacitance jump. The total current is as follows:

$$I = I_1 + I_2 \tag{1}$$

where I_1 is the microcurrent generated by electrical breakdown and I_2 is the microcurrent generated by the capacitance jump.

2.2.1. Model of Capacitance of a Poultry Egg

An electrostatic field with the medium is produced jointly by the bound charge and free charge. In order to represent the electric field, which is under the joint action of both charges, another field vector–electric flux density $\overrightarrow{D}$, also known as electric displacement, is introduced, which is defined in Table 1, where $\overrightarrow{E}$ is the electric field intensity, $\overrightarrow{P}$ is the electric polarization intensity, and ε_0 is the vacuum dielectric constant.

Table 1. Formula table.

Formula Name	Formula
The field vector–electric flux density	$\overrightarrow{D} = \varepsilon_0 \overrightarrow{E} + \overrightarrow{P}$
The total spatial electrostatic field	$\overrightarrow{E} = \overrightarrow{E_0} + \overrightarrow{E'}$
The electric polarization intensity	$\overrightarrow{P} = \varepsilon_0 \mathcal{X}_e \overrightarrow{E}$
The Gauss theorem in the medium	$\oint_S \overrightarrow{D} \cdot \overrightarrow{S} = \sum q$

As shown in Figure 2, when there are poultry eggs in the electric field, the properties of the spatial electrostatic field are related to the free charge (q_0) and the distribution of the dielectric. The macroscopic electrical properties of the dielectric can be replaced by a polarized charge (q'), and then the total spatial electrostatic field consists of $\overrightarrow{E_0}$ and $\overrightarrow{E'}$, as shown in Table 1. Here, $\overrightarrow{E_0}$ represents the applied electric field formed by a free charge, and $\overrightarrow{E'}$ represents the electrolyte polarization electric field formed by a polarized charge.

In a linear isotropic dielectric, the electric polarization intensity $\overrightarrow{P}$ is defined as $\varepsilon_0 \mathcal{X}_e \overrightarrow{E}$, which can be seen in Table 1, where $\mathcal{X}_e$ is the electric polarizability rate. Therefore, we have

$$\overrightarrow{D} = \varepsilon_0(1+\mathcal{X}_e)\overrightarrow{E} = \varepsilon_0\varepsilon_r\overrightarrow{E} \quad (2)$$

In the above formula, $\varepsilon_r = (1+\mathcal{X}_e)$ stands for relative permittivity, which is a physical parameter characterizing the dielectricity or polarization of dielectric materials, also known as relative permittivity. After the electric displacement vector $\overrightarrow{D}$ is obtained, the Gauss theorem in the medium can be formulated, which is defined in Table 1, where $\overrightarrow{S}$ denotes any closed surface in the medium and q denotes a free charge.

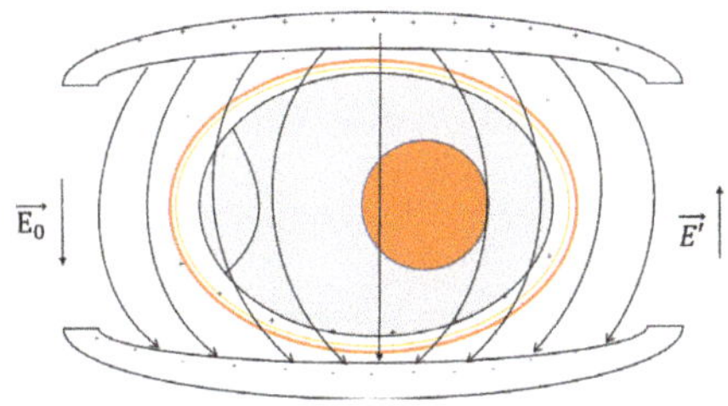

Figure 2. Schematic diagram of surface polarization of eggs in electric field.

We can think of the two electrodes and the egg in the middle as one capacitor, as shown in Figure 3a, where the eggshell is an insulator and the egg liquid is approximately a conductor due to a low resistance value. In an equilibrium state, there is no current in the circuit. The egg liquid has a certain conductivity, so the dielectric constant ε_L of the egg liquid is large. If the egg liquid is approximated as a good conductor, according to the position of the upper and lower electrodes and the poor conductivity of the eggshell, the electrical characteristic model under this connection mode can be approximated as the series of two plate capacitors, as is shown in Figure 3b, and then the electric field distribution under the intact eggshell is $U = E_1 d_1 + E_2 d_2$. Therefore, according to the plate capacitance formula, the equivalent capacitance C_1 is ($d \ll L, d \ll W$, where L is the length of the electrode and W is the the width of the electrode):

$$C_1 = \frac{4\pi\varepsilon_l\varepsilon_r LW}{d_1 + d_2} \quad (3)$$

where d_1 and d_2 are the thickness of the upper and lower layers of eggshell, respectively.

text

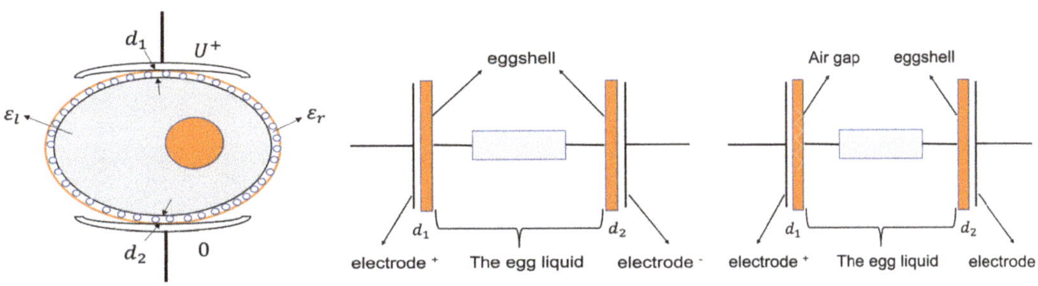

Figure 3. Capacitance system diagram. (**a**) Schematic diagram of the capacitor system, composed of the electrode and egg body. (**b**) Schematic diagram of equivalent capacitance of system when the electrode is not at the crack. (**c**) Schematic diagram of equivalent capacitance of the system when electrode is at the crack.

When a crack exists in an eggshell, the electrical characteristics model of the egg change as shown in Figure 3c, and then

$$U = U_1 + U_2 = \frac{Q}{C_1} + \frac{Q}{C_2} \tag{4}$$

$$U_1 = \frac{Q}{C_1} = \frac{UC_1C_2}{C_1 + C_2} = \frac{UC_2}{C_1 + C_2} = \frac{U\frac{4\pi\varepsilon_l\varepsilon_r S}{d_2}}{\frac{4\pi\varepsilon_l S}{d_1} + \frac{4\pi\varepsilon_l\varepsilon_r S}{d_2}} = \frac{d_1\varepsilon_r}{d_2 + \varepsilon_r d_1}U \tag{5}$$

The electric field at a crack can be defined as

$$E_1' = \frac{U_1}{d_1} = \frac{\varepsilon_r}{d_2 + \varepsilon_r d_1}U \tag{6}$$

When the air breakdown electric field is $E_{1p}' = 30$ KV/cm, and $d_1 = d_2 \approx d = 350$ μm, then the breakdown voltage U_p is

$$U_p = \frac{d_2 + \varepsilon_r d_1}{\varepsilon_r}E_{1p}' \approx dE_{1p}' = 3.5 \times 10^{-4} \times 3 \times 10^4 \times 10^2 = 1050 \text{ V} \tag{7}$$

At this time, the plate capacitance C_2 is

$$C_2 = \frac{4\pi\varepsilon_l\varepsilon_r LW}{\varepsilon_r d_1 + d_2} \tag{8}$$

The experimental results show that if there is no crack in the eggshell of the egg rotating in the middle of two electrodes, the equivalent capacitance value would stay basically stable at C_1 in the whole process. However, if there is a crack in the eggshell, the equivalent capacitance will jump between C_1 and C_2 when the electrode passes the cracks of the rotating egg, resulting in a transient current. Setting the egg rotation as an angular velocity of α, the time to rotate the width of W is $\frac{W}{\alpha R}$, where R is the radius of the egg. Therefore, when the egg rotates from a no crack zone to a crack zone, the current generated is

$$\begin{aligned} I &= \frac{\Delta Q}{\Delta t} = \frac{U\Delta C}{\frac{W}{\alpha R}} = \frac{U\alpha R}{W} \cdot 4\pi\varepsilon_l\varepsilon_r LW\left(\frac{1}{d_1 + d_2} - \frac{1}{\varepsilon_r d_1 + d_2}\right) \\ &= U\alpha R \cdot 4\pi\varepsilon_l\varepsilon_r L \cdot \frac{(\varepsilon_r - 1)d_1}{(d_1 + d_2)(\varepsilon_r d_1 + d_2)} \end{aligned} \tag{9}$$

If $d_1 \approx d_2 = d$, then

$$I \approx \frac{U\alpha R \cdot 4\pi\varepsilon_l\varepsilon_r^2 L}{2(\varepsilon_l + 1)d} \tag{10}$$

The following data were obtained in the experiment: the angular velocity was 2 cycles/SEC, $\alpha = 4\pi$, the radius of the shell $R = 3 \times 10^{-2}$ m, $\varepsilon_l = 8.85 \times 10^{-12}$, the C_aCO_3 dielectric constant of the eggshell $\varepsilon_r \approx 8.8$, the length of the electrode $L = 4 \times 10^{-2}$ m, and the shell thickness was 350 μm. Then, we have

$$I \approx \frac{4\pi \times 3 \times 10^{-2} \times 4\pi \times 8.85 \times 10^{-12} \times 8.8^2 \times 4 \times 10^{-2} \times U}{2 \times (8.8 + 1) \times 3.5 \times 10^{-4}} = 1.894 \times 10^{-8} \times U \tag{11}$$

where when $U = 1500$ V, $I \approx 28.4$ μA.

Figure 4 shows the current curves collected when detecting intact eggs and cracked eggs under the above electrode shapes and experimental parameters. The blue line represents intact eggs, and the red line represents cracked eggs. It is clear that there was a peak in the data for cracked eggs. In the online detection system, the detected current value may be the microcurrent generated by a capacitance jump or microcurrent superposed with that produced in the electric breakdown.

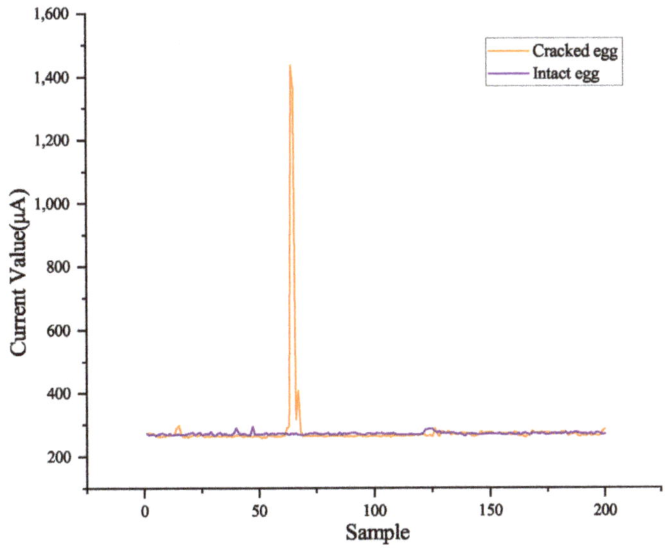

Figure 4. Comparison diagram of current measurement curves without cracks or cracked eggs under discharge electric field.

2.2.2. Electric Breakdown Model of Poultry Eggs

According to the basic principle of electric breakdown, if the voltage applied to an insulator is increased, the number of charge carriers in the material will increase sharply under a certain electric field, and its resistivity will decrease, resulting in producing a strong current. For poultry eggs, an intact one is not conductive under normal conditions, but when there is a crack in the eggshell, an air interlayer with low insulation may occur in the eggshell. Because the breakdown voltage of the air dielectric is much less than that of a solid dielectric, when high voltage is applied on both sides of the egg body, an egg with cracks is more likely to cause electrical breakdown, and there will be a significant difference in the current.

Since the width of the crack is much smaller than the size of the eggshell or the electrode, it can be approximated that the electric field in the crack area is uniform. The gap breakdown voltage is subject to Paschen's law when the air pressure is below 1 standard atmosphere (about 0.1 mpa):

$$V = f(pd) \tag{12}$$

where p is the air pressure and d is the distance between the electrodes.

The breakdown voltage U_b can be calculated according to the empirical formula:

$$U_b = \frac{B_{pd}}{\ln(\frac{A_{pd}}{\ln\frac{1}{\gamma}})} \tag{13}$$

where γ is the ionization coefficient and A and B are constants related to the composition of the air. At standard atmosphere pressure, $A = 43.66$ and $B = 12.8$.

For a static, intact egg, a sudden change in current occurs when solid dielectric breakdown occurs. The breakdown voltage of a solid dielectric is much higher than that of an air dielectric, so if we keep the voltage at both electrodes stable and only allow air dielectric breakdown, we can identify cracked eggs according to the change in current signals. Therefore, the key to the problem is to apply a stable electric field at the crack that can break down the air but not the eggshell. This problem is solved by analyzing the electrode shape and simulation experiments under different voltages. As is shown in Figure 5, there was a tiny crack in the Z direction on top of the egg. U-shaped linear electrodes were applied to the upper and lower sides of the egg to wrap the eggshell to the maximum extent and make the electric field uniform. By adjusting the electrode shape and voltage, the current detection system was optimized in the simulation environment and verified by experiments in the real scene.

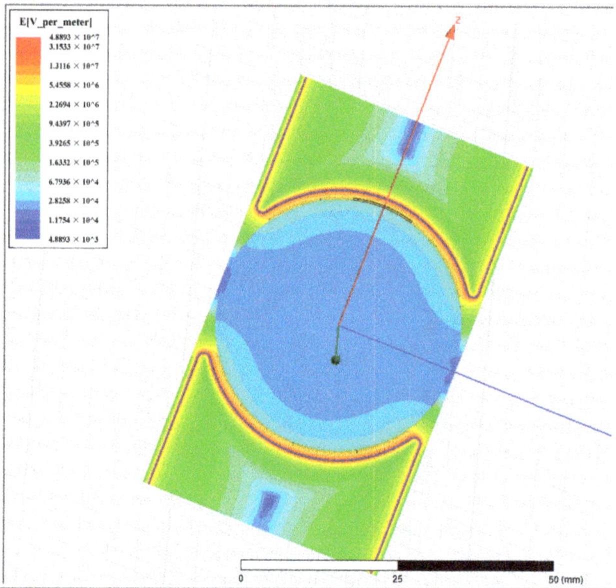

Figure 5. Simulation analysis of egg electric field distribution with cracks.

At the same time, the conditions of the air in the crack gap, such as the temperature, humidity, and other factors, will affect the ionization tendency of the air and correspondingly affect the breakdown voltage or discharge voltage in the crack gap. When the temperature decreases, the density of the air increases, the mean free path of free electrons in the air is shortened, and it is not easy to cause collision ionization, thus causing the breakdown voltage of the air to increase. As an electronegative gas, water vapor easily captures free electrons and transforms them into negative ions when the humidity of the air increases, which weakens the ionization and decreases the breakdown voltage of the air. Given the potential influence of the high temperature and humidity in the egg production line, special attention should be paid to these factors in the process of the analysis and experiment.

In short, the final current value is usually the superposition value of the current generated by the above two cases. When the electrodes are passing the cracked area of the rotating egg, if the detection voltage is less than the breakdown voltage threshold, the total current in Equation (1) is mainly I_2; otherwise, the total current is mainly I_1.

3. System Design and Analysis Methods

3.1. Design of the Detection System

The experimental platform for poultry egg crack detection mainly consisted of five parts: a detection platform, high-voltage power supply, controller, data acquisition circuit, and industrial personal computer, as is shown in Figure 6. The detection platform was composed of a rotating mechanism, discharge electrodes, electrode adjustment mechanism, and other parts, as is shown in Figure 7a. To ensure perfect contact between the electrode and the surface of the poultry eggshell, the upper electrode was made flexible and egg-like and 10 cm wide, and it had four layers of conductive silica gel with different lengths stacked on top of each other. The lower electrode was initially designed to imitate an egg as well, but that led to uneven contact due to the different sizes of the eggs. The lower electrode was later made into a long bar shape, but this shape still did not work because the exposed part of the electrode outside the fixed seat was too short (2 mm) and required the lower electrode to reach for it flexibly, which gave the egg an upward support force and made it difficult to rotate. After a large number of experiments, we found that when the lower electrode took an arc convex shape with little contact, it provided a stable and reliable contact bottom without affecting the rotation. The rotating mechanism included three parts: a servo drive, saddle-shaped support rollers, and an upper spring roller. The servo drive provides a stable driving force to drive rollers on the left and right of the eggs and ensure that eggs of different sizes can rotate evenly without shifting, while the upper spring roller presses the egg to ensure that the eggs can still rotate evenly in place when they come into contact with the electrode and generate friction. The electrode adjustment mechanism can adjust the electrode position according to the egg so as to adapt to different egg sizes, ensure that the electrode fits the egg surface better, and thus provide stable and reliable surface contact. The data acquisition circuit used an STM32F103 microcomputer and 16-bit A/D converter as the core, and the maximum sampling frequency was 12 MHz, which could meet the requirements of the sampling speed and accuracy. The industrial personal computer was used to record and process the current sampling data. Through the analysis and processing of the current signals, it could identify whether there was a crack in the eggshell and then drive the automatic device to remove the cracked egg. The experimental device is shown in Figure 7b.

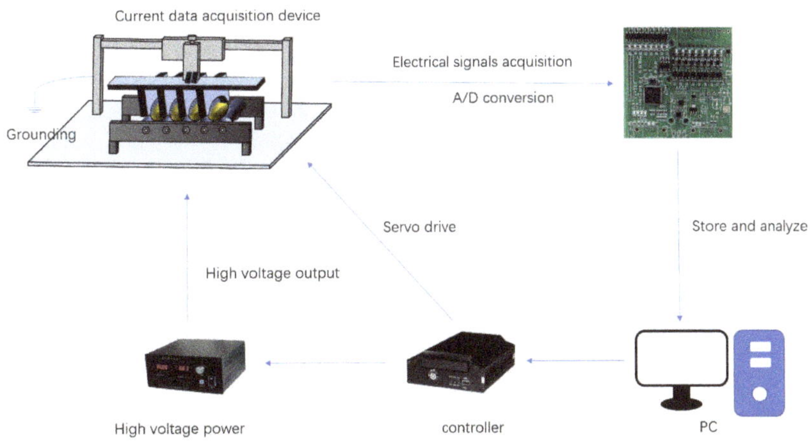

Figure 6. Data acquisition system block diagram.

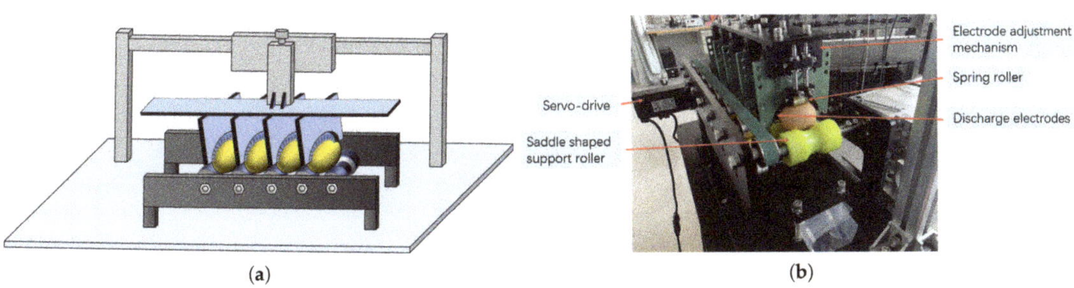

Figure 7. Egg crack detection device. (**a**) Model diagram. (**b**) Physical map.

3.2. Electrode Shape Design

The eggshell is composed of a large amount of calcium carbonate containing tiny pores, and it does not conduct electricity under normal conditions. The inner membrane of the eggshell is a network of organic fibers made of keratin, which together with the egg liquid is a conductor and can conduct electricity under normal conditions. The pores are small in diameter and evenly dispersed. They usually have long and curved air paths extending through the shell toward the inside of the egg, while cracks are characterized by short air paths that extend horizontally on the shell and are concentrated along the crack. Therefore, it is notable to distinguish the pores and cracks in the design of the discharge electrode and ensure that the electric field in the egg body area is uniform. The effective area of detection is another aspect to note. The detection area covered by the electrodes in this paper did not include the tip and blunt end, and only the equatorial part of the egg and the central area between the two ends were covered for crack detection. Moreover, missing out on detection due to gaps between the electrode pieces may have occurred. All these factors mentioned above added difficulty to the design, and they should be carefully dealt with in the design of the electrode.

According to the analysis in Section 2, the charge density is proportional to the curvature of the electrode tip, which means the tip electrode is most likely to produce high-voltage and discharge phenomena. We selected six eggs randomly, made holes at the blunt ends of the eggs, and poured out the inside liquid before we tested the discharge voltages under smooth electrodes, single-tip electrodes, and multi-tip electrodes, as shown in Figure 8. The experimental data shown in Table 2 show that the smooth electrode discharge voltage matched with the polar plate discharge, and the single-tip electrode discharge voltage was slightly higher than that of the smooth electrode. As for the single-tip electrode, it may be difficult to align one end with the other end, which causes the breakdown voltage to increase. On the other hand, this may be because the energy is excessively concentrated in the tip and cannot form a large air column breakdown. The discharge voltage of the multi-tip electrode was close to that of the smooth electrode, which indicates that the multiple tips could reduce the breakdown voltage. Problems were still found in the experiment, such as an increased electrode distance and fewer actual effective tips. The tip electrode had the smallest coverage area on the eggshell surface. When it was in a crack-free area, it could only cover a few pores. When it was in a cracked area, the area ratio of the covered air area changed significantly, so it could effectively distinguish cracks and pores and had a high detection ability. However, the point-shaped tip electrode could only detect eggshells in a very small area near the electrode at one time, and the detection efficiency was low. The spatial electric field generated by the tip electrode was also unevenly distributed, which led to an unstable detection accuracy. Therefore, it is not an ideal electrode shape.

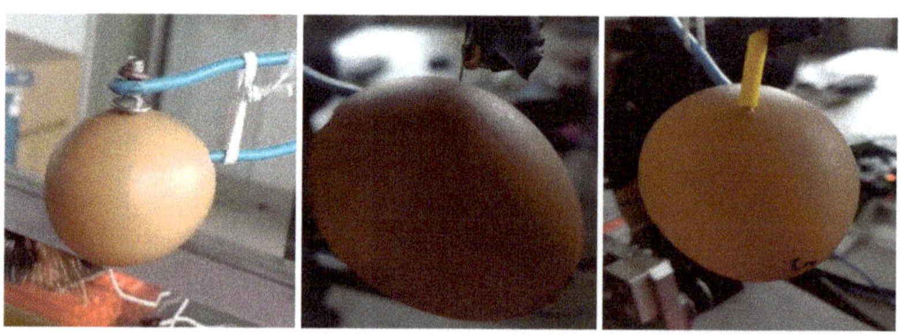

(**a**) (**b**) (**c**)

Figure 8. Tip electrode experimental set-up. (**a**) Smooth electrode. (**b**) Tip electrode. (**c**) Multi-strand tip electrode.

Table 2. Experimental data of tip electrode.

Serial Number	Smooth Electrode Discharge Voltage (V)	Tip Electrode Discharge Voltage (V)	Multi-Strand Tip Electrode Discharge Voltage (V)
No.1	1800	2300	1800
No.2	1700	2500	1900
No.3	1800	2400	1600
No.4	1400	2000	1500
No.5	1400	2200	2100
No.6	1700	2400	1700

Conversely to the point electrodes, planar electrodes offer significant advantages in terms of detection efficiency and spatial distribution of the electric field. However, the shape and size of the egg body vary greatly, and it is difficult to make a flexible electrode that perfectly fits the surface of the egg. The accumulated value of the current generated by too many pores in the non-cracked eggs under the electrode was also close to the current value generated by the cracked egg, resulting in a significant decrease in the detection accuracy, so the planar electrode is also not an ideal shape for electrodes.

The linear electrode combines the advantages of the above two electrodes. It is better in spatial electric field uniformity, more efficient in detection, and more accurate in identification. In addition, the line contact of the conductive material, which can contain the outline of the egg and fit the surface of the eggshell, is an ideal form of contact.

3.3. Electrode Material Analysis

We selected conductive silica gel, conductive rubber, and a conductive brush as the electrode materials for the experiments and found that the egg cracks could be identified with all three materials. The resistivity of the conductive rubber was large, and the current change was not obvious enough when it was used as an electrode. When a conductive brush was used as the electrode, the conductive brushes would fuse after discharge and cause a great loss of electrode material. In contrast, the resistivity of the conductive silica gel was small and could produce an obvious current change when passing the cracked area. Therefore, conductive silica gel was selected as the electrode material in this paper.

To sum up, the current is not only related to the resistivity of electrode materials but also closely related to the contact area of the conductive materials. However, it is not a case of "the larger the better" for the contact area, as too large a contact area will lead to a large current for non-cracked eggs. The more ideal form of contact is line contact, which is made according to the outline of the egg so as to fit the eggshell perfectly. The actual structure of the electrode is shown in Figure 9.

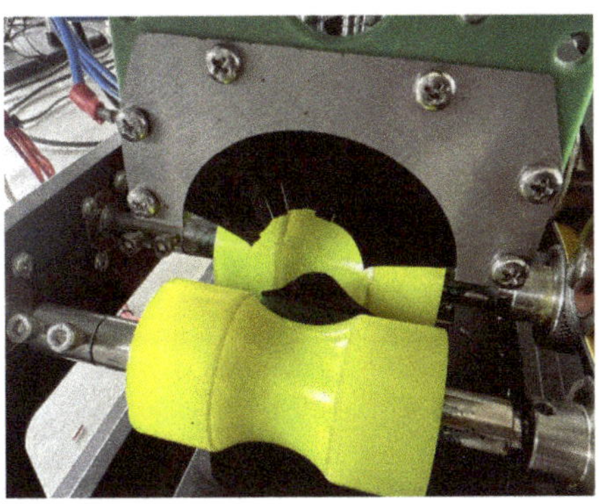

Figure 9. Real figure of electrode.

3.4. The Importance of Multi-Layer Flexible Electrodes

Since eggs vary somewhat in size and shape, the design of a flexible electrode can better fit the eggshell and achieve full coverage of an effective detection area by dynamically adjusting the angle according to the eggs. Although a single-layer flexible electrode can effectively detect cracks, their coverage area is limited. When detecting larger eggs, gaps between the electrode strips may cause omissions during the egg rotation if the cracks are just perpendicular to the gaps. The use of multi-layer flexible electrodes can reduce the chances of missed detection of egg cracks, which plays a significant role in improving the overall detection accuracy and can also further reduce the detection voltage.

3.5. Lab Environment

We selected 10 eggs randomly and put 5 eggs in a group to test the electrical characteristics under different humidity environments. The mean current curve is shown in Figure 10. The experiment found that the measured current value in the environment with a humidity of 72% and voltage of 1500 V was equivalent to that in the environment with a humidity of 54% and voltage of 1800 V, which further proved the conclusion of Section 2 that the detection of egg cracks based on current signals was greatly affected by environmental humidity. Therefore, during the data collection, the humidity and temperature of the experimental environment should be stabilized within a certain range to reduce the influence of the environment on the experimental data.

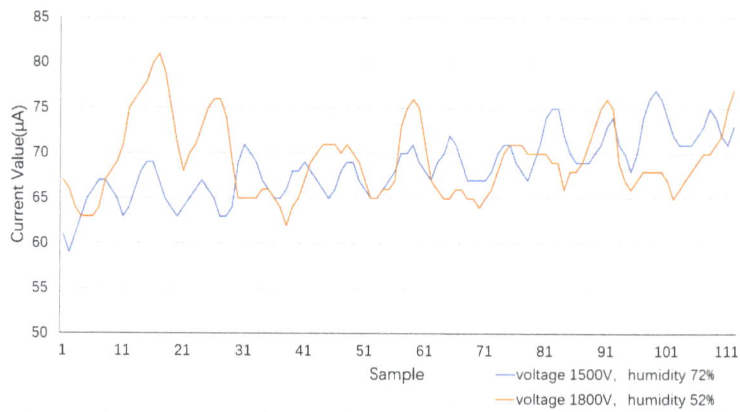

Figure 10. Standard deviation of the current signal of eggs at different voltages.

The voltage value used in HVLD is generally high, even reaching up to tens of thousands of volts at certain times. If it is directly used for the detection of egg cracks, the protein may be denatured. In order to avoid this, we had to choose an appropriate voltage range. All things considered, we finally determined that the experimental environment was perfect at an average temperature of 18.5 °C, a relative humidity of 40%, and a voltage of 1500 V. We strictly controlled the current size, and the system current protection mechanism would be triggered to cut off the power when the current was greater than 1 mA so it would not cause damage to the eggs.

In order to quickly obtain a sufficient number of egg samples with microcracks and avoid the instability of manual striking, we designed an egg crack striking machine to control the size of the artificial cracks and prevent the egg contents from leaking. The machine is shown in Figure 11a. Eggs are fixed at the bottom of the track, and the rollers are released from different heights and strike the egg at the equatorial part to generate controllable microcracks. The width of the artificial microcracks is generally less than 3 microns, which is usually not easy to observe with the human eye. Microcracks are mainly located in the central area between two ends of the egg and only present in the effective detection area. Egg samples with cracks at the tip or blunt end will be discarded. In actual production, there are not only large cracks caused by strong striking but also a large number of microcracks of several microns, which are difficult to detect by traditional methods. The structure of a microcracked egg under an industrial microscope is shown in Figure 11b,c.

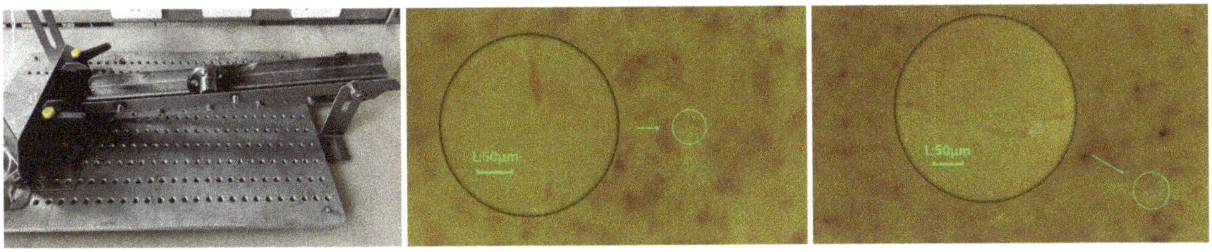

(a) (b) (c)

Figure 11. Egg crack and its generating device. (**a**) Egg crack striking machine. (**b**,**c**) Pictures of cracks of different sizes under the industrial microscope.

3.6. Classification Modeling Methods

3.6.1. Linear Discriminant Analysis

Linear discriminant analysis (LDA) [26] is widely used in the field of high-dimensional data classification as a supervised dimensionality reduction technology. It takes the separability of pattern data as the goal and finds a set of optimal discriminant vectors, which maximizes the between-class scatter measures while minimizing the within-class scatter measures. In this study, the eggshells could be divided into intact eggs and cracked eggs. This was a classification problem. Letting C be the number of categories, where C = 2, x is the n-dimensional features of the training sample, and N is the number of samples, the sample's within-class scatter matrix S_W and between-class scatter matrix S_B are shown below in Equations (14) and (15), respectively:

$$S_W = \frac{1}{N}\sum_{i=1}^{C}\sum_{x\in c_i}(x-\mu_i)(x-\mu_i)^T \tag{14}$$

$$S_B = \sum_{i=1}^{C} p_i(\mu_i-\mu)(\mu_i-\mu)^T \tag{15}$$

where $p_i = N_i/N$ is the prior probability of each class, N_i is the number of training samples of class $C_i(i = 1, 2, \ldots, C)$, μ_i is the mean value of sample C_i, and μ is the mean of all samples.

The goal of LDA is to find the best projection matrix W so that the Fisher criterion is the largest, and its formula is

$$J(W_{opt}) = \arg\max_{W} \frac{|W^T S_b W|}{|W^T S_W W|} \tag{16}$$

3.6.2. K-Means Classification Algorithm

K-means [27] is a common unsupervised learning algorithm that is often used to discover the inherent regularities between datasets. The principle is that K samples are first randomly selected as cluster centers of K categories, and then, the Euclidean distance between the sample data and the k-th centroid is calculated to judge the correlation with this category. Then, it belongs to the category with the highest correlation. Such centroids will also be recalculated with the addition of new samples until the iteration is completed or the preset number of iterations is reached. The Euclidean distance between samples is

$$D(x_i, x_j) = \sqrt{\sum_{n=1}^{N} (x_{i,n} - x_{j,n})^2} \tag{17}$$

where D_{x_i,x_j} is the Euclidean distance between samples x_i and x_j and N is the dimension of the sample data. x_i represents the i-th sample data, and x_j represents the j-th sample data. If the sample has C categories, C_k is used to represent the k-th cluster center, where $k = 1, 2, \ldots, K$. First, K points in the sample are selected as centroids, followed by calculating the similarity between other points and the cluster center points and dividing them into K sets, denoted by C_k. Finally, the new cluster center is recalculated. The formula for C_k is

$$C_k = \frac{1}{m_k} \sum_{x \in C_k} x_k \tag{18}$$

where m_k is the number of k-th category elements. During this process, the K-means clustering algorithm continuously reclassifies and updates the cluster centers, and this ends when the iteration reaches the maximum limit or the objective function is smaller than the threshold. Its objective function is

$$J = \sum_{i=1}^{K} \sum_{x_j \in C_i} D_{x_i,x_j}(x_i, C_k) \tag{19}$$

3.6.3. SVM

A support vector machine (SVM) is based on statistical learning and can solve linear and nonlinear problems at the same time. It shows good performance [28,29], especially in small-sample data when applied in a series of challenging practical problems. The basic idea of SVM is to find the optimal hyperplane that distinguishes the two classes by training the sample set and maximizing the distance between the segmentation plane or hyperplane and the data points in the given dataset.

The current signal obtained in this paper was not linearly separable, so it was necessary to first select an appropriate kernel function to map it to a high-dimensional space and then optimize it. Up to now, there has been no generally accepted selection criterion for the selection of the kernel function. The commonly used kernel functions mainly include Gaussian kernel function, polynomial kernel function, linear kernel function, and sigmoid kernel function. Owing to its advantages of few parameters and fast convergence speed, Gaussian kernel function was used for kernel transformation in this paper. Its mathematical definition is shown in Equation (20) [30]:

$$K(x, y) = e^{-\frac{\|x-y\|^2}{2\sigma^2}} \tag{20}$$

where x and y are the eigenvectors of the current signal.

3.6.4. CART Decision Tree

A decision tree [31] is a supervised machine learning algorithm that can be used to classify or predict unknown objects. The construction of the decision tree is a process of top-down and recursive branching. First, we selected the most effective division method for the samples according to the features, formed a new decision branch, and then pruned the branch to optimize the decision tree. Commonly used decision tree generation algorithms mainly include ID3, C4.5, and CART. We employed the CART model in this study and used the GINI index to select the optimal division points of the optimal features. The basic principle is to form a decision tree structure in the form of a binary tree by cyclic analysis of the training dataset and select the attribute that minimizes the GINI index value of the child nodes as the classification scheme.

3.6.5. Random Forest

A random forest [32] uses a decision tree as the base classifier. It improves the overfitting problem of a decision tree by combining the bagging ensemble learning theory and random subspace method. Based on the idea of multiple decision trees, the random forest generates the training data of each tree by random extraction from the original dataset and then randomly extracts n features from N feature variables before finally selecting the optimal feature variables from these n features as split features to construct multiple decision trees. Finally, each of the decision trees gives a class prediction, and the class with the most votes becomes the model's prediction.

4. Experiments and Results

4.1. Data Acquisition

We purchased 770 eggs at a farmer's market near the laboratory and collected current signals for model training and algorithm verification, including 367 intact eggs and 403 cracked eggs. To avoid the noise introduced by stains on the eggshells, which may have affected the experiment, the cleaning and drying process in the actual egg factory was simulated before data acquisition. As for the impact of cleaning on the test results, we came to the conclusion after small-scale experiments that cleaning could remove the stains on the surface of the eggshell and reduce the interference with the current signal acquisition. Meanwhile, the water molecules during cleaning could wet a part of the crack gaps that were generated and had been blocked for a long time, which contributed to the conductivity of the cracks.

At the initial stage of data acquisition, each egg was used only once for the current signal, which resulted in a lot of waste. In order to improve the utilization rate of the sample eggs and efficiency of data acquisition, the eggs that were detected to be intact would be used again as cracked eggs after being slightly cracked by our crack striking machine. The physical and experimental parameters of the tested eggs are shown in Table 3.

Table 3. Physical and experimental parameters of tested eggs.

	Long Axis Average	Short Axis Average	Weight Average	Voltage	Frequency	Number of Sampling Points	Average Humidity	Average Temperature
Eggs	57.4 mm	44.5 mm	62.7 g	1500 V	100 Hz	450	40% RH	18.5 °C

4.2. Extraction of Data Features

As shown in Figure 12, the current signals of eggs with different sizes, which included three small ones and three large ones, were found to fluctuate significantly. The current signals collected in the experiment were mixed with noise and were easily affected by the environment, reducing the classification accuracy. Therefore, we introduced six common time domain features, three frequency domain features, and wavelet packet coefficients to extract stable and comprehensive feature information from the current signals for the classification models. The six time domain features were the weighted mean, average, standard deviation, range, skewness, kurtosis, and their expressions are listed in Table 4. In the six expressions given in Table 4, x_i $(i = 1,2,\ldots,N)$ is the current data, N is the length of the data, and w is the coefficient. The three frequency domain features were the frequency of the center of gravity, root mean square frequency, and standard deviation of the frequency, and their expressions are described in Table 5. In the three expressions given in Table 5, f is the frequency value and $P(f)$ is the power spectrum.

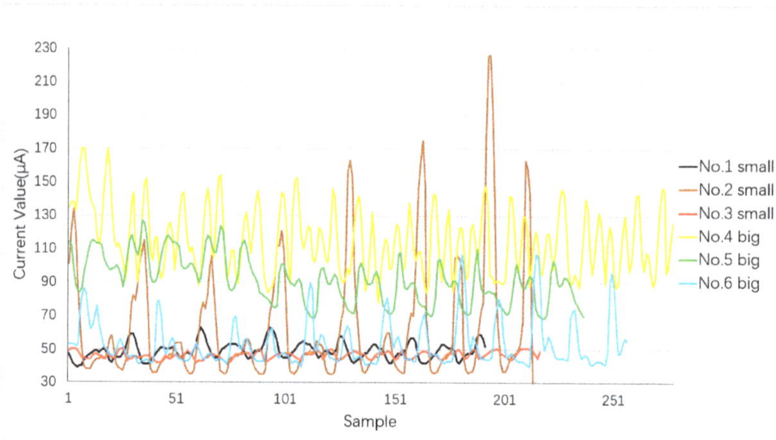

Figure 12. The effect of egg size on current signal.

Table 4. Time domain features.

Time Domain Features	Formula
Weighted mean	$\bar{x} = \frac{\sum_{i=1}^{n} x_i w_i}{\sum_{i=1}^{n} w_i}$
Average	$\mu = \frac{1}{N}\sum_{i=1}^{N} x_i$
Standard deviation	$\sigma = \sqrt{\frac{\sum_{i=1}^{N}(x_i-\mu)^2}{N}}$
Range	$r = max(x) - min(x)$
Skewness	$s = \frac{1}{n}\sum_{i=1}^{n}[(\frac{x_i-\mu}{\sigma})^3]$
Kurtosis	$k = \frac{1}{n}\sum_{i=1}^{n}[(\frac{x_i-\mu}{\sigma})^4]$

Table 5. Frequency domain features.

Frequency Domain Features	Formula
Frequency of center of gravity	$FC = \frac{\int_0^{+\infty} fP(f)df}{\int_0^{+\infty} P(f)df}$
Root mean square frequency	$RMSF = \sqrt{\frac{\int_0^{+\infty} f^2P(f)df}{\int_0^{+\infty} P(f)df}}$
Standard deviation of frequency	$RVF = \sqrt{\frac{\int_0^{+\infty} (f-FC)^2P(f)df}{\int_0^{+\infty} P(f)df}}$

4.3. Analysis of the Results

In the process of acquiring an egg's current signal, there are various discharge phenomena, such as corona discharge, small air gap breakdown, and creeping discharge, which make the current signal mix with a lot of noise. The interference of noise plus the relatively weak current signal at the microcrack cause the current signal to be submerged in the noise. To solve this, the method of wavelet threshold denoising was adopted to remove the high-frequency noise in the signal while retaining the useful high signals. The wavelet threshold denoising was such that, due to the continuity of the real signal $f(t)$, after the discrete wavelet transform, the wavelet coefficients generated at different scales were large, while the wavelet coefficients produced by a corresponding noise signal $e(t)$ were small. Therefore, noise can be effectively suppressed by first selecting appropriate thresholds on different scales to process high-frequency wavelet coefficients, and then performing an inverse wavelet transform on the signal can effectively suppress noise. It is noteworthy that the selection of a wavelet base is of great significance to the effect of wavelet threshold denoising. By analyzing the shape of the current signal at the crack position, the Sym2 wavelet base was finally selected, and it had better symmetry, which could, to a certain extent, reduce the phase distortion when analyzing and reconstructing the signal.

The current signals of two intact eggs and two cracked eggs were randomly selected from the dataset, as shown in Figure 13, where blue represents the signal before denoising and red represents the signal after denoising. The following can be observed from Figure 13: (1) The current signal of the cracked eggs had an evident peak within one cycle, while that of the intact eggs did not. As mentioned in Section 2, when the experimental voltage is smaller than the breakdown voltage, the change in the current curve is mainly dominated by the capacitance jump during the rotation. The experimental voltage in this paper was higher than the breakdown voltage, so the change in the current curve was mainly dominated by the electrical breakdown at the crack. When the crack was small, the experimental voltage may not have reached the breakdown voltage, and the change in the current curve may have also been dominated by a capacitance jump. In addition, we also designed the circuit protection function, where the system would automatically cut off the circuit to protect the safety of the equipment and eggs when the current exceeded the set threshold. (2) The jitter of the current curve was relatively smooth due to the small changes in capacitance of the intact eggs. However, the two wave shapes of the intact eggs were not exactly identical and even had big differences, which may have been related to the different roughnesses of the eggshells.

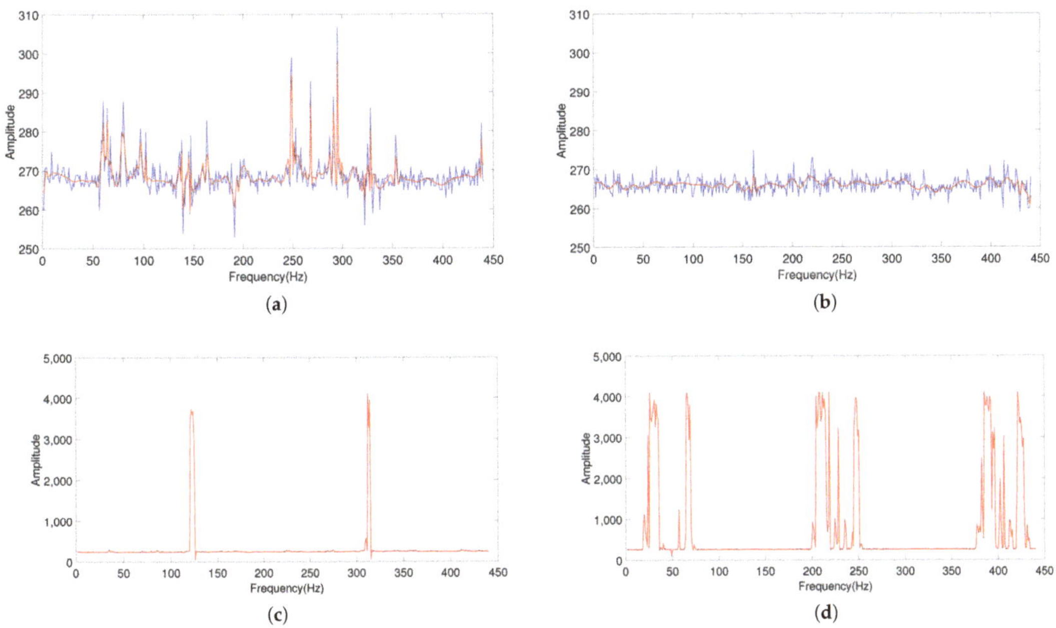

Figure 13. Egg current waveform. (**a**,**b**) Waveforms of intact eggs. (**c**,**d**) Waveforms of cracked eggs.

After the wavelet threshold denoising, the time domain, frequency domain, and wavelet packet coefficients of the current signal were extracted. It can be seen from Figures 14 and 15 that most of the features of the intact eggs and cracked eggs had obvious differences, but some of the differences were not obvious.

We put the time domain, frequency domain, and wavelet packet coefficient features into the SVM model. The experimental results showed that the recognition rate of each feature was different and that the eggs incorrectly recognized by different features were also not the same. This indicates that features in different domains had different classification effects. Therefore, this paper used the multi-domain features to fully reflect the inherent characteristics of the original current signal so as to improve the detection accuracy.

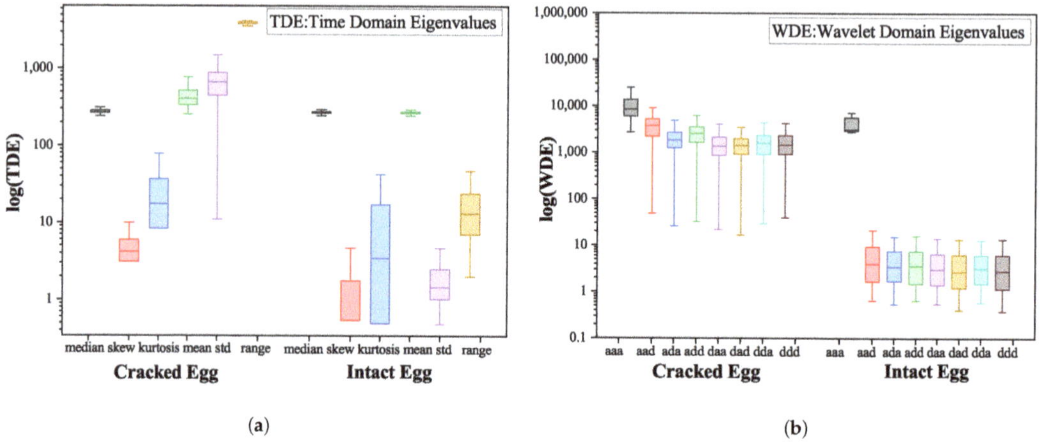

Figure 14. Feature distribution diagram. (**a**) Time domain features. (**b**) Wavelet domain features.

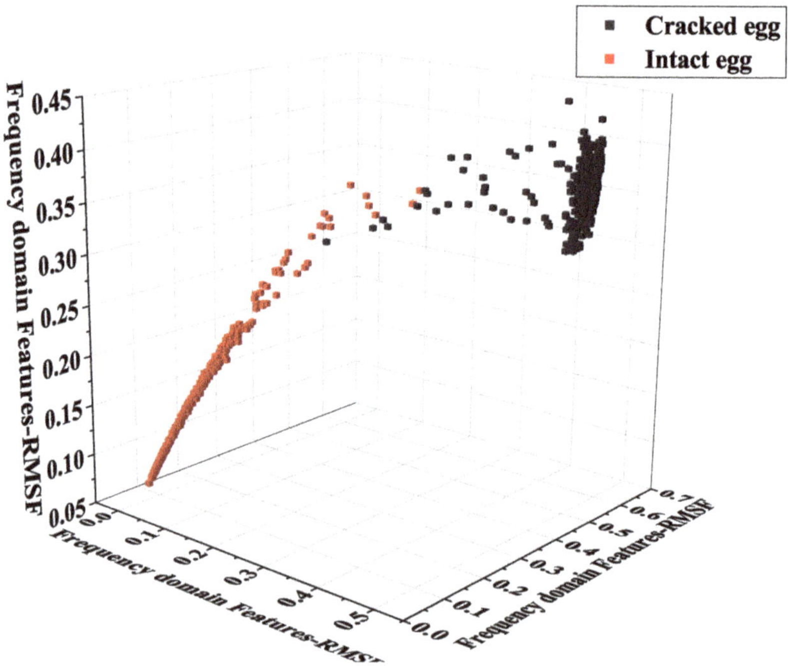

Figure 15. Three-dimensional distribution diagram of frequency domain features.

Finally, we adopted a variety of machine learning methods such as K-means clustering, linear discrimination analysis, and a support vector machine, as mentioned in Section 3.6, for pattern classification, and performance measures such as accuracy, precision, and the recall rate were calculated from the testing data. The experimental results are shown in Table 6.

Table 6. Combination feature classification effect in time domain, frequency domain, and wavelet domain.

	Accuracy	Precision	Recall	F1	AUC
SVM	98.79%	98.27%	99.48%	98.87%	98.75%
LDA	99.31%	99.47%	99.21%	99.34%	99.31%
DT	99.35%	99.29%	99.47%	99.38%	99.36%
KM	99.05%	97.45%	98.73%	99.08%	99.09%
RF	99.44%	99.68%	99.51%	99.59%	99.43%

The following conclusions can be drawn from the experimental results:

1. By selecting a suitable wavelet base for wavelet denoising, the noise in the raw current signal could be effectively suppressed, and thereby, the classification accuracy was improved;
2. By combining the features in various transform domains, more informative and discriminative features could be obtained.

5. Discussion

This paper studied the electric field characteristics of eggs under the action of electrodes on the basis of analyzing the physical properties of the eggshell and established two discharge models. The high-precision detection of eggshell cracks was realized by designing an egg crack detection platform, comparing machine learning classification algorithms, and analysis of the current signal. The most important element of this study is proposing a novel method for crack detection in eggshells based on discharge analysis. The vision-based method has higher requirements for the light source and image processing technology, and the acoustic method has higher requirements for the percussion equipment and environmental noise. However, the method in this paper has high precision, stable results, and less dependence on the environment. It only needs to control the humidity, voltage, and a few other experimental conditions. This section will further discuss the electrical characteristics of poultry eggs and explore the universality and generalization of the method proposed in this paper.

It is worth noting that the classification accuracy did not change significantly under different machine learning methods, which proves that the features extracted based on the current signals were stable. Therefore, the current-based crack detection method is feasible and can be used in actual production, with accuracy rates as high as 99%. In addition, for misclassified eggs, by analyzing the position, condition, and corresponding current signal of the cracks, we found the following problems. Although the cracks were distributed in the effective detection area between the tip and the blunt end, they were blocked by spilled egg liquid and dust due to a long storage time. Therefore, it should be possible to further improve the classification accuracy by improving the design of the brushes.

In addition, we conducted further studies on the electrical properties of the eggs. We randomly selected 10 eggs as samples and recorded the current signals at applied voltages of 800 V, 1000 V, 1200 V, and 1400 V. According to whether there was an obvious discharge that could be directly observed and heard, the eggs could be divided into discharged eggs and undischarged eggs. The current signals of the two kinds of eggs are shown in Figures 16 and 17. Figure 18 compares the current signals of both the discharged and undischarged eggs in the same coordinate system. After analysis, it can be seen that the higher the discharge voltage, the larger the dynamic current of the egg would be. However, the voltage increases would also amplify the current fluctuation, which also indirectly proves that the high voltage will cause breakdown in the eggs. In addition, not all eggs in the discharged samples had cracks, which means it is not reliable for directly identifying whether the eggs had cracks when only using the current signal, and it is very necessary to conduct data analysis on the current signal.

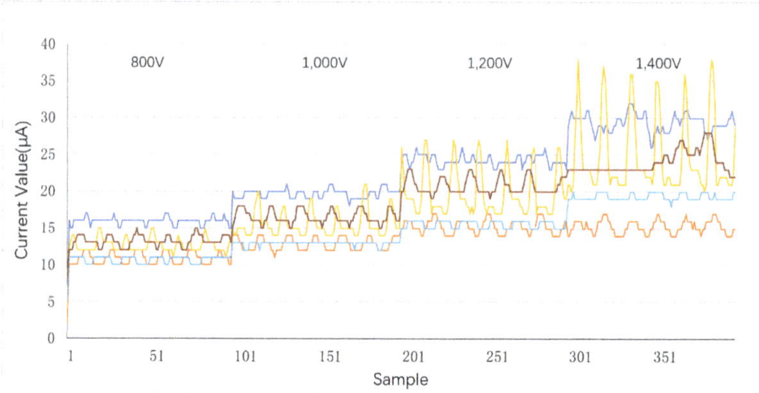

Figure 16. Current signal when the egg had no discharge phenomenon under different voltages.

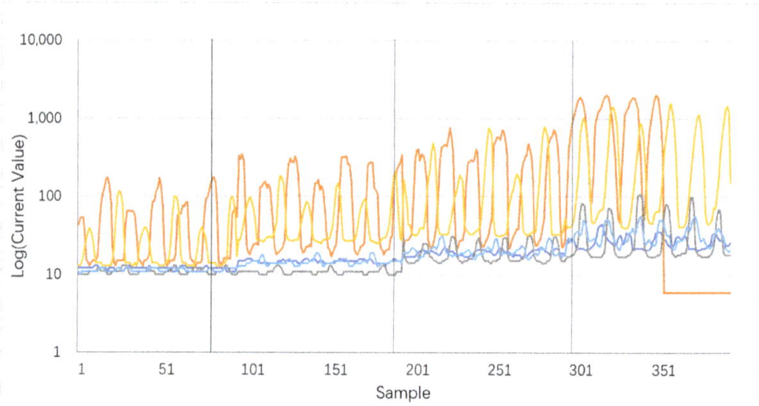

Figure 17. Current signal when the egg produced the discharge phenomenon at different voltages.

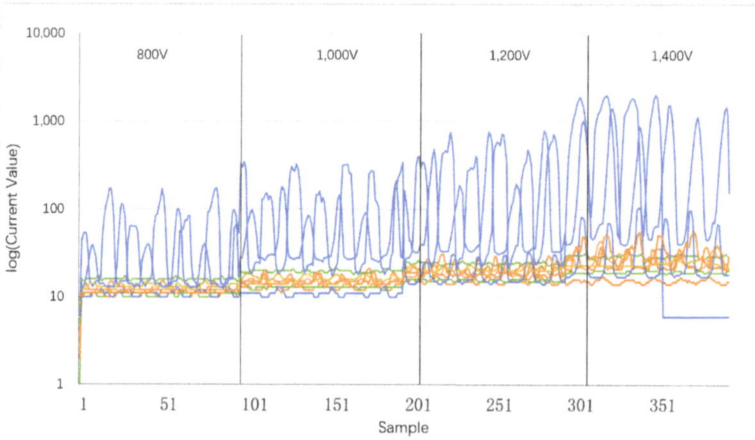

Figure 18. Current signal of the eggs in the voltage range of 800–1400 V. The current signals of 3 eggs with obvious cracks are set to blue, the current signals of 2 eggs with no cracks but obvious discharge are set to green, and the current of the eggs without discharge signal is set to orange.

Crack detection technology based on electrical characteristics is a new research direction for the quality inspection of agricultural products in the future which has great research value and market potential. The method proposed in this paper can not only detect cracks in eggs but also achieve high-precision detection of cracks in duck eggs, among others. It is a universal and generalizable method. We purchased 267 fresh duck eggs from the Dabao Breeding Duck Incubation Base in Xintai Tianbao Town for current signal acquisition, including 130 intact duck eggs and 137 cracked duck eggs. The physical and experimental parameters of the tested duck eggs are shown in Table 7. Based on the analysis in Section 4.3, after the wavelet denoising, the time domain, frequency domain, and wavelet packet coefficient features of the current signal of the duck eggs were extracted and combined, and we selected the RF classifier for training. The results are shown in Table 8. For the duck eggs, the accuracy of the model was slightly reduced but still within a higher accuracy range. We speculate that there are two main reasons for the slight fluctuation of the evaluation index: (1) The number of duck eggs used in verification was quite different from that of the number of eggs.Therefore, according to the equations for the precision rate and recall rate, it can be known that, when the overall base is low, misclassification usually leads to a greater reduction in relevant indicators. (2) Eggs are usually laid in industrialized chicken houses, where the environment is relatively dry and hygienic. While ducks are typical waterfowl, they usually live outdoors and in water, which also leads to a relatively humid and dark environment for duck eggs, and the cracks are easily blocked by impurities such as dust. Although we simulated the cleaning process of the egg factory before testing, the impurities that had been blocked for a long time had solidified, and it was difficult for water molecules to enter the small cracks to wet the blocked substance during flushing, so the conductivity at the cracks would decrease and cause them to be missed during the inspection.

Table 7. Physical and experimental parameters of tested duck eggs.

	Long Axis Average	Short Axis Average	Weight Average	Voltage	Frequency	Number of Sampling Points	Average Humidity	Average Temperature
Duck eggs	67.4 mm	50.3 mm	68.5 g	1500 V	100 Hz	450	55% RH	15 °C

Table 8. Detection results of cracked duck eggs.

	Accuracy	Precision	Recall	F1	AUC
RF	98.16%	98.41%	97.74%	98.04%	98.28%

6. Conclusions

In this study, we established the egg electrical characteristics model and designed a microcrack detection system that has higher accuracy and is more convenient than the traditional methods. Different types of features extracted from the time, frequency, and wavelet domains of the current signals were proven to contain a mass of crack characteristics after reducing the interference of noise in the signal with the sym2 wavelet. Based on the above features, five typical machine learning algorithms were used to divide the eggs into cracked eggs and intact eggs, which verified the proposed model. The experimental results show that the RF had better robustness, and the fusion of multi-domain features can effectively improve the accuracy of classification. It is worth noting that the classification accuracy by different machine learning methods had little variation, with all being around 99%, proving that the model of detecting microcracks by using current signal features has certain stability and reliability. The relevant experiments of duck eggs also confirmed that the method proposed in this paper has a certain universality and generalization. Our research will help relevant enterprises to quickly and accurately detect cracked eggs in the production line, greatly reduce the number of cracked eggs in the end products, improve the quality of related products, and have good practical application prospects. In general, this paper explored a new method for nondestructive testing for egg cracks which lays a foundation for the development of nondestructive testing of egg cracks based on an electrical characteristics model.

Author Contributions: Conceptualization, C.S. and C.Z. (Changsheng Zhu); methodology, C.S., C.Z. (Changsheng Zhu) and Y.W.; software, Y.W., Y.C., B.J. and C.Z. (Changsheng Zhu); validation, C.S., C.Z. (Changsheng Zhu), C.Z. (Chun Zhang) and Y.W.; formal analysis, C.S., C.Z. (Chun Zhang) and J.Y.; investigation, C.S. and J.Y.; writing—original draft preparation, C.S., C.Z. (Changsheng Zhu), Y.W., Y.C. and B.J.; writing—review and editing, C.S., C.Z. (Chun Zhang), C.Z. (Changsheng Zhu), Y.W. and J.Y. All authors have read and agreed to the published version of the manuscript.

Funding: This research was funded by the Tai'an Science and Technology Innovation Development Plan (No.2021GX050 and No.2020GX055).

Institutional Review Board Statement: Not applicable.

Informed Consent Statement: Not applicable.

Data Availability Statement: The data presented in this study are available on demand from the corresponding author at (cs.zhu@sdust.edu.cn).

Conflicts of Interest: The authors declare no conflict of interest.

References

1. Priyadumkol, J.; Kittichaikarn, C.; Thainimit, S. Crack detection on unwashed eggs using image processing. *J. Food Eng.* **2017**, *209*, 76–82. [CrossRef]
2. Xie, C. Vibration Characteristic Analysis of Good Eggs and Cracked Eggs. Ph.D. Thesis, Huazhong Agricultural University, Wuhan, China, 2015.
3. Kertész, I.; Zsom-Muha, V.; András, R.; Horváth, F.; Németh, C.; Felföldi, J. Development of a Novel Acoustic Spectroscopy Method for Detection of Eggshell Cracks. *Molecules* **2021**, *26*, 4693. [CrossRef] [PubMed]
4. Sun, L.; Zhang, P.; Feng, S.; Qiang, M.; Cai, J. Eggshell crack detection based on the transient impact analysis and cross-correlation method. *Curr. Res. Food Sci.* **2021**, *4*, 716–723. [CrossRef] [PubMed]
5. Coucke, P.; Ketelaere, B.D.; Baerdemaeker, J.D. Detection of Eggshell Cracks by Means of Acoustic Frequency Analysis. *IFAC Proc. Vol.* **1997**, *30*, 163–168. [CrossRef]
6. Cho, H.K.; Choi, W.K.; Paek, J.H. Detection of surface cracks in shell eggs by acoustic impulse method. *Trans. Am. Soc. Agric. Eng.* **2000**, *43*, 1921–1926. [CrossRef]
7. Deng, X.; Wang, Q.; Chen, H.; Xie, H. Eggshell crack detection using a wavelet-based support vector machine. *Comput. Electron. Agric.* **2010**, *70*, 135–143. [CrossRef]
8. Sun, L.; Kun Bi, X.; Lin, H.; Wen Zhao, J.; Rong Cai, J. On-line detection of eggshell crack based on acoustic resonance analysis. *J. Food Eng.* **2013**, *116*, 240–245. [CrossRef]
9. Lai, C.C.; Li, C.H.; Huang, K.J.; Cheng, C.W. Duck Eggshell Crack Detection by Nondestructive Sonic Measurement and Analysis. *Sensors* **2021**, *21*, 7299. [CrossRef]
10. Wang, H.; Mao, J.; Zhang, J.; Jiang, H.; Wang, J. Acoustic feature extraction and optimization of crack detection for eggshell. *J. Food Eng.* **2016**, *171*, 240–247. [CrossRef]
11. Wang, J.; Jiang, R.; Yu, Y. Relationship between dynamic resonance frequency and egg physical properties. *Food Res. Int.* **2004**, *37*, 45–50. [CrossRef]
12. Pan, L.; Tu, K.; Zhao, L.; Zhao, Y.; Pan, X. Preliminary research of chicken egg crack detection based on acoustic resonance analysis. *Trans. Chin. Soc. Agric. Eng.* **2005**, *21*, 11–15.
13. Botta, B.; Gattam, S.S.R.; Datta, A.K. Eggshell crack detection using deep convolutional neural networks. *J. Food Eng.* **2022**, *315*, 110798. [CrossRef]
14. Yao, K.; Sun, J.; Chen, C.; Xu, M.; Zhou, X.; Cao, Y.; Tian, Y. Non-destructive detection of egg qualities based on hyperspectral imaging. *J. Food Eng.* **2022**, *325*, 111024. [CrossRef]
15. Elster, R.T.; Goodrum, J.W. Detection of cracks in eggs using machine vision. *Trans. ASAE* **1991**, *34*, 0307–0312. [CrossRef]
16. Goodrum, J.W.; Elster, R.T. Machine vision for crack detection in rotating eggs. *Trans. ASABE* **1992**, *35*, 1323–1328. [CrossRef]
17. Li, Y.; Dhakal, S.; Peng, Y. A machine vision system for identification of micro-crack in egg shell. *J. Food Eng.* **2012**, *109*, 127–134. [CrossRef]
18. Fang, W.; Shu, Z.; Zuojun, T. Non-destructive crack detection of preserved eggs using a machine vision and multivariate analysis. *Wuhan Univ. J. Nat. Sci.* **2017**, *22*, 257–262.
19. Guanjun, B.; Mimi, J.; Yi, X.; Shibo, C.; Qinghua, Y. Cracked egg recognition based on machine vision. *Comput. Electron. Agric.* **2019**, *158*, 159–166. [CrossRef]

20. Turkoglu, M. Defective egg detection based on deep features and Bidirectional Long-Short-Term-Memory. *Comput. Electron. Agric.* **2021**, *185*, 6152. [CrossRef]
21. Song, Y.S.; Gera, M.; Jain, B.; Koontz, J.L. Evaluation of a Non-destructive High-voltage Technique for the Detection of Pinhole Leaks in Flexible and Semi-rigid Packages for Foods. *Packag. Technol. Sci.* **2014**, *27*, 423–436. [CrossRef]
22. Möll, F.; Doyle, D.L.; Haerer, M.; Guazzo, D.M. Validation of a High Voltage Leak Detector for Use With Pharmaceutical Blow-Fill-Seal Containers—A Practical Approach. *PDA J. Pharm. Sci. Technol.* **1998**, *52*, 215–227.
23. Jun, S.; Bin, L.; Hanping, M.; Xiaohong, W.; Hongyan, G.; Ning, Y. Non-Destructive Identification of Different Egg Varieties Based on Dielectric Properties. *Food Sci.* **2017**, *38*, 282–286.
24. Maxwell, T.H. The eggshell: Structure, composition and mineralization. *Front. Biosci.* **2012**, *17*, 1266–1280.
25. Parsons, A. Structure of the eggshell. *Poult. Sci.* **1982**, *61*, 2013–2021. [CrossRef]
26. Zhao, Y.; Wang, J.; Lu, Q.; Jiang, R. Pattern recognition of eggshell crack using PCA and LDA. *Innov. Food Sci. Emerg. Technol.* **2010**, *11*, 520–525. [CrossRef]
27. Wang, L.M.; Shao, Y.M. Crack Fault Classification for Planetary Gearbox Based on Feature Selection Technique and K-means Clustering Method. *Chin. J. Mech. Eng.* **2018**, *31*, 4. [CrossRef]
28. Aljarah, I.; Al-Zoubi, A.; Faris, H.; Hassonah, M.A.; Mirjalili, S.M.; Saadeh, H. Simultaneous Feature Selection and Support Vector Machine Optimization Using the Grasshopper Optimization Algorithm. *Cogn. Comput.* **2017**, *10*, 478–495. [CrossRef]
29. Mohammadi, E.; Khoshab, H.; Kazemnejad, A. Activities of daily living for patients with chronic heart failure: A partnership care model evaluation. *Appl. Nurs. Res.* **2016**, *30*, 261–267. [CrossRef] [PubMed]
30. Gunn, S.R. Support vector machines for classification and regression. *ISIS Tech. Rep.* **1998**, *14*, 5–16.
31. Zhang, L.; Wang, Z.; Wang, L.; Zhang, Z.; Chen, X.; Meng, L. Machine learning-based real-time visible fatigue crack growth detection. *Digit. Commun. Netw.* **2021**, *7*, 551–558. [CrossRef]
32. Ramzan, F.; Klees, S.; Schmitt, A.O.; Cavero, D.; Gültas, M. Identification of Age-Specific and Common Key Regulatory Mechanisms Governing Eggshell Strength in Chicken Using Random Forests. *Genes* **2020**, *11*, 464. [CrossRef]

agriculture

MDPI

Article

Research on Navigation Path Extraction and Obstacle Avoidance Strategy for Pusher Robot in Dairy Farm

Fuyang Tian [1,2], Xinwei Wang [2,3], Sufang Yu [4], Ruixue Wang [5], Zhanhua Song [1,2], Yinfa Yan [1,2], Fade Li [1,2], Zhonghua Wang [6] and Zhenwei Yu [1,2,*]

1 College of Mechanical and Electronic Engineering, Shandong Agricultural University, Tai'an 271018, China; fytian@sdau.edu.cn (F.T.); szh@sdau.edu.cn (Z.S.); yanyinfa@sdau.edu.cn (Y.Y.); lifade@sdau.edu.cn (F.L.)
2 Shandong Provincial Key Laboratory of Horticultural Machineries and Equipment, Tai'an 271018, China; 2021110433@sdau.edu.cn
3 Shandong Provincial Engineering Laboratory of Agricultural Equipment Intelligence, Tai'an 271018, China
4 College of Life Sciences, Shandong Agricultural University, Tai'an 271018, China; sfyu@sdau.edu.cn
5 Chinese Academy of Agricultural Mechanization Sciences, Beijing 100083, China; wangruixue@caams.org.cn
6 College of Animal Science and Technology, Shandong Agricultural University, Tai'an 271018, China; zhwang@sdau.edu.cn
* Correspondence: zhenweiyu615@sdau.edu.cn

Abstract: Existing push robots mainly use magnetic induction technology. These devices are susceptible to external electromagnetic interference and have a low degree of intelligence. To make up for the insufficiency of the existing material pushing robots, and at the same time solve the problems of labor-intensive, labor-intensive, and inability to push material in time at night, etc., in this study, an autonomous navigation pusher robot based on 3D lidar is designed, and an obstacle avoidance strategy based on the improved artificial potential field method is proposed. Firstly, the 3D point cloud data of the barn is collected by the self-designed pushing robot, the point cloud data of the area of interest is extracted using a direct-pass filtering algorithm, and the 3D point cloud of the barn is segmented using a height threshold. Secondly, the Least-Squares Method (LSM) and Random Sample Consensus (RANSAC) were used to extract fence lines, and then the boundary contour features were extracted by projection onto the ground. Finally, a target influence factor is added to the repulsive potential field function to determine the principle of optimal selection of the parameters of the improved artificial potential field method and the repulsive direction, and to clarify the optimal obstacle avoidance strategy for the pusher robot. It can verify the obstacle avoidance effect of the improved algorithm. The experimental results showed that under three different environments: no noise, Gaussian noise, and artificial noise, the fence lines were extracted using RANSAC. Taking the change in the slope as an indicator, the obtained results were about −0.058, 0.058, and −0.061, respectively. The slope obtained by the RANSAC method has less variation compared to the no-noise group. Compared with LSM, the extraction results did not change significantly, indicating that RANSAC has a certain resistance to various noises, but RANSAC performs better in extraction effect and real-time performance. The simulation and actual test results show that the improved artificial potential field method can select reasonable parameters and repulsive force directions. The optimized path increases the shortest distance of the obstacle point cloud from the navigation path from 0.18 to 0.41 m, where the average time is 0.059 s, and the standard deviation is 0.007 s. This shows that the optimization method can optimize the path in real time to avoid obstacles, basically meet the requirements of security and real-time performance, and effectively avoid the local minimum problem. This research will provide corresponding technical references for pusher robots to overcome the problems existing in the process of autonomous navigation and pushing operation in complex open scenarios.

Keywords: dairy farm; pusher robot; path extraction; obstacle avoidance

Citation: Tian, F.; Wang, X.; Yu, S.; Wang, R.; Song, Z.; Yan, Y.; Li, F.; Wang, Z.; Yu, Z. Research on Navigation Path Extraction and Obstacle Avoidance Strategy for Pusher Robot in Dairy Farm. *Agriculture* **2022**, *12*, 1008. https://doi.org/10.3390/agriculture12071008

Academic Editor: Tomas Norton

Received: 23 May 2022
Accepted: 6 July 2022
Published: 12 July 2022

1. Introduction

Dairy farming is an indispensable part of modern agriculture, which occupies a high proportion in the agricultural industry [1,2]. In China, the traditional technology of dairy farming is relatively backward, and most of them adopt the management mode of small-scale scattered farming, which is not conducive to the development of modern agriculture; as a result, the mode is changing toward the direction of large-scale, factory, and standardization. [3,4]. In the past few years, China's dairy industry has developed rapidly, and its output value and scale are at the forefront of the world [5]. According to the data released by the Ministry of Agriculture and Rural Affairs, China's milk production in 2020 was 35.3 million tons, an increase of 7% over 2015, on the other hand, the proportion of farming with more than 100 heads reached 67.2%, an increase of 18.9% compared with 2015. In this situation, it is undeniable that the dairy industry not only meets the residents' consumption demand for milk, but also increases the income of dairy farmers. In addition, it plays a key role in optimizing the rural industrial structure [6].

The continuous prosperity of the social economy makes the public put forward higher requirements for the quality of dairy products, which indirectly promotes the development of the dairy industry [7,8]. However, the rapid development has also exposed new problems, operators gradually found that the existing high-tech aquatic products could not meet the production needs. For example, in the process of cow feeding, part of the feed will be removed from the feeding area due to the cow's activity, resulting in accumulation, which will lead to the deterioration of uneaten feed in the long run. The current solution is to use manual or manual pushing equipment to push the accumulated feedback into the feeding area. In this situation, enterprises need to arrange more labor or equipment to promote feed [9]. Relying on manual labor will make it impossible for the farm to complete the feeding work in a timely and stable manner; as a result, the milk yield of the cow will be reduced. In this case, the robot used to push feed is very practical.

The accuracy and execution efficiency of multimedia target recognition technology have been greatly improved with the development of deep learning (DL) and machine learning, under the circumstances [10–12]; the application of the technology has been extended to the fields of medical imaging [13], video surveillance [14], and robot navigation [15]. In the wave of technological change, traditional agricultural machinery has ushered in a new opportunity for development. Agricultural robots such as feeding robots, transport robots, and picking robots have begun to apply DL and machine learning techniques [16–18]. Among them, the self-propelled robot has been favored by many scholars as a new research hotspot. Some researchers have studied the technical difficulties of navigating the path extraction of agricultural robots based on visual geometry inference and DL [19]. The classical methods to infer visual geometry include simultaneous localization, mapping, and motion structure. This kind of technology obtains parameter values through sensors such as optical detection and ranging (LiDAR), sound navigation and ranging, optical flow, and stereo and monocular cameras, and uses corresponding algorithms for obstacle avoidance and path planning [20]. Among similar sensors, Lidar has the advantages of high-ranging accuracy, good resolution, and a strong anti-jamming ability. It has been widely used in the perception and extraction of agricultural indoor environmental information, and has become a research hotspot for agricultural production robots [21]. In the research field of push robots, new technologies continue to emerge. DeLaval has developed an automatic mixing and pushing robot using magnetic induction technology, which can independently plan the walking route and speed, and is suitable for automatic mixing and the pushing of different types and quantities of feed. Pavkin et al. [22] concentrated on the simulation modeling of a feed pusher robot using Simulink tools in the Matlab environment to facilitate robot modernization or optimize the final cost for artificial testing of typical system elements and reduce production costs. However, the application of Lidar in the bullpen has not been reported, but the research on bullpen path extraction and obstacle avoidance based on Lidar and machine vision has a certain application value.

At present, the existing research at home and abroad has solved the problem of navigation path extraction in some agricultural scenarios, but the working environment of dairy farms was rarely mentioned. In this study, a new type of machine vision system was developed to fill this gap. The system will be used for extraction and tracking control of the working path of the pusher robot. Taking the cowshed environment as the research object, the self-designed pusher robot and 3D lidar were used to collect the point cloud data of the cowshed. The ground point cloud was removed by point cloud preprocessing, and the pass-through filtering algorithm extracted the point cloud data of the region of interest. Then, the least-squares method (LSM) and random sample consensus (RANSAC) were used to extract fence lines, project them and obtain boundary contour features, and extract fence lines and initial paths. At the same time, a robot navigation path optimization and obstacle avoidance method based on the improved artificial potential field method is proposed, which will provide corresponding technical references for pusher robots to overcome the problems existing in autonomous navigation and pushing operations in complex open scenarios. The system designed in this study could autonomously generate accurate navigation paths for robots in a dynamic farm environment, which will provide technical reference for autonomous navigation of farming robots and the development of precision animal husbandry.

This paper is organized as follows: Section 2 details the materials and methods employed to achieve the research objective. In Section 3, experimental results and discussion of the proposed technique are presented. Finally, in Section 4, the conclusion and future work is provided.

2. Materials and Methods

2.1. The Composition of the Pusher Robot System in the Farm

The pusher robot needs to replace the labor for the feeding process, thereby reducing the feed cost and labor intensity of feeding dairy cows. The pusher robot can meet the functions of autonomous walking and pushing. Therefore, the pusher robot was mainly composed of a vehicle navigation hardware system, pusher operation system, and navigation and operation control system. Among them, the vehicle navigation hardware system and the pushing operation system were the specific execution systems of the instructions, which were responsible for receiving and executing the task instructions issued by the control system to complete the autonomous navigation and pushing operation. The navigation and operation control system were responsible for setting the working mode of the vehicle system, issuing target point instructions, displaying the robot position in real time, and controlling the pushing operation system. Through the fusion and analysis of various sensor information, the pusher robot could realize autonomous positioning and navigation in the natural environment.

The hardware device and the control system communicated in real time via a wireless network to complete the autonomous navigation and operation tasks of the pusher robot on the farm together, as shown in Figure 1. The vehicle navigation hardware system mainly included a robot chassis, drive module, control module, environmental information perception module, communication module, and power supply module.

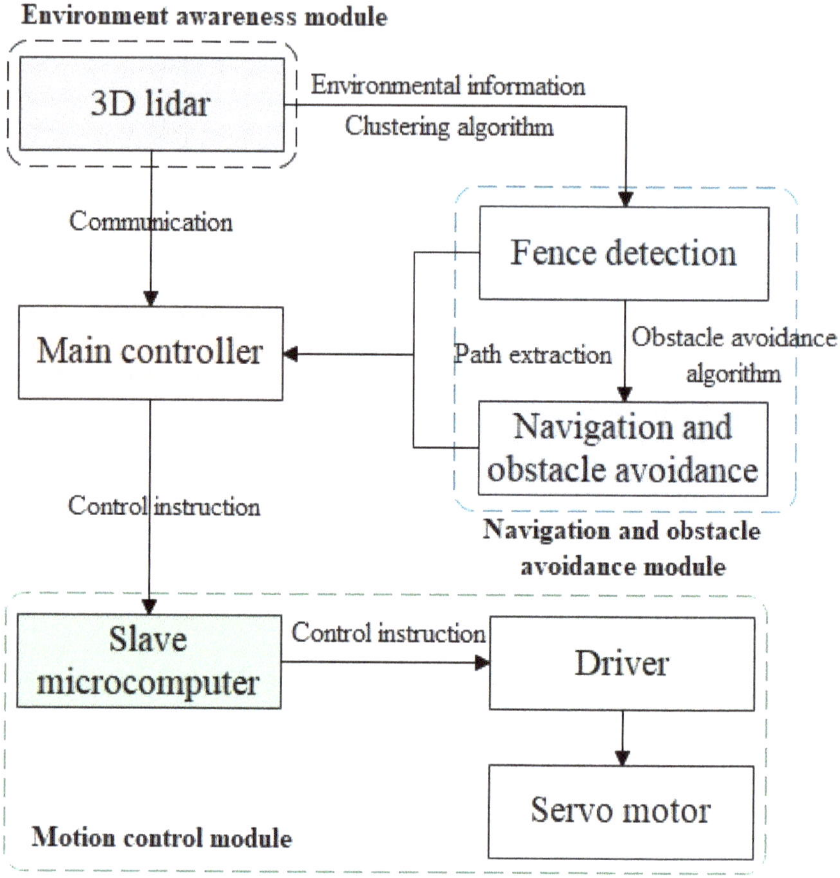

Figure 1. The vehicle navigation hardware system.

According to the task of the farm operation and the needs of the environment, the driving module of the robot adopted a two-wheel differential drive structure, and the steering control of the robot could be realized by setting different speeds for the two driving wheels. This drive system was not only simple in structure and small in turning radius, but also more flexible in movement, which greatly improved the control accuracy of the whole machine. The power system was provided by 60 V lithium battery modules. In order to ensure that the robot had powerful computing functions, the main control unit used Jetson Nano development board (NVIDIA, Shanghai, China), equipped with Tegra X1 heterogeneous SOC (NVIDIA, Shanghai, China), the size of this unit was 100 mm × 80 mm × 29 mm. The basic framework of ROS navigation was built under the Ubuntu 18.04 system, and information was exchanged with the chassis using RS-485 communication. The generated signal was transmitted to the main control unit via USB3.0. The car was equipped with the STM32F103 embedded motherboard (Haoyao, Shenzhen, China) as the underlying controller. According to the speed information provided by the encoder, the odometer data (moving speed, driving distance, and turning angle) of the vehicle system were obtained through kinematics calculation. Finally, the control of the vehicle-mounted system and the pushing operation system was completed through the control algorithm.

The environmental information perception module uses a 16-beam miniature LiDAR (RS-LiDAR-16, Sagitar Juchuang, Shenzhen Sagitar Juchuang Technology Co., Ltd., Shenzhen China). The compact housing of the RS-LiDAR-16, mounted with 16 laser/detector pairs, rapidly spins and sends out high-frequency laser pulses to continuously scan the surrounding environment, collecting real-time 3D point clouds. The 3D space point cloud data and object reflectivity are provided by the distance measurement algorithm, so that the pusher robot can digitally model the cowshed, providing a strong guarantee for its positioning, navigation, and obstacle avoidance. The lidar is installed in the center of the front end of the robot chassis, at a height of 0.6 m from the ground, and its performance parameters are shown in Table 1.

Table 1. Parameters of LiDAR.

Parameters	Values
Wavelength/nm	905
Detecting range/m	1–150
Accuracy/cm	±2
Vertical view/(°)	±15
Horizontal view/(°)	360
Vertical resolution/(°)	2.0
Horizontal resolution/(°)	0.1
Data rate/(pts·s^{-1})	6,000,000
Frame rate/Hz	10
Rotation speed/rpm	600

The 3D schematic diagram and physical map of the installation of each module of the pushing robot are shown in Figure 2. The overall length of the pusher robot is 1.78 m, the width is 1.15 m, the height is 1.40 m, and the rated load is 1 m^3.

Figure 2. The Pusher Robot.

2.2. *Collection and Preprocessing of Point Cloud Data*

2.2.1. Collection of Point Cloud Data

The 3D point cloud data of cowsheds were collected in Jinlan Dairy Farm (Figure 3) in Tai'an City, Shandong Province, China from 16 to 30 October 2021. The cowshed is arranged in a double row with a distance (D) of 6.25 m between the two pens. The point clouds of the cowshed are unevenly distributed, with dense clouds at the near end and gradually sparse ones at the far end (Figure 3b). As shown in Figure 3b, taking the geometric center of the lidar as the origin point O, the forward direction of the robot is the positive direction of the *Y*-axis, the vertical *Y*-axis to the left is the positive direction of the *X*-axis, and the *Z*-axis is determined by the right-hand rule to establish a 3D lidar local coordinate system. This study extracts the 3D point cloud data in the range of *X*-axis (0–10 m), *Y*-axis (−20 to 10 m), and *Z*-axis (0–2 m) from the coordinate system as the region of interest (Region of Interest, ROI). Feeding of dairy cows will lead to the cluttered distribution of some of the far-end feeds, and the collected point cloud data will be messier. Therefore, it is necessary to filter out the ground point cloud to reduce the interference of the ground point cloud data on the initial path extraction.

(**a**) (**b**)

Figure 3. Cattle farm 3D point cloud acquisition: (**a**) Experimental cattle farm, (**b**) 3D point cloud of the original cattle farm.

2.2.2. Preprocessing of Point Cloud Data

There are about 16,000 points in each frame of the collected 3D point cloud data of the cowshed, which is a huge amount of data. In order to reduce the amount of calculation, it is first necessary to preprocess the 3D point cloud data of the cowshed to remove noise and outliers [23,24]. Then, use the pass-through filtering algorithm to extract the ROI point cloud; the centroid of the cube is used to represent all points in the cube, and the voxel filtering algorithm downsamples the point cloud to greatly reduce the number of 3D point clouds while preserving the structural features of the 3D point cloud data. Therefore, this study uses a cube with a side length of 0.1 m to downsample the ROI point cloud. There are still many noise points and outliers in the filtered 3D point cloud, so statistical filtering is used to remove the noise and outliers [25]. In order to reduce the interference of the ground point cloud on the cowshed outline extraction, the ground plane fitting (GPF) algorithm proposed in the literature [26] is used to segment the ground and non-ground point clouds.

2.3. *Fence and Initial Path Extraction*

To make the segmented fence show a better effect, the preprocessed bullpen 3D point cloud was segmented by a high threshold method. The height threshold was determined according to the actual height of the cowshed and empirical methods, and the height threshold here was set to 0.1 m. The fence point cloud has apparent line features. The fence point cloud is projected onto the XY plane, and the fence lines are extracted by LSM and RANSAC, respectively, and the extraction effects of the two are compared. Project the fence point cloud on the XY plane, extract the boundary contour features of the fence point cloud, and calculate the navigation path of the pusher robot according to the fence's boundary outline to improve the mobile robot's accuracy in pushing grass during the operation.

2.3.1. The Least-Squares Method

LSM is a mathematical tool that has been widely used in many disciplines of data processing such as error estimation, system identification and prediction, and forecasting. It finds the best function parameters for point cloud data by minimizing the sum of squared errors. The basic principle is as follows: data $\{(x_i, y_i), i = 1, 2, \ldots, m\}$, obtain the data fitting function $\varphi(x)$. Then, the fitting function $\varphi(x)$ should reflect the changing trend of all data as much as possible, but it is not required to pass all data points; that is to say, there is a certain error between the fitting function $\varphi(x)$ and the actual measured data at x_i. Here, it is represented by ε_i:

$$\varepsilon_i = \varphi(x_i) - f(x_i)(i = 1, \ldots, n)$$

In order to meet the requirement that the fitting function curve can reflect the change trend of all data as much as possible, its 2-norm is required to be a minimum.

$$\| E \|_2 = \left\{ \sum_{i=1}^{n} [\varphi(x_i) - f(x_i)]^2 \right\}^{\frac{1}{2}}$$

where $\| E \|_2$ is the 2 norm of error.

In order to facilitate calculation, analysis, and application, the square of 2 norm is usually calculated, namely:

$$\| E \|_2^2 = \sum_{i=1}^{n} [\varphi(x_i) - f(x_i)]^2$$

This fitting method, which requires the minimum sum of squares of errors, is called the least-squares method.

The Fence Fitting Line was extracted by LSM fitting the point clouds on both sides of the mobile robot's driving direction. When the point cloud coordinates satisfy the minimum value of $F(W)$, W is the parameter matrix of the fitted Fence Fitting Line equation, as shown in Equation (1):

$$F(W) = \min(X'W - Y)^T (X'W - Y) \tag{1}$$

where $W = [k\ d]^T$ is the parameter matrix of the fence line; k is the slope of the fence line; d is the fence line intercept, and $m : X'_{n\times 2} = [X\ I]$ is the matrix composed of the point cloud X-axis coordinate value matrix $X_{n\times 1}$=$[x_1\ x_2 \ldots\ x_n]^T$ and the unit matrix I; $Y_{n\times 1} = [y_1 y_2\ \ldots\ y_n]^T$ is the matrix composed of the Y coordinate values of the point cloud. Taking the derivative of Equation (1), when $X\prime^T X''$ is a positive definite matrix, the parameter matrix W of the fence line equation is shown in Equation (2):

$$W = \left(X'^T X'\right)^{-1} X'^T Y \tag{2}$$

2.3.2. Random Sampling Consistency

The RANSAC method can iteratively estimate the parameters of a mathematical model from a set of observational data sets containing "outliers" [27,28]. The random sampling consensus algorithm can well estimate the model parameters from the data containing a large number of outliers, and can eliminate the interference of outliers on the estimated overall data model, and obtain the global optimal solution. It is an indeterminate algorithm.

It has a certain probability of producing a good result, and to increase the likelihood, the number of iterations must be increased.

The purpose of RANSAC is to find the optimal parameter matrix so that the number of data points that satisfy the matrix is the largest. Usually, h_{33} = 1 is used to normalize the matrix. Since the homography matrix has 8 unknown parameters, at least 8 linear equations are needed to solve, corresponding to the point position information, a set of point pairs can list two equations, and at least 4 sets of matching point pairs are included. The resulting matrix equation is shown in Equation (3):

$$s\begin{bmatrix} x' \\ y' \\ 1 \end{bmatrix} = \begin{bmatrix} h_{11}h_{12}h_{13} \\ h_{21}h_{22}h_{23} \\ h_{31}h_{32}h_{33} \end{bmatrix}\begin{bmatrix} x \\ y \\ 1 \end{bmatrix} \tag{3}$$

where S represents the sample data and h_{ij} represents a single element in the normalized matrix.

The RANSAC algorithm randomly selects 4 samples from the matching data set and ensures that the 4 samples are not collinear, calculates the homography matrix, then uses this model to test all data, and calculates the number and projection of data points that satisfy this model. Error (i.e., cost function), if this model is the optimal model, the corresponding cost function is the smallest. The resulting loss function is shown in Equation (4):

$$L = \sum_{i=1}^{n}\left(x_i', \frac{h_{11}x_i + h_{12}y_i + h_{13}}{h_{31}x_i + h_{32}y_i + h_{33}}\right)^2 + \left(y_i', \frac{h_{21}x_i + h_{22}y_i + h_{23}}{h_{31}x_i + h_{32}y_i + h_{33}}\right)^2 \tag{4}$$

where x_i', y_i' are the elements in the parameter matrix; x_i, y_i are the elements in the surrogate matrix.

A matrix is obtained by random sampling, and using Equation (3), it is verified whether other points conform to the model, and then the conforming points become "internal points", and the nonconforming points become "external points". Next time, extract points from the "new interior point set" to construct a new matrix, and recalculate the error. The final error is the smallest, and the maximum number of points is the final model. The steps of the RANSAC algorithm:

(1) Randomly extract S sample data from the data set, fit multiple models (the 4 samples cannot be collinear), calculate the transformation matrix H, and record it as model M;
(2) Calculate the projection error of all data in the dataset and the model M, if the error is less than the threshold, add the inner point set I;
(3) If the number of elements in the current interior point set I is greater than the optimal interior point set I_best, then update I_ best = I, and update the number of iterations k as shown in Equation (5):

$$k = \frac{\log(1 - p)}{\log(1 - w^m)} \tag{5}$$

where p is the confidence level, which is generally taken as 0.995; w is the proportion of "inner points"; m is the minimum number of samples required to calculate the model;
(4) If the number of iterations is greater than k, exit; otherwise, add 1 to the number of iterations, and repeat the above steps.

RANSAC is used to extract the fence lines on both sides, and a subset is selected from the point clouds of the fences on both sides by random sampling to establish a straight-line model. Then, the number of interior points of the straight-line model is calculated to check the correctness of the straight-line model, and iterate continuously to obtain the optimal straight-line model, which is the extracted fence line (Figure 4).

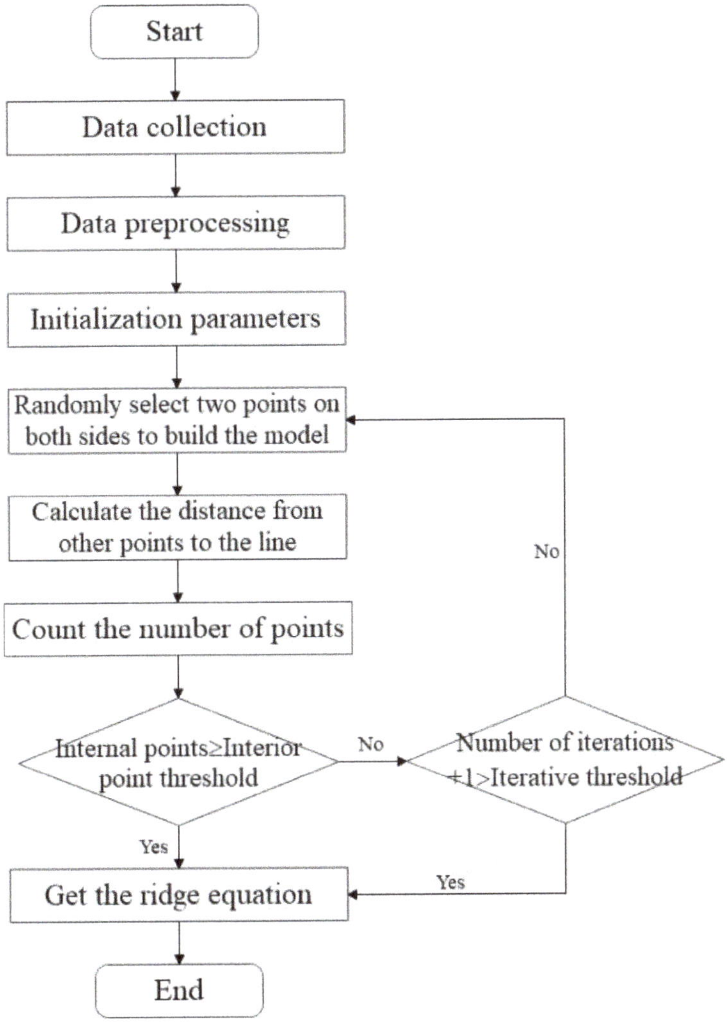

Figure 4. Flow chart of ridgeline extraction based on RANSAC.

The iteration threshold K_{RANSAC} is a key parameter for random sampling fitting. If the value of K_{RANSAC} is set too large, it will take too long, and if the value of K_{RANSAC} is set too small, the fitting effect will be poor. The selection basis of the K value is shown in Equation (6):

$$K_{RANSAC} = \lg(1-\alpha)/\lg(1-\omega^N) \quad (6)$$

where α is the probability that all points selected in the iterative process are interior points, %; ω is the probability that an interior point is selected from the data, %; N is the total number of data points.

2.4. Work Path Extraction

2.4.1. Noise Processing

The real-time performance is evaluated by the processing time of extracting grid lines [29], and the anti-noise ability is the resistance ability of the fence line extraction method to two kinds of noise [30]. This study evaluates the effect of LSM and RANSAC in extracting fence lines from the aspects of real-time performance and anti-noise ability, so that the extraction method can have strong real-time performance and anti-noise ability. Add Gaussian noise and man-made noise to the fence point cloud; among them, Gaussian noise cancels the statistical filtering process, and processes the Gaussian noise with the mean value of 0 and the variance of 0.1 on the point cloud of the fence. The artificial noise is achieved by adding interfering points between two fence point clouds.

2.4.2. Path Extraction

The boundary contour data are extracted from the fence point cloud using the point cloud vector method. First, the fence point cloud and its adjacent points projected on the XY plane are fitted with straight lines. Then, a point P in the fence point cloud is selected, and its adjacent point cloud is set $P_k = \{P_0, P_1 \dots P_{k-1}\}$, a straight line is set $u{\cdot}x + v{\cdot}y = c(u^2 + v^2 = 1)$, and the LSM is used to fit the straight line. At this time, the vector $q(u,\ v)$ is the normal vector of the point P. Then, referring to the content of the literature, the method where the maximum angle between adjacent points is greater than the set threshold is used to extract the fence outline point cloud, and the fence outline point cloud is recorded as the point set $U = \{d_0, d_1 \dots d_{n-1}\}$. Finally, the positional relationship between the fence line and the fence outline data points is judged by Equation (7). After removing the data outside the fence line, the point cloud data inside the fence line are divided into the left point set (U_L) and the right point set (U_R).

$$\begin{cases} d_{li} = \frac{|k_l \cdot x_i + b_l - y_i|}{\sqrt{{k_l}^2 + {b_l}^2}} \\ d_{ri} = \frac{|k_r \cdot x_i + b_r - y_i|}{\sqrt{{k_r}^2 + {b_r}^2}} \\ U_L = \begin{Bmatrix} (k_l \cdot x_i + b_l - y_i) \cdot (k_r \cdot x_i + b_r - y_i) < \\ 0 \cap d_{li} - d_{ri} < 0 | d_i(x_i, y_i) \in U \end{Bmatrix} \\ U_R = \begin{Bmatrix} (k_l \cdot x_i + b_l - y_i) \cdot (k_r \cdot x_i + b_r - y_i) < \\ 0 \cap d_{li} - d_{ri} > 0 | d_i(x_i, y_i) \in U \end{Bmatrix} \end{cases} \tag{7}$$

In the formula, d_{li} and d_{ri} are the distance between the fence outline point $d_i(x_i, y_i)$ and the left and right fence lines, m; i is the index number of the point set, $i = 0, 1, 2, \dots, n-1$; n is the number of point clouds of fence outline.

When the cows have been feeding for a period of time, the cows will push the forage to the outside at will, causing part of the forage to enter the area where they cannot eat, and the shape of the forage pile becomes irregular and the thickness of the pile becomes uneven. Therefore, the pusher robot is operated along the outermost part of the no-eating area, and the forage in the no-eating area is pushed to the eating area. As shown in Figure 5, during the operation of the pushing robot, the sideline of the auger always coincides with the inner boundary of the inaccessible area. The initial path is approximated by the translation transformation of the fence line fitted by the two methods. In this study, the intercept (b_1) of the initial path was used as the index, the width of the edible area was 70 cm, and the length of the auger of the pushing robot was 110 cm.

$$b_1 = b_0 + m_1 + \frac{1}{2} m_2 \tag{8}$$

where b_0 is the intercept of the fence line, m_1 is the width of the edible area, and m_2 is the length of the auger of the pushing robot.

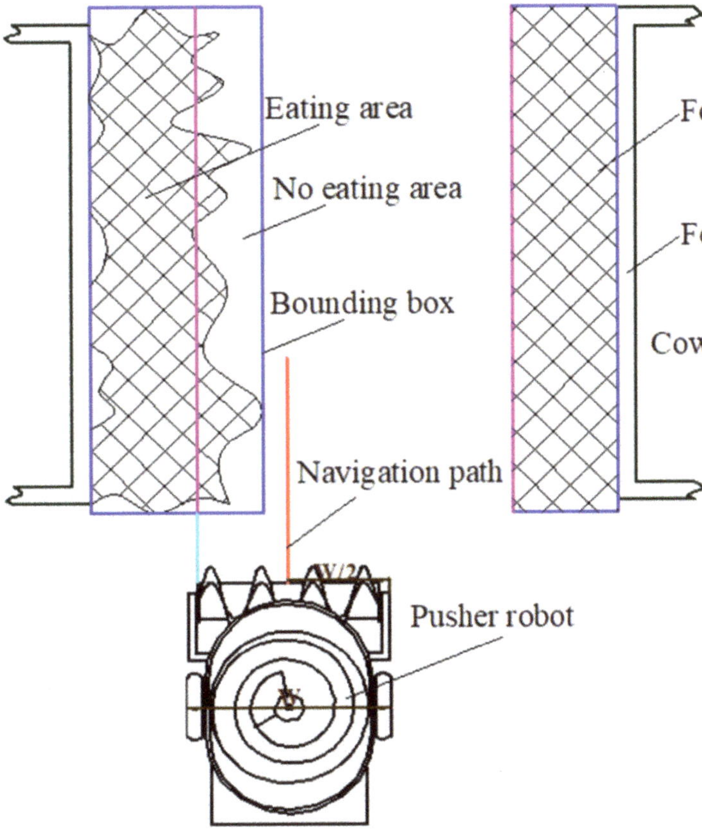

Figure 5. Working diagram of grass pushing robot.

2.5. Improvement of Obstacle Avoidance Strategy

The pusher robot will inevitably encounter obstacles during the operation of the cattle farm. The obstacle avoidance module is an essential part of the pusher robot to ensure that it can pass through obstacles during mobile operations. This paper chooses the artificial potential field method as the basis of the obstacle avoidance algorithm, and improves and analyzes the situation that it falls into the local optimum point.

2.5.1. Artificial Potential Field Method

Artificial Potential Field (APF) is an obstacle avoidance strategy represented by artificially defined virtual forces [31,32]. The mobile robot is assumed to be a point, which moves in a virtual force field, which is composed of the gravitational field of the target point to the robot and the repulsion field of the obstacle to the robot. The gravitational field is generated by the target point, and the repulsive force field is composed of the force field generated by all obstacles [33].

As shown in Figure 6, the repulsive force of the obstacle acting on the mobile machine is denoted as F_{rep}, and the direction is from the obstacle to the mobile robot; the gravitational force of the target point acting on the mobile robot is recorded as F_{att}, and the direction is from the mobile robot to the target point, then the force that the mobile robot receives at this position is the combined force of the repulsion force F_{rep} and the gravitational force F_{att} is F.

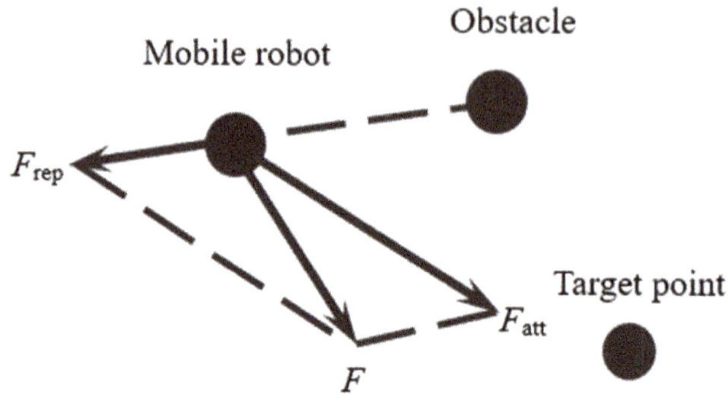

Figure 6. Stress diagram of artificial potential field method.

In the process of path planning, the environment of the unmanned vehicle is treated in a two-dimensional space, but the entire potential field distribution is three-dimensional. As shown in Figure 7, the gravitational potential energy leads to the generation of the third dimension, which is the main force in the process of path planning of the unmanned vehicle. The obstacles in the driving environment form peaks in the potential field distribution map. Under the action of the potential field, the unmanned vehicle can only move from the high potential energy point to the low potential energy point, so that the unmanned vehicle will not hit the obstacles, and it can safely plan the obstacle avoidance route.

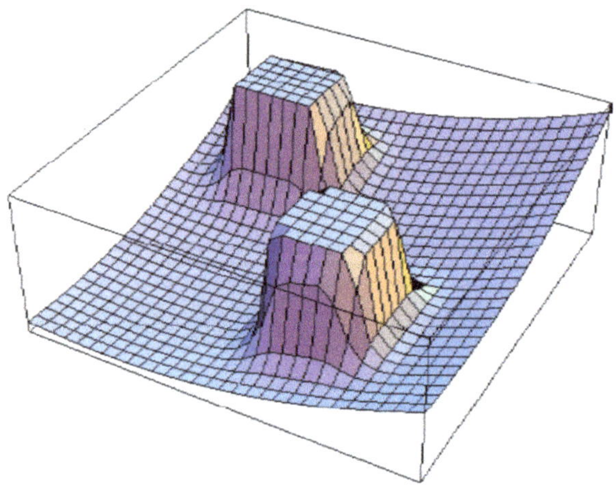

Figure 7. Three-dimensional diagram of artificial potential field obstacle avoidance.

Let the positions of the mobile robot, the target point, and the obstacle, be denoted as $q = (x, y)^T$ and $q_g = (x_g, y_g)$, respectively, and $q_{obs} = (x_{obs}, y_{obs})$ is the gravitational potential field generated by the target point to the mobile robot, and $Uatt(q)$ is the repulsive potential field generated by the obstacle to the mobile robot.

When the mobile robot is far away from the target point, the target point should generate a larger gravitational force for the mobile robot to move the mobile robot towards the target point. At the same time, when the mobile robot is at the target point, the robot should be at the zero-force point, so the gravitational potential field function is expressed as:

$$U_{att}(q) = \frac{1}{2}\xi\rho^2(q, q_g) \tag{9}$$

where ξ is the gain coefficient of the gravitational field, and $\rho(q, q_g)$ represents the distance between the target point and the current position of the mobile robot (expressed in Euclidean distance).

2.5.2. Improvement of Artificial Potential Field Method

The artificial potential field method converts the complex environmental information around the mobile robot into a simple force field model, which can achieve a relatively good obstacle avoidance effect in general [34,35]. However, due to the limitations of the definition of the gravitational potential field function and the repulsive potential field function itself, there may be situations in which the set target cannot be reached as expected and local minima appear before reaching the set target point. The reason for the above situation is mainly due to the defects brought by the definition of the gravitational potential field function and the repulsive potential field function itself. If the gravitational and repulsive forces are zero when the mobile robot reaches the target point, then the target point is the global optimal point. Considering the above problems, the distance between the target point and the robot is introduced into the repulsion function, and the repulsion field function expression is redefined:

$$U_{rep}(q) = \begin{cases} \frac{1}{2}\eta\left[\frac{1}{\rho(q,q_{obs})} - \frac{1}{\rho_0}\right]^2 \left(X - X_{goal}\right)^n, \rho(q,q_{obs}) \leq \rho_0 \\ 0, \qquad \rho(q,q_{obs}) > \rho_0 \end{cases} \tag{10}$$

where $(X - X_{goal})$ is the distance between the robot and the target, and n is a constant and greater than 0. Similarly, the repulsive force on the mobile robot is the negative gradient of the repulsive force field, and the repulsive force F_{rep} (q) is expressed as:

$$F_{rep}(q) = \begin{cases} F_{rep1} + F_{rep2}, \rho(q,q_{obs}) \leq \rho_0 \\ 0 \qquad , \rho(q,q_{obs}) > \rho_0 \end{cases} \tag{11}$$

$$F_{rep1} = \eta\left[\frac{1}{\rho(q,q_{obs})} - \frac{1}{\rho_0}\right]\frac{1}{\rho^2\rho(q,q_{obs})}\left(X - X_{goal}\right)^n \tag{12}$$

$$F_{rep2} = \frac{n}{2}\eta\left[\frac{1}{\rho(q,q_{obs})} - \frac{1}{\rho_0}\right]^2\left(X - X_{goal}\right)^{n-1} \tag{13}$$

In the formula, the direction of F_{rep1} is from the obstacle to the mobile robot, and the direction of F_{rep2} is from the target robot to the target point.

3. Results Analysis

3.1. Cowshed Point Cloud Preprocessing Results

In order to clarify the influence of the preprocessing method of point cloud data on the pusher robot in different motion states, the pusher robot collected 3D point cloud data in static and moving (forward speed is 0.5 m/s) states. From the collected 3D point cloud data, 200 frames of point clouds were selected for preprocessing. The number of preprocessed point clouds and the processing time of the filtering algorithm are shown in Table 2. It can be seen from the table that the pre-processed average point cloud numbers of the data collected by the robot at rest and in motion were 3257 and 3249, and the total average processing time was 0.338 and 0.319 s. There was no significant change in the number of point clouds and the total average processing time, which indicated that the preprocessing method selected in this study was suitable for machines in different motion states. Comparing the processing time of through filtering, downsampling filtering, and statistical filtering, it was found that statistical filtering took the longest time (0.122 s), accounting for 37.5% of the total preprocessing time, which was not conducive to real-time processing.

Table 2. Cowshed point cloud preprocessing results.

Motion States	Preprocessing Methods	Number of Point Clouds	Processing Times/s
Static	Through the filter	9872	0.0763
	Statistical filter	8253	0.125
	Down-sampled filter	4777	0.0684
	Ground plane filter	3257	0.0641
Motion	Through the filter	9865	0.0653
	Statistical filter	8255	0.119
	Down-sampled filter	4765	0.0676
	Ground plane filter	3249	0.0642

The visualization results of the preprocessed 3D point cloud data are shown in Figure 8. The ROI point cloud was extracted by pass-through filtering (Figure 8a), and the number of processed point clouds was 9872, which was reduced by 40%, and significantly reduced the number of point clouds; the number of point clouds after voxel downsampling filtering was 4777, which was reduced by 70%, and still retained the structural features of the original point cloud data (Figure 8b); the number of point clouds after statistical filtering was 8253, and 7% of outliers were removed (Figure 8c); after fitting the ground plane, the number of point clouds was 3257, which reduced the ground point cloud data by 79%, and retained the fence information (Figure 8d).

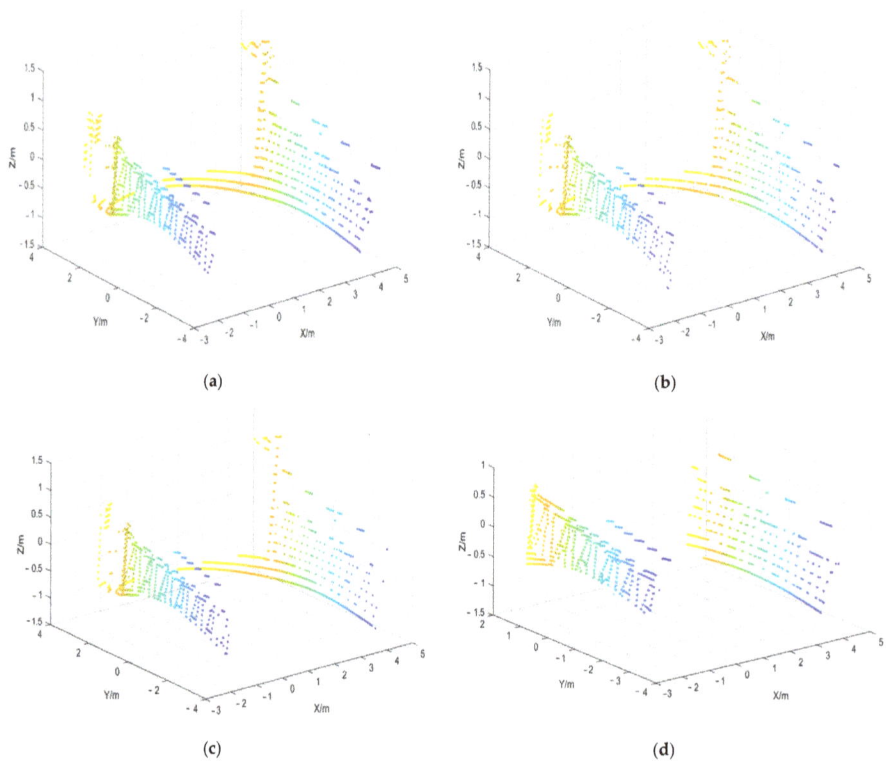

Figure 8. Visualization diagram of point cloud preprocessing: (**a**) Through filtered point cloud, (**b**) Statistical filtered point cloud, (**c**) Down sampling filtered point cloud, (**d**) Point cloud after ground removal.

3.2. Extraction Results of Fence Lines and Paths

In the experiment, Gaussian noise and artificial noise were added to the cowshed point cloud data, and the LSM and RANSAC were used to extract the fence lines and the initial paths. The visualization results were shown in Figure 8 and Table 3. It could be found from Figure 9 that both methods could process the fence point cloud data, and the processing effect was better. From Table 3, the value of the slope of the right fence line extracted by the LSM was −0.095 after adding artificial noise. Compared with no noise added (the value of the slope was about −0.061), there was a larger error and the extraction effect was reduced. The LSM considers the shortest distance from the overall point cloud to the extraction line and the phenomenon that the extracted fence line deviates when many data are deviating from the fence point cloud. The results of the RANSAC extracting fence lines without adding noise, adding Gaussian noise, and adding artificial noise (take the value of the slope of the right fence line as an example) were about −0.058, 0.058, and −0.061, respectively, and there was no significant change, indicating that RANSAC has certain resistance to both Gaussian noise and man-made noise. It could be seen from Table 2 that after adding artificial noise, the intercept value of the initial path extracted by the LSM was about −0.610, and the group without noise (the intercept was about −0.603) was quite different; The intercepts value extracted by the RANSAC in the three groups were −0.602, −0.603, and −0.601, and the intercept changes were not obvious, indicating that the RANSAC was better for initial path extraction.

Table 3. Extraction of bullpen line and initial path analysis by LSM and RANSAC.

Point Cloud Noise	Methods	Intercept of Bullpen b_0/m	Line Slope of Bullpen k_0	Intercept of Initial Path b_1/m	Processing Time/10^{-3} s
Without noise	LSM	-1.853 ± 0.001	0.062 ± 0.002	-0.603 ± 0.001	2.352 ± 0.125
	RANSAC	-1.852 ± 0.001	0.056 ± 0.001	-0.602 ± 0.001	1.157 ± 0.012
Gaussian noise	LSM	-1.853 ± 0.005	0.079 ± 0.012	-0.603 ± 0.005	2.638 ± 0.119
	RANSAC	-1.847 ± 0.005	0.058 ± 0.037	-0.597 ± 0.005	1.162 ± 0.015
Artificial noise	LSM	-1.860 ± 0.007	0.095 ± 0.007	-0.610 ± 0.007	2.931 ± 0.124
	RANSAC	-1.851 ± 0.006	0.062 ± 0.005	-0.601 ± 0.006	1.165 ± 0.013

From the analysis of the processing time of the two methods, it could be found that there was no significant difference in the processing time of RANSAC in the non-noise group, the Gaussian noise group, and the artificial noise group. The processing time (2.352×10^{-3} s) of the LSM without noise group was significantly different from the other two groups, indicating that the processing time of RANSAC was less affected by noise than the LSM method. The running time of RANSAC is significantly lower than that of LSM transform, and RANSAC can obtain better real-time performance by selecting a reasonable number of iterations.

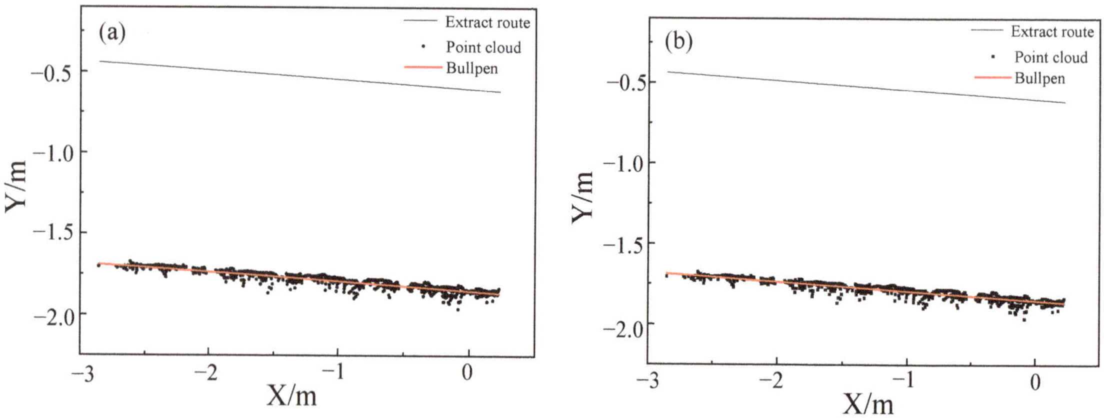

(1) No noise added

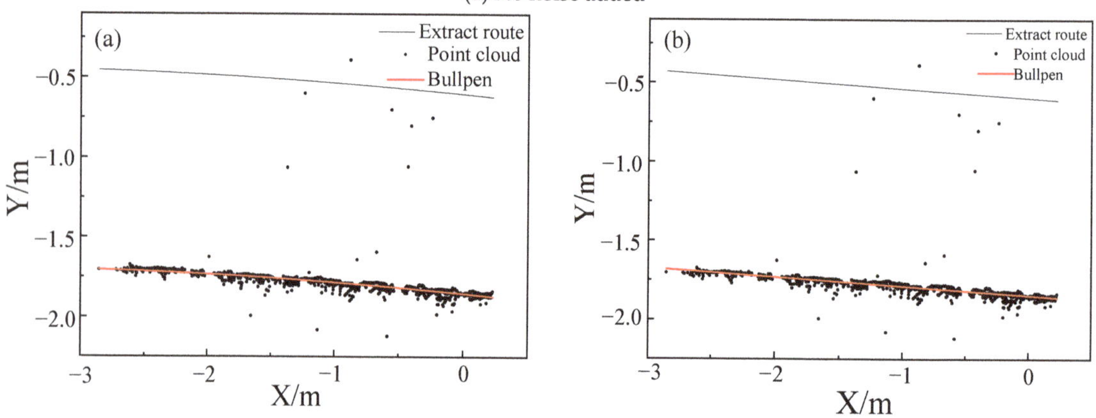

(2) Add Gaussian noise

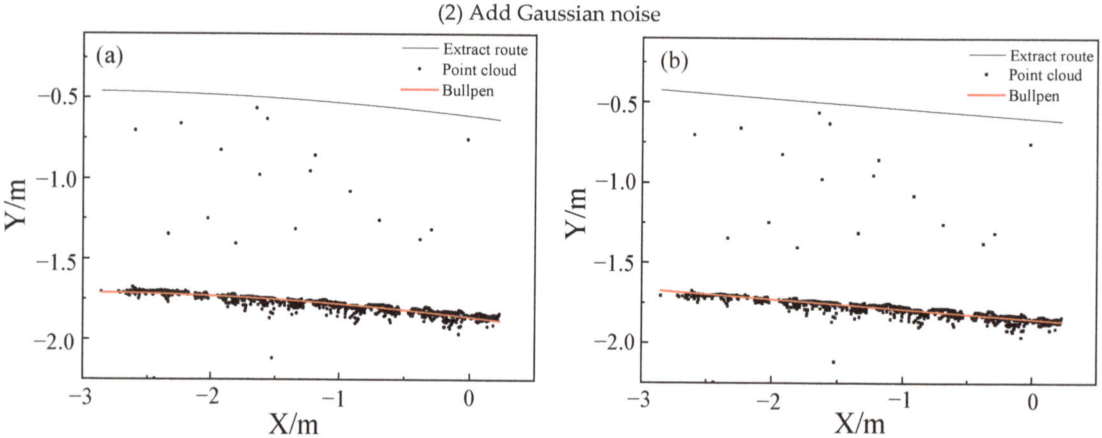

(3) Add man-made noise

Figure 9. Results of extracted ridgeline and initial path by three methods under different noise. (**a**) Least squares method; (**b**) Random sampling consistency.

3.3. Simulation Analysis of Obstacle Avoidance Algorithm

In response to the problem that the traditional artificial potential field method is prone to the defect of falling into minimal values, a new repulsion field function is proposed, and the selection range of the repulsion field gain coefficient is analyzed. In order to verify the effectiveness of the improved artificial potential field method when there are multiple obstacles, and the influence of parameter selection on the obstacle avoidance effect. This paper conducts simulation experiments in MATLAB BR2020a. Since there are cows and feed belts on both sides of the fence, it is only necessary to place obstacles at the proximal and distal ends of the feed belts. The simulation space adopts a 10 × 10 grid map. It sets the coordinates of the starting position of the robot to be (0, 0) marked with a square, and the coordinates of the end point of the target point to be (10, 10), marked with a triangle, and the obstacle coordinate points marked with circles are set between the two to simulate the actual situation.

As shown in Figure 10, five obstacle coordinate points were set, with the coordinates being (1.1, 1.2), (3, 2.4), (5.5, 5.5), (6, 2), and (8, 8.5). Then, the simulation parameters were set: the improved repulsion potential field parameter gain coefficient value $\eta = 5$, the gravitational potential field gain coefficient value $\xi = 15$, the value of the maximum distance that obstacles affect the mobile robot was 1.5 m, and the iterative step size of the mobile robot was 0.1 m. The simulation results showed that the mobile robot could successfully move from the starting point to the target point and achieve the effect of avoiding obstacles.

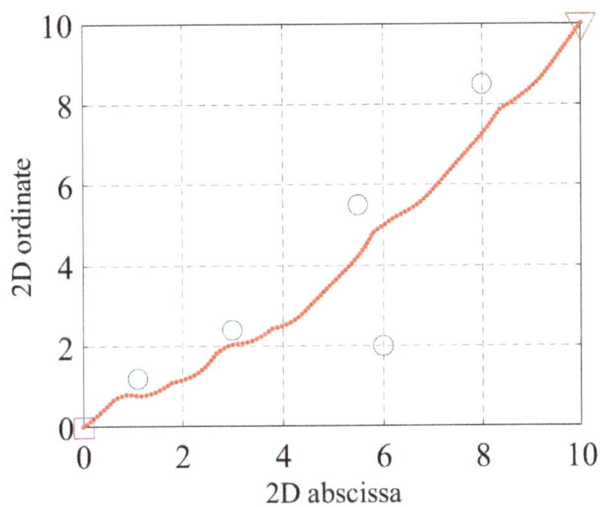

Figure 10. Simulation results of improved artificial potential field method.

To illustrate the influence of the improved artificial potential field algorithm parameter selection on the obstacle avoidance effect, the obstacle avoidance effect of the mobile robot in the moving process is analyzed from the perspective of reducing the repulsion force and the gain coefficient of the gravitational potential field. Figure 11a showed the simulation results when the repulsive potential field gain coefficient was too small. It could be clearly seen that although there would be some collisions with obstacles during the movement, the robot could still move to the target point in the end. Figure 11b shows the simulation results when the gravitational potential field gain coefficient was too small. It can be clearly seen that the mobile robot would oscillate back and forth at certain positions. In addition, the robot cannot reach the set target point.

Figure 12 shows the situation where the obstacle is located on the extension line between the robot and the target point, and the target point is within the repulsive potential field of the obstacle. At this time, the coordinates of the target point are set to (8, 8), the coordinates of the obstacle are set to (9, 9), and the maximum action radius of the repulsive potential field of the obstacle is 1.5 m. The mobile robot reaches the target point smoothly according to the planned path.

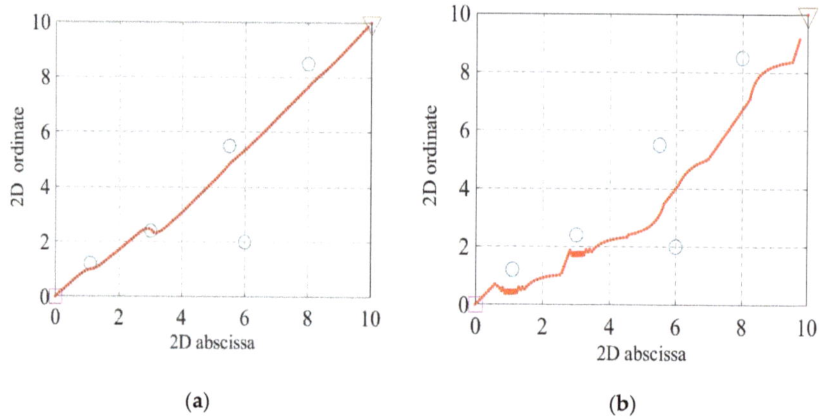

Figure 11. Simulation results when the gain coefficient is not selected: (**a**) The repulsion gain coefficient is too small; (**b**) Gravitational gain coefficient is too small.

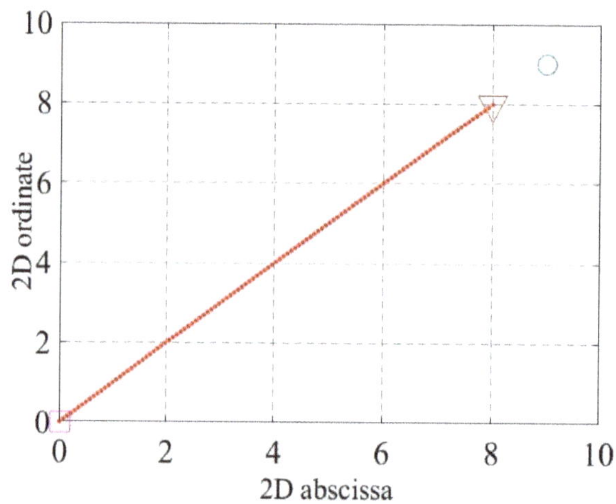

Figure 12. Simulation results when the target point is in the middle.

Figure 13 shows the simulation results when the obstacle is set between the starting point and the target point, and the resultant force is on the connection line between the two, in which the position coordinates of the obstacle in Figure 13a are (5, 5); the position coordinates of the obstacles in Figure 13b are (5, 5), (4.5, 5.5), (4, 6), (3.5, 6.5), and (3, 7). It can be seen that the robot can successfully get rid of the minimum point and avoid obstacles when it falls into a local minimum value during the movement process, and finally can move to the target point.

Take single obstacle and multiple obstacles as examples, the simulation results of the traditional artificial potential field method are shown in Figure 14 below. It can be seen that compared with the improved artificial potential field method, the traditional artificial potential field method will fall into oscillation when the robot is close to the obstacles, and requires more steps when it is away from the obstacles, which is not conducive to the rapid obstacle avoidance of the robot.

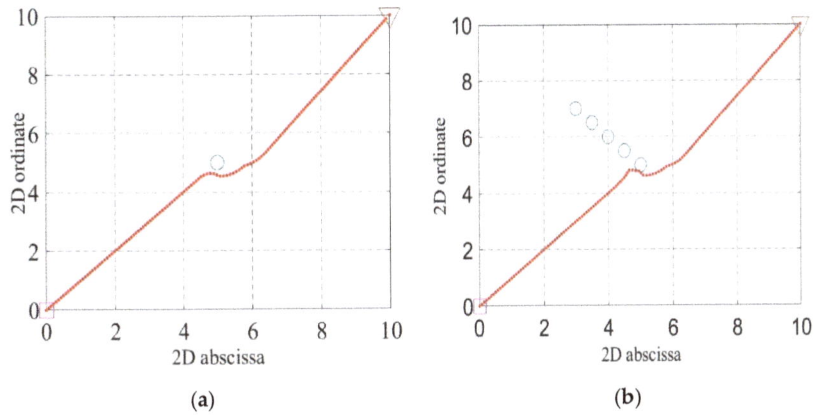

Figure 13. Simulation results when obstacles are in the middle: (**a**) Single obstacle; (**b**) Multiple obstacles.

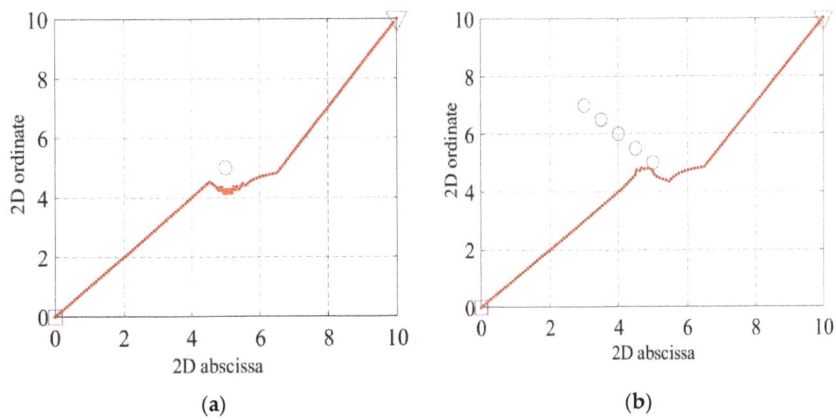

Figure 14. Simulation results under traditional artificial potential field: (**a**) Single obstacle; (**b**) Multiple obstacles.

The above results show that the improved artificial potential field method has better performance under reasonable parameter selection. The forward path is predicted and judged before the robot moves, and simplifies the restricted obstacles; that is, the robot only affects the repulsive force of the obstacles on the target side within the affected range; then, a reasonable virtual target point is set near the simplified obstacle, and the improved repulsion function guides the robot to quickly generate a smooth, stable, and collision-free path in a complex environment. Moreover, the rationality of the selection of the gravitational potential field gain coefficient and the repulsive potential field gain coefficient directly affects the obstacle avoidance effect. The algorithm can realize the obstacle avoidance function of a mobile robot.

3.4. Experimental Research on Obstacle Avoidance

A simple obstacle avoidance test is carried out on the designed pusher robot in this section. The static obstacle is set up in the experiment. In the three scenarios, the maximum speed of the inspection robot is set to 0.5 m/s. Due to the low vehicle speed, the influence radius of obstacles is set to 1.5 m; objects detected within 1.5 m in front of the robot are regarded as obstacles. In the obstacle avoidance experiment, after placing the static obstacle objects in the test area, the robot is initialized to collect and model the surrounding environment information. The robot is instructed to move along the extracted path to verify the robot's response to static obstacles. After the robot completes the map construction and bypasses the obstacles, it quickly moves in the direction of the robot.

The pusher robot performs linear work toward the target until the robot moves to the position shown in Figure 15a. At this time, the robot enters the influence range of the obstacle. Under the combined action of attraction and repulsion, it deflects an angle to the right to drive. Figure 15b shows the position where the robot is closest to the obstacle. Under the action of the resultant force, the robot gradually crosses the obstacle until it successfully reaches the end point (Figure 15c). During the entire driving and obstacle avoidance process, the closest distance to the obstacle is 0.41 m. The optimized path increases the shortest distance value of the obstacle point cloud from the navigation path from 0.18 to 0.41 m, where the average time is 0.059 s and the standard deviation is 0.007 s, which shows that the optimization method can optimize the path in real time to avoid obstacles, basically meeting the requirements of security and real-time performance, and effectively avoiding the local minimum problem The entire obstacle avoidance path is relatively smooth, which can successfully avoid obstacles and reach the destination point. The test proves that the pusher robot can efficiently extract the working path, make timely decisions when detecting static obstacles, avoid collisions with obstacles, and has good stability and reliability.

(a)

Figure 15. *Cont.*

(b)

(c)

Figure 15. Obstacle avoidance scene and process: (**a**) Find the obstacle; (**b**) Get around obstacle; (**c**) Keep working.

4. Conclusions

Based on the research on the existing cowshed fence line extraction method, a robot navigation path optimization method based on the improved artificial potential field method is proposed. This method improves the safety of the unmanned driving of the feeding machine, and provides theoretical support and development basis for the intelligent agricultural equipment in the dairy farm.

(1) The functional requirements of the pushing robot were analyzed and the hardware system was designed. According to the functional requirements of each module, the model was selected and designed, and the research and experimental platform of the pusher robot was built.

(2) The performance of LSM and RANSAC for extracting ridge lines and initial paths was evaluated from the aspects of real-time performance and anti-noise capability. Under three different environments: no noise, Gaussian noise, and artificial noise, the fence lines were extracted using RANSAC, and the obtained results were about −0.058, 0.058, and −0.061, respectively. Compared with LSM, RANSAC was less affected by

noise in processing time (between 1.157×10^{-3} and 1.165×10^{-3} s). It was shown that the running time, anti-noise ability, and extraction effect of RANSAC were better.

(3) The obstacle avoidance method of the pusher robot was optimized. The target point influence factor was introduced into the repulsive potential field function, and the parameter selection strategy of the improved artificial potential field method was analyzed. Finally, the correctness and feasibility of the obstacle avoidance method were verified by simulation experiments.

Author Contributions: Conceptualization, Z.Y. and F.T.; methodology, F.L. and Z.W.; validation, Y.Y.; formal analysis, Z.S.; investigation, R.W.; resources, S.Y.; writing—original draft preparation, X.W.; writing—review and editing, Z.Y. and F.T.; project administration, F.T.; funding acquisition, Z.Y. All authors have read and agreed to the published version of the manuscript.

Funding: This study was funded partly supported by National Natural Science Foundation of China, grant number 32072786, and the Shandong Agricultural University Research Foundation, grant number 030-72195.

Institutional Review Board Statement: Not applicable.

Informed Consent Statement: Not applicable.

Data Availability Statement: The data used to support the findings of this study are included in the article.

Conflicts of Interest: The authors declare no conflict of interest.

References

1. Dong, C.Y.; Ma, B.B.; Lu, C.X. Spatio-temporal dynamics of feed grain demand of dairy cows in China. *Sustainability* **2020**, *12*, 663. [CrossRef]
2. Lovarelli, D.; Brandolese, C.; Leliveld, L.; Finzi, A.; Riva, E.; Grotto, M.; Provolo, G. Development of a new wearable 3D sensor node and innovative open classification system for dairy cows' behavior. *Animals* **2022**, *12*, 1447. [CrossRef] [PubMed]
3. Tan, M.; Hou, Y.; Zhang, L.; Shi, S.; Long, W.; Ma, Y.; Zhang, T.; Li, F.; Oenema, O. Operational costs and neglect of end-users are the main barriers to improving manure treatment in intensive livestock farms. *J. Clean. Prod.* **2021**, *289*, 125149. [CrossRef]
4. Zhang, M.; Wang, X.; Feng, H.; Huang, Q.; Xiao, X.; Zhang, X. Wearable Internet of Things enabled precision livestock farming in smart farms: A review of technical solutions for precise perception, biocompatibility, and sustainability monitoring. *J. Clean. Prod.* **2021**, *312*, 127712. [CrossRef]
5. da Borso, F.; Kic, P.; Kante, J. Analysis of management, labor and economics of milking systems in intensive goat farms. *Agriculture* **2022**, *12*, 513. [CrossRef]
6. Escalante, H.; Castro, L.; Amaya, M.P.; Jaimes, L.; Jaimes-Estevez, J. Anaerobic digestion of cheese whey: Energetic and nutritional potential for the dairy sector in developing countries. *Waste Manag.* **2018**, *71*, 711–718. [CrossRef] [PubMed]
7. Wu, X.; Lu, Y.; Xu, H.; Lv, M.; Hu, D.; He, Z.; Liu, L.; Wang, Z.; Feng, Y. Challenges to improve the safety of dairy products in China. *Trends Food Sci. Technol.* **2018**, *76*, 6–14. [CrossRef]
8. Maitiniyazi, S.; Canavari, M. Exploring Chinese consumers' attitudes toward traceable dairy products: A focus group study. *J. Dairy Sci.* **2020**, *103*, 11257–11267. [CrossRef]
9. Yang, S.; Liang, S.; Zheng, Y.; Tan, Y.; Xiao, Z.; Li, B.; Liu, X. Integrated navigation models of a mobile fodder-pushing robot based on a standardized cow husbandry environment. *Trans. ASABE* **2020**, *63*, 221–230. [CrossRef]
10. Yu, Z.; Liu, Y.; Yu, S.; Wang, R.; Song, Z.; Yan, Y.; Li, F.; Wang, Z.; Tian, F. Automatic Detection Method of Dairy Cow Feeding Behaviour Based on YOLO Improved Model and Edge Computing. *Sensors* **2022**, *22*, 3271. [CrossRef]
11. Rahim, T.; Usman, M.A.; Shin, S.Y. A survey on contemporary computer-aided tumor, polyp, and ulcer detection methods in wireless capsule endoscopy imaging. *Comput. Med. Imaging Graph.* **2020**, *85*, 101767. [CrossRef]
12. Hassan, S.A.; Rahim, T.; Shin, S.Y. Real-time uav detection based on deep learning network. In Proceedings of the 2019 International Conference on Information and Communication Technology Convergence (ICTC), Jeju Island, Korea, 16–18 October 2019.
13. Shen, D.; Wu, G.; Suk, H.I. Deep Learning in medical image analysis. *Annu. Rev. Biomed. Eng.* **2017**, *19*, 221–248. [CrossRef]
14. Wu, X.; Sahoo, D.; Hoi, S.C.H. Recent advances in deep learning for object detection. *Neurocomputing* **2020**, *396*, 39–64. [CrossRef]
15. Luong, M.; Pham, C. Incremental learning for autonomous navigation of mobile robots based on deep reinforcement learning. *J. Intell. Robot. Syst.* **2020**, *101*, 1–11. [CrossRef]
16. Su, D.; Kong, H.; Qiao, Y.; Sukkarieh, S. Data augmentation for deep learning based semantic segmentation and crop-weed classification in agricultural robotics. *Comput. Electron. Agric.* **2021**, *190*, 106418. [CrossRef]

17. Seo, D.; Cho, B.-H.; Kim, K.-C. Development of monitoring robot system for tomato fruits in hydroponic greenhouses. *Agronomy* **2021**, *11*, 2211. [CrossRef]
18. Zheng, Y.Y.; Kong, J.L.; Jin, X.B.; Wang, X.Y.; Zuo, M. CropDeep: The crop vision dataset for deep-learning-based classification and detection in precision agriculture. *Sensors* **2019**, *19*, 1058. [CrossRef] [PubMed]
19. Thombre, S.; Zhao, Z.; Ramm-Schmidt, H.; Vallet Garcia, J.M.; Malkamaki, T.; Nikolskiy, S.; Hammarberg, T.; Nuortie, H.; Bhuiyan, M.Z.; Sarkka, S.; et al. Sensors and AI techniques for situational awareness in autonomous ships: A review. *IEEE Trans. Intell. Transp. Syst.* **2022**, *23*, 64–83. [CrossRef]
20. Chen, M.; Tang, Y.; Zou, X.; Huang, Z.; Zhou, H.; Chen, S. 3D global mapping of large-scale unstructured orchard integrating eye-in-hand stereo vision and SLAM. *Comput. Electron. Agric.* **2021**, *187*, 106237. [CrossRef]
21. Yang, H.; Guo, H.; Dai, W.; Nie, B.; Qiao, B.; Zhu, L. Bathymetric mapping and estimation of water storage in a shallow lake using a remote sensing inversion method based on machine learning. *Int. J. Digit. Earth* **2022**, *15*, 789–812. [CrossRef]
22. Pavkin, D.Y.; Shilin, D.V.; Nikitin, E.A.; Kiryushin, I.A. Designing and simulating the control process of a feed pusher robot used on a dairy farm. *Appl. Sci.* **2021**, *11*, 10665. [CrossRef]
23. Uddin, K.; Jeong, T.H.; Oh, B.T. Incomplete region estimation and restoration of 3D point cloud human face datasets. *Sensors* **2022**, *22*, 723. [CrossRef] [PubMed]
24. Lee, M.-Y.; Lee, S.-H.; Jung, K.-D.; Lee, S.-H.; Kwon, S.-C. A novel preprocessing method for dynamic point-cloud compression. *Appl. Sci.* **2021**, *11*, 5941. [CrossRef]
25. Liu, W.; Chen, R.; Cai, H.; Luo, W. Robust statistical methods for impulse noise suppressing of spread spectrum induced polarization data, with application to a mine site, Gansu province, China. *J. Appl. Geophys.* **2016**, *135*, 397–407. [CrossRef]
26. Zermas, D.; Morellas, V.; Mulla, D.; Papanikolopoulos, N. 3D model processing for high throughput phenotype extraction—The case of corn. *Comput. Electron. Agric.* **2020**, *172*, 105047. [CrossRef]
27. Dal Poz, A.P.; Yano Ywata, M.S. Adaptive random sample consensus approach for segmentation of building roof in airborne laser scanning point cloud. *Int. J. Remote Sens.* **2019**, *41*, 2047–2061. [CrossRef]
28. Liu, M.; Wang, X.; Li, L.; Gao, E. Improved random sampling consistency algorithm employed in three-dimensional point cloud registration. *Laser Optoelectron. Prog.* **2018**, *55*, 165–171.
29. Xia, H.; Yang, B.; Li, Y.; Wang, B. An improved CenterNet model for insulator defect detection using aerial imagery. *Sensors* **2022**, *22*, 2850. [CrossRef]
30. Chen, Y.; Feng, K.; Lu, J.; Hu, Z. Machine vision on the positioning accuracy evaluation of poultry viscera in the automatic evisceration robot system. *Int. J. Food Prop.* **2021**, *24*, 933–943. [CrossRef]
31. Yang, W.; Wu, P.; Zhou, X.; Lv, H.; Liu, X.; Zhang, G.; Hou, Z.; Wang, W. Improved artificial potential field and dynamic window method for amphibious robot fish path planning. *Appl. Sci.* **2021**, *11*, 2114. [CrossRef]
32. Li, L.; Wu, D.; Huang, Y.; Yuan, Z.-M. A path planning strategy unified with a COLREGS collision avoidance function based on deep reinforcement learning and artificial potential field. *Appl. Ocean. Res.* **2021**, *113*, 102759. [CrossRef]
33. Sudhakara, P.; Ganapathy, V.; Priyadharshini, B.; Sundaran, K. Obstacle avoidance and navigation planning of a wheeled mobile robot using amended artificial potential field method. *Procedia Comput. Sci.* **2018**, *133*, 998–1004. [CrossRef]
34. Zhao, T.; Li, H.; Dian, S. Multi-robot path planning based on improved artificial potential field and fuzzy inference system. *J. Intell. Fuzzy Syst.* **2020**, *39*, 7621–7637. [CrossRef]
35. Hou, P.; Pan, H.; Guo, C. Simulation research for mobile robot path planning based on improved artificial potential field method recommended by the AsiaSim. *Int. J. Model. Simul. Sci. Comput.* **2017**, *8*, 1750046. [CrossRef]

agriculture MDPI

Article

Low-Cost Robot for Agricultural Image Data Acquisition

Gustavo José Querino Vasconcelos [1], Gabriel Schubert Ruiz Costa [2], Thiago Vallin Spina [1] and Helio Pedrini [1,*]

1 Institute of Computing, University of Campinas (UNICAMP), Campinas 13083-852, Brazil
2 The Brazilian Center for Research in Energy and Materials (CNPEM), Campinas 13083-100, Brazil
* Correspondence: helio@ic.unicamp.br

Abstract: More sustainable technologies in agriculture are important not only for increasing crop yields, but also for reducing the use of agrochemicals and improving energy efficiency. Recent advances rely on computer vision systems that differentiate between crops, weeds, and soil. However, manual dataset capture and annotation is labor-intensive, expensive, and time-consuming. Agricultural robots provide many benefits in effectively performing repetitive tasks faster and more accurately than humans, and despite the many advantages of using robots in agriculture, the solutions are still often expensive. In this work, we designed and built a low-cost autonomous robot (DARob) in order to facilitate image acquisition in agricultural fields. The total cost to build the robot was estimated to be around $850. A low-cost robot to capture datasets in agriculture offers advantages such as affordability, efficiency, accuracy, security, and access to remote areas. Furthermore, we created a new dataset for the segmentation of plants and weeds in bean crops. In total, 228 RGB images with a resolution of 704×480 pixels were annotated containing 75.10% soil area, 17.30% crop area and 7.58% weed area. The benchmark results were provided by training the dataset using four different deep learning segmentation models.

Keywords: autonomous robots; agriculture; data acquisition; computer vision

Citation: Vasconcelos, G.J.Q.; Costa, G.S.R.; Spina, T.V.; Pedrini, H. Low-Cost Robot for Agricultural Image Data Acquisition. *Agriculture* **2023**, *13*, 413. https://doi.org/10.3390/agriculture13020413

Academic Editors: Jin Yuan, Wei Ji and Qingchun Feng

Received: 31 December 2022
Revised: 5 February 2023
Accepted: 7 February 2023
Published: 10 February 2023

1. Introduction

In recent years, agriculture has been increasingly modernized to address existing problems in agricultural fields, where robots are being used to automate repetitive and tedious tasks for humans, in addition to enabling a more effective application of precision agriculture that uses resources much more efficiently. Advanced robotic systems are used from the process of soil preparation, planting, harvesting, and weed control, to the post-processing of the obtained resources. Although agricultural robots are increasingly common on properties around the world, their use is mainly to directly improve crop productivity, while solutions for scientific applications are still scarce and cost prohibitive for the vast majority of research groups, especially in developing countries [1–4].

The complexity inherent in numerous agricultural tasks; for example, harvesting, weed control, and crop quality analysis, among others; is greatly benefited by the advancement of computer vision and artificial intelligence, due to the greater generalization that this type of technology provides. This offers more robust control solutions for different conditions, such as climate, soil, and vegetation [5–13]. The combination of computer vision with machine learning algorithms in the context of precision agriculture can bring numerous benefits, in addition to the drastic reduction of herbicides; for example, the detailed analysis of the phenotype characteristics of plants by quantitative and individual measurements. For this, a spatial analysis of the images captured by the sensors is necessary. Therefore, the use of semantic segmentation algorithms is of paramount importance for detailing the region of interest.

Semantic segmentation is the task of assigning a class label to every pixel in the image. Figure 1 illustrates an example of a semantic segmentation mask in an agricultural image of the Bean dataset, described in detail in Section 5. All soil class pixels in the image are set to purple, all plant pixels are set to blue, while weeds are set to green. The colors are random and only serve to visually differentiate the classes.

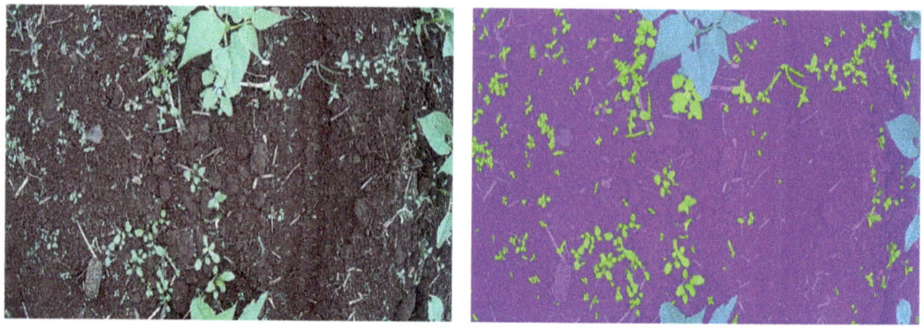

(a) (b)

Figure 1. Example of semantic segmentation in an agricultural image. (**a**) RGB image. (**b**) Segmentation mask.

Currently, the state of the art in semantic segmentation problems is based on deep neural networks, which require a large amount of annotated data in their training process. Several works present adequate solutions to the weed and crop segmentation problem [5–14]. However, due to the large number of parameters found in this type of model, converging to a robust solution requires the use of a lot of data in its training stage. Therefore, the need for large amounts of annotated images of different plant species and at different growth stages is currently one of the biggest challenges in this area.

However, due to the complex task of image acquisition in the field, such as difficulty in location accuracy, high human effort, difficulty in standardizing capture time, angle, height, and lighting, among others, image datasets for training neural networks are still scarce in the literature, especially for crops of regional interest that are not commonly cultivated on a large scale in Europe and North America. While the popularization of drones is reducing this problem for aerial images, for terrestrial images, which are especially important for active weed control, cost and labor are still major factors in the data scarcity.

Since the vast majority of agricultural robots in the literature are designed to perform more work than just image gathering in the field, when we assess the prices of prototype construction or commercial purchase, we often see prohibitive values, especially in the reality of many universities and research centers in developing nations.

A low-cost robot for capturing datasets in agriculture provides several advantages. First, its low fee makes it more affordable for more farms and research groups to purchase, resulting in broader and more affordable data collection. Furthermore, robots are able to collect data more quickly and accurately than humans, increasing the efficiency of data collection. They also avoid human errors such as fatigue or distraction, ensuring more accurate data collection. In addition, robots can be designed to perform dangerous tasks, reducing the risk to human workers. Finally, robots can be sent to remote areas to collect data where it would be difficult or impossible for humans to reach.

This work aims to present the design, construction and use of a low-cost agricultural robot for image acquisition in agricultural fields. The robot prototype was built with simple, economical, ready-to-use components and was developed for the purpose of acquiring images. In order to provide the community with a new dataset to work with domain adaptation, the designed robot was used to collect a new dataset, called the Bean dataset, which used a crop relevant to Brazilian agriculture; in this case, the bean crop.

2. Terrestrial Agriculture Datasets

This section presents the main datasets used in the area of semantic segmentation of plants and weeds that were acquired terrestrially, that is, close to the ground through robots, or acquired manually. Table 1 illustrates the classes, number of images, resolution and sensor type of five different datasets present in the literature. It can be observed that only two terrestrial segmentation datasets have more than 60 images, which limits the advancement of deep learning in this field.

Table 1. Comparison of five different terrestrial agriculture segmentation datasets present in the literature.

Name	Classes	# Images	Resolution	Sensor
Sugar Beets [15]	Sugar Beets and 9 weed species	300	1296 × 966	RGB, NIR, Depth and GPS
Sunflower [16]	Sunflower and weeds	500	1296 × 964	RGB
RGBWD [17]	Carrot and weeds	39	3264 × 2248	RGB
CWFID [18]	Carrot and weeds	60	1269 × 966	RGB
CWDD [19]	Carrot, Onion and weeds	40	2419 × 1986	RGB and NIR

The two main terrestrial image datasets available in the literature and used for evaluation of this work are the Sugar Beets and Sunflower datasets. The other datasets contain few images for use in deep learning models.

Sugar Beets is an agricultural dataset composed of 300 RGB images acquired from a terrestrial robot called BoniRob, illustrated in Figure 2, on a sugar beet farm near Bonn, Germany, over a three-month period in the spring of 2016. The images were recorded three times a week, starting with the emergence of plants and stopping when the robot could no longer access the field without damaging the crops. The robot carried a four-channel multispectral camera and a red, green, blue and depth (RGB-D) sensor to capture detailed information about the crop. Additionally, GPS, LIDAR and wheel encoders are available, resulting in around 5TB of data. Only RGB images and their respective semantic segmentation masks from 2–3 weeks of growth stage after emergence were used in this work. An example image found in the Sugar Beets dataset and its respective segmentation mask can be seen in Figure 3.

Figure 2. Acquisition process of the Sugar Beets dataset using the BoniRob terrestrial robot developed by Bosch DeepField Robotics [15]. BoniRob is equipped with four wheels that can be steered independently of each other, allowing flexible movement and navigation on rough terrain.

(a) (b)

Figure 3. Example of an image found in the Sugar Beets dataset and its respective segmentation mask. (**a**) Image. (**b**) Segmentation mask.

Sunflower is a dataset for weed and crop segmentation that was collected using a custom-built agricultural robot, illustrated in Figure 4, on a sunflower farm in Jesi, Italy. The dataset is composed of 500 images and was recorded in the spring season, over a period of one month, starting from the emergence stage of the crop plants and stopping until the end of the useful period. The images were acquired using a four-channel (RGB + NIR) JAI AD-13 camera, mounted on the robot and facing downwards. The dataset provides RGB and NIR images with pixel-wise annotation of three classes: crop, weed and soil. An example image found in the Sunflower dataset and its respective segmentation mask can be seen in Figure 5.

Figure 4. Acquisition process of the Sunflower dataset using a custom-built manual pushcart. Courtesy of Dr. Alberto Pretto—University of Padova.

(**a**) (**b**)

Figure 5. Example of an image found in the Sunflower dataset and its respective segmentation mask. (**a**) Image. (**b**) Segmentation mask.

3. Agricultural Robots in the Literature

One of the most important features of agricultural robotics is the replacement of humans with field robots or mechanical systems that can perform dangerous and repetitive tasks more precisely and evenly, at lower cost and with more efficiency. The most common applications for agricultural field robots are weed control and accurate spraying. In this context, spot spraying with robots for weed control application has produced satisfactory results, reducing the use of herbicides to less than 10% of total spraying [20]. Several promising weed robot technologies have been introduced and deployed in recent years as a consequence of multidisciplinary collaboration initiatives involving different academic groups and companies, although they have not yet been fully commercialized.

In this section, we present several previous works that have developed robots to operate in agriculture. However, some of them are not only for image acquisition, but also for chemical or mechanical action for weed manipulation. Despite not being the main focus of the robot presented in this work, it would be possible to modify it to accommodate mechanical actuators. Some of the main agricultural robots in the literature that carry cameras for crop analysis are briefly discussed in this section. Table 2 illustrates the price and country of origin of the robots described in this work. N/A means that no value is available.

Table 2. Prices and places of origin of some common agricultural robots described in the literature.

Robot	Price	Country
BoniRob [15]	€175,000	Germany
Terra-Mepp [21]	$20,000	EUA
EcoRobotix [22]	$250,000	Switzerland
Agbot II [23]	€26,000	Australia
Asterix robot [24]	€120,000	Norway
Thorvald II [25]	N/A	Norway
Ladybird [26]	$1 million	Australia
AgriBOT [27]	N/A	Brazil
DARob (Ours)	**$850**	**Brazil**

BoniRob was developed by students at the University of Osnabrück along with Bosch companies and the German agricultural company Amazone [28]. BoniRob is an agricultural robot that detects weeds using camera technology and image recognition, and then drives a screw into the soil to remove the plant. BoniRob was developed for applications in precision agriculture, that is, for mechanical weed control, selective herbicide spraying, as well as for plant and soil monitoring. It provides assemblies to install different tools for these specific tasks. BoniRob is equipped with four wheels that can be steered independently of each other, which allows flexible movement and navigation over rough terrain. They provide visual, depth, 3D laser, GPS and odometry data.

The Terra-Mepp robotic platform is designed to semi-autonomously navigate in a single row and view the crop from above, side and below the canopy to provide a comprehensive set of phenotypic data for each plant [21]. An adjustable mast allows flexible placement of sensors to adapt to changes in canopy height throughout the growing season. This system employs an image-based proximal detection approach to measure plant height using stereo cameras and depth sensors. A stereo camera with a 170° wide-angle lens captured the plant height measurements. The camera was mounted vertically at a specific location on the mast, so that the top of the canopy was centered in the camera's field of view. To measure the width of the plant, the low resolution infrared camera was mounted horizontally on the robot's base. Figure 6a illustrates the Terra-Mepp robot.

(a) (b)

Figure 6. Terra-Mepp (**a**) and EcoRobotix (**b**) robots in the field [22,29].

EcoRobotix, illustrated in Figure 6b, is a four-wheeled robot powered by two electric motors, with wheels designed to ride on off-road surfaces, so it could traverse any farmland with relative ease. It has solar panels on top to generate a continuous source of power for the internal battery, allowing it to run as long as there is daylight, thus removing the need to snap and recharge at the end of the day. It weighs approximately 130 kg. An onboard camera, RTK GPS and a series of sensors allow you to identify crops and keep it on a travel course, as well as detect the presence of weeds between crops.

Agbot II is a two-wheel drive (2WD) system agricultural robot developed at Queensland University of Technology for weed and crop management, as well as horticultural applications [23]. The robot is equipped with a downward facing 1.3 MP global shutter camera with an illuminated field of view using a pulsed lighting system synchronized with data capture. The AgBot II can autonomously navigate and traverse a field performing weeding operations, return to the central station to recharge its batteries, and refill its chemical tank before returning to the field to continue operation.

The Asterix robot was designed with the specific task of applying spot spraying herbicides [24]. The robot has a three-wheel design to maintain maneuverability and stability with the benefits of reduced weight, complexity and cost. The vision unit employs a Nvidia Jetson TK1, with a built-in camera unit using Omnivision 4682 4 MP sensor. Raw images are debayered to RGB (Red–Green–Blue) and HSV (Hue-Saturation-Value) color spaces. The forward facing camera and navigation unit allow the tracking of lines across the field. A combination of vision and GPS location detects the end of a row and aids navigation on the headlands.

Thorvald II is a lightweight, battery-powered autonomous agricultural robot developed from the robotics group at the College of Science and Technology, Norwegian University of Life Sciences (NMBU) [25]. The robot is equipped with two drive modules, two passive wheel modules and no steering modules. As the spacing between the plots is small, approximately 15 cm, the standard wheels of the drive modules are replaced by thinner wheels. The robot is equipped with a RTK-GNSS receiver and an IMU, and navigates through predefined waypoints. It stops and captures images on each plot. The robot is equipped with two pairs of cameras, which consist of an RGB camera and a monochromatic infrared camera. One pair is facing straight down and the other pair is mounted sideways at a 55 degree angle.

The Agricultural Robot (AgriBOT) basically consists of a full-scale four-wheel autonomous robotic tractor equipped with a 4WSD independent steering configuration [27]. The AgriBOT project is a cooperation project between EMBRAPA (Empresa Brasileira de Pesquisa Agropecuária), EESC–USP (Universidade de São Paulo) and Jacto Company. The AgriBOT has two laser sensors attached to its front, pointing downwards at $-30°$, a model inertial measurement unit (IMU), a global positioning system (GPS), real-time kinematic (RTK) and two red, green, and blue (RGB) cameras with charge-coupled device (CCD) sensors of 1600×1200 pixels of resolution.

Researchers at the Australian Centre for Field Robotics (ACFR) of the University of Sydney have developed the Ladybird robot, which is a lightweight, omnidirectional electric vehicle for advancing agricultural robotics technology. The robot is equipped with a forward and backward facing GPS, Light Detection and Ranging (LIDAR) along with the Point Gray Ladybug 3 spherical camera to capture surroundings data to avoid obstacles and detect crop rows. In addition, the robot has a set of three sensors under the structure. The first is a camera to capture RGB images of the crops, whereas the second is a hyperspectral imaging camera to capture infrared and ultraviolet data. It also has a laser sensor to determine the height of crops above the ground.

Because the vast majority of the aforementioned robots are designed to perform more tasks than just image acquisition in the field, when we analyze the costs of construction of the prototype or for commercial purchase, we observe values that are often prohibitive, especially for the reality of many universities and research centers in underdeveloped countries. As an example, the BoniRob robot used to capture the Bonn dataset costs at around $250,000.00, while cheaper versions, such as ecoRobotix and AgBot II, cost around $90,000 and $26,000.00 at outdated prices. At the other end of the spectrum, robots such as Ladybird have an estimated prototype build value of approximately $1 million. In addition, we needed an alternative to the pushcart method used in the Sunflower dataset for image capture that did not require as much manual labor. In this context, the contribution of the development and construction of a low-cost terrestrial robot becomes important for the advancement of computer vision science in agriculture.

4. DARob—Data Acquisition Robot

This section presents the design and construction of the DARob robot that aims to capture images in agricultural fields.

The robot was designed to have an unobstructed region at its center, so that there is a gap for plants to pass through as the robot moves along the crop. The design was intended for strip cropping, in which several crops are planted in alternating rows. Typical combinations include corn, sugar beets, soybeans, and grasses such as hay and wheat. This configuration allows the span to be extended as needed for the plantation rows. The distance between the wheels and their height can be modified, thus facilitating the adaptation of the robot to different configurations of the rows, such as paired or twin rows, as well as the adaptation of the height according to the existence of larger plants or in different growth stages.

The robot is 1530 mm long and 635 mm high in its most compact form. The minimum span width and height are 145 mm and 535 mm, respectively, and the maximum span width is 835 mm. The project was also developed so that the robot is fully modular, facilitating its transport and assembly at the site of use. Figure 7 illustrates the DARob. The batteries and electronic control mechanisms are placed in compartments for easy coupling on the upper part of the robot, which is insulated against dust and moisture.

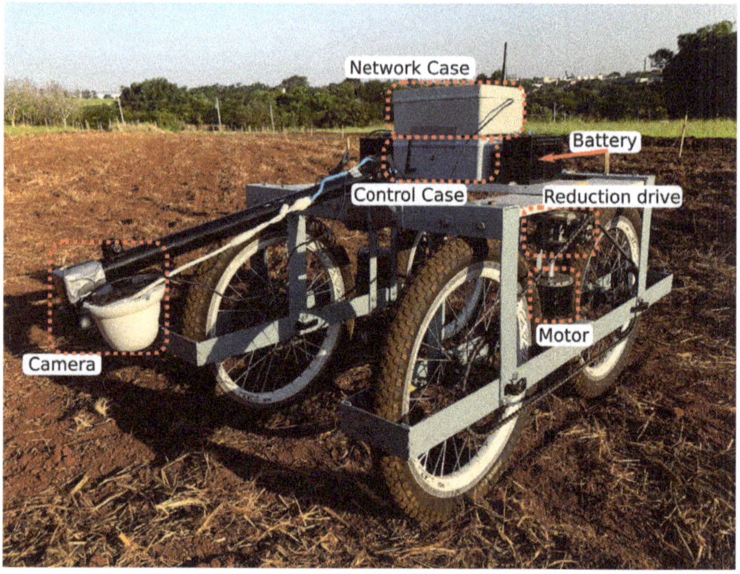

Figure 7. View of the DARob with its main systems.

The robot's maneuvering method is the skid steering type, which is characterized by not having a steering system, but rather changing the direction of the vehicle by controlling the relative speeds of its left and right sides. Although the skid steering mechanism has a simpler construction system, in practice there is a low efficiency for making sharp turns, due to the wheel having to slide sideways to turn. To circumvent this problem, the robot was built to run in either direction (front or back). This means that when the robot reaches the end of the mapped terrain, it can continue driving backwards and does not need to turn 180° to continue, so only minor corrections in the direction are required for it to change planting rows. The robot is controlled by remote control with electric propulsion through the use of two 300 W direct current motors, each controlling the two wheels on its side.

The motors are bolted to the robot's metal frame, and its axes are directly connected with 1:15 transmission reducers, which increases the torque capacity of the motor. A 16-tooth pinion is installed on the output shaft of the gearbox, which is coupled to a chain that connects the 46-tooth chainrings, connected to both wheels (front and rear). This connection also generates a gain in torque of 16/46, totaling approximately a 43-times gain in the motor's rated torque. The wheels are 52 cm in diameter, and the electric motor has 0.8 Nm of nominal torque and 1.7 Nm in stall.

The mechanical structure was built with the goal of modularity. Therefore, the robot, when disassembled, is separated into four parts of equal size (one for each wheel). However, for transport in smaller places, such as car interiors, simply separating the two sides of the robot is sufficient. All electronic components are also easily removable and separable, as they are placed on a plate above the frame. The entire process of assembling the robot and installing the electronics takes about 10 min and requires only one person.

The camera is placed on a vari-angle bar placed in front of the robot above the space where plants can move. The camera chosen for this project was the Intelbras s4020 IP Camera, a low-cost RGB and infrared camera used mainly for security tasks, which, by construction, is designed to work outdoors in extreme temperatures, weather, and unhealthy environmental conditions.

The camera is designed to handle temperatures ranging from −30 to 50 °C, with a 3.6 mm focal length lens and 1/3″ Progressive Scan CMOS sensor. The camera has an IP66 protection rating, making it ideal for working in dusty areas. The transmission of images is carried out via Ethernet cable, its power supply is 12 volts direct current and its consumption is only 4.8 W, with a total weight of 360 g.

The robot's guidance software is controlled by an open hardware platform called Pixhawk, which has magnetometer, accelerometer, barometer, and GPS (Global Positioning System) modules built into the system. These sensors are used as inputs to simultaneous localization and mapping (SLAM), which is calculated from the Ardupilot firmware. Ardupilot is an advanced and reliable open source autopilot software that has been in development since 2010, which from the input of sensors and user commands, controls the current that is delivered to the motors. User control can be performed via remote control or a pre-configured mission via the graphical interface of the QGroundControl ground control station.

QGroundControl is a control station that works with the protocol Micro Air Vehicle Link (MAVLink) compatible with open-source autopilots including the ArduPilot. In this project, QGroundControl was chosen as the ground control station because it provides easy and straightforward use for beginners, as well as offering support to advanced features in mission control and vehicle configurations with autopilot. In addition, QGroundControl is one of the most stable ground control stations, has a simple and efficient interface and is available in different operating systems, such as Windows, Mac OS X, Linux, Android and iOS. Figure 8 illustrates the QGroundControl interface with the waypoints of a planned mission.

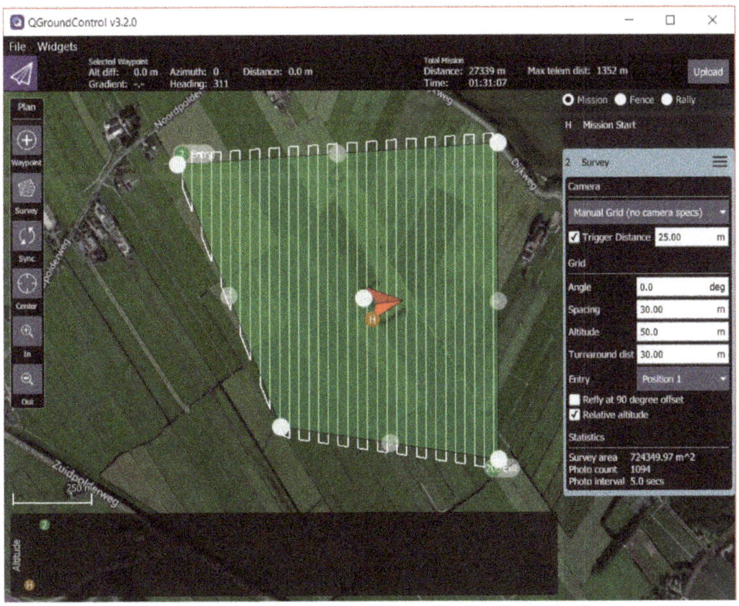

Figure 8. Graphical interface of QGroundControl with an example of the mission's waypoints map. Source: www.qgroundcontrol.com, accessed on 27 March 2022.

In addition to the Pixhawk and the GPS module, the robot's control system also has an H-bridge (model Bts7960) and multiplexers (model CD74HC4067) used to convert the output of the Pixhawk to command the H-bridge. Moreover, the remote control receiver module and the on–off switch button for activating control via Pixhawk are placed in the control system case. Figure 9a illustrates the control system case.

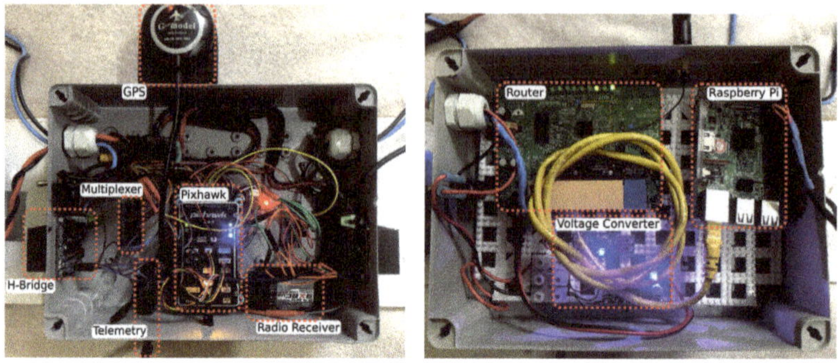

(**a**) (**b**)

Figure 9. Details of the components present in the control and network systems. (**a**) Control System. (**b**) Network System.

In addition to the control system cabinet, the robot has a network system cabinet that includes an embedded microcomputer (Raspberry Pi 3 Model B+) responsible for processing and capturing camera images, and a network router. The images were stored on a 128 GB SD card attached to the Raspberry Pi and, after the missions, the data were transferred to a personal computer. The network router has two different purposes, the first is to allow the raspberry computer to access the IP camera images using a real-time streaming protocol (RTSP), and the second is to allow a ground control computer to access the Raspberry system via Secure Shell (SSH). To carry out this process, the microcomputer runs on Linux (Raspbian OS). The router enables real-time monitoring of collected photos as well as adjusting data collecting settings without having to access or stop the robot. Figure 9b illustrates the network system case. The total cost to build the robot was estimated at around $850. The main components of the robot and their respective costs are shown in Table 3.

Table 3. Main components of the robot and their respective costs.

Item	Price
Raspbeery Pi 3	$127
Pixhawk Kit	$179
Network Router	$19
2xMotor 300W DC 24V	$187
2xBts7960 (Driver)	$13
2xCd74hc4067 (Multiplexer)	$4
Robot Frame	$55
Camera Intelbras s4020	$69
2xBattery 12V	$39
Reduction Drive	$99
Extras	$52
Total	**$843**

5. Bean Dataset Acquisition and Annotation Process

From the analysis of Table 1, it can be observed that only two datasets, Sugar Beets and Sunflower, provide a reasonable amount of images for the application of deep neural network models. The amount of data required to train a deep neural network from scratch varies depending on several factors, such as model complexity, data variety, and degree of overfitting. A deep neural network is typically trained with at least hundreds of pictures [30].

In addition, there are few plant variations available in the datasets, and none of them is a plant of great importance for Brazilian agriculture. In this sense, we chose to create a new dataset to expand the possibilities of evaluating computer vision methods in agriculture and also to create a database more aligned with the interests of Brazilian agriculture, such as plants of the Leguminosae family. Due to the great morphological similarity between the species and the greater availability at the time of planting, we mainly focused on the use of the common bean.

At the end of April 2021, the sowing process of the bean crop began within an experimental area at the School of Agricultural Engineering (FEAGRI) of the University of Campinas (UNICA MP). The area was prepared before sowing by mechanically removing the weeds present in the region. This was the only weed control action in the field, because after the emergence of the plants, no type of control was carried out.

The common bean, the crop used in the proposed dataset, is a perennial herbaceous plant with the scientific name Phaseolus vulgaris L. It is a member of the Leguminosae family. It has two different types of leaves: the simple or primary leaves, which are opposite, and the compound leaves, which are composed of three leaflets and arranged alternately (trifoliolate). Beans can be grown for up to four harvests in a year, as they have a short growing cycle (about 90 days). The best sowing time is determined by the common bean's climatic requirements, which are specific to this variety [31].

One difficulty encountered during the capture of the dataset was the modification of the lighting in the field, due to the robot not having protection from direct sunlight. In some cases, images appear with camera shadows, so priority was given to capturing the dataset in the late afternoon when sunlight was not shining directly on the capture area.

We collected data during one month, a specific period during which we covered various plant growth stages. On average, data was acquired once a week, leading to 4 days of capture in total. When recording on a typical day, the robot drove between two rows of cultivation, each measuring approximately 100 m in length. The robot was radio controlled during the data collection process, maintaining an average speed of 10 cm/s and a capture rate of 0.5 frames per second. Figure 10 illustrates DARob in the acquisition process in the FEAGRI field.

Figure 10. DARob in the acquisition process in the FEAGRI field.

In total, more than 5000 images were captured in RGB and IR format. Even though only a subset was annotated with the segmentation maps, all images will be made publicly available. Although the images are captured from emergence to the adult stage of the plant, all of the images annotated in the Bean dataset are of plants in a growth stage no more than one week apart, 2–3 weeks into the growth stage. Figure 11 illustrates RGB and IR images acquired with DARob. The data collection process covered different growth stages of the bean crop. Our intention was to capture variations over time relevant to weed control. The robot visited various regions of the field in different weather conditions, ranging from sunny to cloudy. Although RGB and IR images were acquired, the applied capture process did not allow a simple capture of the two aligned images.

(**a**) (**b**)

Figure 11. Examples of bean plant images acquired with DARob. (**a**) RGB image. (**b**) Infrared image.

Image annotation was performed using the *supervise.ly* platform [32] (version 6.4.31), which is a Web platform for computer vision data annotation. Many annotation tools are available in the *supervise.ly* interface for effective semantic segmentation, including cuboids, polylines, bitmap brushes, bounding boxes, polygons, and keypoints. Figure 12 illustrates the graphical user interface of the platform and the result of a semantic segmentation case of the annotated images. The process was carried out in stages, where the first 10 images were manually annotated, and later the Deeplab-v3 model [33] with pre-trained ResNet 101 backbone in the COCO dataset [34] was adjusted with the annotated images using 224 × 224 pixel patches. The inferred images were uploaded back to the platform for correction. After 10 corrected images, the process was repeated to increase the inference quality. After three cycles, no further improvement was noticed, so the rest of the images were manually corrected.

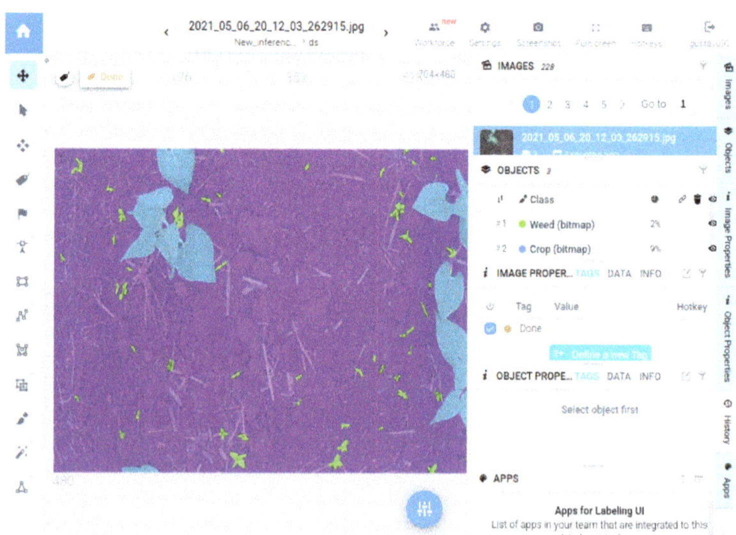

Figure 12. *Supervise.ly* graphical user interface in the process of segmenting an image acquired with DARob.

A total of 228 RGB images with a resolution of 704 × 480 pixels were annotated. The crop growth stage chosen for the annotation was similar to the steps present in the Sugar Beets and Sunflower datasets to enable validation of the domain transfer models. The available dataset was separated into five *k*-fold sets, each one with 183 training images and 45 testing images. The dataset contains 75.10% soil area, 17.30% crop area, and 7.58% weed area. The annotation process lasted approximately 3 months, between November 2021 and January 2022. Each image took approximately 2–3 h to be fully manually annotated. With the iterative process, the average time decreased to 30 min. The dataset is available for download at https://github.com/gustavu92/bean_dataset (accessed on 30 January 2023).

Four different segmentation networks were trained on the dataset to provide benchmark results. For this, we used the BiSeNet [35], DuNet [36], Deeplab-v3 [33], Deeplab-v3+ [37] and PSPNet [38] networks. In all cases, except for BiSeNet, the ResNet50 network was used as the backbone, due to the use of this model in benchmark results in the literature. The BiSeNet network was also evaluated using the ResNet18 backbone. The mIoU results for soil, crop and weed classes are shown in Table 4.

Table 4. Benchmark segmentation results for the acquired Bean dataset.

Model	Bean			
	Soil	Crop	Weed	Mean
BiSeNet (ResNet 18) [35]	0.942	0.920	0.625	0.829
DuNet (ResNet 50) [36]	0.950	0.927	0.662	0.846
Deeplab-v3 (ResNet 50) [33]	0.953	0.935	0.680	0.856
Deeplab-v3+ (ResNet 50) [37]	0.957	0.958	0.682	0.866
PSPNet (ResNet 50) [38]	0.959	0.940	0.708	0.869

The computer used for training is equipped with a GeForce RTX 2070 with 8GB of memory, a tenth-generation Intel i7 CPU and 24GB of RAM. The PyTorch framework was used to implement our methods due to its easy debugging feature and open source license. We also used several libraries for Python programming language, such as NumPy for linear algebra operations, SciPy for scientific computing, OpenCV, PIL and Scikit-Image for computer vision and image processing operations.

Table 4 shows that the tested networks have greater difficulty in segmenting the weed class, which is expected, due to its smaller quantity and smaller size compared to the crop and soil classes. The tests also show good consistency in the results, with little variation due to the complexity of the networks used.

The Sunflower and Sugar Beets datasets were also evaluated with the BiSeNet network, using ResNet18 as the backbone. The results are shown in Table 5. It can be seen that the results for the proposed Bean dataset achieved a higher mIoU in the crop and weed classes. This difference can be explained due to the higher proportion of the weed class in the images. However, it is also observed that the results in the soil class were worse than the results obtained in the other data sets. This can also be explained due to the higher proportion of plants, causing the network to make more errors in the soil class. However, the mIoU in the proposed dataset is slightly higher than the compared datasets.

Table 5. Comparison between the Bean dataset, Sunflower dataset, and Sugar Beet dataset using the BiSeNet network with ResNet 18 as the backbone.

Dataset	Soil	Crop	Weed	Mean
Bean Dataset (Ours)	0.942	0.920	0.625	0.829
Sugar Beets	0.993	0.868	0.430	0.764
Sunflower	0.984	0.840	0.565	0.796

Although Tables 4 and 5 give quantitative values about the results, it is difficult to analyze what would be a good value for the mIoU metric in practice for plant and weed segmentation. This analysis depends on the application in which the result will be used and the specifics of the dataset used. For example, for applications where the result is used for weed control, an average mIoU value of approximately 80 would already be sufficient to detect the main weeds in the analyzed area, but the higher this value, the smaller the size of the detected and segmented weeds.

6. Conclusions

In this work, we designed and constructed a low-cost autonomous robot (DARob) to facilitate the capture of images in agricultural fields. There are some important features to highlight about DARob:

- Low-cost machine: it employed economical and ready-to-use components, which can facilitate the access of other research groups to this type of data acquisition system, increasing the amount of datasets available;
- Automatic operation: the user can program the robot to execute automatically, following the defined mission, which improves the repeatability of the data generated;
- Remote control: it is possible to follow how the data is being acquired during the robot's movement, allowing the operator to correct the acquisition configuration in real time;
- Portability: the robot was designed to be easy to assemble, transport, and also flexible for different types and sizes of crop.

During the operation of the robot, some limiting points were observed:

- Autonomy: the batteries have limited autonomy, reducing the robot's efficiency;
- Bicycle wheels: by using these, the robot has difficulty moving over mud, which reduces the autonomy and disturbs navigability during automatic operation;
- Shadow on images: since it does not have any type of enclosure for the camera, light changing and shadow cause disturbances on the acquired images.

Furthermore, we created a new dataset for segmentation of plants and weeds in bean crops. In total, 228 RGB images with a resolution of 704×480 pixels were annotated containing 75.10% soil area, 17.30% crop area and 7.58% weed area. The benchmark results were provided by training the dataset using four different deep learning segmentation models.

Author Contributions: Data curation, G.J.Q.V. and G.S.R.C.; formal analysis, T.V.S. and H.P.; funding acquisition, H.P.; methodology, G.J.Q.V., G.S.R.C., T.V.S. and H.P.; writing—original draft, G.J.Q.V.; writing—review and editing, T.V.S. and H.P. All authors have read and agreed to the published version of the manuscript.

Funding: This research received no external funding.

Institutional Review Board Statement: Not applicable.

Data Availability Statement: The data presented in this study are available in the main text.

Conflicts of Interest: The authors declare no conflict of interest.

References

1. Shah, T.M.; Nasika, D.P.B.; Otterpohl, R. Plant and Weed Identifier Robot as an Agroecological Tool Using Artificial Neural Networks for Image Identification. *Agriculture* **2021**, *11*, 222. [CrossRef]
2. Ganguly, R.K.; Mukherjee, A.; Chakraborty, S.K.; Verma, J.P. Impact of agrochemical application in sustainable agriculture. In *New and Future Developments in Microbial Biotechnology and Bioengineering*; Elsevier: Amsterdam, The Netherlands, 2021; pp. 15–24.
3. Esposito, M.; Crimaldi, M.; Cirillo, V.; Sarghini, F.; Maggio, A. Drone and Sensor Technology for Sustainable Weed Management: A Review. *Chem. Biol. Technol. Agric.* **2021**, *8*, 1–11. [CrossRef]
4. Groeneveld, D.; Tekinerdogan, B.; Garousi, V.; Catal, C. A Domain-specific Language Framework for Farm Management Information Systems in Precision Agriculture. *Precis. Agric.* **2021**, *22*, 1067–1106. [CrossRef]
5. Picon, A.; San-Emeterio, M.G.; Bereciartua-Perez, A.; Klukas, C.; Eggers, T.; Navarra-Mestre, R. Deep Learning-based Segmentation of Multiple Species of Weeds and Corn Crop using Synthetic and Real Image Datasets. *Comput. Electron. Agric.* **2022**, *194*, 106719. [CrossRef]

6. Mishra, A.M.; Harnal, S.; Gautam, V.; Tiwari, R.; Upadhyay, S. Weed Density Estimation in Soya Bean Crop using Deep Convolutional Neural Networks in Smart Agriculture. *J. Plant Dis. Prot.* **2022**, *129*, 593–604. [CrossRef]
7. Vayssade, J.A.; Jones, G.; Gée, C.; Paoli, J.N. Pixelwise Instance Segmentation of Leaves in Dense Foliage. *Comput. Electron. Agric.* **2022**, *195*, 106797. [CrossRef]
8. Champ, J.; Mora-Fallas, A.; Goëau, H.; Mata-Montero, E.; Bonnet, P.; Joly, A. Instance Segmentation for the Fine Detection of Crop and Weed Plants by Precision Agricultural Robots. *Appl. Plant Sci.* **2020**, *8*, e11373. [CrossRef]
9. Khan, A.; Ilyas, T.; Umraiz, M.; Mannan, Z.I.; Kim, H. CED-Net: Crops and Weeds Segmentation for Smart Farming using a Small Cascaded Encoder-Decoder Architecture. *Electronics* **2020**, *9*, 1602. [CrossRef]
10. Sabzi, S.; Abbaspour-Gilandeh, Y.; Arribas, J.I. An Automatic Visible-Range Video Weed Detection, Segmentation and Classification Prototype in Potato Field. *Heliyon* **2020**, *6*, e03685. [CrossRef]
11. Sheikh, R.; Milioto, A.; Lottes, P.; Stachniss, C.; Bennewitz, M.; Schultz, T. Gradient and Log-based Active Learning for Semantic Segmentation of Crop and Weed for Agricultural Robots. In Proceedings of the IEEE International Conference on Robotics and Automation, Paris, France, 31 May–31 August 2020; pp. 1350–1356.
12. Sodjinou, S.G.; Mohammadi, V.; Mahama, A.T.S.; Gouton, P. A Deep Semantic Segmentation-based Algorithm to Segment Crops and Weeds in Agronomic Color Images. *Inf. Process. Agric.* **2021**, *9*, 355–364. [CrossRef]
13. Zou, K.; Chen, X.; Zhang, F.; Zhou, H.; Zhang, C. A Field Weed Density Evaluation Method Based on UAV Imaging and Modified U-Net. *Remote Sens.* **2021**, *13*, 310. [CrossRef]
14. Vasconcelos, G.; Spina, T.; Pedrini, H. Low-Cost Domain Adaptation for Crop and Weed Segmentation. In Proceedings of the 25th Iberoamerican Congress on Pattern Recognition, Porto, Portugal, 10–13 May 2021; pp. 141–150.
15. Chebrolu, N.; Lottes, P.; Schaefer, A.; Winterhalter, W.; Burgard, W.; Stachniss, C. Agricultural Robot Dataset For Plant Classification, Localization And Mapping On Sugar Beet Fields. *Int. J. Robot. Res.* **2017**, *36*, 1045–1052. [CrossRef]
16. Fawakherji, M.; Potena, C.; Pretto, A.; Bloisi, D.D.; Nardi, D. Multi-Spectral Image Synthesis for Crop/Weed Segmentation in Precision Farming. *Robot. Auton. Syst.* **2021**, *146*, 103861. [CrossRef]
17. Lameski, P.; Zdravevski, E.; Trajkovik, V.; Kulakov, A. Weed Detection Dataset with RGB Images Taken Under Variable Light Conditions. In Proceedings of the ICT Innovations 2017, Skopje, North Macedonia, 18–23 September 2017; Trajanov, D., Bakeva, V., Eds.; Springer International Publishing: Cham, Switzerland, 2017; pp. 112–119.
18. Haug, S.; Ostermann, J. A Crop/Weed Field Image Dataset for the Evaluation of Computer Vision Based Precision Agriculture Tasks. In Proceedings of the European Conference on Computer Vision, Boston, MA, USA, 7–12 June 2015; pp. 105–116.
19. Bosilj, P.; Aptoula, E.; Duckett, T.; Cielniak, G. Transfer Learning between Crop Types for Semantic Segmentation of Crops versus Weeds in Precision Agriculture. *J. Field Robot.* **2019**, *37*, 7–19. [CrossRef]
20. Yeshe, A.; Gourkhede, P.; Vaidya, P. Blue River Technology: Futuristic Approach of Precision Farming. *Just Agric.* **2022**, *2*, 1–14.
21. Kayacan, E.; Kayacan, E.; Ramon, H.; Saeys, W. Towards Agrobots: Identification of the Yaw Dynamics and Trajectory Tracking of an Autonomous Tractor. *Comput. Electron. Agric.* **2015**, *115*, 78–87. [CrossRef]
22. Ecorobotix. The Autonomous Robot Weeder from Ecorobotix. 2017. Available online: https://www.agriculture-xprt.com/products/autonomous-robot-weeder-579629 (accessed on 10 August 2021).
23. Bawden, O.; Ball, D.; Kulk, J.; Perez, T.; Russell, R. A Lightweight, Modular Robotic Vehicle for the Sustainable Intensification of Agriculture. In Proceedings of the 16th Australasian Conference on Robotics and Automation. Australian Robotics and Automation Association (ARAA), Melbourne, VIC, Australia, 2–4 December 2014; pp. 1–9.
24. Utstumo, T.; Urdal, F.; Brevik, A.; Dørum, J.; Netland, J.; Overskeid, O.; Berge, T.; Gravdahl, J. Robotic In-Row Weed Control in Vegetables. *Comput. Electron. Agric.* **2018**, *154*, 36–45. [CrossRef]
25. Grimstad, L.; From, P.J. Thorvald II: A Modular and Re-configurable Agricultural Robot. *IFAC PapersOnLine* **2017**, *50*, 4588–4593. [CrossRef]
26. Underwood, J.; Wendel, A.; Schofield, B.; McMurray, L.; Kimber, R. Efficient in-field plant phenomics for row-crops with an autonomous ground vehicle. *J. Field Robot.* **2017**, *34*, 1061–1083. [CrossRef]
27. Diaz, J.F.A.; Guerrero, H.B.; Tiberti, A.J.; Tabile, R.A.; Tangerino, G.T.; Torres, C.J.; Souza, R. AgriBOT-Mobile Robot to Support of Agricultural Precision Activities. In Proceedings of the 22nd International Congress of Mechanical Engineering, Ribeirao Preto, Brasil, 3–7 November 2013; pp. 6243–6254.
28. Biber, P.; Weiss, U.; Dorna, M.; Albert, A. Navigation System of the Autonomous Agricultural Robot BoniRob. In Proceedings of the Workshop on Agricultural Robotics: Enabling Safe, Efficient, and Affordable Robots for Food Production, Vilamoura, Portugal, 11–12 October 2012.
29. Young, S.N.; Kayacan, E.; Peschel, J.M. Design and Field Evaluation of a Ground Robot for High-Throughput Phenotyping of Energy Sorghum. *Precis. Agric.* **2019**, *20*, 697–722. [CrossRef]
30. Cho, J.; Lee, K.; Shin, E.; Choy, G.; Do, S. How Much Data is Needed to Train a Medical Image Deep Learning System to Achieve Necessary High Accuracy? *arXiv* **2015**, arXiv:1511.06348.
31. Karavidas, I.; Ntatsi, G.; Vougeleka, V.; Karkanis, A.; Ntanasi, T.; Saitanis, C.; Agathokleous, E.; Ropokis, A.; Sabatino, L.; Tran, F. Agronomic Practices to Increase the Yield and Quality of Common Bean (*Phaseolus vulgaris* L.): A Systematic Review. *Agronomy* **2022**, *12*, 271. [CrossRef]
32. Supervisely—Web Platform for Computer Vision. Annotation, Training and Deploy. 2022. Available online: https://supervise.ly/ (accessed on 28 March 2022).

33. Chen, L.; Papandreou, G.; Schroff, F.; Adam, H. Rethinking Atrous Convolution for Semantic Image Segmentation. *arXiv* **2017**, arXiv:1706.05587.
34. Lin, T.Y.; Maire, M.; Belongie, S.; Hays, J.; Perona, P.; Ramanan, D.; Dollár, P.; Zitnick, C.L. Microsoft COCO: Common Objects in Context. In Proceedings of the European Conference on Computer Vision, Zurich, Switzerland, 6–12 September 2014; pp. 740–755.
35. Yu, C.; Wang, J.; Peng, C.; Gao, C.; Yu, G.; Sang, N. BiSeNet: Bilateral Segmentation Network for Real-time Semantic Segmentation. In Proceedings of the European Conference on Computer Vision (ECCV), Munich, Germany, 8–14 September 2018.
36. Jin, Q.; Meng, Z.; Pham, T.D.; Chen, Q.; Wei, L.; Su, R. DUNet: A Deformable Network for Retinal Vessel Segmentation. *Knowl.-Based Syst.* **2019**, *178*, 149–162. [CrossRef]
37. Chen, L.; Zhu, Y.; Papandreou, G.; Schroff, F.; Adam, H. Encoder-Decoder with Atrous Separable Convolution for Semantic Image Segmentation. In Proceedings of the European Conference on Computer Vision (ECCV), Munich, Germany, 8–14 September 2018.
38. Zhao, H.; Shi, J.; Qi, X.; Wang, X.; Jia, J. Pyramid Scene Parsing Network. In Proceedings of the IEEE Conference on Computer Vision and Pattern Recognition, Honolulu, HI, USA, 21–26 July 2017.

agriculture

MDPI

Article

Task Space Model Predictive Control for Vineyard Spraying with a Mobile Manipulator

Ivo Vatavuk *, Goran Vasiljević and Zdenko Kovačić

Faculty of Electrical Engineering and Computing, University of Zagreb, Unska 3, 10000 Zagreb, Croatia; goran.vasiljevic@fer.hr (G.V.); zdenko.kovacic@fer.hr (Z.K.)
* Correspondence: ivo.vatavuk@fer.hr

Abstract: In this paper, a Model Predictive Control (MPC)-based approach for vineyard spraying is presented, able to adapt to different vine row structures and suitable for real-time applications. In the presented approach, the mobile base moves along a row of vines while the robotic arm controls the position and orientation of the spray nozzle. A reference lawnmower pattern trajectory is generated from the vine canopy description, with the aim of minimizing waste while ensuring vine coverage. MPC is used to compute the trajectory of the vehicle along the row and the manipulator tool trajectory, which follow the spray reference, while minimizing vehicle acceleration and tool displacement. The manipulator tool velocity commands provided by the MPC algorithm are tracked using task space control. The presented approach is evaluated in two experiments: a vineyard spraying scenario and an external evaluation scenario in an indoor environment equipped with the Optitrack camera system.

Keywords: mobile manipulation; optimization and optimal control; agricultural robotics; viticulture

Citation: Vatavuk, I.; Vasiljević, G.; Kovačić, Z. Task Space Model Predictive Control for Vineyard Spraying with a Mobile Manipulator. *Agriculture* **2022**, *12*, 381. https://doi.org/10.3390/agriculture12030381

Academic Editors: Jin Yuan, Wei Ji and Qingchun Feng

Received: 8 February 2022
Accepted: 7 March 2022
Published: 9 March 2022

1. Introduction

Agricultural robotics is an exciting, emerging research field that offers a potential solution to the problem of increasing global demand for food production due to exponential population growth and labor shortages [1]. Existing automated agricultural technologies use large and expensive machines that are strictly tied to a specific production process and usually have a significant environmental impact, especially ones used for pesticide, insecticide and herbicide application [2].

The research presented in this paper is a part of the HEKTOR project [3,4], which aims to replace such machines with heterogeneous autonomous robotic systems, capable of jointly performing different tasks in viticulture and mariculture. As a part of the project, a custom flipper-tracked mobile base was developed and equipped with a torque-controlled 7-DoF Kinova Gen3 robotic arm (Figure 1). For the task of vineyard spraying, a spray nozzle is attached to the manipulator end-effector, and its position and orientation are controlled by the robotic arm (Figure 1). This allows for precise control of the spray area, and is suitable for treating the entire vine canopy or specific areas of the plant. The HEKTOR project addresses viticulture activities that need to be carried out on steep terrain, typical of the Mediterranean islands. However, the method presented in this paper is general and was not developed specifically for steep terrains. The design of the mobile robot and the high torques of the track drives allow the storage and transport of a sufficient quantity of protective agent for spraying in conditions that are otherwise difficult for human workers.

This paper attempts to present a solution to the following problem: given a description of a row of grapevines, one must select coordinated mobile vehicle and robot arm commands that result in satisfactory canopy coverage, while aiming to minimize spraying agent waste, and perform the task as quickly as possible. This kind of problem setup calls for a control method that is able to adapt to different row structures, accelerate in areas of the row without grapevines and slow down in areas with the largest foliage heights.

Figure 1. Mobile manipulator developed for the HEKTOR project, with an emphasis on manipulation ability and maneuverability in steep terrain.

1.1. Related Work

Research efforts have already been directed towards the development of robots for vineyard-related tasks [5–22]. The same applies to robots in agricultural spraying tasks [2,6–9,11,13,23]. In [21], Bouloumpasi et al. discuss the possibilities and limitations of autonomous robot technology for performing different tasks in viticulture. Vrochnidou et al. [22] present the system architecture of an autonomous robot for grape harvesting, and discuss the vision system used for the task of grape harvesting.

Monta et al. [5] designed a multipurpose robot for viticulture applications as early as 1995. Approaches for robotic harvesting, berry thinning, spraying and bagging are presented in the paper. At a conceptual level, the authors' approach to vineyard spraying is similar to the one presented in this paper, in which a spray nozzle is mounted on a robotic manipulator that follows a lawnmower pattern path. Monta et al. conclude that this kind of approach results in uniform spraying of the fixed-height foliage, but they do not discuss in detail the control algorithm used to achieve this. In this paper, an expansion of the following idea is given, able to handle different foliage shapes, with controls that allow for spraying where the mobile manipulator is able to adapt to a specific vine row description. Oberti et al. [7,8] mounted a precision spraying end-effector on a robotic manipulator and used it to spray specific disease-affected areas of the vine. In [13], Cantelli et al. present a mobile vehicle developed for autonomous spraying in agriculture, with emphasis on vehicle navigation and mission planning. Berenstein et al. [6] report on grape cluster and foliage detection algorithms for autonomous selective vineyard spraying. The authors present a mobile vehicle with multiple spray nozzles mounted at different heights and use visual feedback to select which nozzles should be active.

Similar work has been done with mobile manipulators for paint spraying [24–26]. Here, a lawnmower path is often used as a reference for the position of the tool, an approach that was adopted in the currently presented work.

Model Predictive Control (MPC) has been a subject of a number of research applications in robotics, such as manipulation, autonomous vehicle control and legged robot control [27–31]. Wieber [27] presents the application of linear MPC to the bipedal walking robot, and discusses the advantages of this kind of controller for this particular task. In [29], Elsisi presents an optimally designed nonlinear model predictive controller, and its application to the robot manipulator. The parameters for the MPC algorithm presented by Elsisi are selected by a modified multitracker optimization algorithm, rather than by trial and error. In later work, Elsisi et al. focus on autonomous vehicle control [30,31]. In [30], an MPC algorithm with a small number of parameters is presented, which is optimally designed using the social ski diver algorithm. In [31], an adaptive MPC algorithm is used for autonomous vehicle control, utilizing a modified grey wolf optimizer, based on opposition-based learning and quasi opposition-based learning. In this paper, MPC parameters are selected experimentally, while the more sophisticated MPC tuning algorithms will be considered as part of future work.

The use of different trajectory optimization algorithms in mobile manipulation is also a well-researched topic [28,32–35]. These trajectory planning algorithms are often designed to be as general as possible, capable of performing multiple tasks while considering obstacles, mobile manipulator dynamics and stability. In [28], Pankert et al. present a nonlinear MPC scheme for continuous mobile manipulation, and show its ability to perform several tasks. The strength of the authors' approach lies in the variety of tasks it can perform, involving position or force control, while avoiding obstacles. One of the experiments presented in the paper is spraying a pattern on a flat surface, which is similar to the presented vision of vineyard spraying with a mobile manipulator. However, the method presented in this paper simplifies the planning problem by using certain insights into this particular task. The complexity of the general mobile manipulation problem includes vehicle path planning while considering obstacles, and controlling the forces or positions of the robot arm end-effector. Grapevines are typically planted in structured rows, the area between the rows being obstacle free. During vineyard spraying, the mobile base is limited to movement between the rows. For this particular application, this paper proposes a task space MPC algorithm that considers only the motion of the mobile base along the row and the two-dimensional motion of the manipulator tool. The result is a low-dimensional linear MPC algorithm that can be solved in real time even with large prediction horizons. Additionally, in the spraying experiment presented in [28], the orientation of the spray nozzle is kept constant throughout the task execution. This greatly reduces the maximum and minimum heights that the spray area can reach, which also depends on the reach of the particular robot arm. Knowing the typical foliage heights and the reach of the Kinova Gen3 robotic arm, it was concluded that constant tool orientation was not appropriate for the vineyard spraying task.

1.2. Contribution

This paper presents a novel method for vineyard spraying with a mobile manipulator that combines the following methodological elements:

- Row-specific reference trajectory generation based on grapevine canopy description;
- Forward mobile base and two-dimensional task space manipulator command generation using linear reference tracking MPC;
- Manipulator joint space velocity command selection using task space control.

Compared to other state-of-the-art autonomous vineyard spraying methods, the presented method includes multiple novelties. The lawnmower spraying reference trajectory is generated based on a specific grapevine row description, aiming to minimize spraying waste and reduce the risk of excessive pollution. The control design based on task space model predictive control allows the spraying agent to follow this reference while optimizing coordinated mobile base and robot arm movements. This results in the desired behavior, where the mobile base keeps constantly moving, accelerating in areas of the row without grapevines and slowing down in areas with foliage of the largest heights. The predictive nature of MPC allows for such behavior, which would be difficult to achieve using standard, instantaneous control methods. Predictive control allows the controller to anticipate any changes in the reference trajectory and optimally select how the mobile base and the robot arm should react to it.

Compared to the state of the art in mobile manipulation, a control algorithm for the specific purpose of vineyard spraying is developed instead of using a general trajectory optimization algorithm. Unlike the general trajectory optimization algorithms, which usually include end-effector pose planning, the presented algorithm combines MPC and task space control in such a way that the MPC plans only the positions of the end-effector while disregarding its orientations, which are handled by an instantaneous task space control algorithm.

This kind of approach results in two quadratic programming problems, for MPC and task space control, which can be solved in real time even for large prediction horizons.

To summarize, our contributions are as follows:

- A novel method for vineyard spraying with mobile manipulators able to adapt to a specific grapevine row description;
- Reference trajectory generation based on grapevine row description;
- Control design based on computationally efficient task space trajectory tracking MPC that exploits the insight into the motion constraints imposed by the specific task of vineyard spraying.

2. Task Space Model Predictive Control Approach

Since the vines are usually planted in structured rows, it is assumed that the vehicle moves in a straight line at a constant distance from the vines, and, because of this, the spray reference for the MPC is generated in a two-dimensional space. As shown in Figure 2, three coordinate frames are defined: a ground-level global frame L_G, a mobile base frame L_B and the spray frame L_S. The spray frame L_S is defined at a fixed distance from the spray nozzle, and its position is computed using a single static transformation from the last link of the robot manipulator. The goal is to control the global position of the L_S frame, which depends on the position of the L_B frame with respect to L_G, and the pose of the robot arm. The y-coordinate of the L_S frame is assumed to be constant at the MPC stage, which is enforced by task space control. The z-coordinate of the L_S frame, $p_{S,z}$, is controlled by the robot arm alone, while the x-coordinate $p_{S,x}$ is a sum of two components

$$p_{S,x} = p_{B,x} + p_{A,x} \tag{1}$$

where $p_{A,x}$ is the x coordinate of the position of the L_S frame with respect to L_B, controlled by the robot arm, and $p_{B,x}$ is the x coordinate of the position of L_B with respect to L_G, controlled by the mobile base.

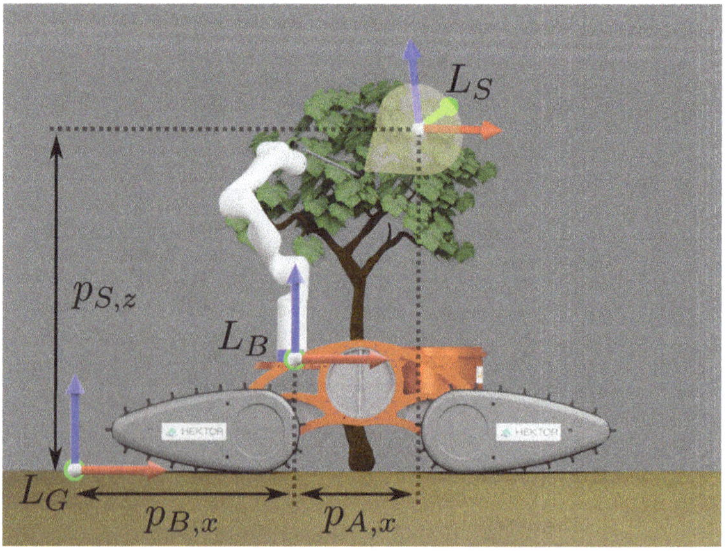

Figure 2. Three coordinate frames are defined: a global frame at the ground level L_G, mobile base frame L_B and the spray frame L_S. The x, y and z axes of the coordinate frames are represented with red, green and blue arrows, respectively.

The overall control diagram of the system is shown in Figure 3. The reference trajectory for the spray frame is computed based on the canopy description, explained in Section 2.1. This reference is used as an input to the MPC solver, which selects optimal mobile base and manipulator task space trajectories. Only the first set of optimized control inputs is sent to the robot, after which the trajectories are recalculated based on the system state feedback and the updated reference trajectory. The task space commands for the manipulator are converted into joint velocities by the task space controller.

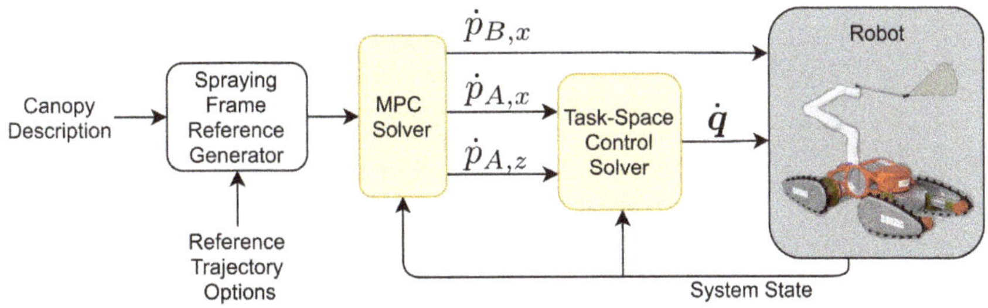

Figure 3. Overall system control diagram. The trajectory of the reference spray frame is generated based on the canopy description and used as input to the MPC solver. The MPC solver provides the velocity of the mobile base along the row and the velocities of the robot arm in the task space. The task space control solver converts the desired velocities in the task space into joint velocity commands $\dot{q}$.

2.1. Reference Spray Frame Trajectory

A reference lawnmower trajectory is generated within the upper and lower boundaries of the foliage, with the aim of providing spray coverage while reducing waste (Figure 4). The foliage canopy description is used to compute the vertices of the lawnmower pattern, along with the Spray Width and Height Offset parameters (Figure 4). These parameters are tuned with respect to the spray pattern of the nozzle mounted at the end-effector of the robot arm. A constant velocity piecewise linear function is used as the reference trajectory between the lawnmower pattern vertices. The reference velocity of the spray frame, together with the flow of the spraying agent through the nozzle, affects the coverage of a fixed area of the canopy. For a fixed amount of spray flow, lower reference velocity will result in more coverage.

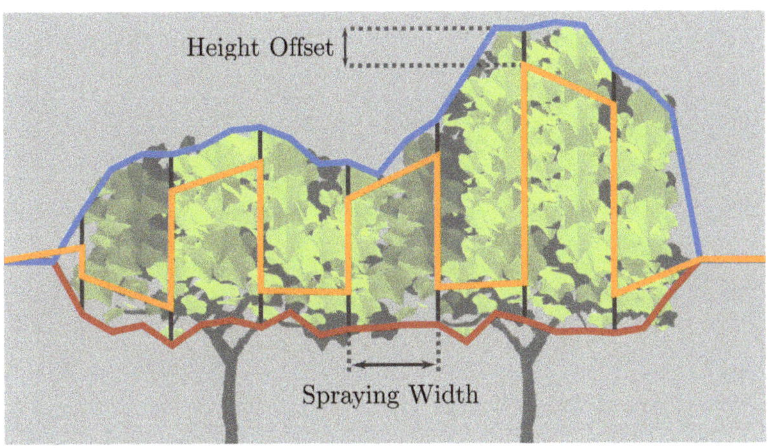

Figure 4. A reference lawnmower trajectory is shown with an orange line. The canopy description is represented by a blue and a red line, representing the upper and lower boundaries of the foliage, respectively.

The canopy description can be created either manually, by an operator, or using foliage detection. In the experiments presented in this paper, the manual approach was used to select the boundaries. In the future work, the plan is to use a foliage detection algorithm for this purpose.

2.2. MPC Algorithm

The MPC algorithm selects trajectories for the mobile base and the robot arm task space that follow a two-dimensional reference, while minimizing an objective function described in this section. The optimization is performed over a prediction horizon N, with fixed discretization time steps T. Although the velocities of the mobile base and the robot arm are used for low-level control (as seen in Figure 3), MPC operates in the acceleration space to achieve the desired behavior. The system state and control inputs in the k-th step, $\boldsymbol{x}(k)$ and $\boldsymbol{u}(k)$, respectively, are

$$\boldsymbol{x}(k) = \begin{bmatrix} p_{B,x}(k) \\ p_{A,x}(k) \\ p_{S,z}(k) \\ \dot{p}_{B,x}(k) \\ \dot{p}_{A,x}(k) \\ \dot{p}_{S,z}(k) \end{bmatrix} \in \mathbb{R}^6 \quad \boldsymbol{u}(k) = \begin{bmatrix} \ddot{p}_{B,x}(k) \\ \ddot{p}_{A,x}(k) \\ \ddot{p}_{S,z}(k) \end{bmatrix} \in \mathbb{R}^3 \tag{2}$$

System state consists of positions $p_{B,x}(k)$, $p_{A,x}(k)$ and $p_{S,z}(k)$, seen in Figure 2, as well as the corresponding velocities. System inputs are the accelerations of the spray frame $\ddot{p}_{B,x}$, $\ddot{p}_{A,x}$ and $\ddot{p}_{S,z}$. The output of the system is a two-dimensional position of the spray frame L_S, relative to the global frame L_G. The output of the system in the k-th step $\boldsymbol{y}(k)$ is

$$\boldsymbol{y}(k) = \begin{bmatrix} p_{B,x}(k) + p_{A,x}(k) \\ p_{S,z}(k) \end{bmatrix} \in \mathbb{R}^2 \tag{3}$$

As already mentioned, the z component of the spray frame position $p_{S,z}$ is controlled solely by the robot arm, and the x component is controlled by both the robot arm and the mobile base (Equation (1)).

The system is therefore linear and its discretization results in the following system dynamics:

$$\boldsymbol{x}(k+1) = \underbrace{\begin{bmatrix} 1 & 0 & 0 & T & 0 & 0 \\ 0 & 1 & 0 & 0 & T & 0 \\ 0 & 0 & 1 & 0 & 0 & T \\ 0 & 0 & 0 & 1 & 0 & 0 \\ 0 & 0 & 0 & 0 & 1 & 0 \\ 0 & 0 & 0 & 0 & 0 & 1 \end{bmatrix}}_{\boldsymbol{A}} \boldsymbol{x}(k) + \underbrace{\begin{bmatrix} \frac{T^2}{2} & 0 & 0 \\ 0 & \frac{T^2}{2} & 0 \\ 0 & 0 & \frac{T^2}{2} \\ T & 0 & 0 \\ 0 & T & 0 \\ 0 & 0 & T \end{bmatrix}}_{\boldsymbol{B}} \boldsymbol{u}(k) \tag{4}$$

$$\boldsymbol{y}(k) = \underbrace{\begin{bmatrix} 1 & 1 & 0 & 0 & 0 & 0 \\ 0 & 0 & 1 & 0 & 0 & 0 \end{bmatrix}}_{\boldsymbol{C}} \boldsymbol{x}(k) \tag{5}$$

MPC optimizes over a system input vector $\boldsymbol{U}$ containing N control inputs,

$$\boldsymbol{U} = \begin{bmatrix} \boldsymbol{u}(0) \\ \boldsymbol{u}(1) \\ \vdots \\ \boldsymbol{u}(N-1) \end{bmatrix} \in \mathbb{R}^{3N} \tag{6}$$

The system state and output vectors $\boldsymbol{X}$ and $\boldsymbol{Y}$ are, respectively,

$$\boldsymbol{X} = \begin{bmatrix} \boldsymbol{x}(0) \\ \boldsymbol{x}(1) \\ \vdots \\ \boldsymbol{x}(N-1) \end{bmatrix} \in \mathbb{R}^{6N}, \ \boldsymbol{Y} = \begin{bmatrix} \boldsymbol{y}(0) \\ \boldsymbol{y}(1) \\ \vdots \\ \boldsymbol{y}(N-1) \end{bmatrix} \in \mathbb{R}^{2N} \tag{7}$$

The MPC optimization problem is defined as the following QP problem:

$$\begin{aligned} \min_{\boldsymbol{U}} \quad & W_y\|\boldsymbol{Y}-\boldsymbol{Y}_d\|^2 + \|\boldsymbol{W}_u\boldsymbol{U}\|^2 + \|\boldsymbol{W}_x\boldsymbol{X}\|^2 \\ \text{s.t.} \quad & \boldsymbol{x}(k+1) = \boldsymbol{A}\boldsymbol{x}(k) + \boldsymbol{B}\boldsymbol{u}(k) \\ & \boldsymbol{y}(k) = \boldsymbol{C}\boldsymbol{x}(k) \\ & \boldsymbol{x}(0) = \boldsymbol{x}_0 \\ & \underline{\boldsymbol{x}} \le \boldsymbol{x}(k) \le \overline{\boldsymbol{x}} \\ & \underline{\boldsymbol{u}} \le \boldsymbol{u}(k) \le \overline{\boldsymbol{u}} \end{aligned} \tag{8}$$

where $\boldsymbol{Y}_d$ is the desired system output and W_y is the reference tracking weight. $\boldsymbol{W}_u$ and $\boldsymbol{W}_x$ are control input and system state weight matrices, respectively.

The reference tracking part of the criterion function ($W_y\|\boldsymbol{Y}-\boldsymbol{Y}_d\|^2$) is used to ensure that the system output follows the reference trajectory for the spray frame. The second part ($\|\boldsymbol{W}_u\boldsymbol{U}\|^2$) minimizes the system inputs: the acceleration of the vehicle along the row and the task space acceleration of the manipulator. The control input weight matrix is

$$\boldsymbol{W}_u = \begin{bmatrix} \boldsymbol{w}_u \\ \boldsymbol{w}_u \\ \vdots \\ \boldsymbol{w}_u \end{bmatrix}^{\mathrm{T}}, \boldsymbol{w}_u = \begin{bmatrix} w_{\ddot{p}_{B,x}} & 0 & 0 \\ 0 & w_{\ddot{p}_{A,x}} & 0 \\ 0 & 0 & w_{\ddot{p}_{S,z}} \end{bmatrix} \tag{9}$$

where the scalars $w_{\ddot{p}_{B,x}}$, $w_{\ddot{p}_{A,x}}$ and $w_{\ddot{p}_{A,z}}$ represent weights with respect to the acceleration of the mobile base along the row, and the acceleration of the robot arm in the x and z directions, respectively.

The third and final part of the criterion function ($\|\boldsymbol{W}_x\boldsymbol{X}\|^2$) minimizes a function of the system state vector. In particular, it minimizes the displacement of the robot arm in the x direction, with a weight matrix of the following form:

$$\boldsymbol{W}_x = \begin{bmatrix} \boldsymbol{w}_x \\ \boldsymbol{w}_x \\ \vdots \\ \boldsymbol{w}_x \end{bmatrix}^{\mathrm{T}}, \ \boldsymbol{w}_x = \begin{bmatrix} 0 & 0 & 0 & 0 & 0 & 0 \\ 0 & w_{p_{A,x}} & 0 & 0 & 0 & 0 \\ 0 & 0 & 0 & 0 & 0 & 0 \\ 0 & 0 & 0 & 0 & 0 & 0 \\ 0 & 0 & 0 & 0 & 0 & 0 \\ 0 & 0 & 0 & 0 & 0 & 0 \end{bmatrix} \tag{10}$$

2.2.1. MPC Parameter Tuning

The main challenge in tuning the parameters of the criterion function (8) is the balance between robot arm and mobile base motion. The x component of the reference trajectory can be tracked by either robot arm or mobile base. The idea behind minimizing arm displacement and mobile base acceleration is to achieve the effect where the mobile base is responsible for slower, global changes in the reference trajectory and the arm is responsible for faster, local changes. Parameters are tuned by trial and error to achieve this effect. Extreme examples of mobile base and robot arm trajectories are shown in Figure 5, with the corresponding optimization weights presented in Table 1.

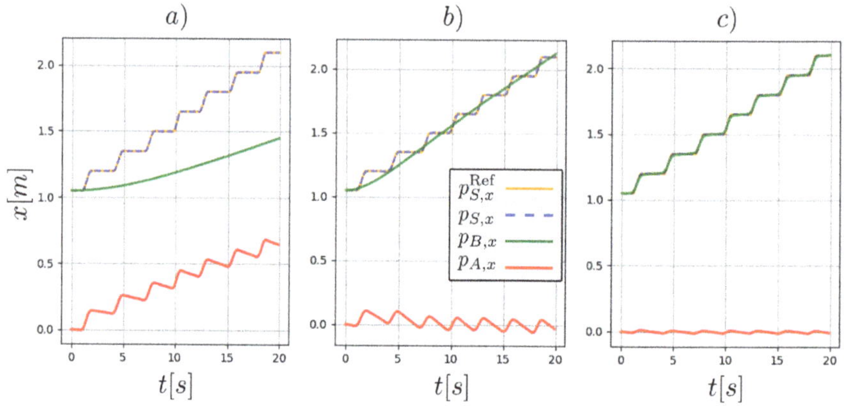

Figure 5. Solutions to the MPC optimization problem in the x direction using different criterion function parameters. Parameters are given in Table 1. Spray frame trajectory $p_{S,x}$ is a sum of $p_{A,x}$ and $p_{B,x}$.

Table 1. MPC criterion function parameters resulting in trajectories shown in Figure 5.

Figure 5.	(a)	(b)	(c)
W_y	800.0	800.0	800.0
$w_{\ddot{p}_{B,x}}$	800.0	80.0	8.0
$w_{\ddot{p}_{A,x}}$, $w_{\ddot{p}_{S,z}}$	4.0	8.0	16.0
$w_{p_{A,x}}$	0.5	1.0	2.0

The behavior of the trajectories selected by the MPC algorithm depends on the ratios between the different optimization weights rather than their exact value. Figure 5 shows solutions to the MPC problem with different ratios between the optimization weights, where the system assumes an initial state with zero velocity. Increasing the weight corresponding to the vehicle acceleration $w_{\ddot{p}_{B,x}}$ too much leads to small accelerations of the mobile base and thus to a large arm displacement (Figure 5a). Lowering the value of $w_{p_{A,x}}$ and $w_{\ddot{p}_{A,x}}$ has a similar effect. Lowering $w_{\ddot{p}_{B,x}}$ or increasing $w_{p_{A,x}}$ and $w_{\ddot{p}_{A,x}}$ too much (Figure 5c) leads to the effect where the mobile base follows the reference with large accelerations, while the robot arm displacement remains close to zero. To achieve the desired behavior (Figure 5b), parameters are tuned following a general tuning rule: a larger weight is used for minimization of the vehicle acceleration than for that of the robot arm. This, in combination with the minimization of the arm displacement, leads to the already discussed effect.

For a set of reference trajectories generated as shown in Section 2.1, it was concluded that different values can be used as weights for the MPC criterion function and still result in satisfactory overall system behavior. The values for the sampling period, horizon and weights of the criterion function used in the experiments can be found in Section 3.

2.2.2. MPC Constraints

The constraints on the system state and the control inputs have the following form:

$$\begin{bmatrix} -p_{B,x}^{\max} \\ -p_{A,x}^{\max} \\ -p_{S,z}^{\max} \\ -\dot{p}_{B,x}^{\max} \\ -\dot{p}_{A,x}^{\max} \\ -\dot{p}_{S,z}^{\max} \end{bmatrix} \leq \boldsymbol{x}(k) \leq \begin{bmatrix} p_{B,x}^{\max} \\ p_{A,x}^{\max} \\ p_{S,z}^{\max} \\ \dot{p}_{B,x}^{\max} \\ \dot{p}_{A,x}^{\max} \\ \dot{p}_{S,z}^{\max} \end{bmatrix} \tag{11}$$

$$\begin{bmatrix} -\ddot{p}_{B,x}^{\max} \\ -\ddot{p}_{A,x}^{\max} \\ -\ddot{p}_{S,z}^{\max} \end{bmatrix} \leq \boldsymbol{u}(k) \leq \begin{bmatrix} \ddot{p}_{B,x}^{\max} \\ \ddot{p}_{A,x}^{\max} \\ \ddot{p}_{S,z}^{\max} \end{bmatrix} \tag{12}$$

The motion of the mobile base is left unconstrained ($p_{B,x}^{\max} = \infty$). The maximum velocity and acceleration of the mobile base, $\dot{p}_{B,x}^{\max}$ and $\ddot{p}_{B,x}^{\max}$, respectively, are known and directly enforced by these constraints. The maximum values of the robot arm velocities and accelerations in the task space ($\dot{p}_{A,x}^{\max}$, $\dot{p}_{S,z}^{\max}$, $\ddot{p}_{A,x}^{\max}$ and $\ddot{p}_{S,z}^{\max}$) must be determined experimentally, as they depend on the current arm configuration and the velocity and acceleration limits of each joint, which are not considered by the MPC algorithm. Instead, the joint space constraints are enforced at the level of the task space control algorithm.

2.3. Manipulator Task Space Control

Joint velocities $\dot{\boldsymbol{q}}$ are selected through task space control and are used to control the robot arm ($\boldsymbol{q}$ represents joint positions). As mentioned earlier, the MPC algorithm only considers the x and z coordinates of the spray frame position. In the planning phase (phase of the MPC algorithm), the position of the spray frame in the y direction ($p_{S,y}$) is assumed to be constant, and its roll, pitch and yaw angles are not considered (ϕ_T, θ_T and ψ_T, respectively, as shown in Figure 6).

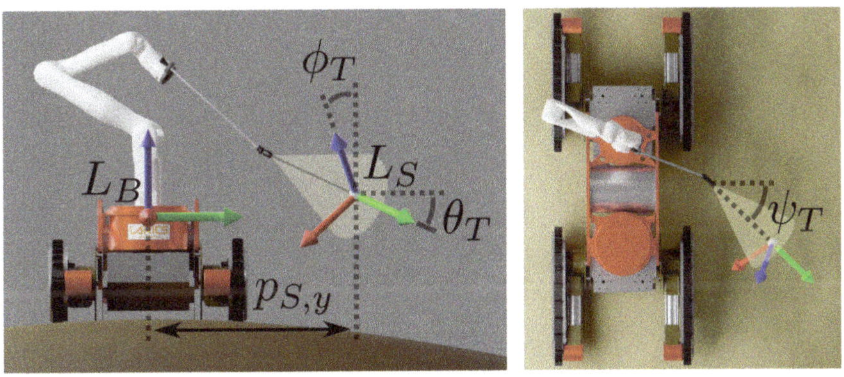

Figure 6. The orientation of the spray frame depends on the joint configuration of the robot arm. Roll, pitch and yaw angles are referred to as ϕ_T, θ_T and ψ_T, respectively.

To achieve the assumption of constant $p_{S,y}$, the desired task space velocity in the y direction, $\dot{p}_{S,y}$, is computed with a proportional controller

$$\dot{p}_{S,y} = K_{P,y}(p_{S,y}^d - p_{S,y}) \tag{13}$$

where $p_{S,y}^d$ is a desired value of $p_{S,y}$ and $K_{P,y}$ is the controller gain.

The desired linear spray frame velocity can be achieved with different angular velocities. To achieve the largest possible linear spray frame velocities, only the roll angle is controlled directly, while the pitch and yaw angles are not considered. The desired roll angle velocity is calculated with a proportional controller

$$\dot{\phi}_T = -K_{P,\phi}\phi_T \tag{14}$$

where $K_{P,\phi}$ is the controller gain and the desired roll angle is zero.

The desired task space velocity is four-dimensional and considers x, y, z position coordinates and the roll angle of the L_S frame,

$$v_d^{4D} = \begin{bmatrix} \dot{p}_{A,x}^M \\ \dot{p}_{S,y} \\ \dot{p}_{S,z}^M \\ \dot{\phi}_T \end{bmatrix} \tag{15}$$

$\dot{p}_{A,x}^M$ and $\dot{p}_{S,z}^M$ are provided by MPC, and $\dot{p}_{S,y}$ and $\dot{\phi}_T$ are provided by proportional controllers (13) and (14), respectively. The four-dimensional task space velocity depends on the joint velocities as follows:

$$v^{4D} = J^{4D}\dot{q} \tag{16}$$

where J^{4D} is the task space Jacobian.

In general, there are multiple solutions for $\dot{q}$ that achieve the desired task space velocities. For this reason, an additional criterion by which the joint commands are selected is introduced. The desired joint velocities $\dot{q}_d$ that drive the robot arm to a desired pose q_d are computed by another proportional controller,

$$\dot{q}_d = K_{P,q}(q_d - q) \tag{17}$$

where $K_{P,q}$ is the controller gain.

Finally, the joint velocity commands are selected by solving the following QP problem:

$$\begin{aligned} \min_{\dot{q}} \quad & \left\| J^{4D}\dot{q} - v_d^{4D} \right\|^2 + w_p \|\dot{q}_d - \dot{q}\|^2 \\ \text{s.t.} \quad & \underline{\dot{q}} \leq \dot{q} \leq \overline{\dot{q}} \end{aligned} \tag{18}$$

where $\underline{\dot{q}}$ and $\overline{\dot{q}}$ are the lower and upper bounds of the joint velocities, respectively, and w_p is the arm pose weight.

3. Results

Two experiments were conducted to evaluate the presented approach: a spraying demonstration in a vineyard and an external validation in an indoor environment equipped with Optitrack cameras. Both experiments were performed with the following optimization weights: $W_y = 800$, $w_{\ddot{p}_{B,x}} = 80$, $w_{\ddot{p}_{A,x}} = 8$, $w_{\ddot{p}_{A,z}} = 8$, $w_{p_{A,x}} = 1$. Parameter values were determined experimentally, as described in Section 2.2.1. Values of proportional controller gains used in the experiments were $K_{P,y} = 0.9$, $K_{P,\phi} = 0.1$ and $K_{P,q} = 1.5$. These values were also determined experimentally. The velocity of the reference lawnmower trajectory used in all experiments was 0.3 m/s. A prediction horizon of 40 steps was used, with a sampling period of $T = 0.1$ s. The optimization problems for the MPC and task space control were both solved using the BPMPD interior point solver for convex QP problems [36]. The average computation times of the MPC and task space control optimization problems were 3.998 ms and 0.201 ms, respectively, for 600 trials. This allows the presented approach to be used in a real-time scenario, even with larger prediction horizons. All experiments were performed on an Intel Core i7-10710U CPU @ 1.60 GHz.

3.1. Equipment

The control equipment used in the experiments is shown in Figure 7. All the software was run on an Intel NUC 10 PC with a Ubuntu 20.04 operating system. The Robot Operating System (ROS) was used as the middleware for controlling the robot arm, the mobile vehicle and the spraying agent pump.

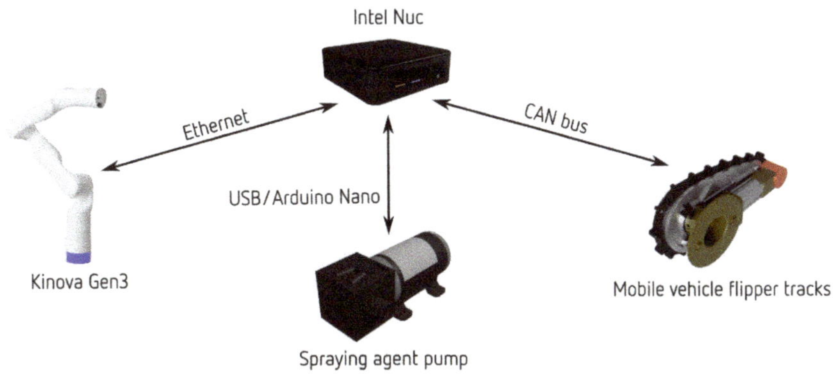

Figure 7. Communication diagram of the equipment used in the vineyard spraying experiments.

The Kinova Gen3 7 DoF robot arm uses an ethernet connection to communicate with ROS drivers running on the PC. As can be seen in Figure 3, low-level control of the robot arm is achieved via joint velocity commands $\dot{\boldsymbol{q}}$. Encoder measurements from the robot arm are used as process feedback. As feedback for the MPC algorithm, the robot arm components of the spray frame position $p_{A,x}$ and $p_{A,z}$ are calculated at every step of the control loop, using forward kinematics and joint position measurements $\boldsymbol{q}$. For the task space control algorithm, joint position data from the encoder are also used as feedback, to calculate the task space Jacobian $\boldsymbol{J}^{4D}$.

Mobile vehicle flipper tracks are actuated by Maxon EC 45 brushless motors, which are controlled using EPOS 70/10 brushless motor drives. EPOS drives communicate with the PC via a CAN bus. ROS drivers for the EPOS drives are used to control the velocities and gather encoder data of each motor. The mobile base component of the spray frame position $p_{B,x}$ is calculated using encoder data odometry and is used as feedback for the MPC algorithm.

Finally, the spraying agent pump is controlled with an Arduino Nano microcontroller board and is either turned on or off based on the current position of the spray frame with respect to the row description.

3.2. *Vineyard Spraying Demonstration*

The first experiment is a demonstration of the presented approach in a vineyard (Figure 8). The canopy description used in the experiment (Figure 9) was selected manually by an operator through a simple graphical user interface (GUI).

A graph showing the reference lawnmower trajectory tracking during the experiment is given in Figure 10. Here, the position of the spray frame $\boldsymbol{p}_S$ is calculated based on the encoder feedback from the robot arm, and the odometry of the mobile base. Since only the odometry feedback is used for mobile base control, its reference tracking is also evaluated with an external sensor, as part of the second experiment in Section 3.3. The MPC criterion function is tuned to follow the reference trajectory imperfectly, to minimize the accelerations of the mobile base and the robot arm end-effector. Figure 11 shows the reference tracking with respect to the actual reference trajectory of the spray frame $\boldsymbol{p}_S^*$, which is formed by accumulating the first control inputs of the optimal control sequences calculated by the MPC algorithm at each control step. There is still some error in the reference tracking caused by the non-ideal following of the joint velocity commands and the error in task space control. The task space Jacobian depends on the joint positions and is computed at the beginning of each control time step. During this time step, the joint positions, and the Jacobian, change. Larger joint velocities and larger time steps result in larger task space control errors. The position tracking errors are shown in Table 2. Spray frame position tracking was evaluated, with a measured root mean square (RMS) error of

4.32 mm, and the maximum error of 22.16 mm. The magnitude of the errors is a result of the fact that the spraying frame is defined at a certain distance to the last link of the robot arm, which makes its position sensitive to small errors in the robot arm joint positions. Based on the imperfections of the spray nozzle and its spraying area, the presented errors are adequate to ensure the effectiveness of the vineyard spraying task. There is a trade-off between reference tracking errors and the velocity of the reference spray frame trajectory. Reducing this velocity would result in slower task execution, but would also reduce the reference tracking errors.

Figure 8. Mobile manipulator performing a spraying experiment in a vineyard.

Figure 9. The canopy description used for the experiment selected by the operator.

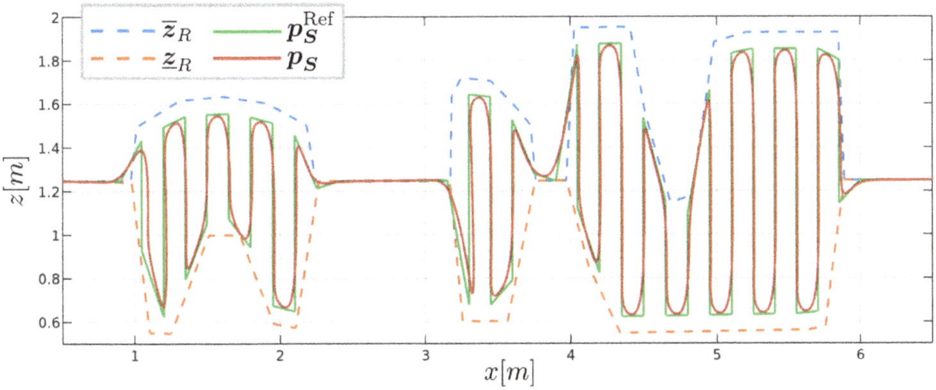

Figure 10. Overall spray frame reference tracking. p_S^{Ref} represents a reference lawnmower trajectory generated based on the row description, where $\bar{z}_R$ and $\underline{z}_R$ represent the upper and lower foliage boundaries, respectively. The spray frame position p_S during the experiment is represented by a red line.

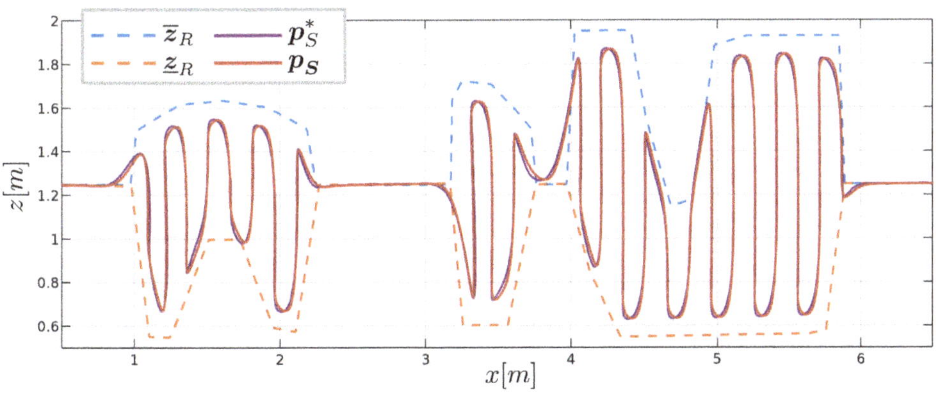

Figure 11. Spray frame tracking with respect to the optimal reference trajectory generated by the MPC algorithm. p_S^* represents the optimal trajectory of the spray frame. This differs from the ideal lawnmower trajectory due to MPC tuning that sacrifices reference tracking to minimize the accelerations of the mobile base and the manipulator end-effector. $\bar{z}_R$ and $\underline{z}_R$ represent the upper and lower foliage boundaries, respectively. The spray frame position p_S during the experiment is represented by a red line.

Table 2. Spray frame position errors during the vineyard spraying experiment.

	p_S	$p_{S,x}$	$p_{S,y}$	$p_{S,z}$
RMS error [mm]	4.32	0.90	3.60	2.20
max error [mm]	22.16	3.92	22.16	18.93

A non-uniform vine row structure allows the demonstration of the adaptability of the presented approach. The velocity of the vehicle adapts to the row description by automatically accelerating in areas without vines and decelerating to the lowest velocities in areas with the highest vines (Figure 12). The footage of the vineyard spraying experiment can be seen in the accompanying video https://youtu.be/BDO7qQldmyQ, accessed on 8 March 2022.

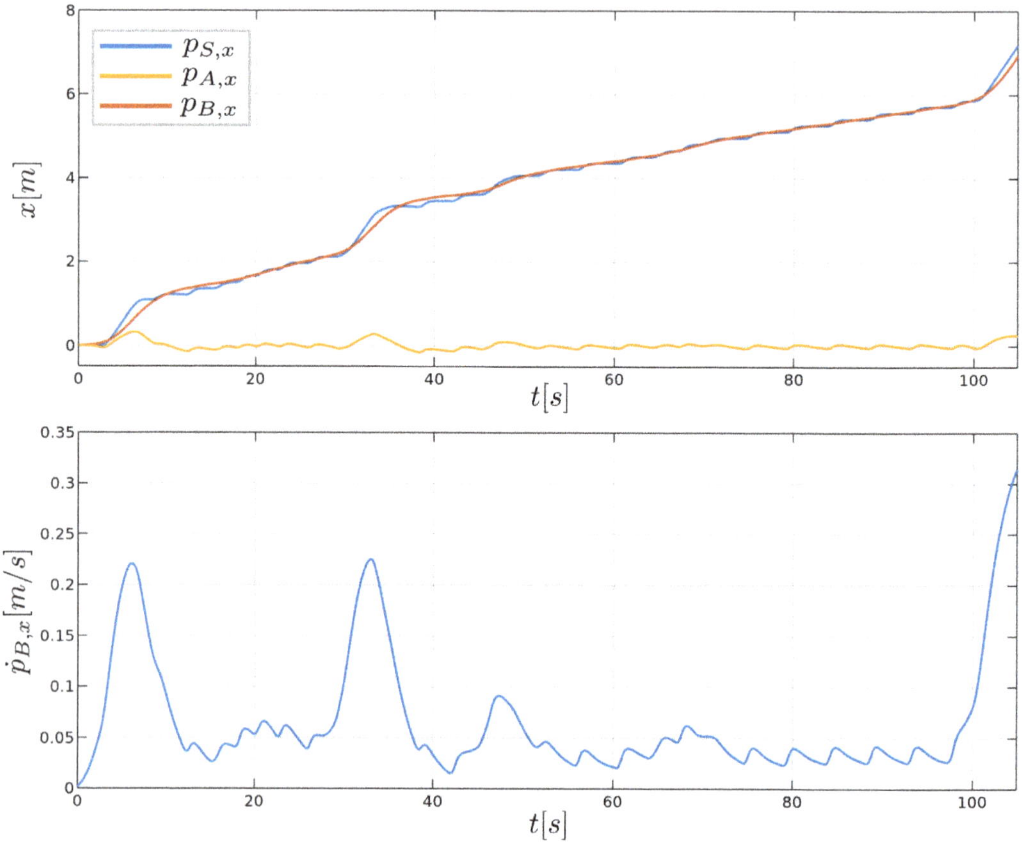

Figure 12. The upper diagram shows the x component of the reference trajectory $p_{S,x}$, along with the robot arm and mobile base components, $p_{A,x}$ and $p_{B,x}$, respectively. The bottom graph shows the forward velocity of the vehicle during the experiment.

3.3. Optitrack Validation

The second experiment was conducted in an indoor environment equipped with the Optitrack camera system, which was used to evaluate the reference tracking with an external sensor. The reference tracking error in this experiment includes the error of the odometry-based mobile base control, the vibrations of the mobile base, the errors present in the robot arm encoder measurements and the Optitrack measurement noise. For practical reasons, the Optitrack markers were placed at the last joint of the robot arm (Figure 13), and the position of the spray frame was calculated using a single static transformation.

Graphs showing the x and z components of the spray frame position during the experiment, along with the tracking errors, are shown in Figures 14 and 15, respectively, and the corresponding error data are given in Table 3. In this experiment, the measured root mean square (RMS) error and the maximum error are equal to 9.76 mm and 52.81 mm, respectively. Errors are larger than the ones in the previous experiment, which is expected due to the already mentioned additional errors that the external sensors are able to capture, and a significant amount of measurement noise. External sensor data confirm that the odometry-based control does not result in a significant drift of the mobile base, as seen in Figure 14.

In Figure 16, the overall spray frame position calculated from the Optitrack data is compared to the position calculated using the joint encoder and vehicle odometry feedback.

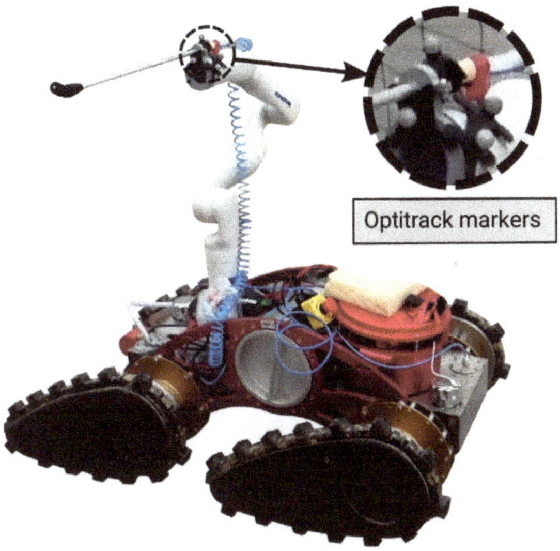

Figure 13. For the second experiment, reference tracking is externally validated using Optitrack cameras to measure the position of the spray frame in the real world. Optitrack markers are attached to the end-effector of the robot arm.

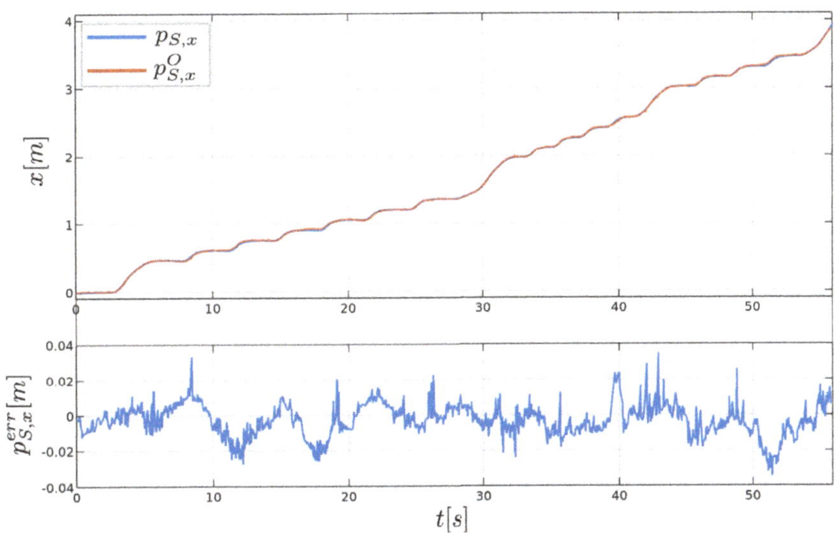

Figure 14. Comparison between the x component of the spray frame position determined by the encoder measurements, and that determined externally via the Optitrack camera system, denoted $p_{S,x}$ and $p^O_{S,x}$, respectively. The bottom plot shows the corresponding error $p^{err}_{S,x}$.

Table 3. Spray frame position errors measured with the Optitrack camera system, during the indoor experiment.

	p_S	$p_{S,x}$	$p_{S,z}$
RMS error [mm]	9.76	7.86	5.79
max error [mm]	52.81	36.59	52.779

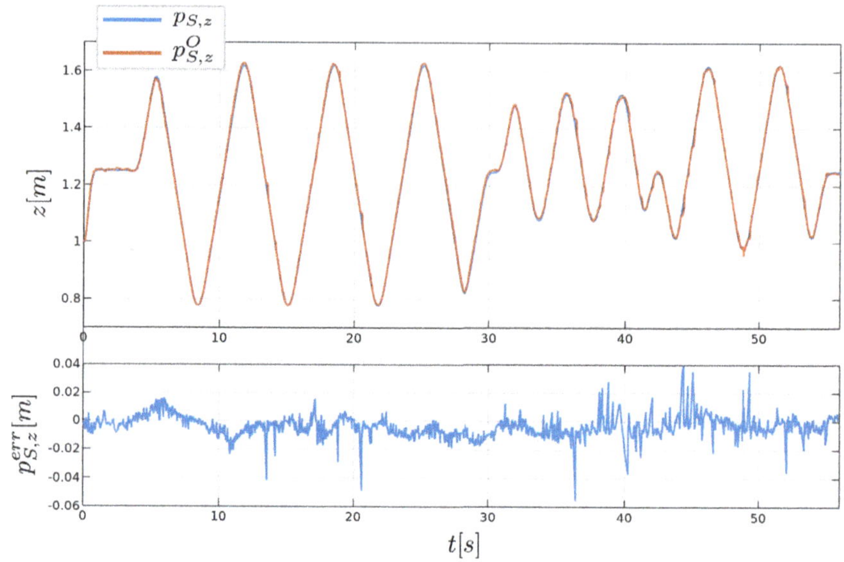

Figure 15. Comparison between the z component of the spray frame position determined by the encoder measurements, and that determined externally via the Optitrack camera system, denoted $p_{S,z}$ and $p^O_{S,z}$, respectively. The bottom plot shows the corresponding error $p^{err}_{S,z}$.

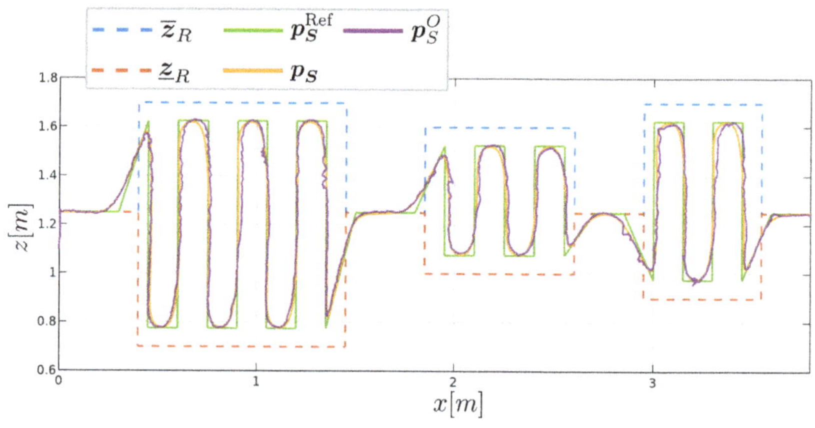

Figure 16. Comparison between the position of the spray frame obtained by the encoder measurements and the position obtained externally via the Optitrack camera system, denoted as $\boldsymbol{p}_S$ and $\boldsymbol{p}^O_S$, respectively. $\boldsymbol{p}^{Ref}_S$ represents the reference lawnmower trajectory, and $\overline{z}_R$ and $\underline{z}_R$ represent the upper and lower foliage boundaries, respectively.

As mentioned earlier, the task space controller selects joint velocities that follow the desired linear and roll spray frame velocities, while attempting to maintain the desired robot arm joint configuration. This results in the yaw and pitch angles of the spray frame shown in Figure 17. This type of control results in a pitch orientation (θ_T) graph similar to the z position, as shown in Figure 15. Similarly, the yaw orientation graph (ψ_T) follows the motion of the robot arm in the x direction.

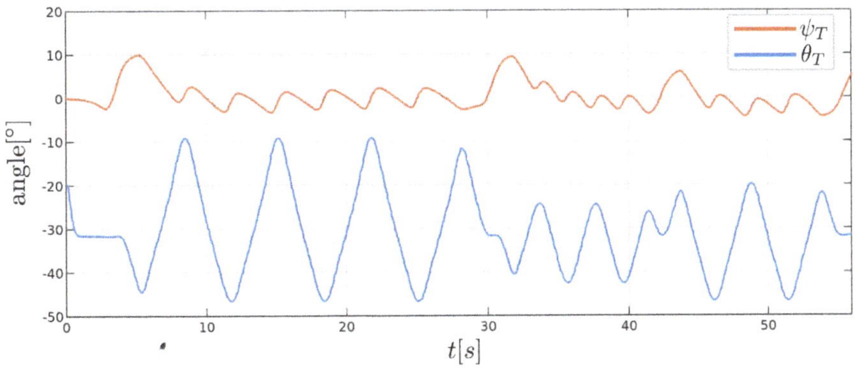

Figure 17. Spray frame orientation during the indoor experiment. The pitch and yaw angles of the spray frame are denoted as θ_T and ψ_T, respectively. These angles are not directly controlled, but are a result of the task space control criterion function.

4. Conclusions and Future Work

In this paper, a vineyard spraying algorithm for mobile manipulators is presented, based on task space model predictive control. The reference is generated based on grapevine canopy description, with the aim of minimizing unnecessary spraying waste and pollution.

There are certain limitations to the presented method. The time required to spray a row of grapevines is limited by the maximum velocity of the vehicle, as well as the maximum joint velocities of the robot arm. Task space control is used to calculate the joint velocity commands for the robot arm, which are not considered in the planning phase (MPC phase) of the algorithm. This could potentially lead to large spray frame velocities that cannot be tracked by the task space controller. Therefore, some experimentation is required to determine the maximum feasible velocity of the lawnmower pattern reference trajectory. Moreover, the task space control algorithm has no direct way of considering joint position constraints of the robot arm. This problem is dealt with indirectly, by allowing different angular velocities of the spraying frame, and constraining the optimization problem in such a way that the solutions moving the joints towards the desired configuration are preferred. No problems were encountered in the experiments regarding joint position constraints. The mobile base is controlled based on odometry feedback, which may lead to certain reference tracking problems since there is no external sensing. The second experiment shows that the open loop control performs well, mainly due to the fact that the vehicle moves in a straight line, which allows precise odometry. In future work, the plan is to close this control loop using a localization algorithm. Moreover, the tilt of the vehicle and other effects of uneven terrain are not taken into account in the current state of the algorithm, which could also be incorporated into future work. In the presented experiments, operator-selected grapevine row description was used. Manual selection of canopy areas proved to be error-prone, tedious and time-consuming. In the future work, a foliage detection algorithm is going to be incorporated for the purpose of generating a grapevine row description. Since the detection algorithm must be robust to changing lighting conditions, it is planned to be based on a combination of deep learning and depth information captured by an RGBD camera. The depth information acquired by the RGBD camera using infrared projection is sensitive to sunlight, so a camera based on pure stereo vision would be suitable for this task.

The presented method was evaluated in a vineyard spraying experiment, demonstrating its ability to adapt to a specific grapevine row structure. Mobile base velocity adapts to the row structure, which can be seen in the accompanying video and the graphs presented in Section 3.2. An additional experiment was performed evaluating the reference tracking with Optitrack cameras as external sensors. Error data show the 4.32 mm and 9.76 mm RMS errors in spray frame position, during the first and second experiment, respectively.

Since the spray frame is located at a certain distance to the last link of the robot arm, its position is sensitive to small joint position errors. The presented error values are sufficiently small for the task of vineyard spraying, while a trade-off exists between reference tracking precision and the time required to execute the task.

The focus of this work was on the control algorithm that sprays a single row of vines. In the future, mission planning and navigation would allow the mobile manipulator to autonomously treat the entire vineyard by entering each row and executing the presented algorithm. Experiments evaluating the spray quality using a water-sensitive paper are planned in the future. Extensive experiments to determine the impact of the presented method on plant health and fruit production and compare it to manual spraying are to be conducted. The presented method will be tested for the task of fruit spraying rather than spraying the entire foliage, which is the focus of this article. Another challenge is the presence of dust in the vineyard, from the influence of which the equipment must be adequately protected. Moreover, while excessive robot arm heating was not noticed during the presented experiments, it could present a potential problem in the case of prolonged robot operation. In this case, some form of active cooling could be used to mitigate the problem. Currently, the spray tank has a volume of 30 L, which will be increased in the future.

Author Contributions: Conceptualization, Z.K., G.V. and I.V.; methodology, I.V.; software, I.V.; validation, I.V. and G.V.; investigation, I.V.; writing—original draft preparation, I.V.; writing—review and editing, G.V. and Z.K.; visualization, I.V.; supervision, Z.K. and G.V.; project administration, Z.K. and G.V.; funding acquisition, Z.K. and G.V. All authors have read and agreed to the published version of the manuscript.

Funding: The research work presented in this article has been supported by the project titled Heterogeneous Autonomous Robotic system in Viticulture and Mariculture (HEKTOR), financed by the European Union through the European Regional Development Fund—The Competitiveness and Cohesion Operational Programme (KK.01.1.1.04.0041).

Institutional Review Board Statement: Not applicable.

Informed Consent Statement: Not applicable.

Conflicts of Interest: The authors declare no conflict of interest.

Abbreviations

The following abbreviations are used in this manuscript:

MPC	Model Predictive Control
QP	Quadratic Programming
CPU	Central Processing Unit
GUI	Graphical User Interface
RMS	Root Mean Square
ROS	Robot Operating System

References

1. Oliveira, L.F.P.; Moreira, A.P.; Silva, M.F. Advances in Agriculture Robotics: A State-of-the-Art Review and Challenges Ahead. *Robotics* **2021**, *10*, 52. [CrossRef]
2. Adamides, G. Agricultural Robots in Targeted Spraying: A mini State-of-the-Art review. *Robot. Autom. Eng. J.* **2017**, *2*, 5581. [CrossRef]
3. Hektor Project Website. Available online: hektor.fer.hr (accessed on 8 March 2022).
4. Goričanec, J.; Kapetanović, N.; Vatavuk, I.; Hrabar, I.; Vasiljević, G.; Gledec, G.; Stuhne, D.; Bogdan, S.; Orsag, M.; Petrović, T.; et al. Heterogeneous autonomous robotic system in viticulture and mariculture-project overview. In Proceedings of the 2021 16th International Conference on Telecommunications (ConTEL), Zagreb, Croatia, 30 June–2 July 2021; pp. 181–188. [CrossRef]
5. Monta, M.; Kondo, N.; Shibano, Y. Agricultural robot in grape production system. In Proceedings of the 1995 IEEE International Conference on Robotics and Automation, Nagoya, Japan, 21–27 May 1995; Volume 3, pp. 2504–2509. [CrossRef]
6. Berenstein, R.; Shahar, O.B.; Shapiro, A.; Edan, Y. Grape clusters and foliage detection algorithms for autonomous selective vineyard sprayer. *Intell. Serv. Robot.* **2010**, *3*, 233–243. [CrossRef]

7. Oberti, R.; Marchi, M.; Tirelli, P.; Calcante, A.; Iriti, M.; Hočevar, M.; Baur, J.; Pfaff, J.; Schütz, C.; Ulbrich, H. Selective spraying of grapevine's diseases by a modular agricultural robot. *J. Agric. Eng.* **2013**, *44*, 149–153. [CrossRef]
8. Oberti, R.; Marchi, M.; Tirelli, P.; Calcante, A.; Iriti, M.; Tona, E.; Hočevar, M.; Baur, J.; Pfaff, J.; Schütz, C.; et al. Selective spraying of grapevines for disease control using a modular agricultural robot. *Biosyst. Eng.* **2016**, *146*, 203–215. [CrossRef]
9. de Soto, M.G.; Emmi, L.; Perez-Ruiz, M.; Aguera, J.; de Santos, P.G. Autonomous systems for precise spraying—Evaluation of a robotised patch sprayer. *Biosyst. Eng.* **2016**, *146*, 165–182. [CrossRef]
10. Botterill, T.; Paulin, S.; Green, R.; Williams, S.; Lin, J.; Saxton, V.; Mills, S.; Chen, X.; Corbett-Davies, S. A Robot System for Pruning Grape Vines. *J. Field Robot.* **2016**, *34*, 1100–1122. [CrossRef]
11. Berenstein, R. The use of agricultural robots in crop spraying/fertilizer applications. In *Robotics and Automation for Improving Agriculture*; Burleigh Dodds Science Publishing: Cambridge, UK, 2019; pp. 109–136. [CrossRef]
12. Santos, L.; Santos, F.; Mendes, J.; Costa, P.; Lima, J.; Reis, R.; Shinde, P. Path Planning Aware of Robot's Center of Mass for Steep Slope Vineyards. *Robotica* **2019**, *38*, 684–698. [CrossRef]
13. Cantelli, L.; Bonaccorso, F.; Longo, D.; Melita, C.D.; Schillaci, G.; Muscato, G. A Small Versatile Electrical Robot for Autonomous Spraying in Agriculture. *AgriEngineering* **2019**, *1*, 391–402. [CrossRef]
14. Santos, L.C.; Aguiar, A.S.; Santos, F.N.; Valente, A.; Ventura, J.B.; Sousa, A.J. Navigation Stack for Robots Working in Steep Slope Vineyard. In *Advances in Intelligent Systems and Computing*; Springer International Publishing: Cham, Switzerland, 2020; pp. 264–285. [CrossRef]
15. Santos, L.C.; Aguiar, A.S.; Santos, F.N.; Valente, A.; Petry, M. Occupancy Grid and Topological Maps Extraction from Satellite Images for Path Planning in Agricultural Robots. *Robotics* **2020**, *9*, 77. [CrossRef]
16. de Aguiar, A.S.P.; dos Santos, F.B.N.; dos Santos, L.C.F.; de Jesus Filipe, V.M.; de Sousa, A.J.M. Vineyard trunk detection using deep learning—An experimental device benchmark. *Comput. Electron. Agric.* **2020**, *175*, 105535. [CrossRef]
17. Sarri, D.; Lombardo, S.; Lisci, R.; Pascale, V.D.; Vieri, M. AgroBot Smash a Robotic Platform for the Sustainable Precision Agriculture. In *Lecture Notes in Civil Engineering*; Springer International Publishing: Cham, Switzerland, 2020; pp. 793–801. [CrossRef]
18. Roure, F.; Moreno, G.; Soler, M.; Faconti, D.; Serrano, D.; Astolfi, P.; Bardaro, G.; Gabrielli, A.; Bascetta, L.; Matteucci, M. GRAPE: Ground Robot for vineyArd Monitoring and ProtEction. In *ROBOT 2017: Third Iberian Robotics Conference*; Springer International Publishing: Cham, Switzerland, 2017; pp. 249–260. [CrossRef]
19. Astolfi, P.; Gabrielli, A.; Bascetta, L.; Matteucci, M. Vineyard Autonomous Navigation in the Echord++ GRAPE Experiment *IFAC-PapersOnLine* **2018**, *51*, 704–709. [CrossRef]
20. Kerkech, M.; Hafiane, A.; Canals, R. Vine disease detection in UAV multispectral images using optimized image registration and deep learning segmentation approach. *Comput. Electron. Agric.* **2020**, *174*, 105446. [CrossRef]
21. Bouloumpasi, E.; Theocharis, S.; Karampatea, A.; Pavlidis, S.; Mamalis, S.; Koundouras, S.; Merou, T.; Vrochidou, E.; Pachidis, T.; Manios, M.; et al. Exploration of viticultural tasks to be performed by an autonomous robot: Possibilities and limitations. In Proceedings of the 11th International Scientific Agriculture Symposium (AGROSYM 2020), Jahorina, Bosnia and Herzegovina, 8–9 October 2020; pp. 56–61.
22. Vrochidou, E.; Tziridis, K.; Nikolaou, A.; Kalampokas, T.; Papakostas, G.A.; Pachidis, T.P.; Mamalis, S.; Koundouras, S.; Kaburlasos, V.G. An Autonomous Grape-Harvester Robot: Integrated System Architecture. *Electronics* **2021**, *10*, 1056. [CrossRef]
23. Adamides, G.; Katsanos, C.; Constantinou, I.; Christou, G.; Xenos, M.; Hadzilacos, T.; Edan, Y. Design and development of a semi-autonomous agricultural vineyard sprayer: Human-robot interaction aspects. *J. Field Robot.* **2017**, *34*, 1407–1426. [CrossRef]
24. Ren, S.; Xie, Y.; Yang, X.; Xu, J.; Wang, G.; Chen, K. A Method for Optimizing the Base Position of Mobile Painting Manipulators. *IEEE Trans. Autom. Sci. Eng.* **2017**, *14*, 370–375. [CrossRef]
25. Liu, G.; Sun, X.; Liu, Y.; Liu, T.; Li, C.; Zhang, X. Automatic spraying motion planning of a shotcrete manipulator. *Intell. Serv. Robot.* **2021**. [CrossRef]
26. Wang, G.; Yu, Q.; Ren, T.; Hua, X.; Chen, K. Task planning for mobile painting manipulators based on manipulating space. *Assem. Autom.* **2018**, *38*, 57–66. [CrossRef]
27. Wieber, P.B. Trajectory Free Linear Model Predictive Control for Stable Walking in the Presence of Strong Perturbations. In Proceedings of the 2006 6th IEEE-RAS International Conference on Humanoid Robots, Genova, Italy, 4–6 December 2006. [CrossRef]
28. Pankert, J.; Hutter, M. Perceptive Model Predictive Control for Continuous Mobile Manipulation. *IEEE Robot. Autom. Lett.* **2020**, *5*, 6177–6184. [CrossRef]
29. Elsisi, M. Optimal design of nonlinear model predictive controller based on new modified multitracker optimization algorithm. *Int. J. Intell. Syst.* **2020**, *35*, 1857–1878. [CrossRef]
30. Elsisi, M.; Ebrahim, M.A. Optimal design of low computational burden model predictive control based on SSDA towards autonomous vehicle under vision dynamics. *Int. J. Intell. Syst.* **2021**, *36*, 6968–6987. [CrossRef]
31. Elsisi, M. Improved grey wolf optimizer based on opposition and quasi learning approaches for optimization: Case study autonomous vehicle including vision system. *Artif. Intell. Rev.* **2022**. [CrossRef]
32. Furuno, S.; Yamamoto, M.; Mohri, A. Trajectory planning of mobile manipulator with stability considerations. In Proceedings of the 2003 IEEE International Conference on Robotics and Automation (Cat. No.03CH37422), Newport Beach, CA, USA, 14–19 September 2003. [CrossRef]

33. Avanzini, G.B.; Zanchettin, A.M.; Rocco, P. Constrained model predictive control for mobile robotic manipulators. *Robotica* **2017**, *36*, 19–38. [CrossRef]
34. Minniti, M.V.; Farshidian, F.; Grandia, R.; Hutter, M. Whole-Body MPC for a Dynamically Stable Mobile Manipulator. *IEEE Robot. Autom. Lett.* **2019**, *4*, 3687–3694. [CrossRef]
35. Korayem, M.; Nazemizadeh, M.; Azimirad, V. Optimal trajectory planning of wheeled mobile manipulators in cluttered environments using potential functions. *Sci. Iran.* **2011**, *18*, 1138–1147. [CrossRef]
36. Mészáros, C. The BPMPD interior point solver for convex quadratic problems. *Optim. Methods Softw.* **1999**, *11*, 431–449. [CrossRef]

agriculture

MDPI

Article

Precision Agriculture Technologies for Crop and Livestock Production in the Czech Republic

Jaroslav Vrchota [1], Martin Pech [1,*] and Ivona Švepešová [2]

[1] Department of Management, Faculty of Economics, University of South Bohemia in Ceske Budejovice, Studentska 13, 370 05 Ceske Budejovice, Czech Republic; vrchota@ef.jcu.cz

[2] Faculty of Economics, University of South Bohemia in Ceske Budejovice, Studentska 13, 370 05 Ceske Budejovice, Czech Republic; svepei01@ef.jcu.cz

* Correspondence: mpechac@ef.jcu.cz

Abstract: Modern technologies are penetrating all fields of human activity, including agriculture, where they significantly affect the quantity and quality of agricultural production. Precision agriculture can be characterised as an effort to improve the results of practical farming, achieving higher profits by exploiting the existing spatial unevenness of soil properties. We aim to evaluate precision agriculture technologies' practical use in agricultural enterprises in the Czech Republic. The research was based on a questionnaire survey in which 131 farms participated. We validated the hypothesis through a Chi-squared test on the frequency of occurrence of end-use technology. The results showed that precision farming technologies are used more in crop than livestock production. In particular, 58.02% of enterprises use intelligent weather stations, 89.31% use uncrewed vehicles, and 61.83% use navigation and optimisation systems for optimising journeys. These technologies are the most used and closely related to autonomous driving and robotics in agriculture. The results indicate how willing are agricultural enterprises to adopt new technologies. For policy makers, these findings show which precision farming technologies are already implemented. This can make it easier to direct funding towards grants and projects.

Keywords: precision agriculture; Industry 4.0; technology; adoption; unmanned vehicles; smart production; drones; robots

Citation: Vrchota, J.; Pech, M.; Švepešová, I. Precision Agriculture Technologies for Crop and Livestock Production in the Czech Republic. *Agriculture* **2022**, *12*, 1080. https://doi.org/10.3390/agriculture12081080

Academic Editors: Jin Yuan, Wei Ji and Qingchun Feng

Received: 19 June 2022
Accepted: 20 July 2022
Published: 22 July 2022

1. Introduction

Today's turbulent times bring new challenges for everyone every day. Society is constantly evolving, and so are the various technologies. The industrial revolution has proceeded gradually since the emergence of mechanisation. The Fourth Industrial Revolution has come sequentially, bringing radical changes across all industries. One of these industries is agriculture. Until a few decades ago, there were more workers in agriculture than in industry. From ancient ages until the early twentieth century, agriculture has always been very demanding, requiring a great deal of physical effort. Still, the profits from these tasks have not been significant. It used to take an average of two acres of cultivated land to feed one man. With the twentieth century came new industrial agriculture, and productivity rose radically [1,2].

Precision agriculture is the term used to describe the association of changes brought about by the Fourth Industrial Revolution in agriculture. Modern technologies enable the precision of work, efficiency, efficient processing of all data and other aspects that will move agriculture to a new level. Precision agriculture means accuracy and implies correctness or precision in any production [3]. The main objective of precision agriculture is to adapt operations to the actual location conditions with the principle of carrying out interventions in the right place, with the right intensity and at the right time. Precision agriculture is currently the most popular in the USA because of its rustic structure and technological maturity. India and North America have the highest [4] technical capacity to pursue an opportunity for smart agriculture. However, McBratney et al. [4] suggest that crop production has the highest potential for precision agriculture according to the spatial index (Ha of cropland per worker) in Canada, Australia, and the USA. Livestock production is highest according to the environmental index (fertiliser use: kg per ha of cropland) in Ireland, the Netherlands, and Egypt.

In the Czech Republic, the agrarian structure is favourable for precision agriculture, as it is dominated by large farms, large plots of land and a diversity of natural conditions combined with soil variability and rugged terrain [5]. These large farms in the Czech Republic combine modern technology, automation and robotisation with a small number of workers focused primarily on maximising production and sales. In contrast, small, family-run farms that focus mainly on quality, regional products, healthy food, horse breeding, or agro-tourism are also thriving. The overall level of involvement in precision agriculture is at a medium level [6]. The adoption of precision agriculture technologies is relatively high [7]. New technologies, the loss of land ownership, the concentration of land in large blocks and the reluctance of people to work seven days a week from morning to night have led to a significant polarisation. The main issues of Czech agriculture are labour shortages, the unfavourable economic situation of most enterprises, and expensive technology.

The main issues in farming communities without precision agriculture are related to uniform and homogenous land management [8]. In this case, the intensity of cultivation interventions is usually chosen based on the average value of the smallest unit area. The most significant advantage of precision agriculture is the ability to identify and determine variability. The primary input information is passed on from generation to generation as the primary know-how. However, agricultural sustainability depends primarily on progress in the efficient use of nitrogen [9] and other agrochemicals. Data from water, nitrogen and pesticide application during the growing season need to be recorded immediately after sensing. It creates significant advantages over traditional farming along with a reduction in human labour and resource efficiency, as outlined in the results of this paper. Thus, the conventional approach is associated with increased costs, production's economic intensity and environmental pollution risk [10]. The main ecological problem is the excessive application of agrochemicals and poorer traceability of records of soil operations. The economic impacts are mainly in the increased cost of material inputs (fertilisers, pesticides and fuel). It is impossible to achieve lower fuel consumption without navigation and satellite technology due to unfamiliarity with the terrain and the use of the optimal route. Another significant problem with these systems is the more challenging identification of harmful organisms due to incorrect demarcation of application zones. Current problems in agriculture are climate change, soil degradation, food unsafety and diversity loss [11].

For the problems mentioned above in farming communities, precision agriculture technologies may be just the solution to enable targeted local interventions. Our research focused on agricultural enterprises to capture the current trends in Czech agriculture. Previous studies of precision agriculture in the Czech Republic are limited to their narrow focus on specific technologies and timeliness. The most recent comprehensive studies date from about five years ago [12,13]. This research gap needed to be filled with current research into determining what technologies are currently the most used in agriculture. Moreover, it would fortify the perspective of enterprises. The article aims to evaluate the practical use of precision agriculture technologies in agricultural enterprises in the Czech Republic. However, there could be differences between crop and livestock production. Thus, we investigated both of these farming areas and compared the results. We stated some recommendations for policymakers and users of precision agriculture technologies.

We divided the article into the following structure: 1. Introduction with basic information on the topic; 2. Theoretical background focused on precision agriculture and technologies; 3. Materials and methods with the definition of research aim and methods; 4. Results including technologies in crop and livestock production; 5. Discussion of main findings; and 6. Conclusion.

2. Theoretical Background

The theoretical background briefly describes the current state of the research field with the definition of precision agriculture and mainly used technologies.

2.1. Definition of Precision Agriculture

Many authors [14–17] speak of revolutions in the era of industrial agriculture. The introduction of tractors brought about the first revolution, combining harvesters and mechanisation. The second revolution was triggered by the development of biotechnology, including the much-discussed genetic manipulation. Later, computer technology began to be used in agriculture to optimise and introduce new production methods.

In recent years, the term Agriculture 4.0 or Precision Agriculture has emerged. The term Industry 4.0 is derived from Agriculture 4.0 or Precision Agriculture. It refers to modern techniques and technology in agriculture to increase the precision of work, reduce costs, increase efficiency, intelligent processing, data evaluation and other aspects leading to the modernisation of agriculture. Precision agriculture is the application of technologies and principles to manage spatial and temporal variability associated with all aspects of agricultural production to improve crop performance and environmental quality [3]. Precision agriculture is already available to all farms using automatic machine control, operating a large tracked tractor or a compact tractor with a small centrifugal spreader. Furthermore, thanks to Agriculture 4.0, it is possible to save a large amount of natural and monetary resources due to the introduction of automatic section control systems and the use of locally variable nutrient applications [18].

With the global increase in land area and the size of farms, this knowledge could not be efficiently obtained. A location was treated more like a homogeneous area whose potential was not fully exploited. This situation has only been changed by the availability of technology and the necessary technical equipment. It provides the spectrum of data needed from many sources and their comprehensive analysis. The outputs help in decision-making on agronomic activities, adapting variable application of fertilisers and pesticides in the right amount at the right place or predicting the condition and characteristics of the soil or crop [19–22].

Precision technologies in livestock production have also evolved quite rapidly. Precision farming in animal nutrition and breeding is referred to abroad as Precision Livestock Farming (PLF). The aim is to improve the precision of farm operations and help the farmer make decisions immediately. New directions in modern farming focus on selecting forage crop varieties, their cultivation, harvesting, nutritional value, silage or storage, while attention is paid to feeding animals and the quality of production.

2.2. Previous Studies and Research Framework

Previous studies on precision agriculture adoption were considered to determine the research framework and questionnaire survey. This short review is focused on the classification of precision agriculture technologies and a summary of commonly used technologies.

Recent results of the Precision Agriculture Dealership Survey [23] show the importance of on-farm data for hybrid/variety selection and nutrient management. Dealers highlighted several uncrewed aerial vehicles for variable pesticide applications and crop input. Virals are GPS-guided controllers on sprayers and guidance-related technologies, which continue to grow. McKinsey Company [24] summarises precision agriculture technologies and divides technologies into five groups: smart-crop monitoring, drone farming, smart-livestock monitoring, autonomous-farming machinery, smart-building, and equipment management.

Several authors have addressed the issue of the adoption of precision agriculture technologies in the Czech Republic. Research in the Czech Republic focuses on mapping soil and crop variability, creating application maps for crop fertilisation, and determining and optimising differentiated doses of fertilisers and herbicides. Stočes et al. [25] developed the User-Technological Index of Precision Agriculture (UTIPA), which is calculated for each technology from obtained relevant data. It is an exciting application result that could help to compare familiarity or usage of the particular technology. Kasparov [26] studied the links between the nature of agricultural subjects on the perception of the attributes of precision agriculture with its adoption. In this research, respondents were asked about the technologies they use. According to application maps, the most widely used technologies were automatic section control, assisted machine travel control, correction signal payment, and variable rate applications. Farmers may not consider the prospects of this area and therefore prefer to invest less in new technologies. The Czech government has supported the adoption of precision technologies through financial incentives for new machines [27]. Research [12] confirmed that investment in agricultural robots is only around 26%.

In the USA, at the University of Nebraska-Lincoln, a study on precision agriculture technology adoption and opinions was conducted [28]. The main results show that agricultural enterprises mostly adopted technology for soil sampling, computer access to high-speed internet, yield maps, yield monitor and GPS guidance systems. According to the Agricultural Resource Management Survey (ARMS), 72% of cornfields and 70% of wheat fields used precision agriculture technologies [29]. Schimmelpfennig [30] found that large corn farms mostly adopted mapping and guidance systems. According to Maloku, adoption of precision agriculture technologies in the USA varied from one state to another. For example, Alabama and Florida predominantly adopted Lightbar Guidance, variable rate technologies, and GIS mapping software [31]. In Kansas, it was lightbar guidance, section control and variable rate fertility [32]. Scientists in Latin America and the Caribbean focused on recent trends in agriculture, new technologies and their applications [33]. Precision agriculture technologies were classified into: soil analysis and environmental assessment, drones and satellite images, remote sensors and georeferenced monitoring, mobile technology, internet of things, big data, artificial intelligence, blockchain, and robots.

The adoption of precision farming in Germany was part of the project related to personal interviews of farmers. Results show that GPS-based soil sampling, yield mapping, area measurement, auto-tracking, and site-specific basic fertilizing were the most adopted technologies [34]. In another study in Germany and Poland, essential technologies were evaluated to determine how and when they could be used for sustainable agriculture [35]. This research divides technology by type of prevailing production. The first group consists of crop production technologies such as nanotechnology, yield management, soil mapping, drones, sensors, and autonomous vehicles. The second group includes technologies for livestock production such as smart devices (position and health sensors), data and on-time software, nanotechnology, and sensors. The highest ratings were considered for collecting (sensors and drones) technologies or using (soil and yield management) data.

Precision agriculture was, according to [36], used in Denmark and the United Kingdom for about 90% of wheat, barley, oilseed rape, grass seed and peas. The most used technologies in Denmark were GPS yield mapping and grid soil sampling. In addition to these technologies, variable rate fertilisation was used extensively in the United Kingdom. In France and Sweden, yield monitors were widely used [37]. According to Cavallo [38], guidance machinery was used to a large extent in Italy. The dependence of adopting precision farming technology on economic and personal factors was studied in Hungary. According to the rankings, precision fertilisation and precision plant protection were precision farming technology's most commonly used elements. An interesting finding was that tractor guidance was a widely used element, but it was not considered a precision farming technology among farmers [16]. Trends in adopting precision farming technologies in Switzerland show that technologies with driver assistance systems are more frequently used in practice. In particular, these technologies reduce the physical labor involved in working [39].

Our research framework determines the most used precision agriculture technologies in the Czech Republic. Based on the studied literature [24], we divided the technologies according to the primary type of agricultural production. The two groups consist of technologies for crop and livestock production. These groups were further subdivided according to the kind of technology. The first group consisted of technologies focused on sensing and data acquisition (primarily sensors). The second group consisted of modern machines or robots. The research does not include some progressive technologies that fall outside both groups (e.g., genetic modification). In particular, we selected technologies for the research that were investigated in a related study. These technologies are the most cited by authors, experts and researchers. The focus on agriculture and the situation in the Czech Republic also influenced the selection of the Appendix A.

2.3. Precision Agriculture Technologies

Computers, sensors and computing technologies were developed in the 1980s, as well as improved vehicle systems (ultrasonic, optical, mechanical etc.). After the 1990s, GPS systems were viral and were used in agricultural applications. Then in quite a short time, these technologies were prohibited in agriculture due to their cost. Due to that, an alternative of machine vision technologies were used. These technologies were used to analyse which crop row structures could be observed efficiently. In 1987 a dynamic thresholding technique helped to extract information from field images. After a short time, a vision guidance system to steer a tractor relative to crop rows was used so that the tractor could automatically acquire its track in the next row [40]. The leading technologies used in precision agriculture are described below.

Precision agriculture uses the new technologies of the digital age to make farming as efficient as possible on the basis of data collection. Drones, satellite images or sensors placed on farm machinery or animal bodies constantly monitor fields, orchards, greenhouses and livestock. This technology saves fertiliser and costs, and higher yields are achieved with its application. It also prevents the overuse of fertilisers and leads to a more environmentally friendly land use [41–43]. Sensor data is processed using information and communication technologies, improving herd management strategies and the farm's economic, social and environmental performance [44–46].

The sensors are mainly included in satellites, ground-based platforms, etc. Ground-based platforms can be divided into three categories: handheld, free-standing in the field, and mounted on tractors or farm machinery. The sensors are used for spatial, spectral, radiometric applications, etc. [47,48]. Remote sensing is used for yield projection, land use classification, biomass estimation, pH measurements, etc. It can be used as a tool for making decisions (e.g., subplot scale). The level of digital agriculture is rapidly growing, and supra-national monitoring is performed using on-farm management tools [49–51].

Sensor data is processed using information and communication technologies, resulting in improved herd management strategies and economic, social and environmental performance of farms. Due to improving technologies, larger volumes of data need to be processed, analysed, and stored. Big Data are also described as data volumes, which are very difficult to process and manage using analytical tools. Databases and storage systems have been created to save the data in real-time and use them for further analyses. These storages are also very helpful for utilising Big Data for agricultural decision support tools. A PDI system is used to process Big Data and helps to innovate, standardise, automate and integrate the data [28,52–55].

Precision farming has become connected to service-oriented architecture services, which help process raw data and extract useful information. New disciplines such as IoT-based companies, automated industries or businesses have been used. Ontology is applied to make the extraction of valuable data easier. Ontology uses many supporting systems, domains, and knowledge. Other authors have developed support systems such as Plants ontology, SAAONT, AgriOnt, etc. [14].

Augmented reality (AR) is a unique application that provides its users with a direct or indirect view of a natural environment (the real world), parts of which are supplemented—augmented or enriched—with additional digital visual elements. AR has many benefits in agriculture because it is possible to create a relationship with other smart city-based technologies (GPS integration etc.) It is possible to couple AR with IoT data, which is one of the benefits of AR [46]. Especially CCD cameras are beneficial in precision agriculture because they capture two-dimensional colour images from which animal information is captured. All of these images can be used for further analysis. One of the ways to use the images captured by CCD cameras is a specification of pig parameters (weight, circumference, height and other body information). Pig identification could be as follows [40,56,57]: facial recognition, live weight detection, growth patterns and mass calculation, and individual pig identification and tracking.

Robotic systems can be involved based on the used applications. Uncrewed Ground Vehicles (UGV) and Unmanned Aerial Vehicles (UAVs) are used in precision agriculture [58]. The biggest challenges in the case of UAVs are costs. Sensors, flight duration, data analytics or requirements are the most significant part of the paid costs for UAVs. Another challenge is data analytics, which needs to be done periodically. The vast data storage of numerous terabytes must be available to store all the data that needs to be analysed. Weather is also a challenge that makes the results of UAV analysis worse due to weather conditions (rain, snowfall, clouds, etc.) [59].

The technology for monitoring crops and soil from the air uses an uncrewed aerial vehicle—a drone. This device can provide the user with multispectral images of soil blocks and can cover up to hundreds of hectares in one flight. The advantage of these devices is that the resolution of the images is higher than that offered by satellite imagery. The photos are then processed using software installed on the user's computer or in a cloud environment [59–61]. Drone outputs are crucial to increasing revenue, reducing costs, and improving business efficiency. The map can also inform which areas need more detailed scouting for effective planning—meaning less time spent examining soil blocks and more time tending the crops that need it.

Another essential step towards more efficient farming is satellite-guided tractor technology. Autonomous steering and turning or control via a touch panel linked to a central system that controls everything and obtains real-time harvest and position data is also standard in the domestic market. The system also allows variable dosing of fertilisers and products. Yields can also be charted thanks to the information recorded by the machine. The system provides information on area threshed, fuel consumption, or working hours. Entrepreneurs can then use this to analyse the profitability of the land. The data obtained can also be easily used for administration and subsidy applications. International satellite navigation systems are used in precision agriculture and conventional farming, and is helpful especially during lower visibility or in case of fatigue in workers [62].

Self-driving tractors have been around for some time and operate on autopilot. The tractor does most of the work, and the farmer only steps in when needed. The technology works with the help of GPS, and the machine spreads fertiliser or ploughs. There is also a device that works on the principle of solar power and can identify the weeds it kills with a dose of herbicide or lasers. Apps available for smartphones can also be used for precision farming. By configuring a precision farming system integrated into a smartphone, it is possible to monitor all the necessary data via the mobile phone. The applications are easily portable, affordable, and have high computing power [63,64].

Another vital area is crops' highly regulated genetic modification (GM) (soybeans, cotton, canola, etc.). Now there is genome editing (GE), which avoids potential risks to human health. These risks are avoided in the GM crops with their productivity, environmental tolerance, and pest resistance. GM crops are currently superseded by GE. In the case of GM crops, by inserting or removing one of the genes or part, the organism changes its specific traits. The development of GM crops is very regulated, and it also needs licenses and approved isolation procedures for field trials. Due to that, the GE techniques (NGTS/OGTR) are well used primarily in Australia to increase the production and tolerance of abiotic and biotic stresses. These techniques rapidly increase costs and exclude GM/GE research and development by small research organisations [65]. One of the introductory chapters of Precision Agriculture is hydroponic farms, which take the form of now commonly available home-grown boxes where seeds are planted; a mobile app runs a program for different types of plants to help oversee successful growth. Hydroponic farms can take the form and size of shipping containers that offer a harvest equivalent to the production of a two-acre farm.

3. Materials and Methods

Industry 4.0 and new technology are primarily applied in the automotive and manufacturing industries, and has a lot of potential in agriculture. The article's main aim is to evaluate the practical use of precision agriculture technologies in agricultural enterprises in the Czech Republic. We wanted to determine whether particular technologies are used more or less frequently in enterprises.

3.1. Data Sample and Research Design

Our research included a questionnaire survey. We created a questionnaire using an online web platform and then sent it to enterprises' email addresses via a web link. The respondents were managers of enterprises involved in crop and livestock production. In some cases, mixed enterprises used both types of production. Data were collected from January to March 2022. The questionnaire was sent to approximately 1500 enterprises, and the total number of responses received was 131, corresponding to a return rate of roughly 8.7% [6]. According to [66], about 89,320 subjects with recognised activities operate in the agriculture, forestry and fishing industry in the Czech Republic. The sample size margin of error at a 95% confidence level was about 8.56%.

We surveyed the size of enterprises by the number of employees. The most significant percentage (49.62%) is small businesses employing 11–50 employees. Next, 28.24% of medium-sized enterprises operating with 51–250 employees were represented in the sample. A total of 21.37% of the enterprises fell into the group of micro-enterprises and employed no more than ten people. Only one enterprise (0.76%) employed more than 250 employees. By legal form of business, they include limited liability enterprises (32.06%), joint-stock enterprises (30.53%), cooperatives (24.43%), self-employed farmers (7.63%), and finally, independent entrepreneurs (4.58%). Finally, we surveyed the predominant type of production, where 25.19% of enterprises are primarily focused on crop production, 6.87% on livestock production and 67.94% on both types of production.

The survey questions concern information obtained through the literature or publicly available studies. The questionnaire consisted of four areas according to the technologies: sensors in crop production, machines in crop production, sensors and IoT devices in livestock production, robots and mobile technology in livestock production. The questions dealt with individual technologies. Respondents were asked about the frequency of occurrence of end-use technology.

3.2. Research Methods and Hypotheses

The results of the technology-related questions were statistically evaluated. We used the Chi-squared test to prove the agreement of frequency distributions for quantitative attributes of each technology. It assesses the difference between the observed frequencies (f_o) and the relative expected frequencies (f_e) that fit the predicted probability distribution. We chose for the theoretically expected frequencies an equal distribution of "yes" and "no" responses (i.e., a probability ratio $p = 0.5$). It decides whether the difference between the empirical and theoretical frequencies is random and comes from a normal population distribution. We formulated a working hypothesis as follows:

H_1: Precision Agriculture technology is used by more than half of the enterprises.

We used a statistical test to check whether the probability p of technology frequency was equal to (H_0: $p = 0.5$; when observed frequencies f_o are similar to expected frequencies f_e) or higher than 0.5 (H_A: $p > 0.5$; when observed frequencies f_o are higher than anticipated frequencies f_e). We tested the hypothesis separately for each technology listed in the questionnaire.

We evaluated the hypothesis via p-values for a one-sided statistical test. If the null hypothesis H_0 could be rejected at the 0.05 significance level, the observed relative frequencies differed from the theoretical ones. It implies that the technology was used in over half of the observed enterprises. The test statistic follows the Chi-square distribution, designated by χ^2 [67,68]:

$$\chi^2 = \sum\left[\frac{(f_o - f_e)^2}{f_e}\right] \tag{1}$$

with $k - 1$ degree of freedom, where:

k is the number of categories.

f_o is and observed frequency in a particular variety.

f_e is an expected frequency in a specific variety.

4. Results

This section outlines the study's results divided into four parts: sensors and machines in crop production, and sensors and robots in livestock production.

4.1. Sensors in Crop Production

The use of sensors is the first step toward precision agriculture. We examined the reasons for the application of sensors and types of sensors. The questionnaire survey results are summarised in Figure 1. A total of 58.02% of enterprises reported that the sensors detect weather conditions and have a weather station function. This function is essential in determining the local weather forecast for a specific location. It provides farmers with information on rainfall, wind speed, wind direction, humidity and temperature and atmospheric pressure. A complete overview of the conditions in the field from the nearest weather station is available. Equally important was the use of sensors for plant protection and nutrition in 53.44%, which leads to the application of substances in only the necessary places. Modern sensors may have built-in rules and algorithms that create dynamic prediction capabilities for the degree of disease risk. It is followed by the option of using sensors for machine positioning, which covered 50.38%. The fundamental advantage of field automation is the stable position and precise dimensions of each cultivated area, which facilitates the basic orientation of the machines. In addition, the direct visibility of the sky also allows satellite navigation to detect and control the position of automated devices. The less frequent option was the sensor function detecting the immediate technical condition. A total of 40.46% of respondents selected this answer. Farmers do not address monitoring machinery's technical situation. Thus, the use of machinery for agricultural work may be still associated with a higher risk of necessary repairs and maintenance. Sensors are used the least to detect crop anomalies, in 21.37%. The main idea is to apply spray only when unavoidable and choose the right time and product.

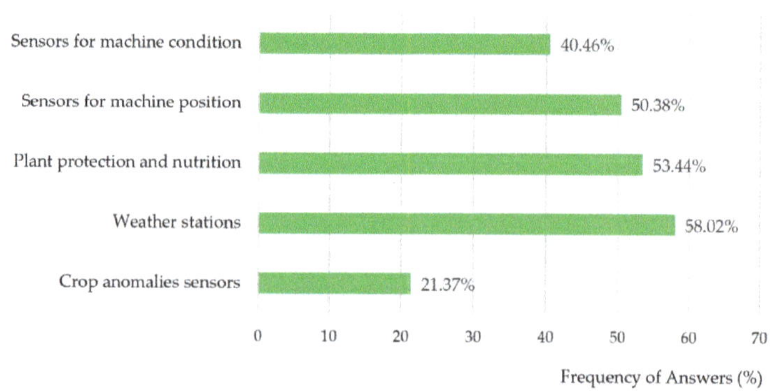

Figure 1. Use of sensors in crop production.

We performed a statistical evaluation of the responses to the survey questions related to sensors in crop production using the Chi-square test. Through working hypothesis H_1, we tried to show statistically significant differences from the mean (see Table 1).

Table 1. Results of statistic evaluation for sensors in crop production.

Technology	χ^2	*p*-Value (H_1)
Machine condition sensors	4.7710	0.9856
Machine position sensors	0.0076	0.4652
Plant protection and nutrition	0.6183	0.2159
Smart weather stations	3.3664	0.0333 *
Crop anomalies sensors	42.9389	1.0000

* the statistically significant differences at the significance level of 5% are marked.

We can prove working hypothesis H_1 that more than half of the enterprises use precision farming technology only for weather stations (*p*-value = 0.0333). It means that the use of this technology is really above average among enterprises. Modern sensors bring new functionalities to mobile applications. Agronomists no longer have to walk miles around the farm every day, checking the current status of the field or stored crop. Sensors provide accurate, updated data online, so they can work much more efficiently and only go where they need to at the time.

4.2. Machines in Crop Production

Drones and self-driving machines are the essential technological contributions of precision agriculture. It can be seen in Figure 2 that 89.31% of enterprises use uncrewed vehicles such as tractors and working machines in crop production. Automatic steering systems are offered by tractor manufacturers already fully integrated into the machine and built-in during its manufacture. The system's control is integrated into the tractor's control terminal. The driver simply enters the machine parameters, records the first pass on the plot, and the autopilot then controls the machine without driver intervention. The human driver only controls the speed of travel and the work of the attachment and monitors obstacles but does not intervene in the steering. The less-used technology of drones for detecting the immediate state of the soil or directly for planting seeds is used by 33.59% of enterprises. Precise mapping of agricultural land would be very time-consuming and technically challenging if it were not for aerial vehicles equipped with specialised sensing technology. Images taken from the air are evaluated and processed into application maps and orthophotos quickly, precisely and efficiently. They can be used to dose fertilisers and sprays accurately, thus exploiting the field's full potential. Instead of uniform tillage, they allow monitoring of soil conditions and dividing areas into several zones that can be approached differently. This technology is probably yet to achieve a "boom" in usage. Drones and drones are not yet used to any significant extent. In both cases, the navigation systems and optimisation software of journeys is a suitable complement to these machines, especially for tractors. This option was indicated by 61.83% of enterprises.

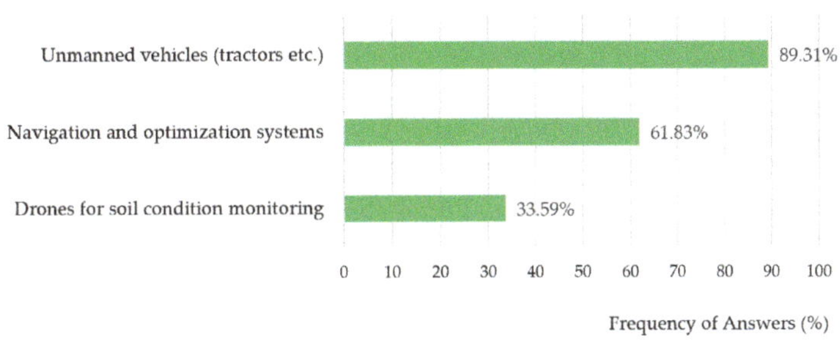

Figure 2. Use of machines in crop production.

The results of the evaluation of working hypothesis H_1 for Machines in Crop production using the Chi-square test are presented in Table 2.

Table 2. The result of statistic evaluation for machines in crop production.

Technology	χ^2	*p*-Value (H_1)
Unmanned vehicles	80.9847	<0.0001 *
Navigation and optimisation systems	7.3359	0.0034 *
Drones for soil condition	14.1145	0.9999

* The statistically significant differences at the significance level of 5% are marked.

We succeed in proving the working hypothesis H_1 that more than half of the enterprises use precision farming technologies for uncrewed vehicles and tractors (p-value < 0.0001) and navigation and optimisation systems for journeys (p-value = 0.0034). According to the above-average results, self-driving tractors and machines can be considered the main benefit of precision agriculture. The self-driving tractor is most often equipped with GPS and terrain mapping technology, thus achieving better efficiency and lower labour costs when cultivating the field. Equally important is route optimisation software, which allows the planning of fieldwork.

4.3. Sensors and IoT Devices in Livestock Production

The next part was dedicated to finding the purpose of using sensors and IoT in livestock production. Smart collars are typical IoT devices using various sensors and performing multiple functions. Figure 3 shows the reasons for using sensors in livestock production. According to the answers, these are most often used as intelligent collars for animals, whose function is to control movement. This answer was selected by 46.56% of enterprises. These smart collars protect grazing animals from theft and help farmers find them quickly if they accidentally escape from the pasture. The second most frequently identified answer is using sensors applied to smart collars with information about animal health. A sensor on the collar senses some of the animal's vital signs. If the animal starts behaving abnormally and the data from the collar deviates from average, it usually means that some health complications are coming. Thanks to the monitoring system, the farmer can react ahead of time and treat the animal earlier or administer vitamins before the disease fully erupts. In this case, 42.75% of enterprises selected this option, followed by intelligent collars with sensors controlling animal nutrition, 2.06%. This technology can help estimate the live weight or health status of animals. The feeding curve can then be modelled accordingly, thus avoiding overfeeding or deterioration of animal health. The last was the possibility of using sensors that can handle the microclimate in the stables in 32.06% of enterprises. Farmers have a system installed in the barn to control the barn microclimate and help maintain it at the necessary values. Sensors check the temperature and humidity of the air or the content of certain gases and adjust the covering of the side walls, the opening of vents, and the running of fans or showers to cool the animals as needed. However, nowadays, it is more typical to use classical recommendations or best practices to create optimal conditions for livestock.

Furthermore, we evaluated working hypotheses H_1 for sensors and IoT devices in livestock production. The results of the Chi-square tests are summarised in Table 3.

We cannot prove hypothesis H_1 that more than half of the enterprises use precision farming technologies for one of the sensors and IoT devices in livestock production. Intelligent collars and microclimate sensors are not yet widely used, and the occurrence of smart collars for health and nutrition monitoring is less frequent in enterprises.

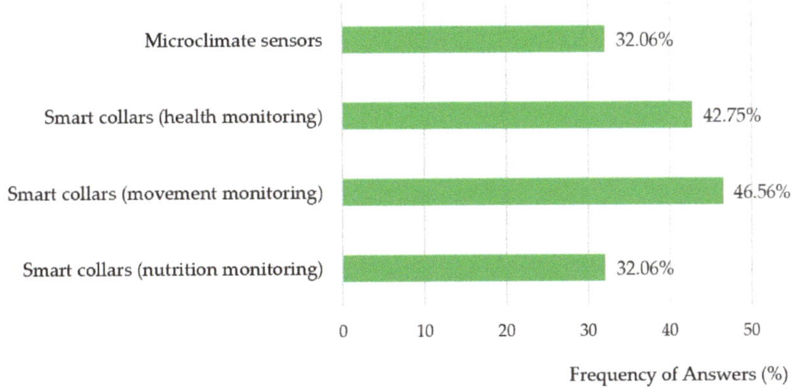

Figure 3. Use of sensors and IoT devices in livestock production.

Table 3. The results of statistical evaluation for sensors and IoT in livestock production.

Technology	χ^2	p-Value (H_1)
Microclimate sensors	16.8626	1.0000
Smart collars (health monitoring)	2.7557	0.9516
Smart collars (movement monitoring)	0.6183	0.7842
Smart collars (nutrition monitoring)	16.8626	1.0000

The statistically significant differences at the significance level of 5% are marked.

4.4. Robots in Livestock Production

The last part is devoted to finding the use of robots in agriculture enterprises. In the future, the automation and robotisation of agriculture are considered one of the most dynamic developments, not only in processing crops, which is already quite common today but also in the cultivation of the fields themselves. Indeed, an "army" of new, more accurate and robust monitoring sensors are set to come into play in a major way, which, in conjunction with more powerful control units, will enable existing types of agricultural machinery to be controlled automatically or semi-automatically. About 19.85% of livestock farmers use robots to feed their animals, and 12.98% of enterprises use milking robots in their business. The animals have freedom of movement, and no one chases them to milk. When they need to be milked, they walk to the robot. When they need to be fed, they walk to the gutter, and when they need to rest, they lie down. These results show that robot technology is not yet widespread in agriculture. The use of robots is, therefore, still very much in the future (see Figure 4).

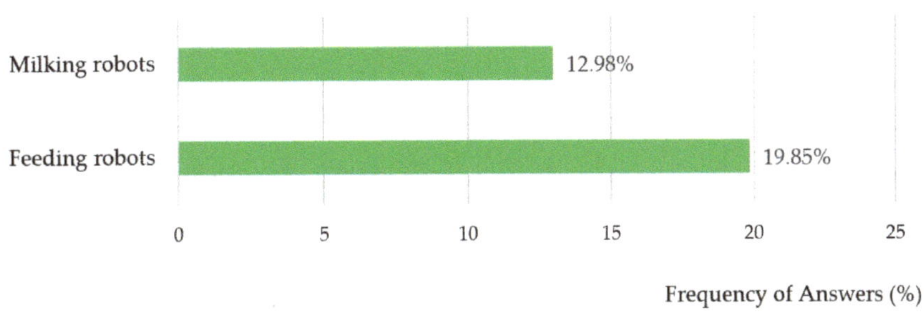

Figure 4. Use of Robots in Livestock Production.

In Table 4, we present the results of the evaluation of working hypothesis H_1 for using robots in livestock production.

Table 4. The results of statistical evaluation for robots in livestock production.

Technology	χ^2	p-Value (H_1)
Milking robots	71.8244	1.0000
Feeding robots	47.6412	1.0000

The statistically significant differences at the significance level of 5% are marked.

In the statistical evaluation of hypothesis H_1, we could not confirm that more than half of the enterprises use the technology of precision agriculture robots in livestock production. The deployment of robots in agriculture is not yet at a high level. The results showed shallow usage values for both milking robots and feeding robots. It means that enterprises still do not favour the advantages of automatic processes.

5. Discussion

This section discusses the results from the perspective of previous studies and the working hypotheses. Future research directions and limitations of our study are highlighted.

5.1. Technology of Precision Agriculture

Furthermore, we focus on technology used by enterprises. Discussion is divided into two parts according to technology usage by enterprises.

5.1.1. The Most Used Technology

First, we focus on technologies more widely used by enterprises. In our case, this confirms working hypothesis H_1, where we hypothesised that more than half of the enterprises would use precision agriculture technology. Our research showed that precision agriculture technologies are predominantly used in crop production. After all, most definitions of the term [3] refer to activities and operations on cultivated land. Moreover, the results showed that most farms do not use sensors, IoT devices and robots. Thus, working hypothesis H_1 was confirmed only for some of the technologies used in crop production. The most widely used precision agriculture technologies are intelligent weather stations, uncrewed vehicles, and navigation and optimisation systems for journeys. Similarly, according to technology expert evaluations [18], the most promising precision agriculture technologies are robots, autonomous machines, sensors, and global navigation satellite systems.

In our research, 89.31% of enterprises used driverless vehicles, such as tractors and machines. These vehicles not only move on the ground but also receive weather data via an internet connection and can also make decisions based on it. According to Kasparov [26], the most widely used technologies in agriculture are those that facilitate machine control and navigation, i.e., automatic section control by 30% of enterprises, and assisted machine travel control by 21% of enterprises. We can conclude that companies have learned to work with these technologies, and the share is gradually increasing. In 2015, USA auto-steer technology was used in about 70% of farms [28] to improve operator performance and reduce excess input usage.

GPS navigation and mapping are the technologies that farmers usually start with and are the most widely used [34]. Our research found that 61.83% of enterprises use navigation systems. Similar results were reported in research [12], where the investment of agricultural enterprises in the Czech Republic in navigation systems is about 70%. In Hungary, 12% of farmers used GPS only for field navigation, not site-specific measures [16]. In Latin America and the Caribbean, sensors for geolocation are used by 36% of farmers [33]. The use of navigation is related to optimising the route and land travel. It is done by special software that records the boundaries of the plot and then can optimise routes for the farmer according to the shape of the property to minimise the number of journeys. Other research that reports on navigation systems use is the Precision Agriculture Dealership Survey [23]. These results show that GPS guidance systems with automatic control are utilised for fertiliser/chemical application in 81%, satellite/aerial imagery in 67%, and GPS to manage vehicle logistics and track locations of vehicles and guide them in 47%. It is becoming apparent from the results of these studies that navigation systems are the most widely used precision agriculture technology overall.

We found that 58.02% of enterprises used weather stations for monitoring and detecting weather conditions. We do not have a direct comparison in this area. However, for example, in Latin America and the Caribbean, 41% mainly used remote sensors [33]. The advantage of these sensors is access to up-to-date data on weather and conditions without the need to walk around the plot and record readings manually. These technologies are prevalent and have utility in precision irrigation, field monitoring and spraying [69].

5.1.2. The Less Used Technologies

Finally, we investigated which precision agriculture technologies are less used by enterprises. We divided the results into two parts: crop production and livestock production.

We found that crop production has an intermediate usage of machine position sensors with 50.38% and plant protection and nutrition sensors with 53.44%. The least used technologies in crop production are machine condition sensors, crop anomaly sensors, and drones. However, some research shows that these technologies could have higher potential. For example, Germany and Poland's highest readiness levels include technology drones, sensors, and soil management systems [35]. Smart-crop monitoring included corresponding sensor data and imagery analysis to optimise resource usage based on location. McKinsey Company estimated the highest range of new global GDP value potential in smart-crop monitoring [24].

In livestock production, 46.56% of enterprises indicated intermediate level usage of smart collars for animal movement monitoring. For livestock production, the least used technologies are microclimate sensors, smart collars for health and nutrition monitoring, and milking and feeding robots. Similarly, research [12] showed that 52% of sensors are used in livestock production to detect newborn calves, peak estrus, health problems, etc. The intelligent tracking collar uses mostly modern GPS technology through which the farmer receives accurate information about the current location of the animals. Monitoring is done through communication between the tracking collar worn by the animal and the base station. Some more sophisticated models communicate with a mobile phone.

It was evident that fewer farmers used leading technologies in livestock production. We, therefore, tried to find an explanation for this situation. One of the reasons why the area of livestock production is not very well developed is that animal breeding in the Czech Republic has been declining recently [70]. Farmers may not consider the prospects of this area and therefore prefer to invest less in new technologies. Research [12] confirmed that investment in robots is only around 26%. Market conditions are conducive to this, and it is questionable whether the current situation is sustainable. Unlike crop production, livestock production is year-round. It, therefore, requires deploying technological and human resources throughout the year, which is a disadvantage for personnel requirements. Livestock production takes place in less variable environmental conditions, unlike crop production. It means that there is higher variability in crop production. Therefore, there is a greater need for modern technologies to cope with this variability. In livestock production, animal nutrition is easily adjustable. Feeding and aftercare needs can be easily predicted. For example, controlled and automated feeding for cows is necessary for above 8000 litres of milk production. For plants, it is more complicated, as nutrient levels depend on soil conditions and fertiliser. In addition, some fertilisers (nitrogen) are easily leachable, affecting fertilisation's overall efficiency. In crop production, the progress and development of plant growth take place in a short time compared to livestock production.

5.2. Future Research

We have identified research gaps that could be further developed from the results of our research and that of other authors. The first challenge is to compare the precision industry's overall level in each country. From the available sources, it has become apparent that this assessment has been done to a limited extent. It is unsuitable for comparison due to differing methodologies across countries. The second challenge is identifying factors that help or hinder the adoption of precision agriculture. Various analyses have already been undertaken in this area, mainly concerning agricultural policy and subsidies. A third research area could be the development of specific case studies of technologies and their use in agriculture. These case studies highlight the advantages of this approach for farmers who have not yet decided to deploy new technologies. Some other streams of research include, for example, a study by McBratney et al. [4] focusing on employers training to acquire knowledge about new technologies, environmental damage costs or economic assessment of precision agriculture.

5.3. Limitations

The limitation of this research may be the sample composition and size, created based on a non-probability sampling method. We used purposive sampling based on our knowledge about the population and the study research aims. Another problem could be the relatively high sample size margin of error of 8.56%. It indicated less likelihood of relying on the results of a survey. Therefore, the confidence in the results will be lower to represent a population. However, the results of other studies show that usage of technologies is very similar to their effects. Therefore, we believe that the results are consistent with the conclusions of the other authors.

Some technologies were not part of the research, such as nanotechnology in livestock production [71], genetic modification [72], automatic planting of seeds in the field, mapping technologies, camera-based imaging, data analysis, and evaluation technologies [73]. The questionnaire was based on what was generally known about agriculture in the Czech Republic. For the selected technologies, we confirmed their use on farms. However, some technologies have been applied in practice only marginally. An overview and description of other precision agriculture technologies include a Smart Farming Platform database (smart-akis.com, accessed on 5 April 2022).

6. Conclusions

Precision agriculture, supported by modern technology, is looking for ways to optimise management. Farmers can better determine what is efficient, cost-effective and time-saving from the knowledge gained. New technology and modern machinery should therefore be thoroughly fostered. However, emphasis should be placed on promoting farming characterised by a broader understanding of local conditions. Digital advances and their implementation are occurring in both livestock and crop production. Automation and electronic data transmission help eliminate the human factor deficit. In our article, we aimed to determine the usage of precision agriculture technologies in agricultural enterprises.

We summarised the results of the technology usage in crop production. In that case, we can conclude that the most used technologies are intelligent weather stations, unnamed vehicles, and navigation and optimisation systems for optimising journeys. We showed that more than half of the enterprises surveyed use these technologies. These technologies can be introduced gradually and create synergies. Thus, we can say that they are more widespread, and the enterprises are solving their daily issues with them. The advantage of autonomous machines in agriculture is to increase productivity and quality and reduce land management costs. Their application is therefore justified for farmers and is already changing the face of agriculture today. Agriculture can consequently be very promising using the latest technological solutions.

Summarising the results of the use of technology in livestock production, we can conclude that precision agriculture principles are not yet widespread in livestock production. Instead, existing animal management practices are used, and only a small number of farms are trying to introduce new technologies. Of these, smart collars for movement are currently the most widely used, often to protect animals from theft, loss or straying. Gradually, with the development of 5G networks and the use of robotics in manufacturing, this situation will change in the future.

Given the anticipated focus of European Union agricultural policy on reducing environmental impact, we consider the use of modern information technology inevitable. Knowing which technologies make sense to support and have future applicability is essential. On the other hand, it is clear that in livestock production, the benefits of technologies still need to be further monitored and communicated to potential farmers. Modern technology should be available to large businesses and small entrepreneurs. Farmers want subsidies and less bureaucracy. Technology can help and benefit everyone. For this reason, it is necessary to educate about information technology so that even older farmers can start to use the new systems.

Author Contributions: Conceptualisation, J.V., M.P. and I.Š.; methodology, M.P. and I.Š.; software, M.P.; validation, M.P.; formal analysis, M.P. and I.Š.; investigation, I.Š.; resources, I.Š. and J.V.; data curation, M.P.; writing—original draft preparation, J.V., M.P. and I.Š.; writing—review and editing, J.V. and I.Š.; visualisation, M.P.; supervision, M.P.; project administration, J.V.; funding acquisition, J.V. All authors have read and agreed to the published version of the manuscript.

Funding: This research was funded by "EF-150-GAJU 047/2019/S", supported by the University of South Bohemia in České Budějovice.

Institutional Review Board Statement: Not applicable.

Informed Consent Statement: Not applicable.

Data Availability Statement: Data is available in a publicly accessible repository.

Acknowledgments: The authors thank the agricultural enterprises for participating in the research.

Conflicts of Interest: The authors declare no conflict of interest.

Appendix A

The questionnaire survey is listed below:
The name of the company (optional):
The legal form of business (select one option)

- Limited liability enterprise
- Joint-stock enterprise
- Cooperatives
- Self-employed farmer
- Independent entrepreneur
- Other

Number of employees (select one option)

- Less than ten employees
- 11–50 employees
- 51–250 employees
- More than 250 employees

The predominant type of production (select one option)

- Crop production
- Livestock production
- Both types (mixed) production

Do you use sensors in crop production? (select one or more options)

- For detecting the instantaneous technical condition of machinery.
- To detect the instantaneous position of the machine.
- For plant protection and nutrition: application only at necessary points on the plot.
- For detecting weather conditions (smart weather stations).
- For the ability to detect anomalies in crops.
- Others

Do you use machines in crop production? (select one or more options)

- Unmanned vehicles in crop production (tractors, work machines)
- Navigation and optimisation systems for optimising journeys around the field
- Drones for the detection of the instantaneous state of the soil condition.
- Others

Do you use of sensors in livestock production? (select one or more options)

- Sensors to control the microclimate in the stables
- Smart collars for animals, controlling their health
- Smart collars for animals, controlling their nutrition
- Smart collars for animals, controlling their movement
- Others

Do you use robots or mobile technology in livestock production? (select one or more options)

- We use robots to feed animals
- We use robots for milking
- Others

References

1. Yang, J.; Guo, X.; Li, Y.; Marinello, F.; Ercisli, S.; Zhang, Z. A Survey of Few-Shot Learning in Smart Agriculture: Developments, Applications, and Challenges. *Plant Methods* **2022**, *18*, 28. [CrossRef] [PubMed]
2. Alkan, A.; Abdullah, U.M.; Abdullah, O.H.; Assaf, M.; Zhou, H. A Smart Agricultural Application: Automated Detection of Diseases in Vine Leaves Usinghybrid Deep Learning. *Turk. J. Agric. For.* **2021**, *45*, 717–729. [CrossRef]
3. Pierce, F.J.; Nowak, P. Aspects of Precision Agriculture. In *Advances in Agronomy*; Elsevier: Amsterdam, The Netherlands, 1999; Volume 67, pp. 1–85. ISBN 978-0-12-000767-7.
4. McBratney, A.; Whelan, B.; Ancev, T.; Bouma, J. Future Directions of Precision Agriculture. *Precis. Agric.* **2005**, *6*, 7–23. [CrossRef]
5. Lukas, V.; Neudert, L.; Křen, J. Precizní Zemědělství a Jeho Přínosy [Precision Agriculture and Its Benefits]. Available online: https://zemedelec.cz/precizni-zemedelstvi-a-jeho-prinosy/ (accessed on 1 January 2022).
6. Svepesova, I. *Průmysl 4.0 v Zemědělství [Industry 4.0 in Agriculture]*; University of South Bohemia in Ceské Budejovice, Faculty of Economics: Ceske Budejovice, Czech Republic, 2022.
7. Kušová, D.; Těšitel, J.; Boukalová, Z. Willingness to Adopt Technologies of Precision Agriculture: A Case Study of Tech Czech Republic. *WIT Trans. Ecol. Environ.* **2017**, *220*, 109–117. [CrossRef]
8. Li, Q.; Hu, G.; Jubery, T.Z.; Ganapathysubramanian, B. A Farm-Level Precision Land Management Framework Based on Integer Programming. *PLoS ONE* **2017**, *12*, e0174680. [CrossRef]
9. Späti, K.; Huber, R.; Finger, R. Benefits of Increasing Information Accuracy in Variable Rate Technologies. *Ecol. Econ.* **2021**, *185*, 107047. [CrossRef]
10. Pánková, L.; Aulová, R.; Jarolímek, J. Economic Aspects of Precision Agriculture Systems. *AOL* **2020**, *12*, 59–67. [CrossRef]
11. Zhang, Y.; Min, Q.; Li, H.; He, L.; Zhang, C.; Yang, L. A Conservation Approach of Globally Important Agricultural Heritage Systems (GIAHS): Improving Traditional Agricultural Patterns and Promoting Scale-Production. *Sustainability* **2017**, *9*, 295. [CrossRef]
12. Duspivová, K.; Nesrstová, M.; Miklová, M.; Doleželová, P. *Vliv Digitalizace a Robotizace Na Charakter Práce a Roli Sociálního Dialogu v Zemědělství [The Impact of Digitalization and Robotization on the Nature of Work and the Role of Social Dialogue in Agriculture]*; Trexima: Zlín, Czech Republic, 2018.
13. Novak, R.; Hrtusova, T. Precizní Zemědělství v Praxi [Precision Agriculture in Practice]. Available online: https://cdn0.erstegroup.com/content/dam/cz/csas/business_csas_cz/precizni-zemedelstvi/Precizni_zemedelstvi_v_praxi_2018_02.pdf (accessed on 10 February 2022).
14. Fahad, M.; Javid, T.; Beenish, H.; Siddiqui, A.A.; Ahmed, G. Extending ONTAgri with Service-Oriented Architecture towards Precision Farming Application. *Sustainability* **2021**, *13*, 9801. [CrossRef]
15. Lee, C.-L.; Strong, R.; Dooley, K.E. Analyzing Precision Agriculture Adoption across the Globe: A Systematic Review of Scholarship from 1999–2020. *Sustainability* **2021**, *13*, 10295. [CrossRef]
16. Lencsés, E.; Takács, I.; Takács-György, K. Farmers' Perception of Precision Farming Technology among Hungarian Farmers. *Sustainability* **2014**, *6*, 8452–8465. [CrossRef]
17. Mulla, D.J. Twenty Five Years of Remote Sensing in Precision Agriculture: Key Advances and Remaining Knowledge Gaps. *Biosyst. Eng.* **2013**, *114*, 358–371. [CrossRef]
18. Fuka, V. Zemědělství 4.0–Co to Vlastně Je? [Agriculture 4.0–What Is It?]. Available online: https://mechanizaceweb.cz/zemedelstvi-4-0-co-to-vlastne-je/ (accessed on 12 February 2022).

19. Balafoutis, A.T.; Evert, F.K.V.; Fountas, S. Smart Farming Technology Trends: Economic and Environmental Effects, Labor Impact, and Adoption Readiness. *Agronomy* **2020**, *10*, 743. [CrossRef]
20. Mendes, J.; Pinho, T.M.; Neves dos Santos, F.; Sousa, J.J.; Peres, E.; Boaventura-Cunha, J.; Cunha, M.; Morais, R. Smartphone Applications Targeting Precision Agriculture Practices—A Systematic Review. *Agronomy* **2020**, *10*, 855. [CrossRef]
21. Bhatnagar, V.; Poonia, R.C.; Sunda, S. State of the Art and Gap Analysis of Precision Agriculture: A Case Study of Indian Farmers. *Int. J. Agric. Environ. Inf. Syst.* **2019**, *10*, 72–92. [CrossRef]
22. Gagliardi, G.; Lupia, M.; Cario, G.; Cicchello Gaccio, F.; D'Angelo, V.; Cosma, A.I.M.; Casavola, A. An Internet of Things Solution for Smart Agriculture. *Agronomy* **2021**, *11*, 2140. [CrossRef]
23. Erickson, B.; Lowenberg-DeBoer, J. *Precision Agriculture Dealership Survey*; Purdue University: West Lafayette, Indiana, 2020; Volume 2020.
24. Goedde, L.; Ménard, A.; Revellat, J. Agriculture's Connected Future: How Technology Can Yield New Growth. Available online: https://www.mckinsey.com/industries/agriculture/our-insights/agricultures-connected-future-how-technology-can-yield-new-growth (accessed on 9 June 2022).
25. Stočes, M.; Jarolímek, J.; Charvát, K.; Masner, J.; Pavlík, J.; Vaněk, J. *User-Technological Index of Precision Agriculture*; Precision Agriculture Association New Zealand: Hamilton, New Zealand, 2017; pp. 1–7.
26. Kasperová, M. Precizní Zemědělství v Česku: Kvantitativní Analýza Faktorů Ovlivňujících Jeho Osvojení. In *Precision Agriculture in the Czech Republic: A Quantitative Analysis of Factors Influencing Its Adoption*; Masaryk University: Brno, Czech Republic, 2018.
27. Mrnuštík Konečná, M.; Sutherland, L.-A. Digital Innovations in the Czech Republic: Developing the Inner Circle of the Triggering Change Model. *J. Agric. Educ. Ext.* **2022**, *28*, 1–24. [CrossRef]
28. Castle, M.; Lubben, D.B.; Luck, J. Precision Agriculture Usage and Big Agriculture Data. *Cornhusker Econ.* **2015**, *2015*, 1–3.
29. U.S. Department of Agriculture Tailored Reports: Precision Agriculture. Available online: https://data.ers.usda.gov/reports.aspx?ID=17883 (accessed on 15 February 2022).
30. Schimmelpfennig, D. *Farm Profts and Adoption of Precision Agriculture*; USDA Economic Research Service: Washington, DC, USA, 2016.
31. Winstead, A.T.; Norwood, S.H.; Griffin, T.W.; Runge, M.; Adrian, A.M.; Fulton, J. *Adoption and Use of Precision Agriculture Technologies by Practitioners*; The International Society of Precision Agriculture (ISPA): Monticello, IL, USA, 2010; pp. 18–21.
32. Griffin, T.W.; Yeager, E.A. How quickly do farmers adopt technology? A duration analysis. In Proceedings of the Precision agriculture '19, Montpellier, France, 8 July 2019; Wageningen Academic Publishers: Montpellier, France, 2019; pp. 843–849.
33. Vitón, R.; Castillo, A.; Lopes Tixeira, T. *AGTECH: Agtech Innovation Map in Latin America and the Caribbean*; Inter-American Development Bank: Washington, DC, USA, 2019.
34. Reichardt, M.; Jürgens, C.; Klöble, U.; Hüter, J.; Moser, K. Dissemination of Precision Farming in Germany: Acceptance, Adoption, Obstacles, Knowledge Transfer and Training Activities. *Precis. Agric.* **2009**, *10*, 525–545. [CrossRef]
35. Maciejczak, M.; Faltmann, J. Assessing Readiness Levels of Production Technologies for Sustainable Intensification of Agriculture. *Apstract* **2018**, *12*, 47–52. [CrossRef]
36. Pedersen, M.S.; Ferguson, B.R.; Lark, A.M. Comparison of Producer Adoption of Precision Agriculture Practices in Denmark, the United Kingdom and the United States. *SJFI–Working Paper No. 2* **2001**, *17*, 1–38.
37. Lowenberg-DeBoer, J.; Erickson, B. Setting the Record Straight on Precision Agriculture Adoption. *Agron. J.* **2019**, *111*, 1552–1569. [CrossRef]
38. Cavallo, E.; Ferrari, E.; Bollani, L.; Coccia, M. Attitudes and Behaviour of Adopters of Technological Innovations in Agricultural Tractors: A Case Study in Italian Agricultural System. *Agric. Syst.* **2014**, *130*, 44–54. [CrossRef]
39. Groher, T.; Heitkämper, K.; Walter, A.; Liebisch, F.; Umstätter, C. Status Quo of Adoption of Precision Agriculture Enabling Technologies in Swiss Plant Production. *Precis. Agric.* **2020**, *21*, 1327–1350. [CrossRef]
40. Vrochidou, E.; Oustadakis, D.; Kefalas, A.; Papakostas, G.A. Computer Vision in Self-Steering Tractors. *Machines* **2022**, *10*, 129. [CrossRef]
41. Fruhwirtova, E. Žádné Sci-Fi, Zemědělství 4.0 Se Stává Realitou [No More Sci-Fi, Agriculture 4.0 Is Becoming a Reality]. Available online: https://zazijzemedelstvi.cz/clanky/zadne-sci-fi-zemedelstvi-40-se-stava-realitou:18/ (accessed on 5 February 2022).
42. Schütze, A.; Helwig, N.; Schneider, T. Sensors 4.0–Smart Sensors and Measurement Technology Enable Industry 4.0. *J. Sens. Sens. Syst.* **2018**, *7*, 359–371. [CrossRef]
43. Vlasov, A.I.; Grigoriev, P.V.; Krivoshein, A.I.; Shakhnov, V.A.; Filin, S.S.; Migalin, V.S. Smart Management of Technologies: Predictive Maintenance of Industrial Equipment Using Wireless Sensor Networks. *JESI* **2018**, *6*, 489–502. [CrossRef]
44. Pedersen, S.R. Smart Farming. Available online: https://goexplorer.org/smart-farming/ (accessed on 10 February 2022).
45. Vrchota, J.; Rehor, P. *Influence of Strategic Management on the Importance of Crises in Farms in the Czech Republic*; Kapounek, S., Krutilova, V., Eds.; Mendel Univ Brno: Brno, Czech Republic, 2017; ISBN 978-80-7509-499-5.
46. Hurst, W.; Mendoza, F.R.; Tekinerdogan, B. Augmented Reality in Precision Farming: Concepts and Applications. *Smart Cities* **2021**, *4*, 1454–1468. [CrossRef]
47. Sishodia, R.P.; Ray, R.L.; Singh, S.K. Applications of Remote Sensing in Precision Agriculture: A Review. *Remote Sensing* **2020**, *12*, 3136. [CrossRef]
48. Pech, M.; Vrchota, J.; Bednář, J. Predictive Maintenance and Intelligent Sensors in Smart Factory: Review. *Sensors* **2021**, *21*, 1470. [CrossRef]

49. Teucher, M.; Thürkow, D.; Alb, P.; Conrad, C. Digital In Situ Data Collection in Earth Observation, Monitoring and Agriculture—Progress towards Digital Agriculture. *Remote Sens.* **2022**, *14*, 393. [CrossRef]
50. Temmen, N.; Schilling, J. *Smart Farming Technology in Japan and Opportunities for EU Companies*; EU-Japan Centre for Industrial Cooperation: Brussel, Belgium, 2021.
51. Vrchota, J.; Řehoř, P. Project Management and Innovation in the Manufacturing Industry in Czech Republic. *Procedia Comput. Sci.* **2019**, *164*, 457–462. [CrossRef]
52. Kharel, T.P.; Ashworth, A.J.; Owens, P.R. Linking and Sharing Technology: Partnerships for Data Innovations for Management of Agricultural Big Data. *Data* **2022**, *7*, 12. [CrossRef]
53. Chumnumporn, K.; Jeenanunta, C.; Komolavanij, S.; Saenluang, N.; Onsri, K.; Fairat, K.; Itthidechakhachon, K. The Impact of IT Knowledge Capability and Big Data and Analytics on Firm's Industry 4.0 Capability. *Proceedings* **2020**, *39*, 22. [CrossRef]
54. McAfee, A.; Brynjolfsson, E. Big Data: The Management Revolution. *Harv. Bus. Rev.* **2012**, *90*, 60–68.
55. Pech, M.; Vrchota, J. The Product Customisation Process in Relation to Industry 4.0 and Digitalization. *Processes* **2022**, *10*, 539. [CrossRef]
56. Abad, I.; Cerrada, C.; Cerrada, J.A.; Heradio, R.; Valero, E. Managing RFID Sensors Networks with a General Purpose RFID Middleware. *Sensors* **2012**, *12*, 7719–7737. [CrossRef]
57. Huang, G.Q.; Zhang, Y.F.; Chen, X.; Newman, S.T. RFID-Enabled Real-Time Wireless Manufacturing for Adaptive Assembly Planning and Control. *J. Intell. Manuf.* **2008**, *19*, 701–713. [CrossRef]
58. Aslan, M.F.; Durdu, A.; Sabanci, K.; Ropelewska, E.; Gültekin, S.S. A Comprehensive Survey of the Recent Studies with UAV for Precision Agriculture in Open Fields and Greenhouses. *Appl. Sci.* **2022**, *12*, 1047. [CrossRef]
59. Velusamy, P.; Rajendran, S.; Mahendran, R.K.; Naseer, S.; Shafiq, M.; Choi, J.-G. Unmanned Aerial Vehicles (UAV) in Precision Agriculture: Applications and Challenges. *Energies* **2022**, *15*, 217. [CrossRef]
60. Nakrošienė, A.; Bučiūnienė, I.; Goštautaitė, B. Working from Home: Characteristics and Outcomes of Telework. *IJM* **2019**, *40*, 87–101. [CrossRef]
61. Fernández-Caramés, T.M.; Blanco-Novoa, O.; Froiz-Míguez, I.; Fraga-Lamas, P. Towards an Autonomous Industry 4.0 Warehouse: A UAV and Blockchain-Based System for Inventory and Traceability Applications in Big Data-Driven Supply Chain Management. *Sensors* **2019**, *19*, 2394. [CrossRef] [PubMed]
62. Radočaj, D.; Plaščak, I.; Heffer, G.; Jurišić, M. A Low-Cost Global Navigation Satellite System Positioning Accuracy Assessment Method for Agricultural Machinery. *Appl. Sci.* **2022**, *12*, 693. [CrossRef]
63. Modrak, V.; Soltysova, Z.; Poklemba, R. Mapping Requirements and Roadmap Definition for Introducing I 4.0 in SME Environment. In *Advances in Manufacturing Engineering and Materials*; Hloch, S., Klichová, D., Krolczyk, G.M., Chattopadhyaya, S., Ruppenthalová, L., Eds.; Springer International Publishing: Cham, Germany, 2019; pp. 183–194. ISBN 978-3-319-99352-2.
64. Shoaib, M.; Bosch, S.; Incel, O.; Scholten, H.; Havinga, P. Fusion of Smartphone Motion Sensors for Physical Activity Recognition. *Sensors* **2014**, *14*, 10146–10176. [CrossRef] [PubMed]
65. Redden, R. Genetic Modification for Agriculture—Proposed Revision of GMO Regulation in Australia. *Plants* **2021**, *10*, 747. [CrossRef] [PubMed]
66. Czech Statistical Office Businesses by Principal Activity. Available online: https://vdb.czso.cz/vdbvo2/faces/en/shortUrl?su=2f1468c8 (accessed on 10 February 2022).
67. Witte, R.S.; Witte, J.S. *Statistics*, 12th ed.; Wiley: Hoboken, Hudson, 2016.
68. Lind, A.D.; Marchal, G.W.; Wathen, A.S. *Basic Statistics for Business & Economics*, 9th ed.; McGraw-Hill Education: New York, NY, USA, 2018.
69. European Commission The Digitisation of the European Agricultural Sector. Available online: https://digital-strategy.ec.europa.eu/en/policies/digitisation-agriculture (accessed on 10 February 2022).
70. Institute of Agricultural Economics and Information. *Zpráva o Stavu Zemědělství ČR Za Rok 2020: Zelená Zpráva [State of Agriculture Report 2020: Green Report]*; Ministry of Agriculture of the Czech Republic: Prague, Czech Republic, 2020.
71. Smolkova, B.; Dusinska, M.; Gabelova, A. Epigenetic Effects of Nanomaterials. In *Encyclopedia of Environmental Health*; Elsevier: Amsterdam, The Netherlands, 2019; pp. 678–685. ISBN 978-0-444-63952-3.
72. Jedlicka, M. Precizní Zemědělství v Chovech Zvířat [Precision Agriculture in Animal Farming]. Available online: https://naschov.cz/precizni-zemedelstvi-v-chovech-zvirat/ (accessed on 14 February 2022).
73. Balafoutis, A.T.; Beck, B.; Fountas, S.; Tsiropoulos, Z.; Vangeyte, J.; van der Wal, T.; Soto-Embodas, I.; Gómez-Barbero, M.; Pedersen, S.M. Smart Farming Technologies–Description, Taxonomy and Economic Impact. In *Precision Agriculture: Technology and Economic Perspectives*; Pedersen, S.M., Lind, K.M., Eds.; Progress in Precision Agriculture; Springer International Publishing: Cham, Germany, 2017; pp. 21–77. ISBN 978-3-319-68713-1.

agriculture

Article

Information Perception Method for Fruit Trees Based on 2D LiDAR Sensor

Yong Wang, Changxing Geng *, Guofeng Zhu, Renyuan Shen, Haiyang Gu and Wanfu Liu

Robotics and Microsystems Centre, Soochow University, Suzhou 215000, China; 20195229071@stu.suda.edu.cn (Y.W.); 20195229074@stu.suda.edu.cn (G.Z.); 20205229012@stu.suda.edu.cn (R.S.); 20215229104@stu.suda.edu.cn (H.G.); 20215229086@stu.suda.edu.cn (W.L.)

* Correspondence: chxgeng@suda.edu.cn; Tel.: +86-186-6229-9889

Abstract: To solve the problem of orchard environmental perception, a 2D LiDAR sensor was used to scan fruit trees on both sides of a test platform to obtain their position. Firstly, the two-dimensional iterative closest point (2D-ICP) algorithm was used to obtain the complete point cloud data of fruit trees on both sides. Then, combining the lightning connection algorithm (LAPO) and the density-based clustering algorithm (DBSCAN), a fruit tree detection method based on density-based lightning connection clustering (LAPO-DBSCAN) was proposed. After obtaining the point cloud data of fruit trees on both sides of the test platform using the 2D-ICP algorithm, the LAPO-DBSCAN algorithm was used to obtain the position of fruit trees. The experimental results show that the positive detection rate was 96.69%, the false detection rate was 3.31%, and the average processing time was 1.14 s, verifying the reliability of the algorithm. Therefore, this algorithm can be used to accurately find the position of fruit trees, meaning that it can be applied to orchard navigation in a later stage.

Keywords: point cloud registration of fruit trees; lightning attachment procedure optimization; density-based spatial clustering of applications with noise; information perception of fruit trees

Citation: Wang, Y.; Geng, C.; Zhu, G.; Shen, R.; Gu, H.; Liu, W. Information Perception Method for Fruit Trees Based on 2D LiDAR Sensor. *Agriculture* **2022**, *12*, 914. https://doi.org/10.3390/agriculture12070914

Academic Editor: Francesco Marinello

Received: 14 April 2022
Accepted: 18 June 2022
Published: 23 June 2022

1. Introduction

To accelerate the development of smart agriculture, agricultural vehicle navigation technology has been developed rapidly. Agricultural machinery autonomous navigation systems based on machine vision, GPS, and LiDAR sensors have emerged [1]. Machine vision is greatly affected by the operating environment and lighting conditions. The application of GPS is affected by satellite signals. A LiDAR sensor can provide a large amount of accurate distance information at a higher frequency, reliably provide the position and depth information of surrounding objects [2], and provide more comprehensive information.

There are many ways to identify fruit trees in orchards. Judging from the existing research results, LiDAR sensors, cameras, or multisensor fusion can be used to detect fruit trees. Since the overall characteristics of trees are obvious, the trunks of fruit trees can be regarded as circles which can be detected by LiDAR sensors [3]. Due to the different installation methods and types of LiDAR sensors used, the data obtained are also different. (1) A LiDAR sensor can be installed vertically to extract the contour information of fruit trees [4,5]. Although this method can obtain the information of the trunks of fruit trees, as the LiDAR sensor is installed vertically, it can only extract the information of one tree at a time. This perception method is usually used to find the specific growth information of a fruit tree, such as fruit trees contour reconstruction. (2) A ground LiDAR sensor can be used to scan the environment to obtain fruit tree information [6,7]. (3) A mobile ground LiDAR sensor has been used to identify Fuji apples [8]. (4) An airborne LiDAR sensor has been used to obtain the scan data of fruit tree trunks [9,10].

Some scholars have also obtained tree information by analyzing LiDAR sensor data found from scanning. Using the same distance between the positions of fruit trees in an orchard, the data points in the arithmetic sequence of the concave points in the LiDAR sensor scan data can be extracted as the trunk points to obtain data [11]. LiDAR sensors can be used to scan woodland environments to obtain woodland data [12,13]. Since the data type of LiDAR sensors can be approximated by a point set, a clustering algorithm can be used to obtain fruit tree trunk information. Two-dimensional LiDAR sensors can be used to scan orchard environments and perform data clustering to extract the arc information of trunks [14,15]. Besides obtaining fruit tree information from clustering, 2D LiDAR sensors can be used to extract the central feature point data of tree trunks using the Euclidean clustering algorithm and the important geometric theorem of three-point collinearity [16]. Three-dimensional LiDAR sensor data are more abundant than 2D LiDAR sensor data, so many people use 3D LiDAR sensors for tree detection [17,18]. Although machine vision is greatly affected by the operating environment and lighting conditions, there have been many studies on the use of cameras for fruit tree inspection in orchards [19]. Due to the complex environment of orchards, a variety of sensor fusion methods can be used for research [20–23].

In previous studies, various sensors have been used to obtain orchard environmental information for orchard intelligent equipment. Usually, the information of fruit trees is used to pave the way for the application of intelligent equipment in orchard navigation.

The main purpose of this article is to obtain the position information of fruit trees using a 2D LiDAR sensor. After obtaining the position information of fruit trees with the algorithm proposed in this paper, it can be used for positioning, fitting navigation lines, and the navigation of orchard intelligent equipment in later stages. For the complex environment of orchards, this environment perception method is studied. Firstly, a fruit tree information acquisition method based on 2D-ICP is proposed. After the iterative registration of the point cloud data of both sides of fruit trees obtained by the 2D LiDAR sensor, the point cloud data of each fruit tree in the orchard are obtained. Then, by improving the LAPO and DBSCAN algorithms, a new method based on LAPO-DBSCAN is used to obtain the position of each fruit tree and realize their detection. Finally, the accuracy of the algorithm is verified by a field test.

2. Materials and Methods

2.1. Experimental Equipment

In this research, a differential test platform with a maximum speed of 1 m/s was built, as shown in Figure 1. The LiDAR sensor scans the surrounding data in real time, and the obtained LiDAR sensor data are transmitted to the industrial computer. The industrial computer runs a self-made software system to analyze the LiDAR sensor data. The LiDAR sensor is Rashen N30103B and it adopts the horizontal installation method, which is located in the front and middle of the orchard transportation robot. The installation height is 0.65 m and the parameters are shown in Table 1.

Figure 1. Test platform.

Table 1. Two-dimensional LiDAR sensor parameters.

2D LiDAR Sensor Specifications	Parameter Index
Detection range (m)	30
Ranging accuracy (mm)	±30
Scanning angle (°)	360
Angle resolution (°)	0.18
Scanning frequency (Hz)	10

2.2. *Fruit Tree Information Perception Method*

In this section, we introduce the complete method of fruit tree information perception; the specific process is shown in Figure 2. When the test platform runs in the orchard, the surrounding fruit tree information can be obtained by 2D LiDAR sensor scanning, and the point cloud data of the fruit tree will be preprocessed. Preprocessing is used so as to only retain fruit trees on both sides of the test platform for point cloud registration and clustering. The most important is point cloud registration and clustering. Firstly, complete fruit tree data on both sides can be obtained by point cloud registration. Then, the position of each fruit tree can be obtained by a clustering algorithm.

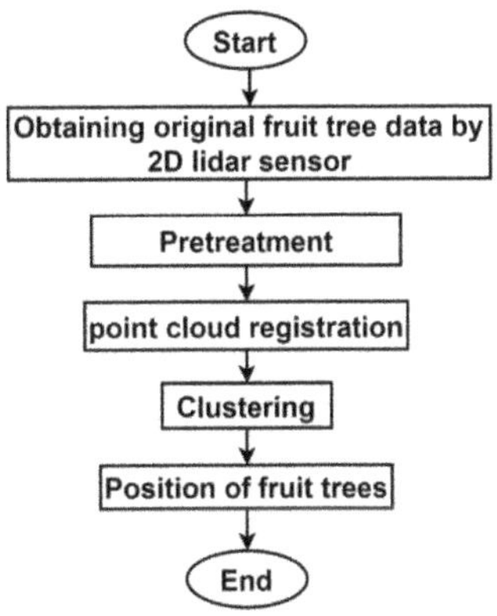

Figure 2. The process of fruit tree perception.

2.3. *Fruit Tree Point Cloud Data Collection Method Based on 2D-ICP*

With the increase in distance, the amount of LiDAR sensor data collected will become less and less. At the same time, in the process of acquisition, the fruit tree rows on both sides of the non-test platform will be blocked and the information will be incomplete, resulting in a poor iterative effect. Therefore, for the preprocessing of the collected LiDAR point cloud data, only the fruit tree row data on both sides of the test platform will be retained. After obtaining the LiDAR sensor data of fruit trees on both sides of the test platform, the 2D-ICP algorithm can be used for registration. The specific algorithm steps are as follows. The preprocessed target point cloud and source point cloud are $P^k = \left\{P_i^k\right\}$ and $P^{k+1} = \left\{P_i^{k+1}\right\}, i = 1,2,3\ldots n$. According to the 2D-ICP algorithm, Equation (1) gives the objective function.

$$Dist(R,T)_{min} = \frac{1}{n}\sum_{i=1}^{n}\left|P_i^k - \left(RP_i^{k+1} + T\right)\right|^2 \tag{1}$$

There are two variables in Equation (1) that can be regularized by considering only the rotation matrix R—that is, the centers of two frame point clouds are $C_{pk} = \frac{1}{n}\sum_{i=1}^{n} P_i^k$ and $C_{pk+1} = \frac{1}{n}\sum_{i=1}^{n} P_i^{k+1}$, where $P_i^k = P_i^k - C_{pk}$ and $P_i^{k+1} = P_i^{k+1} - C_{pk+1}$.

Then, Equation (1) becomes Equation (2).

$$Dist(R,T)_{min} = \frac{1}{n}\sum_{i=1}^{n}\left|P_i^k - RP_i^{k+1}\right|^2 \tag{2}$$

Decomposing Equation (2) into Equation (3) gives us:

$$Dist(R,T)_{min} = \frac{1}{n}\left(\sum_{i=1}^{n}\left|P_i^k\right|^2 + \sum_{i=1}^{n}\left|RP_i^{k+1}\right|^2 - 2\sum_{i=1}^{n}\left|P_i^k RP_i^{k+1}\right|^2\right) \tag{3}$$

If the objective function is minimized, Equation (4) is maximized:

$$F(R)_{max} = \frac{1}{n}\sum_{i=1}^{n}\left|P_i^k RP_i^{k+1}\right|^2 \tag{4}$$

Since the fruit tree data are two-dimensional, the rotation matrix R is valued as $\begin{bmatrix} cos\theta & -sin\theta \\ sin\theta & cos\theta \end{bmatrix}$, where θ is the rotation angle between the two frame point clouds, which is substituted into Equation (4) to obtain Equation (5).

$$F(R)_{max} = \frac{1}{n}\sum_{i=1}^{n}\left|\begin{bmatrix} Px_i^k & Py_i^k \end{bmatrix}\begin{bmatrix} cos\theta & -sin\theta \\ sin\theta & cos\theta \end{bmatrix}\begin{bmatrix} Px_i^{k+1} \\ Py_i^{k+1} \end{bmatrix}\right|^2 \tag{5}$$

The derivation and extreme value of θ are used to deduce Equation (6).

$$\frac{sin\theta}{cos\theta} = sin\theta\frac{1}{n}\sum_{i=1}^{n}\left(\frac{Py_i^k \times Px_i^{k+1} - Px_i^k \times Py_i^{k+1}}{Px_i^k \times Px_i^{k+1} + Py_i^k \times Py_i^{k+1}}\right) \tag{6}$$

After θ is calculated, R can be obtained. T can be obtained by Equation (7), and then iterated until the threshold is satisfied.

$$T = \begin{bmatrix} \Delta x \\ \Delta y \end{bmatrix} = \begin{bmatrix} C_{pk}^x \\ C_{pk}^y \end{bmatrix} - R\begin{bmatrix} C_{pk+1}^x \\ C_{pk+1}^y \end{bmatrix} \tag{7}$$

2.4. Fruit Tree Position Detection Algorithm

2.4.1. Introduction to LAPO and DBSCAN Algorithms

The LAPO algorithm [24] has four important stages, including the cloud surface penetrating the air phase, the lightning channel moving downward, the upward pilot starting to spread from the ground (or grounded object), and the last fight back stage. The LAPO algorithm has a strong optimization ability in many engineering problems, and no additional parameters need to be set, which can help to avoid subjective factors influencing the results of the algorithm. Due to the influence of randomness and other factors, the standard LAPO algorithm may also fall into a local optimum, which makes it impossible to obtain a better solution every time. There is room for further improvement in its stability. Therefore, this algorithm has also been applied and improved in various clustering algorithms [25,26].

The DBSCAN algorithm [27] is based on a certain distance measurement criterion, which clusters closely related data points based on their criteria into one category. The following two parameters are set before clustering. The first one is *Eps* (the radius of the given object is the neighborhood). The second one is *MinPts* (the minimum number of components that make up a class). The traditional DBSCAN clustering algorithm is affected by *Eps* and *MinPts*. These two parameters are global and fixed so that only the data in the data set that meet the threshold condition can be effectively clustered, meaning that data of other densities may be treated as noise. In addition, the traditional DBSCAN algorithm needs to traverse each data point. When the data scale is large, the algorithm execution efficiency will be low, and the processing time will be long, which is not conducive to the realization of the algorithm. In view of the shortcomings of the traditional DBSCAN algorithm, our predecessors in this area have carried out a considerable amount of research and improved the DBSCAN algorithm [28–30].

2.4.2. Fruit Tree Detection Algorithm Based on LAPO-DBSCAN

Due to the shortcomings of the two algorithms, this paper proposes a fruit tree position detection algorithm based on LAPO-DBSCAN. This algorithm is mainly used to obtain the position of fruit trees. This process includes preparation and detailed steps, and its flow chart is shown in Figure 3.

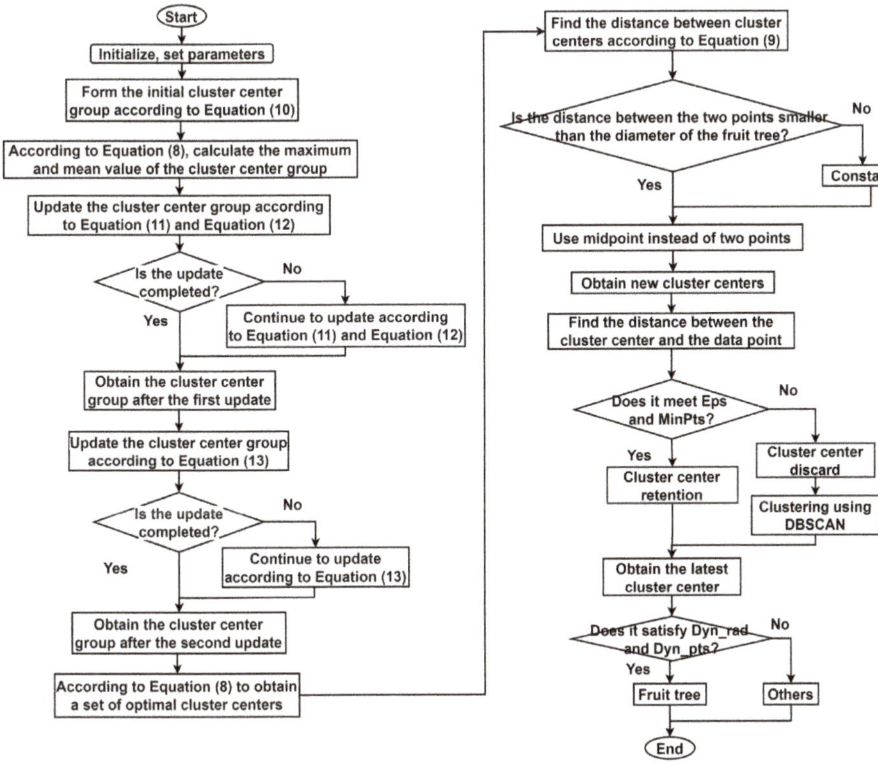

Figure 3. Fruit tree detection based on LAPO-DBSCAN.

A. Preparation.

Parameter settings. 1. Setting the parameters of the LAPO algorithm. (1) n groups of initial clustering centers are used as an initial clustering center group for each group of k clustering centers (k is set randomly). (2) t is the number of iterations. (3) t_{max} is the maximum number of iterations. 2. Setting the DBSCAN algorithm parameters. (1) The traditional DBSCAN algorithm: *Eps* and *MinPts* are set according to the actual situation. (2) The dynamic DBSCAN algorithm: the radius range of the dynamic fruit tree is set according to the radius of the fruit tree, which is $Dyn\_rad \in [rad_{min}, rad_{max}]$. According to the LiDAR sensor parameters and distance, different neighborhood density thresholds are obtained as $Dyn\_pts \in [Pts\_num_{min}, Pts\_num_{max}]$.

Setting various functions. 1. Equation (8) is the objective function.

$$F_{min} = \sum_{i=1}^{n}\sum_{j=1}^{k} v_{i,j} \parallel x_i - c_j \parallel^2 \tag{8}$$

where $v_{i,j} = \begin{cases} 1\ if\ arg\ min_j dis(x_i, c_j) \\ 0\ else \end{cases}$ is the confidence function, which means that each datum in the data set can only belong to one class.

2. Equation (9) is the Euclidean distance between two points, where $i \neq j$.

$$dis(X_i, X_j) = \sqrt{\sum_{i=1}^{n}\sum_{j=1}^{k} (X_i - X_j)^2} \tag{9}$$

B. Detailed steps.

1. Initializing the LiDAR data and setting the corresponding parameters.

2. Randomly selecting an initial cluster center group (n groups of cluster centers for each group of k) to form a matrix C of n rows and k columns:

$$C = \begin{bmatrix} \{c_{1,1}, c_{1,2}, \cdots c_{1,k-1}, c_{1,k}\} \\ \{c_{2,1}, c_{2,2}, \cdots c_{2,k-1}, c_{2,k}\} \\ \vdots \\ \{c_{n-1,1}, c_{n-1,2}, \cdots c_{n-1,k-1}, c_{n-1,k}\} \\ \{c_{n,1}, c_{n,2}, \cdots c_{n,k-1}, c_{n,k}\} \end{bmatrix}.$$

The cluster center of the row is expressed by Equation (10).

$$C_i = Data_{min} + rand \times (Data_{max} - Data_{min}) \tag{10}$$

where *rand* is a random number in the range [0, 1]. $Data_{max}$ and $Data_{min}$ are the maximum and minimum values of the radar data, respectively.

3. According to Equation (8), the fitness of cluster center group (C) is calculated. The optimal value, worst value, and average value of the cluster center group are C_{best}, C_{low}, and C_{ave}, respectively. According to the objective function, if $F_{low} > F_{ave}$, we can assign the value of C_{ave} to C_{low}.

4. If the cluster center (C_i) of a certain row is updated, a group of cluster centers (C_j) is randomly selected from the population, where $i \neq j$. If $F_{ave} > F_j$, Equation (11) can be used to iterate.

$$C_i = C_i + rand \times (C_{ave} - rand \times (C_i)) \tag{11}$$

If $F_j > F_{ave}$, Equation (12) can be used to iterate.

$$C_i = C_i - rand \times (C_{ave} - rand \times (C_i)) \tag{12}$$

After the above process is complete, the updated cluster center group (C_{new}) can be obtained.

5. Return to step (3). Updating the cluster center group (C_{new}) and obtaining the optimal value, the worst value and mean value of the cluster center group will be C_{new_best}, C_{new_low}, and C_{new_ave}, respectively.

6. C_{new} can be iterated through Equation (13) to obtain a new cluster center group C_{new}.

$$C_{new} = C_{new} + rand \times S \times (C_{new\_ave} + rand \times (C_{new\_low} - C_{new\_best})) \tag{13}$$

where $S = 1 - \left(\frac{t}{t_{max}}\right) \times exp\left(-\frac{t}{t_{max}}\right)$.

7. Return to step (3). The cluster center group (C_{new}) needs to be updated and the optimal value (C_{new_best}), worst value (C_{new_low}), and mean value (C_{new_ave}) of the cluster center group can be obtained.

8. With step (2) to step (7), the optimal cluster center group (C_{best}) can be obtained and the fitness can be calculated according to the objective function to obtain a set of optimal cluster centers (c_{best}), namely, $\{c_{1\_best}, c_{2\_best}, \cdots c_{k-1\_best}, c_{k\_best}\}$.

9. According to the obtained cluster centers, Equation (9) can be used to find the distance between each cluster center. When the distance between two points is less than the diameter of the fruit tree (the diameter of the fruit tree at the height of the radar installation), the distance between the two points is used. The two cluster centers can be replaced with the midpoint (when encountering bifurcated fruit trees, the midpoint is also used to replace the two points). Otherwise, the cluster center remains unchanged. Finally, the updated cluster center (c_{best}) can be obtained.

10. Equation (9) can be used to calculate the distance from each point (x_i) to the cluster center (c_{best}) in the data set and dividing each data point into each cluster center according to the confidence function, which is $Dist(x_i, c_{n\_best}) = \sqrt{(x_i - c_{n\_best})^2}$.

11. According to the actual situation, when the *Eps* and *MinPts* conditions are met, the cluster center is retained; otherwise, the cluster center is discarded.

12. After discarding some of the cluster centers that do not meet the criteria, according to the DBSCAN algorithm (still must meet the *Eps* and *MinPts* conditions), clustering is performed to obtain the corresponding cluster, and the mean value of the corresponding cluster is used to represent the cluster center (c_i).

13. When the distance between two cluster centers is less than the diameter of the fruit tree (the diameter of the fruit tree at the radar installation position), return to step (9). Otherwise, the final cluster center (c_{final}) should be obtained to form a cluster.

14. According to the final cluster center (c_{final}), the following two conditions need to be met to determine whether something is a fruit tree. Firstly, if the fruit tree dynamic radius (*Dyn_rad*), which is $rad_{min} \leq Dist(X_i, C_{n\_best}) \leq rad_{max}$, is satisfied, an object is a fruit tree. At the same time, if it meets the dynamic neighborhood density threshold ($Dyn\_pts \in [Pts\_num_{min}, Pts\_num_{max}]$), the object is a fruit tree.

15. Finally, the position of each fruit tree can be detected.

2.4.3. Algorithm Improvement

The algorithm solves the problem of local optimal solutions in the LAPO algorithm and parameter globality in the DBSCAN algorithm. In the LAPO algorithm, n initialized cluster centers are randomly selected, as shown in step (2). As the selected n cluster centers are random, there will be local optimal solutions in the calculation process for LAPO (from step (3) to step (8)). In the actual test, the local optimal solution may appear in the following situations. The results of the algorithm identify a set of clustering centers, but there is obvious deviation (the obvious deviation here refers to the situation where the clustering center is not on the data point and cannot correspond to the relevant data point) and two clustering centers are together. At the same time, the LAPO algorithm also has the problem of missing perception. To solve the above problems, the LAPO algorithm is improved. When the two clustering centers are together, the two clustering centers can be combined into one by taking the midpoint in step (9). In the cases of obvious deviation and missing detection, the final cluster center can be obtained by the DBSCAN algorithm (from step (10) to step (12)), but fruit trees may still not be detected. As the DBSCAN algorithm is affected by *Eps* and *MinPts*, in the step (14) *Eps* and *MinPts* are used to distinguish fruit trees by the dynamic threshold method. Through three-layer detection, the positive detection rate of fruit tree detection can be greatly improved.

2.4.4. Simulation Data Verification

Both the LAPO and DBSCAN algorithms have shortcomings, so they need to be improved to adapt them to more scenes. Using the 2D LiDAR sensor to scan the contours of fruit trees at different heights, the point cloud data will include two kinds of point cloud data, which are the point cloud data of the main tree trunk and the point cloud data of the canopy. Usually, the more data there are, the better the algorithm will be. To verify that the LAPO-DBSCAN algorithm used in this paper is better than the LAPO and DBSCAN algorithms, simulation data similar to the trunk of fruit trees are used for verification according to the point cloud data of the trunks of fruit trees scanned by 2D LiDAR sensors, as shown in Figure 4.

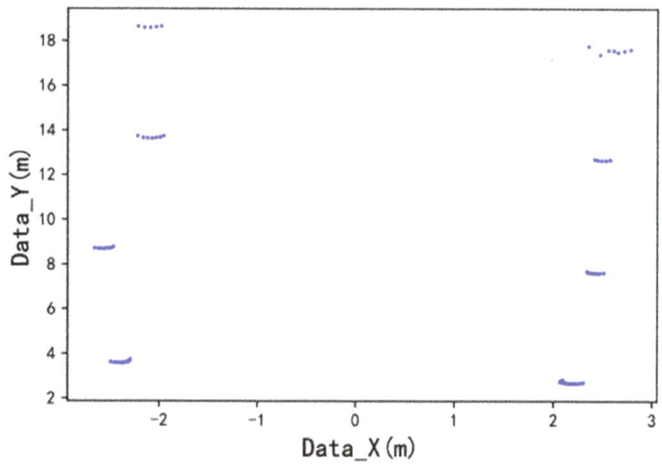

Figure 4. Simulation data.

To prove that the method based on the LAPO-DBSCAN algorithm is able to detect more characteristic information than the method based on the LAPO algorithm, this paper uses simulated fruit tree data to test the above two methods for one hundred frames. The results obtained from the method based on the LAPO-DBSCAN algorithm and the method based on the LAPO algorithm are shown in Table 2. The simulation data used in this paper are similar to the trunk of fruit trees. Scholars [30] have used the improved DBSCAN algorithm to detect the trunk of fruit trees, and the accuracy can reach 95.5%. Compared with previous algorithms, this algorithm increases the accuracy by 3.92%. Therefore, the algorithm used in this paper will no longer be compared with the DBSCAN algorithm for detecting the trunk of fruit trees.

Table 2. Actual scene test results.

Algorithm Type	Times	Results (%)		Average Handling Time (s)
LAPO	100	Positive detection rate False detection rate	97.00% 3.00%	0.41
LAPO-DBSCAN	100	Positive detection rate False detection rate	99.42% 0.58%	0.07

In Table 2, the positive detection rate of the LAPO-DBSCAN algorithm is better than that of the LAPO algorithm, and the detection result is 2.42% higher. In terms of the false detection rate, the LAPO-DBSCAN algorithm is better than the LAPO algorithm, and the difference between the detection results is 2.42%. In terms of the average processing time, the LAPO-DBSCAN algorithm consumes 82.92% less time than the LAPO algorithm. Therefore, the simulation results show that the LAPO-DBSCAN algorithm is superior to the LAPO algorithm and has a better detection effect.

3. Results

3.1. Experimental Scene

Due to the different planting mode and row spacing of each fruit tree, the data obtained by 2D LiDAR sensor scans of different fruit trees are also different, which will affect the accuracy and stability of the algorithm. The test site used in this paper was selected from the orchard of Nijiawan water field in Xiangcheng District of Suzhou, as shown in Figure 1. The distance from the ground to the main trunk of the fruit tree selected in this paper was about 0.5 m, and the area above 0.5 m was the canopy. According to the installation height of the 2D LiDAR sensor, the collected point cloud data were all the point cloud data of fruit tree crowns. As shown in Table 3, data on two rows of fruit trees used in the experiment were obtained.

Table 3. Fruit tree data.

Fruit Tree Information							
The outline length of the fruit tree on the left (m)	3.83	5.03	4.08	2.82	2.67	2.98	3.92
The outline length of the fruit tree on the right (m)	2.83	4.40	4.24	2.98	3.61	3.30	4.46
Distance between left and right fruit trees (m)				4			
Distance between adjacent fruit trees on the same side (m)				3			

3.2. Algorithm Verification

In the previous chapter, we introduced the method of fruit tree information perception. Next, the point cloud registration based on the 2D-ICP algorithm and the fruit tree position detection based on the LAPO-DBSCAN algorithm are tested.

3.2.1. Experiment on Fruit Tree Information Acquisition

Firstly, the test platform is controlled to drive slowly from the beginning of the fruit tree to the end of the fruit tree to collect the point cloud data of the fruit trees. Then, all the initial point cloud data of the fruit trees are preprocessed, and the final result only retains the nearest point cloud data of the fruit tree on both sides of the test platform. The original point cloud data, as shown in Figure 5, show that the data in the untreated orchard have many interference points, such as the data collected from non-bilateral fruit trees and the "zero-points" generated by the 2D LiDAR sensor at a certain angle. During the preprocessing, we keep the data in the red box in Figure 5. The preprocessed data are shown in Figure 6.

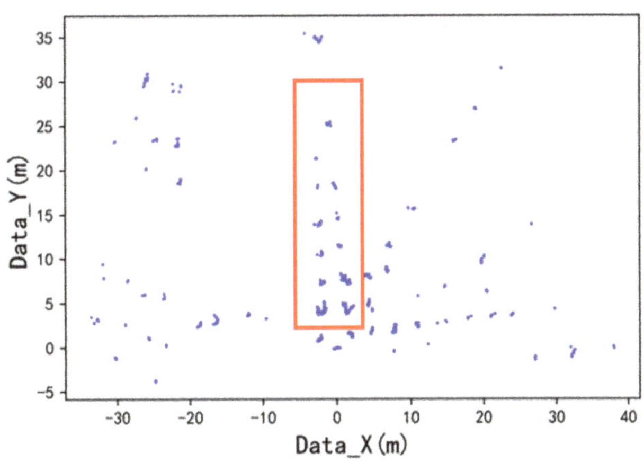

Figure 5. A frame of original data.

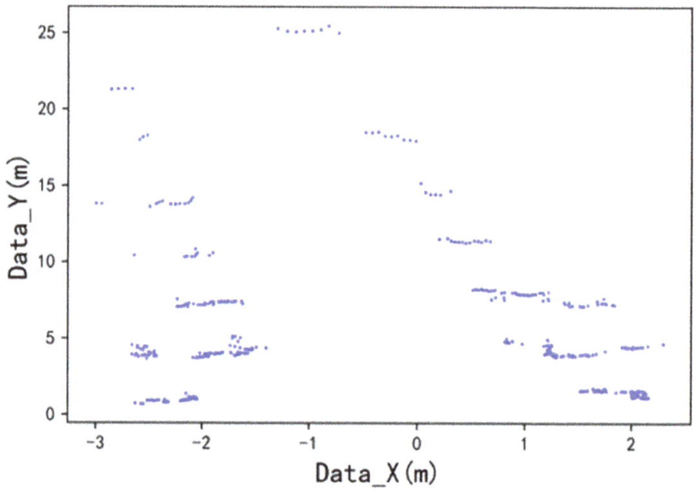

Figure 6. A frame of preprocessed data.

After preprocessing, registration is performed based on the 2D-ICP algorithm to obtain the completed fruit tree row point cloud data. This paper takes the registration of two frame point cloud data as an example to illustrate the registration process of fruit tree point cloud data. The registration process is shown in Figure 7. In Figure 7a–e, the iterative registration processes of two point clouds are shown. The R and T from Figure 7a–e are calculated using Equations (6) and (7). The R and T are $\begin{bmatrix} 0.999948 & -0.010145 \\ 0.010145 & 0.999948 \end{bmatrix}$ and $\begin{bmatrix} 0.027403 & 0.139726 \end{bmatrix}$ in Figure 7a. The R and T are $\begin{bmatrix} 0.999992 & -0.003975 \\ 0.003975 & 0.999992 \end{bmatrix}$ and $\begin{bmatrix} 0.010213 & 0.062397 \end{bmatrix}$ in Figure 7b. The R and T are $\begin{bmatrix} 0.999997 & -0.002030 \\ 0.002030 & 0.999997 \end{bmatrix}$ and $\begin{bmatrix} 0.007544 & 0.030963 \end{bmatrix}$ in Figure 7c. The R and T are $\begin{bmatrix} 0.999999 & -0.000753 \\ 0.000753 & 0.9999996 \end{bmatrix}$ and $\begin{bmatrix} 0.001915 & 0.018437 \end{bmatrix}$ in Figure 7d. The R and T are $\begin{bmatrix} 0.999999 & -0.000435 \\ 0.000435 & 0.999999 \end{bmatrix}$ and $\begin{bmatrix} 0.002288 & 0.000870 \end{bmatrix}$ in Figure 7e. Figure 7f is the result of using a two-frame point cloud registration as the next target point cloud. All the preprocessed point cloud data of fruit trees on both sides of the test platform can be iteratively registered by the 2D-ICP algorithm to obtain complete the point cloud data of fruit trees, including the point cloud information of each fruit tree position, as shown in Figure 8. The fruit trees on both sides of the test platform in Figure 8a show the point cloud data in Figure 8b.

When using the 2D-ICP algorithm to construct point cloud data of fruit trees on both sides of the test platform, we need to consider two problems. The first question is how to obtain as much complete point cloud data as possible. The second problem is how to prevent the oscillation of the collected point cloud data due to the uneven ground of the orchard during movement. Therefore, on the one hand, we move the test platform as slowly as possible, so the vibration amplitude of the vehicle is not large in the process of moving. On the other hand, we choose a 2D LiDAR sensor with a high frequency, as shown in Table 1, to obtain more fruit point cloud data within a short period of time. Through these two measures, the adverse effects caused by the vibration of the test platform can be compensated for to a certain extent, and the obtained fruit tree point cloud information can be enriched, which is conducive to obtaining better point cloud registration results.

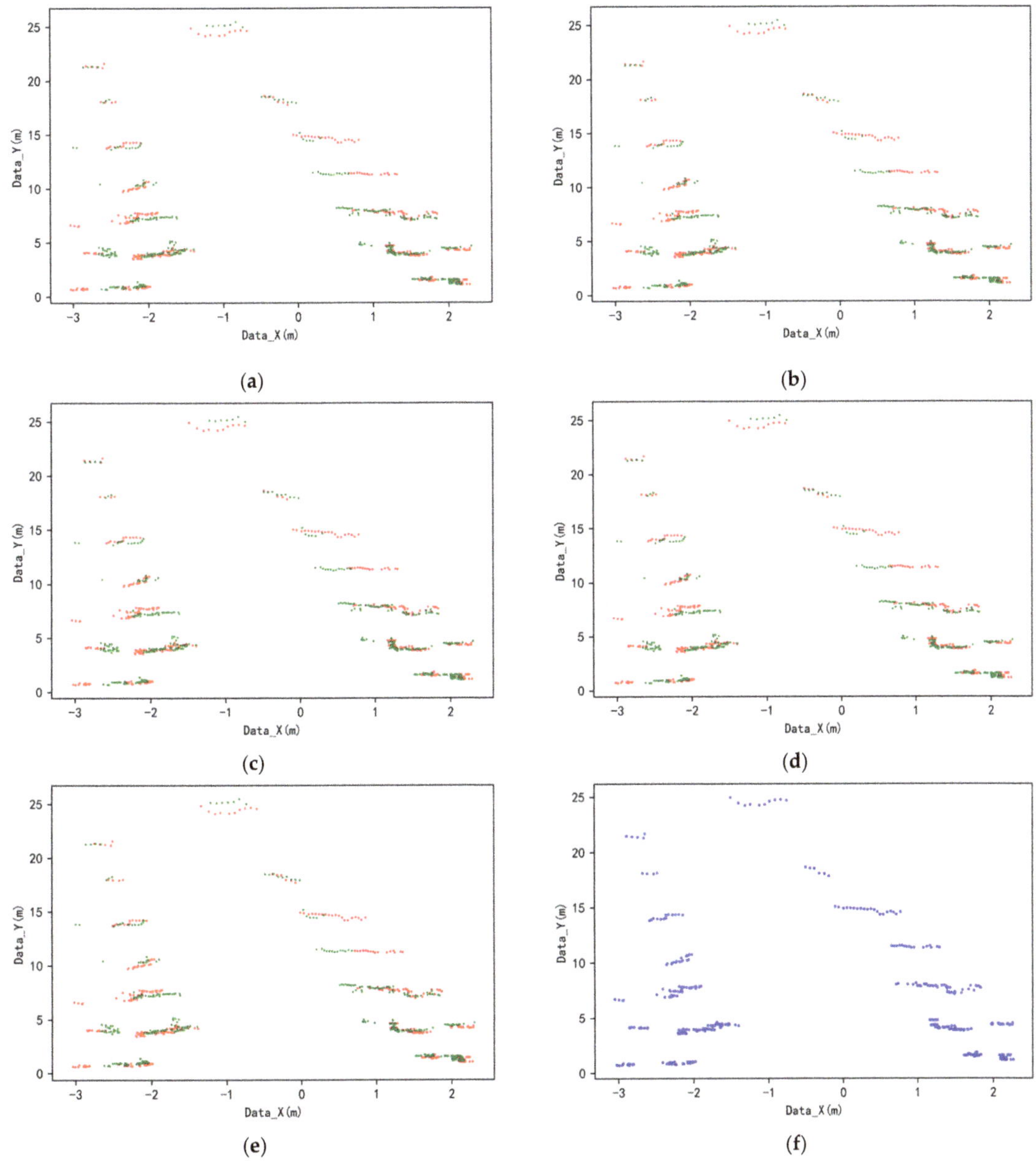

Figure 7. The 2D-ICP algorithm registration process. (**a**) First point cloud registration; (**b**) second point cloud registration; (**c**) third point cloud registration; (**d**) fourth point cloud registration; (**e**) last point cloud registration; (**f**) results after two-frame point cloud alignment.

Figure 8. Complete fruit tree row point cloud data. (**a**) Fruit trees on both sides of the test platform in the actual scene; (**b**) point cloud data corresponding to fruit trees on both sides of the test platform.

3.2.2. Experiment on Fruit Tree Position Acquisition

After obtaining the point cloud data of the fruit trees on both sides of the test platform, we need to obtain the position of the fruit trees. The first task is to set relevant parameters. The number of initial cluster centers are set to four groups with six cluster centers in each group, and the maximum number of iterations is 10. Although the above parameters are set randomly, they should also be designed according to the actual situation. The value of the parameters should not be too large or too small. In the traditional DBSCAN algorithm, *Eps* is set to 0.58 m and *MinPts* is set to 3. *Eps* depends on the average radius of fruit trees on both sides. In the dynamic DBSCAN algorithm, *Dyn_rad* depends on the minimum radius and maximum radius of fruit trees on both sides—namely, $Dyn\_rad \in [0.42, 0.80]$. *Dyn_pts* depends on the quotient of the number of point clouds per frame of the 2D LiDAR sensor and the number of cluster centers in each group, namely, $Dyn\_pts \in [3, 333]$. The radius of fruit trees depends on their contours, as shown in Table 3. We regard the outline of the fruit tree as a circle and use the formula of the circumference of a circle to calculate the radius of the fruit tree.

The implementation process based on the LAPO-DBSCAN algorithm is as follows. According to the LAPO-DBSCAN algorithm, a set of clustering centers are obtained from step (3) to step (8). In three situations of the missing detection of some fruit trees, two clustering centers appear on one fruit tree and the clustering center obviously deviates from the fruit tree data. Here, we take two clustering centers together as examples, as shown in Figure 9a. Then, we obtain the complete cluster center through step (9) to step (13). Finally, the test results of fruit trees are obtained through step (14), as shown in Figure 9b, where the blue point is the data point and the red "×" is the clustering center. The corresponding cluster center coordinates are shown in Table 4. According to Figure 8b and Table 4, this algorithm can accurately detect the position of fruit trees.

Table 4. Two-dimensional LiDAR sensor parameters.

The Coordinates of the Left Fruit Tree (m)	The Coordinates of the Right Fruit Tree (m)
(−2.34, 0.95)	(1.86, 1.61)
(−2.27, 4.08)	(1.94, 4.51)
(−2.23, 7.59)	(1.37, 7.18)
(−2.63, 10.57)	(1.28, 11.16)
(−2.71, 14.23)	(1.88, 15.32)
(−2.11, 17.84)	(1.76, 18.05)
(−2.55, 20.29)	(2.21, 20.68)

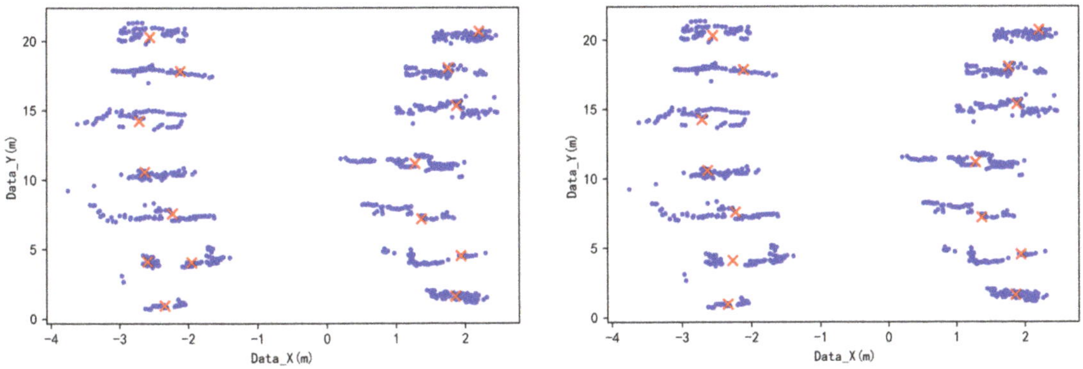

Figure 9. Algorithm detection results: (**a**) two cluster centers on a fruit tree; (**b**) results processed using this algorithm.

In Figure 8, the point cloud information of fruit trees on both sides was stored by the 2D-ICP algorithm. Therefore, the point cloud data were directly tested 100 times by the LAPO-DBSCAN algorithm, as shown in Table 5. One hundred iterations of the LAPO-DBSCAN algorithm are shown in Figure 10.

Table 5. Actual scene test results.

Algorithm Type	Times	Results (%)		Average Handling Time (s)
LAPO-DBSCAN	100	Positive detection rate False detection rate	96.69% 3.31%	1.14

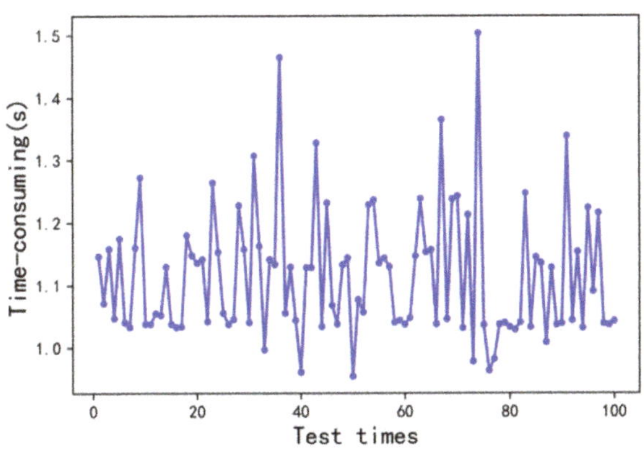

Figure 10. Actual scene, 100 iterations.

In Table 5, the positive detection rate of the algorithm is 96.69%, the false detection rate is 3.31%, the accuracy is maintained at more than 95%, and the average processing time is 1.14 s, which meets the accuracy requirements for the actual scene. Therefore, this algorithm can be used for the detection of fruit trees in orchards. This has certain practical significance for future navigation in orchards.

4. Discussion

In this research, the contour information of surrounding fruit trees was collected by a 2D LiDAR sensor mounted on an experimental platform, and the point cloud registration of fruit trees on both sides of the transportation robot was completed by the 2D-ICP algorithm. Then, the point cloud data were analyzed using the LAPO-DBSCAN clustering method to obtain the coordinate position points of each fruit tree. The most important thing is to propose a fruit tree position detection algorithm based on LAPO-DBSCAN. This algorithm has obvious advantages over those proposed in previous studies. Compared with K-means clustering, this algorithm does not need to set the number of clustering centers to be detected in advance, which makes it more convenient. Compared with the DBSCAN, this algorithm is more adaptable and can classify fruit trees more accurately. Compared with LAPO, this algorithm takes less time and is more accurate. Comparing the results of the simulation data (Table 2) with the results of the actual scene (Table 5), it can be seen that with the increase in environmental characteristics (from the detection of fruit tree trunks to the detection of fruit tree crowns), although the accuracy of the algorithm is reduced and the processing time is prolonged, the accuracy remains above 95% and the average processing time is 1.14 s, which generally meets the accuracy requirements of actual scenes. Therefore, this method can be used for the detection of fruit trees in orchards.

5. Conclusions

In this paper, a fruit tree position information perception method based on a 2D LiDAR sensor was proposed and verified on an experimental platform. According to the actual detection effect, the positive detection rate of the algorithm could reach 96.69%, the false detection rate was as low as 3.31%, and the average processing time was 1.14 s, indicating that the algorithm can be used in fruit tree detection to obtain the position of fruit trees. Although the algorithm has a good perception effect, there are also shortcomings. In the process of the experiment, because of the limitations of the 2D LiDAR sensor itself, the fruit tree information obtained was limited. When the algorithm is used for verification, there will be false detection and missed detection. From the detection of fruit tree trunks to the detection of fruit tree crowns, the amount of point cloud data for fruit trees will increase, resulting in a decrease in the positive detection rate of the algorithm. However, overall, the algorithm can still meet the requirements for the detection of fruit trees. In the future, the positions of fruit trees obtained by this algorithm could play a role in orchard navigation.

Author Contributions: Conceptualization, Y.W. and C.G.; Data curation, Y.W., G.Z. and H.G.; Formal analysis, R.S. and W.L.; Investigation, C.G.; Methodology, Y.W.; Software, Y.W.; Validation, G.Z., R.S. and H.G.; Writing—original draft, Y.W.; Writing—review and editing, C.G. and G.Z. All authors have read and agreed to the published version of the manuscript.

Funding: This research was funded by the National Key Research and Development Program of China (No. 2016YFD0700600).

Institutional Review Board Statement: Not applicable.

Informed Consent Statement: Not applicable.

Data Availability Statement: Not applicable.

Conflicts of Interest: The authors declare no conflict of interest.

References

1. Dworak, V.; Selbeck, J.; Ehlert, D. Ranging sensors for vehicle-based measurement of crop stand and orchard parameters: A review. *Trans. ASABE* **2011**, *54*, 1497–1510. [CrossRef]
2. Qiu, Q.; Han, J. 2.5-dimensional angle potential field algorithm for the real-time autonomous navigation of outdoor mobile robots. *Sci. China Inf. Sci.* **2011**, *54*, 2100. [CrossRef]
3. Lehtomäki, M.; Jaakkola, A.; Hyyppä, J.; Kukko, A.; Kaartinen, H. Detection of Vertical Pole-Like Objects in a Road Environment Using Vehicle-Based Laser Scanning Data. *Remote Sens.* **2010**, *2*, 641–664. [CrossRef]
4. Bargoti, S.; Underwood, J.P.; Nieto, J.I.; Sukkarieh, S. A Pipeline for Trunk Localisation Using LiDAR in Trellis Structured Orchards. In *Field and Service Robotics*; Springer International Publishing: Cham, Switzerland, 2015; Volume 105.
5. Underwood, J.P.; Jagbrant, G.; Nieto, J.I.; Sukkarieh, S. Lidar-Based Tree Recognition and Platform Localization in Orchards. *J. Field Robot.* **2015**, *32*, 1056–1074. [CrossRef]

6. Cabo, C.; Ordóñez, C.; López-Sánchez, C.A.; Armesto, J. Automatic dendrometry: Tree detection, tree height and diameter estimation using terrestrial laser scanning. *Int. J. Appl. Earth Obs.* **2018**, *69*, 164. [CrossRef]
7. De Conto, T.; Olofsson, K.; Görgens, E.B.; Rodriguez, L.C.E.; Almeida, G. Performance of stem denoising and stem modelling algorithms on single tree point clouds from terrestrial laser scanning. *Comput. Electron. Agric.* **2017**, *143*, 165–176. [CrossRef]
8. Gené-Mola, J.; Gregorio, E.; Guevara, J.; Auat, F.; Sanz-Cortiella, R.; Escolà, A.; Llorens, J.; Morros, J.-R.; Ruiz-Hidalgo, J.; Vilaplana, V. Fruit detection in an apple orchard using a mobile terrestrial laser scanner. *Biosyst. Eng.* **2019**, *187*, 171–184. [CrossRef]
9. Antonarakis, A.S.; Richards, K.S.; Brasington, J.; Bithell, M.; Muller, E. Retrieval of vegetative fluid resistance terms for rigid stems using airborne lidar. *J. Geogr. Res. Biogeosci.* **2015**, *113*, 96. [CrossRef]
10. Liu, L.; Lim, S.; Shen, X.; Yebra, M. A hybrid method for segmenting individual trees from airborne lidar data. *Comput. Electron. Agric.* **2019**, *163*, 104871. [CrossRef]
11. Liu, P.; Chen, J.; Zhang, M. Automatic control system of orchard tractor based on laser navigation. *Trans. CSAE* **2011**, *27*, 196–199.
12. Wang, D.; Liu, J.; Wang, J. Identification and classification of scanned target in forest based on hierarchical cluster. *Trans. CSAE* **2011**, *27*, 173–177.
13. Pierzchała, M.; Giguère, P.; Astrup, R. Mapping forests using an unmanned ground vehicle with 3D LiDAR and graph-SLAM. *Comput. Electron. Agric.* **2018**, *145*, 217–225. [CrossRef]
14. Zhou, J.; Hu, C. Inter-row Localization Method for Agricultural Robot Working in Close Planting Orchard. *Trans. CSAE* **2015**, *46*, 22–28.
15. Zhang, C.; Yong, L.; Chen, Y.; Zhang, S.; Ge, L.; Wang, S.; Li, W. A Rubber-Tapping Robot Forest Navigation and Information Collection System Based on 2D LiDAR and a Gyroscope. *Sensors* **2019**, *19*, 2136. [CrossRef] [PubMed]
16. Zhang, S.; Guo, C.; Gao, Z.; Sugirbay, A.; Chen, J.; Chen, Y. Research on 2D Laser Automatic Navigation Control for Standardized Orchard. *Appl. Sci.* **2020**, *10*, 2763. [CrossRef]
17. Monnier, F.; Vallet, B.; Soheilian, B. Trees detection from laser point clouds acquired in dense urban areas by a mobile mapping system. *ISPRS Ann. Photogramm. Remote Sens. Spat. Inf. Sci.* **2012**, *3*, 245–250. [CrossRef]
18. Chakraborty, M.; Khot, L.R.; Sankaran, S.; Jacoby, P.W. Evaluation of mobile 3D light detection and ranging based canopy mapping system for tree fruit crops. *Comput. Electron. Agric.* **2019**, *158*, 284–293. [CrossRef]
19. Zhang, X.; Li, X.; Zhang, B.; Zhou, J.; Tian, G.; Xiong, Y.; Gu, B. Automated robust crop-row detection in maize fields based on position clustering algorithm and shortest path method. *Comput. Electron. Agric.* **2018**, *154*, 165–175. [CrossRef]
20. Underwood, J.P.; Hung, C.; Whelan, B.; Sukkarieh, S. Mapping almond orchard canopy volume, flowers, fruit and yield using lidar and vision sensors. *Comput. Electron. Agric.* **2016**, *130*, 83–96. [CrossRef]
21. Narváez, F.J.Y.; del Pedregal, J.S.; Prieto, P.A.; Torres-Torriti, M.; Cheein, F.A.A. LiDAR and thermal images fusion for ground-based 3D characterisation of fruit trees. *Biosyst. Eng.* **2016**, *151*, 479–494. [CrossRef]
22. Astolfi, P.; Gabrielli, A.; Bascetta, L.; Matteucci, M. Vineyard Autonomous Navigation in the Echord++ GRAPE Experiment. *IFAC-PapersOnLine* **2018**, *51*, 704–709. [CrossRef]
23. Chen, X.; Wang, S.A.; Zhang, B.; Luo, L. Multi-feature fusion tree trunk detection and orchard mobile robot localization using camera/ultrasonic sensors. *Comput. Electron. Agric.* **2018**, *147*, 71–108. [CrossRef]
24. Nematollahi, A.F.; Rahiminejad, A.; Vahidi, B. A Novel Physical Based Meta-Heuristic Optimization Method Known as Lightning Attachment Procedure Optimization. *Appl. Soft Comput.* **2017**, *59*, 596–621. [CrossRef]
25. Zheng, T.; Luo, W. An Enhanced Lightning Attachment Procedure Optimization with Quasi-Opposition-Based Learning and Dimensional Search Strategies. *Comput. Intell. Neurosci.* **2019**, *2019*, 1589303. [CrossRef] [PubMed]
26. Wei, L.; Yang, S.; Ye, Z.W.; Huang, Q.; Huang, Y.K. An Image Segmentation Method Based on Two-Dimensional Entropy and Chaotic Lighting Attachment Procedure Optimization Algorithm. *Int. J. Pattern Recognit. Artif. Intell.* **2020**, *34*, 2054030.
27. Ester, M.; Kriegel, H.P.; Sander, J.; Xu, X. A Density-Based Algorithm for Discovering Clusters in Large Spatial Databases with Noise. In *Proceedings of the 2nd International Conference on Knowledge Discovery and Data Mining*; AAAI Press: Portland, OR, USA, 1996; pp. 226–231.
28. Guo, C.L.; Liu, G.; Zhang, W.J.; Feng, J. Apple tree canopy leaf spatial location automated extraction based on point cloud data. *Comput. Electron. Agric.* **2019**, *166*, 104975. [CrossRef]
29. Liu, X.; Hu, C.H.; Li, P.P. Automatic segmentation of overlapped poplar seedling leaves combining Mask R-CNN and DBSCAN. *Comput. Electron. Agric.* **2020**, *178*, 105753. [CrossRef]
30. Niu, R.X.; Zhang, X.Y.; Wang, J.; Zhu, H.; Huang, J.; Chen, Z.W. Orchard Trunk Detection Algorithm for Agricultural Robot Based on Laser Radar. *Trans. CSAE* **2020**, *51*, 21–27.

agriculture

MDPI

Article

Apple Grading Method Design and Implementation for Automatic Grader Based on Improved YOLOv5

Bo Xu [1], Xiang Cui [1], Wei Ji [1,*], Hao Yuan [2] and Juncheng Wang [1]

[1] School of Electrical and Information Engineering, Jiangsu University, Zhenjiang 212013, China
[2] School of Mechanical Engineering, Jiangsu University, Zhenjiang 212013, China
* Correspondence: jiwei@ujs.edu.cn

Abstract: Apple grading is an essential part of the apple marketing process to achieve high profits. In this paper, an improved YOLOv5 apple grading method is proposed to address the problems of low grading accuracy and slow grading speed in the apple grading process and is experimentally verified by the designed automatic apple grading machine. Firstly, the Mish activation function is used instead of the original YOLOv5 activation function, which allows the apple feature information to flow in the deep network and improves the generalization ability of the model. Secondly, the distance intersection overUnion loss function (DIoU_Loss) is used to speed up the border regression rate and improve the model convergence speed. In order to refine the model to focus on apple feature information, a channel attention module (Squeeze Excitation) was added to the YOLOv5 backbone network to enhance information propagation between features and improve the model's ability to extract fruit features. The experimental results show that the improved YOLOv5 algorithm achieves an average accuracy of 90.6% for apple grading under the test set, which is 14.8%, 11.1%, and 3.7% better than the SSD, YOLOv4, and YOLOv5s models, respectively, with a real-time grading frame rate of 59.63 FPS. Finally, the improved YOLOv5 apple grading algorithm is experimentally validated on the developed apple auto-grader. The improved YOLOv5 apple grading algorithm was experimentally validated on the developed apple auto grader. The experimental results showed that the grading accuracy of the automatic apple grader reached 93%, and the grading speed was four apples/sec, indicating that this method has a high grading speed and accuracy for apples, which is of practical significance for advancing the development of automatic apple grading.

Keywords: apple grader; YOLOv5; attention mechanism SE; DIoU_Loss; mish

Citation: Xu, B.; Cui, X.; Ji, W.; Yuan, H.; Wang, J. Apple Grading Method Design and Implementation for Automatic Grader Based on Improved YOLOv5. *Agriculture* **2023**, *13*, 124. https://doi.org/10.3390/agriculture13010124

Academic Editor: Xiuliang Jin

Received: 22 December 2022
Revised: 29 December 2022
Accepted: 30 December 2022
Published: 2 January 2023

1. Introduction

Today, labour on farms and orchards relies heavily on manual labour by skilled farmers, which can lead to increased time and production costs. Smart farming has become a popular concept with the development of precision farming and information technology [1]. China is a major apple-producing country globally, and apple sorting has a high economic application value [2]. With increased economic development, people have higher requirements for fruit quality [3,4]. As a critical element in improving apple quality and liberating orchard labour, apple grading technology is of great significance in increasing the added value of products, improving market competitiveness, and alleviating labour shortages in orchards. Therefore, a high precision and speed grading method is needed for the effective and objective grading of apples.

In the research of fruit grading based on traditional machine learning, Abdullah et al. [5] detected the quality features of poppy peaches by machine learning, the features considered mainly included fruit surface color and fruit shape, and developed automatic machine vision detection software to detect the ripeness grade of poppy peaches by linear discriminant analysis and multilayer neural network. Marchant et al. [6] studied the method of automatic potato detection and grading based on a computer vision system. Moallem et al. [7]

proposed a computer vision-based grading algorithm for golden crown apples where texture and geometric features were extracted from the defective areas. Finally, a support vector machine (SVM), a multilayer perception (Muti-Layer Perception), and a K-Nearest Neighbor classifier were used to classify the apples into first-class, second-class, and out-of-class fruits. Gui et al. [8] proposed a wavelet rectangle-based apple classification method based on apple shape, which classified apples into normal fruit shape, mild deformity, and severe deformity with a classification accuracy of 86.2%, 85.8%, and 90.8%, respectively. In the above machine learning classification methods, preprocessing of images is often required, and the classification relies on single features, which has the problems of poor real-time performance and low robustness.

In the research of fruit grading based on deep learning, Fan et al. [9] used a convolutional neural network (CNN) architecture for apple quality recognition, trained a convolutional neural network, and achieved an accuracy of 96.5% in the test set, designed classification software for CNN-based convolutional neural networks, and used a computer vision module to sort at a rate of 5/s on a four-threaded fruit sorter. The classification accuracy reached 92%. However, the model was large, and the computational efficiency was relatively low. Raikar et al. [10] studied the quality grade of okra and used three deep learning models, AlexNet, GoogLeNet, and ResNet50, to classify okra into four types based on length: small, medium, large and extra large, where the accuracy of the ResNet deep learning model reached over 99%. Luna et al. [11] proposed a deep learning-based method for single tomato defect area detection, implemented through the OpenCV library and Python programming. He collected 1200 tomato images of different qualities using an image capture box and used the images for training VGG16, InceptionV3, and ResNet50 deep learning models, respectively, compared the experimental results and found that VGG16 was the best deep learning model for defect recognition. However, there are still problems of insufficient model optimization and poor real-time performance in the above deep learning model grading methods.

In terms of research on automatic fruit grading equipment, Cubero et al. [12] designed a computer vision-based automated citrus sorting device. The sorting device was deployed on a mobile platform, and the low-power industrial camera image acquisition and powerful lighting system enabled the device to work better in the field. Experiments showed that the sorting device could achieve a sorting speed of up to eight per second. Baigvand et al. [13] developed a machine learning-based fig sorting system, which first uses a feeder and a belt. The figs were first transported under a CCD camera by a feeder and belt conveyor. The figs were classified into five categories by extracting fig characteristics from the pictures taken by the CCD camera, including size, colour, segmentation size and fig centre position, etc. The experiments verified that the grading system was 95% accurate in recognizing the five categories of figs. However, the designed automatic fruit grader tends to be large and more suitable for large assembly line working modes and is not suitable for the needs of small and medium-sized farmers for detection and grading.

Although the above methods have achieved specific results in terms of fruit feature detection and equipment implementation, there are still problems, such as insufficient model optimization and equipment implementation. Based on this, this paper takes red Fuji apples as the research object. It provides an in-depth discussion on the grading detection of apple features and the implementation of automatic apple sorting equipment. An apple grading algorithm based on the improved YOLOv5 is proposed, using the Mish activation function instead of the original Relu activation function to improve the model generalization ability. A loss function (DIou_Loss) is introduced to speed up the rate of edge regression and improve localization accuracy. The attention mechanism squeeze excitation (SE) module is embedded into the backbone feature network to improve the feature extraction ability of the model. Experimental results show that the improved method can improve the model detection without increasing the model training cost. Finally, the automatic apple grader designed based on this paper was experimentally validated, and some conclusions were obtained.

2. Materials and Methods

2.1. Automatic Apple Grader Design

The structure of the automatic apple grader designed in this paper is shown in Figure 1. It consists of Feeding and material handling lifting mechanism, turnover detection conveyor, visual inspection and automatic grading control system, and graded actuators. The design is based on a two-level layout to reduce the space required.

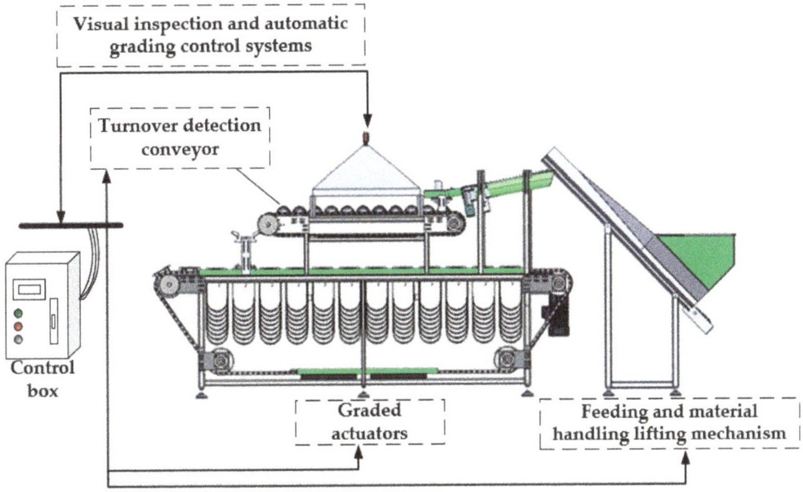

Figure 1. Structure of the automatic apple grader.

(1) Feeding and material handling lifting mechanism. The Feeding and material handling lifting mechanism is a scraper elevator, as shown in Figure 2. The scraper elevator consists of a funnel-shaped storage tank and a vertical conveyor belt, where the funnel-shaped storage tank includes the back plate of the hopper and the support plate, the three-dimensional conveyor belt includes the guide plate and the curved scraper, and the whole mechanism is placed at an inclination of 45°. The scraper elevator moves the conveyor belt by means of an AC motor driven by a frequency converter, which organizes the disordered apples into an orderly quadruple queue, transporting them from the bottom upwards and conveying them into the Turnover detection conveyor.

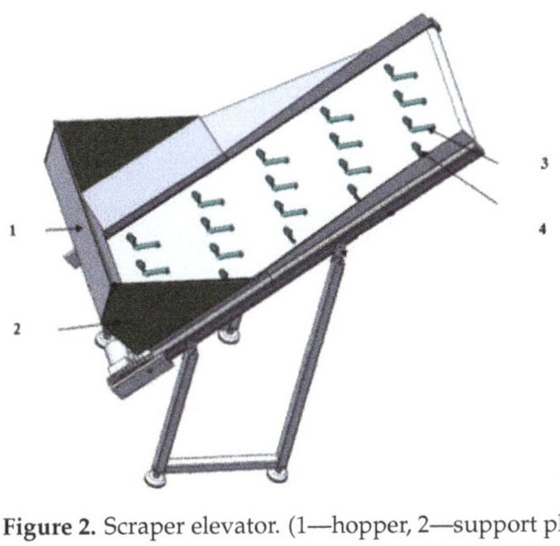

Figure 2. Scraper elevator. (1—hopper, 2—support plate, 3—guide plate, 4—curved scraper).

(2) Turnover detection conveyor. The Turnover detection conveyor is shown in Figure 3 and consists of sprockets, chains, sponge rollers, and motors. The apples are lifted by the scraper elevator into the turnover detection conveyor. The turnover detection conveyor uses pairs of double-tapered rollers to turn the apples axially, and a CCD industrial camera mounted on top of the lampshade collects images of the tumbling apples several times to obtain complete surface information about the apples in a moving position.

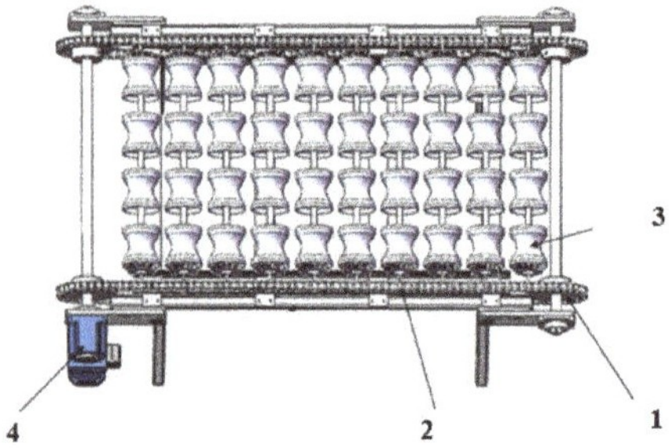

Figure 3. Turnover detection conveyor. (1—sprockets, 2—chains, 3—sponge rollers, 4—motor).

(3) Visual inspection and automatic grading control system. The visual inspection and automatic grading control system are shown in Figure 4 and consist of a CCD industrial camera and automatic grading control system. The visual inspection and automatic grading control system determines the grading of apples according to the information collected by the CCD industrial camera on the whole surface of the apples and finally sends the grading results to the graded actuators.

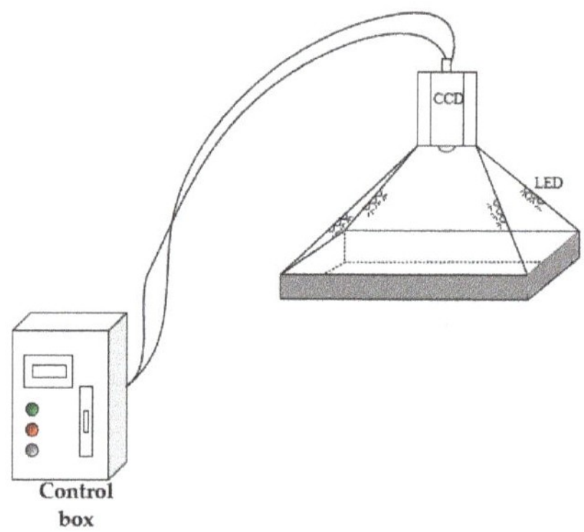

Figure 4. Visual inspection and automatic grading control system structure diagram.

(4) Graded actuators. The graded actuator is shown in Figure 5 and consists of a Trigger grading mechanism, sprocket chain drive, grading fruit cup, grading channel, and a motor. The grading fruit cup is shown in Figure 5b and consists of a cup body, a drop door, and rollers. The grading actuator receives the grading results from the image detection and automatic grading system and allows the apples to reach the corresponding grade position and then open the cups and fall into the corresponding grade storage bin.

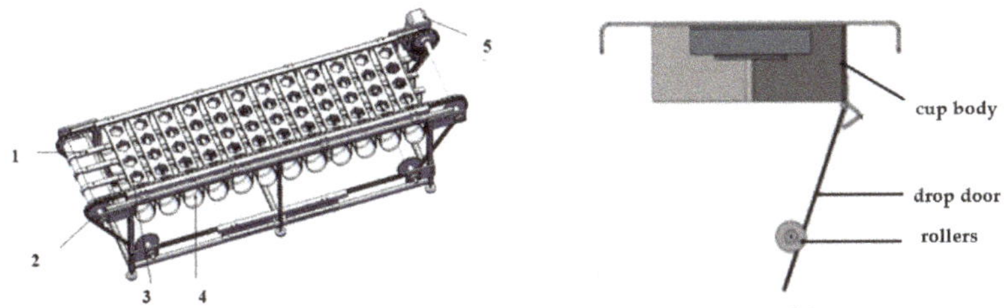

(a) (b)

Figure 5. Graded actuators. (**a**) Grading actuator; (**b**) Detail of grading fruit cup. (1-Trigger grading mechanism; 2-Sprocket chain drive; 3-Grading fruit cup; 4-Grading channel; 5-Motor).

2.2. Apple Image Acquisition and Data Augmentation

2.2.1. Image Acquisition

The apples used in this dataset are "Fengxian apples" from Xuzhou City, Jiangsu Province, and "Yantai apples" from Yantai City, Shandong Province, which are among the representative brands of red Fuji. The sources of apples include purchases from apple markets and picking from orchards. The image acquisition equipment used in the experiments is a CCD industrial camera, MER2-G. The camera was mounted on a bracket above a lampshade with a 90° angle of view directly above the flip mechanism and a fixed height of 70 cm. The lampshade was illuminated with a LED strip as a fill light source to capture images of apples under diffuse lighting. A final dataset of 2000 apple images was obtained, including grade-1, grade-2, and grade-3 apples. The grade-1 and grade-2 apples were mainly purchased from the market (differentiated by price), while grade-3 apples were marketed in smaller quantities, mainly from orchard picking. The CCD industrial camera uses a GigE interface for data transmission and acquisition with an industrial computer. The image resolution is 1280 × 1024, the pixel size is 4.5 × 4.5 μm, and the operating temperature range is 0° to 45°. The camera is installed according to the position of the flipping mechanism, the shooting angle is 90° directly above the flipping mechanism, the shooting height is a fixed value of 70 cm, and the image acquisition of the apple is carried out on the flipping mechanism, the image acquisition method and the image acquisition effect are shown in Figure 6.

(a) (b)

Figure 6. Data set production. (**a**) Image acquisition devices; (**b**) Image acquisition.

2.2.2. Apple Grading Criteria

In this paper, based on the Red Fuji GB/T 10651-2008 national standard [14], as shown in Table 1, ripeness, fruit shape, defects, and fruit diameter were selected as grading features to classify Red Fuji apples into 3 grades for this dataset.

Table 1. Red Fuji GB/T 10651-2008 national standard.

Index / Projects	Quality Grade		
	Grade-1	**Grade-2**	**Grade-3**
Ripeness	Bright red or dark red	Greenish red	Greenish yellow
Fruit shape	No deformities	No deformities	deformities
defects	NO	NO	Area not exceeding 4 cm^2
diameter(Maximum cross-sectional diameter)/mm	≥70	≥70	≥65

2.2.3. Dataset Annotation and Expansion

LabelImg was used to annotate the apple images, saving the image categories and target rectangle boxes according to the PASCAL VOC dataset format, generating an annotation file in XML format. As the height of the industrial camera is a fixed value, the longest side of the rectangular box calibrated in the dataset is used as the criterion for fruit diameter; the ratio of the long side to the short side of the rectangular box is used as the criterion for fruit shape; apples with poor ripeness and defects are not carefully classified and are judged to be grade-3 apples. The collected apple images were expanded using MATLAB (2019) to make the training model more robust. The expansion methods included horizontal mirroring, vertical mirroring, multi-angle rotation (90.180.270), and image tiling. The expanded dataset is shown in Figure 7. The extended dataset has 6000 images with a uniform image size of 1280 × 1024, with a high number of grade-1 and grade-2 apples, each accounting for 40%, and a low number of grade-3 apples, accounting for 20%. The extended dataset was allocated to the training, test, and validation sets in a ratio of 7:2:1.

Figure 7. The expanded dataset.

2.3. Design of Apple Grading Method Based on Improved YOLOv5

YOLOv5 is an algorithm proposed by Glenn-Jocher with high real-time performance in terms of algorithmic efficiency [15,16]. The YOLOv5 network has four main components, which are the input side, the backbone network (backbone), the Neck network part, and the Output part. The YOLO family of algorithms has promising results on open-source datasets, but there is no comprehensive and mature method for grading different state fruits [17].

Therefore, this paper proposes an improved YOLOv5 model structure for apple grading based on the lightweight network YOLOv5s shown in Figure 8. Using the Mish activation function instead of Leaky-ReLU, the distance intersection overUnion (DIoU_Loss) loss function is used at the output of the model. Finally, a simple and efficient channel attention module (Squeeze Excitation) is introduced, which allows the model to focus on apple refinement features without increasing the computational effort of the model.

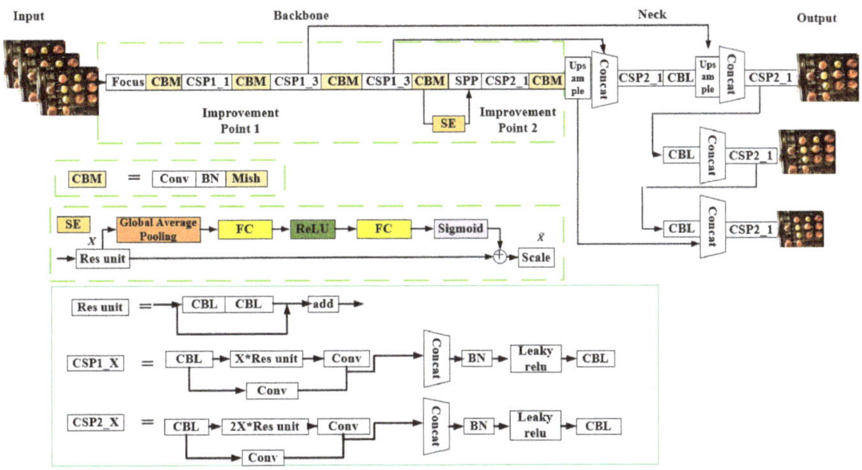

Figure 8. Diagram of the improved network structure of YOLOv5.

2.3.1. Improvement of the Activation Function

The role of the activation function in a convolutional neural network structure is to combine features thoroughly. The activation functions commonly used in YOLOv5 neural networks are Leaky ReLU, Sigmoid, etc. Leaky ReLU (see Equation (1)) can handle the gradient disappearance problem, but it suffers from neuron necrosis due to data sparsity, while Sigmoid (see Equation (2)) can map real numbers to a specified interval, and his curve is smooth and easy to find derivatives for, but it suffers from the problem of gradient disappearance. The Mish (see Equation (3)) activation function has outperformed the Leaky ReLU and other standard activation functions in many deep-learning models [18,19]. The depth of the model in this paper is deeper, and the apple features are more abstract, so this study uses the Mish activation function in the backbone of the YOLOv5 model to achieve better feature extraction results. The CBM module in the backbone network consists of a convolutional layer, a normalization layer, and the Mish activation function. The rest of the model still uses the Leaky ReLU activation function.

$$f_1(x) = \begin{cases} x & x > 0 \\ \alpha x & others \end{cases} \tag{1}$$

$$f_2(x) = \frac{1}{1+e^{-x}} \tag{2}$$

$$f_3(x) = \frac{x}{1+e^{-x}} \tag{3}$$

As can be seen from Figure 9, the Mish activation function can output arbitrarily large positive values while allowing slight negative gradient values, which avoids gradient saturation due to the gradient being close to zero. The Mish function is non-monotonic and continuously differentiable, which allows the deep neural network to achieve better accuracy and generalization, and facilitates the optimization of gradient updates [20,21].

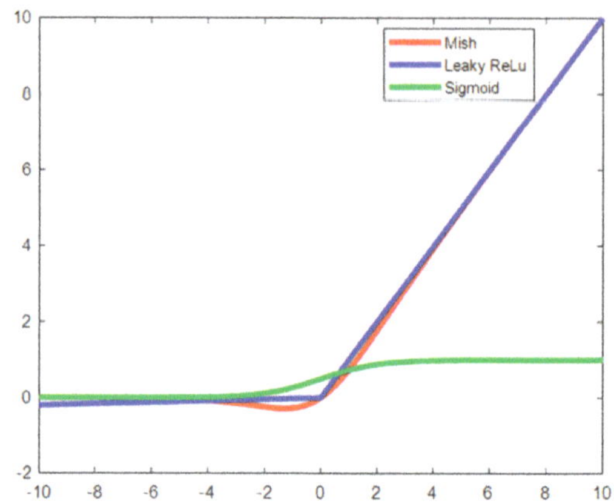

Figure 9. Comparison of Mish, Leaky ReLu, and Sigmoid function images.

2.3.2. Improvement of the Loss Function

Deep learning networks adjust the weights between the layers of the network during the training process through optimization algorithms, and they can reduce the loss so that the predicted frames and the actual frames overlap as much as possible. The loss function is the key to adjusting the weights [22,23]. GIoU has scale invariance. When the target is enlarged or reduced, the loss value can remain the same magnitude, and it considers both the overlapping and non-overlapping parts between the detection frame and the target frame. When IoU = 0, the distance of the bounding box does not affect the loss value, GIoU overcomes this shortcoming and can make the corresponding loss expression according to the distance of the two bounding boxes. GIoU expressions are as follows:

$$\begin{cases} GIOU = IOU - \frac{C-(A \cup B)}{C} \\ GIOU = -1 + \frac{A \cup B}{C} \quad (IOU = 0) \end{cases} \tag{4}$$

As shown in Equation (4), when there is an intersection between predicted frame A and actual frame B, convergence is slower in the horizontal and vertical directions. When there is an inclusion relation C between the predicted and actual frames (when C is the smallest closed frame containing A and B), the *GIoU* degrades to an *IoU* and does not work. In this paper, the apples in the flip turnover detection conveyor are relatively dense, and the apples rotate in all directions with the sponge rollers, which makes it impossible to accurately distinguish the actual region from the background region in the grading work of the prediction frame. Therefore, in this paper, DIoU_Loss is chosen as the boundary loss function in the output layer instead of GIoU_Loss to speed up the target grading accuracy and detection speed.

DIoU inherits the advantages of GIoU and adds the centroid distance geometric information [24,25]. As shown in Figure 10, which takes into account both the overlapping area and the distance between the two centroids, DIoU can provide the accurate gradient direction for the model when the prediction frame and the actual frame have crossed or overlapped. The introduction of the distance penalty makes DIoU converge faster than GIoU. The equation is shown in Equation (5).

$$DIoU = 1 - IoU + \frac{p^2(b, b^{gt})}{c^2} \tag{5}$$

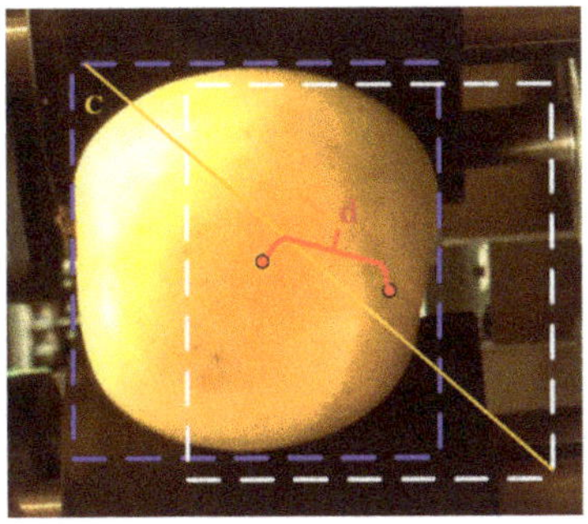

Figure 10. DIoU schematic.

In the above equation, b, b^{gt} represents the target and prediction box centroids, p(*) represents the Euclidean distance, and c is the diagonal length of the minimum enclosing box covering the target and prediction boxes.

2.3.3. Integration of Attentional Mechanisms

Attention is one of the most critical mechanisms in human perception. The human eye is adept at recognizing key image features from complex images and ignoring irrelevant information, which is where the attention mechanism excels. With the booming development of deep learning, the attention mechanism can be used for machine vision. Apples have characteristics such as many features and small sizes, which can easily lead to wrong and missed detection, thus making the grading accuracy of apple features low [26]. By introducing the attention mechanism in the convolutional layer, the learning representation can be enhanced autonomously, and the method is highly operational and effective [27,28]. The Backbone module in YOLOv5 adds the Focus structure, which improves the computational speed by slicing the feature map, but may have an impact on the features. In order to improve the target feature extraction effect of the Backbone module, this paper introduces the channel focus mechanism squeeze excitation (SE) module [29], which is embedded into the last layer of the Backbone module to improve the accuracy of apple grading without increasing the model size.

The SE module can effectively capture the channel and position information of the image, which in turn can improve the grading accuracy of the model. Figure 11 shows the working principle of the SE module, which consists of two main parts, Squeeze and Excitation. The SE module first obtains a global description of the input through Squeeze, which enables a wider perceptual field of view, and then obtains the weights of each channel in the Feature Map through Excitation's two-layer fully connected bottleneck structure as input to the lower layer network.

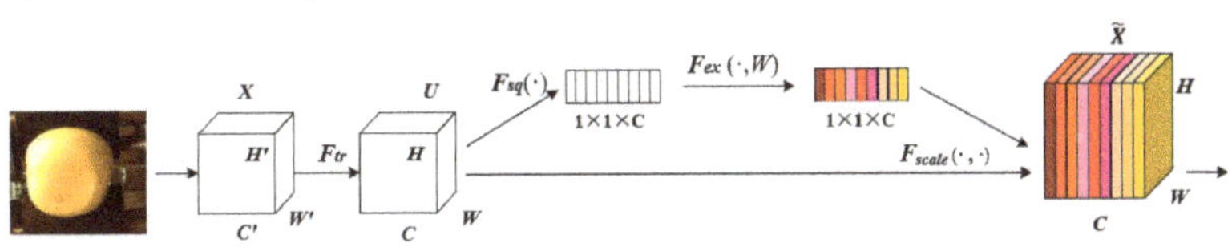

Figure 11. Squeeze and excitation.

In Figure 11, the squeeze operation first encodes the entire spatial feature on the channel as a local feature by global averaging pooling. Then the operation of the connected channel is performed through two fully connected layers and a non-linear activation function (see Equation (6)), followed by a Sigmoid activation function to obtain the weight of each channel, and finally, a multiplicative weighted multiplication to each channel to complete the recalibration of the attention mechanism. The calculation results are shown in Equations (7) and (8). A correlation between channels was established through global average pooling, two fully connected layers, and a non-linear activation function.

$$Z_c = \frac{1}{H \times W} \sum_{i=1}^{H} \sum_{j=1}^{W} u_c(i,j) \tag{6}$$

where Zc represents the Cth element in the statistic, H, W the space dimension, and the subscripts i, j the number of channels. After the Squeeze operation has obtained the channel information, it uses two fully connected layers to form the gate mechanism and activates it with Sigmod. The calculation is as follows:

$$s = F_{ex}(z, W) = \sigma(g(z, W) = \sigma(W_2 \delta(W_1 z) \tag{7}$$

where δ is the ReLu activation function, σ is the Sigmoid function, W_1 and W_2 are two fully connected layers equal to C/r×C and C×C/r, respectively, r is the scaling parameter that limits the complexity of the model and increases its capability, and s represents the set of weights of the feature maps obtained through the fully connected and non-linear layers. Finally, the output weights are assigned to the original features. The calculation formula is as follows.

$$\widetilde{X_c} = s_c \times u_c \tag{8}$$

where $\widetilde{X_c}$ is the feature map of the featured channel X, S_c is the weight, and U_c is a two-dimensional matrix.

3. Result and Discussion

3.1. Experimental Validation and Analysis of Results

3.1.1. Experimental Environment

The experimental models in this paper were constructed, trained, and the results were tested based on the Windows 10-x64-bit operating system. The experimental programming environment is Python 3.7, using Cudnn for GPU acceleration, and Apple hierarchical model training is implemented under the PyTorch 1.7 deep learning framework. The experimental environment configuration is shown in Table 2. The number of iterations of the training process was set to 150, the weight decay coefficient was 0.001, the learning rate was 0.917, and the maximum training batch was eight. An IOU threshold of 0.5 was taken as the standard.

Table 2. Experimental environment.

Computer Configuration	Specific Parameters
CPU	Intel i7-9750k
GPU	NVIDIA GTX1660Ti(16G)
Operating system	Windows 10-x64
Random Access Memory	DDR4 32G (8G*4)
CUDA	CUDA 10.3

In order to better calculate the classification accuracy and reliability of this model, this paper selects loss function curve (Loss), Precision, Average Precision (AP), Recall, Mean Average Precision(mAP), and frames per second (Fps) as the algorithm performance evaluation indexes [30]. The relevant evaluation indexes are calculated as shown in Equations (9)–(12).

$$Precision = \frac{TP(TruePositive)}{TP + FP(FalsePositive)} \tag{9}$$

$$Reacll = \frac{TP(TruePositive)}{TP + FN(FalseNegative)} \tag{10}$$

$$Average\ Precesion = \int_0^1 (P(R)dR) \tag{11}$$

$$Mean\ AveragePrecesion = \frac{\sum Average\ Precesion}{n(Class)} \tag{12}$$

In the above equation, *TP* represents the number of apple samples correctly identified by the model, *FP* represents the number of apples incorrectly identified by the model, *FN* represents the number of apple samples not identified by the model, and n represents the number of categories.

3.1.2. Analysis of Experimental Results

(1) Experiments related to the improved algorithm

The loss function can visually reflect whether the model can converge stably or not [31]. In the process of network training, the following three models were selected for comparison, taking into account the comparative effects of different algorithmic models: the YOLOv5 algorithm using the Mish activation function to optimize the backbone network, denoted by YOLOv5-M; the YOLOv5 algorithm using the DIoU optimization loss function, denoted by YOLOv5-D; and the simultaneous use of the Mish activation function and DIoU optimization method, denoted by Im-YOLOv5. The resulting loss function curve after training is shown in Figure 12.

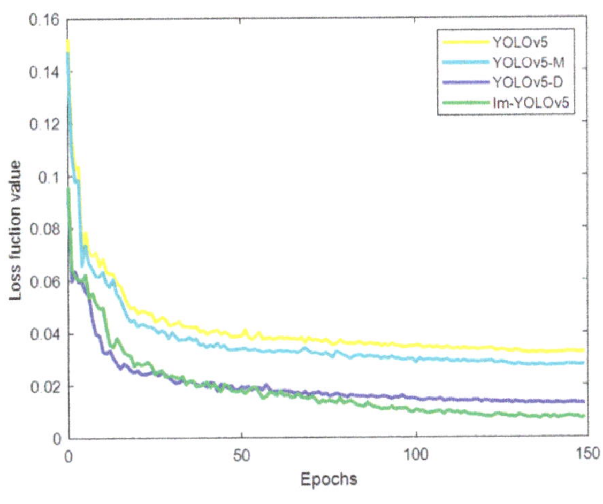

Figure 12. Loss value curve changes with epochs.

As shown in Figure 12, the overall trend of the loss values of the four models in training is the same. They decrease rapidly in training and eventually stabilize. The YOLOv5-M and YOLOv5-D loss values and convergence rates are significantly faster than the original YOLOv5 algorithm, and the degree of fluctuation is less, which proves that the localization accuracy and convergence rate of the models can be increased when using complete loss and activation functions [31]. From Im-YOLOv5, the loss value and convergence speed are slightly lower than YOLOv5-D for the first 50 iterations of the model, but after 50 iterations, the loss value and convergence speed are due to the rest of the models. This indicates that the Im-YOLOv5 algorithm can improve the convergence speed and localization accuracy of the model, which helps to obtain a more accurate resultant model, which proves the effectiveness of the improved model.

In order to verify the effectiveness of the improved method in this paper for apple grading, this study trained the YOLOv5 and the Im-YOLOv5 models under the same dataset and training set. The PR curve represents the relationship between accuracy and recall, which can measure the model's generalization ability. The PR curves of the two models after the training was completed are shown in Figure 13. The area between the PR curve and the coordinate axes of the Im-YOLOv5 is larger than that of the original YOLOv5 model, which indicates that the improved model has better overall performance.

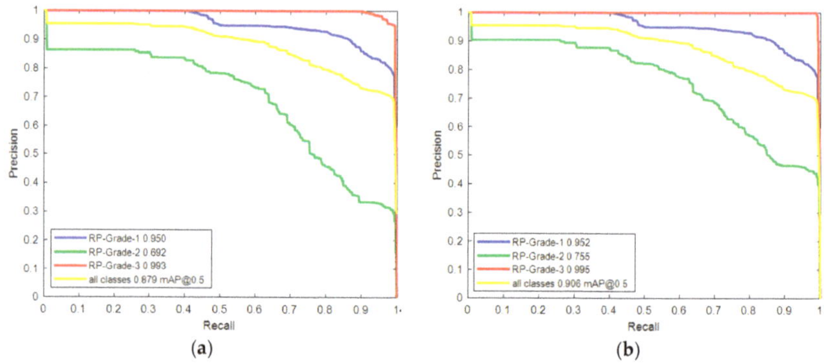

Figure 13. Comparison of YOLOv5 and Im-YOLOv5 PR curve. (**a**) YOLOv5; (**b**) Im-YOLOv5.

As can be seen from Figure 13, the Im-YOLOv5 model has improved the grading accuracy for different apple quality levels, with higher mAP of over 95% for Grade-1 and Grade-3 apples. The mAP for Grade-2 apples reached 0.755, an improvement of 9.1% over the original model. The average accuracy for all apple grades was 0.906, an increase of 3.1% compared to the original model.

The Im-YOLOv5 model and YOLOv5s model trained in this paper were used to grade apples of different qualities in an automatic apple grader.

Figure 14a shows the grading results before the improvement of the YOLOv5 model, and Figure 14b shows the results of the Im-YOLOv5 model grading. The accuracy of the apple grading in Figure 14a is low, where the apples in the first and second images appear to have duplicate detection frames, and the second image shows incorrect grading of the grade-1 and grade-2 apples, marking the grade-1 apples as grade-2 apples. The third panel shows no duplicate detection frames but incorrectly marks three grade-1 apples with low accuracy. In contrast, there is an improvement in grading accuracy for all grades of apples in Figure 14b, with no duplicate boxes. The improved model was able to pay more attention to apple feature information, which improved the robustness of the model while increasing the grading accuracy. Therefore, the Im-YOLOv5 model can satisfy apple grading in actual production environments.

In order to explore the effectiveness of visual attention mechanisms in convolutional networks and to enhance the interpretability of the apple grading model in this paper, a part of the improved YOLOv5 feature extraction layer in this paper was visualized [32]. The results of feature extraction from the convolutional layer of the backbone network are shown in Figure 15. As shown in Figure 15a, the initial feature size of the convolutional layer of the backbone network is large, the feature extraction is more fine-grained, and the apple features are extracted while containing complex background information; as the network deepens, the extracted features are gradually blurred and sparse and more semantic. As can be seen in Figure 15b, after the attention SE module, there are some highlighted areas in the figure, and the location of the apples is highlighted in the spatial pyramid pooling (SPP) output feature map, which indicates that after adding the SE module, the deep network layer of the Im-YOLOv5 model in this paper filtered the extracted features, which helped to highlight the target apples as well as filter the background information in the grading stage and improved the network model accuracy.

(a) (b)

Figure 14. Grading results. (**a**) YOLOv5s model; (**b**) Im-YOLOv5 model.

(2) Comparison experiments between different models

In order to further verify the superiority of the proposed algorithm, the improved YOLOv5 algorithm was compared with several classical algorithms commonly used in the current deep learning field, including a single shot multibox detector (SSD) [33], a fast and superior generalization among One-Stage detectors, and YOLOv4 and YOLOv5s [34], which have better comprehensive performance. The comparison experiments selected accuracy, recall, mAP, and Fps as the evaluation metrics of each algorithm, and the models were trained and tested under the same initial conditions. The apple grading results are shown in Table 3 below.

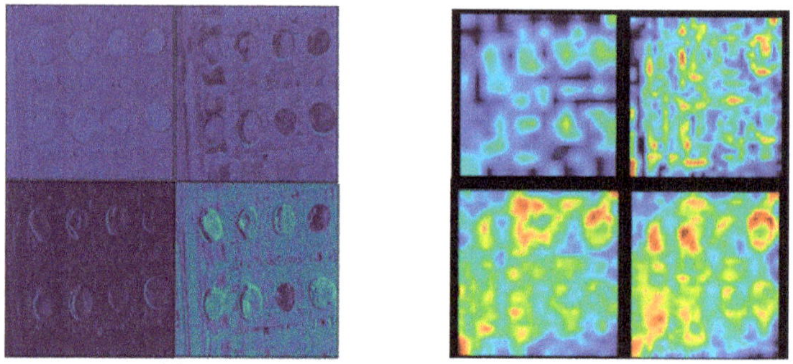

(a) (b)

Figure 15. Feature map visualization of the Im-YOLOv5 model. (**a**) CSP network output 56 × 56 feature map; (**b**) SPP network output 14 × 14 feature map.

Table 3. Comparison of different models.

Models \ Index	Precision			Recall			mAP @0.5	FPS(f/s)
	Grade-1	Grade-2	Grade-3	Grade-1	Grade-2	Grade-3		
SSD	0.812	0.612	0.884	0.926	0.645	0.895	0.789	34.78
YOLOv4	0.821	0.656	0.892	0.862	0.609	0.923	0.815	50.42
YOLOv5s	0.938	0.692	0.991	0.950	0.655	0.993	0.879	56.64
Im-YOLOv5	0.951	0.806	0.992	0.952	0.751	0.995	0.906	59.63

As can be seen from the results in Table 3, the SSD model has lower accuracy and recall, with an average accuracy mAP of 0.789 and a real-time frame rate FPS of 34.78 for the Apple classification. As the model improves, its accuracy, recall, mAP, and FPS gradually increase, with the Im-YOLOv5 model having the highest mAP of 0.906, compared to the YOLOv5, YOLOv4, and SSD models by 14.8%, 11.1%, and 3.7%, respectively. The accuracy and recall of the grade-2 apple reached 0.806 and 0.751, respectively, which were 16.4% and 14.6% higher than the original YOLOv5 method. On the other hand, the real-time image frame rate of the Im-YOLOv5 method in this paper was improved, and the FPS of the improved model reached a maximum of 59.63, which has better real-time performance compared with the lightweight model YOLOv5s. The results show that the grading effect and real-time performance of the Im-YOLOv5 model proposed in this paper are better than those of the traditional deep learning model, proving the effectiveness of the proposed method.

3.2. System Solution Validation

3.2.1. Automatic Apple Grader Control System Set Up

The automatic apple grader designed and developed in this paper is shown in Figure 16, and its workflow is shown in Figure 17. When the automatic apple grader is started, the apples are lifted by a feeding and material handling lifting mechanism to the turnover detection conveyor. The turnover conveyor uses pairs of double conical sponge rollers to turn the apples. At this point, the automatic grading control system uses the improved YOLOv5 algorithm to grade the apples based on the surface information collected by the image acquisition device and sends the grading decision to the grading execution device [35,36]. The grading actuator automatically places the apples in the appropriate storage bin when they reach the appropriate grade based on the grading results assessed by the grading control system. The bins are equipped with cushioning material to reduce the impact of falling apples.

Figure 16. Physical view of the automatic apple grader.

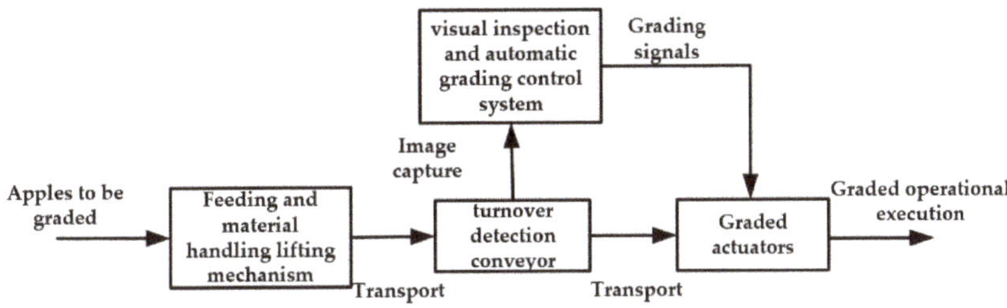

Figure 17. Workflow diagram for automatic apple grading.

The hardware of the automatic apple grading control system includes a CCD industrial camera, IPC-610L industrial computer, PLC-1212 controller, AC contactor, inverter, and AC motor. The CCD industrial camera uses a GigE interface for data transmission and acquisition with an IPC-610L industrial computer. The industrial computer and PLC-1212 controller use the snap7 library to transmit information via a network cable. PLC controls the AC motor to drive a grading actuator through the AC contactor. The processor CPU of the IPC-610L is the same as that of the training computer, an Intel i7-9750k, an Intel i7-9750k with two graphics cards GTX1660Ti (6G), the operating system is Windows 10-x64, and the software environment is Python3.7, CUDA10.3, TIA Portal V15.1.

In order to facilitate debugging and observe the improvement effect of the model algorithm, the PyQt-based apple automatic grading control system software developed in this study is shown in Figure 18, which implements local video detection and real-time grading functions to achieve fast and accurate apple grading. The software designed in this paper sends the processed apple grade and location information to the TIA Portal V15.1 software through the snap7 library. After the grading actuator receives the apples in order, the grading operation is finally completed by the PLC controller in the corresponding grading lane [36,37].

3.2.2. Results of the Grading Experiment

In order to verify the feasibility of the algorithm and the grading scheme of the apple automatic grading platform system in this paper, the designed and developed automatic apple grader was experimentally verified. One hundred apples of each quality grade were manually selected as samples, and the apple grades were determined based on the red Fuji GB/T 10651-2008 grading standard mentioned in Section 3.2. The experimental results are shown in Table 4.

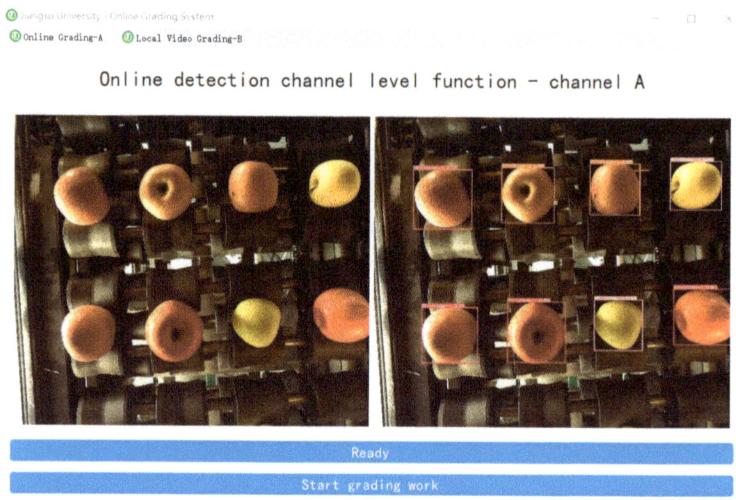

Figure 18. Apple auto-grading software.

Table 4. Grading experimental data.

Grade	Manual Grading Results	Equipment Grading Results	Consistency Rates	Completions Time(/s)
RP-grade-1	100	92	92%	27
RP-grade-2	100	88	88%	27
RP-grade-3	100	100	100%	27
Accuracy			93%	81

As can be seen from Table 4 of the experimental results, there was some grading error for Grade-1 and Grade-2 apples. Grade-1 apple sorting was 92% accurate, grade-2 apple sorting was 88% accurate, and grade-3 apple sorting was 100% accurate, with an average accuracy of 93%. The average classification accuracy for the three apple grades was 93% for the three apple grades, with an average classification speed of four apples/second. Both the real-time and accuracy rates are high enough to meet the grading requirements of small and medium-sized fruit farmers and to verify the effectiveness of the algorithm.

4. Conclusions

This paper proposes an improved apple grading model of YOLOv5, which better balances the grading accuracy and speed of apples, and also carries out experimental verification on the automatic apple grader designed and developed in this paper. The main conclusions of this study are as follows.

(1) In order to achieve more accurate apple grading and better real-time performance, the DIoU loss function and Mish loss function were chosen to replace the GIoU function and Relu activation function of the original algorithm model in terms of algorithm optimization, which improved the feature extraction capability and convergence speed of the model. The attention SE module is embedded in the Backbone structure to discard unnecessary features, which improves the training accuracy of the model without burdening the model. The experimental results show that the improved YOLOv5 has improved the average accuracy rate mAP by 3.1% compared to YOLOv5, 11% compared to YOLOv4, and 15% compared to SSD, and the real-time grading speed has reached 59.63 FPS, which is a large improvement in both the apple-grade grading accuracy rate and real-time performance. A portion of the improved YOLOv5 feature extraction layer was visualized to show the features extracted by different

convolutional layers, enhancing the interpretability of the apple grading model in this paper.

(2) An automatic apple grader was developed and designed, and the grading method in this paper was experimentally verified on an automatic apple grading machine platform. The experimental results showed that the grading accuracy of the grading method on the automatic apple grader reached 93%, with an average grading speed of four apples/sec. It has high accuracy and real-time performance, which can meet the grading needs of farmers and small and medium-sized enterprises in the field and has practical application in the apple grading industry.

Author Contributions: Conceptualization, B.X. and X.C.; methodology, W.J. and X.C.; software, X.C. and J.W.; validation, J.W.; formal analysis, X.C.; investigation, B.X.; data curation, X.C.; resources, B.X. and H.Y.; writing—original draft preparation, X.C.; writing—review and editing, W.J. and J.W.; visualization, B.X.; supervision, W.J.; project administration, W.J. and H.Y.; funding acquisition, B.X. and W.J. All authors have read and agreed to the published version of the manuscript.

Funding: This research was funded by the National Natural Science Foundation of China (No. 61973141 and 62173162), the Jiangsu agriculture science and technology innovation fund (No. CX(20)3059), and A Project Funded by the Priority Academic Program Development of Jiangsu Higher Education Institutions (No. PAPD-2018-87).

Institutional Review Board Statement: Not applicable.

Data Availability Statement: Not applicable.

Conflicts of Interest: The authors declare no conflict of interest.

References

1. Tian, Y.N.; Yang, G.D.; Wang, Z.; Wang, H.; Li, E.; Liang, Z.Z. Apple detection during different growth stages in orchards using the improved YOLO-V3 model. *Comput. Electron. Agric.* **2019**, *157*, 417–426. [CrossRef]
2. Zhu, D.S.; Ren, X.J.; Wei, L.; Cao, X.H.; Ge, Y.; Liu, H.; Li, J.R. Collaborative analysis on difference of apple fruits flavour using electronic nose and electronic tongue. *Sci. Hortic.* **2020**, *260*, 108879. [CrossRef]
3. Musacchi, S.; Serra, S. Apple fruit quality: Overview on pre-harvest factors. *Sci. Hortic.* **2018**, *234*, 409–430. [CrossRef]
4. Courtemanche, B.; Montpetit, B.; Royer, A.; Roy, A. Creation of a Lambertian microwave surface for retrieving the downwelling contribution in ground-based radiometric measurements. *IEEE Geosci. Remote Sens. Lett.* **2015**, *12*, 462–466. [CrossRef]
5. Abdullah, M.Z.; Mohamad-Saleh, J.; Fathinul-Syahir, A.S.; Mohd-Azemi, B.M.N. Discrimination and classification of fresh-cut starfruits using automated machine vision system. *J. Food Eng.* **2006**, *76*, 506–523. [CrossRef]
6. Marchant, J.A.; Onyango, C.M.; Street, M.J. High speed sorting of potatoes using computer vision. *ASAE Pap.* **1988**, *88*, 3540.
7. Moallem, P.; Serajoddin, A.; Pourghassem, H. Computer vision-based apple grading for golden delicious apples based on surface features. *Inf. Processing Agric.* **2017**, *4*, 33–40. [CrossRef]
8. Gui, J.S.; Zhang, Q.; Hao, L. Apple shape classification method based on wavelet moment. *Sens. Transducers* **2014**, *178*, 182–187.
9. Fan, S.X.; Li, J.B.; Zhang, Y.H.; Tian, X.; Wang, Q.Y.; He, X.; Zhang, C.; Huang, W.Q. On line detection of defective apples using computer vision system combined with deep learning methods. *J. Food Eng.* **2020**, *286*, 110102. [CrossRef]
10. Raikar, M.M.; Meena, S.M.; Kuchanur, C.; Girraddi, S.; Benagi, P. Classification and grading of Okra-ladies finger using deep learning. *Procedia Comput. Sci.* **2020**, *171*, 2380–2389. [CrossRef]
11. Luna, R.; Dadios, E.P.; Bandala, A.A.; Vicer, R.R.P. Tomato fruit image dataset for deep transfer learning-based defect detection. In Proceedings of the 2019 IEEE International Conference on Cybernetics and Intelligent Systems and IEEE Conference on Robotics, Automation and Mechatronics, Bangkok, Thailand, 18–20 November 2019; pp. 356–361.
12. Cubero, S.; Aleixos, N.; Albert, F.; Torregrosa, A.; Ortiz, C.; Garcia-Navarrete, O.; Blasco, J. Optimised computer vision system for automatic pre-grading of citrus fruit in the field using a mobile platform. *Precis. Agric.* **2014**, *15*, 80–94. [CrossRef]
13. Baigvand, M.; Banakar, A.; Minaei, S.; Khodaei, J.; Behroozi-Khazaei, N. Machine vision system for grading of dried figs. *Comput. Electron. Agric.* **2015**, *119*, 158–165. [CrossRef]
14. *GB/T 10651—2008*; Fresh apple General Administration of Quality Supervision, Inspection and Quarantine of the People's Republic of China. Standardization Administration of PRC (SAC): Beijing, China, 2008.
15. Chen, Z.Y.; Wu, R.H.; Lin, Y.Y.; Li, C.Y.; Chen, S.Y.; Yuan, Z.E.; Chen, S.W.; Zou, X.J. Plant disease recognition model based on improved YOLOv5. *Agronomy* **2022**, *12*, 365. [CrossRef]
16. Xu, X.W.; Zhang, X.L.; Zhang, T.W. Lite-YOLOv5: A lightweight deep learning detector for on-board ship detection in large-scene sentinel-1 SAR images. *Remote Sens.* **2022**, *14*, 1018. [CrossRef]

17. Lv, J.D.; Xu, H.; Han, Y.; Lu, W.B.; Xu, L.M.; Rong, H.L.; Yang, B.A.; Zou, L.; Ma, Z.H. A visual identification method for the apple growth forms in the orchard. *Comput. Electron. Agric.* **2022**, *197*, 106954. [CrossRef]
18. Jagtap, A.D.; Kawaguchi, K.; Karniadakis, G.E. Adaptive activation functions accelerate convergence in deep and physics-informed neural networks. *J. Comput. Phys.* **2020**, *404*, 109136. [CrossRef]
19. Li, S.; Wang, Y.H.; Feng, C.Y.; Zhang, D.; Li, H.Z.; Huang, W.; Shi, L. A thermal imaging flame-detection model for firefighting robot based on YOLOv4-F model. *Fire* **2022**, *5*, 172. [CrossRef]
20. Li, Y.M.; Zhang, J.; Hu, Y.; Zhao, Y.N.; Cao, Y. Real-time safety helmet-wearing detection based on improved YOLOv5. *Comput. Syst. Sci. Eng.* **2022**, *43*, 1219–1230. [CrossRef]
21. Wang, Y.N.; Fu, G.Q. A novel object recognition algorithm based on improved YOLOv5 model for patient care robots. *Int. J. Hum. Robot.* **2022**, *19*, 2250010. [CrossRef]
22. Lv, H.F.; Lu, H.C. Research on traffic sign recognition technology based on YOLOv5 algorithm. *J. Electron. Meas. Instrum.* **2021**, *35*, 137–144.
23. Ji, W.; Gao, X.X.; Xu, B.; Pan, Y.; Zhang, Z.; Zhao, D. Apple target recognition method in complex environment based on improved YOLOv4. *J. Food Process Eng.* **2021**, *44*, e13866. [CrossRef]
24. Ji, W.; Peng, J.Q.; Xu, B.o.; Zhang, T. Real-time detection of underwater river crab based on multi-scale pyramid fusion image enhancement and MobileCenterNet model. *Comput. Electron. Agric.* **2023**, *204*, 107522. [CrossRef]
25. Zheng, Z.; Wang, P.; Liu, W.; Li, J.Z.; Ye, R.G.; Ren, D.W. Distance-IoU Loss: Faster and better learning for bounding box re–gression. In Proceedings of the 34th AAAI Conference on Artificial Intelligence, New York, NY, USA, 7–12 December 2020.
26. Chen, W.; Zhang, J.F.; Guo, B.Y.; Wei, Q.Y.; Zhu, Z.Y. An apple detection method based on des-YOLO v4 algorithm for harvesting robots in complex environment. *Math. Probl. Eng.* **2021**, *2021*, 7351470. [CrossRef]
27. Guo, S.Y.; Li, L.L.; Guo, T.Y.; Cao, Y.Y.; Li, Y.L. Research on mask-wearing detection algorithm based on improved YOLOv5. *Sensors* **2022**, *22*, 4933. [CrossRef] [PubMed]
28. Ji, W.; Pan, Y.; Xu, B.; Wang, J.C. A real-time apple targets detection method for picking robot based on ShufflenetV2-YOLOX. *Agriculture* **2022**, *12*, 856. [CrossRef]
29. Yao, J.; Qi, J.M.; Zhang, J.; Shao, H.M.; Yang, J.; Li, X. A real-time detection algorithm for kiwifruit defects based on YOLOv5. *Electronics* **2021**, *10*, 1711. [CrossRef]
30. Jung, H.K.; Choi, G.S. Improved YOLOv5: Efficient object detection using drone images under various conditions. *Appl. Sci.* **2022**, *12*, 7255. [CrossRef]
31. Wang, C.; Zhang, X.F.; Liu, C.; Zhang, W.; Tang, Y. Detection method of wheel hub weld defects based on the improved YOLOv3. *Opt. Precis. Eng.* **2021**, *29*, 1942–1954. [CrossRef]
32. Shin, H.C.; Roth, H.R.; Gao, M.C.; Lu, L.; Xu, Z.Y.; Nogues, I.; Yao, J.H.; Mollura, D.; Summers, R.M. Deep convolutional neural networks for computer-aided detection: CNN architectures, dataset characteristics and transfer learning. *IEEE Trans. Med. Imaging* **2016**, *35*, 1285–1298. [CrossRef]
33. Zhou, T.; Yu, Z.T.; Cao, Y.; Bai, H.Y.; Su, Y. Study on an infrared multi-target detection method based on the pseudo-two-stage model. *Infrared Phys. Technol.* **2021**, *118*, 103883. [CrossRef]
34. Chang, Y.H.; Zhang, Y.Y. Deep learning for clothing style recognition using YOLOv5. *Micromachines* **2022**, *13*, 1678. [CrossRef]
35. Ji, W.; Cui, X.; Xu, B.; Ding, S.H.; Ding, Y.; Peng, J.Q. Cross-coupled control for contour tracking error of free-form curve based on fuzzy PID optimized by improved PSO algorithm. *Meas. Control.* **2022**, *55*, 807–820. [CrossRef]
36. Raza, T.Z.; Lang, W.; Jedermann, R. Integration of wireless sensor networks into industrial control systems. In *Dynamics in Logistics*; Springer: Cham, Switzerland, 2017; pp. 209–218.
37. Cai, W.Z.; Tian, J.Y.; Wang, S.Y.; Li, J.F.; Yang, S.Q. Research of joint virtual commissioning of robotic grinding based on NX MCD and TIA. *Mod. Manuf. Eng.* **2022**, *7*, 37–42.

agriculture

MDPI

Article

Research on Multiobjective Optimization Algorithm for Cooperative Harvesting Trajectory Optimization of an Intelligent Multiarm Straw-Rotting Fungus Harvesting Robot

Shuzhen Yang [1,2,*], Bocai Jia [2], Tao Yu [2] and Jin Yuan [3]

1 School of Intelligent Manufacturing and Control Engineering, Shanghai Polytechnic University, Shanghai 201209, China
2 School of Mechatronic Engineering and Automation, Shanghai University, Shanghai 200444, China; 18800205923@163.com (B.J.); yutao@shu.edu.cn (T.Y.)
3 College of Mechanical and Electronic Engineering, Shandong Agricultural University, Tai'an 271018, China; jyuan@sdau.edu.cn
* Correspondence: szyang@sspu.edu.cn; Tel.: +86-21-5021-6899

Abstract: In view of the difficulties of fruit cluster identification, the specific harvesting sequence constraints of aggregated fruits, and the balanced harvesting task assignment for the multiple arms with a series-increasing symmetric shared (SISS) region, this paper proposes a multi-objective optimization algorithm, which combines genetic algorithm (GA) and ant colony optimization (ACO) stepwise, to optimize the multiarm cooperative harvesting trajectory of straw-rotting fungus to effectively improve the harvesting efficiency and the success rate of non-destructive harvesting. In this approach, firstly, the multiarm trajectory optimization problem is abstracted as a multiple travelling salesman problem (MTSP). Secondly, an improved local density clustering algorithm is designed to identify the cluster fruits to prepare data for harvesting aggregated fruits in a specific order later. Thirdly, the MTSP has been decomposed into M independent TSP (traveling salesman problem) problems by using GA, in which a new DNA (deoxyribonucleic acid) assignment rule is designed to resolve the problem of the average distribution of multiarm harvesting tasks with the SISS region. Then, the improved ant colony algorithm, combined with the auction mechanism, is adopted to achieve the shortest trajectory of each arm, which settles the difficulty that the clustered mature fruits should be harvested in a specified order. The experiments show that it can search for a relatively stable optimal solution in a relatively short time. The average harvesting efficiency is up to 1183 pcs/h and the average harvesting success rate is about 97%. Therefore, the proposed algorithm can better plan the harvesting trajectory for multiarm intelligent harvesting, especially for areas with many aggregated fruits.

Keywords: straw-rotting fungus; multiarm harvesting trajectory optimization; multiobjective optimization; cluster fruit; genetic ant colony stepwise algorithm

Citation: Yang, S.; Jia, B.; Yu, T.; Yuan, J. Research on Multiobjective Optimization Algorithm for Cooperative Harvesting Trajectory Optimization of an Intelligent Multiarm Straw-Rotting Fungus Harvesting Robot. *Agriculture* **2022**, *12*, 986. https://doi.org/10.3390/agriculture12070986

Academic Editor: Francesco Marinello

Received: 30 May 2022
Accepted: 6 July 2022
Published: 8 July 2022

1. Introduction

Straw rotting fungi are fungi that absorb the decayed humus of grass straws (such as straw and wheat straw) as the main source of nutrition [1], mainly including Agaricus bisporus, Agaricus blazei, straw mushrooms, capsule mushrooms, etc. Its factory production usually adopts the bed planting method. The growth of its fruit is similar to a round or spherical shape, easy to cluster, the combination of fruit and culture medium (soil) is relatively rigid and compact, the ripening time of each fruit is different, and the fruit is tender and vulnerable. This kind of fruit usually needs to be harvested by selective harvesting, and the aggregated mature fruits need to be harvested in the specified order; otherwise, the fruit is easy to be damaged. Therefore, it is difficult to achieve effective automatic nondestructive harvesting [2]. Under the trend of serious labor shortages, although the cultivation of such crops can achieve factory-like and intensive production, the labor-intensive harvesting process [3] still relies on manual labor, which has become the main bottleneck affecting further production and efficiency improvement. Therefore, there is an urgent need for harvesting robots that can adapt to the factory and intensive cultivation environment for intelligent and efficient harvesting.

In the 1990s, Reed, J. N. and Tillett, R.D. [4] proposed the first selective harvesting robot for Agaricus bisporus, which made selective autonomous harvesting of straw-rotting fungus feasible. Since then, scholars have carried out many researches to on the edible fungus harvesting robot technology but mainly focused on visual recognition and end-effectors to improve the harvesting success rate [5–13], and few studies on trajectory planning. Yang [14] aimed at the issue of mushroom harvesting path planning. GA was used to optimize the harvesting path to improve the harvesting efficiency after obtaining the location coordinates of all the mushrooms that can be harvested. In order to improve the harvesting efficiency of the mushroom harvesting robot, Hu et al. [15] proposed an improved simulated annealing algorithm to find the optimal path, which can increase the harvesting efficiency by 14–18%. However, both of the above studies are all optimized for the path of the single-arm harvesting robot, which can improve the harvesting efficiency to a certain extent, but far from manual efficiency. In addition, these methods do not consider the specific harvesting order of the aggregated fruits and are merely suitable for harvesting the fruits, which are relatively sparse. When the fruits grow densely, take Agaricus bisporus as an example, as shown in Figure 1, for the clustered fruits in the red circled area, the height of fruit body A is higher than B and C. If B or C is harvested before A, B and C will be damaged or even be harvested unsuccessfully because their cap is covered by A. Moreover, a may be pushed down, which will cause its cup center to deviate greatly, resulting in failure when picking A. So, the aggregated fruits should be harvested in order of height. Otherwise, the success rate of non-destructive harvesting and the harvesting quality will be reduced. So, it is significant to greatly improve harvesting efficiency; meanwhile, to take into account the harvesting order of aggregated fruits to improve the success rate of non-destructive harvesting further.

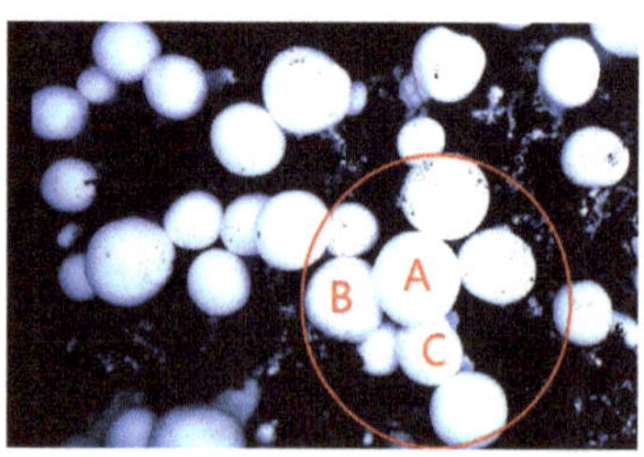

Figure 1. Schematic diagram of aggregated fruits.

Because the fruit is delicate and vulnerable and the harvesting environment is complex, the harvesting operation is usually limited to slow and time-consuming [16]. So, the harvesting efficiency is normally much lower than manual efficiency, which leads to the situation that the harvesting robot cannot be widely used in actual production. To solve this issue, using multiple harvesting arms is a typical approach.

A cotton harvesting robot with multiple robotic arms has been developed to achieve multiple plucking of crops, which increases the yield by 20–25% [17].

Zion [18] developed a melon harvesting robot with multiple Cartesian arms to accelerate the speed. The robot travels along a two-dimensional field at a constant velocity. The multiarm assignment is modeled as a k-colorable sub-graph problem and uses a greedy algorithm to achieve an optimal solution. Because the greedy algorithm focuses on local optimization, the effect of global optimization may not be very good.

A strawberry harvesting robot with dual cartesian arms was developed to reduce the cost and optimize the harvesting efficiency. The fruits were partitioned into several subsections equally for each arm to harvest. The harvesting speed can be reduced to 4.6 s [19].

An oyster mushroom harvesting robot with four harvesting arms connected in parallel on a common mobile platform is provided to improve efficiency [20]. Each arm is allocated to harvest in the divided area independently, which is divided from left to right with an approximately equal number of target fruits, and the mature fruit closest to the end actuator is picked first in each area.

A multiple robot arm system for kiwifruit harvest has been designed to reduce the harvesting cycle time, thereby increasing the efficiency to meet the requirements of commercial applications [21]. The multiarm harvesting robot sorts and partitions target fruits according to their x coordinates and assigns them to the harvesting arms so that the amount of the harvested fruit is approximately equal for each arm. However, for fruits growing in clusters, the clustered fruits are completely allocated to the same arm to be harvested from low to high to avoid touching or moving other fruits' positions. Although this method can improve the picking success rate of fruit clusters, it will affect the distribution uniformity of each arm, thereby reducing the picking efficiency.

Most of the above multiarm harvesting trajectory planning uses the traditional method, which partitions target fruits approximately evenly according to the arrangement direction of the harvesting arms and allocates them to the robot arms, and each arm harvests fruits from left to right or from right to left. Normally, this method can achieve a good result. However, the trajectory planning effect of this method will be greatly reduced when the fruits are distributed seriously ununiformly and even with many fruit clusters. Moreover, this method assumes that each fruit can be picked by each arm; that is, the accessibility of the task to be executed to the individual is consistent. While the fact is that not every fruit can be picked by each arm due to the size of the arm, which indicates that the accessibility of the task to be executed in the instance is inconsistent. This shortcoming will also increase the difficulty of equal allocation among each arm. Therefore, a more global and flexible optimization approach is required to resolve the above issues. In order to effectively improve harvesting efficiency and adapt to the environment of straw-rotting fungus factory bed planting, the harvesting robot can be designed as a highly cost-effective Cartesian coordinate harvesting robot as shown in Figure 2, equipped with multiple harvesting arms [22]. Here, because the mature fruits of straw-rotting fungus are little difference in height, it can be assumed that the time spent in the height direction (Z-axis) of each fruit during harvesting is the same, so that the dimension of the harvesting trajectory planning of such fruits can be reduced as a two-dimensional trajectory planning problem in the XY plane. In addition, since the shape of the straw-rotting fungus is basically spherical, the projection of the shape of the fruit on the XY plane is further approximated as a circle. Just as the harvesting sequence planning problem of a single-arm harvesting robot can be regarded as a TSP problem [23]. Given the above assumptions, the multiarm cooperative harvesting trajectory planning problem studied in this paper can be simplified as a typical MTSP as well.

Similar to most fruit and vegetable picking robots mentioned above, the harvesting time for a straw-rotting fungus fruit is also time-consuming (about 5 s) due to the operation of grasping the fruit and detaching it from the culture medium or soil cannot be too fast to avoid damaging the fruit. Compared with the harvesting time (time required for harvesting at the target fruit position), the moving time (time used to move from the picked fruit to the next target fruit) is much shorter. Therefore, it is much more significant to allocate the harvesting tasks to each arm as uniformly as possible before optimizing the shortest path for each arm [24–26]. In the meantime, it is also necessary to take into account the harvesting sequence of aggregated fruits to improve the success rate of non-destructive harvesting.

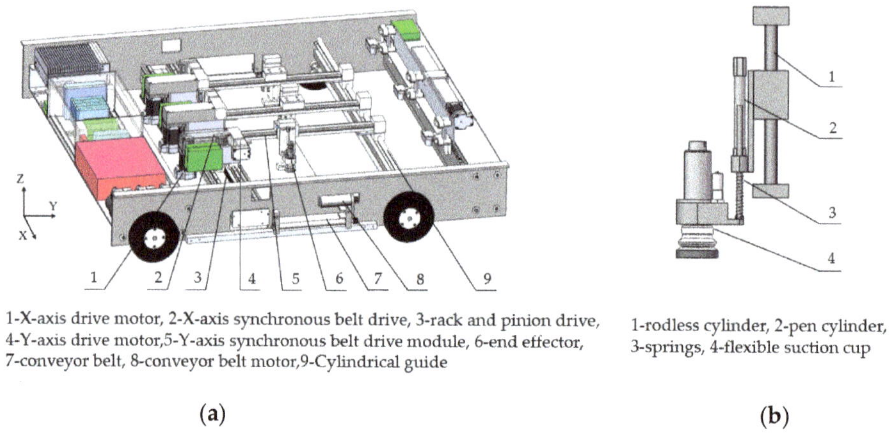

Figure 2. Detailed mechanical design of MFA. (**a**) Detailed mechanical structure of MFA; (**b**) detailed mechanical structure of end-effector.

For the sake of the uniform assignment of tasks for each harvesting arm, in addition to the region segmentation method described in the previous literature analysis, the stepwise algorithm used for MTSP is more appropriate to solve the issue, which is also superior to the general heuristic algorithm [27–30]. A stepwise algorithm is proposed to resolve the MTSP of multi-UAV cooperative airport bird repelling, which adopts a genetic algorithm to divide the MTSP into M independent TSP. This paper shows that GA is suitable to solve the task balance assignment of MTSP [31]. Lu et al. [32] combined the K-means clustering algorithm with GA to solve multiobjective MTSP, although a better-balanced task assignment is obtained, but it still has poor trajectory searchability.

The ant colony algorithm is widely used in combinatorial optimization problems due to its strong search ability and fast convergence speed [33]. Necula et al. [34] used ACO as the bi-standard surface for solving the multitraveling salesman problem. Changdar et al. [35] adopted ACO to resolve the multi-stop multitraveling salesman problem with non-random parameters. Although both of them achieve a good result, it is easier to fall into the local optimal solution prematurely in MTSP with high complexity problems.

In addition to solving the problem of equal allocation of multiarm tasks under SSIS restrictions to improve efficiency, it is also necessary to take into account the harvesting order of aggregated fruits to improve the success rate of non-destructive harvesting mentioned above. To achieve this aim, the above algorithm for solving similar MTSP should be further improved by combining a multiobjective optimization method. More than this, another challenge is how to improve the algorithm so that the fruit clusters that need to be picked in a specific order can be split and allocated to multiple different arms, to overcome the shortcoming of the whole fruits in the same cluster being merely allocated to the same arm, so as to further increase the efficiency even though the mature fruits are seriously unevenly distributed on a culture medium (soil) and with many fruit clusters.

Through the above analysis, this paper proposes an improved genetic ant colony multiobjective optimization algorithm, which makes comprehensive use of the advantages of both the genetic algorithm and the ant colony algorithm to resolve the difficulties of multiarm cooperative harvesting of straw-rotting fungus and achieve both high harvesting efficiency and a high success rate. The main contributions of this approach are as follows:

(1) The trajectory planning problem of multiarm cooperative harvesting of straw-rotting fungus is transformed into an MTSP problem;
(2) To resolve the difficulty of accurately recognizing the fruit clusters of straw-rotting fungus due to their different shapes and uncertain density, a density-based clustering

algorithm is improved by designing a new method for calculating local density, which can better meet the clustering analysis of straw-rotting fungus;

(3) A multiobjective optimization model is built for the trajectory optimization of an intelligent multiarm straw rotting fungus harvesting robot;

(4) The improved ant colony algorithm combined with the auction mechanism is used to achieve the shortest trajectory of each TSP problem. Meanwhile, the fruit clusters that are required to be picked in a specific order can be allocated to different arms instead of being allocated to a single arm by combining the auction mechanism with an ant colony.

2. Description of Multiarm Cooperative Harvesting Trajectory Planning Problem

2.1. Intelligent Multiarm Straw-Rotting Fungus Harvesting Robot

As shown in Figure 2, an intelligent multiarm straw rotting fungus harvesting robot consists of the following four main parts: (i) a mobile platform (MP) with a conveyor belt; (ii) a visual position system (VPS), and (iii) multiple flexible arms (MFA); (iv) control system (CS). The MP is placed on the rails of the multistory shelves to move along the rail with the help of the MP, the VPS can recognize and locate all mature fruits. The CS can plan the harvesting trajectory and control MFA to harvest the mature fruits and send them to the convey for blanking.

MFA consists of at least two Cartesian arms with three DOF. As shown in Figure 2a, the X-axis adopts the gear rack motor traverse multi-axis mechanism, and each is mounted on a guide rail. The X-axis motor drives the gears in the gear rack through a synchronous belt drive structure to move the end-effector along the X-direction. The Y-axis uses the synchronous belt drive module to move the end-effector in the Y direction. The end-effector is designed into a two-stage driving structure, as shown in Figure 2b, to adapt to the narrow layer height of the edible fungus culture rack (available design height is only about 250 mm while travel should be 160 mm). The first stage is driven by a rodless cylinder, and the second stage is driven by a pen cylinder. The suction cup is connected to the pen cylinder, which is connected to the rodless cylinder. During harvesting, the rodless cylinder drives the pen cylinder lower. Firstly, the rodless cylinder inflates to drive the pen cylinder down. Then, the pen cylinder deflates to cause the suction cup to drop under the action of gravity until it touches the surface of the target fruit. Finally, vacuum to grasp the target fruit, rotate or wobble, and pull up to detach the fruit from the soil. The structure can better adapt to the large height difference in fruits and realize picking action by wobble or rotation motion.

The working flow of the proposed robot is shown in Figure 3, as follows: First, the camera traverses all fruit images within the current visual area; Secondly, the image is transferred to the host computer, and the positions of the mature fruits are identified by image processing; Thirdly, all of the coordinates of fruits to be picked are scheduled and allocated to multiple harvesting arms; Fourth, the controller drives the arms to harvest fruits after receiving the assigned task; Finally, the robot moves forward as a whole by MP to start the next harvesting cycle until the harvest of one layer is finished. In addition to the use of three arms, in order to improve efficiency, the following are also performed: (i) Multiple economical depth cameras are used to shorten the photographing and identification time; (ii) Visual processing and harvesting operation work in parallel rather than in series. In the current cycle, the multiarm harvester harvests the fruits identified in the previous cycle.

As shown in Figure 4, the control system takes the motion controller (TRIO MC4N) as the core. The motion controller communicates with the PC through the Ethercat bus to obtain the picking task for each arm, then drives the motor of each axis and controls the cylinder, sucker, and other actuators of the end-effector to work monitor the working status of each actuator in real-time and returns it to the upper computer. The economical and compact stepping servo motor is used for each axis motor, and the communication between the motor and the controller is also via the Ethercat bus.

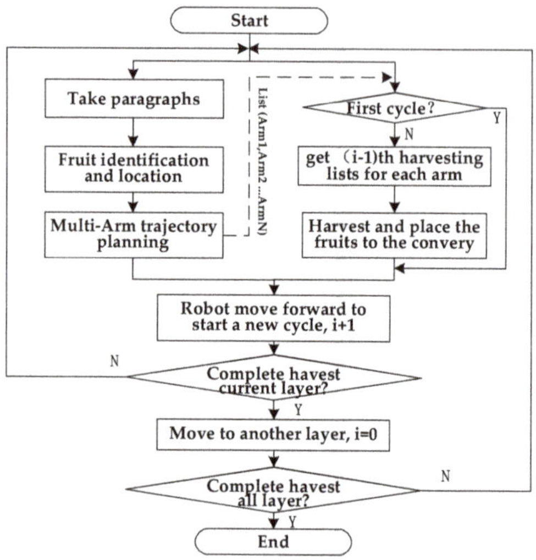

Figure 3. Workflow of the multiarm harvesting robot.

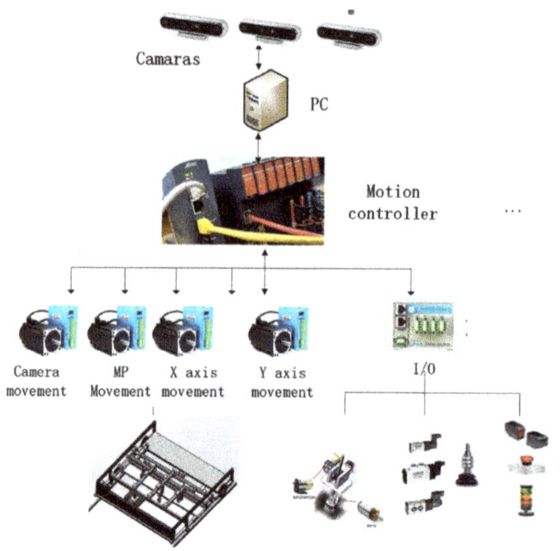

Figure 4. Hardware composition diagram of control system.

2.2. Accessibility Analysis of Multiple Harvesting Arms

Generally, the MTSP problem assumes that all cities can be visited by each traveler, but this is not the case with the multiple harvesting arm structure in this paper. As shown in Figure 2a, multiple arms are arranged in a series along the X-direction. Due to the unignorable width of the end-effector, each arm has a certain inaccessible area, and the reachable range of each arm is different.

The accessibility of the picking arm in this paper is characterized by exclusive area, partially shared area, and fully shared area, and they are symmetrically distributed as shown in Figure 5, which is called the serial increasing symmetric shared (SISS) area. The definition of the SSIS is described in detail in [36]. Such an SSIS area makes it more difficult to assign the harvesting tasks equally among each arm.

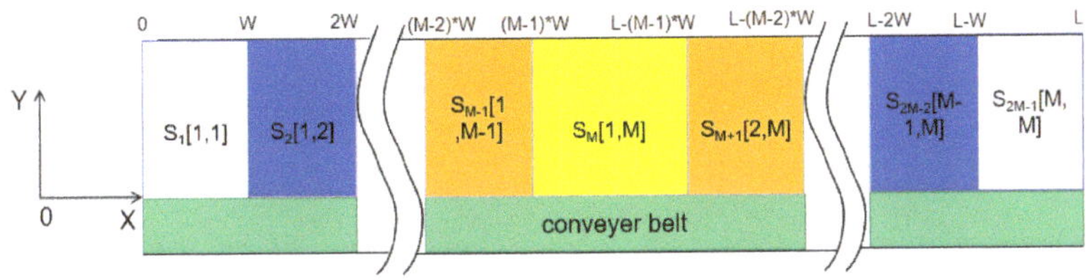

Figure 5. Distribution of accessible area.

The range of each harvesting region can be described as follows:

$$S_k = (X_{k-1}, X_k),\ k \in \{1,\ 2,\ \ldots, 2M-1\} \quad (1)$$

$$X_k = \begin{cases} W \times k, & k \in \{1,\ 2, \ldots, M\} \\ L - (2M-1-k) \times W\,, & k \in \{M+1, M+2, \ldots, 2M-1\} \end{cases} \quad (2)$$

where L is the length of the harvesting region, W is the width of arm and end-effector, M is the amount of harvesting arm, S_k is the k-th harvesting region, and X_k is the coordinate of the end point in the X-direction of the k-th harvesting area.

Equation (3) shows the accessible harvesting arms per area as follows:

$$S_k(j1, j2) = \begin{cases} S_k(1,k)\,, & k \in \{1, 2\,, \ldots, M\} \\ S_k(k-M+1, M)\,, & k \in \{M+1, M+2, \ldots, 2M-1\} \end{cases} \quad (3)$$

where ($j1$,$j2$) indicate harvesting arms $j1$ to $j2$ can reach the k-th area, $1 \le j1 \le j2 \le M$.

2.3. Description of Trajectory Planning for Multiarm Cooperative Harvesting

As described in Section 1, the multiarm trajectory planning problem of this paper can be abstracted to an MTSP problem with an SSIS Area. In which case, the aggregated fruits should be harvested in the order from high to low in the Z direction. The goal of trajectory optimization is to achieve both high harvesting efficiency and a high success rate with the following constraints:

(1) Each harvesting arm's accessible region is limited by the SISS region shown in the above section;
(2) To avoid collision between the adjacent harvesting arms, a safety distance Dist is required;
(3) After completing the harvesting tasks, each harvesting arm should go back to its start point independently, which is (0,0), (W,0), … ,((M − 1) × W,0).

3. Mathematical Model of Multiarm Cooperative Harvesting Trajectory Planning Problem

It can be known from Section 2.3 that the trajectory optimization problem of an intelligent multiarm straw-rotting fungus harvesting robot is regarded as a multiobjective optimization problem. Researchers mainly use multiobjective optimization methods such as the weighted coefficient method, multiobjective genetic algorithm, multiobjective particle swarm optimization algorithm, etc. [37,38]. Among them, because the optimization problem in this paper is a two-dimensional, that is, a low-dimensional objective optimization problem, and to reduce the complexity of the problem, the simple and easy-to-use weighted coefficient method is adopted, which decomposes multiobjective into a single objective and then optimizes this single objective.

Suppose the robot has M arms and N fruits to be harvested. The average harvesting efficiency of the robot is C pcs/h and the average manual working efficiency is M pcs/h; R is the ratio of the average working efficiency of the robot and man. In order to obtain the optimal solution to double the objectives of harvesting efficiency and success rate, take

R and success rate P to solve the objective function E by the objective weighting method as follows:

$$\text{Max } E = R \times K_1 + P \times K_2 \tag{4}$$

$$R = \frac{C}{M} = \frac{N}{T \times M} \tag{5}$$

$$P = \frac{N - N_{lose}}{N} \times 100\% \tag{6}$$

$$T = [\max(t\_ARM_i) + T_c]/3600 \tag{7}$$

$$t\_ARM_i = \sum_{j=1}^{N}\sum_{k=1}^{N} t_{jk}x_{ijk} + \sum_{j=1}^{N} t_{d_ij}x_{id_ij} + \sum_{j=1}^{N} t_{jd_i}x_{ijd_i}, \forall i \in M \tag{8}$$

$$x_{ijk} = \begin{cases} 1, \text{ when Arm i from fruit j to k} \\ 0, \text{ other} \end{cases} \tag{9}$$

where,

K_1 and K_2 are weighting coefficients, K_1 and $K_2 > 0$ and $K_1 + K_2 = 1$;

N_{lose} is the number of fruits in the clusters that failed to be harvested in the order from high to low in the Z direction when the fruits were gathered;

T is the total harvesting time required for harvesting all mature fruits;

T_C is the avoidance time;

$x_{id_{ij}}$ indicates that Arm i from the starting point d_i to fruit j;

x_{ijd_i} indicates that Arm i from fruit j to the starting point d_i

t_{jk} is the time taken by the harvesting arm to perform the task from fruit j to k;

The objective function must satisfy the following constraints:

Given the mature fruits set is V = {1,2, ... ,N}; U is the set of harvesting arms ($1 \leq M \leq N$), U = {1,2, ... ,M}; Since the working range of each harvesting arm is limited, d_i is the starting point of Arm i.

$$y_{ij} = \begin{cases} 1, j \in I_i, \text{ when Arm i harvests fruit j} \\ 0, \text{ other} \end{cases} \tag{10}$$

where I_i means the mature fruits subset must be harvested by Arm i, $I_i \in V$

The sum of fruits picked by each arm should be equal to N as follows:

$$\sum_{j=1}^{n} y_{ij} = Q_i, \forall i \in M, \tag{11}$$

$$\sum_{i=1}^{m} Q_i = n \tag{12}$$

where Q_i indicates the number of fruits allocated to Arm i, $i \in U$;

Each fruit can only be harvested by one arm as follows:

$$\sum_{i=1}^{m} y_{ij} = 1, \forall j \in V \tag{13}$$

Each arm should start from and come back to its own start point after harvesting all assigned fruits as follows:

$$\sum_{j=1}^{n} x_{id_ij} = \sum_{j=1}^{n} x_{ijd_i} = 1, \forall i \in M \tag{14}$$

4. Multiobjective Optimization Algorithm for Multiarm Cooperative Harvesting Trajectory

As discussed in Section 1, there are the following two main difficulties required to overcome: (1) The SSIS region restricts the working area of the harvesting arm as mentioned in Section 2.2; (2) The allocation of cluster fruits to ensure the non-destructive success rate.

For the first problem, the stepwise algorithm, which can divide the MTSP into m-independent TSPs to reduce the complexity of the problem, is an appropriate approach.

For the second sequence planning problem, the normal approach is to regard the clustered fruits as a whole and assign them to a certain arm. However, when the number of fruits in the same cluster is large and the distribution density of mature fruits in the harvesting area is uneven (for example, most of the mature fruits are located on the left and a few on the right), and if the fruits of a cluster can only be assigned to one arm, there will be more harvesting tasks for one arm and fewer for the other arms, which will greatly affect the harvesting efficiency. Therefore, for clustered fruits, it is necessary not only to ensure that they can be harvested in order but also to be assigned to different arms to help achieve the balanced allocation of the total harvesting task to each arm.

Therefore, an improved genetic algorithm and ant colony stepwise multiobjective optimization algorithm (IGAACMO) is proposed. The algorithm flow is shown in Figure 6.

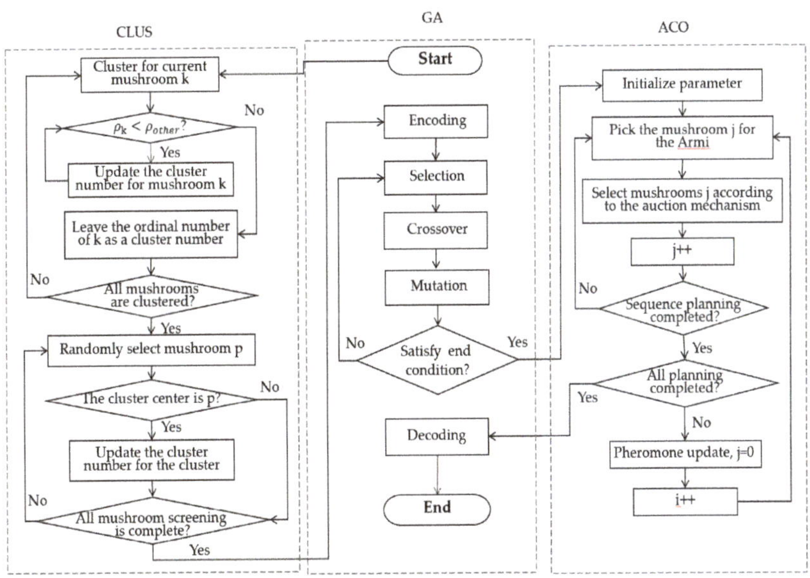

Figure 6. The flow chart of the multitarget optimization algorithm proposed in this paper.

In the first step, an improved local density bi-directional clustering algorithm is designed to identify the clustered fruits to provide preparation for harvesting the clustered fruits in the specified order; Then, in the second step, the MTSP problem is decomposed into m independent TSP problems by a genetic algorithm with strong global optimization ability, so as to settle uneven task assignment of the MTSP problem with SSIS region; The third step is to use the fast convergence speed of the ant colony algorithm to plan the trajectory of the above M-independent TSP respectively, and combined it with the auction mechanism to resolve the allocation issue under the restriction of clustering fruit sequence planning.

4.1. Clustering Algorithm Optimization for Fruits of Straw-Rotting Fungus

The clustering algorithms in the current research can be roughly divided into the following five categories: partitional-based, hierarchical-based, grid-based model-based, and density-based [39]. The partitional-based algorithm is suitable for identifying datasets with small sample sizes and spherical cluster shapes. However, it depends on the user to specify the number of clusters in advance, and the processing for large-scale datasets and clusters with complex shapes still needs to be improved further [40]. The hierarchical-based clustering algorithm is sensitive to the noise and abnormal data points in the data and cannot be rolled back after the upward or downward iteration [41]. The grid-based algorithm runs at a high speed because its processing time is only related to the number of cells and has nothing to do with the number of objects. However, the grid-based division method may also reduce the clustering accuracy [42,43]. The advantage of a model-based clustering algorithm is that it can find noise and isolated data points and can automatically identify the number of classes. The disadvantage is that it is not suitable for clustering with a large amount of data [44]. The density-based clustering algorithm can identify clusters with different shapes. It can effectively eliminate abnormal data points or isolated data points in the dataset, and has good noise resistance, but are sensitive to the density of adjacent data points [45,46].

The fruit clustering state of straw-rotting fungus is relatively complex. Taking Agaricus bisporus as an example, as shown in Figure 7a, it has the characteristics of complex and different cluster shapes, and the number of clusters is unpredictable in advance, which makes the partitional clustering algorithms and hierarchical clustering algorithms unsuitable to discriminate against it. In addition, as shown in Figure 7b, it also has the characteristics of many small clusters and many discrete values globally, which makes grid clustering algorithms and model clustering algorithms less suitable.

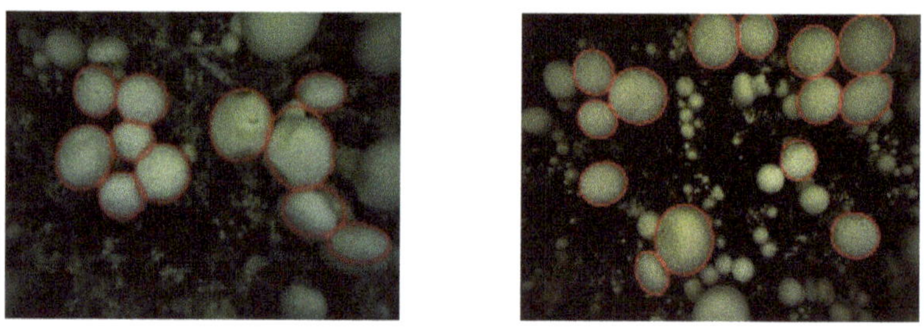

(a) (b)

Figure 7. Cluster shape of Agaricus bisporus (**a**) clusters with different shapes and densities; (**b**) clusters with many small clusters and many discrete values.

The density-based clustering algorithm can identify clusters with different shapes. However, due to the different diameters of each fruit of the straw-rotting fungus, the density of the fruit clusters of the straw-rotting fungus is uncertain, while the general density-based algorithm is not effective in solving such clusters with variable density. Therefore, an improved local density bi-directional clustering algorithm is designed in this paper. The designed local density calculation method can better adapt to the problems of complex and different cluster shapes, especially for uncertain density, so that the algorithm can better meet the requirements of fruit cluster analysis of straw-rotting fungus.

4.1.1. Fruit Clustering Definition

In order to determine which cluster set should a fruit belong to; the following definitions are given:

(1) If D_{ij}, the center distance between fruit i and j, is less than or equal to the sum of their radius, as Equation (15), fruits i and j belong to the same cluster;

$$D_{ij} \leq r_i + r_j \tag{15}$$

where,

r_i, r_j represent the radius of fruit i and j, respectively.

(2) If fruits i and j are in the same cluster, while fruits j and k are in the same cluster, then fruits i and k are also in the same cluster;
(3) If the center distance between fruit i and any other fruit fails to satisfy Equation (15), then the fruit i does not belong to any cluster, which is called discrete fruit.

4.1.2. Local Density Calculation

Let N be the set of all mature fruits and n be the set of the number of fruits. The local density ρ_i represents the number of fruits that belong to the same cluster (that is, meet the definition (1) in Section 4.1). The larger the local density, the more likely the fruit is the center of the cluster center; when the local density is 0, the fruit is a discrete fruit. The calculation process of local density is as follows:

$$\rho_i = \sum_{j=1}^{n} \mathrm{Clu}_{ij},\ i \neq j,\ i,j \in N, \tag{16}$$

$$\mathrm{Clu}_{ij} = \begin{cases} 1, & D_{ij} \le r_i + r_j \\ 0, & D_{ij} > r_i + r_j \end{cases},\ i \neq j,\ i,j \in N, \tag{17}$$

where Clu_{ij} is used to determine if fruit i and j belong to the same cluster.

4.1.3. Improved Density-Based Clustering Algorithm

The improved clustering algorithm is divided into an ascending process and a descending process, as shown in Figure 8.

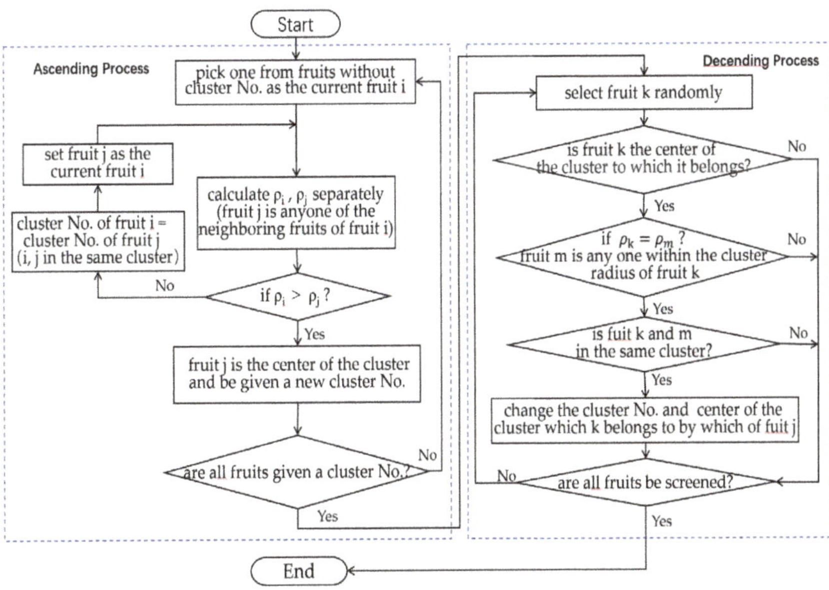

Figure 8. The flow chart of the improved local density-based clustering algorithm.

In the ascending process, calculate the local density of different fruit points, find the high local density point closest to the fruit point, form a data chain from the data points of low local density to high local density, and find the cluster center of the ascending process for all fruits point. The codes in details are shown in Table 1.

Table 1. The codes of ascending process of the improved clustering algorithm.

Algorithm 1.

```
Input: N        // set of all mature fruits
Output: Unit[]  //set of cluster centers corresponding to each fruit
Initialization();
UpProcess(a,b); // a is the data coordinate and b is the data serial number.
  c = b;         // If no other point within the cutoff distance has a higher cluster
                    density than it, its cluster center is itself
  while i < n do;
    if i! = b and density[i]> = density[b];   // The local density of i is greater
                                                  than the input fruit
      Distance = math.sqrt((N[i][0]-a [0])^2+(N[i][1]-a [1])^2);
      rc = Clu[i][2]+a [2]                 // Cluster radius
      if x1<=x2:                 // Two fruits are clustered
        c=i;
        break;
  return c;
while i < n do;  #Ascending process, computing the cluster center for each fruit
  Unit[i]=UpProcess(N[i],i)
return Solution;
```

In the descending process, the data point with the highest local density is used as the cluster center, and then the data chain is merged. After all data points are traversed, and finally, clustering is performed to complete the unified operation of all fruit clustering centers in the same cluster, the codes in detail are shown in Table 2. In addition, consolidation operations were added to the descent. Because the growth characteristics of straw-rotting fungus easily lead to the highest local density points within the same cluster, which may not be unique, they need to be integrated into the same cluster. For example, the local density values of A and B in Figure 9 are both equal to 3, which are both the highest local density points in the cluster. In this case, fruit A and B may be the cluster centers of each other, so it is necessary to integrate Fruit A and B into the same cluster. Its processing method is shown in lines 9–10 in Table 2.

Table 2. The codes of descending process of the improved clustering algorithm.

Algorithm 2.

```
Input: Unit[]     // The set of cluster centers obtained during the ascent
Output: Unit[ ]   // The final set of cluster centers corresponding to each fruit
1 Initialization();
2 DownProcess(a,b);
3 if a == b;             // If the cluster center of the data point is itself
4   return b;
5 else;
6   a = DownProcess(Unit[a],a);
7   return a;
8 for i in range(n);    // Descending process, if the cluster center of a data is another
                           point, it will be merged into its subclass.
9   if i == Unit[Unit[i]];    // If it is the cluster center with another point, select one
                                 of the points as the cluster center and merge the two.
10     Unit[i] = i;
11  if Unit[Unit[i]]! = Unit[i];   // If the final cluster center corresponding to the
                                      point is found
Unit[i] = DownProcess(Unit[i],i);
12 return Solution;
```

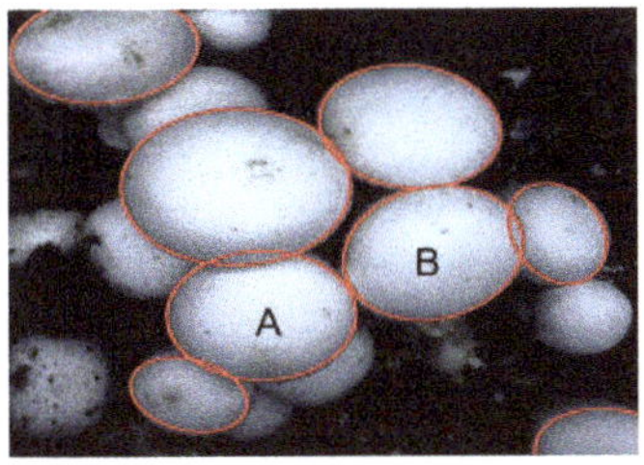

Figure 9. The highest local density point within the cluster is not unique.

4.2. Genetic Algorithm

A genetic algorithm is used to divide the MTSP into 2M-1 independent TSPs for SSIS region constraint by designing an appropriate encoding so that the MTSP can be decomposed into 2M-1 TSPs appropriately.

(1) Encoding

The gene sequence of GA is divided into 2M-1 corresponding segments to the 2M-1 different accessible regions in the SSIS region, as shown in Figure 10 so that the MTSP can be decomposed the MTSP into 2M-1 TSPs. The gene indicates which arm the fruit should be allocated to. The DNA fragment corresponds to the picking region S_k in Figure 5 one by one.

S_1			...	S_M				...	S_{2M-1}		
1	1	...	...	1	2	M	...	...	M	M	...

Figure 10. Diagram of DNA sequence proposed in this paper.

Each ripe fruit in the reachable region corresponds to each element in the corresponding DNA segment. $DNA_k(j)$ indicates the picking arm allocated to the j-th fruit in the k-th area. Therefore, in order to be consistent with the harvesting arm allowed to enter each accessible region of Equation (3), the assignment rule of $DNA_k(j)$ in the initialization of the corresponding population is as follows:

$$DNA_k(j) = \begin{cases} random(1,k)\,, & k \in \{1,2,\ldots,M-1\} \\ random(k-M+1,M), & k \in \{M,M+1,\ldots,2M-1\} \end{cases} \tag{18}$$

where random (1,*k*) represents any integer in the randomly assigned closed interval 1 to *k*.

(2) Selection operator

Roulette is adopted as the selection operator to improve the optimization ability of the algorithm. In this method, two individuals are selected at a time, and then the individual with the better fitness of the two individuals is selected by the probability of survival.

(3) Crossover operator

To increase the global search ability, a multipoint crossover is used to randomly select multiple segments in the gene sequence for crossover.

(4) Mutation operator

Different mutation rules are required for each DNA segment, and it can merely mutate into the code for the harvesting arm accessible to the corresponding reachable area of the segment.

All of the mature mushrooms in the current cycle were divided into M groups based on the DNA sequences of the best individuals in the population.

4.3. Improved Ant Colony Algorithm

In order to solve the problems of sequence harvesting of cluster fruits and collision avoidance when M harvesting arms work together, the respective trajectory planning of each harvesting arm should be carried out in parallel, so that it can be judged in real-time whether there the clustered fruits are harvested in the specified order and whether will be collisions between the arms.

The ant colony algorithm has good parallelism and late convergence of the algorithm, so this paper adopts the ant colony algorithm to solve the trajectory planning problem of each of the M harvesting arms and combines the auction mechanism to deal with the sequence harvesting of cluster fruits when the M harvesting arms work together.

In actual harvesting, in addition to harvesting efficiency, the harvesting success rate is also a very important indicator. According to actual harvesting requirements and experiments, the trajectory planning algorithm designed in this paper needs to ensure that the harvesting success rate is more than 95%. Therefore, the following approach is designed so that in the early stage of the evolution of the algorithm, the success rate is the main guide, while after the success rate meets the requirements, the pheromone concentration of the current fruit to be harvested should be temporarily increased to increase the probability of its selection.

The specific calculation process for the success rate is shown in Figure 11.

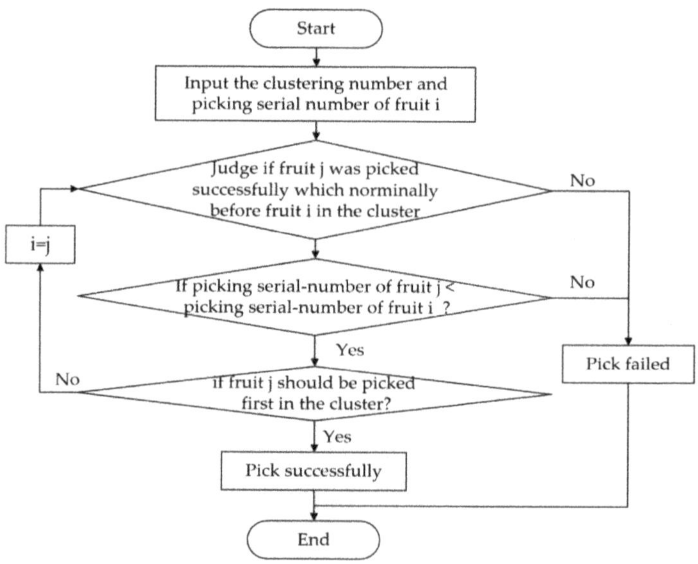

Figure 11. Flow chart for judging whether fruit in the cluster be harvested in specific order.

The following are the detailed steps:

Step 1 Initialize.

Initialize the pheromone matrix, the path taboo table, the set containing the nominal harvesting order of clustered fruits (for calculating the success rate), the set of coordinates of the fruit to be harvested for each arm, and the matrix containing the information corresponding to the time axis and displacement of the X-axis.

Step 2 Build trajectory.

Ants construct m-picking arms in parallel. First, the path taboo table is used to remove the picked fruits and generate a preliminary candidate fruit set. Additionally, then, the auction mechanism is used to determine the current candidate fruit set for each ant and choose the fruit to be picked next from the set according to the pheromone concentration until all ants have completed the trajectory construction.

Step 3 Evaluation.

The objective function E, which can be calculated by Equation (4), is used for evaluation. To make the success rate of harvesting meet the requirement of more than 95%, the K_1 and K_2 coefficients in Equation (4) are dynamically adjusted. When the success rate is less than 95%, set $K_1 = 0.4$, $K_2 = 0.6$; after the success rate is greater than or equal to 95%, set $K_1 = 0.6$, $K_2 = 0.4$.

Step 4 Update the pheromone matrix.

The trajectory with maximum E in Equation (3) is selected to update the pheromone matrix.

Step 5 Determine the number of iterations.

If the maximum number of iterations is reached, turn to End, otherwise go to step 1.

Step 6 End.

Output the final optimal trajectory.

5. Experiments and Analysis

5.1. Experiments of Clustering Algorithm

To verify the effect of the improved local density clustering algorithm (ILDCA) in this paper, the agaricus bisporus was taken as an example to test, and the test data were all from the site of the planting factory. The data is shown in Figure 12a. The pictures were taken on the spot by the harvesting robot, and the mature fruits recognized by visual are marked with red circles. The mature fruit data obtained from the image identification are processed by the clustering algorithm proposed in this paper, and the obtained clustering result is shown in Figure 12b, in which the fruits belonging to the same cluster are marked with the same color. Comparing a and b of Figure 12, it can be seen that the success rate of clustering is close to 97%, which fully meets the requirement of clustering identification in robotic harvesting.

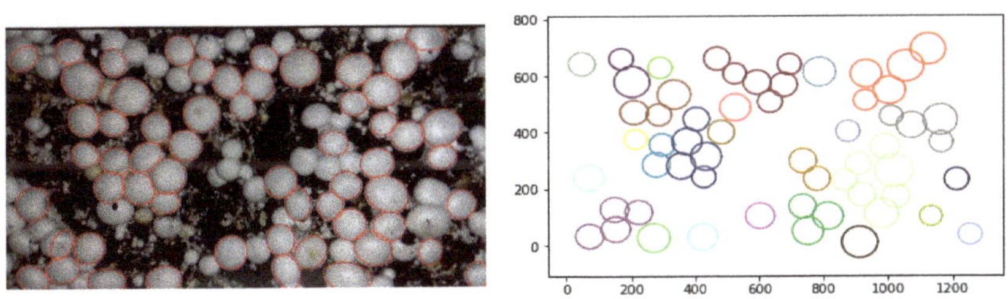

(a) (b)

Figure 12. The processing result of clustering algorithm processing. (**a**) visual identity map of ripe fruits; (**b**) The processing result of the improved clustering algorithm.

In order to further verify the effectiveness of the algorithm, this paper selects much more samples with different fruit numbers, cluster numbers, and discrete point numbers to conduct multiple sets of experiments and compares them with the commonly used clustering algorithm K-means algorithm and Gaussian mixture algorithm. The results are shown in Table 3. It shows that the K-means algorithm has the worst effect in processing the clustering of the fruits in this paper, whose average success rate is only 68%. Compared with the K-means algorithm, the Gaussian mixture algorithm is more flexible in the shape of the clustering, but it is more difficult to adapt to the characteristics of this paper with many small clusters and many discrete values, and the average success rate is merely 78%. However, the effect of the improved algorithm is much better than the other algorithm, with its average success rate is up to 97%. Additionally, as the number and complexity of clusters increase, the superiority of the improved clustering algorithm remains stable. So, the improved clustering algorithm is suitable to solve the clustering problem of straw-rotting fungus.

Table 3. Clustering algorithm comparison of many groups of samples.

Group	Num of Fruits	Num of Cluster	Num of Discrete Points	The Recognition Success Rate of Clustered Fruits		
				K-Means	Gaussian Mixture	Improved Clustering Algorithm
1	40	18	12	82%	88%	99%
2	40	25	20	78%	85%	99%
3	55	27	16	72%	80%	98%
4	55	38	27	68%	78%	96%
5	70	36	28	63%	73%	96%
6	70	45	36	57%	66%	95%
AVG	55	189	139	68%	78%	97%

5.2. Experiments of Multiobjective Optimization Algorithm for Multiarm Cooperative Harvesting Trajectory

Take the three-arm Agaricus bisporus harvesting robot as an example to verify the effect of the proposed approach, as shown in Figure 13.

Figure 13. The multiarm intelligent harvesting robot working in the multistory shelf trays in the factory environment.

Three sets of data containing 40, 55, and 70 fruits, respectively, are selected as the first experimental data, which are shown in Figure 14. The detailed harvesting information for the fruits to be harvested in Figure 14 is shown in Appendix A, where (X, Y, Z, C) is used to express the harvesting information for fruits. X, Y, and Z represent the coordinates of the center point of the fruit to be harvested. C indicates the cluster number the fruit should belong to, which can be obtained by the clustering algorithm proposed in this paper.

The proposed IGAACMO algorithm is used to optimize the harvesting trajectory of the real fruit data (Figure 14). Furthermore, the two-chromosome genetic algorithm (DCGA) and the genetic stepwise algorithm (GAGA) are also used to plan the trajectory of the three-arm robot with the experimental data to compare with the processing results of the algorithm proposed in this paper.

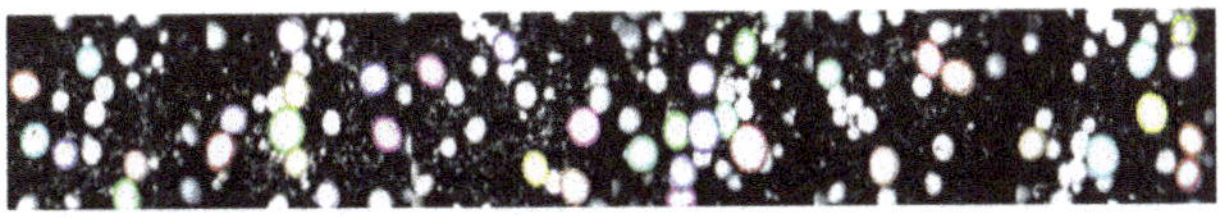

(a) Contains 40 fruits

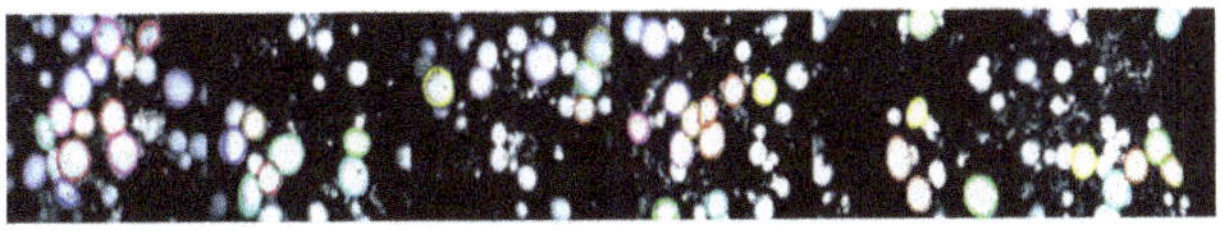

(b) Contains 55 fruits

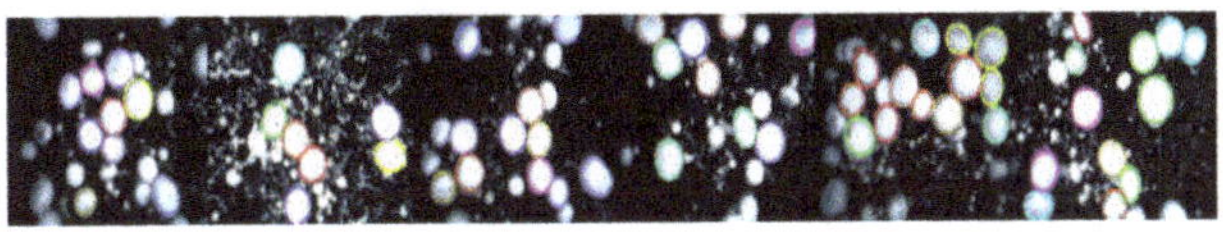

(c) Contains 70 fruits

Figure 14. Pictures of fruit harvesting area from planting factory. (**a**) contain 40 fruits; (**b**) contain 55 fruits; (**c**) contain 70 fruits.

The parameter settings are as follows: (1) the crossover probability is set to 0.15, the population is set to 30, the mutation probability is set to 0.015, and the maximum iteration number is set to 500. (2) In the ant colony algorithm, let 1 be the set of the important factors of pheromone, let 30 be the set of the number of ants, let 5 be the set of the intensity of pheromone, and let 10 be the set of the important factors of heuristic pheromone, let 0.1 be the set of the volatile factors and let 500 be the set of the maximum iteration number; (3) the moving speed of the harvesting arm (V) is given as 100 mm/s and the harvesting execution time (t1) is given as 5 s.

The convergence performance of the algorithm is shown in Figure 15. It indicates that when the picking scale is 40 (i.e., 40 fruits need to be picked), the iteration number of the proposed algorithm is about 50% less than that of DCGA, 67% less than that of GAGA, and the optimal harvesting time of GAAC is 14% better than DCGA and 11% better than GAGA; When the picking scale is 55, the iteration number of GAAC is about 67% less than that of DCGA, and 75% less than that of GAGA, and the optimal harvesting time of GAAC is 22% better than DCGA, and 15% better than GAGA; When the picking scale is 70, the iteration number of the proposed algorithms is about 28% less than that of GAGA, about 22% less than that of GAGA, and the optimal harvesting time of the proposed algorithm is 26% better than DCGA, and 19% better than GAGA. Therefore, compared with the other two methods, the convergence speed and optimization ability of the algorithm proposed in this paper are better.

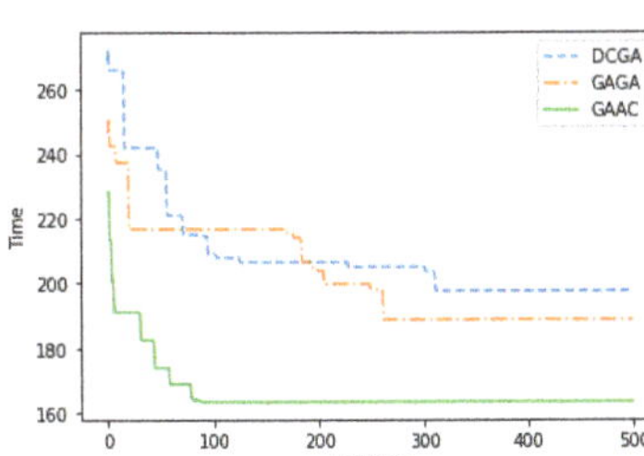

(**a**)

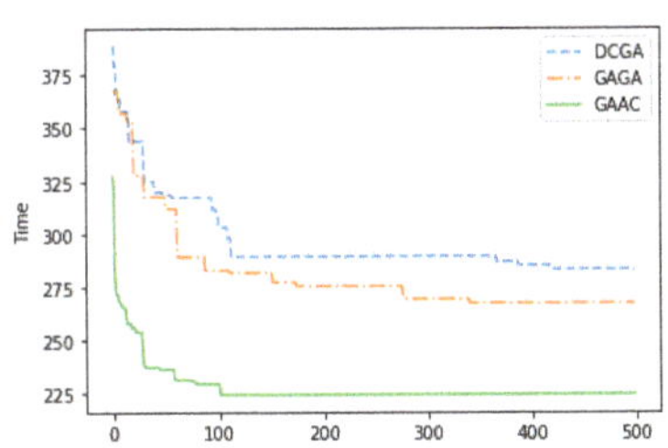

(**b**)

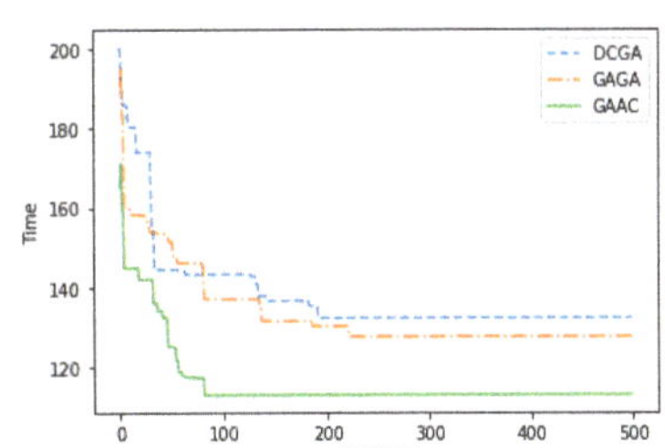

(**c**)

Figure 15. Comparison of convergence performance of three algorithms. (**a**) contain 40 fruits; (**b**) contain 55 fruits; (**c**) contain 70 fruits.

The other results obtained by the three algorithms and the important parameters are shown in Table 4. By comparing these parameters, the following can be seen: (1) The algorithm proposed in this paper has the best multiarm task distribution uniformity and the highest utilization of multiarm cooperation. The greater the number of fruits to be harvested, the more obvious the advantages compared with the other two algorithms; (2) The harvesting success rate after using the improved algorithm in this paper can always be guaranteed to be above 95%.

Table 4. Comparison of experimental results of the three algorithms.

Num of Fruits	Algorithm	Avoid Times	Harvesting Time(s)	Harvesting Efficiency(pcs/h)	Harvesting Success Rate
	DCGA	2	130.21	1106	96%
40	GAGA	1	126.54	1138	98%
	IGAACMO	0	113.81	1265	100%
	DCGA	5	196.16	1009	90%
55	GAGA	3	185.42	1068	94%
	IGAACMO	1	161.31	1227	98%
	DCGA	9	283.29	890	82%
70	GAGA	6	267.95	940	87%
	IGAACMO	2	224.64	1122	96%

The harvesting trajectory optimized by the IGAACMO algorithm is presented in Figure 16. The harvesting assignment task of each harvesting arm is relative balance, and there is basically no redundancy in the trajectories.

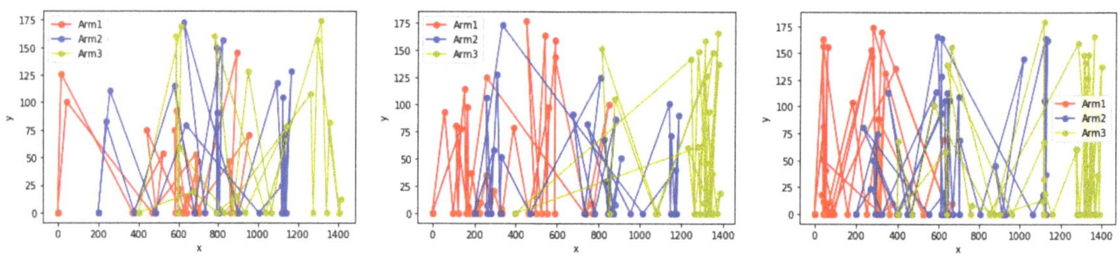

(a) (b) (c)

Figure 16. Trajectory diagram optimized by IGAACMO algorithm. (**a**) contain 40 fruits; (**b**) contain 55 fruits; (**c**) contain 70 fruits.

Because the larger the ratio of the number of clusters to the total number of fruits is, the more it will affect the performance of the algorithm. Another 10 more experiments are added to verify the stability of the proposed algorithm further. The number of ripe fruits ranges from 40 to 70, with the proportion of clusters ranging from 20% to 60% as well.

The results of the ten group experiments are shown in Table 5. The average harvesting efficiency optimized by the proposed algorithm is 1183 pcs/h, which is about 21% higher than that of the DCGA algorithm and about 15% higher than that of the GAGA algorithm. In the meantime, the average harvesting success rate is 97%, much better than the other two algorithms as well. All of the group results are basically consistent with Table 4. This indicates that the algorithm designed in this paper can achieve a better harvesting trajectory for the multiarm intelligent harvesting robot for fruits with different distributions.

Table 5. Comparison of the three algorithms with 10 groups data.

Group	Num of Fruits	Ratio of Cluster	Harvesting Efficiency (pcs/h)			Harvesting Success Rate		
			GAGA	DCGA	IGAACMO	GAGA	DCGA	IGAACMO
1	40	20%	1162	1182	1308	98%	99%	100%
2	42	38%	1112	1141	1258	97%	98%	100%
3	45	58%	1046	1093	1114	92%	95%	98%
4	49	21%	1068	1114	1238	95%	98%	100%
5	52	30%	1027	1074	1232	92%	95%	98%
6	57	45%	992	1034	1205	90%	94%	98%
7	60	60%	931	983	1161	85%	90%	95%
8	62	23%	945	1008	1186	86%	91%	97%
9	67	46%	892	947	1132	81%	87%	96%
10	70	57%	849	896	1106	78%	85%	95%
AVG	54.4	41%	977	1025	1183	88%	92%	97%

6. Discussion

The harvesting trajectory planning of a multiarm straw-rotting fungus harvesting robot is a typical NP-hard problem. It can be better optimized by the IGAACMO algorithm, which is proposed in this paper.

In terms of running speed, the IGAACMO is obviously superior to the other two methods (DCGA and GAGA). Moreover, the larger the processing scale (the more fruits to be picked), the greater the convergence advantage.

In terms of the optimization results, the amount and distribution of ripe fruits have an impact on the results. The algorithm is sensitive to the distribution density of the fruit to be picked. With the increase in fruit density, the picking efficiency will decrease. This shows that the closer the fruit distribution is, the more difficult it is to avoid a collision, which makes some picking arms have to wait and reduces the picking efficiency. However, compared with the other two algorithms, the optimization effect of the proposed algorithm is better under the same conditions, especially in the case of the fruit distribution with high density.

In particular, there is another important issue with the fruit cluster that needs to be harvested in a specific order. There are two ways to deal with this issue. One is to regard the fruits in the same cluster as a whole and assign them to the same arm to harvest them in a specific order, which is mostly adopted at present. The other is to allocate them to multiple different arms on the premise of ensuring the required harvesting order, which is an improved method proposed in this paper. The latter method is superior to the former one, especially when the distribution of fruits in each accessible area is seriously uneven, with large fruit clusters stretching across two different accessible areas as well, as shown in Figure 17. The red circle represents the fruit to be picked, and the black circle represents the immature fruit. The picking robot has three arms, and the working area is divided into five accessible areas, where S_1 (1,1) represents the exclusive area for Arm1, and the fruits in this area can only be harvested by arm1, S_2 (1,2) is the partial shared area that can only be harvested by Arm 1 and Arm 2, S_3(1,3) is the fully shared area that can be harvested by all three arms, S_4 (2,3) is the partial shared area that can be harvested by Arm 2 and Arm 3, and S_5(3,3) is the exclusive area that can only be harvested by arm 3. Most of the fruits to be picked are distributed in the exclusive area S_1 (1,1) and the partial shared area S_2 (1,2), and there is a large fruit cluster C1 over the two areas, meanwhile. The comparison of the results of the above two methods is shown in Table 6. All of the fruits in the fruit cluster C1 are allocated to Arm2 and the number of fruits allocated to Arm1 is very few by using the GAAC algorithm, which greatly increases the cycle time. However, by comparison, C1 is split and assigned to Arm1 and Arm2 respectively, resulting in a more uniform harvesting task among each arm, thereby improving the harvesting efficiency further.

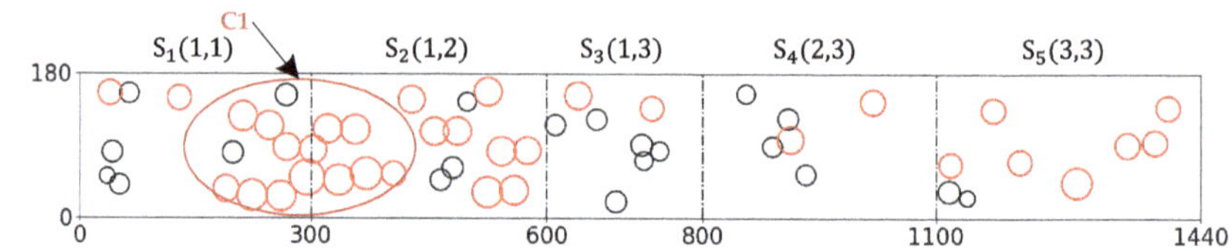

Figure 17. Diagram with serious uneven distribution of fruits.

Table 6. Comparison results of different allocation methods for fruit clusters.

Algorithm	Avoid Times	Harvesting Time(s)	Harvesting Success Rate	N1:N2:N3
GAAC (assigned to the same arm)	1	126.29	100%	03:17:14
IGAACMO (assigned to different arms)	1	105.25	100%	13:12:09

It can be shown that when the distribution of mature fruits is seriously ununiformly, with some fruit clusters across multiple accessible areas as well, it is easy to cause the uneven task assigned to each arm by assigning the total fruits in a cluster merely to the same arm, which results in some arms waiting for no picking tasks meanwhile other arms have too many picking tasks. This will greatly increase the cycle time of harvesting, thereby seriously reducing the picking efficiency. The algorithm proposed in this paper, combined with the auction mechanism, can allocate the fruits in a cluster to different arms on the premise of ensuring the required harvesting order instead of allocating them to a single arm, which can resolve this issue appropriately. Therefore, it can be concluded from all the above discussions that the algorithm proposed in this paper has strong optimization ability and good stability. For fruits with different densities, the picking tasks for each arm can be evenly distributed even though the fruits are not uniformly distributed on the culture medium or soil, with some fruit clusters across multiple accessible areas as well; thereby, it can not only achieve higher harvesting efficiency but also a higher success rate. The algorithm can better adapt to the issues of dense and uneven distribution of fruits caused by the natural growth of straw-rotting fungus.

7. Conclusions

This paper takes a straw-rotting fungus multiarm harvesting robot as the research object. Aiming at the problem of uniform task allocation and sequential harvesting for clustered mature fruits in multiarm cooperative harvesting trajectory optimization, an improved multiobjective optimization algorithm, IGAACMO, is proposed. The multiarm cooperative harvesting trajectory planning is abstracted to an MTSP problem. We use an improved local density bi-directional clustering algorithm to identify the clustered fruits to provide preparation for harvesting the clustered mature fruits in the specified order; Then, GA is adopted to decompose the MTSP into m independent TSP problems, where a new DNA coding method is designed to make the harvesting task of each harvesting arm evenly distributed under the constraining of the SSIS area. Subsequently, we use the ant colony algorithm to plan the trajectory of the above M-independent TSP, respectively; Here, by combining with the auction mechanism, the clustered fruits can be planned to be harvested in their specified order.

From all the above experiments and discussion, it can be shown that the optimization ability of the proposed algorithm, IGAACMO, is significantly stronger than the other two methods. The average harvesting efficiency optimized by the proposed algorithm is up to 1183 pcs/h, and the average harvesting success rate is 97%.

In addition, since the hourly harvesting efficiency of the multiarm robot has reached the manual efficiency, the daily harvesting efficiency of the robot will be significantly higher than the manual, even if it can be up to at least twice that of the manual. Because the robot can work for at least 16 h per day (considering battery replacement, layer change, and other auxiliary work), while people generally work for 8 h per day. This efficiency greatly increases the feasibility of the robot applied to the actual harvesting of straw-rotting fungus instead of manual.

However, the operation time of the algorithm is not faster enough. In future research, the algorithm needs to be improved to increase its efficiency of the algorithm to improve its real-time control further.

Author Contributions: S.Y.: Writing—original draft, writing—review and editing, conceptualization, methodology, investigation, data curation, formal analysis, software, validation, production of related equipment, funding acquisition. B.J.: writing—original draft, writing—review and editing, data curation, validation, software, production of related equipment. T.Y.: investigation, conceptualization, project administration. J.Y.: investigation, writing-review and editing, conceptualization. All authors have read and agreed to the published version of the manuscript.

Funding: This research was funded by the Shanghai Science and Technology Innovation Action Plan-Agriculture, Grant No. 21N21900600; Shanghai Agriculture Applied Technology Development Program, Grant No. 2019-02-08-00-10-F01123; Major Scientific and Technological Innovation Project of Shandong Province, Grant No. 2022CXGC010609-7; Mechanical Engineering (Intelligent Manufacturing) Plateau Discipline of Shanghai Polytechnic University.

Institutional Review Board Statement: Not applicable.

Informed Consent Statement: Not applicable.

Data Availability Statement: Anyone can access our data by sending an email to szyang@sspu.edu.cn.

Acknowledgments: We acknowledge the Shanghai Lianzhong edible fungus professional cooperative for the use of their mushrooms and facilities.

Conflicts of Interest: The authors declare no conflict of interest.

Appendix A

Table A1. The detail data with the information of the fruits to be harvested in Figure 14.

Group	Number of Fruits	Detail Information of Fruits
a	40	(168.43, 136.53, 163.29, 0.0), (72.03, 129.46, 170.47, 0.0),(130.93,121.4,165.45,0.0), (183.02,162.07,149.93,0),(75.41,10.54,170.04,0),(33.48,35.3,190.66,0),(11.47,98.3,212.67,0), (84.48,168.46,173.35,0),(62.15,75.78,183.99,0),(13.53,133.29,188.81,0),(124.38,8.55,159.71,0), (406.7,39.63,143.94,0),(431.28,77.04,134.9,0),(288.84,157.6,169.56,0),(296.91,43.28,170.86,0), (399.99,87.99,158.08,0),(332.95,7.87,162.36,0),(262.74,170.84,160.23,0),(325.74,175.71,168.02,0), (514.44,43.76,156.76,0),(702.29,35.59,172.61,0),(646.05,34.69,171.18,0), (763.04,101.5,154.61,0), (874.45,137.36,156.47,0),(892.59,65.13,159.9,0),(825.83,162.81,161.57,0),(741.06,76.6,167.01,0), (1087.35,28.05,152.2,0),(1049.77,102.33,159.7,0),(1128.99,53.9,166.76,0),(1165.21,13.61,173.77,0), (1013.76,10.89,175.49,0),(1167.15,171.3,161.97,1),(1054.98,64.92,169.91,1),(1385.3,155.13,173.38,2), (1312.73, 160.07, 175.86, 2.0), (1239.82, 172.1, 176.6, 3.0), (1274.88, 117.37, 179.35, 3.0), (1236.71, 9.1, 168.9, 4.0), (1347.87, 167.54, 160.21, 4.0)
b	55	(56.71,93.06,163.55,0.0),(167.53,97.5,155.25,0.0),(132.8,114.33,156.23,0.0),(134.06,77.44,135.51,0.0), (229.94,9.72,159.53,0),(111.97,80.42,142.6,0),(172.3,6.49,160.79,0),(186.14,36.68,153.34,0), (309.16,127.35,170.77,0),(257.31,124.79,170.59,0),(258.59,34.41,172.91,0),(294.92,20.13,170.75,0), (293.61,58.17,178.57,0),(387.83,78.1,139.3,0),(326.2,142.62,171.57,0),(332.02,172.98,173.62,0), (331.68,4.6,174.6,0),(449.56,176.44,179.32,0),(260.4,6.76,165.98,0),(256.9,106.31,250.58,0), (330.21,51.61,159.84,0),(558.02,96.99,154.04,0),(675.93,90.7,158.23,0),(538.99,162.97,171.9,0), (590.67,158.46,147.43,2),(587.62,143.37,160.68,2),(846.49,100.02,145.07,0),(813.59,151.48,152.09,0), (823.82,67.78,155.26,0),(807.23,124.87,148.15,0),(877.89,104.9,155.6,4),(878.81,86.16,155.87,4), (832.51,29.5,159.6,0),(908.39,50.74,169.97,0),(879.47,8.38,160.7,0),(741.88,81.64,184.95,0),

Table A1. *Cont.*

Group	Number of Fruits	Detail Information of Fruits
c	70	(1186.72,90.04,162.32,3),(1140.91,100.87,159.13,3),(1172.68,40.46,169.81,3),(1150.34,71.37,160.56,3), (1347,148.08,151.3,0),(1315.87,126.89,158.6,0),(1282.68,61.11,149.35,0),(1278.79,148.58,163.01,0), (1312.53,159.13,147.72,0),(1316.57,45.91,154.21,0),(1231.16,60.71,167.48,0),(1350.77,36.87,165.88,0), (1308.44,78.13,163.01,0),(1245.3,141.76,174.2,0),(1373.8,166.04,151.64,0),(1384.82, 18.79, 172.77, 4.0), (1378.14,137.58,162.09,4.0),(1329.02, 93.26, 136.63, 5.0), (1309.99, 9.79, 173.65, 5.0) (36.54,51.33,173.4,0.0),(34.66,18.28,172.73,0)(38.94,163.27,161.22,0.0),(39.91,156.28,169.74,0.0), (43.31,81.47,159.44,0.0),(43.32,12.06,168.63,0.0),(52.59,40.87,167.5,0),(61.23,155.52,153.22,0), (62.45,9.62,183.5,0),(181.64,103.87,144.54,0),(230.97,80.64,171.97,0),(268.83,153.14,161.5,0), (270.25,23.83,174.2,0),(273.84,146.9,172.97,0),(284.45,49.35,157.29,0),(291.88,60.67,169.3,0), (299.04,74.55,181.48,0),(351.08,113.39,162.1,0),(372.31,9.62,153.22,0),(386.24,135.54,167.47,0), (402.12,68.05,207.25,0),(574.22,101.72,166.15,0),(586.16,114.01,158.9,0),(590.18,165.77,157.53,0), (611.49,163.4,160.1,0),(626.03,69.38,172.7,0),(628.9,19.62,175.28,0),(631.23,105.9,183.47,0), (633.71,59.05,175.13,0),(635.86,18.45,164.45,0),(640.67,139.46,150.91,0),(641.93,113.39,152.45,0), (652.2,105.64,172.64,0),(661.64,155.52,169.74,0),(669.45,9.62,178.28,0),(699.63,109.9,172.31,0), (705.25,69.05,267.25,0),(765.4,8.45,166.08,0),(878.4,45.15,168.06,0),(1018.96,145.14,172.04,0), (1114.33,12.86,174.04,0),(1115.94,179,177.31,0),(1116.67,32.84,154.18,0),(1117.28,105.43,159.69,0), (1118.38,66.59,162.1,0),(1122.56,18.75,153.22,0),(1122.93,163.7,183.5,0),(1130.49,162.3,239.41,0), (1274.22,60.15,153.1,0),(1280.38,60.58,168.15,0),(1283.99,159.01,147.51,0),(1320.04,45.58,155.63,0), (1331.13,126.89,158.6,0),(1334.07,148.09,151.35,0),(1341.03,78.13,163.01,0),(1363.14,166.09,151.68,0), (1378.89,36.51,167.16,0),(1380.81,94.34,174.28,0),(1397.66,137.94,163.47,0),(277.82,174.17,165.13,1), (318.43,169.18,164.42,1),(1127.94,37.02,146.94,2),(1115.54,18.48,164.17,2),(1311.72,148.48,163.4,3), (1317.97, 141.62, 173.72, 3.0), (1325.74, 8.28, 177.16, 3.0), (336.15, 131.55, 249.33, 4.0), (305.25, 88.58, 157.51, 4.0), (611.4, 128.39, 144.99, 5.0), (612.86, 94.02, 152.85, 5.0)

References

1. Guo, H.; Ren, P.; Huang, C. Green prevention and control technology of straw rotting fungus. *Edible Med. Mushrooms* **2019**, *3*, 218–221.
2. Noble, R.; Reed, J.N.; Miles, S.; Jackson, A.F.; Butler, J. Influence of mushroom strains and population density on the performance of a robotic harvester. *J. Agric. Eng. Res.* **1997**, *52*, 215–222. [CrossRef]
3. Schiau, H.G. Considerations on the evolution of mushrooms harvesting systems. *Res. Sci. Today* **2013**, *7*, 170–182.
4. Reed, J.N.; Tillett, R.D. Initial experiments in robotic mushroom harvesting. *Mechatronics* **1994**, *4*, 265–279. [CrossRef]
5. Reed, J.N.; Mrilles, S.J.; Butler, J.; Baldwin, M.; Noble, R. Automatic mushroom harvester development. *J. Agric. Eng. Res.* **2001**, *78*, 15–23. [CrossRef]
6. Yu, G.H.; Luo, J.M.; Zhao, Y. Region labeling technology and mushroom image segmentation method based on sequential scanning algorithm. *J. Agric. Eng.* **2006**, *4*, 139–142.
7. Lu, C.P.; Liaw, J.J.; Wu, T.C.; Hung, T.F. Development of a Mushroom Growth Measurement System Applying Deep Learning for Image Recognition. *Agronomy* **2019**, *9*, 32. [CrossRef]
8. Kuchinskiy, N.A. Development of an Autonomous Robotic Mushroom Harvester. Master's Thesis, The University of Western Ontario, London, ON, Canada, 2016.
9. Huang, M.; He, L.; Choi, D.; Pecchia, J.; Li, Y. Picking dynamic analysis for robotic harvesting of *Agaricus bisporus* mushrooms. *Comput. Electron. Agric.* **2021**, *32*, 185–256. [CrossRef]
10. Lu, W.; Wang, P.; Wang, L.; Deng, Y. Design and experiment of flexible gripper for mushroom non-destructive picking. *Trans. Chin. Soc. Agric. Mach.* **2020**, *51*, 28–36.
11. Yang, S.; Ni, B.; Du, W.; Yu, T. Research on an improved segmentation recognition algorithm of overlapping *Agaricus bisporus*. *Sensors* **2022**, *22*, 3946. [CrossRef]
12. Huang, M.; Jiang, X.; He, L.; Choi, D.; Pecchia, J.; Li, Y. Development of a robotic harvesting mechanism for button mushroom. *Trans. ASABE* **2021**, *64*, 565–575. [CrossRef]
13. Mohanan, M.G.; Salgaonkar, A. Robotic mushroom harvesting by employing probabilistic road map and inverse kinematics. *BOHR Int. J. Future Robot. Artif. Intell.* **2021**, *1*, 1–10. [CrossRef]
14. Yang, Y. Key Techniques for Automatic Mushroom Picking System Based on Machine Vision. Master's Thesis, Nanjing University of Aeronautics and Astronautics, Nanjing, China, 2019.
15. Hu, X.M.; Wang, C.; Yu, T. Design and application of visual system in the *Agaricus bisporus* picking robot. *JOP Conf. Ser.* **2019**, *1187*, 032034.
16. Tinoco, V.; Silva, M.F.; Santos, F.N.; Rocha, L.F.; Santos, L.C. A review of pruning and harvesting manipulators. In Proceedings of the 2021 IEEE International Conference on Autonomous Robot Systems and Competitions, ICARSC 2021, Santa Maria da Feira, Beja, Portugal, 28–29 April 2021; pp. 155–160.

17. Nas, B.; Nrv, A.; Vsh, A.; Vd, A. Cotton harvester through the application of machine learning and image processing techniques. *Mater. Today Proc.* **2021**, *47*, 2200–2205.
18. Zion, B.; Mann, M.; Levin, D.; Shilo, A.; Rubinstein, D.; Shmulevich, I. Harvest-order planning for a multiarm robotic harvester. *Comput. Electron. Agric.* **2014**, *103*, 75–80. [CrossRef]
19. Xiong, Y.; Ge, Y.; Grimstad, L.; From, P.J. An autonomous strawberry-harvesting robot: Design, development, integration, and field evaluation. *J. Field Robot.* **2020**, *37*, 202–224. [CrossRef]
20. Rong, J.; Wang, P.; Yang, Q.; Huang, F. A field-tested harvesting robot for oyster mushroom in greenhouse. *Agronomy* **2021**, *11*, 1210. [CrossRef]
21. Barnett, J.; Duke, M.; Chi, K.A.; Shen, H.L. Work distribution of multiple cartesian robot arms for kiwifruit harvesting. *Comput. Electron. Agric.* **2020**, *169*, 105202. [CrossRef]
22. Jia, B.C.; Yang, S.Z.; Yu, T. Research on three picking arm avoidance algorithms for agaricus mushroom picking robot. In Proceedings of the 2020 IEEE International Conference on Advances in Electrical Engineering and Compute Applications, AEECA 2020, Dalian, China, 25–27 August 2020; pp. 325–328.
23. Kurtser, P.; Edan, Y. Planning the sequence of tasks for harvesting robots. *Robot. Auton. Syst.* **2020**, *131*, 103591. [CrossRef]
24. Venkatesh, P.; Singh, A. Two metaheuristic approaches for the multiple traveling salesperson problem. *Appl. Soft Comput.* **2015**, *26*, 74–89. [CrossRef]
25. Zhang, H. The Area Divison of Express Delivery Terminal and Vehicle Routing Problem with Balance. Master's Thesis, Beijing Jiaotong University, Beijing, China, 2019.
26. Yin, Y. Research on Path Optimization of Warehouse Robot Based on Improved Genetic Algorithm. Master's Thesis, Chengdu University of Technology, Chengdu, China, 2019.
27. Xu, X.; Yuan, H.; Mark, L.; Marcello, T. Two phase heuristic algorithm for the multiple-travelling salesman problem. *Soft Comput.* **2018**, *22*, 6567–6581. [CrossRef]
28. Liu, W.; Wang, Y. Path planning of multi-UAV cooperative search for multiple targets. *Electron. Opt. Control.* **2019**, *26*, 35–38.
29. Dong, X.; Cai, Y. A novel genetic algorithm for large scale colored balanced traveling salesman problem. *Future Gener. Comput. Syst.* **2019**, *95*, 727–742. [CrossRef]
30. Luo, Z.; Feng, S.; Liu, X.; Che, J.; Wang, R. Method of area coverage path planning of multi-unmanned cleaning vehicles based on step by step genetic algorithm. *J. Electr. Measur. Instr.* **2020**, *34*, 43–50.
31. Wang, M. Cooperative Task Assignment and Path Planning for Multi-Uavs Bird-Driving System at the Airport. Master's Thesis, Nanjing University of Aeronautics and Astronautics, Nanjing, China, 2019.
32. Lu, Z.; Zhang, K.; He, J.; Niu, Y. Applying K-means clustering and genetic algorithm for solving MTSP. Proceedings of Bio-inspired Computing–Theories and Applications-11th International Conference, BIC-TA 2016, Xi'an, China, 28–30 October 2016; pp. 278–284.
33. Kencana, E.N.; Harini, I.; Mayuliana, K. The performance of ant system in solving multi traveling salesmen problem. *Proc. Comput. Sci.* **2017**, *124*, 46–52. [CrossRef]
34. Necula, R.; Breaban, M.; Raschip, M. Tackling the bi-criteria facet of multiple traveling salesman problem with ant colony systems. In Proceedings of the IEEE 27th International Conference on Tools with Artificial Intelligence, Vietrisul Mare, Italy, 9–11 November 2015; pp. 873–880.
35. Changdar, C.; Pal, R.K.; Mahapatra, G.S. A genetic ant colony optimization based algorithm for solid multiple travelling salesmen problem in fuzzy rough environment. *Soft Comput.* **2017**, *21*, 4661–4675. [CrossRef]
36. Yang, S.Z.; Jia, B.C.; Yu, T. Research on Path Planning Algorithm of Multi-Arm Collaborative Picking for *Agaricus Bisporus* Intelligent Picking Robot. In Proceedings of the 5th IEEE/IFToMM International Conference on Reconfigurable Mechanisms and Robots, Toronto, ON, Canada, 12–14 August 2021; pp. 776–796.
37. Cao, X.; Yan, H.; Huang, Z.; Ai, S.; Xu, Y.; Fu, R.; Zou, X. A Multi-Objective Particle Swarm Optimization for Trajectory Planning of Fruit Picking Manipulator. *Agronomy* **2021**, *11*, 2286. [CrossRef]
38. Xu, H.; Hu, Z.; Zhang, P.; Gu, F.; Wu, F.; Song, W.; Wang, C. Optimization and Experiment of Straw Back-Throwing Device of No-Tillage Drill Using Multi-Objective QPSO Algorithm. *Agriculture* **2021**, *11*, 986. [CrossRef]
39. Saxena, A.; Prasad, M.; Gupta, A.; Bharill, N.; Patel, O.P.; Tiwari, A.; Er, M.J.; Ding, W.; Lin, C.T. A review of clustering techniques and developments. *Neurocomputing* **2017**, *267*, 664–681. [CrossRef]
40. Sisodia, D.; Sisodia, S.; Saxena, K. Clustering techniques: A brief survey of different clustering algorithms. *Int. J. Latest Trends Eng. Technol.* **2012**, *1*, 82–87.
41. Karna, A.; Gibert, K. Automatic identification of the number of clusters in hierarchical clustering. *Neural. Comput. Appl.* **2022**, *34*, 119–134. [CrossRef]
42. Boonchoo, T.; Ao, X.; Liu, Y.; Zhao, W.; He, Q. Grid-based dbscan: Indexing and inference. *Pattern Recognit.* **2019**, *90*, 271–284. [CrossRef]
43. Liu, F.; Wen, P.; Zhu, E. Efficient grid-based clustering algorithm with leaping search and merge neighbors method. *JOP Conf. Ser.* **2017**, *1*, 12122–12132. [CrossRef]
44. Melnykov, V. Challenges in model-based clustering. *WIREs Comput. Stat.* **2013**, *5*, 135–148. [CrossRef]

45. Bechini, A.; Marcelloni, F.; Renda, A. Tsf-dbscan: A novel fuzzy density-based approach for clustering unbounded data streams. *IEEE Trans. Fuzzy Syst.* **2022**, *30*, 623–637. [CrossRef]
46. Zhu, Y.; Ting, K.M.; Jin, Y.; Angelova, M. Hierarchical clustering that takes advantage of both density-peak and density-connectivity. *Inform. Syst.* **2022**, *103*, 101871. [CrossRef]

agriculture

Article

Design and Performance Test of a Jujube Pruning Manipulator

Bin Zhang [1,2], Xuegeng Chen [3], Huiming Zhang [2], Congju Shen [3,4] and Wei Fu [1,2,*]

1 School of Information and Communication Engineering, Hainan University, Haikou 570228, China; 21110810000031@hainanu.edu.cn
2 Mechanical and Electrical Engineering College, Hainan University, Haikou 570228, China; 994330@hainanu.edu.cn
3 College of Mechanical and Electrical Engineering, Shihezi University, Shihezi 832003, China; chenxg130@shzu.edu.cn (X.C.); shencongju@stu.shzu.edu.cn (C.S.)
4 Xinjiang Academy of Agricultural and Reclamation Science, Shihezi 832003, China
* Correspondence: 994026@hainanu.edu.cn

Abstract: To solve the problems of poor working conditions and high labor intensity for artificially pruning jujube trees, a pruning scheme using a manipulator is put forward in the present paper. A pruning manipulator with five degrees of freedom for jujube trees is designed. The key components of the manipulator are designed and the dimension parameters of each joint component are determined. The homogeneous transformation of the DH parameter method is used to solve the kinematic equation of the jujube pruning manipulator, and the kinematic theoretical model of the manipulator is established. Finally, the relative position and attitude relationship among the coordinate systems is obtained. A three-dimensional mathematical simulation model of the jujube pruning manipulator is established, based on MATLAB Robotics Toolbox. The Monte Carlo method is used to carry out the manipulator workspace simulation, and the results of the simulation analysis show that the working space of the manipulator is −600~800 mm, −800~800 mm, and −200~1800 mm in the X, Y, and Z direction, respectively. It can be concluded that the geometric size of the jujube pruning manipulator meets the needs of jujube pruning in a dwarf and densely planted jujube garden. Then, based on the high-speed camera technology, the performance test of the manipulator is carried out. The results show that the positioning error of the manipulator at different pruning points of jujube trees is less than 10 mm, and the pruning success rate of a single jujube tree is higher than 85.16%. This study provides a theoretical basis and technical support for the intelligent pruning of jujube trees in an orchard.

Keywords: jujube pruning; manipulator; kinematic analysis; high-speed photography technology; performance test

Citation: Zhang, B.; Chen, X.; Zhang, H.; Shen, C.; Fu, W. Design and Performance Test of a Jujube Pruning Manipulator. *Agriculture* **2022**, *12*, 552. https://doi.org/10.3390/agriculture12040552

Academic Editor: Francesco Marinello

Received: 6 March 2022
Accepted: 8 April 2022
Published: 12 April 2022

1. Introduction

As one of the endemic tree species in China, jujube ranks first in the world in terms of the planting area and yield [1]. With its unique geographical and climatic conditions, Xinjiang has become the main production area in China [2]. By 2020, the planting area for jujube in Xinjiang is about 445,225 ha, and the output is up to 3,727,729 t [3]. The pruning of jujube trees is an important part of jujube orchard management, because it improves nutrient digestion and absorption, adjusts the tree's structure, extends the tree's life, and improves the yield and quality of the jujube tree [4,5]. At present, the pruning of jujube trees is mainly carried out manually, which causes significant problems, such as poor operating conditions, high labor intensity, low work efficiency, and high labor costs [6]. Therefore, it is an inevitable trend to develop a high degree of automation for pruning manipulators to replace manual pruning.

Recently, the manipulators were widely used in the field of agricultural picking, plant protection, and other orchard management links [7–10]. Li et al. designed a multi-terminal manipulator for apple picking, which cut off the fruit's stem via blade rotation and a toothed fruit collector, and the position error of the manipulator end was less than 9 mm [11]. Zhao et al. developed an apple harvesting robot that adopted a 5-DOF manipulator with a PRRRP structure and an end-effector with a spoon-shaped pneumatic gripper, for which the harvesting success rate was 77% [12,13]. Henten et al. designed a 7-DOF manipulator for cucumber picking, and the cutting device of the end-effector used medical thermal cutting technology to pick the cucumbers, with a picking success rate of 74% [14,15]. Bac et al. developed a 9-DOF manipulator system for picking sweet peppers, and the picking success rate reached 84% [16–19]. In the field of pruning, shaping and pruning machinery is mostly studied [20,21]. Domestic and foreign researches on intelligent pruning robots are basically in the laboratory research stage [22,23]. The typical foreign research cases are as follows: Kawasaki et al. developed a new robot for climbing pruning that could perform climbing pruning quickly [24]. Soni et al. designed a 9-DOF pruning robot for climbing areca, and the 5-DOF PUMA manipulator was able to complete the pruning of areca branches with different diameters [25]. Botterill et al. developed a pruning robot for grape trees that took approximately 2 min to prune a single grape tree, and the target estimation error was within 1% [26]. Zahid et al. designed a pruning robot for apple trees planted within a hedge. The 3-DOF end-effector was integrated into the Cartesian mechanical arm, which could cut 25 mm fruit-tree branches [27,28]. Zahid et al. studied the obstacle avoidance trajectory planning of the developed 6-DOF apple pruning manipulator, which provided the research foundation and technical support for the pruning robot to realize intelligent pruning [29]. Van Marrewijk et al. developed a new pruning robot, which could prune spherical, cylindrical, and rectangular shapes of horticultural plants [30]. The typical domestic studies mainly include the following: Chai et al. designed a pruning robot for green fences with a 14-DOF body structure based on the exoskeleton [31]. Luo et al. conducted a study on obstacle avoidance by the arm of a pruning robot for green fences [32]. Li and Chen et al. studied the motion characteristics of a pruning robot for green fences [33,34]. Huang et al. designed a cylindrical coordinate pruning robot for loquat, for which the average pruning and crushing times of a single branch was approximately 55 s [35,36]. Wu et al. designed a high-branch pruning manipulator with a pruning height of 5–20 m, a maximum pruning radius of 5 m, and a maximum pruning diameter of 12 cm [37]. To sum up, the manipulators are mainly used for agricultural fruit and vegetable harvesting, and in the field of agricultural pruning, the pruning robots are mainly studied for the single pruning way of forest trees and green fences. However, due to the great diversity of fruit-tree growth information, different regional pruning requirements, and the unstructured orchard working environment, there are few studies on the technology of orchard pruning robots. More specifically, the research on pruning robots for jujube is rarely reported.

Consequently, a jujube pruning manipulator is designed in this paper; the theoretical model of kinematics for the manipulator is established; the three-dimensional simulation model of the jujube pruning manipulator is generated based on the MATLAB Robotics Toolbox; the Monte Carlo method is used to verify the workspace simulation of the manipulator; and, finally, the performance test of the manipulator prototype is carried out. The results provide a foundation for the research and technical support for the intelligent pruning of the trees in jujube orchards.

2. The Design of the Jujube Pruning Manipulator

2.1. Structure Composition and Working Principle

2.1.1. Structure Composition

The body structure of the pruning manipulator for jujube is mainly composed of a machine arm with 5 degrees of freedom (5-DOF), an end-effector, and a control system. Among them, the 5-DOF manipulator is mainly composed of the foundation support, the rotary joint of the foundation support, the machine body, the mobile joint of the machine body, the shoulder joint, the big arm, the elbow joint, the rotary joint of the forearm, and the forearm. The shear end-effector is mainly composed of a moving cutter and a stationary cutter. The control system is mainly composed of a lower control system and an upper man–machine interface. The diagram for the structure of the overall machine is shown in Figure 1.

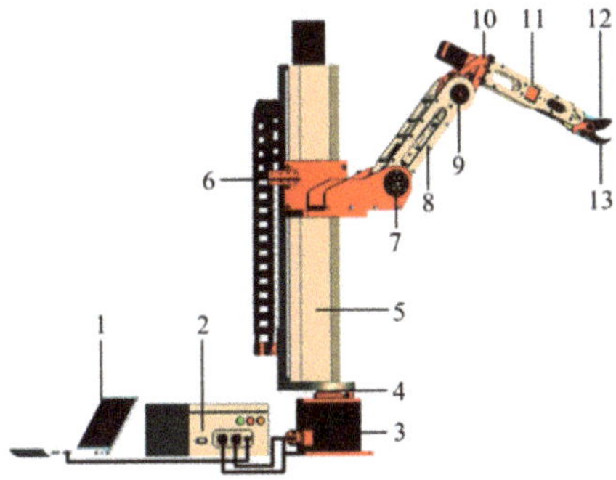

Figure 1. Schematic diagram of the structure composition for the manipulator. 1. PC machine; 2. Control box; 3. Foundation support; 4. Rotary joint of the foundation support; 5. Machine body; 6. Mobile joint of the machine body; 7. Shoulder joint; 8. Big arm; 9. Elbow joint; 10. Rotary joint of the forearm; 11. Forearm; 12. Moving cutter; and 13. Stationary cutter.

2.1.2. Working Principle

When the manipulator is working, the upper computer of the control system in the teaching mode obtains the coordinate information of the pruning points for the jujube, according to the experience and knowledge obtained from the jujube farmers, and sends them to the lower computer of the manipulator control system. After the lower controller of the control system receives the location information instruction for the coordinates of the jujube branches that need to be pruned, the motor of each joint of the manipulator arm is controlled to rotate correspondingly, according to the forward and inverse kinematics analysis data, so that the manipulator reaches the target pruning point for pruning. According to the pruning point information recorded in the man–machine teaching mode of the upper computer, the manipulator is controlled to arrive at each target pruning point in turn for pruning. After the pruning of all the pruning branches has been completed, the manipulator is reset.

2.2. The Design of the Mechanical Arm

2.2.1. Structure Design

The structural forms of the manipulator mainly include the type of cylindrical coordinate, polar coordinate, rectangular coordinate, and joint coordinate [38]. The joint coordinate manipulator is similar to the human arm, and it has the advantages of a compact structure, flexible movement, large working space, and small occupation area. To simulate the manual pruning process, the joint coordinates were selected to design the pruning manipulator for dwarf and densely planted jujube trees in Xinjiang.

When manually pruning jujube trees, farmers hold pruning scissors through the coordination and cooperation of each joint for pruning. Therefore, when designing the manipulator, three rotating joints were used to determine the position of the target pruned branches. To meet the standards for pruning jujube trees at different heights and branches at different positions, manual pruning needs to be supplemented by a long ladder. Therefore, a movement joint was used to realize the function of moving up and down. In addition, a rotary joint should be added at the end of the manipulator to adjust the attitude of the end-effector to facilitate the pruning. Finally, the 5-DOF mechanical arm can meet the pruning requirements of jujube trees. The designed manipulator consists of four rotary joints and one mobile joint. The four rotary joints are the rotary joint of the foundation support, shoulder, elbow, and forearm, and one mobile joint is the mobile joint of the machine body. The structure and motion direction of each joint for the manipulator are shown in Figure 2. The rotary joint of the foundation support can rotate left and right about the Z axis, and it drives all other joint movements along with it when it turns. The Z axis is perpendicular to the horizontal plane (XOY plane) and moves upwards vertically. The mobile joint of the machine body moves up and down the Z axis, and it drives the rotary joints of the shoulder, elbow, and forearm movement along with it when it moves. The shoulder joint can rotate up and down around the l_1 axis, parallel to the horizontal plane (XOY plane). When it rotates, it will drives the elbow and forearm movements together. The elbow joint can rotate up and down about the l_2 axis, which is parallel to the l_1 axis, and when it moves, it drives the rotation joint of the forearm movement together. The rotation joint of the forearm rotates around the m axis, and the m and l_2 axes are perpendicular to each other on different planes. When it moves, it drives the attitude of the end-effector to change.

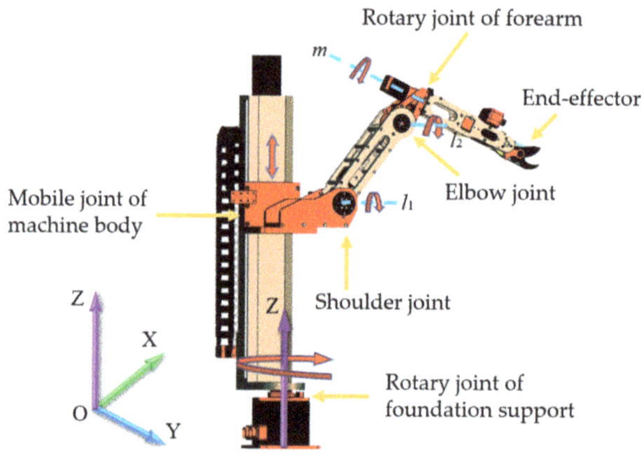

Figure 2. The structure and motion direction of each joint for the manipulator.

The rotating joint of the base drives the overall machine to realize the azimuth adjustment and expand the target working area in the horizontal direction (XOY plane). The mobile joint of the body adjusts the manipulator at different heights by moving up and down to expand the target working area in the vertical direction (Z axis), and adapt to the pruning of jujube trees at different heights. The shoulder and elbow joints coordinate with the base joint, and the body joint is used to locate the branches at different positions and adjust the end-effector pruning posture in real time through the forearm rotation joint to adapt to jujube branches with different growth postures.

2.2.2. The Parameters Design of the Links Dimension

The link size parameters of each joint for the manipulator were determined by the size information of the jujube tree before and after pruning. Therefore, a field investigation was carried out on jujube trees from 2 to 8 years old in dwarf and densely planted jujube gardens in Xinjiang, and the size information of the jujube trees before and after pruning was obtained by actual measurements. The specific size parameters of the jujube tree growth information are as follows: the row space of jujube trees is generally 3000 mm; the plant space is 800~1000 mm; the height of the jujube trees is generally 1500~2500 mm; the diameter of the canopy is 1000~1800 mm; the height of the canopy is 1200~2000 mm; and the height range of the main branch is 300~500 mm. According to the agronomic requirements of jujube pruning, the height of the canopy after pruning is between 800~1600 mm and the diameter of the canopy is between 600~1400 mm. The area formed by the maximum diameter and height of the jujube canopy is rotated around the direction of its trunk to form a cylinder, which envelopes all of the branches of the jujube tree. In combination with the growth information of the jujube trees, the target pruning space of the jujube trees is analyzed. The manipulator is placed on a mobile chassis with a height of 400 mm, and the horizontal distance between the main stem of the jujube tree and the base of the manipulator is 1000 mm. The analysis for the target pruning space of the manipulator is shown in Figure 3.

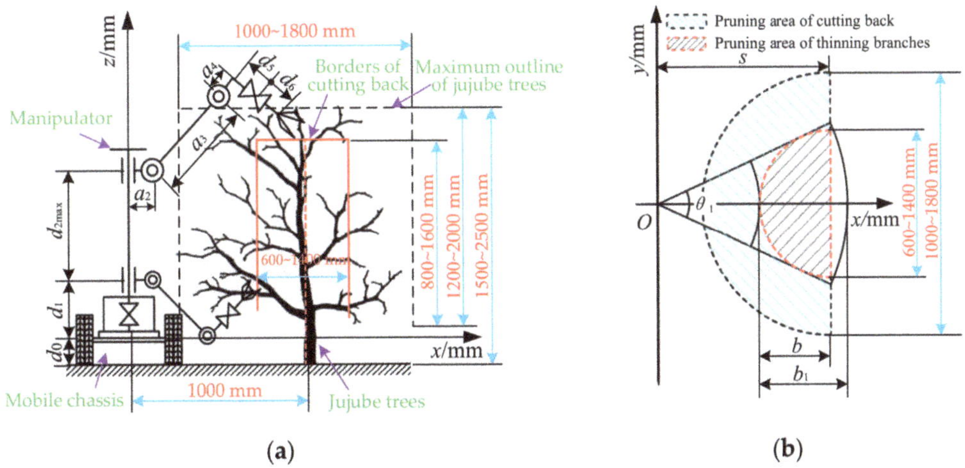

Figure 3. Schematic diagram of the pruning space analysis for the manipulator. (**a**) Main view; (**b**) top view. Note: d_0 is the distance between the base of the manipulator and the ground, mm; d_1 is the height of the base, mm; d_{2max} is the maximum travel of the machine body, mm; a_2 is the offset of the shoulder joint, mm; a_3 is the length of the big arm, mm; a_4 is the offset for the rotatory joint of the forearm, mm; d_5, d_6 is the length of the forearm, mm; θ_1 is the rotation angle of the base, degree; b is the radius of the jujube canopy after pruning by shortening the branches, mm; and b_1 is the operating width of the manipulator for pruning by thinning the branches, mm.

The pruning of jujube trees in winter mainly involves shortening and thinning the branches. Additionally, the range of shortening and thinning branches on one side is shown in Figure 3b. When the end of the manipulator reaches the junction of the shortening and thinning branches area, the shortening of the branches can be completed. At the same time, to meet the space requirements of the operation of thinning the branches, the rotation angle of the base should correspond to Equation (1).

$$\begin{cases} \theta_1 \geq 2\arcsin(b_{max}/s) \\ b_{max} = d_{max}/2 \end{cases} \tag{1}$$

where b_{max} is the maximum radius of the jujube canopy after pruning, mm, d_{max} is the maximum diameter of the jujube canopy after pruning, mm.

The maximum diameter of the canopy for 2–8-year-old jujube trees after pruning is 1400 mm, which can be substituted into Equation (1) to obtain $\theta_1 \geq 88.9$ degrees. At the same time, when the geometric dimensions of each joint meet Equation (2), the manipulator can complete the unilateral pruning requirements of jujube trees in any horizontal region (*xoy*). When the rotation angle θ_1 of the base is 180 degrees, its travel range is −90~+90 degrees, and the problem of satisfying the three-dimensional space pruning can be simplified as the problem of satisfying the rectangle $b_1 \times h$ in the longitudinal plane (*xoz*). When $b_1 \times h$ is satisfied in the longitudinal plane, the base joint of the manipulator is used to rotate the corresponding angle θ_1 around the z axis to achieve the required pruning space.

$$\begin{cases} a_2 + a_3 + d_5 + d_6 \geq \sqrt{s^2 + b_{min}^2} \\ a_2 + a_3 \leq s - b_{min} \end{cases} \tag{2}$$

where b_{min} is the minimum radius of the jujube canopy after pruning, mm.

According to the structural layout requirements of the manipulator, when the offset of the shoulder joint a_2 is 100 mm and the offset of the forearm rotary joint a_4 is 100 mm, the interference between the shoulder joint and forearm rotary joint in the actual assembly and movement can be avoided. To reduce the load arm of the manipulator, the rotary motor of the forearm is arranged at the tail of the forearm, and d_5 is 0 mm, which can be obtained by substituting it into Equation (2).

$$\begin{cases} a_3 \leq 600 \\ d_6 \geq 344 \end{cases} \quad (3)$$

The mechanical arm is a key component of the manipulator. The longer the moment arm of the manipulator, the lower its performance. In the process of movement, if the structure size of the big arm and forearm is larger, the performance of the pruning manipulator is reduced. Therefore, on the premise for meeting the requirements of the pruning space, the design of the big arm and forearm should achieve a compact structure and harmonious proportion. According to Equation (3), the big arm a_3 of the mechanical arm designed in this paper is 550 mm, and the forearm d_6 is 350 mm. According to the height of the canopy before and after pruning, base d_1 is 200 mm, the maximum travel of the machine body $d_{2\max}$ of the machine body is 700 mm. By analyzing the pruning space of the manipulator, the dimension parameters of each link of the manipulator are shown in Table 1.

Table 1. The dimension parameters of the manipulator links.

a_2	a_3	a_4	d_1	$d_{2\max}$	d_5	d_6
100 mm	350 mm	100 mm	200 mm	700 mm	0 mm	350 mm

2.3. *The Design of the End-Effector*

2.3.1. Structure Design

The common pruning methods for fruit trees are shear and saw cutting. As the method of supported pruning, the operation process of shear pruning is stable. In combination with the structural characteristics of the articulated manipulator, the shear structure was selected as the end-effector of the jujube pruning manipulator. It is mainly composed of an executive motor, planetary reducer, gear transmission mechanism, moving cutter, stationary cutter, diagonal photoelectric sensor, mounting plate, and a fixed support. During the operation, the mechanical arm drives the end-effector installed on the forearm to reach the target branch position, and the moving cutter is closed under the action of the executive motor when the diagonal photoelectric sensor detects that the branch has entered the scissor mouth. When the moving and stationary cutters are completely closed, the motor of the end-effector is reversed to make the moving cutter and the stationary cutter open automatically. To enable the pruned branches to effectively enter the cutting mouth of end-effector, the diameter of the pruned jujube branches was 5–20 mm, the opening angle of the moving and fixed cutters was 40 degrees, the maximum vertical distance of the scissor's mouth was 35 mm, and the distance between the cutting position of the jujube branch and the rotating axis of the moving cutter was 50 mm. The specific structure diagram of the end-effector is shown in Figure 4.

2.3.2. The Design of the Moving Cutter

As a key part of the end-effector, the moving cutter completes the cutting of the branches. To achieve the purpose of saving labor and improving the incision quality, it is necessary to ensure that the cutting angle α of each cutting edge point is equal to the friction angle φ between the moving cutter and the branch during the cutting process. Therefore, the design of the cutting edge curve can achieve the stable and sliding pruning of the jujube branches.

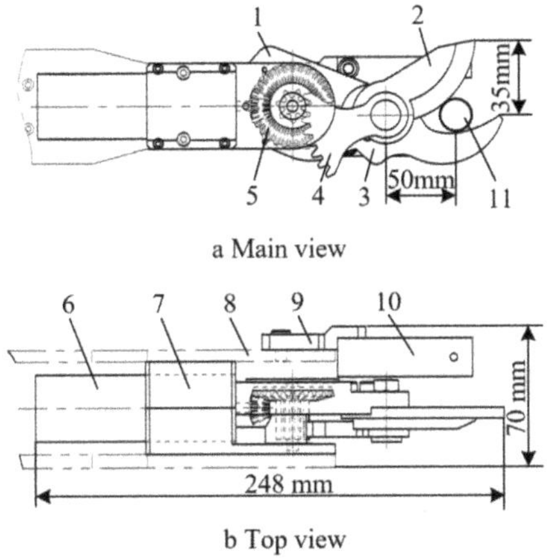

Figure 4. The structural diagram of the end-effector. 1. Fixed base; 2. Moving cutter; 3. Stationary cutter; 4. Incomplete gear mechanism; 5. Bevel gear mechanism; 6. Force motor; 7. Planetary reducer; 8. Forearm; 9. Mounting plate; 10. Diagonal photoelectric sensor; and 11. Branch of the jujube tree.

The moving cutter rotated around the hinge point O to shear the jujube branch during the operation. Suppose the blade curve is $A\hat{B}C$, the cutting angle α of any point on the curve is equal to the friction angle φ, and the hinge point O is taken as the origin of the coordinates; a coordinate system is established to analyze the blade curve of the moving cutter, as shown in Figure 5.

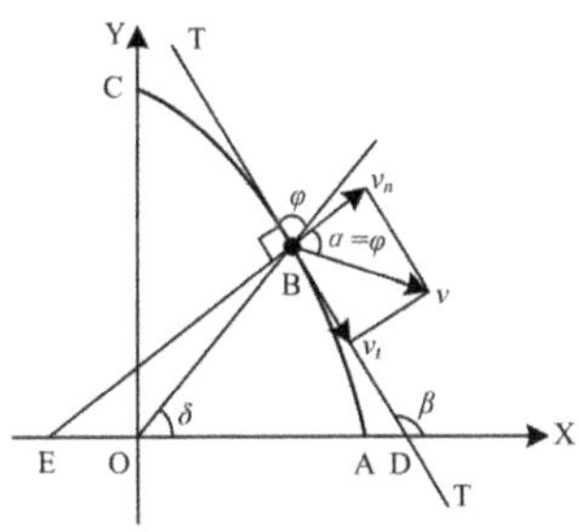

Figure 5. The analysis of the cutting edge curve of the moving cutter. Note: T–T is the tangent line to point B; EB is the normal line at point B; OB is the rotation radius of point B, mm; v is the sliding cutting speed at point B, m/s; v_t is the tangential velocity at point B, m/s; and v_n is the normal velocity at point B, m/s.

According to the geometric relation of ΔOBD, $\beta = \delta + \varphi$. There are:

$$\tan\beta = \frac{\tan\delta + \tan\varphi}{1 - \tan\delta \cdot \tan\varphi} \tag{4}$$

where $\tan\beta = \frac{dy}{dx}$, $\tan\delta = \frac{y}{x} = u$. Substitute them into Equation (4) to obtain:

$$\frac{1 - u\tan\varphi}{\tan\varphi + u^2\tan\varphi}du = \frac{1}{x}dx \tag{5}$$

Integrate both sides of Equation (5) to obtain:

$$\frac{1}{2}\ln\left(x^2 + y^2\right) = \frac{1}{\tan\varphi}arctan\frac{y}{x} + C \tag{6}$$

By substituting $x = \rho \cdot \cos\delta$, $y = \rho \cdot \sin\delta$, and substitute them into Equation (6); the polar coordinate equation for the blade curve of the moving cutter is:

$$\rho = Ce^{\frac{\delta}{\tan\varphi}} \tag{7}$$

where ρ is the polar diameter, mm; δ is the polar angle; and C is the integration constant.

According to Equation (7), when the blade curve of the moving cutter is a logarithmic spiral, stable and sliding pruning can be realized. When the polar angle changes from δ_1 to δ_2, the required cutting edge arc length l is the following:

$$l = \int_{\delta_2}^{\delta_1} dl = \int_{\delta_2}^{\delta_1} \sqrt{\rho_1^2 + \rho_2^2}\, d\delta = \frac{\rho_2 - \rho_1}{\cos\varphi} \tag{8}$$

According to Equation (8), $\rho_2 - \rho_1 \geq d$ must be satisfied when cutting the jujube branch with diameter d. Additionally, the actual diameter range of pruning the jujube branches is 5–20 mm, so the length of the designed moving cutter is 80 mm. According to the relevant design research of the cutting tools, the slide angle was designed to be 35 degrees and the edge inclination angle was 20 degrees. Figure 6a shows the blade curve of the moving cutter established by MATLAB, and Figure 6b shows the structure of the moving cutter designed by using the blade curve.

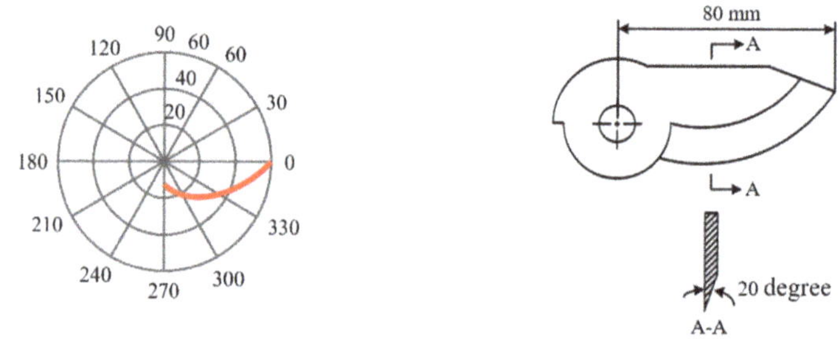

(a) (b)

Figure 6. The model of the moving cutter. (**a**) The blade curve of the moving cutter, and (**b**) the structure of the moving cutter.

2.4. The Design of the Control System

The design and construction of the control system for the jujube pruning manipulator are very important for its pruning function. The main function of the control system of the jujube pruning manipulator designed in this paper is to realize the delivery, processing, and execution of the control instructions, so as to realize the data communication between the upper and lower computers. The diagram for the overall control scheme of the jujube pruning manipulator is shown in Figure 7.

The control system of the manipulator adopts a two-layer structure control, including the upper and lower computers. The lower computer control system adopts a six-axis off-line motion controller (YJ-CTRL-A601; Shenzhen Yijia Technology Co., Ltd.; Shenzhen; China). The driving motor of each joint is an integrated closed-loop stepper motor (ESS60-P; Shenzhen YAKO Automation Technology Co., Ltd.; Shenzhen; China). The controller is connected to each joint motor of the manipulator through the signal output port, pulse output port, direction port, servo enable port, servo alarm, and alarm clearing port of the encoder, and the driver of each joint motor is controlled by the sending direction and pulse signal. In addition, the motion controller is connected to the solid-state relay by the switching output. The relay signal is used as the input signal of the end-effector controller to control the moving cutter. Finally, the switch signal output by the sensor of the end-effector controls the power-on or power-off of the relay coil to play the role of system protection or automatic control. The diagram for the electrical schematic of the jujube pruning manipulator is shown in Figure 8.

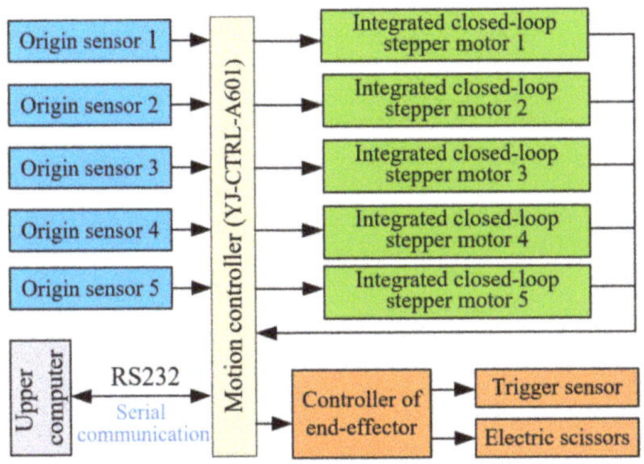

Figure 7. The diagram for the overall control scheme of the jujube pruning manipulator.

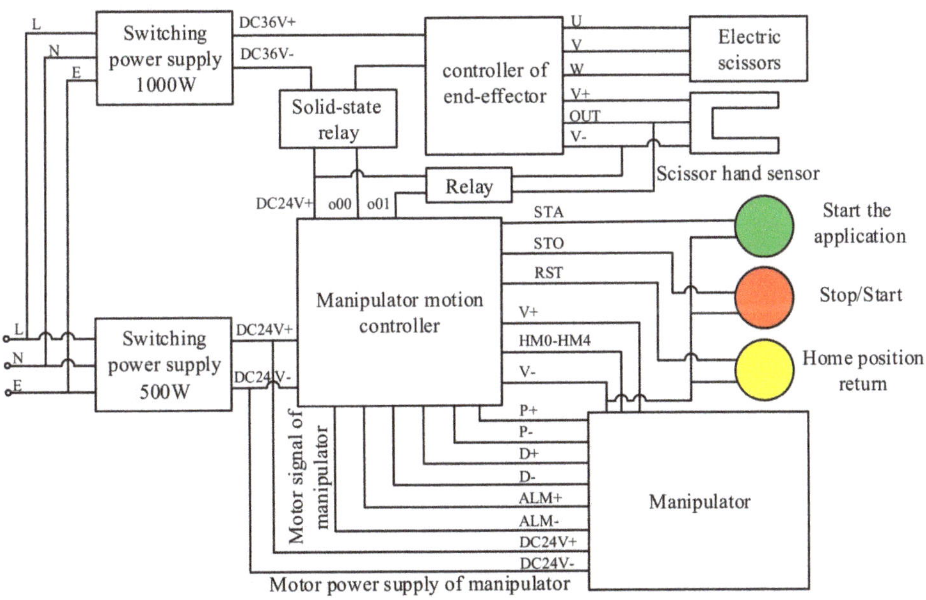

Figure 8. The diagram for the electrical schematic of the jujube pruning manipulator.

The upper computer adopts a PC machine (Lenovo Y9000P; Lenovo Group; Beijing; China), of which the basic frequency is 2.30 GHz, the development environment is Visual Studio, and the development language is C#. The lower computer communicates with the upper computer through a serial port, for which the serial port communication protocol is RS232, the serial port parameter's baud rate is 115,200, the data bit is 8, and the stop bit is 1.

2.5. The Kinematics Analysis of the Manipulator

Based on the kinematics analysis of the manipulator designed in this paper, the relationship between the pose of the end-effector and the joint variables of the manipulator was established, and the workspace simulation was carried out based on the kinematics model to verify whether the workspace of the manipulator met the requirements of the pruning space. The coordinate system of the link of the jujube pruning manipulator is shown in Figure 9. The parameters of the link of the jujube pruning manipulator are shown in Table 2.

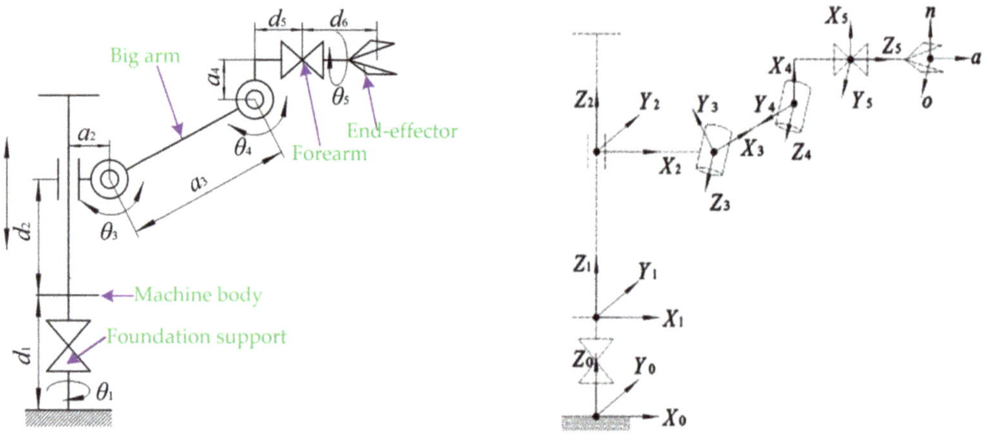

(a) **(b)**

Figure 9. The coordinate system for the link of the jujube pruning manipulator. (**a**) The schematic diagram of the manipulator structure and (**b**) the coordinate system for the manipulator link. Note: θ_1 is the rotation angle of the base, degree; θ_3 is the rotation angle of the shoulder joint, degree; θ_4 is the rotation angle of the elbow joint, degree; θ_5 is the rotation angle of the forearm, degree; $X_0Y_0Z_0$ is the base coordinate system; $X_1Y_1Z_1$ is the coordinate system at the top of the base; $X_2Y_2Z_2$ is the coordinate system for the mobile joint of the machine body; $X_3Y_3Z_3$ is the coordinate system of the shoulder joint; $X_4Y_4Z_4$ is the coordinate system of the elbow joint; $X_5Y_5Z_5$ is the coordinate system for the rotary joint of the forearm; and *noa* is the coordinate system of the end-effector.

Table 2. The parameters for the link of the jujube pruning manipulator.

Link *i*	θ_i/Degree	α_{i-1}/Degree	a_{i-1}/mm	d_i/mm	Range of Variables
1	θ_1	0	0	d_1 (200)	−90~+90 degree
2	0	0	0	d_2	0~700 mm
3	θ_3	90	a_2 (100)	0	−30~+180 degree
4	θ_4	0	a_3 (350)	0	−90~+90 degree
5	θ_5	90	a_4 (100)	d_5 (0)	−160~+160 degree
6	0	0	0	d_6 (350)	-

2.5.1. Forward Kinematics Analysis

The DH parameter method [11,39] was used for the kinematic analysis, and a kinematic model of the manipulator was established to describe the relative position and attitude among the coordinate systems. According to the kinematics theory of the robot, the general formula ${}^{i-1}_{i}T$ of the transformation matrix under the DH parameters of the adjacent link of the manipulator is:

$$
{}^{i-1}_{i}T = \begin{bmatrix} \cos\theta_i & -\sin\theta_i & 0 & a_{i-1} \\ \sin\theta_i\cos\alpha_{i-1} & \cos\theta_i\cos\alpha_{i-1} & -\sin\alpha_{i-1} & -d_i\sin\alpha_{i-1} \\ \sin\theta_i\sin\alpha_{i-1} & \cos\theta_i\sin\alpha_{i-1} & \cos\alpha_{i-1} & d_i\cos\alpha_{i-1} \\ 0 & 0 & 0 & 1 \end{bmatrix} \tag{9}
$$

where θ_i is the joint angle, degree; d_i is the horizontal distance, mm; a_{i-1} is the distance of the common normal, i.e., the length of the rod, mm; and a_{i-1} is the torsion angle, degree.

According to Equation (9) and the parameters of the link presented in Table 2, the transformation matrix ${}^0_6T$ for the end pose of the jujube pruning manipulator can be obtained:

$$ {}^0_6T = {}^0_1T{}^1_2T{}^2_3T{}^3_4T{}^4_5T{}^5_6T = \begin{bmatrix} n_x & o_x & a_x & p_x \\ n_y & o_y & a_y & p_y \\ n_z & o_z & a_z & p_z \\ 0 & 0 & 0 & 1 \end{bmatrix} \tag{10} $$

where

$$ \left.\begin{array}{r} n_x = c1c5(c3c4 - s3s4) + s1s5 \\ n_y = s1c5(c3c4 - s3s4) - c1s5 \\ n_z = c5(c3s4 + s3c4) \\ o_x = s1c5 - c1s5(c3s4 - s3s4) \\ o_y = -s1s5(c3c4 - s3s4) - c1c5 \\ o_z = -s5(c3c4 + s3c4) \\ a_x = c1(c3s4 + s3c4) \\ a_y = s1(c3s4 + s3c4) \\ a_z = s3s4 - c3c4 \\ p_x = c1[a_2 + a_3c3 + a_4(c3c4 - s3s4) + (d_5 + d_6)(c3s4 + s3c4) \\ p_y = s1[a_2 + a_3c3 + a_4(c3c4 - s3s4) + (d_5 + d_6)(c3s4 + s3c4) \\ p_z = a_3s3 + a_4(c3s4 + s3c4) + d_1 + d_2 + (d_5 + d_6)s3s4 - c3c4 \end{array}\right\} \tag{11} $$

In Equation (11), $ci = \cos\theta_i$, $si = \sin\theta_i$, where i is 1, 3, 4, and 5, respectively. The same is expressed below.

The transformation matrix ${}^0_6T$ represented by Equation (10), describes the pose of the base coordinate system {0} relative to the end-effector coordinate system {6} of the pruning manipulator. To test the correctness of the model ${}^0_6T$, the initial positions ($\theta_1 = 0$ degree, $\theta_3 = 90$ degree, $\theta_4 = 0$ degree, $\theta_5 = 0$ degree) of the manipulator were obtained for checking and calculation; substituting them into Equation (11), the result of calculating the arm transformation matrix ${}^0_6T$ is:

$$ {}^0_6T_{Initial\ position} = \begin{bmatrix} 0 & 0 & 1 & 450 \\ 0 & -1 & 0 & 0 \\ 1 & 0 & 0 & 650 \\ 0 & 0 & 0 & 1 \end{bmatrix} \tag{12} $$

The test results of Equation (12) are consistent with the initial position parameters of the designed manipulator, indicating that the established mathematical model of the manipulator kinematics is correct.

2.5.2. Inverse Kinematic Analysis

Before the manipulator is driven to the desired position, all the joint variables related to the position must be obtained. Therefore, it is necessary to carry out the inverse kinematics analysis of the manipulator.

The desired pose coordinate of the end-effector of the manipulator is assumed as [n, o, a, p]. Firstly, multiply ${}^0_1T^{-1}$ at both sides of Equation (10) by the inverse transformation method. After simplification, it can be determined that ${}^0_1T^{-1}{}^0_6T = {}^1_2T{}^2_3T{}^3_4T{}^4_5T{}^5_6T$. According to the equal elements of the matrices at both sides, it can be determined that $c1 \times p_y = s1 \times c1 \times p_x$. Finally, the rotation angle of the base joint is shown in Equation (13):

$$ \theta_1 = arctan\frac{p_y}{p_x} \tag{13} $$

Similarly, multiply ${}^{1}_{2}T^{-1}$, ${}^{2}_{3}T^{-1}$, ${}^{3}_{4}T^{-1}$, ${}^{4}_{5}T^{-1}$, ${}^{5}_{6}T^{-1}$ at both sides of Equation (12), and, according to the elements at both sides, which are equal, the general expressions of the revolute joint variables θ_3, θ_4, θ_5 are obtained, as shown in Equation (14)~(16):

$$\theta_3 = arctan\frac{t_2 - (d_5 + d_6)a_z - a_4(c1a_x + s1a_y)}{t_1 - (d_5 + d_6)(c1a_x + s1a_y) + a_4a_z} \tag{14}$$

$$\theta_4 = arctan\frac{a_4(c3t_2 - s3t_1) + (d_5 + d_6)(s3t_2 + c3t_1 - a_2)}{a_4(s3t_2 - c3t_1 - a_3) - (d_5 + d_6)(c3t_2 + s3t_1)} \tag{15}$$

$$\theta_5 = arctan\frac{n_x\cdot \sin\theta_1 - n_y\cdot \cos\theta_1}{o_x\cdot \sin\theta_1 - o_y\cdot \cos\theta_1} \tag{16}$$

where $t_1 = c1p_x + s1p_y - a_2, t_2 = p_z - d_1 - d_2$.

In conclusion, the DH parameter method was used to establish the theoretical model of the manipulator kinematics, and the relative position and pose relationship between the coordinate systems of each joint were obtained. Meanwhile, the inverse kinematic analysis of the manipulator was carried out to obtain the general expressions for the joint angles of the manipulator, which provides the theoretical basis for the simulation analysis of the manipulator workspace.

3. The Performance Test Method of the Manipulator

To further verify whether the designed manipulator meets the performance requirements of jujube pruning, the performance test of the jujube pruning manipulator prototype was carried out, based on high-speed camera technology [40–42].

3.1. The Analysis of the Agronomic Pruning Point for Jujube Trees

Jujube pruning agronomy mainly consists of cutting back and thinning the branches. Cutting back mainly consists of cutting off part of the lateral branches of the current year's growth along the height of the jujube trees, which can inhibit the excessive growth of the lateral branches and promote the main branches to produce flowers and fruit. The thinning of the branches mainly entails the cutting off of the dense or dead branches along the depth of the jujube trees, which can improve ventilation and light, and promote the rejuvenation of dead branches. By analyzing the process of using the manipulator to prune the jujube trees, it can be concluded that the manipulator needs to reach the different heights of the jujube tree canopy for pruning when cutting back the branches, and the manipulator needs to complete the pruning of jujube branches at different depths when thinning the branches. The schematic diagram of the agronomic pruning analysis for a jujube tree is shown in Figure 10. In the actual operation, the manipulator is installed on the mobile chassis. In order to be convenient for analysis, the jujube tree height corresponding to the installation height of the manipulator base is taken as zero. A field investigation was carried out on the growth information of the jujube trees before and after pruning; it was found that the cutting back points were mainly distributed in the range of 200~1000 mm in the height direction of the jujube trees, and the points of the thinning branches were mainly distributed in the range of 100~700 mm, in the depth direction of the jujube trees.

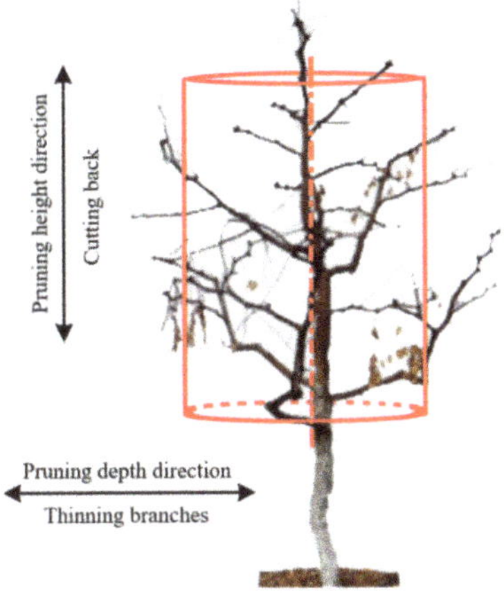

Figure 10. The schematic diagram for the agronomic analysis of jujube tree pruning.

3.2. The Workspace Simulation of the Manipulator

Based on the MATLAB Robotics Toolbox, a 3D mathematical simulation model of the jujube pruning manipulator was established. The Monte Carlo method [43] was used to simulate the workspace of the manipulator to verify whether the theoretical design of the manipulator met the requirements of the jujube pruning space. According to the kinematic theoretical analysis of the manipulator, in combination with the parameters and variable ranges of each joint size of the manipulator presented in Tables 1 and 2, the Rand function in MATLAB was used to program the manipulator workspace for the calculation and simulation. The random values of each joint variable generated by the Rand function are shown in Equation (17):

$$\theta_i = \theta_i^{min} + \left(\theta_i^{max} - \theta_i^{min}\right) \times Rand(N, 1) \tag{17}$$

where θ_i^{min} is the minimum value of the angle range of joint i, degree; θ_i^{max} is the maximum angle range of joint i, degree; and N is the number of cycles, N = 10,000.

3.3. The Platform Construction and Test of the Prototype

The self-made prototype for the jujube pruning manipulator was used to build its performance test platform, as shown in Figure 11. The test results were recorded by a 3D high-speed camera system. The test equipment mainly includes a pruning manipulator prototype, a 3D high-speed camera (FASTECIMAGING-TS4; Fastec Imaging Corporation; San Diego, CA, USA), a graduated scale (accuracy: 1 mm), and a calibration plate.

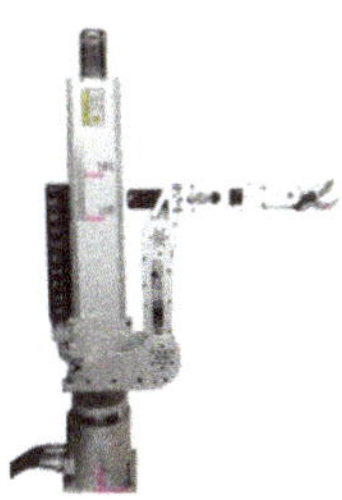

(a)

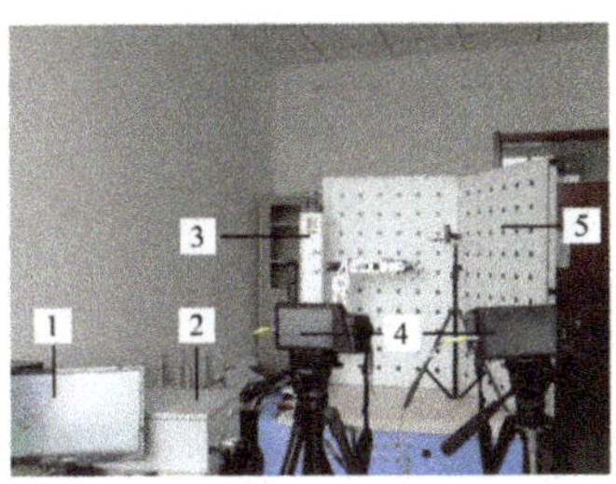

(b)

Figure 11. The platform for the manipulator performance test. (**a**) Prototype and (**b**) test platform. 1. PC machine; 2. Control box; 3. Manipulator; 4. 3D high-speed camera; and 5. Calibration plate.

3.3.1. The Scheme for the Positioning Accuracy Test

The end-effector was driven by the manipulator to move to the target pruning point of the branch when pruning the jujube trees, and the branches triggered sensors to complete the pruning operation. The positioning accuracy of the end-effector to the pruning point is one of the key factors for completing the pruning operation. Therefore, the positioning error of the end for the manipulator was taken as the evaluation index to verify the positioning accuracy of the manipulator moving to the pruning points of the jujube trees. The calculation of the positioning error is shown in Equation (18):

$$D = \sqrt{(X - X_0)^2 + (Y - Y_0)^2 + (Z - Z_0)^2} \tag{18}$$

where $P_0(X_0,Y_0,Z_0)$ are the theoretical coordinates of the pruning points, mm, and $P(X,Y,Z)$ are the measured coordinates of the pruning points, mm.

By taking the base of the manipulator as the origin, the positions for the end-effector of the manipulator to the 9 pruning points with the horizontal distance of 600 mm and the height of 200~1000 mm were recorded, and the positioning accuracy was tested. Similarly, the positions of the end of the manipulator to the 5 pruning points with equal spacing ranging from 100~700 mm in the depth direction were recorded, and the positioning accuracy of the end-effector to the pruning points with different depths was tested. The video data analysis software ProAnalyst was used to analyze the test results for the positioning accuracy of the manipulator end-effector. Firstly, the manipulator in the video was calibrated. The ruler placed in advance on the manipulator was marked, and the actual size of the ruler was input in the software; then, the manipulator in the video was restored to the actual size after calibration. Secondly, the center position of the base of the manipulator in the video data analysis software was set as the base coordinate system of the manipulator. Thirdly, the position of the end-effector in the video data analysis software was marked as the tracking point. The motion track of the manipulator along different height directions and different depth directions was automatically tracked. Finally, the coordinates for the tracked trajectory of the manipulator end-effector in the video data analysis software were output and recorded.

3.3.2. The Scheme for the Pruning Test

The test subjects were five two-year-old jujube trees from the Science and Technology Park of Shihezi University. The average height of the jujube trees was 1.8 m, and the average width of the canopy was 1.4 m. The jujube tree was fixed on the performance test platform of the manipulator to conduct the pruning test, as shown in Figure 12.

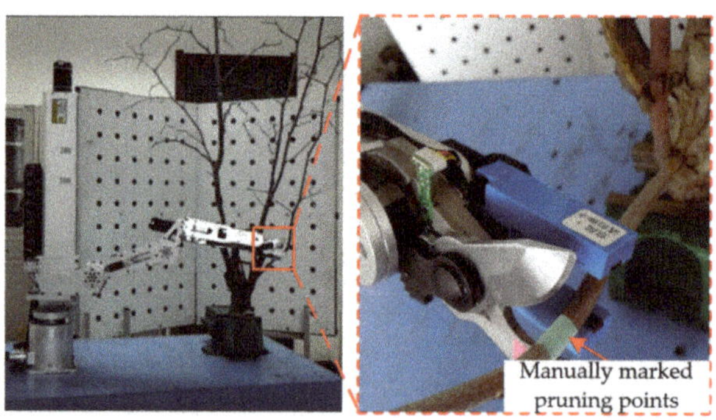

Figure 12. The pruning test.

The specific operational procedures of the jujube pruning test are as follows: firstly, according to the artificial pruning of the jujube agronomic knowledge and the experience of the jujube farmers, the branches that needed to be cut and the location of the pruning points were identified manually, and each pruning point was marked with green tape. Secondly, the manipulator was set to teaching mode (when the manipulator was in teaching mode, the sensor of the end-effector was in the closed state, and the pruning function of the end-effector could not be triggered when the manipulator reached the pruning point), and the manipulator was controlled manually to reach the pruning point of the jujube tree. Additionally, the coordinate information of the current pruning point was obtained and recorded by the upper computer. The above operations were repeated to obtain and record the coordinate information of each pruning point. Finally, the manipulator was reset to the initial state and set to working mode (when the manipulator was in working mode, the sensor of the end-effector was in an open state. When the manipulator reached the pruning point of the jujube tree, the branch entered the detection area of the sensor, which could trigger the pruning function of the end-effector). The coordinates of the pruning point were manually input into the upper computer, the manipulator was controlled to automatically reach the pruning point of the jujube tree, and the pruning test was carried out. The 3D high-speed camera was used to record the real-time video data of the motion position and pose for the manipulator in the pruning process. The video data analysis software ProAnalyst was used to extract the pruning time and judge the effect of pruning.

The main purpose of the pruning manipulator is to complete the pruning task in a short period of time. Therefore, the success rate of pruning R and the pruning time T are taken as the evaluation indexes of the pruning performance for the manipulator. The success rate of pruning (R) and the pruning time (T) were calculated as follows:

$$R = \frac{\sum_{i=1}^{n} L_i}{\sum L} \times 100\% \tag{19}$$

$$T = \sum_{i=1}^{n} T_i \tag{20}$$

where $\sum L$ is the total pruning time of a single jujube tree; n is the number of successful pruning attempts of a single jujube tree; and T_i is the time taken to complete the i-th pruning, min.

4. Results and Discussion

4.1. The Simulation Results and Analysis of the Manipulator Workspace

The simulation results of the manipulator workspace are presented in Figure 13. The workspace of the manipulator is −600~800 mm in the X direction, −800~800 mm in the Y direction, and −200~1800 mm in the Z direction. Additionally, the pruning points are more dense in the range of 0~600 mm in the X direction, −600~600 mm in the Y direction, and 0~1700 mm in the Z direction. The simulation results show that the geometric size of the jujube pruning manipulator can meet the requirements of the pruning space of the jujube trees in the dwarf and densely planted jujube garden.

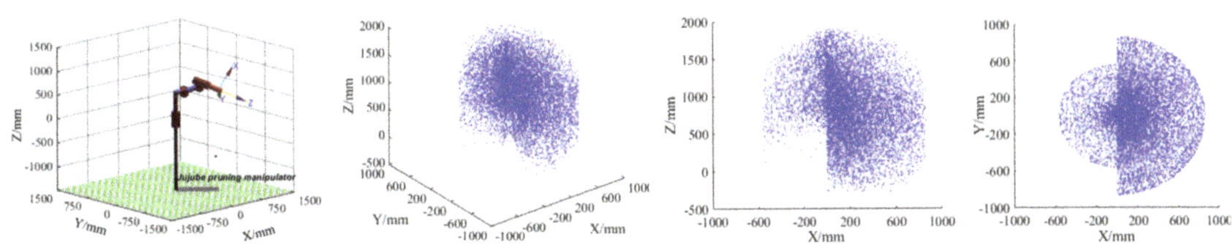

Figure 13. The simulation results of the manipulator workspace. (**a**) The three-dimensional mathematical simulation model of the manipulator; (**b**) the three-dimensional manipulator workspace; (**c**) the projection of the workspace onto the XOZ plane; and (**d**) The projection of the workspace onto the XOY plane.

4.2. The Results and Discussion of the Positioning Accuracy

The test results for the positioning error of the manipulator at different pruning points are presented in Table 3. The schematic diagram of the positioning error trend for the manipulator end-effector is shown in Figure 14.

Table 3. The test results for the positioning error of the manipulator at different pruning points.

Number	Theoretical Coordinates of the Pruning Points/mm			Measured Coordinates of the Pruning Points/mm			Absolute Value of the Positioning Error/mm			
	X_0	Y_0	Z_0	X	Y	Z	D_x	D_y	D_z	D
1	600	0	200	601.59	-	208.78	1.59	-	8.78	8.92
2	600	0	300	605.32	-	301.12	5.32	-	1.12	5.44
3	600	0	400	601.10	-	395.93	1.10	-	4.07	4.21
4	600	0	500	599.12	-	507.36	0.88	-	7.36	7.41
5	600	0	600	597.03	-	598.68	2.97	-	1.32	3.25
6	600	0	700	597.47	-	697.91	2.53	-	2.19	3.34
7	600	0	800	598.87	-	800.69	1.13	-	0.69	1.32
8	600	0	900	598.21	-	897.62	1.79	-	2.38	2.97
9	600	0	1000	602.35	-	998.46	2.35	-	1.54	2.81
10	150	0	600	148.27	-	603.08	1.73	-	3.08	3.53
11	250	0	600	249.56	-	603.91	0.44	-	3.91	3.93
12	350	0	600	351.28	-	600.79	1.28	-	0.79	1.50
13	450	0	600	453.37	-	602.48	3.37	-	2.48	4.18
14	550	0	600	554.95	-	598.93	4.95	-	1.07	5.06
15	650	0	600	658.25	-	596.09	8.25	-	3.91	9.13

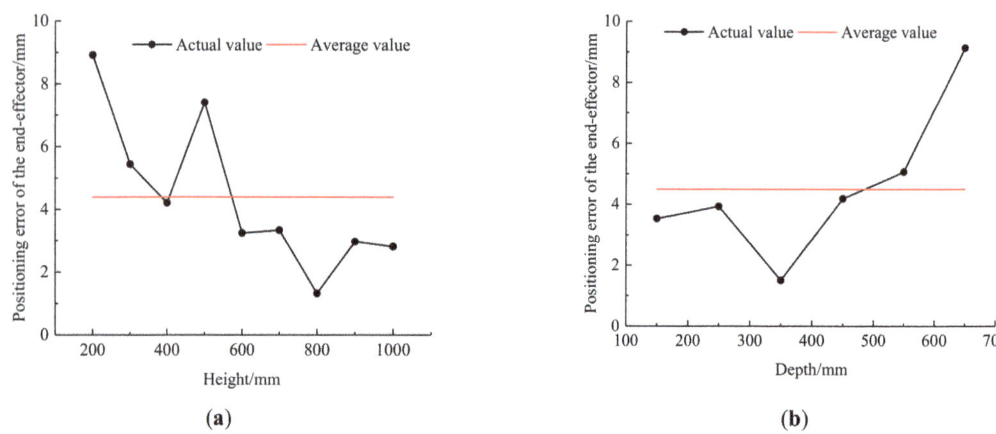

Figure 14. The schematic diagram of the positioning error trend for the manipulator end-effector. (**a**) Positioning error at different heights, and (**b**) positioning error at different depths.

Table 3 and Figure 14 show that, in the height directions, the positioning error of the end-effector tends to decrease as the height increases when the manipulator moves from the initial position (450, 0, 650) to different height positions (600, 0, 200~1000), and the average error value is 4.4 mm. The maximum error occurs at the lowest position (Z = 200 mm), which is 8.92 mm. The main reason for this phenomenon is that, when the manipulator end-effector moves from the initial position to different heights below 650 mm, the moment arm of the machine arm gradually increases with the decrease in the height of the pruning position, and the direction of the manipulator movement is consistent with the gravity direction of the center-of-mass gravity of the manipulator body, resulting in the positioning error of the end-effector increasing with the decrease in the height of the pruning position. When the manipulator is at the lowest position (Z = 200 mm), the motion inertia force reaches its maximum, resulting in the maximum positioning error occurring in this position. When the end-effector moves from the initial position to different heights above 650 mm, the moment arm of the machine arm gradually increases with the increase in the height of the pruning position. However, the direction of the manipulator movement is opposite to the gravity direction of the center-of-mass gravity of the manipulator body, resulting in the positioning error of the end-effector decreasing with the increase in the height of the pruning position. Therefore, the positioning error of the end-effector tends to decrease as the height of the pruning position increases.

In the depth directions, when the manipulator moves from the initial position (450, 0, 650) to different depth positions (150~650, 0, 600), the positioning error of the end-effector tends to increase as the depth of the pruning position increases, and the average error value is 4.5 mm. The maximum positioning error occurs at the farthest position of pruning point (X = 650 mm), which is 9.13 mm. The main reason for this phenomenon is that the moment arm of the machine arm increases as the moving distance of the manipulator end-effector increases in the direction of the depth. Therefore, the positioning error of the end-effector increases with the increase in the depth of the pruning position.

In conclusion, the positioning errors of the end-effector of the pruning manipulator at different heights and depths are all less than 10 mm. There are two main reasons for the positioning error of the manipulator in the process of the test. On the one hand, there are errors in the manufacturing and assembly of all the parts of the manipulator, and a mechanical vibration occurs in the process of operation. On the other hand, the center of gravity for the machine arm changes in real time during the operation of the manipulator. In the follow-up study, the positioning error is improved by improving the manufacturing and assembly accuracy of the manipulator parts and further optimizing the control system.

4.3. The Results and Discussion of the Pruning Test

The results of the pruning test are shown in Table 4. Due to the unstructured natural growth of jujube tree canopy, the number of branches, which were identified to be pruned, and the location of the pruning point vary from tree to tree. Therefore, the number and position of the pruning points were different for each jujube tree in the test. Among them, when the first jujube tree was pruned, a total of 36 pruning points were determined, and 33 points were successfully pruned. The success rate of pruning was 91.67%, which was the highest among the 5 jujube trees. Additionally, the pruning time was about 29.3 min. When the 5th jujube tree was pruned, 23 of the 27 pruning points were successfully pruned, and the success rate of a single jujube tree was 85.16%, which was the lowest among the 5 jujube trees. The pruning time was about 25.6 min.

Table 4. The results of the pruning test.

Number	Total Number of Pruning Points	Number of Successfully Pruned Points	Success Rate/%	Pruning Time/min
1	36	33	91.67	29.3
2	30	26	86.67	27.6
3	33	30	90.91	28.8
4	30	27	90.00	27.2
5	27	23	85.16	25.6
Total	156	139	89.10	27.7

A total of 156 pruning points were determined in the 5 tests, and the results show that 139 points were successfully pruned. The average success rate of pruning a single jujube tree was about 89.10%, and the average time was about 27.7 min. Additionally, the manipulator ran smoothly in each pruning process. The test verified the reasonableness and feasibility of the designed pruning manipulator.

Figure 15 shows the whole process of the failed pruning of the manipulator. The main reason for the failure was that the branches deviated from the sensor detection area in the pruning process of the moving cutter. Figure 16 shows the whole process of the successful pruning performed by the manipulator. The main reason for the success was the branches in the pruning process of the moving cutter; the branches are always in the sensor detection area. To summarize, the remote and small branches on the side of the jujube tree were easy to fail pruning in the test. The main reason is that the mechanical arm is in a state when the manipulator runs lateral to the jujube trees in a remote location. According to the results of the positioning accuracy of the positioning error of the manipulator, the end-effector is large at this time. In addition, small branches are easy to bend when touching the moving cutter, leading to branches deviating from the sensor detection area. The next study is to optimize and improve the sensitivity and detection range of the end-effector, so that the improved sensor can effectively avoid the jujube branches from breaking away from the detection area after bending.

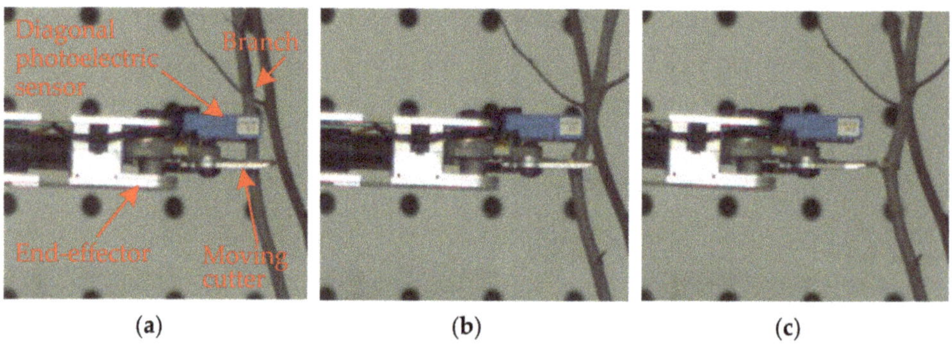

Figure 15. The whole process of failed pruning of the manipulator. (**a**) The start of the pruning process; (**b**) during the pruning process; and (**c**) the branch has not been pruned.

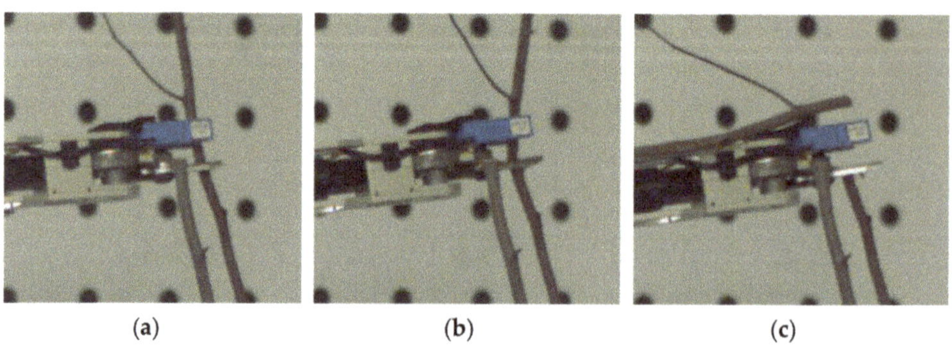

Figure 16. The whole process of the successful pruning of the manipulator. (**a**) The start of the pruning process; (**b**) during the pruning process; and (**c**) the branch has been pruned.

At present, there are many researches on the robot technology in agricultural fields, such as orchard picking, plant protection, and fruit-tree pruning. For different agricultural production links and different operation objects, each form of research has put forward different strategies. In many cases, it is difficult to compare and evaluate the performance of different machines, because the operating objects and operating conditions greatly vary. At present, typical researches in the field of orchard pruning, such as the grape-pruning robot designed by Botterill et al. [26], the apple-tree pruning robot designed by Zahid et al. [27–29], the loquat-pruning robot designed by Huang et al. [35,36], and the high-branch pruning manipulator designed by Wu et al. [37], are still in the initial stage of research. Because the growth information and agronomic pruning requirements of jujube trees are different from other fruit trees, it is necessary to design special pruning equipment for jujube trees, according to the growth characteristics and agronomic pruning requirements of dwarf and densely planted jujube trees in Xinjiang. Fu et al. developed a shaping and pruning machine for dwarf and densely planted jujube trees [20]. This machine can realize the rapid shortening pruning function of large-scale jujube trees, with a high pruning efficiency. However, it cannot realize the thinning branch pruning function of jujube trees, and the internal ventilation and light transmission of jujube trees after pruning are poor. Therefore, on the basis of this research, we propose the manipulator pruning jujube tree program. According to the characteristics of the artificial pruning of jujube trees, a 5-DOF jujube pruning manipulator was designed by choosing a joint manipulator structure to realize the function of the selective pruning of jujube trees.

5. Conclusions and Future Work

A 5-DOF pruning manipulator was designed, and the relative position and attitude of each coordinate system were obtained by establishing the theoretical model of manipulator kinematics. The workspace of the manipulator was obtained through the simulation analysis of the workspace of the manipulator (−600~800 mm in the X direction, −800~800 mm in the Y direction, and −200~1800 mm in the Z direction). It was verified that the geometric size of the manipulator met the requirements of the pruning space of jujube trees in the dwarf and densely planted jujube garden. Finally, a prototype manipulator was developed, and the positioning accuracy test of the end-effector and pruning performance test of the manipulator were carried out, based on high-speed camera technology. The results show that the positioning errors of the manipulator at different pruning points were all less than 10 mm, the average pruning success rate of the manipulator was about 89.10%, and the average pruning time of a single jujube tree was 27.7 min. It was verified that the structure and control system of the pruning manipulator was reasonable and feasible. This study can provide a theoretical basis and technical support for the intelligent pruning of a jujube garden.

This paper mainly studied the mechanical structure and control system of the manipulator, but there are still pruning failures in the pruning tests of jujube trees. The aim of the subsequent study is to optimize the structure of the manipulator body and improve the control system to further improve the success rate of manipulator pruning. At the same time, the machine vision system will be equipped on the manipulator to realize the intelligent recognition and positioning of pruning points. Additionally, the mobile chassis and manipulator were integrated to carry out the experiment research of jujube-tree pruning in a natural environment, so as to realize the intelligent pruning of a jujube garden.

Author Contributions: Conceptualization, B.Z. and X.C.; Data curation, B.Z. and H.Z.; Formal analysis, B.Z. and C.S.; Funding acquisition, W.F. and C.S.; Investigation, B.Z., H.Z. and C.S.; Methodology, B.Z. and X.C.; Project administration, W.F. and C.S.; Resources, W.F. and C.S.; Software, B.Z. and H.Z.; Supervision, W.F.; Validation, B.Z., W.F., X.C., H.Z. and C.S.; Visualization, B.Z. and C.S.; Writing—original draft, B.Z. and H.Z.; Writing—review and editing, B.Z. and W.F. All authors have read and agreed to the published version of the manuscript.

Funding: This work was supported by the National Natural Science Foundation of China (NSFC), No. 51765058, Regional Innovation Guidance Project of the Xinjiang Production and Construction Group (No.2021BB020), Scientific and Technological Innovation Talents Program of the Xinjiang Production and Construction Group (No.2020CB008).

Institutional Review Board Statement: Not applicable.

Informed Consent Statement: Not applicable.

Data Availability Statement: The data presented in this study are available on-demand from the first author at (994026@hainanu.edu.cn).

Acknowledgments: The authors would like to thank their schools and colleges, as well as the funding of the project. All support and assistance are sincerely appreciated. Additionally, we sincerely appreciate the work of the editor and the reviewers of the present paper.

Conflicts of Interest: The authors declare no conflict of interest.

References

1. Zhang, B.; Fu, W.; Wang, X.F.; Lou, Z.X.; Liu, Y.D.; Fu, Y.X.; Chen, Y.Y. Bench test and parameter optimization of jujube pruning tools. *IAEJ* **2020**, *29*, 58–68.
2. Fu, W.; Zhang, Z.Y.; Ding, K.; Cao, W.B.; Kan, Z.; Pan, J.B.; Liu, Y.D. Design and test of 4ZZ-4A2 full-hydraulic self-propelled jujube harvester. *Int. J. Agric. Biol. Eng.* **2018**, *11*, 104–110. [CrossRef]
3. Gao, W.H.; Han, R. *Xinjiang Statistical Yearbook*; China Statistics Press: Beijing, China, 2020.
4. Wang, Y.K.; Hui, Q.; Wang, X.; Ma, J.P.; Zhang, W.F. Growth and Water Use Efficiency of Water saving Type Pruning Jujube Tree in Dry Soil. *Trans. Chin. Soc. Agric. Mach.* **2017**, *48*, 247–254.
5. Fu, W.; Liu, Y.D.; Kan, Z.; Pan, J.B.; Cui, J.; Zhang, H.M. The Situation and Expectation of Fruit Tree Pruning Machine. *J. Agric. Mech. Res.* **2017**, *39*, 7–11.
6. Ge, Y.; Fang, J.; Wang, S.L.; Liu, R.; Chen, Z.Y. Status of Mechanized Pruning Techniques Zaoyuan Close Planting Dwarf. *J. Agric. Mech. Res.* **2013**, *35*, 249–252.
7. Kootstra, G.; Wang, X.; Blok, P.M.; Hemming, J.; Van Henten, E. Selective harvesting robotics: Current research, trends, and future directions. *Curr. Robot. Rep.* **2021**, *2*, 95–104. [CrossRef]
8. Wang, T.; Xu, X.; Wang, C.; Li, Z.; Li, D. From Smart Farming towards Unmanned Farms: A New Mode of Agricultural Production. *Agriculture* **2021**, *11*, 145. [CrossRef]
9. Vatavuk, I.; Vasiljević, G.; Kovačić, Z. Task Space Model Predictive Control for Vineyard Spraying with a Mobile Manipulator. *Agriculture* **2022**, *12*, 381. [CrossRef]
10. Vrochidou, E.; Tziridis, K.; Nikolaou, A.; Kalampokas, T.; Papakostas, G.A.; Pachidis, T.P.; Mamalis, S.; Koundouras, S.; Kaburlasos, V.G. An Autonomous Grape-Harvester Robot: Integrated System Architecture. *Electronics* **2021**, *10*, 1056. [CrossRef]
11. Li, G.L.; Ji, C.Y.; Gu, B.X.; Xu, W.Y.; Dong, M. Kinematics Analysis and Experiment of Apple Harvesting Robot Manipulator with Multiple End-effectors. *Trans. Chin. Soc. Agric. Mach.* **2016**, *47*, 14–21, 29.
12. Zhao, D.A.; Lv, J.D.; Ji, W.; Zhang, Y.; Chen, Y. Design and control of an apple harvesting robot. *Biosyst. Eng.* **2011**, *110*, 112–122.
13. Ji, W.; Zhao, D.A.; Cheng, F.Y.; Xu, B.; Zhang, Y.; Wang, J.J. Automatic recognition vision system guided for apple harvesting robot. *Comput. Electr. Eng.* **2012**, *38*, 1186–1195. [CrossRef]
14. Van Henten, E.J.; Tuijl, B.A.J.; Hemming, J.; Kornet, J.G.; Bontsema, J.; Van Os, E.A. Field Test of an Autonomous Cucumber Picking Robot. *Biosyst. Eng.* **2003**, *86*, 305–313. [CrossRef]
15. Van Henten, E.J.; Hemming, J.; Tuijl, B.A.J.; Kornet, J.G.; Bontsema, J. Collision-free Motion Planning for a Cucumber Picking Robot. *Biosyst. Eng.* **2003**, *86*, 135–144. [CrossRef]
16. Hemming, J.; Bac, C.W.; Tuijl, B.A.; Barth, R.; Bontsema, J.; Pekkeriet, E.; Henten, E.J. A robot for harvesting sweet-pepper in greenhouses. In Proceedings of the International Conference of Agricultural Engineering, Zurich, Switzerland, 28–29 July 2022.
17. Bac, C.W.; Hemming, J.; Henten, E.J. Robust pixel-based classification of obstacles for robotic harvesting of sweet-pepper. *Comput. Electron. Agric.* **2013**, *96*, 148–162. [CrossRef]
18. Bac, C.W.; Hemming, J.; Henten, E.J. Stem localization of sweet-pepper plants using the support wire as a visual cue. *Comput. Electron. Agric.* **2014**, *105*, 111–120. [CrossRef]
19. Bac, C.W.; Hemming, J.; Tuijl, B.A.J.; Barth, R.; Wais, E.; Henten, E.J. Performance Evaluation of a Harvesting Robot for Sweet Pepper. *J. Field. Robot.* **2017**, *34*, 1123–1139. [CrossRef]
20. Zhang, B.; Liu, Y.; Zhang, H.; Shen, C.; Fu, W. Design and Evaluation of a Shaping and Pruning Machine for Dwarf and Densely Planted Jujube Trees. *Appl. Sci.* **2022**, *12*, 2699. [CrossRef]
21. Li, M.; Ma, L.; Zong, W.; Luo, C.; Huang, M.; Song, Y. Design and Experimental Evaluation of a Form Trimming Machine for Horticultural Plants. *Appl. Sci.* **2021**, *11*, 2230. [CrossRef]
22. Zahid, A.; Mahmud, M.S.; He, L.; Heinemann, P.; Choi, D.; Schupp, J. Technological advancements towards developing a robotic pruner for apple trees: A review. *Comput. Electron. Agric.* **2021**, *189*, 106383. [CrossRef]
23. Tinoco, V.; Silva, M.F.; Santos, F.N.; Rocha, L.F.; Santos, L.C. A Review of Pruning and Harvesting Manipulators. In Proceedings of the 2021 IEEE ICARSC, Santa Maria da Feira, Portugal, 28–29 April 2021.
24. Kawasaki, H.; Murakami, S.; Kachi, H.; Ueki, S. Novel Climbing Method of Pruning Robot. In Proceedings of the 2008 SICE Annual Conference, Tokyo, Japan, 20–22 August 2008; pp. 160–163.
25. Soni, D.P.; Ranjana, M.; Gokul, N.A.; Swaminathan, S.; Binoy, B.N. Autonomous arecanut tree climbing and pruning robot. In Proceedings of the INTERACT-2010 IEEE, Santa Maria da Feira, Portugal, 28–29 April 2021; pp. 278–282.
26. Botterill, T.; Paulin, S.; Green, R.; Williams, S.; Lin, J.; Saxton, V.; Mills, S.; Chen, X.Q.; Corbett-Davies, S. A Robot System for Pruning Grape Vines. *J. Field. Robot.* **2017**, *34*, 1100–1122. [CrossRef]
27. Zahid, A.; Mahmud, M.S.; He, L.; Choi, D.; Heinemann, P.; Schupp, J. Development of an integrated 3R end-effector with a cartesian manipulator for pruning apple trees. *Comput. Electron. Agric.* **2020**, *179*, 105837–105847. [CrossRef]

28. Zahid, A.; He, L.; Zeng, L.H.; Choi, D.; Schupp, J.; Heinemann, P. Development of a Robotic End-Effector for Apple Tree Pruning. *Trans. ASABE* **2020**, *63*, 847–856. [CrossRef]
29. Zahid, A.; He, L.; Choi, D.; Schupp, J.; Heinemann, P. Investigation of branch accessibility with a robotic pruner for pruning apple trees. *Trans. ASABE* **2021**, *64*, 1459–1474. [CrossRef]
30. Van Marrewijk, B.M.; Vroegindeweij, B.A.; Gené-Mola, J.; Mencarelli, A.; Hemming, J.; Mayer, N.; Kootstra, G. Evaluation of a boxwood topiary trimming robot. *Biosyst. Eng.* **2022**, *214*, 11–27. [CrossRef]
31. Chai, Y.Q. Development of Hedge Pruning Robot Based on Exoskeleton. Master's Thesis, Zhengzhou University, Zhengzhou, China, 2019.
32. Luo, T.H.; Tang, G.; Ma, X.Y.; Zhou, J.C. Obstacle avoidance path planning for expressway hedgerow pruning robot manipulator. *Chin. J. Eng.* **2019**, *41*, 134–142.
33. Li, M.D. Dynamics Simulation Research on the Urban Landscape Automatic Pruning Robot. *J. Agric. Mech. Res.* **2019**, *41*, 61–65.
34. Chen, F.F.; Wu, X.F. The kinematics simulation of the forestry special robot for the hedge pruning. *Mech. Electr. Eng. Technol.* **2009**, *38*, 30–31, 59, 117.
35. Huang, B.; Shao, M.; Song, L. Vision Recognition and Frameworks Extraction of Loquat Branch Pruning Robot. *J. South China Univ. Technol.* **2015**, *43*, 114–119, 126.
36. Huang, B.; Shao, M.; Chen, W.J. Design and Research on End Effector of a Pruning Robot. *Int. J. Simul. Syst. Sci. Technol.* **2016**, *17*, 1–5.
37. Wu, X.; Shen, Y.; Yang, H.; Guo, J.; Teng, F. Design and Test of High Branch Pruning Manipulator in Forest and Fruit Industry. *Trans. Chin. Soc. Agric. Eng.* **2020**, *42*, 70–75.
38. Wang, Y.; Yang, Q.H.; Bao, G.J.; Xun, Y.; Zhang, L.B. Optimization Design and Experiment of Fruit and Vegetable Picking Manipulator. *Trans. Chin. Soc. Agric. Mach.* **2011**, *42*, 191–195.
39. Chen, Y.; Zhang, B.; Fu, Y.; Fu, W.; Shen, C.; Jin, X. Path Planning of Jujube Pruning Manipulator. *Trans. Chin. Soc. Agric. Eng.* **2021**, *43*, 37–41.
40. Quan, L.Z.; Peng, T.; Shen, L.Y.; An, S.Y.; Ji, Z.L.; Sun, T. Parameter optimization and experiment of manipulator for three-dimensional seedling tray management robot. *Trans. Chin. Soc. Agric. Eng.* **2017**, *33*, 10–19.
41. Cui, T.; Liu, J.; Yang, L.; Zhang, D.X.; Zhang, R.; Lan, W. Experiment and simulation of rolling friction characteristic of corn seed based on high-speed photography. *Trans. Chin. Soc. Agric. Eng.* **2013**, *29*, 34–41.
42. Islam, M.N.; Iqbal, M.Z.; Ali, M.; Chowdhury, M.; Kabir, M.S.N.; Park, T.; Kim, Y.-J.; Chung, S.-O. Kinematic Analysis of a Clamp-Type Picking Device for an Automatic Pepper Transplanter. *Agriculture* **2020**, *10*, 627. [CrossRef]
43. Tian, H.B.; Ma, H.W.; Wei, J. Workspace and Structural Parameters Analysis for Manipulator of Serial Robot. *Trans. Chin. Soc. Agric. Mach.* **2013**, *44*, 196–201.

agriculture

MDPI

Article

A Deep-Learning Extraction Method for Orchard Visual Navigation Lines

Jianjun Zhou [1], Siyuan Geng [2], Quan Qiu [3,*], Yang Shao [1] and Man Zhang [4,*]

1 College of Information Engineering, Beijing Institute of Petrochemical Technology, Beijing 102617, China
2 Beijing Electro-Mechanical Engineering Institute, Beijing 100074, China
3 Academy of Artificial Intelligence, Beijing Institute of Petrochemical Technology, Beijing 102617, China
4 Key Laboratory of Smart Agriculture System Integration Research, Ministry of Education, China Agricultural University, Beijing 100083, China
* Correspondence: qiuquan0110@ustc.edu (Q.Q.); cauzm@cau.edu.cn (M.Z.); Tel.: +86-10-8129-3195 (Q.Q.); +86-10-6273-7188 (M.Z.)

Abstract: Orchard machinery autonomous navigation is helpful for improving the efficiency of fruit production and reducing labor costs. Path planning is one of the core technologies of autonomous navigation for orchard machinery. As normally planted in straight and parallel rows, fruit trees are natural landmarks that can provide suitable cues for orchard intelligent machinery. This paper presents a novel method to realize path planning based on computer vision technologies. We combine deep learning and the least-square (DL-LS) algorithm to carry out a new navigation line extraction algorithm for orchard scenarios. First, a large number of actual orchard images are collected and processed for training the YOLO V3 model. After the training, the mean average precision (MAP) of the model for trunk and tree detection can reach 92.11%. Secondly, the reference point coordinates of the fruit trees are calculated with the coordinates of the bounding box of trunks. Thirdly, the reference lines of fruit trees growing on both sides are fitted by the least-square method and the navigation line for the orchard machinery is determined by the two reference lines. Experimental results show that the trained YOLO V3 network can identify the tree trunk and the fruit tree accurately and that the new navigation line of fruit tree rows can be extracted effectively. The accuracy of orchard centerline extraction is 90.00%.

Keywords: autonomous navigation; navigation line extraction; orchard machinery; deep learning; least-square

Citation: Zhou, J.; Geng, S.; Qiu, Q.; Shao, Y.; Zhang, M. A Deep-Learning Extraction Method for Orchard Visual Navigation Lines. *Agriculture* **2022**, *12*, 1650. https://doi.org/10.3390/agriculture12101650

Academic Editor: Michele Pisante

Received: 7 September 2022
Accepted: 6 October 2022
Published: 9 October 2022

1. Introduction

In recent years, the new orchard intelligent machinery has shown great advantages in improving agricultural production efficiency and solving the labor shortage problem. First, such machinery has the ability to avoid direct contact between people and their working environments [1]. For example, there are some toxic or high-temperature scenarios which are not conducive to the human body in some operations. Moreover, the repetitive and monotonous nature of some phases of the orchard fruit production process, such as fruit picking, can be tiring and lead to missed operations or accidents. How to achieve autonomous navigation is one of the hot research topics in the field of intelligent machinery for orchards. With its advantages of wide range of detection information and comprehensive information acquisition, visual navigation has become the most widely used robotic navigation method throughout the world. The key aspect of visual navigation is its accurate and reliable extraction of the navigation baseline through image processing technology [2–4].

For the autonomous navigation problem, research ideas are focused on two aspects: road- or sky-based navigation line generation and crop detection–based fitting of navigation lines. Road- or sky-based navigation methods are highly robust to plant species, shape, and height and therefore constitute a hot research topic for scholars throughout the world. Crop detection–based navigation methods require accurate identification of crop trunks and are highly robust to complex road environments, thus requiring high adaptability.

Using the features shown in orchard images, He et al. proposed a horizontal projection method to recognize the main trunk area dynamically [5]. The color-difference R-B and two-dimensional Otsu algorithm were employed to segment the trunk from the background. A morphological method was adopted to eliminate noises from tiny branches and fading fallen leaves. Similarly, an optimal path extraction method proposed by Li also adopted color-model and segmentation methods [6]. The least-square and Hough transform methods are the most generally used line fitting methods. Based on the least-square method, both studies fit the reference lines of the fruit trees on both sides. The experimental results showed that the path generation method can provide a theoretical basis and technical support for the walking of a kiwi fruit–picking robot [6].

To achieve a better result, Ali et al. proposed a classification-based tree detection algorithm [7]. Color and texture cues were combined to yield better performance than individual cues could accomplish. Lyu et al. applied the Naive Bayesian classification (Artificial Neural Networks (ANN) and K-nearest neighbor (KNN) in [7]) to detect the boundary between trunk and ground and proposed a method to determine the centerline of orchard rows [8]. The advantage of the Bayesian classification is that it requires a small number of samples and a simple training process. In addition, it can effectively reduce impact from branches, soil, weeds, or tree shadows on the ground. In orchard navigation tests, the steering angle deviations generated by the proposed algorithm were much smaller than those generated from manual decisions. This showed that the orchard navigation method is more stable than a method that determines the centerline extraction manually.

Thus far, most researchers have developed algorithms that take advantage of the ground structures of orchards. These studies use the segmented sky from the tree canopy background and the centroid features of the segmented object as the process variables to guide the unmanned ground vehicle moving in the tree rows [1]. Experiments have shown that these approaches have the potential to guide utility vehicles.

Light detection and ranging (LiDAR) technology is also widely used in orchard navigation. Zhou et al. proposed a method for calculating the center point of the trunk with LiDAR sensory data [9]. LiDARs were used to scan the trunks on both sides of the fruit tree row. Point clusters with approximately circular arc shapes were formed. The central coordinate position and the radius of the trunk could be determined through geometric derivation. As the robot moved, its position and posture were corrected in real time by comparing the detected coordinates of the center point of the trunk with those obtained previously. Blok et al. paid more attention to the robot's self-positioning [3]. This research validated the applicability of two probabilistic localization algorithms that used a single 2D LiDAR scanner for in-row robot navigation in orchards. The first localization algorithm was a particle filter (PF) with a laser beam model, and the second was a Kalman filter (KF) with a line detection algorithm. Experiments were designed to test the navigation accuracy and robustness of the two methods, and the results showed that PF with a laser beam model was preferred over a line-based KF for in-row navigation.

Shalal et al. combined LiDAR and cameras in their research [10,11]. The LiDAR was used to detect edge points to determine the width of trunks and of non-trunk objects. The color and parallel edges of the trunks and non-trunk objects were verified by camera images.

Traditional image processing methods are easily affected by sunlight, canopy occlusion, and weeds. With the development of artificial intelligence, Zhang et al. tried to apply deep learning image processing in orchard management [12]. A multi-class object detection algorithm was proposed on the basis of a region convolutional neural network (R-CNN) model to detect branches, trunks, and apples in the orchard environment. VGG16 and VGG19 (the highest MAP of 82.4%) both achieved higher detection accuracy than Alexnet for the skeleton fitting of branches and trunks [13–15]; this study provided a foundation and possibility for developing a fully automated shake-and-catch apple harvesting system.

According to the above analysis of orchard autonomous navigation research results, the limitations of current orchard navigation are reflected in the following three points: ① In orchards with large tree canopies, it is more difficult to extract the vanishing point, and the application of generating navigation lines based on roads or skies will be limited. ② The use of traditional image processing methods based on tree trunk detection to fit the navigation path is susceptible to light intensity, shadows, and other factors. ③ Using radar data to improve the midpoint of fruit tree trunks provides a method for fruit tree row extraction, and image sensors have the advantage of low cost.

To address the limitations of the existing methods, we provide a DL_LS method that uses a deep learning model to extract the trunks of fruit trees near the ground and calculate the fruit tree reference points, fit the fruit tree row lines through the reference points, and calculate the centerlines through the row lines on both sides. In our method, we employ the YOLO V3 network to detect trunks of fruit trees in contact with the ground area, which can be basically independent of light intensity, shade, and disturbances. Furthermore, we use the detected trunk bounding box to determine the key points or reference points of the tree row, which are the middle points of the bottom lines of the bounding boxes, and then extract the tree row lines by the least-square method in order to improve the accuracy of the tree row line extraction.

Our method consists of four steps: detection of the fruit tree trunks using the deep learning method, determination of the fruit tree reference points, fitting of the fruit tree reference row lines, and generation of the orchard centerlines. The deep convolution neural network, which replaces the traditional feature extraction methods, can automatically detect the target after training with enough sampled learning data. The algorithm of the fruit tree row line fitting is put forward using a least-square algorithm, which can effectively extract the orchard machinery walking route.

2. Materials and Methods

The DL-LS algorithm proposed in this study can carry out path planning tasks for autonomous orchard machinery by combining deep learning methods with fruit tree line fitting algorithms. We selected the YOLO V3 network to accurately identify tree trunks with a bounding box, determine key or reference points with the middle points of the bottom lines of the bounding boxes, and fit the tree row reference lines with the least-square algorithm, which can carry out tree row line detection with higher accuracy under different disturbances in orchard scenarios. We collected a large quantity of actual orchard image data. These images were employed to train the YOLO V3 network after the sorting and labeling. Then the coordinates of the bounding box were generated after the tree trucks were detected. The reference point coordinates of the fruit tree can be calculated with these coordinates. The reference lines of the fruit tree rows were fitted by the least-square method. Finally, the centerline of the fruit tree rows was fitted with two reference lines. The principle is shown in Figure 1. This centerline is regarded as the tracking or moving path for the orchard machinery. Figure 2 is a flowchart of the deep learning-based tree/trunk extraction method. In the training stage, images of fruit tree rows in orchards are collected to form a dataset. The dataset is divided into a training set and a test set, and the manual labeling includes two types of tree trunks and fruit trees. The YOLO V3 network is trained using the training set to generate weight files. While testing, the trunk and fruit tree rectangular boxes are generated by the trained network; then fruit tree row reference point coordinates can be obtained by using trunk rectangular box coordinates calculation, and the fruit tree row lines are generated by means of least-squares fitting. Finally, the centerline of the fruit tree rows is obtained using the algorithm.

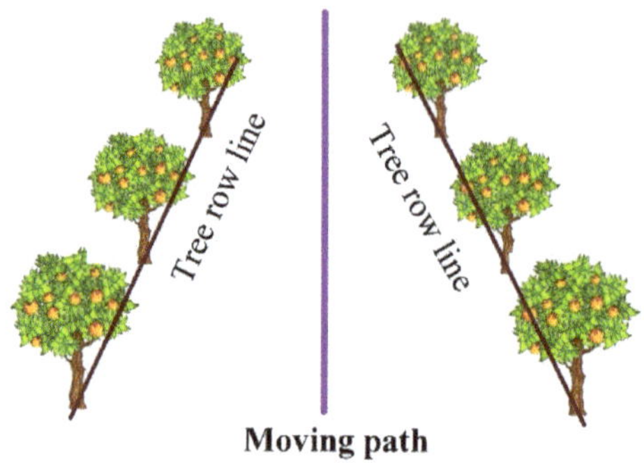

Figure 1. Schematic diagram of orchard navigation line extraction.

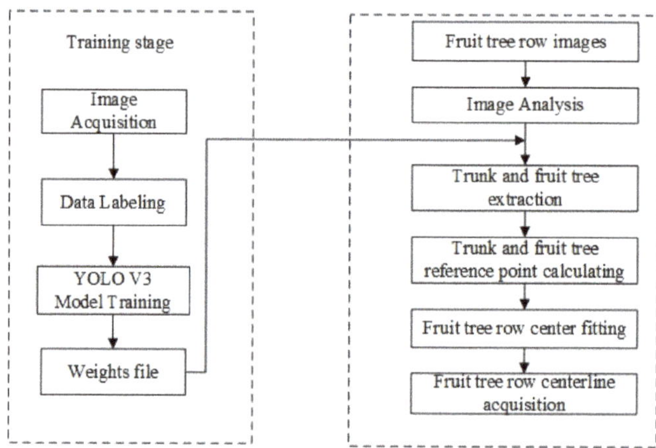

Figure 2. Flowchart of the deep-learning extraction method of orchard visual navigation line.

2.1. Detection of Fruit Tree and Trunk

Traditional target recognition methods are strongly dependent on specific images and are susceptible to light intensity, shade, etc. In this thesis, the YOLO V3 network is used to identify fruit trees and the trunk of fruit trees in contact with the ground area.

2.1.1. Network Structure of YOLO V3

YOLO V3 uses the residual module to improve the phenomenon of gradient disappearance or gradient explosion, and YOLO V3 borrows the idea of the feature pyramid networks (FPN) algorithm, which has excellent performance for small-target detection. The YOLO v3 network is based on a regression approach to feature extraction, enabling end-to-end object detection. Thus, it is more suitable for field application environments as it can quickly predict and classify targets while ensuring high accuracy.

The backbone network of YOLO V3 is Darknet-53. There are 53 layers of convolutional neural networks. The last layer is the fully-connected layer, and the other 52 layers appear as the layers for feature extraction [16]. The structure is as shown in Figure 3. Moreover, the residual module is widely used in the Darknet-53 network [13]. The gradient will disappear or explode if there are too many layers in the network. The residual module can improve this situation. YOLO V3 adopts the mechanism of multiscale fusion and multiscale prediction. YOLO V3's excellent performance for small-target detection is highly suitable for the task of trunk detection. It uses both the rich detail and location information of the low-level feature map and the rich semantic information of the high-level feature map to improve the detection precision and detect small targets better [17–21].

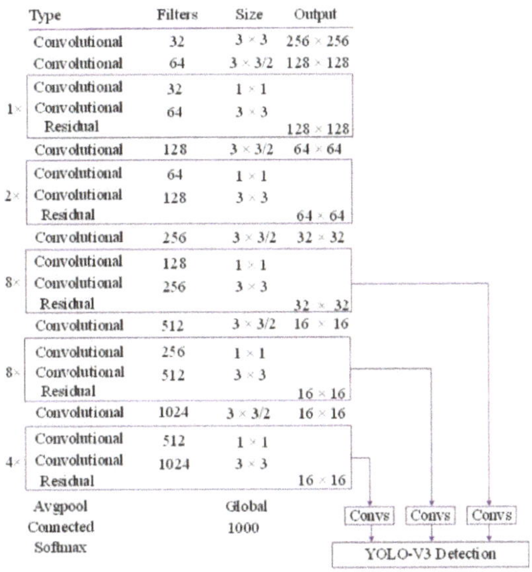

Figure 3. The structure of YOLO V3.

2.1.2. Image Datasets

The training of deep neural networks requires a great amount of data. The image dataset in this study was acquired from a pear orchard in the Daxing District, Beijing, which contains fruit trees of different ages, including young and adult trees. A large number of images of fruit trees were taken under different angles and illumination. The data collection scenarios are shown in Figure 4. In order to improve the training and prediction speed, the resolution of the input side of the image is uniformly converted to 512 × 512 pixels during image pre-processing. To improve the robustness of the model and suppress overfitting, random perturbations are added to expand the amount of data during training, such as random adjustment of contrast, saturation, brightness, etc. Finally, 971 images are obtained. In each sample image, the position and category of trunks and fruit trees are marked by a rectangle box, and the marked data are saved in a particular format. We chose LabelMe V3.16 installed on Anaconda for image labeling.

Figure 4. Some examples of the image datasets.

2.1.3. Model Training

The experiments in this study were conducted on a computer with Intel i7, 64-bit and a GTX 1080Ti GPU. The dataset was split into 70% for training and 30% for testing. In the training and testing processes, the unit of the images was pixel. In the process of model training, there are many hyperparameters that need to be set manually, and the difference in parameters seriously affects the quality of the model, such as the learning rate and batch size. In our model we set the initial learning rate to 0.001 and the batch size to 8. The learning rate is an important hyperparameter in the deep-learning optimizer which determines the speed of the weight updating. If the learning rate is too high, the training result will exceed the optimal value; if the learning rate is too low, the model will converge too slowly. The batch size depends on the size of the computer memory, and the larger the batch, the better the model training effect. After many parameter adjustments, we trained a model with relatively high accuracy which can accurately identify the trunk and fruit trees in the image. After the training, the loss value curve was drawn, as shown in Figure 5. The line reflects the relationship between the loss value and the number of epochs in the training process. The detection error of YOLO V3 dropped rapidly after the first 10 iterations. And the loss value hardly changed after 50 epochs.

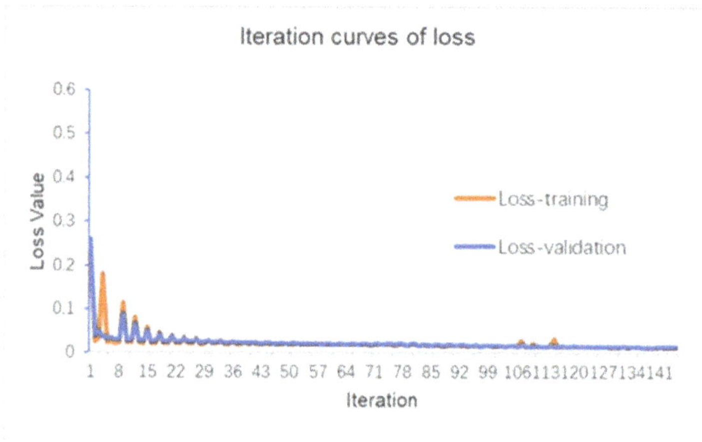

Figure 5. Loss curves of the YOLO V3 model.

2.2. *Path Extraction of Orchard Machinery Navigation*

The previous section extracts the information on tree trunk position coordinates from orchard images taken in the row. In this section, the centerline of the fruit tree rows is extracted on the basis of the trunk box coordinates.

2.2.1. Reference Point Generation

The coordinate value of the bounding-box border can be read clearly by generating the position information of the trunk, which contains the coordinate value of the points in the upper-left and lower-right corner. The coordinates of the points in the upper-left and lower-right corner are P_l (x_l, y_l) and P_r (x_r, y_r), respectively. The reference point of this trunk is $(\frac{x_r - x_l}{2} + x_l, y_r)$. The algorithm's pseudocodes are shown in Algorithm 1.

Algorithm 1 Obtain available coordinate points

```
Input: Acquired raw image
            [r c] = size(img)
Imghalfwidth = c/3/2
            A = importdata (txt)
            [m,n] = size (A.data)
for I = 1:m
  if textdata including "trunk"
      if Second data < imghalfwidth
            y = The fifth data value in A
          x = 0.5 (fourth data value - second data value) + second data value
        else
            y = The fifth data value in A
          x = 0.5 (fourth data value - second data value) + second data value
      end
    end
end
```

2.2.2. Line Fitting of the Tree Rows

The reference points of the fruit trees are fitted into the reference lines of the fruit trees on both sides of the row by the least-square method. If there are fewer than three available tree trunks extracted in case of missing fruit trees, we simply connect the nearest two reference points. The process is shown in Algorithm 2.

Algorithm 2 Obtain the reference lines

```
Input: Sorting the coordinates of the reference points of the left- and right-side fruit trees,
respectively
if the number of points on the left is equal to or greater than 3
      least-square method
else if less than 3 points on the left
      Connect two points
end
The right-fitting line is the same as above
if the number of points on the right is equal to or greater than 3
  Fit a straight line using the least-square method
else if less than 3 points on the left
Connect two points line 11
      k = (ycord (1) - ycord (2))/(xcord (1) - xcord (2))
      b = ycord (1)
end
```

2.2.3. Obtaining the Centerline

The centerline of the previously obtained two reference lines of the fruit tree rows on both sides is the reference line of the orchard machinery, and its detailed principle is shown in Figure 6. We denote point P_{l1} as the farthest reference point on the left reference line in the image. Its corresponding point on the right reference line is $P_{r1.}$ We connect the segment P_{l1} P_{r1} and calculate the midpoint P_{m1}. Similarly, we denote point P_{l2} as the nearest reference point and connect the segment P_{l2} P_{r2} to determine the point P_{m2}. Currently, the straight line passing through P_{m1} and P_{m2} is the reference line for the orchard machinery. The algorithm flow is shown in Algorithm 3.

Algorithm 3 Obtain the centerline

Input: the left and right reference lines
sort coordinate for the left rectangle label
search the nearest point P_{l2}corresponding Point $P_{r2}(x_{r2}, y_{r2})$
middle point coordinate (x_{m2}, y_{m2})
search the furthest point P_{l1} corresponding Point P_{r1} (x_{r1}, y_{r1})
calculate the coordinates of point P_{m1} (x_{m1}, y_{m1}) by points P_{l1} and P_{r1}
calculate the coordinates of point P_{m2} (x_{m2}, y_{m2}) by points P_{l2} and P_{r2}
line connecting points P_{m1} and P_{m2}
end

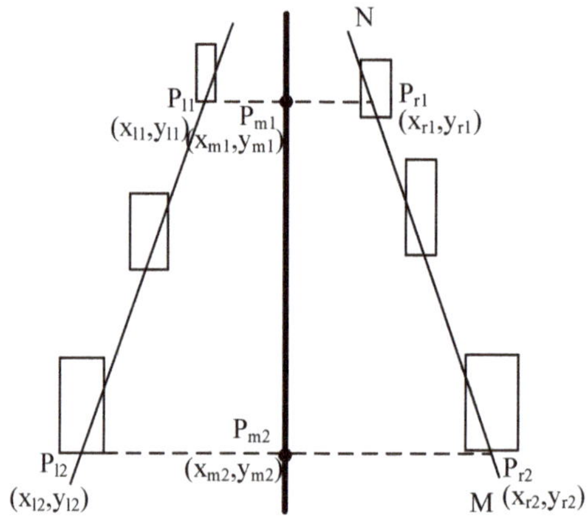

Figure 6. Centerline acquisition for orchard machinery.

3. Results and Discussion

3.1. Tree and Trunk Detection Results

The trained network can identify the tree trunk and fruit tree accurately. The detection accuracy is shown in Table 1. The average precision (AP) of the trees is 92.7%, and the AP of the trunks is 91.51%. The MAP of detection can reach 92.11%, which is not easily affected by sunlight. The trunk of the same fruit tree can be accurately detected under normal sunlight and strong sunlight, as shown in Figure 7. This method has a stronger anti-interference ability compared with traditional methods, especially in the morning and afternoon when the lighting condition changes. Furthermore, weeds easily affect the results of the interference; this is because the color and shape of weeds and leaves are very similar and because weeds occasionally become entangled with the tree trunks. Figure 8 shows the detection result under strong sunlight. The recognition result of the trunks and fruit trees obtained by this network in weed-rich environments shows it to be helpful in alleviating the interference caused by weeds. As shown in Figure 9, the effect of trunk extraction on both sides of the fruit tree rows is excellent under normal sunlight, which is an important basis of this study. Figure 10 shows the result of weak sunlight.

Table 1. Detection accuracy.

Type	AP/%
Tree	92.70
Trunk	91.51

Figure 7. Detection results under different sunlight conditions.

Figure 8. Detection results of tree and trunk under strong sunlight.

Figure 9. Detection results of trunk under normal sunlight.

Figure 10. Detection results of trunk under weak sunlight.

3.2. Results of Reference Point Generation

The accuracy of the algorithm is tested by comparing the coordinate points manually marked with those extracted by our algorithm. The error is calculated by the distance between two pixels. Referring to Figure 10, the error of Figure 10b is larger than errors of the other Figure 10a,c because the height of the tree trunks buried in the soil is irregular, and the error of manually marked points is lower than those obtained in our method. As shown in Table 2, there are five points in three sub-figures. The average error is 1.93 pixels.

Table 2. Error analysis of the trunk reference point.

Sub-Figure	Original Coordinates	Reference Point Coordinates	Manual Marking Coordinates	Error (Pixel)
(a)	(232,425) (308,512)	(270,512)	(268,512)	2.00
(b)	(180,439) (345,632)	(262.5,632)	(260,631)	2.69
	(171,532) (302,685)	(236.5,685)	(235,685)	1.50
(c)	(29,328) (53,373)	(41,373)	(40,372)	1.41
	(534,352) (581,411)	(557.5,411)	(558,413)	2.06

3.3. Results of Tree-Row Line Fitting

Determining the position of each fruit tree in the image is the basis of orchard mechanical operation, which can obtain the straight line of fruit trees on both row sides. The fitting results of reference lines on both sides of the fruit tree row under different sunlight are shown in Figure 11, including weed environment, strong sunlight, weak sunlight, and normal sunlight. The analysis of the line fitting of fruit tree rows is shown in Table 3.

(**a**) corridor with weed (**b**) strong sunlight (**c**) weak sunlight

(**d**) normal sunlight on a corridor with leaves (**e**) normal sunlight

Figure 11. Reference line of a fruit tree row under different sunlight.

Table 3. Accuracy of line fitting of the fruit tree rows.

	Weed Environment			Weak Sunlight			Strong Sunlight			Normal Sunlight		
	to the left	to the right	correct	to the left	to the right	correct	to the left	to the right	correct	to the left	to the right	correct
Left row line	1	0	6	1	0	6	0	0	7	1	0	8
Right row line	0	0	7	0	0	7	0	1	6	1	0	8
Total	1	0	13	1	0	13	0	1	13	2	0	16

Thirty images were selected to test the accuracy of the fruit tree line fit, all of which included both left and right fruit tree rows. The four environments were weeded, low light, high light, and normal light. There were 7 images of the weedy environment, 7 images of the low light environment, 7 images of the high light environment, and 9 images of the normal light environment. When performing this study, the fitted lines should describe the fruit tree rows evenly and accurately, with errors categorized as right-leaning or left-leaning on the left side and right-leaning or left-leaning on the left side. Sixty lines were included in the 30 images, divided into 30 lines on the left and 30 lines on the right side; the specific fruit tree lines fitted are shown in Table 3. In the weed environment, one line on the left side of the tree was fitted to the left, while the other 13 lines were fitted correctly. In the weak-light environment, one line on the left side was fitted to the left, while the other 13 lines were fitted correctly. The right fruit tree row line was fitted to the right in the strong-light environment, while the other 13 were fitted correctly. Under normal light conditions, one line on the left side of the tree was fitted to the left and one line on the right side of the tree was fitted to the right, while the remaining 16 lines were fitted correctly. A total of 55 lines were fitted correctly and 5 lines were fitted inaccurately. The average accuracy of the fruit tree line fits was calculated to be 91.67%.

3.4. Results of Centerline Extraction

As shown in Figure 12, the green lines are the centerlines of the orchard machinery. The combination of deep learning and least-square yields a great improvement in efficiency and accuracy compared with traditional methods.

Figure 12. Centerlines of fruit rows calculation in the orchard.

In order to evaluate the accuracy of the centerline generation, the benchmark line is selected manually; the difference between the algorithm-generated centerline and the best navigation line is then analyzed. Table 4 shows the fitting results of the centerline in the fruit rows. The accuracy of orchard centerline extraction is 90.00% according to 27 extracted proper centerlines out of 30 images.

Table 4. The fitting of the centerline in fruit rows.

	Weed Environment		**Weak Sunlight**		**Strong Sunlight**		**Normal Sunlight**	
type	Little deviation	correct	Little deviation	correct	Little deviation	correct	Little deviation	correct
amount	1	6	1	6	0	7	1	8

Han, et al. proposed a U-Net network-based approach for visual navigation path recognition in orchards [22]. Table 5 gives a comparative analysis of the maximum and mean value pixel error of the centerline of the fruit tree rows calculated by both U-Net and DL_LS. Under weak light, the maximum pixel error of the centerline is 19 pixels for U-Net and 8 pixels for DL_LS, and the mean value pixel error of the centerline is 11.8 pixels for U-Net and 5.2 pixels for DL_LS; under normal light, the maximum pixel error of the centerline extracted by U-Net is 10 pixels and 5 pixels for DL_LS, and the mean value pixel error of the centerline extracted by U-Net is 6.5 pixels and 3.4 pixels for DL_LS; under strong light, the maximum pixel error of the centerline extracted by U-Net is 7 pixels and 4 pixels for DL_LS, and the mean value pixel error of the centerline extracted by U-Net is 2.9 pixels and 2.1 pixels for DL_LS. From Table 5, we can infer that our DL_LS can give higher centerline extraction results than those of U-Net.

Table 5. Comparison of centerline maximal pixel errors of different methods.

Method	Weak Sunlight		Normal Sunlight		Strong Sunlight	
	Maximum	Mean Value	Maximum	Mean Value	Maximum	Mean Value
U-Net [22]	19	11.8	10	6.5	7	2.9
DL_LS	8	5.2	5	3.4	4	2.1

3.5. Discussion

Although our method can extract the centerline of two adjacent orchard tree rows with high accuracy, there are still some drawbacks or limitations in our method. First, some of the trunks detected by the deep learning algorithm are side views or parts of the whole trunks, which introduces pixel error while determining the reference points. As a result, the centerline extraction accuracy could be improved if a smart reference point selection strategy is designed. Second, fruit tree trunks of other rows may be captured into the images, so that the extracted feature points are distributed in a zigzag shape, which affects the accurate generation of fruit tree row centerlines. Therefore, a reference or feature point selection or filtering strategy should be proposed to improve our algorithm.

The trained network can identify the tree trunk and fruit tree accurately. The single-target average accuracies for trees and trunks are 92.7% and 91.51% respectively. Trunks and fruit trees are well identified in different sunlight and weed-rich environments. The model has strong robustness, and it takes about 50 milliseconds to process an image, which meets the reliability of the algorithm in real-time mode.

4. Conclusions

A centerline extraction algorithm of orchard rows was proposed based on the YOLO V3 network, which can detect fruit trees and trunks in contact with the ground area independent of light intensity, shade, and disturbances. The average detection accuracy of the tree trunks and fruit trees was 92.11% by outputting the coordinate text file of the bounding box at the same time.

With the trunk bounding box, the reference points of the fruit tree trunks were extracted and the least-squares method was applied to fit the fruit tree rows on both sides of the walking routine of the agricultural machinery. According to the experimental results, the centerline of the orchard line was finally fitted. The average accuracy of the fruit tree line extraction was calculated to be 90%.

In the future, our research will consider the fusion of multiple sensors which can acquire richer environmental information and enable automated navigation in complex and changing orchard environments.

Author Contributions: Conceptualization, J.Z.; methodology, J.Z and S.G.; validation, J.Z, S.G and Q.Q.; formal analysis, Q.Q., S.G. and J.Z; writing—original draft preparation, J.Z and S.G.; writing—review and editing, J.Z., M.Z. and Q.Q.; visualization, Y.S., S.G. and J.Z.; supervision, M.Z.; funding acquisition, Q.Q. All authors have read and agreed to the published version of the manuscript.

Funding: This study was supported by National Natural Science Foundation of China (No.61973040, No. 31101088).

Institutional Review Board Statement: Not applicable.

Data Availability Statement: The data that support the findings of this study are available from the first author and the second author, upon reasonable request.

Acknowledgments: The authors would like to thank Senliu Chen, for his instructive suggestions and kind helps in experiment implementation and paper writing.

Conflicts of Interest: The authors declare no conflict of interest.

References

1. Radcliffe, J.; Cox, J.; Bulanon, D.M. Machine vision for orchard navigation. *Comput. Ind.* **2018**, *98*, 165–171. [CrossRef]
2. Guo, S.Q. Research on Identifying and Locating Apple in Orchard Based on Neural Network and 3D Vision. Master's Thesis, Beijing Jiaotong University, Beijing, China, 2019.
3. Blok, P.M.; van Boheemen, K.; van Evert, F.K.; IJsselmuiden, J.; Kim, G.H. Robot navigation in orchards with localization based on Particle filter and Kalman filter. *Comput. Electron. Agric.* **2019**, *157*, 261–269. [CrossRef]
4. Feng, J.; Liu, G.; Si, Y.S.; Wang, S.; He, B.; Ren, W. Algorithm based on image processing technology to generate navigation directrix in orchard. *Trans. Chin. Soc. Agric. Mach.* **2012**, *43*, 184–189. [CrossRef]
5. He, B.; Liu, G.; Ji, Y.; Si, Y.S.; Gao, R. Auto recognition of navigation path for harvest robot based on machine vision. In Proceedings of the International Conference on Computer and Computing Technologies in Agriculture, Nanchang, China, 23–24 October 2010. [CrossRef]
6. Li, W.Y. Research on the Method of Generating Visual Navigation Path of Kiwi Picking Robot. Master's Thesis, North West Agriculture and Forestry University, Yangling, China, 2017.
7. Ali, W.; Georgsson, F.; Hellstrom, T. Visual tree detection for autonomous navigation in forest environment. In Proceedings of the IEEE Intelligent Vehicles Symposium, Eindhoven, The Netherlands, 4–6 June 2008. [CrossRef]
8. Lyu, H.K.; Park, C.H.; Han, D.H.; Kwak, S.W.; Choi, B. Orchard free space and center line estimation using naive bayesian classifier for unmanned ground self-driving vehicle. *Symmetry* **2018**, *10*, 355. [CrossRef]
9. Zhou, J.J.; Hu, C. Inter-row localization method for agricultural robot working in close planting orchard. *Trans. Chin. Soc. Agric. Mach.* **2015**, *46*, 22–28.
10. Shalal, N.; Low, T.; McCarthy, C.; Hancock, N. Orchard mapping and mobile robot localisation using on-board camera and laser scanner data fusion–Part A: Tree detection. *Comput. Electron. Agric.* **2015**, *119*, 254–266. [CrossRef]
11. Shalal, N.; Low, T.; McCarthy, C.; Hancock, N. Orchard mapping and mobile robot localisation using on-board camera and laser scanner data fusion–Part B: Mapping and localisation. *Comput. Electron. Agric.* **2015**, *119*, 267–278. [CrossRef]
12. Zhang, J.; Karkee, M.; Zhang, Q.; Zhang, X.; Yaqoob, M.; Fu, L.S.; Wang, S.M. Multi-class object detection using faster R-CNN and estimation of shaking locations for automated shake-and-catch apple harvesting. *Comput. Electron. Agric.* **2020**, *173*, 105384. [CrossRef]
13. Redmon, J.; Divvala, S.; Girshick, R.; Farhadi, A. You Only Look Once: Unified, real-time object detection. In Proceedings of the IEEE Conference on Computer Vision and Pattern Recognition, Las Vegas, NV, USA, 27–30 June 2016. [CrossRef]
14. Liu, J.; Wang, X.W. Early recognition of tomato gray leaf spot disease based on MobileNetv2-YOLOv3 model. *Plant Methods* **2020**, *16*, 83. [CrossRef] [PubMed]
15. Cenggoro, T.W.; Aslamiah, A.H.; Yunanto, A. Feature pyramid networks for crowd counting. *Procedia Comput. Sci.* **2019**, *157*, 175–182. [CrossRef]
16. Luo, Z.; Yu, H.; Zhang, Y. Pine Cone Detection Using Boundary Equilibrium Generative Adversarial Networks and Improved YOLOv3 Model. *Sensors* **2020**, *20*, 4430. [CrossRef] [PubMed]
17. He, K.M.; Zhang, X.Y.; Ren, S.Q.; Sun, J. Deep residual learning for image recognition. In Proceedings of the IEEE Conference on Computer Vision and Pattern Recognition, Las Vegas, NV, USA, 27–30 June 2016. [CrossRef]
18. Redmon, J.; Farhadi, A. YOLOv3: An incremental improvement. *arXiv* **2018**, arXiv:1804.02767.
19. Wu, D.H.; Wu, Q.; Yin, X.Q.; Jiang, B.; Wang, H.; He, D.J.; Song, H.B. Lameness detection of dairy cows based on the YOLOv3 deep learning algorithm and a relative step size characteristic vector. *Biosyst. Eng.* **2020**, *189*, 150–163. [CrossRef]
20. Tian, Y.N.; Yang, G.D.; Wang, Z.; Wang, H.; Li, E.; Liang, Z.Z. Apple detection during different growth stages in orchards using the improved YOLO-V3 model. *Comput. Electron. Agric.* **2019**, *157*, 417–426. [CrossRef]
21. Liu, G.; Nouaze, J.C.; Mbouembe, P.L.T.; Kim, J.H. YOLO-Tomato: A robust algorithm for tomato detection based on YOLOv3. *Sensors* **2020**, *20*, 2145. [CrossRef] [PubMed]
22. Han, Z.H.; Li, J.; Yuan, Y.W.; Fang, X.F.; Zhao, B.; Zhu, L.C. Path Recognition of Orchard Visual Navigation Based on U-Net. *Trans. Chin. Soc. Agric. Mach.* **2021**, *52*, 30–39. [CrossRef]

MDPI
St. Alban-Anlage 66
4052 Basel
Switzerland
Tel. +41 61 683 77 34
Fax +41 61 302 89 18
www.mdpi.com

Agriculture Editorial Office
E-mail: agriculture@mdpi.com
www.mdpi.com/journal/agriculture

www.ingramcontent.com/pod-product-compliance
Lightning Source LLC
Chambersburg PA
CBHW061501220926
51850CB00004B/145
* 9 7 8 3 0 3 6 5 8 3 7 6 1 *